Communications
in Computer and Information Science 204

Dhinaharan Nagamalai Eric Renault
Murugan Dhanuskodi (Eds.)

# Trends in Computer Science, Engineering and Information Technology

First International Conference on Computer Science, Engineering and Information Technology, CCSEIT 2011
Tirunelveli, Tamil Nadu, India, September 23-25, 2011
Proceedings

Springer

Volume Editors

Dhinaharan Nagamalai
Wireilla Net Solutions PTY Ltd
Melbourne, VIC, Australia
E-mail: dhinthia@yahoo.com

Eric Renault
Institut Telecom/Telecom SudParis (ex. GET-INT)
Departement Reseaux et Services Multimedia Mobiles (RS2M)
Samovar UMR INT-CNRS 5157
9, rue Charles Fourier, 91011 Evry Cedex, France
E-mail: eric.renault@it-sudparis.eu

Murugan Dhanuskodi
Manonmaniam Sundaranar University
Department of Computer Science and Engineering
Tirunelveli, Tamil Nadu, India
E-mail: dhanushkodim@yahoo.com

ISSN 1865-0929 e-ISSN 1865-0937
ISBN 978-3-642-24042-3 ISBN 978-3-642-24043-0 (eBook)
DOI 10.1007/978-3-642-24043-0
Springer Heidelberg Dordrecht London New York

Library of Congress Control Number: Applied for

CR Subject Classification (1998): I.2, C.2, H.4, H.3, D.2, J.1



*Typesetting:* Camera-ready by author, data conversion by Scientific Publishing Services, Chennai, India

Printed on acid-free paper

Springer is part of Springer Science+Business Media (www.springer.com)

# Preface

The First International Conference on Computer Science, Engineering and Information Technology (CCSEIT-2011), The First International Conference on Parallel, Distributed Computing Technologies and Applications (PDCTA-2011), and The First International Conference on Digital Image Processing and Pattern Recognition (DPPR-2011) were held in Tirunelveli - Tamil Nadu, India, during September 23–25, 2011. The events attracted many local and international delegates, presenting a balanced mixture of intellects from all over the world. The goal of this conference series is to bring together researchers and practitioners from academia and industry to focus on understanding parallel, distributed computing technologies, digital image processing and pattern recognition and all areas of computer science, information technology, and to establish new collaborations in these areas.

The CCSEIT 2011, PDCTA 2011 and DPPR 2011 committees invited original submissions from researchers, scientists, engineers, and students that illustrate research results, projects, survey works, and industrial experiences describing significant advances in the areas related to the relevant themes and tracks of the conferences. This effort guaranteed submissions from an unparalleled number of internationally recognized top-level researchers. All the submissions underwent a strenuous peer-review process which comprised expert reviewers. Besides the members of the Technical Program Committee, external reviewers were invited on the basis of their specialization and expertise. The papers were reviewed based on their technical content, originality, and clarity. The entire process, which includes the submission, review, and acceptance processes, was done electronically. All these efforts undertaken by the Organizing and Technical Program Committees led to an exciting, rich, and high-quality technical conference program, which featured high-impact presentations for all attendees to enjoy and to expand their expertise in the latest developments in this field.

There were a total 1,256 submissions to the conference, and the Technical Program Committee selected 185 papers for presentation at the conference and subsequent publication in the proceedings. This small introduction would be incomplete without expressing our gratitude and thanks to the General and Program Chairs, members of the Technical Program Committees, and external reviewers for their excellent and diligent work. Thanks to Springer for the strong support. Finally, we thank all the authors who contributed to the success of the conference. We also sincerely wish that all attendees benefited academically from the conference and wish them every success in their research.

Dhinaharan Nagamalai
Eric Renault
Murugan Dhanushkodi

# Organization

## General Chairs

| | |
|---|---|
| David C. Wyld | Southeastern Louisiana University, USA |
| Michal Wozniak | Wroclaw University of Technology, Poland |

## Steering Committee Chairs

| | |
|---|---|
| Murugan Dhanuskodi | Manonmaniam Sundaranar University, India |
| Jan Zizka | SoNet/DI, FBE, Mendel University in Brno, Czech Republic |
| John Karamitsos | University of the Aegean, Samos, Greece |
| Khoa N. Le | University of Western Sydney, Australia |
| Nabendu Chaki | University of Calcutta, India |
| Salah S. Al-Majeed | University of Essex, UK |
| Dhinaharan Nagamalai | Wireilla Net Solutions, Australia |

## Publicity Chairs

| | |
|---|---|
| Roberts Masillamani | Hindustan University, India |
| Chih-Lin Hu | National Central University, Taiwan |

## Program Committee

| | |
|---|---|
| A. Arokiasamy | Eastern Mediterranean University, Cyprus |
| A.P. Sathish Kumar | PSG Institute of Advanced Studies, India |
| Abdul Aziz | University of Central Punjab, Pakistan |
| Abdul Kadir Ozcan | The American University, Cyprus |
| Al-Sakib Khan Pathan | Kyung Hee University, South Korea |
| Andreas Riener | Johannes Kepler University Linz, Austria |
| Andy Seddon | Asia Pacific Institute of Information Technology, Malaysia |
| Antelin Vijila | Manonmaniam Sundaranar University, India |
| Arvinth Kumar | M.S. University, India |
| Athanasios Vasilakos | University of Western Macedonia, Greece |
| Atilla Elci | Eastern Mediterranean University, Cyprus |
| B. Srinivasan | Monash University, Australia |

| | |
|---|---|
| Balasubramaniam | Manonmaniam Sundaranar University, India |
| Bong-Han Kim | Chongju University , South Korea |
| Boo-Hyung Lee | KongJu National University, South Korea |
| Brajesh Kumar Kaushik | Indian Institute of Technology, India |
| Charalampos Z. Patrikakis | National Technical University of Athens, Greece |
| Chih-Lin Hu | National Central University, Taiwan |
| Chin-Chih Chang | Chung Hua University, Taiwan |
| Cho Han Jin | Far East University, South Korea |
| Cynthia Dhinakaran | Hannam University, South Korea |
| Danda B. Rawat | Old Dominion University, USA |
| David W. Deeds | Shingu College, South Korea |
| Debasis Giri | Haldia Institute of Technology, India |
| Deepak Garg | Thapar University, India |
| Dhinaharan Nagamalai | Wireilla Net Solutions Pty Ltd., Australia |
| Dimitris Kotzinos | Technical Educational Institution of Serres, Greece |
| Dong Seong Kim | Duke University, USA |
| Emmanuel Bouix | iKlax Media, France |
| Eric Renault | Institut Telecom – Telecom SudParis, France |
| Farhat Anwar | International Islamic University , Malaysia |
| Firkhan Ali Bin Hamid Ali | Universiti Tun Hussein Onn Malaysia, Malaysia |
| Ford Lumban Gaol | University of Indonesia |
| Geuk Lee | Hannam University, South Korea |
| Girija Chetty | University of Canberra, Australia |
| H.V. Ramakrishnan | Dr. MGR University, India |
| Henrique Joao Lopes Domingos | University of Lisbon, Portugal |
| Ho Dac Tu | Waseda University, Japan |
| Hoang, Huu Hanh | Hue University, Vietnam |
| Hwangjun Song | Pohang University of Science and Technology, South Korea |
| J.Arunadevi | Thiagarajar College, Madurai, India |
| Jacques Demerjian | Communication & Systems, Homeland Security, France |
| Jae Kwang Lee | Hannam University, South Korea |
| Jan Zizka | SoNet/DI, FBE, Mendel University in Brno, Czech Republic |
| Jansirani | Manonmaniam Sundaranar University, India |
| Jeong-Hyun Park | Electronics Telecommunication Research Institute, South Korea |
| Jivesh Govil | Cisco Systems Inc. - CA, USA |
| Johann Groschd | University of Bristol, UK |

| | |
|---|---|
| John Karamitsos | University of the Aegean, Samos, Greece |
| Johnson Kuruvila | Dalhousie University, Halifax, Canada |
| Jose Enrique Armendariz-Inigo | Universidad Publica de Navarra, Spain |
| Jung-Gil Song | Hannam University, South Korea |
| Jungwook Song | Konkuk University, South Korea |
| K.P. Thooyamani | Bharath University, India |
| Kamaljit I Lakhtaria | Atmiya Institute of Technology & Science, India |
| Kannan | Anna University, Chennai, India |
| Khamish Malhotra | University of Glamorgan, UK |
| Khoa N. Le | Griffith University, Australia |
| Krzysztof Walkowiak | Wroclaw University of Technology, Poland |
| L. Ganesan | Alagappa University, India |
| Lu S. Veiga | Technical University of Lisbon, Portugal |
| Lu Yan | University of Hertfordshire, UK |
| Maode Ma | Nanyang Technological University, Singapore |
| Marco Roccetti | Universty of Bologna, Italy |
| Michael Peterson | University of Hawaii at Hilo, USA |
| Michal Wozniak | Wroclaw University of Technology, Poland |
| Mohsen Sharifi | Iran University of Science and Technology, Iran |
| Murugan D. | Manonmaniam Sundaranar University, India |
| Murugeswari | M.S. University, India |
| Muthulakshmi | M.S. University, India |
| N. Krishnan | Manonmaniam Sundaranar University, India |
| Nabendu Chaki | University of Calcutta, India |
| Natarajan Meghanathan | Jackson State University, USA |
| Neerajkumar | Madha Vaisnavi Devi University, India |
| Nicolas Sklavos | Technological Educational Institute of Patras, Greece |
| Nidaa Abdual Muhsin Abbas | University of Babylon, Iraq |
| Paul D. Manuel | Kuwait University, Kuwait |
| Phan Cong Vinh | London South Bank University, UK |
| Ponpit Wongthongtham | Curtin University of Technology, Australia |
| Prabu Dorairaj | NetApp Inc., India |
| Rajalakshmi | Manonmaniam Sundaranar University, India |
| Rajendra Akerkar | Technomathematics Research Foundation, India |
| Rajesh Kumar P. | The Best International, Australia |
| Rajesh | Manonmaniam Sundaranar University, India |
| Rajesh Bawa | Punjabi University, India |

| | |
|---|---|
| Rajkumar Kannan | Bishop Heber College, India |
| Rakhesh Singh Kshetrimayum | Indian Institute of Technology-Guwahati, India |
| Ramayah Thurasamy | Universiti Sains Malaysia, Malaysia |
| Rituparna Chaki | West Bengal University of Technology, India |
| Roberts Masillamani | Hindustan University, India |
| S. Arumugam | Nandha Engineering College, India |
| S. Hariharan | B.S.Abdur Rahman University, India |
| Sadasivam | Manonmaniam Sundaranar University, India |
| Sagarmay Deb | Central Queensland University, Australia |
| Sajid Hussain | Acadia University, Canada |
| Sajid Hussain | Fisk University, USA |
| Salah S. Al-Majeed | University of Essex, UK |
| Sanguthevar Rajasekaran | University of Connecticut, USA |
| Sarmistha Neogy | Jadavpur University, India |
| Sattar B. Sadkhan | University of Babylon, Iraq |
| Seemabawa | Thapar University, India |
| Sergio Ilarri | University of Zaragoza, Spain |
| Serguei A. Mokhov | Concordia University, Canada |
| Seungmin Rho | Carnegie Mellon University, USA |
| Shivan Haran | Arizona State University, USA |
| Somitra Sanadhya | IIT-Delhi, India |
| Soodamani, ramalingam | Uuniversity of Hertfordshire,UK |
| Sriman Narayana Iyengar | VIT University, India |
| Subha | M.S. University, India |
| Sudip Misra | Indian Institute of Technology-Kharagpur, India |
| Sundarapandian Viadyanathan | Vel Tech Dr.RR & Dr.SR Technical University, India |
| Sundaresan | Manonmaniam Sundaranar University, India |
| SunYoung Han | Konkuk University, South Korea |
| Suruliandi | Manonmaniam Sundaranar University, India |
| Susana Sargento | University of Aveiro, Portugal |
| Syed Rahman | University of Hawaii-Hilo, USA |
| Syed Rizvi | University of Bridgeport, USA |
| Taruna S. | Banasthali University, India |
| Thamaraiselvi | Madras Institute of Technology,India |
| Thambidurai | Pondicherry University, India |
| Velmurugan Ayyadurai | Center for Communication Systems, UK |
| Vishal Sharma | Metanoia Inc., USA |
| Wei Jie | University of Manchester, UK |

| | |
|---|---|
| Yan Luo | University of Massachusetts Lowell, USA |
| Yannick Le Moullec | Aalborg University, Denmark |
| Yao-Nan Lien | National Chengchi University, Taiwan |
| Yeong Deok Kim | Woosong University, South Korea |
| Yuh-Shyan Chen | National Taipei University, Taiwan |
| Yung-Fa Huang | Chaoyang University of Technology, Taiwan |

## External Reviewers

| | |
|---|---|
| Alejandro Regalado Mendez | Universidad del Mar, Mexico |
| Alireza Mahini | Islamic Azad University-Gorgan, Iran |
| Amandeep Singh Thethi | Guru Nanak Dev University Amritsar, India |
| Ashok Kumar Sharma | YMCA Institute of Engineering, India |
| Ayman Khalil | Institute of Electronics and Telecommunications of Rennes (IETR), France |
| Buket Barkana | University of Bridgeport, USA |
| Christos Politis | Kingston University, UK |
| Hao Shi | Victoria University, Australia |
| Indrajit Bhattacharya | Kalyani Government Engineering College, India |
| Jyotirmay Gadewadikar | Alcorn State University, USA |
| Khoa N. Le | University of Western Sydney, Australia |
| Laiali Almazaydeh | University of Bridgeport, USA |
| Lakshmi Rajamani | Osmania University, India |
| Michel Owayjan | American University of Science & Technology - AUST, Lebanon |
| Mohamed Hassan | American University of Sharjah, UAE |
| Monika Verma | Punjab Technical University, India |
| N.K. Choudhari | Bhagwati Chaturvedi College of Engineering, India |
| Nitiket N. Mhala | B.D. College of Engineering - Sewagram, India |
| Nour Eldin Elmadany | Arab Acadmy for Science and Technology, Egypt |
| Premanand K. Kadbe | Vidya Pratishthan's College of Engineering, India |
| R. Murali | Dr. Ambedkar Institute of Technology, Bangalore, India |
| Raman Maini | Punjabi University, India |
| Rushed Kanawati | LIPN - Université Paris 13, France |
| S.A.V. Satyamurty | Indira Gandhi Centre for Atomic Research, India |
| Shrikant K. Bodhe | Bosh Technologies, India |

| | |
|---|---|
| Sridharan | CEG Campus - Anna University, India |
| Utpal Biswas | University of Kalyani, India |
| Wichian Sittiprapaporn | Mahasarakham University, Thailand |

## Technically Supported By

Networks & Communications Community (NCC)
Digital Signal & Image Processing Community (DSIPC)
Computer Science & Information Technology Community (CSITC)

## Organized By

ACADEMY & INDUSTRY RESEARCH COLLABORATION CENTER (AIRCC)

# Table of Contents

## Computer Science, Engineering and Information Technology

# Hybrid Chaos Synchronization of Liu and Lü Systems by Active Nonlinear Control

Sundarapandian Vaidyanathan

Research and Development Centre,
Vel Tech Dr. RR & Dr. SR Technical University,
Avadi-Alamathi Road, Avadi, Chennai-600 062, India
sundarvtu@gmail.com
http://www.vel-tech.org/

**Abstract.** This paper investigates the hybrid chaos synchronization of identical Liu systems, identical Lü systems, and non-identical Liu and Lü systems by active nonlinear control. Liu system (Liu *et al.* 2004) and Lü system (Lü and Chen, 2002) are important models of three-dimensional chaotic systems. Hybrid synchronization of the three-dimensional chaotic systems considered in this paper are achieved through the synchronization of the first and last pairs of states and anti-synchronization of the middle pairs of the two systems. Sufficient conditions for hybrid synchronization of identical Liu, identical Lü, and non-identical Liu and Lü systems are derived using active nonlinear control and Lyapunov stability theory. Since the Lyapunov exponents are not needed for these calculations, the active nonlinear control is an effective and convenient method for the hybrid synchronization of the chaotic systems addressed in this paper. Numerical simulations are shown to illustrate the effectiveness of the proposed synchronization schemes.

**Keywords:** Chaos, hybrid synchronization, Liu system, Lü system, nonlinear control.

## 1 Introduction

Chaotic systems are dynamical systems that are highly sensitive to initial conditions. This sensitivity is popularly referred to as the *butterfly effect* [1].

Chaos is an interesting nonlinear phenomenon and has been extensively and intensively studied in the last two decades ([1]-[17]). In 1990, Pecora and Carroll [2] introduced a method to synchronize two identical chaotic systems and showed that it was possible for some chaotic systems to be completely synchronized. From then on, chaos synchronization has been widely explored in a variety of fields including physical [3], chemical [4], ecological [5] systems, secure communications ([6]-[7]) etc.

In most of the chaos synchronization approaches, the master-slave or drive-response formalism is used. If a particular chaotic system is called the *master* or *drive system* and another chaotic system is called the *slave* or *response system*, then the idea of synchronization is to use the output of the master system to control the slave system so that the output of the slave system tracks the output of the master system asymptotically.

D. Nagamalai, E. Renault, M. Dhanushkodi (Eds.): CCSEIT 2011, CCIS 204, pp. 1–10, 2011.

Since the seminal work by Pecora and Carroll [2], a variety of impressive approaches have been proposed for the synchronization of the chaotic systems such as the PC method [2], sampled-data feedback synchronization method [8], OGY method [9], time-delay feedback method [10], backstepping method [11], adaptive design method [12], sliding mode control method [13], etc.

So far, many types of synchronization phenomenon have been studied such as complete synchronization [2], phase synchronization [5], generalized synchronization [7], anti-synchronization [14], hybrid synchronization [15] etc.

Complete synchronization (CS) is characterized by the equality of state variables evolving in time, while anti-synchronization (AS) is characterized by the disappearance of the sum of relevant state variables evolving in time.

In hybrid synchronization of chaotic systems [15], one part of the systems is synchronized and the other part is anti-synchronized so that complete synchronization (CS) and anti-synchronization (AS) co-exist in the systems. The co-existence of CS and AS is very useful in secure communication and chaotic encryption schemes.

This paper is organized as follows. In Section 2, we derive results for the hybrid synchronization of identical Liu systems ([16], 2004). In Section 3, we derive results for the hybrid synchronization of two identical Lü systems ([17], 2002). In Section 4, we derive results for the hybrid synchronization of non-identical Liu and Lü systems. In Section 5, we present the conclusions of this paper.

## 2 Hybrid Synchronization of Identical Liu Systems

In this section, we consider the hybrid synchronization of identical Liu systems [16]. Thus, we consider the *master* system as the Liu dynamics, which is described by the dynamics

$$\begin{aligned} \dot{x}_1 &= a(x_2 - x_1) \\ \dot{x}_2 &= bx_1 - x_1 x_3 \\ \dot{x}_3 &= -cx_3 + dx_1^2 \end{aligned} \tag{1}$$

where $x_i (i = 1, 2, 3)$ are the *state* variables and $a, b, c, d$ are positive parameters.

We consider the slave system as the controlled Liu system, which is described by the dynamics

$$\begin{aligned} \dot{y}_1 &= a(y_2 - y_1) + u_1 \\ \dot{y}_2 &= by_1 - y_1 y_3 + u_2 \\ \dot{y}_3 &= -cy_3 + dy_1^2 + u_3 \end{aligned} \tag{2}$$

where $y_i (i = 1, 2, 3)$ are the states of the system and $u_i (i = 1, 2, 3)$ are the active controls.

The Liu system ([16], 2004) is one of the paradigms of the three dimensional chaotic systems and this system is chaotic when the parameters take the values

$$a = 10, \quad b = 40, \quad c = 2.5 \quad \text{and} \quad d = 4.$$

Figure 1 illustrates the chaotic portrait of the Liu system (1).

For the hybrid synchronization of the identical Liu systems (1) and (2), the *errors* are defined as

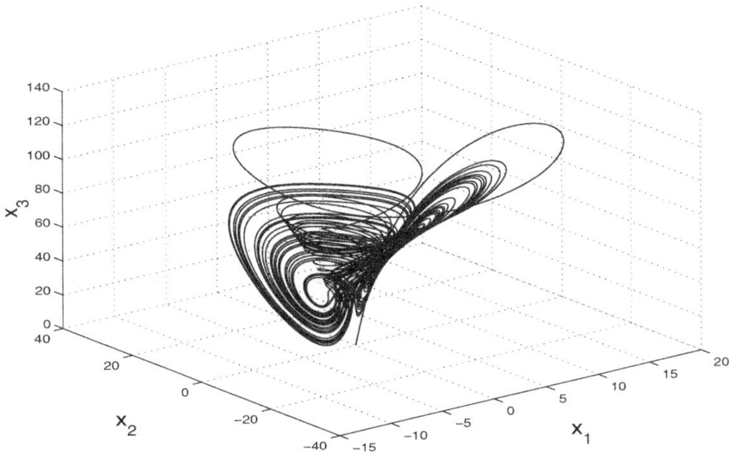


**Fig. 1.** State Orbits of the Liu Chaotic System

$$\begin{aligned} e_1 &= y_1 - x_1 \\ e_2 &= y_2 + x_2 \\ e_3 &= y_3 - x_3 \end{aligned} \tag{3}$$

A simple calculation yields the error dynamics as

$$\begin{aligned} \dot{e}_1 &= a(e_2 - e_1) - 2ax_2 + u_1 \\ \dot{e}_2 &= be_1 + 2bx_1 - (y_1y_3 + x_1x_3) + u_2 \\ \dot{e}_3 &= -ce_3 + d(y_1^2 - x_1^2) + u_3 \end{aligned} \tag{4}$$

We consider the nonlinear controller defined by

$$\begin{aligned} u_1 &= -(a+b)e_2 + 2ax_2 \\ u_2 &= -e_2 - 2bx_1 + y_1y_3 + x_1x_3 \\ u_3 &= -d(y_1^2 - x_1^2) \end{aligned} \tag{5}$$

Substitution of (5) into (4) yields the linear error dynamics

$$\begin{aligned} \dot{e}_1 &= -ae_1 - be_2 \\ \dot{e}_2 &= be_1 - e_2 \\ \dot{e}_3 &= -ce_3 \end{aligned} \tag{6}$$

We consider the candidate Lyapunov function defined by

$$V(e) = \frac{1}{2}\, e^T e = \frac{1}{2}\, \left(e_1^2 + e_2^2 + e_3^2\right) \tag{7}$$

Differentiating (7) along the trajectories of the system (6), we get

$$\dot{V}(e) = -ae_1^2 - e_2^2 - ce_3^2 \tag{8}$$

which is a negative definite function on $\mathbb{R}^3$, since $a$ and $c$ are positive constants.

Thus, by Lyapunov stability theory [18], the error dynamics (6) is globally exponentially stable. Hence, we obtain the following result.

**Theorem 1.** *The identical hyperchaotic Liu systems (1) and (2) are globally and exponentially hybrid synchronized with the active nonlinear controller (5).* ■

**Numerical Simulations.** For the numerical simulations, the fourth order Runge-Kutta method with initial step $h = 10^{-6}$ is used to solve the two systems of differential equations (1) and (2) with the nonlinear controller (5).

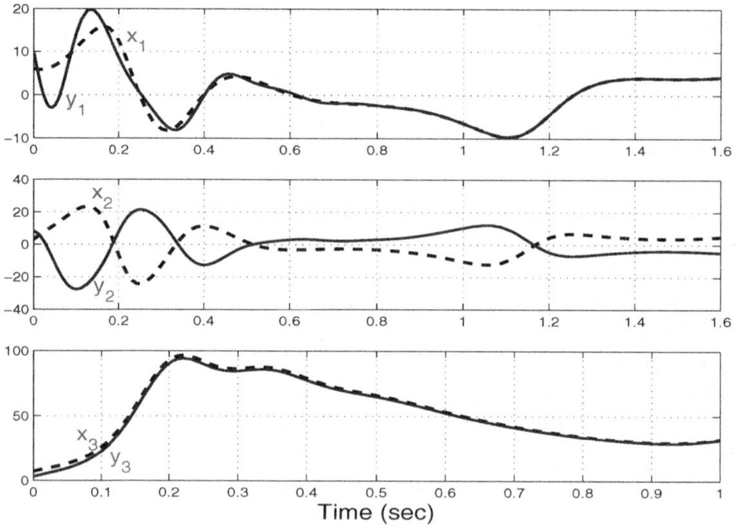


**Fig. 2.** Hybrid Synchronization of Identical Liu Systems

The parameters of the identical Liu systems (1) and (2) are taken as those which result in the chaotic behaviour of the systems, *viz.* $a = 10,\ b = 40,\ c = 2.5$ and $d = 4$.

The initial values for the master system (1) are taken as

$$x_1(0) = 6,\ x_2(0) = 3,\ x_3(0) = 7$$

and the initial values for the slave system (2) are taken as

$$y_1(0) = 10,\ y_2(0) = 8,\ y_3(0) = 3$$

Figure 2 depicts the hybrid synchronization of the identical Liu systems (1) and (2).

## 3 Hybrid Synchronization of Identical Lü Systems

In this section, we consider the hybrid synchronization of identical Lü systems [17]. Thus, we consider the *master* system as the Lü dynamics, which is described by the dynamics

$$\begin{aligned} \dot{x}_1 &= \alpha(x_2 - x_1) \\ \dot{x}_2 &= \gamma x_2 - x_1 x_3 \\ \dot{x}_3 &= -\beta x_3 + x_1 x_2 \end{aligned} \tag{9}$$

where $x_i (i = 1, 2, 3)$ are the *state* variables and $\alpha, \beta, \gamma$ are positive constants.

We consider the slave system as the controlled Lü system, which is described by the dynamics

$$\begin{aligned} \dot{y}_1 &= \alpha(y_2 - y_1) + u_1 \\ \dot{y}_2 &= \gamma y_2 - y_1 y_3 + u_2 \\ \dot{y}_3 &= -\beta y_3 + y_1 y_2 + u_3 \end{aligned} \tag{10}$$

where $y_i (i = 1, 2, 3)$ are the states of the system and $u_i (i = 1, 2, 3)$ are the active controls.

The Lü system ([17], 2002) is one of the paradigms of the three dimensional chaotic systems and this system is chaotic when the parameters take the values

$$\alpha = 36, \quad \beta = 3 \quad \text{and} \quad \gamma = 20.$$

Figure 3 illustrates the chaotic portrait of the Lü system (9).

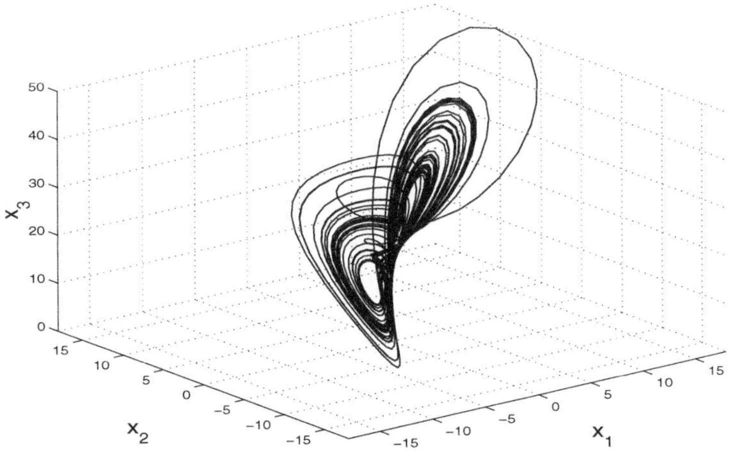


**Fig. 3.** State Orbits of the Lü Chaotic System

For the hybrid synchronization of the identical Lü systems (9) and (10), the *errors* are defined as

$$\begin{aligned} e_1 &= y_1 - x_1 \\ e_2 &= y_2 + x_2 \\ e_3 &= y_3 - x_3 \end{aligned} \tag{11}$$

A simple calculation yields the error dynamics as

$$\begin{aligned} \dot{e}_1 &= \alpha(e_2 - e_1) - 2\alpha x_2 + u_1 \\ \dot{e}_2 &= \gamma e_2 - (y_1 y_3 + x_1 x_3) + u_2 \\ \dot{e}_3 &= -\beta e_3 + y_1 y_2 - x_1 x_2 + u_3 \end{aligned} \tag{12}$$

We consider the nonlinear controller defined by

$$\begin{aligned} u_1 &= -\alpha e_2 + 2\alpha x_2 \\ u_2 &= -(\gamma + 1) e_2 + y_1 y_3 + x_1 x_3 \\ u_3 &= -y_1 y_2 + x_1 x_2 \end{aligned} \tag{13}$$

Substitution of (13) into (12) yields the linear error dynamics

$$\begin{aligned} \dot{e}_1 &= -\alpha e_1 \\ \dot{e}_2 &= -e_2 \\ \dot{e}_3 &= -\beta e_3 \end{aligned} \tag{14}$$

We consider the candidate Lyapunov function defined by

$$V(e) = \frac{1}{2} e^T e = \frac{1}{2} \left( e_1^2 + e_2^2 + e_3^2 \right) \tag{15}$$

Differentiating (15) along the trajectories of the system (14), we get

$$\dot{V}(e) = -\alpha e_1^2 - e_2^2 - \beta e_3^2 \tag{16}$$

which is a negative definite function on $\mathbb{R}^3$, since $\alpha$ and $\beta$ are positive constants.

Thus, by Lyapunov stability theory [18], the error dynamics (14) is globally exponentially stable. Hence, we obtain the following result.

**Theorem 2.** *The identical Lü systems (9) and (10) are globally and exponentially hybrid synchronized with the active nonlinear controller (13).* ∎

**Numerical Simulations.** For the numerical simulations, the fourth order Runge-Kutta method with initial step $h = 10^{-6}$ is used to solve the two systems of differential equations (9) and (10) with the nonlinear controller (13).

The parameters of the identical Lü systems (9) and (10) are taken as those which result in the chaotic behaviour of the systems, *viz.* $\alpha = 36$, $\beta = 3$ and $\gamma = 20$.

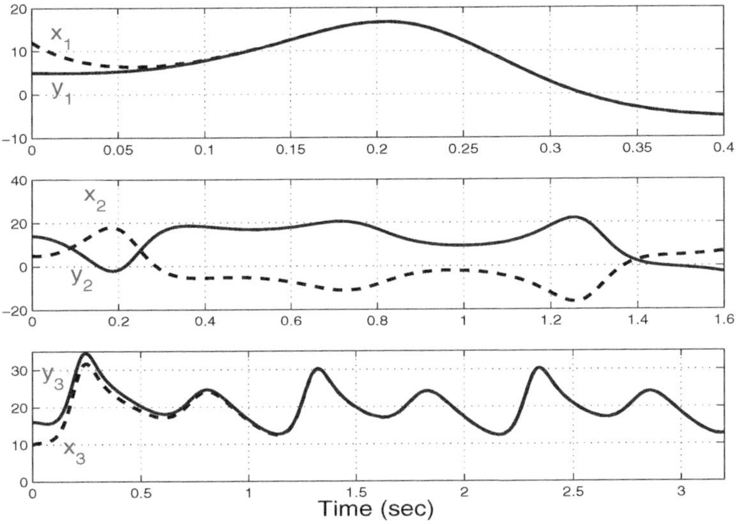


**Fig. 4.** Hybrid Synchronization of Identical Lü Systems

The initial values for the master system (9) are taken as

$$x_1(0) = 12, \; x_2(0) = 5, \; x_3(0) = 10$$

and the initial values for the slave system (10) are taken as

$$y_1(0) = 5, \; y_2(0) = 14, \; y_3(0) = 16$$

Figure 4 depicts the hybrid synchronization of the identical Lü systems (9) and (10).

# 4 Hybrid Synchronization of Liu and Lü Systems

In this section, we consider the hybrid synchronization of the non-identical Liu and Lü systems.

Thus, we consider the master system as the Lü system (2002), which is described by the dynamics

$$\begin{aligned} \dot{x}_1 &= \alpha(x_2 - x_1) \\ \dot{x}_2 &= \gamma x_2 - x_1 x_3 \\ \dot{x}_3 &= -\beta x_3 + x_1 x_2 \end{aligned} \tag{17}$$

where $x_i (i = 1, 2, 3)$ are the states of the system and $\alpha, \beta, \gamma$ are positive parameters of the system.

Next, we consider the slave system as the controlled Liu system (2004), which is described by the dynamics

$$\begin{aligned} \dot{y}_1 &= a(y_2 - y_1) + u_1 \\ \dot{y}_2 &= by_1 - y_1 y_3 + u_2 \\ \dot{y}_3 &= -cy_3 + dy_1^2 + u_3 \end{aligned} \tag{18}$$

where $y_i (i = 1, 2, 3)$ are the states of the system and $u_i (i = 1, 2, 3)$ are the active controls.

For the hybrid synchronization of the Lü system (17) and Liu system (18), the *errors* are defined as

$$\begin{aligned} e_1 &= y_1 - x_1 \\ e_2 &= y_2 + x_2 \\ e_3 &= y_3 - x_3 \end{aligned} \tag{19}$$

A simple calculation yields the error dynamics as

$$\begin{aligned} \dot{e}_1 &= a(e_2 - e_1) - (a - \alpha)x_1 - (a + \alpha)x_2 + u_1 \\ \dot{e}_2 &= be_1 + bx_1 + \gamma x_2 - (y_1 y_3 + x_1 x_3) + u_2 \\ \dot{e}_3 &= -ce_3 + (\beta - c)x_3 - x_1 x_2 + dy_1^2 + u_3 \end{aligned} \tag{20}$$

We consider the nonlinear controller defined by

$$\begin{aligned} u_1 &= -(a + b)e_2 + (a - \alpha)x_1 + (a + \alpha)x_2 \\ u_2 &= -e_2 - bx_1 - \gamma x_2 + y_1 y_3 + x_1 x_3 \\ u_3 &= -(\beta - c)x_3 + x_1 x_2 - dy_1^2 \end{aligned} \tag{21}$$

Substitution of (21) into (20) yields the linear error dynamics

$$\begin{aligned} \dot{e}_1 &= -ae_1 - be_2 \\ \dot{e}_2 &= be_1 - e_2 \\ \dot{e}_3 &= -ce_3 \end{aligned} \tag{22}$$

We consider the candidate Lyapunov function defined by

$$V(e) = \frac{1}{2} e^T e = \frac{1}{2} \left( e_1^2 + e_2^2 + e_3^2 \right) \tag{23}$$

Differentiating (23) along the trajectories of the system (22), we get

$$\dot{V}(e) = -ae_1^2 - e_2^2 - ce_3^2 \tag{24}$$

which is a negative definite function on $\mathbb{R}^3$, since $a$ and $c$ are positive constants.

Thus, by Lyapunov stability theory [18], the error dynamics (22) is globally exponentially stable. Hence, we obtain the following result.

**Theorem 3.** *The Lü system (17) and the Liu system (18) are globally and exponentially hybrid synchronized with the active nonlinear controller (21).* ■

**Numerical Simulations.** For the numerical simulations, the fourth order Runge-Kutta method with initial step $h = 10^{-6}$ is used to solve the two systems of differential equations (17) and (18) with the nonlinear controller (21).

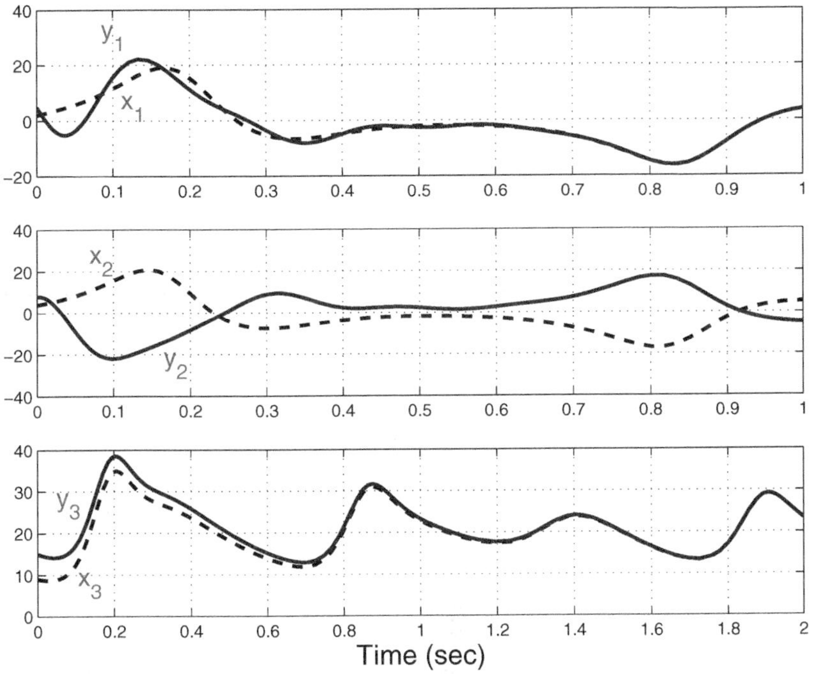


**Fig. 5.** Hybrid Synchronization of the Liu and Lü Systems

For the Lü and Liu systems, the parameter values are taken so that the systems are chaotic. Hence, we take

$$a = 10,\ b = 40,\ c = 2.5,\ d = 4,\ \alpha = 36,\ \beta = 3 \text{ and } \gamma = 20.$$

The initial values for the master system (17) are taken as

$$x_1(0) = 2,\ x_2(0) = 4,\ x_3(0) = 9$$

and the initial values for the slave system (18) are taken as

$$y_1(0) = 5,\ y_2(0) = 8,\ y_3(0) = 15$$

Figure 5 depicts the hybrid synchronization of the non-identical Lü system (17) and Liu system (18).

## 5 Conclusions

In this paper, nonlinear control method based on Lyapunov stability theory has been deployed to globally and exponentially hybrid synchronize identical Liu systems (2004), identical Lü systems (2002) and non-identical Liu and Lü systems. Numerical simulations are also shown to illustrate the effectiveness of the proposed hybrid synchronization schemes. Since Lyapunov exponents are not required for these calculations, the proposed active control method is effective and convenient to achieve global hybrid chaos synchronization of the chaotic systems addressed in this paper.

## References


1. Alligood, K.T., Sauer, T., Yorke, J.A.: Chaos: An Introduction to Dynamical Systems. Springer, New York (1997)
2. Pecora, L.M., Carroll, T.L.: Synchronization in chaotic systems. Phys. Rev. Lett. 64, 821–824 (1990)
3. Lakshmanan, M., Murali, K.: Chaos in Nonlinear Oscillators: Controlling and Synchronization. World Scientific, Singapore (1996)
4. Han, S.K., Kerrer, C., Kuramoto, Y.: Dephasing and bursting in coupled neural oscillators. Phys. Rev. Lett. 75, 3190–3193 (1995)
5. Blasius, B., Huppert, A., Stone, L.: Complex dynamics and phase synchronization in spatially extended ecological system. Nature 399, 354–359 (1999)
6. Feki, M.: An adaptive chaos synchronization scheme applied to secure communication. Chaos, Solit. Fract. 18, 141–148 (2003)
7. Murali, K., Lakshmanan, M.: Secure communication using a compound signal from generalized synchronizable chaotic systems. Phys. Rev. Lett. A 241, 303–310 (1998)
8. Yang, T., Chua, L.O.: Control of chaos using sampled-data feedback control. Internat. J. Bifurcat. Chaos. 9, 215–219 (1999)
9. Ott, E., Grebogi, C., Yorke, J.A.: Controlling chaos. Phys. Rev. Lett. 64, 1196–1199 (1990)
10. Park, J.H., Kwon, O.M.: A novel criterion for delayed feedback control of time-delay chaotic systems. Chaos, Solit. Fract. 17, 709–716 (2003)
11. Yu, Y.G., Zhang, S.C.: Adaptive backstepping synchronization of uncertain chaotic systems. Chaos, Solit. Fract. 27, 1369–1375 (2006)
12. Liao, T.L., Tsai, S.H.: Adaptive synchronization of chaotic systems and its applications to secure communications. Chaos, Solit. Fract. 11, 1387–1396 (2000)
13. Konishi, K., Hirai, M., Kokame, H.: Sliding mode control for a class of chaotic systems. Phys. Lett. A. 245, 511–517 (1998)
14. Jian, X.: Anti-synchronization of Lienard chaotic systems via feedback control. Internat. J. Phys. Sci. 5, 2757–2761 (2010)
15. Sudheer, K.S., Sabir, M.: Hybrid synchronization of hyperchaotic Lü system. Pramana 73, 1387–1396 (2000)
16. Liu, C., Liu, T., Liu, L., Liu, K.: A new chaotic attractor. Chaos, Solit. Fract. 22, 1031–1038 (2004)
17. Lü, J., Chen, G.: A new chaotic attractor coined. Internat. J. Bifur. Chaos. 12, 659–661 (2002)
18. Hahn, W.: The Stability of Motion. Springer, New York (1967)

# Methods for Preventing Search Engine-Based Web Server Attacks

Keerthiram Murugesan and Mukesh Singhal

Department of Computer Science,
University of Kentucky,
Lexington, Kentucky - 40508, USA

**Abstract.** Many Internet users make use of search engines to locate a web page or information they want. These search engines have become a fundamental tool in the World Wide Web. Since the capability of search engines to reach any part of the web has increased, security loopholes of any websites face the risk of being exposed globally. This poses serious security threats on the web. In this paper, we discuss various methods to prevent web server attacks through search engines.

**Keywords:** Search engine security, information security.

## 1 Introduction

The number of websites in the Internet is growing at a fast rate. This increase in number of websites forces Internet users to turn to search engines to find specific information. A user can search for any content available in the Internet with the help of search engines.

Due to the power of search engines to reach any part of the web, vulnerabilities in a website is accessible to anyone who actively searches for compromised websites. These vulnerabilities occur due to known and/or unknown software bugs, improper or insufficient security measures, etc. in the websites. If a website is unaware of these vulnerabilities, search engine might accidently reveal these information in their search results. Attackers can now use search engines as a tool for identifying websites with potential vulnerabilities and search for the queries that expose these vulnerabilities in search results without revealing the attackers identity. This results in certain users having an unauthorized access to restricted information, posing a serious threat to web security. There are several cases reported on web server attacks using search engine as a tool [1,2].

The rest of the paper is organized as follows: The next section provides a background on search engine architecture. In Section 3, we examine the usage of malicious queries in search engines. Section 4 presents web server attacks that have been mounted successfully using search engines. We propose several methods in Section 5 to prevent web server attacks through search engines.

D. Nagamalai, E. Renault, M. Dhanushkodi (Eds.): CCSEIT 2011, CCIS 204, pp. 11–18, 2011.

## 2 General Search Engine Architecture

In this section, we provide a brief description of general search engine architecture [3]. We focus on three main components of search engine architecture; namely, crawler, indexer and query engine that help us to understand various security attacks through search engine. Figure 1 shows the general search architecture with all the main components.

The crawler is an important component of the search engine architecture and runs with a basic set of URLs to start with. There are one or more crawlers run concurrently to maximize the searching for information. Crawlers hop from one web page to another in the Internet in search of any useful information.

When a crawler visits a web page, it extracts all the URLs in the web page and gives them to the crawler controller. The crawler controller organizes the URLs received from each crawlers based on the feedback information received from the indexer and query engine components. The web page is then added to the queue, for further processing, by the Indexer component. Optionally, a copy of the web page is stored in the page repository for future uses.

The indexer uses a parser to extract all the important keywords or terms from each page in the queue. These keywords are stored along with their link information in the indexes using the index controller. The index controller links these indexed web pages together based on their contents. These indexes, which may be distributed, form the basic data structure of the search engine and can be used for further processing like classification, text summarization, etc.

When a user sends a query to the query engine, it extracts all the important terms from the query using the query parser. This query parser is similar in functionality to the index parser. The parsed query is then sent to the index controller, which uses the indexes to generate the list of URLs that are relevant to the user query. The query engine may use a specific ranking algorithm (PageRank, HITS, etc.) to sort the retrieved list.

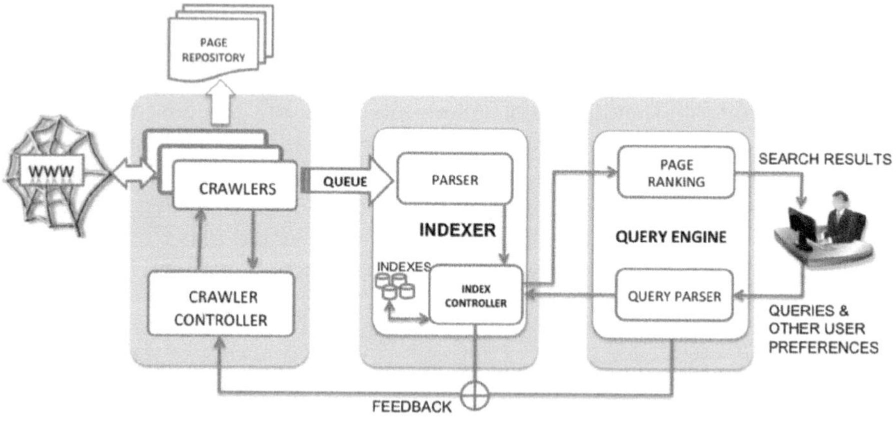


**Fig. 1.** General Search Engine Architecture

As mentioned earlier, the indexer and query engine components generate feedback information such as usage patterns, index last modified details and ranking and importance of a web page, with respect to the search engine model and users. These feedbacks are used by the crawler controller for computing refresh policies for each web page in the Internet. Refresh policies help the crawlers to identify the next important web page to crawl and the frequency with which they need to fetch each web page in the Internet.

Although there are many other supporting modules in a search engine architecture, these three components determine the search result returned for a user-supplied query. In short, crawlers work together to extract information about a web page, whereas the indexer and query engine (along with ranking module) work together to provide list of URLs relevant to a user query.

There are two main metrics (cost functions) that determine the efficiency and quality of search engines:

- **Freshness:** This metric tells whether the local copy is accurate or not.
- **Age:** This metric tells how outdated the local copy is.

A good search engine should maintain the freshness of a collection of web pages as high as possible and the age as low as possible. Readers are referred to [3] for the complete description of the search engine architecture.

## 3 Malicious Queries

A search for vulnerabilities is easier if you know what you are looking for. An attacker can search queries for website vulnerabilities that expose this information in the search results, without revealing the attackers identity.

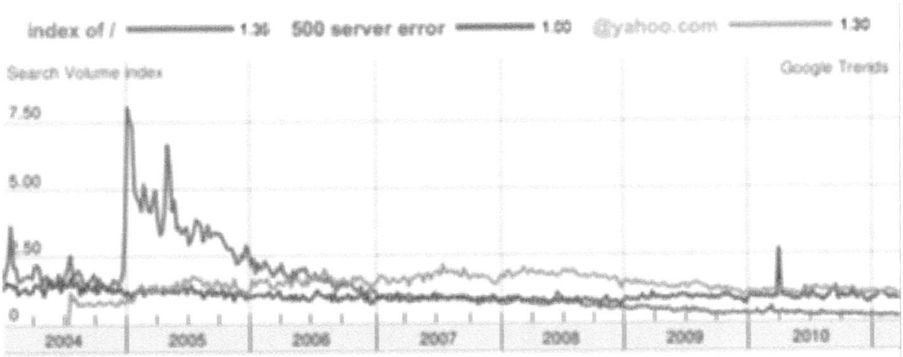


**Fig. 2.** Search engine query statistics for queries such as *index of/*, *500 server error* and *@yahoo.com* from the Google search engine since 2004 (Google Trends [4]). The data is scaled based on the average search traffic of the given terms over the period (Search Volume Index).

For example, searching *index of /* can show the list of websites with a directory listing enabled. Sometimes, it could be easy to find sensitive files or even system password files like */etc/passwd*, *.htpasswd*, *sam* password files, etc. with this search, if the websites have accidently enabled a directory listing. A search for the query *500 server error* shows a list of vulnerable websites that can provide useful information like operating system version, port numbers, sql queries, source code, etc. for an attacker. Similarly, a query *@yahoo.com* reveals a list of yahoo mail users, which can be used by a spammer for phishing or advertising. Figure 2 shows the Google search engine statistics for the above queries.

## 4 Search Engine-Based Web Server Attacks

In this section, we will discuss about web server attacks that involves search engines to find vulnerable servers and to propagate its attack.

In July 2004, a variant of *MYDOOM* worm used SMTP mail service to spread and steal email addresses from the address book of target machines [5]. When an email address (*mail@example.com*) is found, the worm uses the domain name (*example.com*) as a query in search engines such as Google, Yahoo, Lycos, and AltaVista to find email addresses of that domain in the search result and propagates further.

In December 2004, PhpBB Bulletin Board System, an open source php-based web forum application, has been reported with a remote PHP script injection and input validation vulnerabilities due to an improper use of *urldecode()* on user-supplied input for *highlight* parameter in *viewtopic.php* phpBB script file [6]. The *urldecode()* function, which is used to filter potentially malicious content from user-supplied input, failed to properly translate input. Using these vulnerabilities, a remote attacker could have executed arbitrary commands on a vulnerable server without sufficient privileges, which could have been exploited to trigger SQL injection attacks.

After this issue was posted on the Internet, a *Santy.A* worm was created to exploit these vulnerabilities using the Google search engine as a tool to search for the keyword *viewtopic.php* in order to find the potentially vulnerable servers [7]. When a victim site was found, the worm would attack the system by exploiting the *highlight* parameter vulnerability in the phpBB software. This worm would execute itself by using the PERL environment of the compromised server and propagate further. When Google stopped the worm from using its search engine for finding the vulnerable sites, variants of the worm (*Santy.B*) started using AOL and Yahoo search engines to locate the vulnerable websites [8]. Thus, if one of the several search engines is able to stop a threat, there are several other search engines that attackers can use to perform their malicious actions.

In May 2008, *Asprox* trojan, a spam botnet for sending phishing emails, with *Danmec*, a password-stealing trojan, searches Google for web servers that use asp, aspx, or php scripts to dynamically generate web pages from SQL databases [9,10]. With the list of URLs from search engine, it launches SQL-Injection attack to inject an iframe into the website source which will redirect visitors of that website to *direct84.com* to download a malicious javascript file. This javascript file may redirects the visitor to another website for downloading additional malicious javascript files which tries one or more exploits to crack the visitor's computer. When successful, it builds a data stream directly into the visitor's file system and installs this trojan on the computer, which in turn repeats the above steps on other set of potential URLs returned by a search engine.

In August 2008, a directory listing vulnerability was reported in TWiki, an open source enterprise wiki application. Due to a failure in filtering the malicious shell meta-characters from the user-supplied string for the TWiki query image parameter of the *open()* function, an attacker could access any files if the *twiki/bin/configure* script was not properly secured and the web server had sufficient permissions to view them. For example, to get the */etc/passwd* file of the vulnerable server, an attacker could go to: `http://www.example.org/twiki/bin/configure?action=image;image=../../../../../../etc/passwd;type=text/plain` [11]. This could also be used to execute arbitrary commands via shell meta-characters. Since the TWiki application uses predictable URLs, the vulnerable servers could be found using the search engines.

Although these attacks were fixed, many web security threats are still exposed through search engines everyday.

## 5 Proposed Methods

In this section, we propose several methods to address the growing problem of web attacks through search engines. The main purpose of this section is to stimulate research ideas in this problem domain.

### 5.1 IN:SITE Crawlers

IN:SITE crawlers or local crawlers are a simple extension to the crawler component shown in Figure 1. This crawler may be specific to a web server or a network of servers that belongs to an organization. Unlike a crawler in Figure 1, this local crawler is deployed in each web server in the Internet and is completely under the control of web administrator. Together they provide a distributed approach to the search engine architecture as shown in Figure 3. The effort needed to generate a powerful but secured search engine are distributed throughout the Internet. Each web server in the Internet now acts as a resource pool, which constantly provides information to the search engine to make it effective and secure.

IN:SITE crawler can crawl all the web pages or publicly accessible contents in a web server or network and send this information to a search engine. This crawler can keep track of last modified information of all publicly accessible web pages of that web server. When a web page is modified, it notifies the crawler controller about the modified content. Crawler controller updates the refresh policy of the web page based on its importance and its freshness and age metrics.

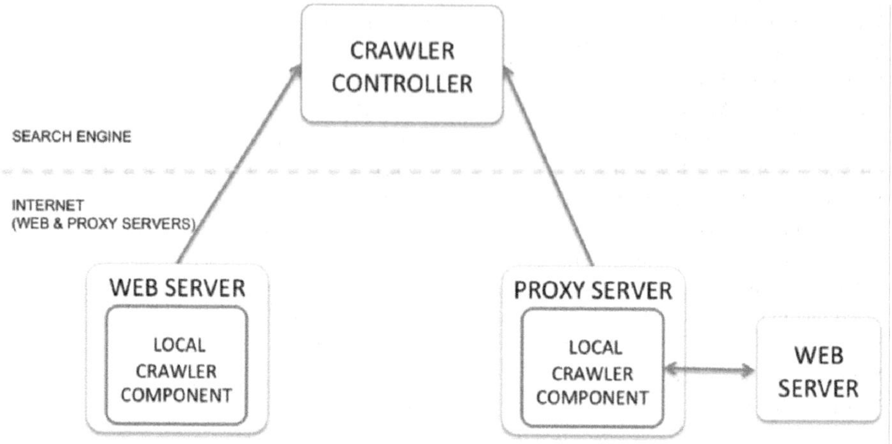


**Fig. 3.** IN:SITE crawler. All the information exchanged between a search engine and web servers are encrypted.

Since a crawler is local to a web server or a network, it may have an access to some restricted information that cannot be accessed from outside their computer or network. For instance, employee details of an organization, which can be accessed within the organization network, may be open to a local crawler of that network. This can be avoided by placing this local crawler in a proxy server of that network or web server or by providing a set of judgments or rules on what to access, to a local crawler, specified by the web administrator.

IN:SITE crawler has several advantages:

- These local crawlers reduce the unnecessary check performed by a search engine crawler to identify any recent modifications in a web page since its last crawl. This reduces the bandwidth wastages caused by a search engine in the Internet.
- This increases the freshness metric and considerably reduces the age metric of each web page in the search engine.
- The complete control of information sent to the crawler controller is with the web administrator.
- When sensitive information is accidently released into the Internet, this crawler can be configured to inform a search engine to block an access to a specific web page or an entire web site immediately.

## 5.2 Intelligent Crawlers

Intelligent crawlers measure the amount of security issues exposed in a website. From this knowledge, intelligent crawlers can now avoid indexing vulnerable websites or block an access to those websites from search results to reduce further security risks.

In general, an intelligent crawler performs two steps: fetching a web page from the Internet and risk analysis. The risk analysis step analyzes each web page for any vulnerabilities and sensitive information before adding it to the queue. The risk analysis step uses patterns to identify the sensitive information. These patterns are based on simple regular expressions. For instance, patterns for SSN and email address can be given as:

```
SSN: ^(?!000)([0-6]\d{2}|7([0-6]\d|7[012]))
([-]?)(?!00)\d\d\3(?!0000)\d{4}$

Email Address: ^([\w\-\.]+)@((\[([0-9]{1,3}
\.){3}[0-9]{1,3}\])|(([\w\-]+\.)+)([a-zA-Z]{2,4}))$
```

In addition, the risk analysis step scans a web page for any known vulnerabilities such as SQL injection, directory listing, etc. using vulnerability signatures [12,13]. If this step finds any security issues in a web page, the search engine will blacklist the web page or reduce the ranking based on the concentration of security issues in a web site.

### 5.3 Feedback Information

Search engines can use feedback information to improve search results returned by the query engine. In addition, this can be used to identify vulnerable websites to avoid listing them in a search result [14].

Explicit feedback is the feedback information given by a user to search engine about vulnerabilities in a web site and update refresh policies or rank web pages. These feedbacks are based on ratings given by users for a set of web pages. Since these feedbacks are based on the users judgment, it may have false positive or false negative errors. Modules in query engine and indexer components should analyze this feedback information again to reduce these errors. These feedbacks are more effective if security experts rate web pages listed in search results for popular queries. In order to ensure the security of the web, search engines can use these feedbacks to avoid listing vulnerable websites or reduce their ranking. This feedback mechanism can be used by search engine to prevent web attacks through search engine in the long run.

### 5.4 Query Filters

When a user sends a query to the query engine, the query filter checks for any malicious content in the user-supplied query using known regular expressions, keywords and/or phrases. In addition, malicious queries can be identified based on the percentage of blacklisted web pages listed in search results returned for a user-supplied query.

## 6 Conclusion

Search engines can be used as a tool in locating vulnerable servers in the Internet anonymously. Vulnerabilities, such as SQL injection, Cross-Site Scripting, etc. in a website can be found easily if an attacker knows what to look for. In this paper, we discussed several methods to address the problem of search engine-based web server attacks. These proposed methods can only prevent an attacker from using a search engine to locate a vulnerable server. A web administrator should take certain precautionary measures such as keeping the server up-to-date, blocking unwanted ports and services, using firewall and antivirus programs, etc. to protect a web server completely from these attacks.

## References


1. Hernandez, J.C., Sierra, J.M., Ribagorda, A., Ramos, B.: Search engines as a security threat. Computer 34, 25–30 (2001)
2. Newman, A.C.: Search Engines used to attack databases. White Paper, Application Security, Inc.
3. Arasu, A., Cho, J., Molina, H.G., Paepcke, A., Raghavan, S.: Searching the Web. ACM Trans. Internet Technol. 1, 2–43 (2001)
4. Google Trends, https://www.google.com/trends
5. Symantec Threats and Risks, http://www.symantec.com/security_response/writeup.jsp?docid=2004-072615-3527-99
6. United States Computer Emergency Readiness Team, http://www.us-cert.gov/cas/techalerts/TA04-356A.html
7. McAfee Virus Profile and Definition, http://home.mcafee.com/VirusInfo/VirusProfile.aspx?key=130471
8. Symantec Threats and Risks, http://www.symantec.com/security_response/writeup.jsp?docid=2004-122610-5904-99
9. Dell SecureWorks,
http://www.secureworks.com/research/threats/danmecasprox
10. Shin, Y., Myers, S., Gupta, M.: A Case Study on Asprox Infection Dynamics. In: Flegel, U., Bruschi, D. (eds.) DIMVA 2009. LNCS, vol. 5587, pp. 1–20. Springer, Heidelberg (2009)
11. TWiki Security Alert,
http://twiki.org/cgi-bin/view/Codev/SecurityAlert-CVE-2008-3195
12. Xie, Y., Yu, F., Achan, K., Panigrahy, R., Hulten, G., Osipkov, I.: Spamming botnets: signatures and characteristics. In: ACM SIGCOMM Conference on Data Communication (SIGCOMM 2008), pp. 171–182. ACM, New York (2008)
13. Brumley, D., Newsome, J., Song, D., Wang, H., Jha, S.: Towards Automatic Generation of Vulnerability-Based Signatures. In: IEEE Symposium on Security and Privacy (SP 2006), pp. 2–16. IEEE Computer Society, Washington, DC (2006)
14. White, R., Jose, J., Ruthven, I.: Comparing explicit and implicit feedback techniques for web retrieval: Trec-10 interactive track report. In: Text Retrieval Conference, TREC (2001)

# Sea Object Detection Using Shape and Hybrid Color Texture Classification

M. Uma Selvi and S. Suresh Kumar

Vivekanadha College of Engineering for Women,
{Mumaselvi,nice.ssk}@gmail.com

**Abstract.** Sea target detection from remote sensing imagery is very important, with a wide array of applications in areas such as fishery management, vessel traffic services, and naval warfare. This paper focuses on the issue of ship detection from spaceborne optical images (SDSOI). Although advantages of synthetic aperture radar (SAR) result in that most of current ship detection approaches are based on SAR images. But disadvantages of SAR still exist. Such as the limited number of SAR sensors, the relatively long revisit cycle, and the relatively lower resolution. To overcome these disadvantages a new classification algorithm using color and texture is introduced for Ship detection. Color information is computationally cheap to learn and process. However in many cases, color alone does not provide enough information for classification. Texture information also can improve classification performance. This algorithm uses both color and texture features. In this approach for the construction of a hybrid color-texture space we are using mutual information and three aspects: 1) Classifies ship candidates 2) The relevant classes are automatically built by the samples' Appearances and 3) Shape and Texture features. Experimental results of SDSOI on a large image set captured by optical sensors from multiple satellites show that our approach is effective in distinguishing between ships and non ships, and obtains a satisfactory ship detection performance.. Feature extraction is done by the co-occurrence matrix with SVM (Support Vectors Machine) as a classifier. Therefore this algorithm may attain a very good classification rate.

**Keywords:** Ship detection from spaceborne optical images, synthetic aperture radar, Support vector machines classifier, Co- Occurrence Matrix, Hybrid Color, Hybrid color-texture.We would like to encourage you to list your keywords in this section.

## 1 Introduction

Ship detection from remote sensing imagery is very important and has a wide array of applications such as fishery management, vessel traffic services, and naval warfare. In particular, in recent years, because of the decrease in fishery resources in the world, ship detection has become much more important for effective and efficient ship monitoring to prohibit illegal fishing activities in time.

D. Nagamalai, E. Renault, M. Dhanushkodi (Eds.): CCSEIT 2011, CCIS 204, pp. 19–31, 2011.

However, ship detection based on SAR has limitations. First, with a limited number of SAR satellites, the revisit cycle is relatively long and, then, cannot meet the needs of the application of real-time ship monitoring. Second, the resolution of most satellite SAR images is often not high enough to extract detailed ship information.

Ship detection based on satellite optical images can partly overcome the aforementioned shortcomings of ship detection based on SAR and is complementary to SAR-based ship detection. Thus, it is advantageous to investigate SDSOI to better satisfy the requirements of ship monitoring. There are not many researches about target detection in sea through direct image processing and among them papers that utilize color features are even fewer. For this reason almost all of the existing approaches suffer from one or more drawbacks like sensitivity to different sensors or invalidity for low SNR images . Many papers have worked on ship detection based on SAR2 images [5]. A new method based on combinatorial improved PNN3 model for ship detection in SAR imagery and [4] proposes a method to reduce speckle noise for SAR images and to improve the detected ratio for SAR ship targets from the SAR imaging mechanism. A specific technique for automatic spot detection, based on the Wavelet Transform is presented and justified in [10]. Marivi et al [11] proposed a ship target detection method of SAR images by means of the discrete wavelet transform, taking advantage of the difference of statistical behavior of the vessels and the sea. All above methods are dependent on SAR images; they expend largely and can only obtain target points, which cannot be used to recognize targets. There are also some papers that use remote sensing images for ship detection [6]. They present a method based on cumulative projection curve (CPC) to estimate the number of ships of small size, which is only efficient on especial images from stationary ships along coastline in a harbor. One of the few researches that uses color feature, from Lab color coordinate system, for sea target detection is [9]. They presented a definition on the degree of overlap between two clusters and developed an algorithm for calculating the overlap rate. Using this theory, they also developed a new hierarchical cluster merging algorithm for image segmentation and applied it to the ship detection in high resolution images is one of the several papers that worked on IR images. They used PCA, Bayes classification and wavelet-denoising to classify the sea targets, but in several papers, limitations and disadvantages of the methods based on statistical analysis are pointed out [8]. One of the papers that have more superiority to previous works in visible images domain is [7]. Their work is based on calculating different chaos by obtaining largest Lyapunov exponent of target and sea background which is not appropriate for images that contain seaside or some low chaos objects. Also the authors have proposed [8] based on the natural measure feature. Although the results of the method are considerable for some images but it still suffers from previously mentioned imperfection and needs analyzing several frames for exact results. So it is still important to find new methods of detecting target from sea background. Except [9] all above researches have worked on grey level images and for sure our method is one of the few one that utilizes color feature for sea target detection.

We want to further investigate how we can resolve issues in SDSOI. SDSOI includes two stages: 1) sea detection and 2) ship detection in sea. This paper mainly focuses on how we can detect a ship in sea, assuming that sea regions have been detected by prior geographic information or algorithms. In addition, in this paper, we are only interested in ship detection on a panchromatic band or on a band of a multispectral image and not on multispectral images.In SDSOI, several factors such as clouds, ocean waves, and small islands are often detected as false ship candidates due to similar characteristics, which affect the performance of ship detection.

In this paper, a novel hierarchical complete and operational SDSOI approach with multiple features.

*Stage 1–Extraction of Ship Candidates:* This stage includes image segmentation and simple shape analysis. On the one hand, image segmentation with global and local information, including the characteristics of gray and edge features, is applied to obtain possible ship candidate regions. On the other hand, because ships are generally thin and long, simple shape features are extracted to eliminate obvious false candidates. In this stage, the main aim is to extract ship candidates with missing alarms as low as possible, and false alarms that are due to objects such as small clouds and islands may be extracted as ship candidates. *Stage 2–Classification of Ship Candidates:* In this stage, a novel semisupervised hierarchical classification of ship candidates based on shape and texture features is further conducted to remove most false alarms and obtain the results of ship detection.

Some differences between the feature distributions of ships and nonships can be noticed; therefore, we can remove nonship candidates and detect ships by pattern classification. Furthermore, a significant difference usually exists between the feature distributions of different samples within one class, which often has a negative influence on the classification performance. Thus, all candidates are divided into several subclasses according to their appearances, and the subclasses are divided into miniclasses by feature clustering (so-called semisupervised) to improve the performance. In addition, a hierarchical ensemble classification approach based on the miniclass distributions, rather than on the directed extracted features, is presented for candidate recognition. Moreover, in feature extraction, besides commonly used shape and gray distribution features, a new texture operator, local multiple patterns (LMP), for enhancing the representation ability of the feature set is first introduced. The basic idea of LMP is that it extends binary patterns to multiple patterns. In addition, the LMP preserves more structural information and is more suitable for image analysis than local binary patterns (LBP).

## 2 Learning and Extraction of Ship Candidates

### 2.1 Image Segmentation

In spaceborne optical images, a stationary gray distribution with a slow variation in the sea region usually exists, whereas usually, an obvious edge between the ship and sea due to a high ship hull is observed. Thus, image segmentation with edge information can be applied to extract possible candidate regions, which may include many false alarms such as islands, clouds, and ocean waves due to their similar edge characteristics with those of ships. On the other hand, there is usually a strong wake round steering ship, and it is sometimes difficult to distinguish the ship and the wake in optical images with low resolutions. When the ship and its wake are segmented into one region, the area is expanded, which often helps us find the small ships. Thus, they are considered a whole part in the following processing. In the experiments that involve extraction of ship candidates, only the original regions of more than ten pixels in size and more than four pixels in length are considered for analysis. The algorithm is described as follows.

**Image preprocessing:** There is sometimes a dark gray distribution of ships, which has a negative effect on ship segmentation. However, there is usually obvious edge information around the ship. To enhance the distinction between the ship and sea to extract the whole ship, a new mixed image is proposed. The new mixed image is computed with the original gray value and edge magnitude as follows. Suppose that $f(i,j)$ is the gray value of a point $(i,j)$ in an image. Then, the Sobel operator is applied to compute the edge magnitude $Mag(i,j)$. The mixed image is defined as

$$Mixf(i,j) = f(i,j) + a \times Mag(i,j) \tag{1}$$

where $a$ is a coefficient that controls the proportion of the edge magnitude to gray value, which was experientially set to equal to 1 in our experiments.

*Coarse image segmentation:* The whole ship region is relatively salient in the mixed image, ship candidates can be obtained by coarse image segmentation of the mixed image with a proper threshold. Here, a simple and effective segmentation method based on the optimal principle of maximum between-class variance and minimum within-class variance [14] is used to compute a global adaptive threshold. Furthermore, a region-filling algorithm [15] is adopted to delete the holes within regions, and morphological open and close operators with a three-pixel-diameter circle element [15] are adopted to eliminate very thin lines and random noise regions, which are perhaps generated by ocean waves for example.

*Refined image segmentation:* Due to the usage of only a global threshold, the coarse segmentation results may have edge localization errors. This processing aims at refining every region's contour with the local characteristics for the following feature extraction. A level set by the Chan–Vese model [16] is adopted to refine image segmentation. Chan and Vese (2001) proposed an algorithm based on the Mumford–Shah model, which can provide an optimal partition of two classes. The simple energy function is given as follows:

$$F(C) = F_o(C) + F_b(C) = \int_{inside(C)} |u(x,y) - c_o|2\, dx\, dy + \int_{outside(C)} |u(x,y) - c_b|2\, dx\, dy \tag{2}$$

where $c_o$ is the mean value inside of the curve $C$, and $c_b$ is the mean value outside of the curve $C$. The minimum energy given by the Chan–Vese model will be an optimal piecewise smooth approximation of the edge. A level set is done on the corresponding subimages that are cut from the original images, with ranges depending on the ranges of segmented regions with adaptive thresholds. The initial curve $C$ is set as the edge of the coarse segmentation.

### 2.2 Simple Shape Analysis

After image segmentation, simple shape analyses can be applied to eliminate obvious false candidates. First, ships have a limited area, length, and width range. According to this constraint, false candidates such as very large or very small islands and clouds can be eliminated with proper thresholds, which should take the wake imaging into account.

Second, ships are commonly long and thin. Thus, the ratio of the length to the width of the region ounding rectangle [15], [17] is larger than a given threshold. According to this condition, obvious false alarms, including islands and clouds with very small ratios, are eliminated. The relatively low threshold aims at keeping us from eliminating ships that may be either amid a curved wake or too small to exactly extract the ratio value. It aims at having a low missing alarm rate. Ship candidates are detected by image segmentation and simple shape analysis, as previously discussed. In our experiments, to reduce the time consumption, the level set is done after the simple shape analysis: only ship candidates need refined image segmentation with level set. Fig. 3 gives the intermediate results of one typical image sample. It is easily concluded that the refined segmentation results with level sets are closer to the true ship edges than those with an adaptive threshold due to the use of local gray characteristics.

## 3 Feature Extraction of Ship Candidates

It is important to extract effective features to distinguish ships from other objects, which mainly comprise clouds and ocean waves. Both shape and appearance attributes are crucial for each object class. In our approach, various features, including shape and texture, are extracted from a ship candidate and are concatenated as a feature vector for classification.

### 3.1 Shape Feature Extraction of Ship Candidates

Binary images, which are obtained from image segmentation, readily provide simple geometric properties such as perimeter and area. In our approach, the following shape descriptors are adopted.

*Compactness:* Compactness is given as [17] Compactness = $(\text{Perimeter})^2$ /Area The most compact region in a Euclidean space is a circle.

*Convexness:* Let $S$ represent a set of contour points obtained from level set and $CH(S)$ [15], [17] be defined as its convex hull. The convexity measure is defined as

$$CM(S) = Area(S)/Area(CH(S))\ [15], [17].$$

*Rectangularity and Eccentricity:* The simplest eccentricity [17] is the ratio of the major to the minor axes of an object approximated by its best fit ellipse. Ships are commonly long and thin; therefore, it can be adopted as a shape feature. Rectangularity [17] is the maximum ratio of region area to the area of a bounding rectangle according to its different directions. It assumes values in the interval (0, 1], with 1 representing a perfectly rectangular region.

*Moment invariants:* Moments are extensively used for shape representation, pattern recognition [18], and image reconstruction, which makes them a very useful feature set to include. Here, we adopt the first seven moment invariants introduced by Hu [19].

### 3.2 Texture Feature Extraction of Ship Candidates

After taking a close look at these candidates that have been expanded to more than 100 pixels, we consider that most candidates are large enough so that some stable features can be computed from the gray distribution. Furthermore, little difference between the gray distribution of a ship and that of a nonship can be observed. Texture analysis can be applied to eliminate false alarms.

In our approach, besides commonly used features such as simple texture features, wavelet-based features, and multiscale Gaussian differential features (MGDFs), a new texture operator, LMP, is introduced to enhance the representation ability of the feature set. Simple texture features include mean, variance, moments, and entropy of gray value . Other features are described briefly as follows.

*Wavelet-Based Features:* Wavelet-based features aim at extracting the information in the frequency domain, and their capability of undergoing texture analysis has been shown [20]. By applying the wavelet transform to the candidate image, a number of subbands are generated. For each subband $I_{(l,s)}$, the following features are calculated [20]:

$$e_1^{(l,q)} = 1/A \sum_j \sum_j \left| I_{i,j}^{(l,q)} \right|$$

$$e_2^{(l,q)} = -1/\log A \sum_j \sum_j \left( \left| I_{i,j}^{(l,q)} \right| / anorm) \log( \left| I_{i,j}^{(l,q)} \right| / anorm \tag{3}$$

$$\text{Where } anorm = \sum_{i,j} \left| I_{i,j}^{(l,q)} \right| )$$

*A* is the area of each subband, *l* is the decomposition level, and *q* represents the subband number within the decomposition level *l*. The feature $e_1^{(l,q)}$ shows the amount of signal energy at a specific resolution, whereas $e_2^{(l,q)}$ shows the nonuniformity of the subband values.

## 4 Learning Colour and Texture Features

### 4.1 Color Feature

Color information is readily available as input from a color camera so no extra processing is required. Color of a pixel is represented as a vector in the Red, Green, Blue (RGB) color space or Hue, Saturation, Value (HSV). Real world surfaces often have more than a single color. They are either white, pink, gray, or black. To learn the color appearance of a real world surface, the histogram method is often used. A normalized histogram H approximates the distribution of colors in an image patch P, each bin in H is calculated by

$$Hi = ni/N \tag{1}$$

where $ni$ is the number of pixels whose color falls into the region defined by bin $i$ and $N$ is the total number of pixels in P. When histogram is used for classification, during the training stage, a set of model histograms of a class is learned. In classification stage, a sample histogram is compared with the model histograms. The sample's class is assigned with the best matched model's class.

### 4.2 Texture Feature

Although there is no formal definition of texture, the texture of a visual surface can be thought of as spatial distribution of intensity/color variations. In the case of co-occurrence matrix, the performance depends on the number of gray levels, displacement sets are used. Texture is used to improve the performance of a color based classifier but with a consideration of computational speed and memory requirement. A good tradeoff between performance improvement and added complexity is desired. For this reason, two texture learning methods were chosen: the Local Binary Pattern (LBP) method and a statistical method using 8 histograms of the intensity difference between two pixels at 2, 4, 6, and 8 pixels apart at horizontal and vertical directions. These two methods are significantly less complex than other texture methods. The LBP contains local intensity pattern of a pixel. The LBP of pixel $n$ with intensity $x$ is defined as:

$$lbp\ (n) = \sum_{i=1...8} 2^{bi(x-x1)} \tag{2}$$

$$b_i(x) = \{1 \text{ if } x \geq 0, 0 \text{ if } x < 0. \tag{3}$$

where $x1...8$ are the intensity values of the 8 pixels neighboring $n$.

## 5 Combining Colour and Texture Features

### 5.1 Partitioning Training Set Using Color Histograms

From initial tests of the color based classifier, it was found that the average performance was relatively good. The Color information is rich enough in most cases. Classification errors arise when the sample images were taken under bad illumination condition when there were targets with very similar color appearance to the path (even to a human eye). It is noted that extracting texture features is several magnitude more complex than extracting color feature. From these observations, algorithm was built that uses color as the primary feature and only consider texture feature in special cases where it is difficult to use color to discriminate. Instead of having one threshold for each model color histogram, two thresholds are learned. One high threshold is used to determine if a sample is matched with a model, one low threshold used to determine if it is not. If the intersection value falls in between these two thresholds, the texture feature is used. The samples that fall in this region are the difficult cases for the color classifier.

### 5.2 Classification Using Color and Texture

During classification stage, the algorithm maps a pixel patch of unknown class label to either targets or non-targets. A patch is labeled as path if it's color histogram is matched with any of the h color histograms saved from the learning stage using the high threshold. If none of the model histograms matches with the sample, the algorithm iterates through the h models again. This time the intersection value between the sample and the model color histogram is compared with the low threshold. If a matched model is found, the sample's texture histogram is compared with the model's texture histograms. A pseudo-code of the algorithm is listed below

for *i* = 1 to h do
if *M(sco,mco,i)* > *high thresholdi* then
*s* is path
return
end if
end for
for *i* = 1 to h do
if *M(sco,mco,i)* > *low thresholdi* AND
    *M(stx,mtx, i)* > *texture thresholdi* then
*s* is path
return
end if
end for
*s* is non path

*sco, stx* are the sample's color and texture histogram, *mco,i* and *mtx,i* are color and texture histogram of the *ith* model.

## 6 Hybrid Color-Texture Space Construction

### 6.1 Color Spaces

The construction of the hybrid color-texture space starts with transforming a classified image, initially represented in the RGB standard system, into different color spaces. Color can be divided into four families [4] namely:

1. The primary spaces which are based on the trichromatic theory assuming that it is possible to match any color by mixing appropriate amounts of three primary colors.
2. The perceptual spaces which try to quantify the subjective human color perception using the intensity, the hue and the saturation.
3. The luminance–chrominance spaces where one component represents the luminance and the two others the chrominance.
4. The independent axis spaces resulting from different statistical methods which provide as less correlated components as possible.

a global color space *Cs* gathers the components chosen for our approach, with *NCs* =30 vectors:

*Cs={R,G,B,r,g,b,X,Y1,Z,x,y,z,L,a,b,u1,v1,c,h,H,S,V,Y,i,q,u2,v2,I1,I2,I3}*

The color study is associated with texture analysis, by calculating 2 statistical parameters and 5 parameters of Haralick of the co-occurrence matrix.

## 6.2 Gray-Level Co-occurrence Matrices

Let *E* denote the space of candidates attributes. The number of attributes of *E* is *7 x 30 =210*, where *7* represent the five parameters of Haralick + the two statistical parameters, and *30* is the number of colors components. From *E* we select the attributes that will Compose the hybrid color-texture space using mutual information.

## 6.3 Mutual Information

Mutual information (*MI*) is the measure of the dependence between random variables. The *MI* of two discrete variables *x* and *y* is defined based on their joint probabilistic distribution *p(x, y)* and the respective marginal probabilities *p(x) and p(y):*

$$\mathbf{MI}\,(x,y)=\sum_{i,j} a\log(a/b) \qquad (4)$$

$$a= p(xi,yj),\ b=p(xi)p(yj)$$

The estimation of the joint probabilistic distribution p(x, y) represents the principal difficulty for the use of *MI*. Various methods have been studied in the literature to estimate such joint distribution (histograms and kernel-based methods). Authors in propose a fast method to estimate the *MI*. One of the advantages of this method is that it based in the estimation of the entropy. The relation between *MI* and the entropy can be defined as :

$$MI(x, y) = H(x) - H(y \mid x) \qquad (5)$$

Using the properties of the entropy, the mutual information can be rewritten into

$$MI(x, y) = H(x) + H(y) - H(x, y) \qquad (6)$$

This entropy is calculated using *k*-nearest neighbor method. This method has a second advantage. It allows the calculation between a set of features *x* and the output variable *y*.

## 6.4 Hybrid Color-Texture Space

In this part, an algorithm will select the attributes composing the hybrid color-texture space. These attributes are chosen from the space of candidates attributes *E*. The algorithm based on the mutual information is developed. It works in two steps. In the first step, the forward phase, attributes are adding one by one. At each iteration, the attribute selected to integrate the current subspace is the one that most increases the mutual information with the output variable. The forward phase is stopped when adding any new attribute decreases the mutual information. The second step is the backward phase. Attributes are eliminated one at time. The attribute that most increase the mutual information when it is discarded is eliminated from the subset of features. As in the forward step, the backward phase is stopped when discarding any other attributes decreases the mutual information of the subspace with the output variable. The final subspace is the hybrid color-texture space. It contains the attributes that are the most discriminate.

## 7 Support Machine Vector

Among the methods with kernels based on the statistical learning theory of Vladimir Vapnik, SVM are the most known. SVM is a method of binary classification by supervised learning, it was introduced by Vapnik in 1995 . This method is an alternative to the recent classification. This method relies on the existence of a linear classifier in an appropriate space.

### 7.1 SVM Classifier

Support Vector Machine Algorithm are attractive, especially for non-linearly separable problems, where the problem is high-dimensional. A classification task usually involves with training and testing data which consists of some data instances. Each instance in training set contains one “target value”(class lables) and several “attributes”(features).The goal of SVM is to produce a model which predicts target value of data instances in the test set which are given only the attributes. It provides a Novel means of classification using the principles of structural risk minimization (SRM). It is one of the most sophisticated nonparametric supervised classifiers available today, with many different configurations depending upon the functions used for generating the transform space in which the decision surface is constructed. Here the classification problem is restricted to consideration of the two-class problem without loss of generality. In a binary classification task, the aim is to find an optimal separating hyper-plane. Considering the example shown in fig.1,there are many possible linear classifiers that can separate the data, but there is only one that maximizes the margin(maximizes the distance between the hyper-plane and the nearest data of every other class).

Only those vectors that are on the perimeter of the data cloud are used to determine the boundary. In other words, those data vectors, which are closest to the separating margins in the transformed space, are called the support vectors(SV).

Since this is a problem of classification of two classes, this method uses a set of learning data to learn the parameters of the model. It is based on the use of so-called kernel function that allows an optimal separation of the data. It was successfully evaluated on pattern recognition problem . For example in two classes problem(positive and negative sets of simples), the basic form of linear SVM classifier try to find an optimal hyperplane that separates the set of positive samples from the set of negative samples. The LIBSVM package that support multi-class problem available on to classify the different textures is used in our work.

## 8 Experiments

Our experiments were conducted using a PC with a Pentium 4 CPU 1.8 G with 1-GB memory, and they involve the following two image data sets.

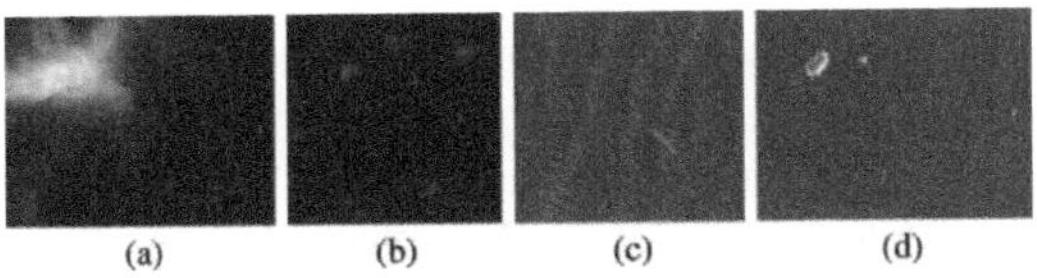


**Fig. 1.** Typical sea images. (a) Images with large clouds. (b) Images withsmall clouds. (c) Images with strong ocean waves. (d) Images with small islands.

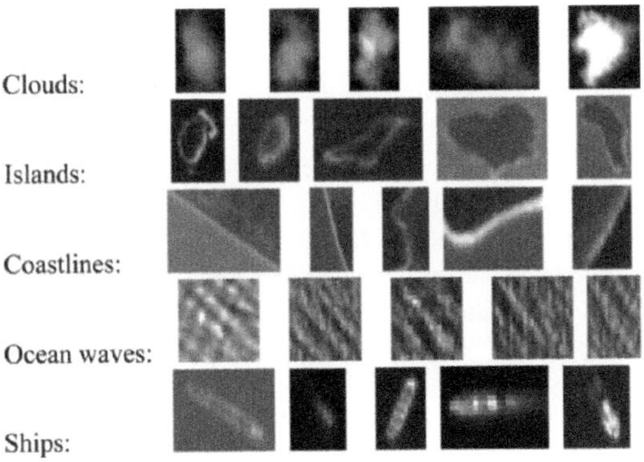


**Fig. 2.** Typical samples in Data set

*Data Set 1:* This data set consists of spaceborne optical images, about 2000 $*$ 2000 pixels in size, with a total of 232 images and a resolution of 5–20 m. Typical samples are shown in Fig. 1. It is used to test our proposed approach of ship detection. For simplicity, all of the images were scaled to the spatial resolution of 10 m to train only one classifier in our experiments.

*Data Set 2:* This data set includes more than 600 typical ship candidate sub images obtained by our ship candidate extraction from the spaceborne optical images in Data set 1. Some typical samples are shown in Fig. 2. The aim is to test the performance of the presented hierarchical classification approach, which is a very important section of the ship detection approach.

## 9 Conclusion

Because of the security problems sea target databases, unlike face and skin images, are not published. Lack of famous color images for sea target detection, forced us to create our own database. I am going to test the method on 3 different databases. First database is about 2000 frames of the sea targets from several mainstreams movies. The second database is collected from different military websites and the third one is database from low quality movies filmed by one of famous military companies for sea target detection purposes. It is obvious that because of observing framing rules the third database has the best result.

An efficient classification algorithm for ship detection that combines color and texture features were presented to detect ship in the sea . The Hybrid color-texture space was constructed for classified image using mutual information. Thus the algorithm may achieve good classification rate when compared to existing algorithms.

## References

1. Zhu, C., Zhou, H., Wang, R., Guo, J.: A Novel Hierarchical Method of Ship Detection from Spaceborne Optical Image Based on Shape and Texture Features. IEEE Transactions On Geoscience And Remote Sensing 48, 3446–3456 (2010)
2. Mirghasemi, S., Yazdi, H.S., Lotfizad, M.: Linear and Quadratic PSO Based Color Space Conversion for Sea Target Detection. International Journal of Computer and Electrical Engineering 2, 1793–8163 (2010)
3. Mirghasemi, S.: Sea Target Detection Based on SVM Method Using HSV Color Space. In: IEEE Students Conference on Research and Development (November 2009)
4. Huang, S.-q., Liu, D.-z., Gao, G.-q., Guo, X.-j.: A novel method for speckle noise reduction and ship target detection in SAR images. In: Science Direct, Pattern Recognition, pp. 1533–1542 (October 2010)
5. Du, Z., Liu, R., Liu, N., Chen, P.: A New Method for Ship Detection in SAR Imagery based on Combinatorial PNN model. In: IEEE First Int. Conf. on Intelligent Networks and Intelligent Systems, pp. 531–534 (2008)
6. Hu, Y., Wu, Y.: Number Estimation of Small-Sized Ships in Remote Sensing Image Based on Cumulative Projection Curve. In: IEEE Int. Conf. on Audio, Language and Image Processing (ICALIP), pp. 1522–1526 (July 2008)
7. Yang, S., He, S., Lin, H.: Video Image Targets Detection Based on the Largest Lyapunov Exponent. In: IEEE The 9th Int. Conf. for Young Computer Scientists, pp. 2973–2977 (November 2008)
8. He, S., Yang, S., Shi, A., Li, T.: A Novel Image Moving Sea Targets Detection Method Based on Natural Measure Feature. In: IEEE Int. Symposium on Information Science and Engineering, pp. 397–400 (December 2008)
9. Hong, Z., Jiang, Q., Guan, H., Weng, F.: Measuring overlap-rate in hierarchical cluster merging for image segmentation and ship detection. In: Fourth Int. Conf. on Fuzzy Systems and Knowledge Discovery (FSKD), vol. 4, pp. 420–425 (2007)
10. Tello, M., Lopez-Martinez, C., Mallorqui, J.J., Bonastre, R.: Automatic Detection of Spots and Extraction of Frontiers in SAR Images by Means of the Wavelet Transform: Application to Ship and Coastline Detection. In: IEEE Int. Con. on Geoscience and Remote Sensing Symposium, IGARSS, pp. 383–386 (August 2006)
11. Marivi, T., Carlos, L.M., Jordi, J.M.: A novel algorithm for ship detection in SAR imagery based on the wavelet transform. IEEE Geoscience And Remote Sensing Letters 2, 201–205 (2005)
12. Maia, H.E., Aboutajdine, D., Hammouch, A.: Color-texture analysis by mutual information for multispectral image classification. In: IEEE Pacific Rim Conference on Communications, Computers and Signal Processing, pp. 359–364 (August 2009)
13. El Maia, H., Hammouch, A., Aboutajdine, D.: Hybridization of the color information with the texture by using the mutual information for the classification of satellite images. In: Int. Conference on Multimedia Computing and Systems (ICMCS), pp. 372–376 (April 2009)
14. Otsu, N.: A threshold selection method from gray-level histogram. IEEE Trans. Syst., Man, Cybern. SMC-9(1), 62–66 (1979)

15. Gonzalez, R.C., Woods, R.E.: Digital Image Processing, 2nd edn. Prentice-Hall, Upper Saddle River (2002)
16. Chan, F.T., Vese, L.: Active contours without edges. IEEE Trans. Image Process. 10(2), 266–277 (2001)
17. Sonka, M., Hlavac, V., Boyle, R.: Image Processing, Analysis, and Machine Vision, 2nd edn. Brooks/Cole, Pacific Grove (2002)
18. Dudani, S.A., Breeding, K., McGhee, R.: Aircraft identification by moment invariants. IEEE Trans. Comput. C-26(1), 39–46 (1977)
19. Hu, M.: Visual pattern recognition by moment invariants. IEEE Trans. Inf. Theory IT-8(2), 179–187 (1962)
20. Kourosh, J.K., Hamid, S.Z.: Radon transform orientation estimation for rotation invariant texture analysis. IEEE Trans. Pattern Anal. Mach. Intell. 27(6), 1004–1008 (2005)

# Multi-layer Logon Verification System: A Case Study of Indian Banks

Mohammad Asim

Dept.of CSSE
University of Hail, KSA
m.asim@uoh.edu.sa

**Abstract.** Internet is only the medium which connects billions of people globally and finally has been removed all physical barriers. Bank is also one of the sector which greatly influenced with the invent of Internet. However, Internet banking is one of the service of the bank which provides many services online to the customers without being physically visit the bank.On the one end Internet banking bring many services to the customer but on the contrary also pose a risk of fraud, theft and vandalism.

The key aspect of any online services to authenticate the customer without any face-to-face communication.which ultimately challenge for the technology to validate them without any compromise. In the whole scenario security can be compromised at three level ;at the client ,in transit and at the server.Presently there are number of authentications methods proposed by the researcher worldwide for the financial institutions but my study is particularly focus on Indian Banks. Moreover, a special focus given on online services providing by the banks to their customers. Study reveals the authentication methods is a layered approach to achieve high degree of privacy and security in online services and which is a combination of two approach such as mobile and Token based authentication methods.

I have conducted a study of 25 banks of India to collect the data regarding present Authentication methods adopted presently. The paper describe in detail working of the said authentication methods and future implementation of these technologies.

**Keywords:** Internet banking, Protection and Security, Mobile Authentication, Token Authentication.

## 1 Introduction

In the current scenario, the Internet has become extensively used gizmo for corporate organizations, education and bank is using Internet Technology to provide various online services to serve their customers with ease and comfort. However, with the growing application of the Internet, security has become an important area of attention for various levels of users and however, a big challenge for the organization to implement security methods for safe and hassle free services for their customers to gain their confidence. There are several factors that may be affecting the adoption of internet banking. However, an important factor is the level of security and the risk associated with it[1].

D. Nagamalai, E. Renault, M. Dhanushkodi (Eds.): CCSEIT 2011, CCIS 204, pp. 32–41, 2011.

The paper mainly discusses and analysis the current procedures adopted by major banks in India.however, it is important to note that whenever there is trust, there is risk. Thus, trust is accepting risk. Trust, risk, security and reliability are fundamental to internet banking.I investigated the security of the internet banking systems in Banks in India including Public,private and foreign banks through visiting their sites and security guidelines/manual they have online.

The rest of paper is outlined as follows. In section 2, I describe current Internet Banking Authentication System adopted by Indian Banks. In section 3, I have purposed a approach for authentication, In section 4 is concern the case study of the banks.In section 5,There are some recommendations for the banks and finally I concluded in section 6.

## 2 Existing Internet Banking Authentication Methods in India

I have conducted the study of 25 major banks in India through visit their official sites to obtain the information for their security methods adopted for authentication process:

SSL + HTTP + Virtual Keyboard

Authentication is a method to verify the credentials of the customer before using any online services without any paper work or physical identification system. However, now the question arises here for validating the credentials of the customer[2].After a thoroughly studied of the banks sites, I have found that all the banks are using method of authentication based on HTTPS. Which allow a secure session between the clients machine and the bank's server.

Presently the authentication system of the Internet banking in India is based on public key cryptography system and the digital signature technology for authenticating, thus the safe and smooth electronic transaction can be ensured. The Certificate based authentication system is a certificate center system based on PKI. The security authentication system of internet banking is composed of Central Bank Certification Authority (CA) and the Commercial Bank internal management CA. These two systems work independently and cooperatively. The tasks are divided as follows:

The Central Bank CA issues the digital certificates to the clients of the internet banking, and the internal management CA issues management digital certificates to commercial bank employees. These digital management certificates are mainly used in the Registration Authority (RA) system. The main purpose of Commercial bank employees to use digital certificate to authenticate ID, encrypt transmission and resist repudiation, also audit clients' certificate requests and create the clients' certificates.

Security proxy permit a web-based transaction interface embedded with functionalities such as authentication, integrity, non-repudiation, encryption, authorization and security enhancements. The main function of Digital signature is to ensure the integrity and non-repudiation of any transaction [3].

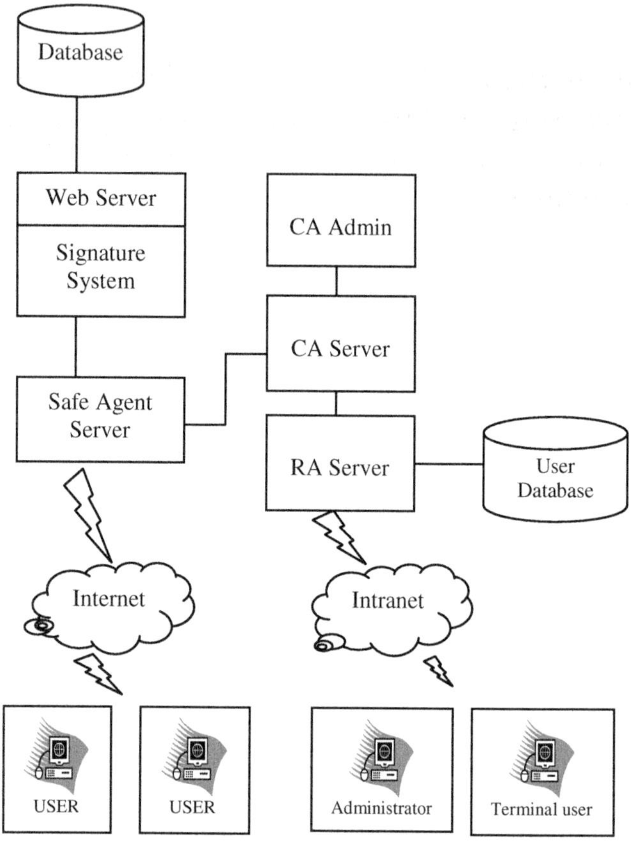


**Fig. 1.** Internet Banking Security Architecture (Source: IEEE 2006,p.1570 )

### 2.1 Virtual Keyboard for Internet Banking

Virtual Keyboard is an online application to enter password with the help of a mouse.

### 2.2 Advantage of a Virtual Keyboard

The Virtual Keyboard is designed to protect your password from malicious "Spyware" and "Trojan Programs". Use of Virtual keyboard will reduce the risk of password theft.

**Process to Use Virtual Keyboard**

Steps to use Virtual Keyboard are as follows:

- Enter your Login Id using Physical Keyboard.
- Select the check box "Use Virtual Keyboard".
- Use the Virtual Keyboard to enter the login password.
- Once you have entered your password, click "Log-in"[4].

# 3 Proposed Approach

In today's Internet-driven society, most financial institutions offer online banking services to their customers. However, with the fast growth of online fraud, Internet banking comes with risks. A majority of users are worried about the improper use of their bank account details on the Internet.

Therefore, banks are looking for ways to protect their brand image and preserve the trust they have with their customers [5].

My approach is based on the concept that encryption of the user name/password is not enough that is why we introduce a second layer of security which would be highly safe and secure for online banking services. That is called multi-layer logon verification system.

1. Layer is concern with the level and type of security and authentication method.In this layer customer is suppose to enter pair of information username/password like traditional method which would be secure and protected by using HTTPS technology.
2. Second layer allow authenticating the user further to enter some sort of code generated by bank server and send it to the registered mobile number of the user or the token generated by hardware devices. The two methods are:

   a. Mobile Authentication and
   b. Token Authentication

I will discuss in detail, the working of these two methods in the next two sections.

## 3.1 Mobile Authentication

**The First Method: Using the mobile phone**

One of new security enhancements is the new authentication method that has been developed for our individual customers. This method allows the customer to use his/her mobile phone as an authentication tool.

**How does the authentication using the mobile phone work?**

- This relates to when the user chooses to authenticate some transactions/payments through his/her mobile phone, while using Internet Banking.
- When the customer attempts to make a transaction/payment, the banking system generates a special password for this transaction, and sends it as an SMS message to their mobile phone.
- The user then enters this password in order to authenticate the requested transaction, thereby adding another layer of security.
- The banking system checks the entered password against the system-generated password, and if it matches the transaction is completed.

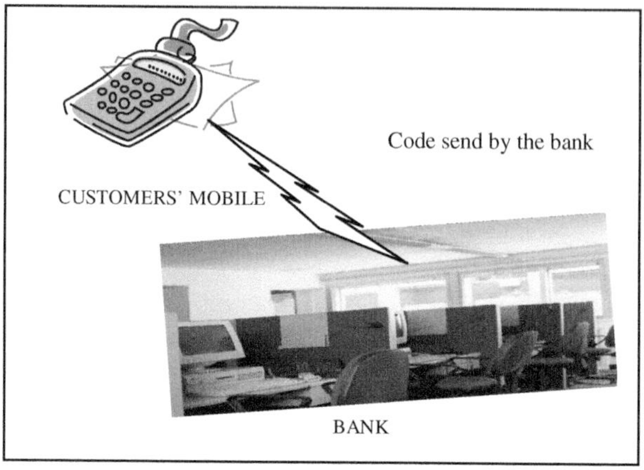


**Fig. 2.** Architecture of Mobile Authentication

**How to register a mobile phone number of the customer on the banks site to allow it to be used as an authentication tool?**

The user enters the mobile phone number, on Internet banking service after which the system will give the user the following options for authenticating the mobile phone number:

**Through the ATM:**

- When using this method the system creates an application code and sends it as an SMS message to the entered mobile number.
- The user needs to authenticate his/her ID and confirm the entered mobile number using the application code. This can be done by visiting a any Bank ATM and entering the Order Number in the appropriate service menu.
- Upon entry, the system checks the entered application code against the system-generated application code, and if it matches, the system then registers the mobile number for use as an authentication tool.

**Through Internet Banking Module – Security Token**

- When the user chooses this method to authenticate the mobile phone number, the system then asks for a security token code to be entered.
- When successfully entered the system then generates an application code and then sends it as an SMS message to the entered mobile number.
- The user then need to authenticate his/her ID and confirm the entered mobile number using the application code received, by entering the application code on Internet banking module.
- Upon entry, the system checks the application code against the system-generated application code, and if it matches, the system then registers the mobile number for use as an authentication tool.

### 3.2 Token Authentication

This is a method for authentication that will be expanded to cover all third party transactions as well as any Internet banking services. This option applies using a small electronic device supplied by the Bank to the customers; the device generates a one off password, for every transaction. To date this security token device has only been used for some transactions (i.e. transactions exceeding a certain amounts). However, my idea to introduce this method of transaction authentication with all transactions/payments made from the customer's account to third party, as well when updating certain essential information.

The small devices are typically of the size of small key chains, calculators or credit cards. Usually an authentication token has the following features:

1. Processors
2. Liquid Crystal Display (LCD) for displaying outputs.
3. Battery
4. (Optionally) a small keypad for entering information.
5. (Optionally) a real time clock.

Each authentication token is pre-programmed with a unique number, called as a random seed. The seed forms the basis for ensuring the uniqueness of the output produced by the token.

Let we understand. How does it work step by step?

**Step1: Creation of a Token:** Whenever a authentication token is created, the corresponding random seed is generated for the token by the authentication server. This seed is stored or pre-programmed inside the token, as well as its entry is made

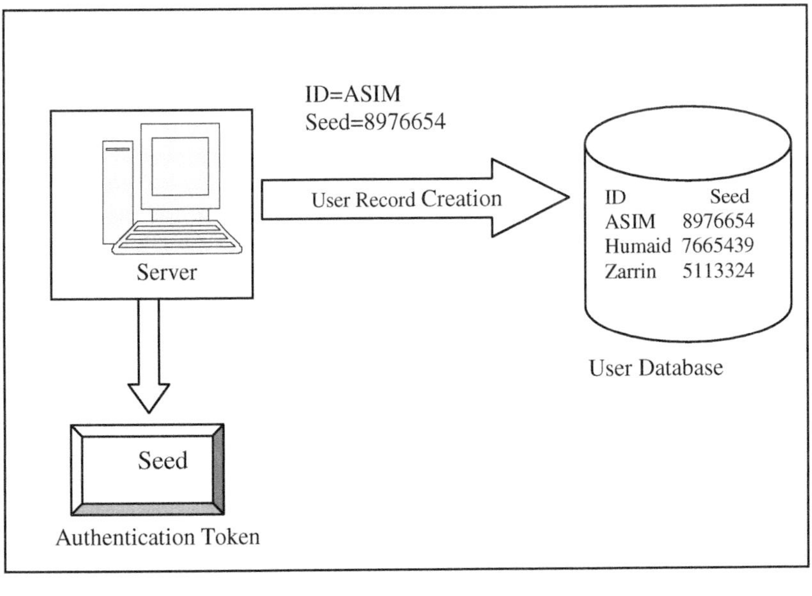


**Fig. 3.** Random seed storage in the database and the authentication token

against that user's record in the user database. Conceptually it's like a password but not exactly a password because the user know about the password but the user does not know about the value of the seed unlike password.

**Step 2: Use of Token**

An authentication token automatically generates pseudorandom numbers ,called as one time passwords.One time passwords are generated randomly by an authentication token, based on the seed value that they are programmed with.They are one-time because they are generated ,used once ,and discarded forever. When a user wants to be authenticated, the user will get a screen to enter the user id and the latest one time password obtained from the authentication token.The user id and password travel to the server as a part of the login request.The server obtains the seed corresponding to the user id from the user database,using a *seed retrieval program*.It then calls another program called as password validation program,to which the server gives the seed and the one time password .This program knows how to establish the relationship between the seed and the one time password.Now we are explaining in simple way,the program uses synchronization techniques,to generate the same one-time password as was done by the authentication token.However,the main point to be noted here is that the authentication server can use this program to determine if a particular seed value relates to a particular one-time password or not.This is shown in the figure 4.A question at this stage could be ,what would happen if a user loses an authentication token? Can another user simply grab it and use it?To deal such situations,the authentication token is generally protected by a password or a 4 digit PIN. Only when this PIN is entered can the one-time password be generated. This is also the basis for what is called as multi-layer logon verification system.What are these factors?There are three most common factors:

- Something that you know,e.g a password or PIN
- Something you have,e.g a credit card or an identity card
- Something you are,e.g you voice or finger print

Based on these principles,we can see that a password is a single layer authentication,because it is only something that you know.In contrast ,authentication token are examples of second layer of authentication,here you must have something (the authentication token itself) and you must also know something(the PIN is used to protect it).Someone only knowing the PIN or only having the token cannot use it. Both the layers of authentication are required that is why it is called multi -layer authentication.

***Step 3:* Server return an appropriate message back to the user**

Finally ,the server sends an appropriate message back to the user,depending on whether the previous operations yielded success or failure[6].This is shown in figure 5:

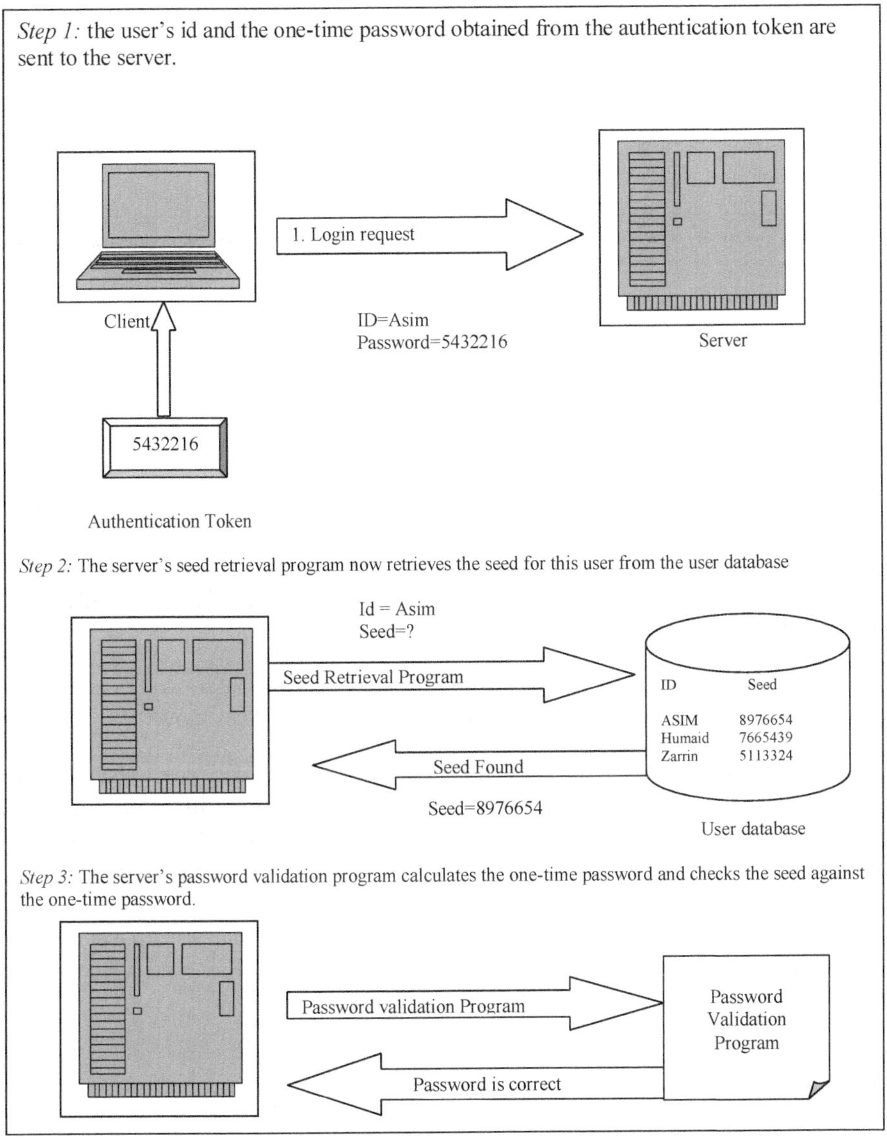


**Fig. 4.** Server validate the one-time password

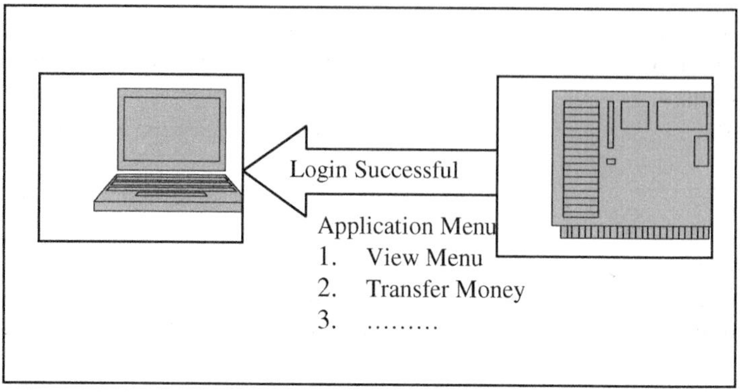


**Fig. 5.** Server sends an appropriate message back to the user

# 4 Case Study of Indian Banks

The following table shows the current methods by the banks in India and I have collected and analysis the data after visiting the official site of the banks.

| S.No | Name of the Bank | URL | Security Methods for online Banking | |
|---|---|---|---|---|
| 1 | State Bank of India | https://www.onlinesbi.com | PKI/Digital Signature | Virtual Keyboard |
| 2 | Union Bank of India | http://www.unionbankonline.co.in | PKI/Digital Signature | Virtual Keyboard |
| 3 | Bank of Baroda | http://bankofbaroda.com | PKI/Digital Signature | Virtual Keyboard |
| 4 | Andhra Bank | https://www.onlineandhrabank.net.in | PKI/Digital Signature | Virtual Keyboard |
| 5 | Allahabad Bank | https://www.allbankonline.in | PKI/Digital Signature | Virtual Keyboard |
| 6 | ICICI Bank | http://icicibank.com | PKI/Digital Signature | Virtual Keyboard |
| 7 | HDFC Bank | http://www.hdfcbank.com | PKI/Digital Signature | Virtual Keyboard |
| 8 | Bank of India | http://www.bankofindia.com | PKI/Digital Signature | Virtual Keyboard |
| 9 | Axis Bank | http://www.axisbank.com | PKI/Digital Signature | NO |
| 10 | Karnataka Bank | http://www.karnatakabank.com | PKI/Digital Signature | Virtual Keyboard |
| 11 | Kotak Mahindra Bank | http://www.kotak.com | PKI/Digital Signature | Virtual Keyboard |
| 12 | Yes bank | http://www.yesbank.co.in | PKI/Digital Signature | NO |
| 13 | American Express | https://home.americanexpress.com | PKI/Digital Signature | NO |
| 14 | Corporation Bank | http://www.corpbank.com/ | PKI/Digital Signature | Virtual Keyboard |
| 15 | Indian bank | https://www.indianbank.net.in | PKI/Digital Signature | NO |
| 16 | Punjab National Bank | http://www.pnbindia.in | PKI/Digital Signature | Virtual Keyboard |
| 17 | Syndicate Bank | http://www.syndicatebank.in | PKI/Digital Signature | Virtual Keyboard |
| 18 | Vijaya bank | http://vijayabank.com/vijaya/vijaya | PKI/Digital Signature | Virtual Keyboard |
| 19 | UCO Bank | http://www.ucobank.com/default.htm | PKI/Digital Signature | Virtual Keyboard |
| 20 | Induslnd bank | http://www.indusind.com/indusind/wcms/en/home/personal-banking | PKI/Digital Signature | Virtual Keyboard |
| 21 | INGVysya | http://www.ingvysyabank.com | PKI/Digital Signature | Virtual Keyboard |
| 22 | State Bank of Hyderabad | http://www.sbhyd.com/default.asp | PKI/Digital Signature | Virtual Keyboard |
| 23 | Oriental Bank of Commerce | https://www.obcindia.co.in/obcnew/site/index.aspx | PKI/Digital Signature | Virtual Keyboard |
| 24 | State Bank of Bikaner & Jaipur | http://www.sbbjbank.com | PKI/Digital Signature | Virtual Keyboard |
| 25 | Indian Overseas Bank | http://www.iob.in | PKI/Digital Signature | Virtual Keyboard |

## 5 Recommendations for the Banks

1. Winning the faith of the customers is the primary goal for any financial institutions. But customers are not merely looking towards the services as well as they are worried about the protection and security of their personal and financial data.
2. Study revealed, that most of the banks in India, they are using traditional approach to authenticate the customers which is not enough. However, increasing cyber attacks demands more sophisticated and advance methods.
3. Study proposed a method of authentication. Which is basically broken into two separate layers and each one is performing individual function to achieve high degree of privacy and security for online services.

## 6 Conclusion

This paper is based on the case study of Indian banks and particularly focuses on authentication methods in Internet banking system. I have also explored the current method of authentications by the banks and proposing the new techniques for authentication addition to the existing methods. The implementation of these new techniques would be greatly influence the security and create confidence in customers to adopt Internet banking. Still Internet banking is less popular due to lack of knowledge of using Internet banking, privacy issues and risks involved among the customers. My research definitely proved to be good idea for banking people to understand the role of technology to win the faith of the customer. The recognition of any banks depends on two factors, the services they are providing to their customers and the technology implementation especially at security side.

## References

[1] Sanayei, A., Noroozi, A.: Security of Internet Banking Services and its linkage with Users' Trust. In: International Conference on Information Management and Engineering 2009, pp. 3–7 (2009)

[2] Xing, F., Zhan, J.: Online banking Authentication Using Mobile Phones.Future Information Technology (FutureTech). In: 2010 5th International Conference, pp. 1–5 (2010)

[3] Lin, S., Liu, Y.: An Improved Security Authentication Description and Implementation in Internet Banking System. In: Proceedings of International Conference in Communications, Circuits and Systems, vol. 3, pp. 1570–1572 (2006)

[4] http://www.icicibank.com/pfsuser/webnews/virtualkeyboad.htm

[5] http://www.gemalto.com/brochures/download/ezio_mobile_auth.pdf

[6] Kahate, A.: Cryptography and Network Security. Tata McGraww-Hill Publishing Company Limited, New York

# Image Compression Using Two Dimensional Spherical Coder in Wavelet Lifting

S. Panimalar, R. Anisha, and M. Gomathi

Manonmaniam Sundaranar University,
Tirunelveli
Snowb4ufriends@yahoo.co.in
anisha.mtech@gmail.com

**Abstract.** In recent years, many wavelet coders that use various spatially adaptive coding technique to compress the image. Level of flexibility and the coding efficiency are two crucial issues in spatially adaptive methods. So in this paper "spherical coder" is introduced. The objective of this paper is to combine the spherical tree with the wavelet lifting technique and compare the performance of the spherical tree between the different coding technique such as arithmetic, Huffman and run length. The Comparison is made by using PSNR and Compression Ratio (CR). It is shown the Spherical tree in wavelet lifting with the arithmetic coder gives high CR value.

**Keywords:** Discrete Wavelet Transform, Wavelet Lifting, Shperical Coder, Arithmetic Coding, Run length Coding, Huffman Coding.

## 1 Introduction

Image compression is the application of Data Compression on Digital Images. When a digital image is transmitted through a communication medium like a cable or a simple telephone line, the cost of the transmission depends on the size of the file and is a very important factor. To reduce the transmission cost the file has to be compressed. The objective of the compression is to reduce redundancy of the image data in order to be able to store or data in an efficient form.

Image compression may be lossy or lossless. Lossless compression is mostly preferred for artificial images such as icons, comics and technical drawings. Because of using low bit rate lossy compression produce some compression artifacts. Lossy methods are suitable for natural images. This is because the minor loss of information is acceptable to achieve a substantial reduction in bit rate.

Lossy compression is most commonly used to compress the multimedia data such as audio, video, still images especially in applications such as internet telephony and streaming media. Lossless compression is most commonly used to compress the text and data files such as bank records, text articles, etc.

To compress the image any one of the coder is used. All image coders are based on some statistical model for natural images. The coder is explicitly or implicitly optimized for the specific model and applied to sample images. Wavelet domain provides a good match to the space-frequency characteristics of natural images.

D. Nagamalai, E. Renault, M. Dhanushkodi (Eds.): CCSEIT 2011, CCIS 204, pp. 42–54, 2011.

Hence, it is much easier to build a realistic image model in the wavelet domain than in the pixel domain. That is because a simple coder in the wavelet domain could outperform a complex coder in pixel domain.

Wavelet subbands are best characterized by spatially varying nonhomogeneous processes. The "spherical representation" that has been introduced in [1] as a novel way of representing image information in wavelet domain. Based on the spherical representation, we develop a coding algorithm that handles this nonhomogeneity in an effective and nonparametric way.

Successful coding schemes used in DCT-based algorithms, such as run-length coding, and vector quantization [2], [3] were tried in wavelet image coding, but demonstrated modest coding gains over standard transform-based algorithms. The breakthrough in wavelet image coding arrived with coders using hierarchical wavelet trees, such as EZW [4], SPIHT [5], SFQ [6]–[9]. Grouping wavelet coefficients that belong to the same spatial region under a tree-structure, these coders were able to adapt to the properties of different regions in the image. Other coders, such as EQ [10], classification-based algorithms [11], EBCOT [12], etc., achieved improved coding efficiency by introducing more spatial adaptivity in modeling the subbands.

The rest of this paper is organized as follows: in Section II we review the Discrete Wavelet Transform (DWT). Wavelet Lifting Scheme and features are discussed in Section 3. Section 4 describes the details of the spherical representation and spherical coding algorithm. In Section 5 the different coding techniques are discussed. The performance of the spherical representation with various coders and various wavelet domain are compared in Section 6 in a series of computational experiments. In Section 7 we summarise the results of the study . The conclusion of the paper is in Section 8.

## 2 Discrete Wavelet Transform

The discrete wavelet transform (DWT) is an implementation of the wavelet transform using a discrete set of the wavelet scales and translations obeying some defined rules. In other words, this transform decomposes the signal into mutually orthogonal set of wavelets, which is the main difference from the continuous wavelet transform (CWT), or its implementation for the discrete time series sometimes called discrete-time continuous wavelet transform (DT-CWT).

The discrete wavelet transform is a hierarchical subband technique. The subbands are created by applying decomposition on original image. To start the decomposition, the image is filtered in both the horizontal and vertical directions using separable filters. This creates four subbands as follows: Subband LL1 represents the horizontal and vertical low frequency components of the image. Subband HH1 represents the horizontal and vertical high frequency components of the image. Subband LH1 represents the horizontal low and vertical high frequency components. Subband HL1 represents the horizontal high and vertical low frequency components.

To obtain the next level in the decomposition the two filters are again applied, but only to the LL1 subband. Again we obtain four subbands labeled LL2, HH2, LH2, and HL2 with representations similar to the first level subbands. The following example illustrates the discrete wavelet transform.

**Fig. 1.** Original Lena Image

The two level decomposition image and corresponding decomposition map of above figure is show below.

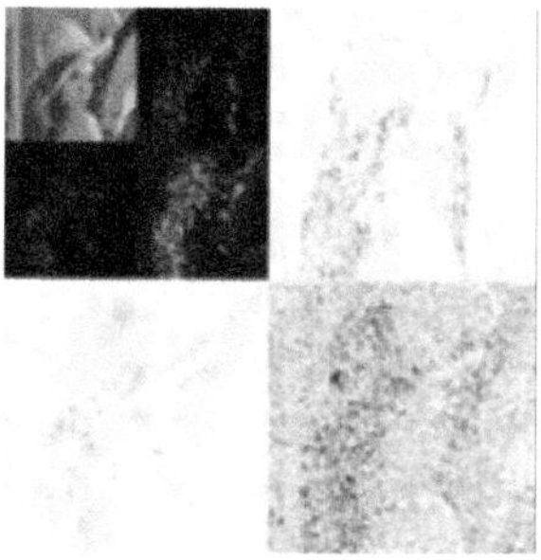

**Fig. 2.** Two Level Decomposed Image

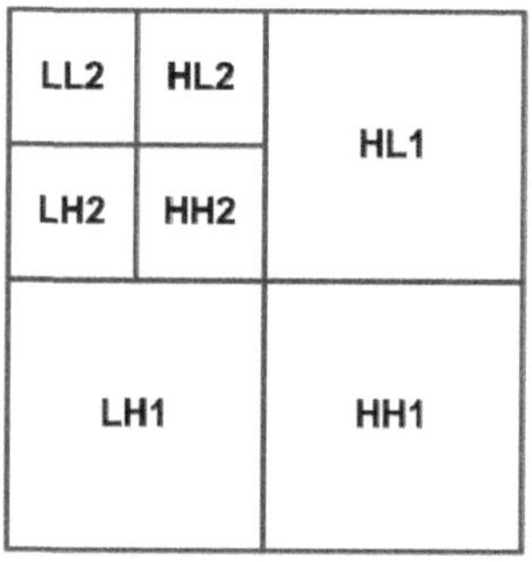


**Fig. 3.** Two Level Decomposition Map

## 3 Wavelet Lifting Sceme

The wavelet lifting scheme divides the wavelet transform into a set of steps. One of the elegant qualities of wavelet algorithms expressed via the listing scheme is the fact that the inverse transform is a mirror of the forward transform. lifting scheme wavelets the predict wavelet transform starts with a split step, which divides the data set into odd and even elements. The predict step uses a function that approximates the data set. The difference between the approximation and the actual data replaces the odd elements of the data set. The even elements are left unchanged and become the input for the next step in the transform. The predict step, where the odd value is "predicted" from the even value is described by the equation

$$odd_{j+1,i} = odd_{j,i} - P( even_{j,i} ) \quad (1)$$

The inverse predict transform adds the prediction value to the odd element (reversing the prediction step of the forward transform). In the inverse transform the predict step is followed by a merge step which interleaves the odd and even elements back into a single data stream.

The simple predict wavelets are not useful for most wavelet applications. The even elements that are used to "predict" the odd elements result from sampling the original data set by powers of two (e.g., 2, 4, 8...). Viewed from a compression point of view this can result in large changes in the differences stored in the odd elements (and less compression). One of the most powerful applications for wavelets is in the construction of filters. The "down sampling" in the predict wavelets does not provide an approximation of the data at each step, which is one of the requirements for filter design.

The update step replaces the even elements with an average. This results in a smoother input for the next step of the next step of the wavelet transform. The odd elements also represent an approximation of the original data set, which allows filters to be constructed. A simple lifting scheme forward transform is diagrammed in below figure.

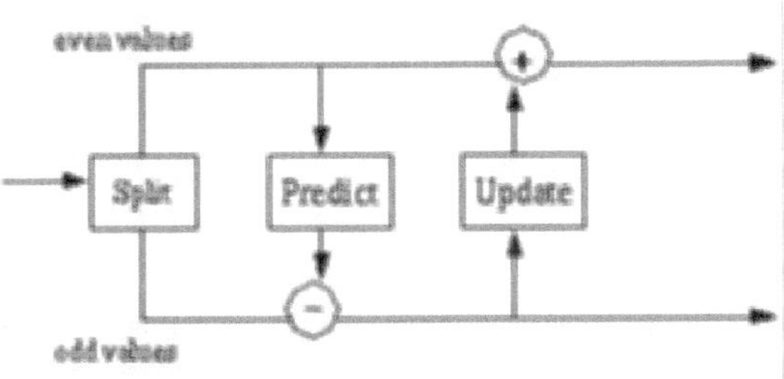


**Fig. 4.** Lifting Scheme Forward Wavelet Transform

The update phase follows the predict phase. The original value of the odd elements has been overwritten by the difference between the odd element and its even "predictor". So in calculating an average the update phase must operate on the differences that are stored in the odd elements:

$$evenj+1,i = evenj,i + U( oddj+1,i ) \quad (2)$$

In the lifting scheme version of the Haar transform, the prediction step predicts that the odd element will be equal to the even element. The difference between the predicted value (the even element) and the actual value of the odd element replaces the odd element. For the forward transform iteration j and element i, the new odd element, j+1,i would be

$$oddj+1,i = oddj,i - evenj,I \tag{3}$$

In the lifting scheme version of the Haar transform the update step replaces an even element with the average of the even/odd pair (e.g., the even element si and its odd successor, si+1):

$$even_{j+1,i=(}even_{ji}+odd_{ji})/2 \tag{4}$$

The original value of the oddj,i element has been replaced by the difference between this element and its even predecessor. Simple algebra lets us recover the original value:

$$oddj,i = evenj,i + oddj+1,I \tag{5}$$

Substituting this into the average, we get

$$even_{j+1,i}=(even_{j,i}+even_{j,i}+odd_{j+1,i})/2$$

$$even_{j+1,i}= even_{j,i}+(odd_{j+1,i})/2 \tag{6}$$

The averages (even elements) become the input for the next recursive step of the forward transform. This is shown in Figure 5, below

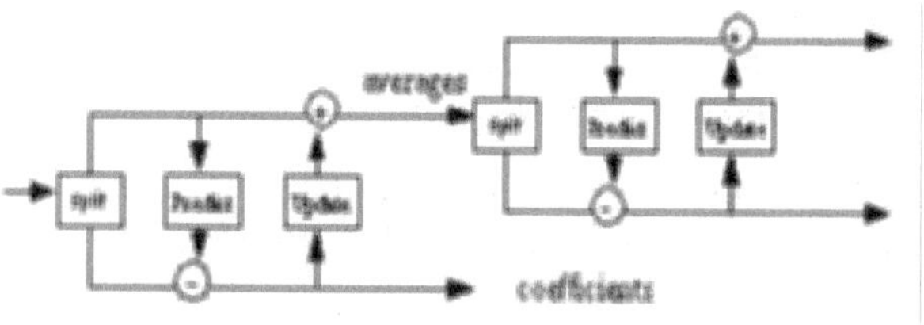


**Fig. 5.** Two Steps in Wavelet Lifting Scheme Forward Scheme

One of the elegent features of the lifting scheme is that the inverse transform is a mirror of the forward transform. In the case of the Haar transform, additions are substituted for subtractions and subtractions for additions. The merge step replaces the split step. This is shown in Figure 6, below

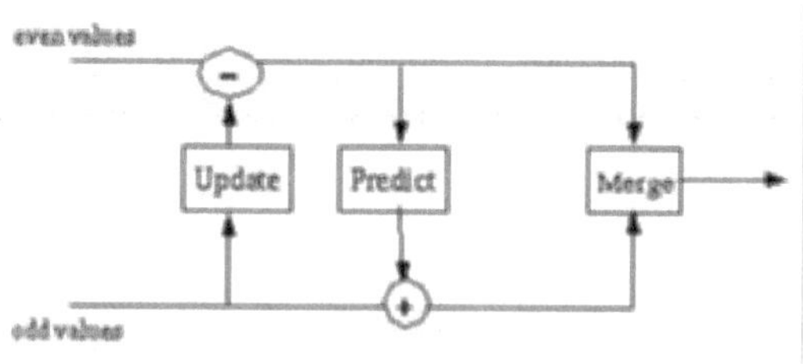


**Fig. 6.** Lifting Scheme Inverse Transform

# 4 Spherical Representation

The "spherical representation" is a hierarchical description of how total coefficient energy gets distributed within each wavelet subband. A hierarchical tree of subband energy is formed by summing up the squared coefficients. Phase variables are defined that describe how the energy in a given region is split into energies of two sub-regions. Phase variables are coded based on a simple and effective model. The nonhomogeneity of wavelet subbands is handled through this nonparametric model of the hierarchy.

We use the following hierarchical structure to represent a random process **X** (see Fig.7): In 1-D for $\mathbf{X}=\{x_i\}_{1\leq i\leq 2k}$ and for $0<m \leq k$, $0 \leq n < 2^{k-m}$.

$$E_m(n)=\sum_{i=2^m (n+1)}^{2^m (n+1)} x_i^2 \tag{7}$$

$$\psi_m (n) = arctan\left(\sqrt{\frac{E_{m-1}(2n)}{E_{m-1}(2n+1)}}\right) \tag{8}$$

where $0 \leq \Psi_m(n) \leq \Pi/2$. Here, **X** could be seen as one of the wavelet subbands of a 1-D signal. In the next section, this formulation is easily extended to 2-D subbands.

The variables $E_m(n)$ provide local energy information at different resolution levels m. The phase variables $\Psi_m(n)$ indicate how the local energy gets split between the two neighboring regions

$$E_{m-1}(2n) = E_m(n)\sin(\psi_m(n))^2 \tag{9}$$

$$E_{m-1}(2n+1) = E_m(n)\cos(\psi_m(n))^2 \tag{10}$$

The phase variables in a sense represent the difference in information content between the two regions. Going from the top level (m=k) to the bottom level (m=1) of the hierarchy, the phase values provide a refinement of the available information in each region of the subband.

When the total energy, $E_k(0)=\sum_{i=1}^{2k} x_i^2$ and $2^k$-1the phases at all levels of the hierarchy are given, the coefficients are easily determined up to a sign bit; i.e.,$x_i = sign(x_i)\sqrt{E_0}(i)$. The sign bits could also be defined as part of the epresentation if the phase values at the bottom of the hierarchy cover full range;i.e., $-\pi<\psi_1(i)\leq\pi$, and

$$X_{2i} = \sqrt{E_1}\,(i)\sin(\varphi(i)) \tag{11}$$

$$X_{2i+1} = \sqrt{E_1}\,(i)\cos(\varphi(i)) \tag{12}$$

In this type of representation, we are able to use local energy not only to differentiate between statistically distinct parts of the process but also to provide direct information about the underlying coefficient values. Coding $E_k(0)$ and $_m(n)$ can be seen as an alternative to coding $x_i$. We might say that, instead of cartesian coordinates, spherical coordinate system is used in representing the process; hence, the name spherical representation.

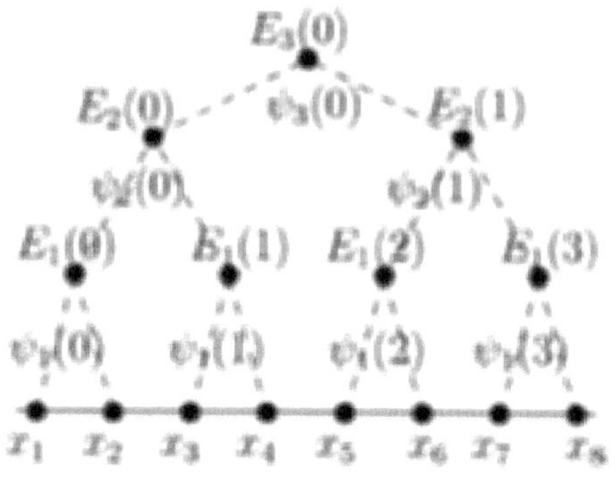

Fig. 7. Spherical Representation in 1-D

Spherical representation could be easily extended to 2-D to be used in wavelet image coding. Partial squared sums need to be defined in both vertical and horizontal directions in an alternating fashion (see Fig. 8). Let us represent the phase variables $\psi_{u,v}(s,t)$ and the local energies $\Gamma_{u,v}(s,t)$ as and respectively.

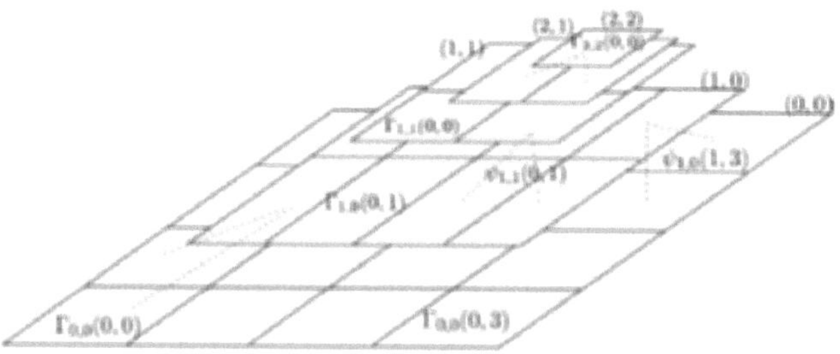

Fig. 8. Spherical Representation in 2-D

## 5 Encoding

Encoding is the process of transforming information from one format into another.

*A. Run Length Encoding(RLE)*

One relatively simple way to compress an image is called Run Length Encoding (RLE), which describes the image as a list of "runs", where a run is a sequence of horizontally adjacent pixels of the same color. It codes the data by measuring the length of runs of the values

The simplest form of compression technique which is widely supported by most bitmap file formats such as TIFF, BMP, and PCX. RLE performs compression regardless of the type of information stored, but the content of the information does affect its efficiency in compressing the information.

*B. Huffman Encoding*

We use this coding to encode/compress a file. Steps:

1. Read the file.
2. Calculate the probability of each symbol (since we use an ASCII file, there are 256 possible symbols). Instead of actual probability we can use instance count.
3. Use the Huffman coding algorithm to find the coding for each symbol. Need to apply only on subset of symbols that actually appear in the file.
4. Encode the file, and write it out.

The out file must also include the encoding table, so as to permit decoding.

*C. Arithmetic Encoding*

The principles of arithmetic coding describe an arithmetic coding engine that will produce a compliant bit stream when used in conjunction with the correct methods for binarization and context selection (described below).

Arithmetic coding is employed on integer-valued wavelet coefficients and uses a two-stage process as shown in figure 9. First, integer values are binarized into a sequence of bits or boolean values. At the same time a "context" is generated for each of these bits. These boolean values, together with the corresponding context, are then coded by the binary arithmetic coding engine.

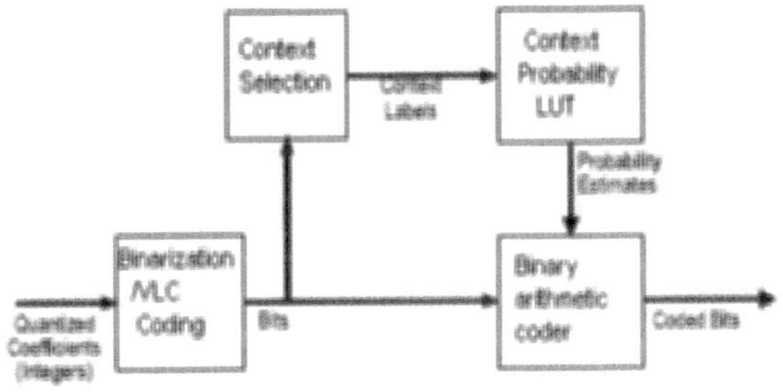


**Fig. 9.** Arithmetic Encoding Process

Arithmetic coding is a lossless coding technique. In arithmetic coding, a data is encoded as a real number in an interval from one to zero. Arithmetic coding typically has a better compression ratio than Huffman coding. Arithmetic coding produces a single symbol rather than several separate codeword.

## 6 Spherical Coding Scheme

The spherical coding algorithm is given (for each wavelet subband at different scales and in different orientations) as follows

1. Apply Wavelet Lifting to input image
2. Use soft-thresholding to estimate zero-quantized wavelet coefficients

$$C^\wedge(m,n)= c(m,n) \quad \text{if } |c(m,n)| > T$$

$$C^\wedge(m,n)= 0 \quad \text{if } |c(m,n)| < T \tag{13}$$

3. Create a spherical representation

$$\Gamma_{u,v}(s,t) = \sum_{m=2^u s}^{^u(s+1)-1} \sum_{n=2^v t}^{2^v(t+1)-1} (m,n)^2$$

$$0\leq S<2^{(J-u)},\ 0\leq 2^{(J-v)} \text{ and for}, 0\leq u<J$$

$$\Psi_u+1,\ u+1(s,t)=\arctan\left(\frac{\sqrt{\Gamma_{u+1,u(s,2t)}}}{\sqrt{\Gamma_{u+1,u(s,2t+1)}}}\right)$$

$$0\leq s,t<2^{(J-u-1)}$$

$$\Psi_u+1,\ u(s,t)=\arctan\left(\frac{\sqrt{\Gamma_{u,u(2s,t)}}}{\sqrt{\Gamma_{u,u(s,2s+1,t)}}}\right) \tag{14}$$

$$0\leq s<2^{(J-u-1)}, 0\leq t<2^{(J-u)}$$

4. Find the Lagrangian cost by using the below forumla where I(m,n) represents the sign bit for coefficient c(m,n)

$$L_{0,0}(m,n)=(c(m,n)-c(m,n))^2+I(m,n) \tag{15}$$

where I(m,n) represents the sign bit for coefficient c(m,n), i.e.,

$$I(m,n)=0 \text{ if } c(m,n)=0$$

$$I(m,n)=1 \text{ else} \tag{16}$$

5. Compare the Lagrangian cost to previous cost. If the latter is smaller then quantize to zero
6. Coding the local energies and phase variables using coder(arithmetic coder, runlength coder, Huffman coder)
7. Decode the local energies and phase variables using arithmetic decoder
8. Decode the wavelet coefficients

$$\dot{c}(m,n) = sign(c(m,n))\sqrt{\hat{\Gamma}_{0,0}(m,n)} \tag{17}$$

Apply inverse Wavelet Lifting finally we get compressed image

The block diagram of the proposed spherical coding scheme is shown below

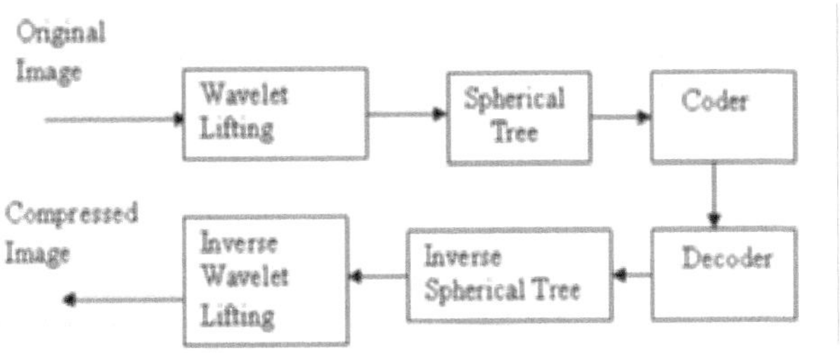


**Fig. 10.** Proposed Spherical Coding Process

## 7 Performance Analysis of Spherical Coding Scheme

In this performance analysis of proposed method we have to analyse the spherical coding with various wavelet domain such as Discrete Wavelet Domain and Wavelet Lifting as well as with various coder such as arithmetic coder, run length coder and the Huffman coder and find out which one is the best. We carried out computational experiments to conform the effectiveness of the above techniques.

To evaluate the performance of the compression techniques several performance metrics are available. we use the PSNR and CR to analyses the performance.

**PSNR** means **Peak Signal to Noise Ratio (PSNR)**. This performance metric gives the signal to noise ratio between the original image and the compressed image. PSNR is calculated by using the following formula.

$$PSNR = 20log\frac{2^n - 1}{RMS}$$

$$RMS = \sqrt{\frac{1}{n}}\ \sum_{i=1}^{N}(x_i - x_i')^2$$

CR means Compression Ratio. This performance metric gives compression ratio. This ratio gives an indication of how much compression is achieved for a particular image.

The performance of the algorithm is evaluated on Lena, Peppergray and Sailboat All the images are of 512x512 pixels. These pictures are the most widely used standard test images used for compression algorithms. The image contains a nice mixture of detail, flat regions, shading, and texture that do a good job of testing various image processing algorithms.

We can evaluate these images by the CR value and PSNR value.

(a) (b) (c)

**Fig. 11.** Experimental Images (a) Lena (b) Pepper (c) Sailboat

The performance analysis of spherical coding on PSNR value is given below

**Table 1.** Performance analysis on PSNR value

| Image | Wavelet Lifting | | |
|---|---|---|---|
| | RLC | HMC | AMC |
| Lena | 34.468 | 34.468 | 34.468 |
| Pepper | 35.046 | 35.046 | 35.046 |
| Sailboat | 32.848 | 32.848 | 32.848 |

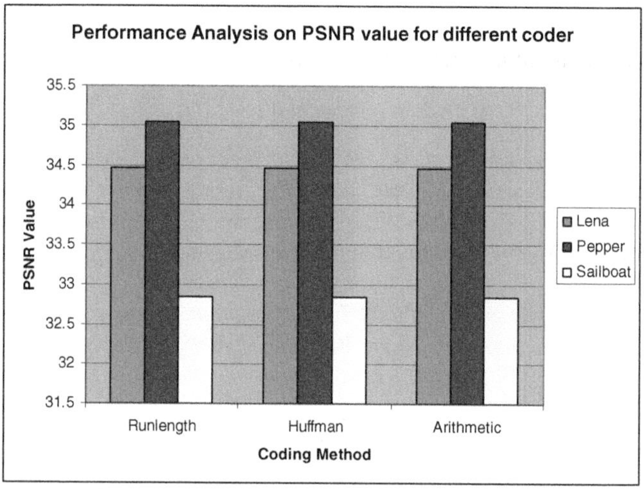


**Fig. 12.** Performance Analysis on PSNR Value of different coder

In the above graph that comparing is made with different coding techniques such as run length , Huffman and arithmetic for lena, pepper gray and sailboat images. From that graph the psnr value for Pepper image in all the coding technique is very high.

**Table 2.** Performance analysis on CR value

| Image | Wavelet Lifting | | |
|---|---|---|---|
| | RLC | HMC | AMC |
| Lena | 16.20 | 33.33 | 66.12 |
| Pepper | 12.54 | 34.44 | 66.13 |
| Boat | 26.84 | 33.14 | 66.72 |

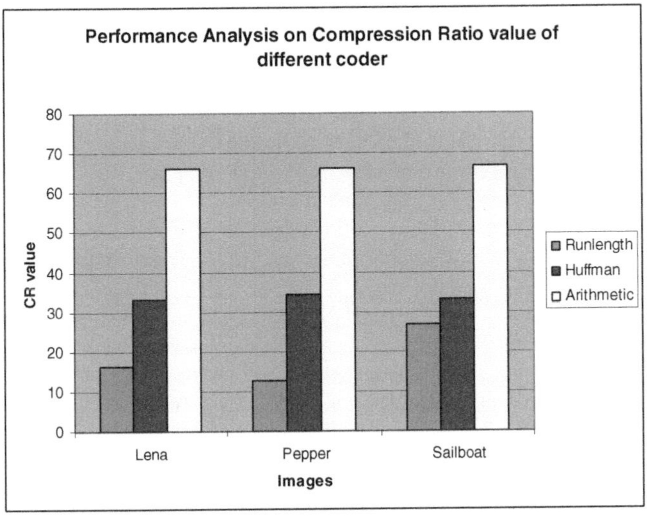


**Fig. 13.** Performance Analysis on CR Value of different coder

From the above graph we found that comparing with arithmetic coder, run length coder and Huffman coder on lena pepper gray and sailboat images the arithmetic coder gives higher CR value than the other coders.

## 8 Conclusions

In this paper the Spherical coding in the wavelet lifting with various wavelets domain with various encoders have discussed. From that discussion the wavelet lifting scheme and arithmetic coding gives best reconstructed image and also better CR and PSNR is concluded. Arithmetic coder gives better CR as well as gives better PSNR.

Wavelet compression is a lossy method. To make it lossless and more efficient, we go in for adaptive schemes in various factors for the transform. For example, we have different algorithms like directional adaptive algorithm, Directional Qucinix Lifting, Adaptive filters etc for future enhancement.

## References

[1] Ates, H., Orchard, M.: Wavelet image coding using the spherical representation. In: Proc. IEEE Int. Conf. Image Processing, Genova, Italy, vol. 1, pp. 89–92 (September 2005)

[2] Antonini, M., Barlaud, M., Mathieu, P., Daubechies, I.: Image coding using wavelet transform. IEEE Trans. Image Process. 1(2), 205–220 (1992)

[3] Antonini, M., Barlaud, M., Mathieu, P., Daubechies, I.: Image coding using vector quantization in the wavelet transform domain. In: Proc. IEEE Int. Conf. Acoustics, Speech, Signal Processing, Albuquerque, NM, vol. 4, pp. 2297–2300 (April 1990)

[4] Shapiro, J.: Embedded image coding using zerotrees of wavelet coefficients. IEEE Trans. Signal Process. 41(12), 3445–3462 (1993)

[5] Said, A., Pearlman, W.: A new fast and efficient image codec based on set partitioning in hierarchical trees. IEEE Trans. Circuit Syst. Video Technol. 6(3), 243–250 (1996)
[6] Xiong, Z., Ramchandran, K., Orchard, M.: Space-frequency quantization for wavelet image coding. IEEE Trans. Image Process. 6(5), 677–693 (1997)
[7] Xiong, Z., Ramchandran, K., Orchard, M.: Joint optimization of scalar and tree-structured quantization of wavelet image decomposition. In: Proc. Conf. Rec. 33th Asilomar, Pacific Grove, CA, vol. 2, pp. 891–895 (November 1993)
[8] Ramchandran, K., Orchard, M.T.: An investigation of waveletbased image coding using an entropy-constrained quantization framework. IEEE Trans. Signal Process. 46(2), 342–353 (1998)
[9] Xiong, Z., Galatsanos, N., Orchard, M.: Marginal analysis prioritization for image compression based on a hierarchical wavelet decomposition. In: Proc. IEEE Int. Conf. Acoustics, Speech, Signal Processing, Minneapolis, MN, vol. 5, pp. 546–549 (April 1993)
[10] LoPresto, S.M., Ramchandran, K., Orchard, M.T.: Image coding based on mixture modeling of wavelet coefficients and a fast estimation-quantization framework. In: Proc. Data Compression Conf., Snowbird, UT, pp. 221–230 (March 1997)
[11] Joshi, R., Jafarkhani, H., et al.: Comparison of different methods of classification in subband coding of images. IEEE Trans. Image Process. 6(11), 1473–1486 (1997)
[12] Taubman, D.: High performance scalable image compression with EBCOT. IEEE Trans. Image Process. 9(7), 1219–1235

# An Efficient Visible Watermarking for Copyright Protection Using Discrete Wavelet Transform

R. Anisha, S. Panimalar, and M. Gomathi

Manonmaniam Sundaranar University,
Tirunelveli
anisha.mtech@gmail.com,
Snowb4ufriends@yahoo.co.in

**Abstract.** The objective of this project is to use a novel method for generic visible watermarking with a capability of lossless image recovery. This method is based on one-to-one compound mapping algorithm. The compound mappings are proved to be reversible, which allows for lossless recovery of original images from watermarked images. The mappings may be adjusted to yield pixel values close to those of desired visible watermarks. Different types of visible watermarks, including opaque monochrome and translucent full color ones, are embedded as applications of the proposed generic approach. To increase the visual perception Discrete Wavelet Transforms is used. Experimental results show that our method provides a higher embedding capacity compared to the other algorithms proposed in the literature.

**Index Terms:** Lossless reversible visible, watermarking, one-to-one compound mapping, Discrete Wavelet Transforms, Translucent watermark.

## 1 Introduction

Digital watermarking is defined as a process of embedding data (watermark) into a multimedia object to help to protect the owner's right to that object. The embedded data (watermark) may be either visible or invisible .In visible watermarking , a secondary image ie. a watermark is embedded in a primary (source) image such that watermark is intentionally visible to a human observer whereas in the case of invisible watermarking the embedded data is not visible, but may be extracted/detected by a computer program.

Some of the desired characteristics of visible watermarks are listed below [23][24].

1. A visible watermark should be obvious in both color and monochrome images.

2. The watermark should be spread in a large or important area of the image in order to prevent its deletion by clipping.

D. Nagamalai, E. Renault, M. Dhanushkodi (Eds.): CCSEIT 2011, CCIS 204, pp. 55–64, 2011.

3. The watermark should be visible yet must not significantly obscure the image details beneath it.

4. The watermark must be difficult to remove; removing a watermark should be more costly and labor intensive than purchasing the image from the owner.

5. The watermark should be applied automatically with little human intervention and labor.

Embedding of watermarks, either visible or invisible, affect the quality of the source image in general. A group of techniques, named *reversible* watermarking [8]–[19], allow users to remove the embedded watermark and restore the original content as needed. However, not all reversible watermarking techniques guarantee *lossless image recovery*, which means that the recovered image is identical to the original, pixel by pixel. Lossless recovery is important in many applications where serious concerns about image quality arise. Some examples include forensics, medical image analysis, historical art imaging, or military applications.

Compared with their invisible counterparts, there are relatively few mentions of lossless visible watermarking in the literature. Several lossless invisible watermarking techniques have been proposed in the past. The most common approach is to compress a portion of the original host and then embed the compressed data together with the intended payload into the host [5], [13]–[15]. Another approach is to superimpose the spread-spectrum signal of the payload on the host so that the signal is detectable and removable [3]. A third approach is to manipulate a group of pixels as a unit to embed a bit of information [16], [17]. Although one may use lossless invisible techniques to embed removable visible watermarks [11], [18], the low embedding capacities of these techniques hinder the possibility of implanting large-sized visible watermarks into host media.

As to lossless visible watermarking, the most common approach is to embed a monochrome watermark using deterministic and reversible mappings of pixel values or DCT coefficients in the watermark region [6], [9], [19]. Another approach is to rotate consecutive watermark pixels to embed a visible watermark [19]. One advantage of these approaches is that watermarks of arbitrary sizes can be embedded into any host image. However, only *binary* visible watermarks can be embedded using these approaches, which is too restrictive since most company logos are colorful.

In this paper, a method for lossless visible watermarking using an appropriate *compound mappings* that allow mapped values to be controllable is used. The mappings are proved to be *reversible* for lossless original image recovery . The approach is *generic*, leading to the possibility of embedding different types of visible watermarks into cover images. Two applications of the proposed method are demonstrated, where opaque monochrome watermarks and non uniformly translucent full-color ones are respectively embedded into color images. More specific compound mappings are also created and proved to be able to yield visually more distinctive visible watermarks in the watermarked image. Embedding *removable translucent full-color watermarks* provide better advertising effects than traditional monochrome ones.

In the remainder of this paper, DWT is described in SectionII. The one-to-one compound mappings is described in Section III. Related lemmas and theorems are also proved and security protection measures described. Applications of the proposed method for embedding opaque monochrome and translucent color watermarks into color images are described in Sections 3 and 4, respectively. In Section 5, the specific compound mapping for yielding more distinctive visible watermarks is described. In Section 6, experimental results are presented to demonstrate the effectiveness of the proposed method. Finally, a conclusion with some suggestions for future work is included in Section 7.

## 2 Discrete Wavelet Transform

The discrete wavelet transform (DWT) is an implementation of the wavelet transform using a discrete set of the wavelet scales and translations obeying some defined rules. In other words, this transform decomposes the signal into mutually orthogonal set of wavelets, which is the main difference from the continuous wavelet transform (CWT), or its implementation for the discrete time series sometimes called discrete-time continuous wavelet transform (DT-CWT).

The discrete wavelet transform is a hierarchical subband technique. The subbands are created by applying decomposition on original image. To start the decomposition, the image is filtered in both the horizontal and vertical directions using separable filters. This creates four subbands as follows: Subband LL1 represents the horizontal and vertical low frequency components of the image. Subband HH1 represents the horizontal and vertical high frequency components of the image. Subband LH1 represents the horizontal low and vertical high frequency components. Subband HL1 represents the horizontal high and vertical low frequency components.

To obtain the next level in the decomposition the two filters are again applied, but only to the LL1 subband. Again we obtain four subbands labeled LL2, HH2, LH2, and HL2 with representations similar to the first level subbands. The following example illustrates the discrete wavelet transform.

**Fig. 1.** Original Lena Image

The two level decomposition image and corresponding decomposition map of above figure is show below.

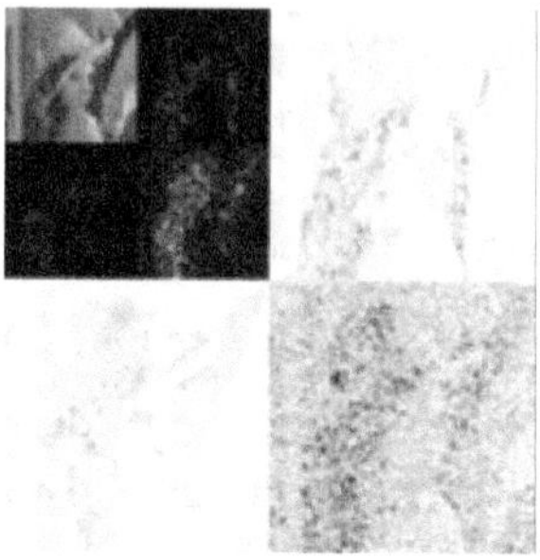

**Fig. 2.** Two Level Decomposed Image

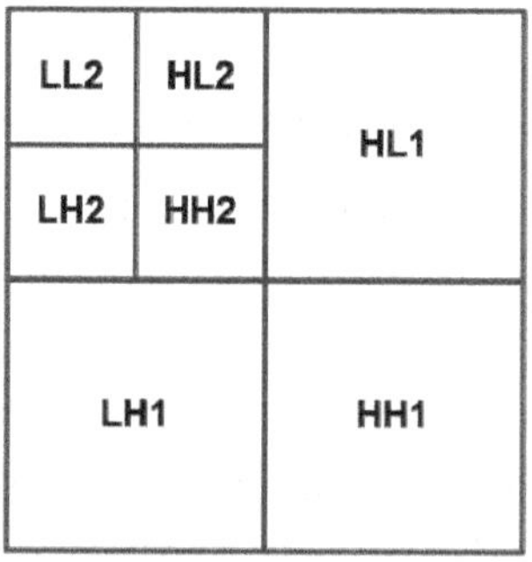


## 3 Reversible One-to-One Compound Mapping

First, we propose a generic *one-to-one compound mapping* for converting a set of numerical values $P = \{p1, p2 \ldots pM\}$ to another set Q={q1,q2...qM}, such that the respective mapping from $p_i$ to $q_i$ for all i=1,2...,M is *reversible*. Here, for the copyright protection applications investigated in this study, all the values $p_i$ and $q_i$ are image pixel values (grayscale or color values). The compound mapping f is governed by a one-to-one function $F_x$ with one parameter x=a or b in the following way:

$$q=f(p)= F_b^{-1}(F_a(p)) \tag{1}$$

where $F_x^{-1}$ is the inverse of $F_x$ which, by the one-to-one property, leads to the fact that if $F_a(p)=p'$ , then $F_a^{-1}(p')=p$ for all values of a and b. On the other hand, $F_a(p)$ and $F_b(p)$ generally are set to be *unequal* if a≠b.

The compound mapping described by (1) is indeed *reversible*, that is, p can be derived exactly from q using the following formula:

$$p= f^{-1}(q)=F_a^{-1}(F_b(q)) \tag{2}$$

# 4 Generic Visible Watermark Embedding and Extraction

*Theorem 1 (Lossless Reversible Visible Watermarking):*

There exist one-to-one compound mappings for use to embed into a given imageI a visible watermark Q whose pixel values are close to those of a given watermark L , such that the original image can be recovered from Q losslessly.

*Proof:* This is a consequence of Lemmas 1 and 2 after regarding the individual pixel values in I ,L and Q respectively as those of p,l and q mentioned in Lemma 2. And it is clear by Lemma 1 that the value p can be recovered losslessly from the mapped value q which is derived in Lemma 2.

The above discussions are valid for embedding a watermark in a grayscale image. If color images are used both as the cover image and the watermark, we can apply the mappings to each of the color channels to get multiple independent results. The resulting visible watermark is the composite result of the color channels.

Based on Theorem 1, the proposed generic lossless reversible visible watermarking scheme with a given image and a watermark as input is described as an algorithm as follows.

-------------------------------------------------------------

**Algorithm 1: Generic Visible Watermark Embedding**

-------------------------------------------------------------

***Input***: an image and a watermark L .

***Output***: watermarked image W.

***Steps***:

1) Select a set P of pixels from I where L is to be embedded, and call P a *watermarking area*.
2) Denote the set of pixels corresponding to P in W by Q in by .
3) For each pixel X with value p in P , denote the corresponding pixel in Q as Z as and the value of the corresponding pixel Y in L as l , and conduct the following steps.

a) Apply an estimation technique to derive a to be a value close to p , using the values of the neighboring pixels of X (excluding X itself).
b) Set b to be the value l .
c) Map p to a new value $p=F_b^{-1}(F_a(p))$ .
d) Set the value of Z to be q .

4) Set the value of each remaining pixel in W, which is outside the region P, to be equal to that of the corresponding pixel in I .

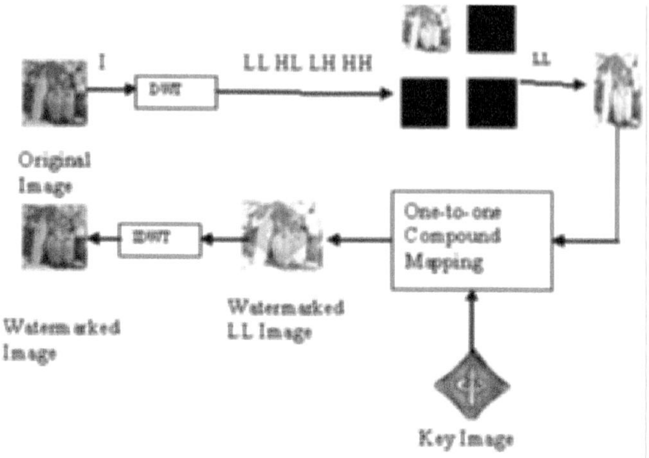


The corresponding watermark removal process for a watermarked Image W generated by Algorithm 1 is described as an algorithm as follows.

---

**Algorithm 2: Generic Watermark Removal for Lossless Image Recovery**

-------------------------------------------------------------

***Input*:** a watermarked image W and a watermark L.

***Output*:** the original image R recovered from W .

***Steps:***

1) Select the same watermarking area Q in W as that selected in Algorithm 1.
2) Set the value of each pixel in R , which is outside the region Q , to be equal to that of the corresponding pixel In W .
3) For each pixel Z with value q inQ , denote the corresponding pixel in the recovered image R as X and the value of the corresponding pixel Y in L as l, and conduct the following steps.

a) Obtain the same value a as that derived in Step 3a of Algorithm 1 by applying the same estimatdion technique used there.
b) Set b to be the value l .
c) Restore p from q by setting $p=F_a^{-1}(F_b(q))$ .
d) Set the value of X to be p .

---

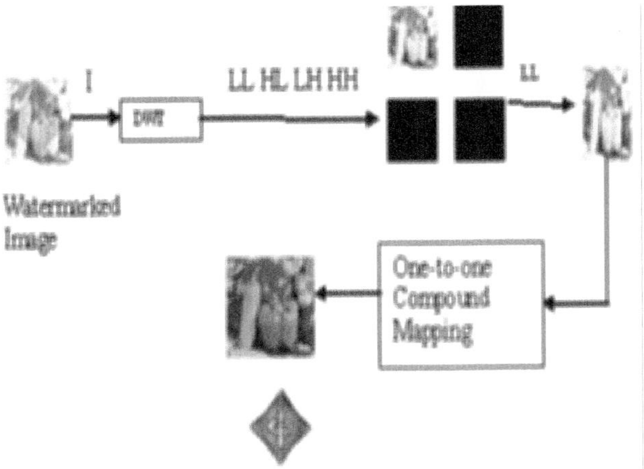


## 5 Experimental Results

The following four images are taken for this experiment.

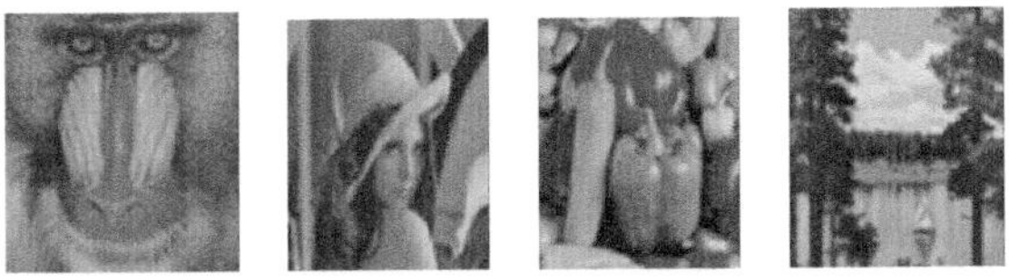

To quantitatively measure the effectiveness of this method, we define a set of performance metrics here. First, the quality of a watermarked Image W is measured by the peak signal-to-noise ratio (PSNR) of W with respect to the non recoverable watermarked image B in the following way:

$$PSNR_W = 20 \times \log_{10}\left(255 \Big/ \sqrt{\frac{1}{w \times h}\sum_{y=1}^{h}\sum_{x=1}^{w}[W(x,y)-B(x,y)]^2}\right)$$

Also, the quality of a *recovered* image R is measured by the PSNR of R with respect to the original image I in a similar way

$$PSNR_R = 20 \times \log_{10}\left(255 \Big/ \sqrt{\frac{1}{w \times h}\sum_{y=1}^{h}\sum_{x=1}^{w}[R(x,y)-I(x,y)]^2}\right)$$

**PSNR analysis after watermark embedding**

| IMAGES | PSNR |
|---|---|
| Leena | 45.02 |
| Pepper | 43.01 |
| Boat | 42.23 |
| Baboon | 47.22 |

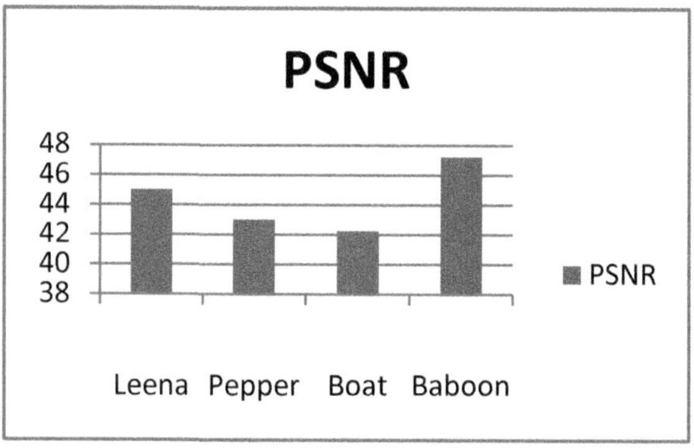


**PSNR analysis after watermark extraction**

| IMAGES | PSNR |
|---|---|
| Leena | 48.34 |
| Pepper | 46.24 |
| Boat | 45.01 |
| Baboon | 50.02 |

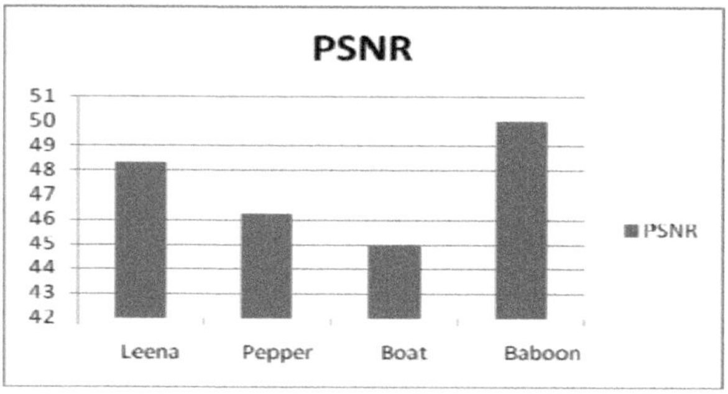


## 6 Conclusions

In this paper, a new method for reversible visible watermarking with lossless image recovery capability has been developed. The method uses one-to-one compound mappings that can map image pixel values to those of the desired visible watermarks. The compound mappings allow different types of visible watermarks to be embedded. To increase the visual perception Discrete Wavelet Transform is used. PSNR values are increased.

## References


[1] Petitcolas, F.A.P., Anderson, R.J., Kuhn, M.G.: Information hiding—A survey. Proc. IEEE 87(7), 1062–1078 (1999)

[2] Johnson, N.F., Duric, Z., Jajodia, S.: Information Hiding. In: Steganography and Watermarking—Attacks and Countermeasures. Kluwer, Boston (2001)

[3] Cox, I.J., Kilian, J., Leighton, F.T., Shamoon, T.: Secure spread spectrum watermarking for multimedia. IEEE Trans. Image Process. 6(12), 1673–1687 (1997)

[4] Kankanhalli, M.S., Rajmohan, Ramakrishnan, K.R.: Adaptive visible watermarking of images. In: Proc. IEEE Int. Conf. Multimedia Computing and Systems, vol. 1, pp. 568–573 (1999)

[5] Hu, Y., Kwong, S.: Wavelet domain adaptive visiblewatermarking. Electron. Lett. 37(20), 1219–1220 (2001)

[6] Mohanty, S.P., Ramakrishnan, K.R., Kankanhalli, M.S.: A DCT domain visible watermarking technique for images. In: Proc. IEEE Int. Conf. Multimedia and Expo, vol. 2, pp. 1029–1032 (July 2000)

[7] Braudaway, G., Magerlein, K.A., Mintzer, F.: Protecting publicly available images with a visible image watermark. In: Proc. SPIE Int. Conf. Electronic Imaging, vol. 2659, pp. 126–133 (February 1996)

[8] Cheng, Y.J., Tsai, W.H.: A new method for copyright and integrity protection for bitmap images by removable visible watermarks and irremovable invisible watermarks. Presented at the Int. Computer Symp.—Workshop on Cryptology and Information Security, Hualien, Taiwan, R.O.C. (December 2002)

[9] Huang, P.M., Tsai, W.H.: Copyright protection and authentication of grayscale images by removable visible watermarking and invisible signal embedding techniques: A new approach. Presented at the Conf. Computer Vision, Graphics and Image Processing, Kinmen, Taiwan, R.O.C. (August 2003)

[10] Hu, Y., Kwong, S., Huang, J.: An algorithm for removable visible watermarking. IEEE Trans. Circuits Syst. Video Technol. 16(1), 129–133 (2006)

[11] Hu, Y., Jeon, B.: Reversible visible watermarking and lossless recovery of original images. IEEE Trans. Circuits Syst. Video Technol. 16(11), 1423–1429 (2006)

[12] Macq, B.: Lossless multiresolution transform for image authenticating watermarking. Presented at the European Signal Processing Conf., Tampere, Finland (September 2000)

[13] Fridrich, J., Goljan, M., Du, R.: Lossless data embedding—New paradigm in digital watermarking. J. Appl. Signal Process. (2), 185–196 (2002)

[14] Awrangjeb, M., Kankanhalli, M.S.: Lossless watermarking considering the human visual system. Presented at the Int. Workshop on Digital Watermarking, Seoul, Korea (October 2003)

[15] Awrangjeb, M., Kankanhalli, M.S.: Reversible watermarking using a perceptual model. J. Electron. Imag. 14(013014) (March 2005)

[16] de Vleeschouwer, C., Delaigle, J.F., Macq, B.: Circular interpretation of bijective transformations in lossless watermarking for media asset management. IEEE Trans. Multimedia 5(1), 97–105 (2003)

[17] Tian, J.: Reversible data embedding using a difference expansion. IEEE Trans. Circuits Syst. Video Technol. 13(8), 890–896 (2003)
[18] Tsai, H.M., Chang, L.W.: A high secure reversible visible watermarking scheme. In: Proc. IEEE Int. Conf. Multimedia and Expo, Beijing, China, pp. 2106–2109 (July 2007)
[19] Yip, S.K., Au, O.C., Ho, C.W., Wong, H.M.: Lossless visible watermarking. In: Proc. IEEE Int. Conf. Multimedia and Expo, pp. 853–856 (July 2006)
[20] Lumini, A., Maio, D.: Adaptive positioning of a visible watermark in a digital image. In: Proc. Int. Conf. Multimedia and Expo, Taipei, Taiwan, R.O.C., pp. 967–970 (June 2004)
[21] Li, X., Orchard, M.T.: Edge-directed prediction for lossless compression of natural images. IEEE Trans. Image Process. 10(6), 813–817 (2001)
[22] Liu, T.-Y., Tsai, W.-H.: Generic Lossless Visible Watermarking—A New Approach. IEEE Transactions on Image Processing 19(5) (May 2010)
[23] Yeung, M.M., et al.: Digital Watermarking for High- Quality Imaging. In: Proc. of IEEE First Workshop on Multimedia Signal Processing, Princeton, NJ, pp. 357–362 (June 1997)
[24] Mintzer, F., et al.: Effective and Ineffective Digital Watermarks. In: Proc. of IEEE International Conference on Image Processing, ICIP 1997, vol. 3, pp. 9–12 (1997)

# Evolving Test Patterns for Use Case Maps

P.G. Sapna[1], Hrushikesha Mohanty[2], and Arunkumar Balakrishnan[1]

[1] Coimbatore Institute of Technology, Coimbatore, India
[2] University of Hyderabad, Hyderabad, India
sapna@cit.edu.in, {mohanty.hcu,arunkumar.cta.cit}@gmail.com

**Abstract.** An important aspect of software testing today is the ability to seamlessly move from analysis models to generating test scenarios and test cases. Use Case Maps(UCMs) are a scenario-based software engineering technique used at early stages of software development. The notation is most suited to capturing and eliciting use cases as well as high-level architectural design and for test case generation. Test patterns provide established solutions for designing tests or for supporting the testing process. In this work, we define test patterns based on the the primitives of the Use Case Maps to aid the process of test scenario generation.

## 1 Introduction

There is a lot of effort to standardize the User Requirements Notation(URN). Use Case Maps(UCMs) are used in this direction as it is a scenario based notation that helps in capturing functional requirements of a system. Also, UCMS bridge the modeling gap between requirements(represented as use cases) and design by linking behaviour and structure in an explicit way. UCMs elaborate one or more paths describing the causal relationships between responsibilities of one or more use cases i.e. a UCM shows the progression of scenarios along use cases. Thus, UCMs integrate many scenarios enabling reasoning about interactions between scenarios.

A pattern is a named description of a problem and solution that can be applied to new contexts. It suggests how to apply in varying novel circumstances, considering the trade-offs [6]. Patterns exists that address design and architecture issues. Many well-known patterns address design, architecture, or process issues. Testing patterns help give solutions for designing tests. For example, Binder [4] suggests a collection of test design patterns for object-oriented systems. Patterns for artifacts like classes, methods and scenarios as well as for test automation and test oracles are discussed. Testing patterns help bridge the divide between manual test generation and automated test case generation.

Individual scenarios help in the understanding of functionalities as well as guide the definition of details scenarios for design and testing phase. Amyot and Eberlein [7] in their work handle the issue of extracting individual scenarios from UCM specifications thereby enabling the validation of requirements and facilitating the transition from the requirements to the design phase. The objective of

D. Nagamalai, E. Renault, M. Dhanushkodi (Eds.): CCSEIT 2011, CCIS 204, pp. 65–74, 2011.

this work is to develop test patterns from Use Case Map patterns. The number of patterns and their unstructured nature(e.g. OR-forks and OR-joins in UCM need not be nested) require that test patterns be defined to aid in the process of automated test case generation. For this, we consider that UCMs consisting of multiple scenarios embedded in one have been converted to individual scenarios. The work of [2] looks at building test patterns for basic UCM patterns, namely, alternatives, concurrency, loops and stubs. Mussbacher's [5] work uses new notational elements and defines 18 additional patterns of UCMs in addition to work in [2]. This work studies [5] and aims at building test patterns using basic testing techniques, namely, equivalence class partitioning, boundary value testing, loop testing and orthogonal array testing techniques.

Section 2 introduces UCMS and discusses related work in the area. Section 3 details the approach used and then discusses the UCM patterns. Test patterns corresponding to each UCM patterns is given. Section 4 gives concludes the work and identifies areas for future work.

## 2 Survey

### 2.1 Use Case Maps

Use Case Maps (UCMs) are used as a visual notation for describing causal relationships between responsibilities of one or more use cases [1]. Responsibilities relate to processing like actions, activities, operations and tasks. Software entities like objects, processes and databases are represented using components. The relationships in UCMs are called causal as they link causes represented using preconditions and triggering events to effects represented using postconditions and

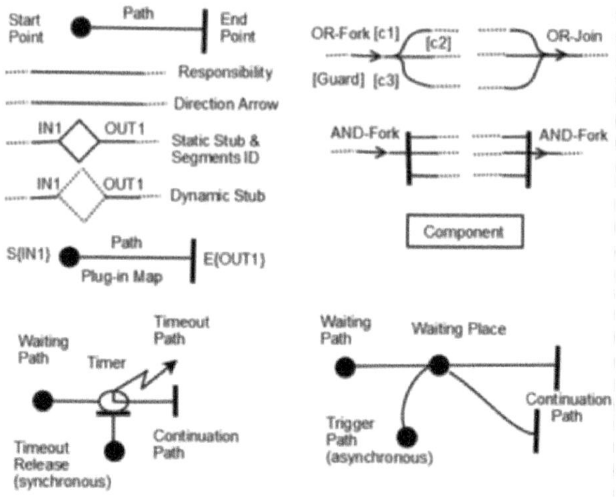


**Fig. 1.** Basic Elements of UCM Notation

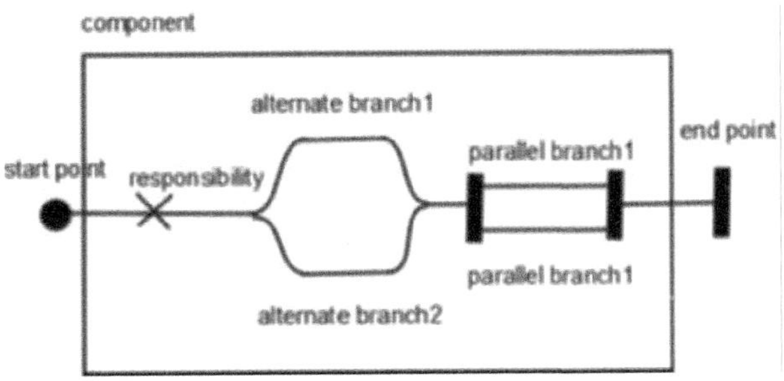


**Fig. 2.** Example Use Case Map

resulting events. Thus, UCMs show use cases and use case maps scenarios model functional requirements of a system.

The basic elements of the UCM notation are shown in Figure 1. A map can contain any number of paths and structural elements(components). Actions to fulfil a scenario are represented using *responsibilities*. Paths show causal sequences. Alternatives are represented using OR-forks and OR-joins whereas concurrency is represented using AND-forks and AND-joins. Loops can be modeled using OR-forks and OR-joins. UCM models can be decomposed using stubs which contain sub-maps called *plug-ins*, which are reusable units of behaviour and structure. There are two types of stubs: static and dynamic. A static stub can have atmost one plug-in, whereas a dynamic stub may have many plug-ins that are selected during runtime. The choice of a plug-in in the case of dynamic stub is dependent on a selection policy. Figure 2 shows a sample use case map.

### 2.2 Defining Scenarios Using UCM

UCMs are useful for describing multiple scenarios abstractly in a single view. The advantage of such a representation is that it helps understanding and reasoning about the system as a whole as well as early detection of conflicting or inconsistent scenarios [7]. However, this same advantage causes difficulty in extracting individual scenarios.

Scenarios begin at the start point representing some precondition and triggering event and finish at end points, which describe postconditions and resulting events [8].

For example, the scenarios extracted from the sample UCM in Figure 2 include:

1. <start point, responsibility, alternate branch1, parallel branch1, parallel branch2, end point>
2. <start point, responsibility, alternate branch2, parallel branch1, parallel branch2, end point>
3. <start point, responsibility, alternate branch1, parallel branch2, parallel branch1, end point>
4. <start point, responsibility, alternate branch2, parallel branch2, parallel branch1, end point>

### 2.3 Related Work

Amyot et al [1] handle the problem of automated scenario generation from UCMs in their work. Scenario definitions and traversal algorithms are used to extract individual scenarios from UCMs, which are stored as XML files. Transformation to UML diagrams(viz. sequence diagrams) is achieved through the use of XSLT.

Roy et al [10] in their work built a new Eclipse-based editing tool for UCMs, jUCMNav. The tool helps in creating, maintaining, analyzing and transforming UCM models. Kealey et al [11] in their work propose enhancement to the jUCMNav Eclipse plug-in by introducing new scenario traversal semantics as well as transformation to Message Sequence Charts(MSC). The objective is to use UCMs for scenario generation as well as translating UCMs to design diagrams like MSC.

Mussbacher [5] study the applicability of UCMs for workflow description with reference to scenarios and study how to evolve the UCM notation. They compare with standards for workflow, business process design and business process execution languages like Business Process Modeling Notation(BPMN), Business Process Execution Language for Web Services(BPEL4WS) and UML Activity diagrams. With the simple traversal mechanism as the basis, Mussbacher shows that UCMs support only eight of the possible twenty one workfloww patterns which is relatively lesser than for the other notations. However, with new notational elements and a new traversal mechanism, the number of patterns UCM supports raises to eighteen. In our work, we concentrate our efforts on developing test patterns for the UCM patterns defined in [5].

In [4], testing patterns are developed that target the coverage of scenarios described in terms of Use Case Maps. The objective is to cover functional scenarios at different levels of completeness, like all results, all causes and all results, all path segments, all end-to-end paths, all plug-ins. The idea is that covering paths in the UCMs leads to coverage of events and responsibilities that together form scenarios. The basis for the patterns is the use of white-box techniques for test case generation applied at the level of scenarios.

## 3 Test Patterns for Use Case Maps

### 3.1 The Approach

UCMs as mentioned before, is a scenario based notation that helps in capturing functional requirements of a system. The number of scenarios generated for even a small system is very large which makes exhaustive testing impossible. Hence, there is a need to test software in an effective way considering cost, time, coverage and customer requirements. Path testing is considered as a criteria to determine adequacy of testing with reference to scenarios. The number of paths tested can be used as a measure to determine the level of testing. Questions that can be asked are [9]:

- What is the maximum number of different paths possible ?
- What is the fewest number of paths possible ?
- How many different paths can be deduced ?

Also, equivalence class partitioning and boundary value analysis is used in case of preconditions and loops respectively. In this work, for each UCM pattern, we define corresponding test patterns. The patterns taken into consideration include:

a. Basic Control patterns - includes, sequence, parallel split, synchronization, exclusive choice and simple merge
b. Advanced branching and synchronization patterns - includes, multiple choice, synchronizing merge, multiple merge.

### 3.2 Basic Control Patterns

Consider two segments 'a' and 'b'. Here, a path that consists of two successive path segments is expressed by the path product(a*b) of the segment names. In case of parallel paths between two nodes, path sum is taken(a+b). In case of loops, an infinite sum is denoted by b* and a loop taken atleast once is denoted by $b^+$. Here, we look at developing test cases for basic control patterns. Test cases for alternatives, concurrency, loops, multiple start points, single stub and plugins and causal stubs have been discussed in [5]. In this work, we look at obtaining test cases from other patterns found in UCMs.

**Sequence:** Figure 3a shows the sequence pattern consisting of a sequence of responsibilities from start to end. The test case is a path product of all the sequences contributing to the path.

**Parallel Split:** Figure 3b indicates that activities can be performed in parallel or in any order. An AND-fork is used to represent parallel split. Hence, we consider that the activities may happen in sequence or parallel.

**Synchronization:** Figure 3c indicates a case where concurrent branches converge at an AND-join. The assumption here is that each concurrent branch is executed only once, and all incoming branches should be executed before control moves from the AND-join forwards.

**Exclusive Choice:** Figure 3d shows an exclusive choice represented using an OR-fork. Here, the branches have to be mutually exclusive.

**Simple Merge:** Figure 3e shows a case where branches merge together into one represented using an OR-join.

### 3.3 Advanced Branching and Synchronization Patterns

**Multiple Choice:** Figure 4a shows a case where more than one alternative branch can be selected and executed at the same time. Here, equivalence class

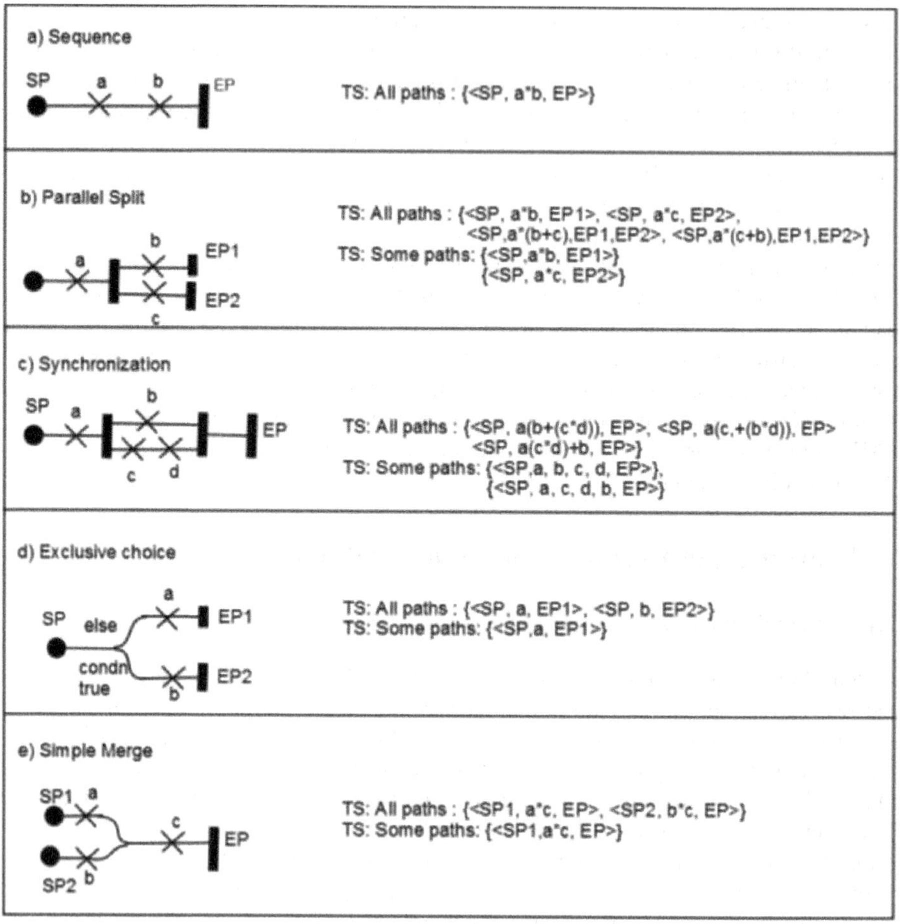


**Fig. 3.** Testing patterns for Basic control patterns

partitioning(ECP) is used to to partition the input domain, thereby eliciting all possible input classes for testing.

**Multiple Merge :** Figure 4b shows the case where multiple branches are merged without synchronization. The assumption is that activities following the merge will be performed once for each active branch. Here, each path from the start to the end is considered.

**Static Stub:** Figure 4c shows the case of static stub which contains only one plug-in and therefore having hierarchical decomposition of the UCM.

**Synchronizing Merge:** Figure 4d shows the case of dynamic stubs represented as dashed diamond containing several plug-ins. Here, the selection of the plug-in can be determined at run-time based on the selection policy defined(otherwise called pre-condition). The traversal waits until all plug-ins have completed their

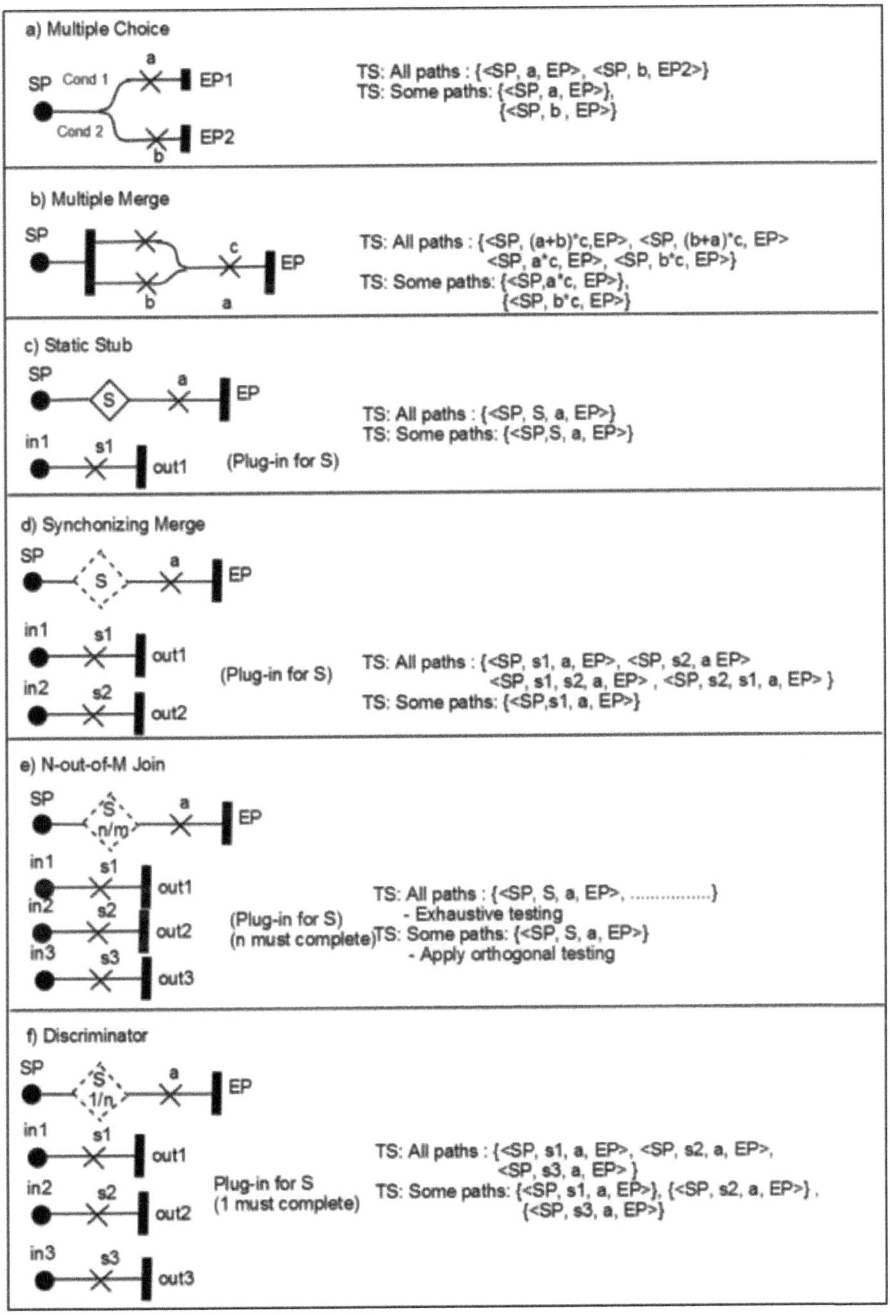


**Fig. 4.** Testing patterns for Advanced control patterns

activities before continuing along the path after the stub. Here, ECP is used taking into consideration valid and invalid input classes, thereby generating test cases where at least one branch is executed and at most all branches are executed.

**N-out-of-M Join:** Figure 4e is a type of merge where multiple concurrent incoming branches merge. Here, the continuation of the workflow happens when $n^{th}$ out of m incoming branches complete and thereby trigger the continuation of the activity. The other plug-ins may complete but do not trigger the continuation of the workflow. In this case, equivalence class partitioning and boundary value analysis is used. Equivalence class partitioning is used for defining the input domain and boundary value analysis used to determine the bounds groups for n out of m branches.

**Discriminator:** Figure 4f is a special case of n-out-of-m join and test cases generated similarly.

### 3.4 Structural Patterns

**Unstructured cycles:** Figure 5 shows an unstructured loop. With UCMs it is possible to create unstructured loops with OR-forks and OR-joins as nesting of the same is not a requirement. Beizer [9] defines four difference classes of loops and a methodology to design test cases for both simple and nested loops which is followed in this work. In cases of unstructured loops, Beizer recommends that unstructured loops be redesigned as structured loops which is one solution. Otherwise, in such a case, a condition that loops must be executed such that a responsibility is executed a maximum of two times can be applied[Linzhang04]. For the UCM in Figure 4, a total of three test cases will ensure that all paths are tested if the maximum condition is taken.

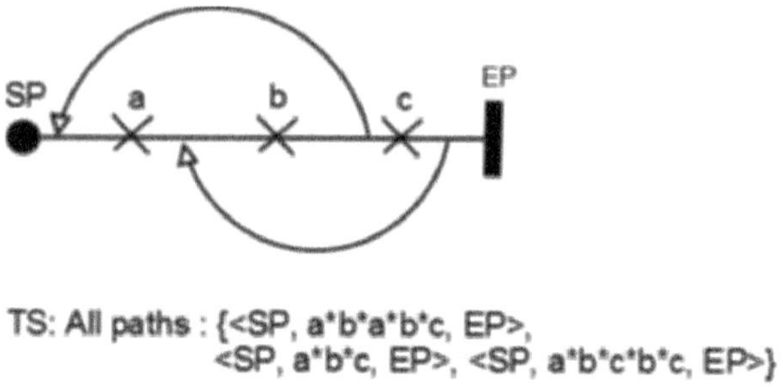


**Fig. 5.** Testing pattern for UCM Structural pattern

### 3.5 Patterns Involving Multiple Instances

Figure 6 shows a pattern involving multiple instances. Three such patterns have been defined by [5]. However, in all these cases, boundary value analysis with

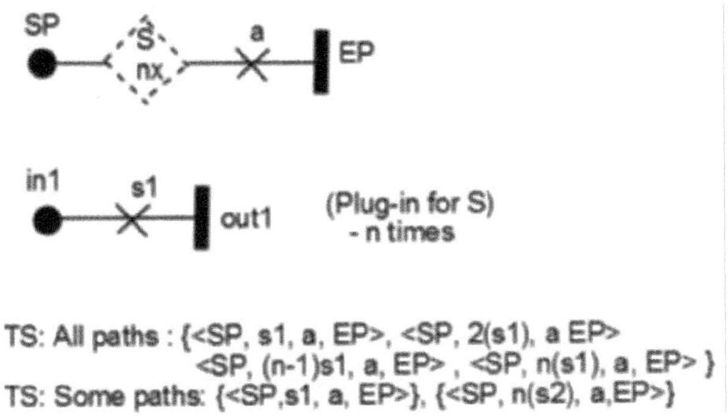


**Fig. 6.** Testing pattern for Multiple Instances

reference to the number of times the instances need to be executed is adopted. Valid test cases for ends of the range and invalid test cases for just above and below the range is included.

## 4 Conclusions and Future Work

The focus of this paper was to create scenarios for testing based on Use Case Maps. Use Case Maps have the advantage of integrating scenarios in a single view which helps designers and test engineers to understand and analyse requirements better than when they are represented individually. However, given the varied patterns possible when using UCMs, there is need to formulate procedures for testing. Currently, UCMs support only eight workflow patterns for which test scenarios have been defined. These patterns are basic in nature. This work looks at additional patterns defined in [5] and defines test patterns for the same. Future work involves building test patterns for 'communication patterns' defined in [5] which is not handled in this work as well as looking at a reduction procedure to reduce the number of scenarios generated.

## References

1. Amyot, D.: Specification and Validation of Telecommunications Systems with Use Case Maps and LOTOS, Ph.D. thesis, SITE, University of Ottawa, Canada (September 2001)
2. Amyot, D., Logrippo, L., Weiss, M.: Generation of test purposes from Use Case Maps. Computer Networks (2005)
3. Wang, L., Yuan, J., Yu, X., Hu, J., Li, X., Zheng, G.: Generating Test Cases from UML Activity Diagram based on Gray-Box Method. In: Proceedings of the 11th Asia-Pacific Software Engineering Conference (APSEC). IEEE Computer Society, Los Alamitos (2004)
4. Binder, R.V.: Testing Object-Oriented Systems Models, Patterns, and Tools. Addison Wesley, Reading (1999)

5. Mussbacher, G.: Evolving Use Case Maps as a Scenario and Workflow Description Language. In: 10th Workshop of Requirement Engineering (WER 2007), pp. 56–67 (2007)
6. Larman, C.: Applying UML and Patterns: An Introduction to Object-Oriented Analysis and Design and the Unified Process, 2nd edn. Prentice Hall, Englewood Cliffs (2001)
7. Amyot, D., Eberlein, A.: An Evaluation of Scenario Notations and Construction Approaches for Telecommunication Systems Development. Telecommunications Systems Journal 24(1), 61–94 (2003)
8. Amyot, D., He, X., He, Y., Cho, D.Y.: Generating Scenarios from Use Case Map Specifications. In: 3rd International Conference on Quality Software (QSIC), pp. 108–115 (2003)
9. Beizer, B.: Software Testing Techniques. Dreamtech Press (2002)
10. Roy, J.-F., Kealey, J., Amyot, D.: Towards Integrated Tool Support for the User Requirements Notation. In: Gotzhein, R., Reed, R. (eds.) SAM 2006. LNCS, vol. 4320, pp. 198–215. Springer, Heidelberg (2006)
11. Kealey, J., Amyot, D.: Enhanced Use Case Map Traversal Semantics. In: Gaudin, E., Najm, E., Reed, R. (eds.) SDL 2007. LNCS, vol. 4745, pp. 133–149. Springer, Heidelberg (2007)

# A Novel Approach for Compression of Encrypted Grayscale Images Using Huffman Coding

S. Saravana Kumar and S. Panimalar

Manonmaniam Sundaranar University, Tirunelveli
ssaravanan.puli@gmail.com,
snowb4ufriends@yahoo.co.in

**Abstract.** Currently information security is becoming more important in data storage and transmission. Images are broadly used in several processes. Therefore, the protection of image data from unauthorized access is essential. Image encryption plays a major role in the field of information hiding. The objective of this paper is to provide an efficient coding technique for compress the encrypted gray scale images without any loss. This is made possible by using Huffman coding. This paper uses resolution-progressive compression (RPC) technique to achieve better result. RPC compresses an encrypted image progressively in resolution, such that the decoder can observe a low-resolution version of the image, study local statistics based on it, and use the statistics to decode the next resolution level. The experimental results show that the method used in this paper for compression gives good result.

**Keywords:** Compression of encrypted images, resolution progressive compression, Huffman coding.

## 1 Introduction

In many convenient scenarios multimedia contents need both to be compressed and protected. Whereas the conventional way to compress and protect data requires that data are first compressed and then encrypted, in some cases it may be desirable to operate the other way around. This is the case, for example of an encrypted content that is transmitted without compression through the internet, but at a certain point needs to be compressed in order to match the characteristics of the transmission link. If the node is not trusted it may not be provided with the encryption key, then compression of the encrypted content is required.

In the literature, solutions have been proposed to treat image and video data as Markov sources and to exploit the Markovian property in the Slepian-Wolf decoder [1], [12].A similarwork is also found in [13] for nonencrypted colored sources. Some good results have been reported for binary images. However, there are some limitations with this approach: first, Markov decoding in a Slepian-Wolf decoder is expensive, especially in dealing with sources with nonbinary alphabets; second, bit-plane based Markov decoding certainly reduces the complexity, but the source dependency that originally defined in the symbol domain is usually not fully utilized when translated to bit-planes; third, since image and video data are known to be

D. Nagamalai, E. Renault, M. Dhanushkodi (Eds.): CCSEIT 2011, CCIS 204, pp. 75–83, 2011.

highly nonstationary, a global Markov model cannot describe its local statistics precisely. As reported in [12], for 8-bit grayscale images, only the first two most significant bit-planes (MSB) are compressible by employing a 2-D Markov model in bit-planes. The challenging issue still remains is, how to effectively exploit the correlation of encrypted image data.

In this paper, lossless compression techniques are discussed and experimented on encrypted gray scale images. This paper provides an efficient way to compress encrypted images through resolution-progressive compression (RPC). The encoder starts by sending a downsampled version of the ciphertext. At the decoder, the corresponding low-resolution image is decoded and decrypted, from which a higher-resolution image is obtained by intraframe prediction. The predicted image, together with the secret encryption key, is used as the side information (SI) to decode the next resolution level. This process is iterated until the whole image is decoded. Huffman technique is used as coder and decoder.

The rest of paper is organized as follows. The next section II briefly describes the resolution-progressive compression technique. Section III describes the context adaptive technique. Section IV describes the Huffman coding technique. Section V briefly describes the compression algorithm by step wise. Experimental setup is discussed in Section V.Finally the conclusion is presented in Section VI.

## 2 Resolution Progressive Compression of Encrypted Images

The input grayscale image I is encrypted by using encryption algorithm. After applying the encryption algorithm the input image is converted into the ciphertext Y. This is shown in below Fig.1.

The encoder takes the ciphertext Y and decomposes it into four subimages. It is called as 00, 01, 10, and 11 sub-images. Each sub-image is a downsampled-by-two version of the encrypted image Y. The name of a sub-image denotes the horizontal and vertical offsets of the downsampling. The 00 sub-image is further downsampled to create multiple resolution levels. An example of the decomposition is illustrated in Fig. 2.

(a) (b)

**Fig. 1.** Encryption Process (a) Pepper Image (b) Encrypted Image Y

(a) (b)

**Fig. 2.** (a) Illustration of a three-level decomposition of the unencrypted "Pepper" image. (b) Layout of the sub-images.

After the downsampling, each sub-image is encoded independently using Huffman codes, and the resulting bits are transmitted from the lowest resolution to the highest. Decoding starts from the 00 sub-image of the lowest-resolution level, say, level $00_n$. This paper suggests transmitting the uncompressed $00_n$ sub-image as the doped bits]. Thus, the $00_n$ sub-image can be known by the decoder without ambiguity, and knowledge about the local statistics will be derived based on it. Next, other sub-images of the same resolution level are interpolated from the decrypted $00_n$ sub-image. Then the Context Adaptive Interpolation technique is applied for SI generation.

# 3 Context Adaptive Interpolation

The SI is generated through interpolation. For simplicity, for any pixel in the target sub-image, this technique only uses the four horizontal and vertical neighbors or the four diagonal neighbors in the known sub-images for the interpolation. For this purpose a simple and effective context adaptive interpolator (CAI) technique is used in this paper. Due to the isomorphism, this technique only describes the horizontal-vertical interpolator illustrated in Fig. 3.

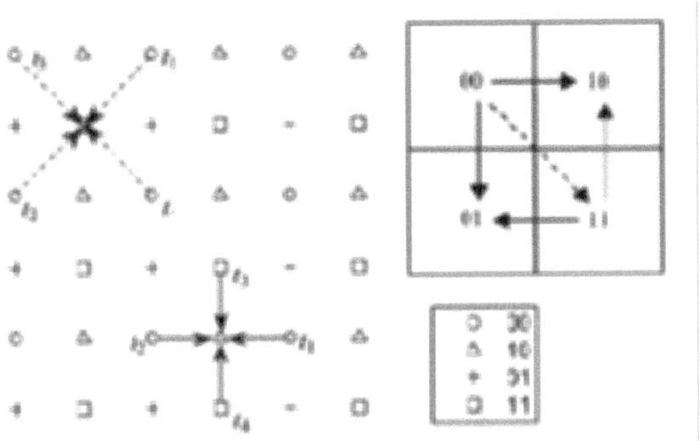

**Fig. 3.** Illustration of the two-step interpolation at the decoder side. The dashed arrows denote the first step interpolation, and the solid arrows denote the second step.

Let s be the pixel value to be interpolated, $t=[t_1,t_2,t_3,t_4]$be the vector of neighboring pixels. The interpolator classifies the local region into four types: smooth, horizontally-edged, vertically-edged and other. In smooth regions, a mean filter is applied; in horizontally/vertically edged regions, the interpolation is done along the edge; otherwise use a median filter. CAI is formulated as

$$\hat{s}=\begin{cases} mean(t), & (\max(t)-\min(t)\leq 20) \\ \frac{(t_1+t_2)}{2} & (|t_3-t_4|-|t_1-t_2|)>20) \\ \frac{(t_3+t_4)}{2} & (|t_1-t_2|-|t_3-t_2|>20) \\ median(t), & (otherwise). \end{cases} \tag{1}$$

## 4 Huffman Encoding

We use this coding to encode/compress a file. Steps:

1. Read the file.
2. Calculate the probability of each symbol (since we use an ASCII file, there are 256 possible symbols). Instead of actual probability we can use instance count.
3. Use the Huffman coding algorithm to find the coding for each symbol. Need to apply only on subset of symbols that actually appear in the file.
4. Encode the file, and write it out.

The out file must also include the encoding table, so as to permit decoding.

## 5 Compression Algorithm

1. Get the input gray scale image
2. Apply the encryption algorithm on to the input image to get the cipher text
3. Then apply the downsampling technique on to the cipher text

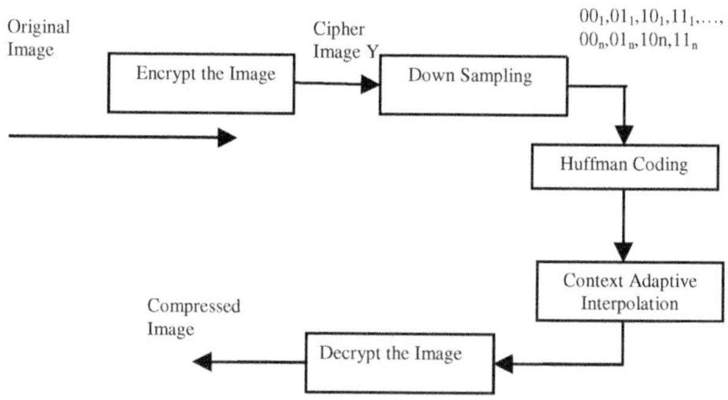


**Fig. 4.** Overall Process Flow Diagram of Compression of Encrypted Gray Scale Images

4. Apply the Huffman coding algorithm on to the each downsampled images.
5. Then the context adaptive interpolation technique to create the Side Information
6. Then apply the decryption algorithm to get the compressed image.

The overall process flow diagram is shown in Fig.4

# 6 Expriements and Performance Analysis

The images Pepper, Lena, Barbara and Plane and are used for the experiments. This is shown in Fig.5.

(a)

(b)

(c)

(d)

**Fig. 5.** Experimental Images (a) Pepper (b) Lena (c) Barbara (d) Plane

First the experimental images are encrypted to create the cipher text. The encrypted images of the experimental images are shown in Fig.6. Then the encrypted images are downsampled to create the sub images. Then the each sub images are encoded using the Huffman techniques.

Then the context adaptive interpolation technique is used for predicts the unknown sub images. Finally the interpolated images are decrypt to get the compressed image. The decrypted images are shown in Fig.7.

**Fig. 6.** Encrypted Images (a) Pepper (b) Lena (c) Barbara (d) Plane

**Fig. 7.** Decrypted Images (a) Pepper (b) Lena (c) Barbara (d) Plane

### A. *Performance Analysis*

The Peak-Signal-to-Noise-Ratio (PSNR) and Compression Ratio (CR) are two metrics used for analysis the performance of the above algorithm. The PSNR and CR values for the images are tabulated in Table1.

PSNR means Peak Signal to Noise Ratio (PSNR). This performance metric gives the signal to noise ratio between the original image and the compressed image. PSNR is calculated by using the following formula.

$$PSNR = 20log\frac{2^n - 1}{RMS}$$

$$RMS = \sqrt{\frac{1}{n}}\ \sum_{i=1}^{N}(x_i - x_i')^2$$

CR means Compression Ratio. This performance metric gives compression ratio. This ratio gives an indication of how much compression is achieved for a particular image. The PSNR & CR analysis of experimental images is shown in Fig.8.

**Table 1.** PSNR & CR Value For expirental images

| Image | PSNR | CR |
|---|---|---|
| Lena | 34.468 | 33.33 |
| Pepper | 35.046 | 34.44 |
| Boat | 32.848 | 33.14 |
| Barbara | 33.235 | 32.15 |

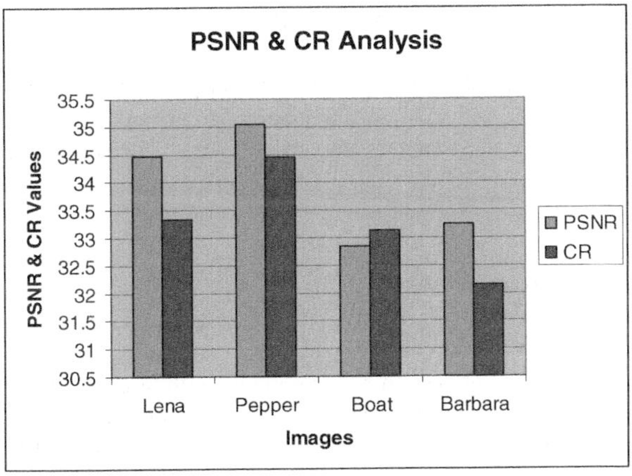


**Fig. 8.** PSNR & CR Analysis of Experimental Images

## 7 Conclusions

This paper provides an efficient coding technique for compress the encrypted gray scale images without any loss. This is made possible by using Huffman coding. This paper uses resolution-progressive compression (RPC) technique to achieve better result. RPC compresses an encrypted image progressively in resolution, such that the decoder can observe a low-resolution version of the image, study local statistics based on it, and use the statistics to decode the next resolution level. The experimental results show that the method used in this paper for compression gives good result.

## References

[1] Johnson, M., Ishwar, P., Prabhakaran, V.M., Schonberg, D., Ramchandran, K.: On compressing encrypted data. IEEE Trans. Signal Process. 52(10), 2992–3006 (2004)

[2] Stallings, W.: Cryptography and Network Security: Principles and Practice, 3rd edn. Prentice-Hall, Upper Saddle River (2003)

[3] Pradhan, S.S., Ramchandran, K.: Distributed source coding using syndromes (DISCUS). IEEE Trans. Inf. Theory 49(3), 626–643 (2003)

[4] Slepian, J.D., Wolf, J.K.: Noiseless coding of correlated information sources. IEEE Trans. Inf. Theory IT-19, 471–480 (1973)

[5] García-Frías, J., Zhao, Y.: Compression of correlated binary sources using turbo codes. IEEE Commun. Lett. 5(10), 417–419 (2001)

[6] Bajcsy, J., Mitran, P.: Coding for the Slepian-Wolf problem with turbo codes. In: Proc. IEEE Global Telecommun. Conf., San Antonio, TX, pp. 1400–1404 (November 2001)

[7] Aaron, A., Girod, B.: Compression with side information using turbo codes. In: Proc. IEEE Data Compression Conf., Snowbird, UT, pp. 252–261 (April 2002)

[8] Liveris, A., Xiong, Z., Georghiades, C.: Compression of binary sources with side information at the decoder using LDPC codes. IEEE Commun. Lett. 6(10), 440–442 (2002)

[9] Varodayan, D., Aaron, A., Girod, B.: Rate-adaptive distributed source coding using low-density parity-check codes. In: Proc. Asilomar Conf. Signals, Systems and Computers, Pacific Grove, CA, pp. 1203–1207 (November 2005)

[10] Xiong, Z., Liveris, A., Cheng, S.: Distributed source coding for sensor networks. IEEE Signal Process. Mag. 21, 80–94 (2004)

[11] Yang, Y., Stankovic, V., Xiong, Z.: Image encryption and data hiding: Duality and code designs. In: Proc. Inf. Theory Workshop, Lake Tahoe, CA, pp. 295–300 (September 2007)

[12] Schonberg, D.: Practical Distributed Source Coding and its Application to the Compression of Encrypted Data. Ph.D. dissertation, Univ. California, Berkeley (2007)

[13] Varodayan, D., Aaron, A., Girod, B.: Exploiting spatial correlation in pixel-domain distributed image compression. In: Proc. Picture Coding Symposium, Beijing, China (April 2006)

[14] Weinberger, M., Seroussi, G., Sapiro, G.: The LOCO-I lossless image compression algorithm: principles and standardization into JPEG-LS. IEEE Trans. Image Process. 9(8), 1309–1324 (2000)

[15] Wu, X., Memon, N.: Context-based adaptive lossless image coding. IEEE Trans. Commun. 45(4), 437–444 (1997)

[16] Aaron, A., Rane, S., Setton, E., Girod, B.: Transform-domain Wyner-Ziv codec for video. In: Proc. SPIE Visual Commun. Image Process., San Jose, CA, pp. 520–528 (January 2004)

[17] Berger, T.: Rate Distortion Theory: A Mathematical Basis for Data Compression. Prentice-Hall, Upper Saddle River (1971)

[18] Gersho, A., Gray, R.: Vector Quantization and Signal Compression. Kluwer, Boston (1992)

[19] Jayant, N., Noll, P.: Digital Coding of Waveforms: Principles and Applications to Speech and Video. Prentice-Hall, Upper Saddle River (1984)

[20] Rowitch, D.N., Milstein, L.B.: On the performance of hybrid FEC/ARQ systems using rate-compatible punctured turbo (RCPT) codes. IEEE Trans. Commun. 48, 948–959 (2000)
[21] Yao, Q., Zeng, W., Liu, W.: Multi-resolution based hybrid spatiotemporal compression of encrypted videos. In: Proc. IEEE Int. Conf. Acous., Speech and Sig. Process., Taipei, Taiwan, R.O.C., pp. 725–728 (April 2009)
[22] Liu, W., Zeng, W., Dong, L., Yao, Q.: Efficient Compression of Encrypted Grayscale Images. IEEE Transactions on Image Processing 19(4) (April 2010)

# Global Chaos Synchronization of Hyperchaotic Pang and Wang Systems by Active Nonlinear Control

Sundarapandian Vaidyanathan[1] and Karthikeyan Rajagopal[2]

[1] R & D Centre, Vel Tech Dr. RR & Dr. SR Technical University
Avadi-Alamathi Road, Avadi, Chennai-600 062, India
sundarvtu@gmail.com
http://www.vel-tech.org/

[2] School of Electronics and Electrical Engineering, Singhania University
Dist. Jhunjhunu, Rajasthan-333 515, India
rkarthiekeyan@gmail.com
http://www.singhaniauniversity.co.in/

**Abstract.** This paper investigates the global chaos synchronization of identical hyperchaotic Pang systems (Pang and Liu, 2011) and synchronization of non-identical hyperchaotic Pang system and Wang system (Wang and Liu, 2006). Active nonlinear feedback control is the method used to achieve the synchronization of the identical and different hyperchaotic Pang and Wang systems addressed in this paper and our results are established using Lyapunov stability theory. Since the Lyapunov exponents are not required for these calculations, the active control method is effective and convenient to synchronize identical and different hyperchaotic Pang and Wang systems. Numerical simulations are given to illustrate the effectiveness of the proposed synchronization schemes for the global chaos synchronization of hyperchaotic systems addressed in this paper.

**Keywords:** Active control, chaos synchronization, nonlinear control, Pang system, Wang system.

## 1 Introduction

Chaos theory has been developed and extensively studied over the past two decades. A chaotic system is a nonlinear dynamical system that is highly sensitive to initial conditions. This sensitivity is popularly referred to as the *butterfly effect* [1]. There has been active research done in the past three decades on the control and synchronization of nonlinear chaotic systems.

Chaos synchronization problem was first described by Fujisaka and Yemada [2] in 1983. This problem did not receive great attention until Pecora and Carroll ([3]-[4]) published their results on chaos synchronization in early 1990s. From then on, chaos synchronization has been extensively and intensively studied in the last three decades ([3]-[21]). Chaos theory has been explored in a variety of fields including physical [5], chemical [6], ecological [7] systems, secure communications ([8]-[10]) etc.

In most of the chaos synchronization approaches, the master-slave or drive-response formalism is used. If a particular chaotic system is called the *master* or *drive* system

D. Nagamalai, E. Renault, M. Dhanushkodi (Eds.): CCSEIT 2011, CCIS 204, pp. 84–93, 2011.

and another chaotic system is called the *slave* or *response* system, then the idea of the synchronization is to use the output of the master system to control the slave system so that the output of the slave system tracks the output of the master system asymptotically.

Since the seminal work by Pecora and Carroll ([3]-[4]), a variety of impressive approaches have been proposed for the synchronization for the chaotic systems such as PC method ([3]-[4]), the sampled-data feedback synchronization method ([10]-[11]), OGY method [12], time-delay feedback approach [13], backstepping design method [14], adaptive design method ([15]-[18]), sliding mode control method [19], active control method ([20]-[21]), etc.

In this paper, we derive new results for the global chaos synchronization for identical and different hyperchaotic Pang and Wang systems. Explicitly, using the active nonlinear control method, we achieve global chaos synchronization for identical hyperchaotic Pang systems ([22], 2011) and non-identical hyperchaotic Pang system and Wang system ([23], 2006).

This paper has been organized as follows. In Section 2, we give the problem statement and our methodology. In Section 3, we derive results for the chaos synchronization of two identical hyperchaotic Pang systems. In Section 4, we discuss the synchronization of non-identical hyperchaotic Pang and Wang systems. In Section 5, we present the conclusions of this paper.

## 2 Problem Statement and Our Methodology

Consider the chaotic system described by the dynamics

$$\dot{x} = Ax + f(x) \tag{1}$$

where $x \in \mathbb{R}^n$ is the state of the system, $A$ is the $n \times n$ matrix of the system parameters and $f$ is the nonlinear part of the system. We consider the system (1) as the *master* or *drive* system.

As the *slave* or *response* system, we consider the following chaotic system described by the dynamics

$$\dot{y} = By + g(y) + u \tag{2}$$

where $y \in \mathbb{R}^n$ is the state of the slave system, $B$ is the $n \times n$ matrix of the system parameters, $g$ is the nonlinear part of the system and $u$ is the controller of the slave system.

If $A = B$ and $f = g$, then $x$ and $y$ are the states of two *identical* chaotic systems. If $A \neq B$ or $f \neq g$, then $x$ and $y$ are the states of two *different* chaotic systems. In the active nonlinear control approach, we design a feedback controller $u$ which synchronizes the states of the master system (1) and the slave system (2) for all initial conditions $x(0), y(0) \in \mathbb{R}^n$.

If we define the synchronization error as

$$e = y - x, \tag{3}$$

then the synchronization error dynamics is obtained as

$$\dot{e} = By - Ax + g(y) - f(x) + u \tag{4}$$

Thus, the global synchronization problem is essentially to find a feedback controller (*active control*) $u$ so as to stabilize the error dynamics (4) for all initial conditions, *i.e.*

$$\lim_{t \to \infty} \|e(t)\| = 0, \quad \forall\, e(0) \in \mathbb{R}^n \tag{5}$$

We use the Lyapunov stability theory as our methodology. We take as a candidate Lyapunov function

$$V(e) = e^T P e,$$

where $P$ is a positive definite matrix. Note that $V$ is a positive definite function by construction. We assume that the parameters of the master and slave systems are known and that the states of both systems (1) and (2) are available for measurement.

If we find a feedback controller $u$ so that

$$\dot{V}(e) = -e^T Q e$$

where $Q$ is a positive definite matrix, then $V$ is a negative definite function on $\mathbb{R}^n$. Thus, by Lyapunov stability theory [24], the error dynamics (4) is globally exponentially stable and hence the states of the master system (1) and slave system (2) are globally exponentially synchronized.

## 3 Synchronization of Identical Hyperchaotic Pang Systems

In this section, we apply the active nonlinear control method for the synchronization of two identical hyperchaotic Pang systems ([22], 2011).

Thus, the master system is described by the Pang dynamics

$$\begin{aligned} \dot{x}_1 &= a(x_2 - x_1) \\ \dot{x}_2 &= cx_2 - x_1 x_3 + x_4 \\ \dot{x}_3 &= -bx_3 + x_1 x_2 \\ \dot{x}_4 &= -k_1 x_1 - k_2 x_2 \end{aligned} \tag{6}$$

where $x_1, x_2, x_3, x_4$ are the state variables and $a, b, c, d, k_1, k_2$ are positive real constants.

When $a = 36, b = 3, c = 20, k_1 = 2$ and $k_2 = 2$, the Pang system (6) is hyperchaotic as illustrated in Figure 1.

The slave system is also described by the Pang dynamics

$$\begin{aligned} \dot{y}_1 &= a(y_2 - y_1) + u_1 \\ \dot{y}_2 &= cy_2 - y_1 y_3 + y_4 + u_2 \\ \dot{y}_3 &= -by_3 + y_1 y_2 + u_3 \\ \dot{y}_4 &= -k_1 y_1 - k_2 y_2 + u_4 \end{aligned} \tag{7}$$

where $y_1, y_2, y_3, y_4$ are the state variables and $u_1, u_2, u_3, u_4$ are the nonlinear controllers to be designed.

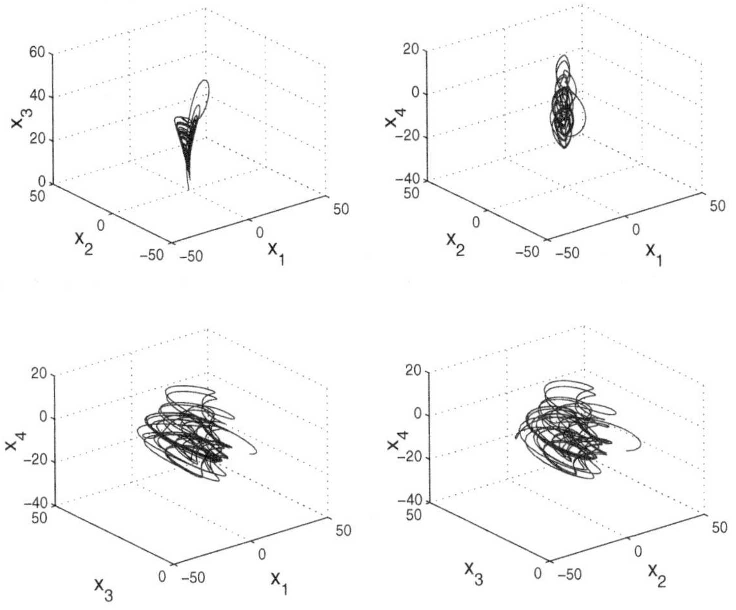


**Fig. 1.** The Hyperchaotic Pang System

The synchronization error is defined by

$$e_i = y_i - x_i, \quad (i = 1, 2, 3, 4) \tag{8}$$

A simple calculation yields the error dynamics as

$$\begin{aligned} \dot{e}_1 &= a(e_2 - e_1) + u_1 \\ \dot{e}_2 &= ce_2 + e_4 - (y_1 y_3 - x_1 x_3) + u_2 \\ \dot{e}_3 &= -be_3 + y_1 y_2 - x_1 x_2 + u_3 \\ \dot{e}_4 &= -k_1 e_1 - k_2 e_2 + u_4 \end{aligned} \tag{9}$$

We choose the nonlinear controller as

$$\begin{aligned} u_1 &= -ae_2 \\ u_2 &= -(c+1)e_2 - e_4 + y_1 y_3 - x_1 x_3 \\ u_3 &= -(y_1 y_2 - x_1 x_2) \\ u_4 &= k_1 e_1 + k_2 e_2 - e_4 \end{aligned} \tag{10}$$

Substituting the controller $u$ defined by (10) into (9), we get

$$\dot{e}_1 = -ae_1, \quad \dot{e}_2 = -e_2, \quad \dot{e}_3 = -be_3, \quad \dot{e_4} = -e_4 \tag{11}$$

We consider the candidate Lyapunov function

$$V(e) = \frac{1}{2} e^T e = \frac{1}{2} \left( e_1^2 + e_2^2 + e_3^2 + e_4^2 \right) \tag{12}$$

which is a positive definite function on $\mathbb{R}^4$.

Differentiating $V$ along the trajectories of (11), we find that

$$\dot{V}(e) = -ae_1^2 - e_2^2 - be_3^2 - e_4^2 \tag{13}$$

which is a negative definite function on $\mathbb{R}^4$ since $a$ and $b$ are positive constants.

Thus, by Lyapunov stability theory [24], the error dynamics (11) is globally exponentially stable. Hence, we have proved the following result.

**Theorem 1.** *The identical hyperchaotic Pang systems (6) and (7) are exponentially and globally synchronized with the active nonlinear controller $u$ defined by (10).* ■

**Numerical Results.** For the numerical simulations, the fourth order Runge-Kutta method with time-step $10^{-6}$ is used to solve the two systems of differential equations (6) and (7). We take the parameter values as $a = 36, b = 3, c = 20, k_1 = 2, k_2 = 2$ so that the two Pang systems (6) and (7) are hyperchaotic. The active controller $u$ is defined by (10).

The initial values of the master system (6) are taken as

$$x_1(0) = 18, \;\; x_2(0) = 12, \;\; x_3(0) = 20, \;\; x_4 = 10$$

and the initial values of the slave system (7) are taken as

$$y_1(0) = 26, \;\; y_2(0) = 30, \;\; y_3(0) = 10, \;\; y_4 = 20.$$

Figure 2 shows that synchronization between the states of the master system (6) and the slave system (7) occur in 4 seconds.

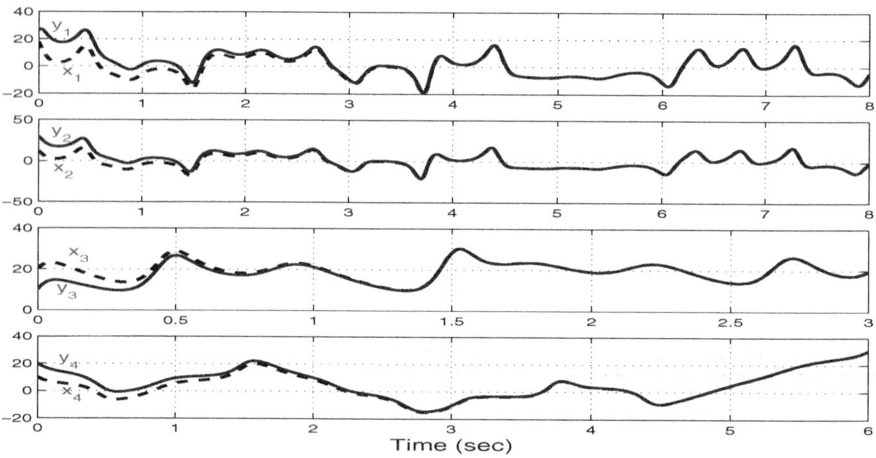


**Fig. 2.** Synchronization of the Identical Hyperchaotic Pang Systems

## 4 Synchronization of Hyperchaotic Pang and Wang Systems

In this section, the nonlinear control method is applied for the synchronization of the non-identical hyperchaotic systems, *viz.* Pang system (2011) and Wang system (2006).

The dynamics of the Wang system (2006) taken as the master system, is described by

$$\begin{aligned}\dot{x}_1 &= \alpha(x_2 - x_1)\\ \dot{x}_2 &= \beta x_1 - \eta x_1 x_3 + x_4\\ \dot{x}_3 &= -\gamma x_3 + \epsilon x_1^2\\ \dot{x}_4 &= -\delta x_1\end{aligned} \tag{14}$$

where $x_1, x_2, x_3, x_4$ are the state variables and $\alpha, \beta, \gamma, \delta, \epsilon, \eta$ are positive real constants.

When $\alpha = 10, \beta = 40, \gamma = 25, \delta = 10.6, \epsilon = 4$ and $\eta = 1$, the Wang dynamics (14) is hyperchaotic as illustrated in Figure 3.

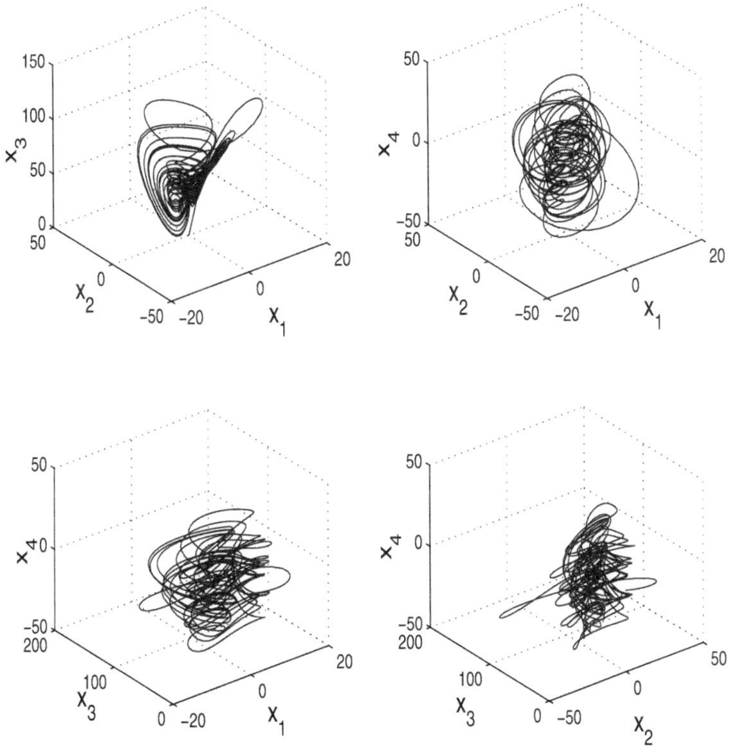


**Fig. 3.** The Hyperchaotic Wang System

The slave system is described by the Wang dynamics

$$\begin{aligned}\dot{y}_1 &= a(y_2 - y_1) + u_1\\ \dot{y}_2 &= cy_2 - y_1y_3 + y_4 + u_2\\ \dot{y}_3 &= -by_3 + y_1y_2 + u_3\\ \dot{y}_4 &= -k_1y_1 - k_2y_2 + u_4\end{aligned} \tag{15}$$

where $y_1, y_2, y_3, y_4$ are the state variables and $u_1, u_2, u_3, u_4$ are the nonlinear controllers to be designed.

A simple calculation yields the error dynamics as

$$\begin{aligned}\dot{e}_1 &= a(e_2 - e_1) + (a - \alpha)(x_2 - x_1) + u_1\\ \dot{e}_2 &= ce_2 + e_4 + cx_2 - \beta x_1 - (y_1y_3 - \eta x_1x_3) + u_2\\ \dot{e}_3 &= -be_3 + (\gamma - b)x_3 + y_1y_2 - \epsilon x_1^2 + u_3\\ \dot{e}_4 &= -k_1e_1 - k_2e_2 + (\delta - k_1)x_1 - k_2x_2 + u_4\end{aligned} \tag{16}$$

We choose the nonlinear controller as

$$\begin{aligned}u_1 &= -ae_2 - (a - \alpha)(x_2 - x_1)\\ u_2 &= -(c+1)e_2 - e_4 - cx_2 + \beta x_1 + (y_1y_3 - \eta x_1x_3)\\ u_3 &= -(\gamma - b)x_3 - y_1y_2 + \epsilon x_1^2\\ u_4 &= k_1e_1 + k_2e_2 - e_4 - (\delta - k_1)x_1 + k_2x_2\end{aligned} \tag{17}$$

Substituting the controller $u$ defined by (17) into (16), we get

$$\begin{aligned}\dot{e}_1 &= -ae_1\\ \dot{e}_2 &= -e_2\\ \dot{e}_3 &= -be_3\\ \dot{e}_4 &= -e_4\end{aligned} \tag{18}$$

We consider the candidate Lyapunov function

$$V(e) = \frac{1}{2}\, e^T e = \frac{1}{2}\, \left(e_1^2 + e_2^2 + e_3^2 + e_4^2\right) \tag{19}$$

which is a positive definite function on $\mathbb{R}^4$.

Differentiating $V$ along the trajectories of (18), we find that

$$\dot{V}(e) = -ae_1^2 - e_2^2 - be_3^2 - e_4^2 \tag{20}$$

which is a negative definite function on $\mathbb{R}^4$ since $a$ and $b$ are positive constants.

Thus, by Lyapunov stability theory [24], the error dynamics (18) is globally exponentially stable. Hence, we have proved the following result.

**Theorem 2.** *The non-identical hyperchaotic Wang system (14) and hyperchaotic Pang system (15) are exponentially and globally synchronized with the active nonlinear controller $u$ defined by (17).* ■

**Numerical Results.** For the numerical simulations, the fourth order Runge-Kutta method with time-step $10^{-6}$ is used to solve the two systems of differential equations (14) and (15) with the active nonlinear controller $u$ defined by (17). The parameter values are chosen so that the master and slave systems exhibit hyperchaotic behaviour, *i.e.*

$$a = 36, b = 3, c = 20, k_1 = 2, k_2 = 2$$

$$\alpha = 10, \beta = 40, \gamma = 25, \delta = 10.6, \epsilon = 4 \text{ and } \eta = 1$$

The initial values of the master system (14) are taken as

$$x_1(0) = 15, \ \ x_2(0) = 20, \ \ x_3(0) = 10, \ \ x_4(0) = 40$$

and the initial values of the slave system (15) are taken as

$$y_1(0) = 25, \ \ y_2(0) = 10, \ \ y_3(0) = 26, \ \ y_4(0) = 18$$

Figure 4 shows that synchronization between the states of the master system (14) and the slave system (15) occur in 3 seconds.

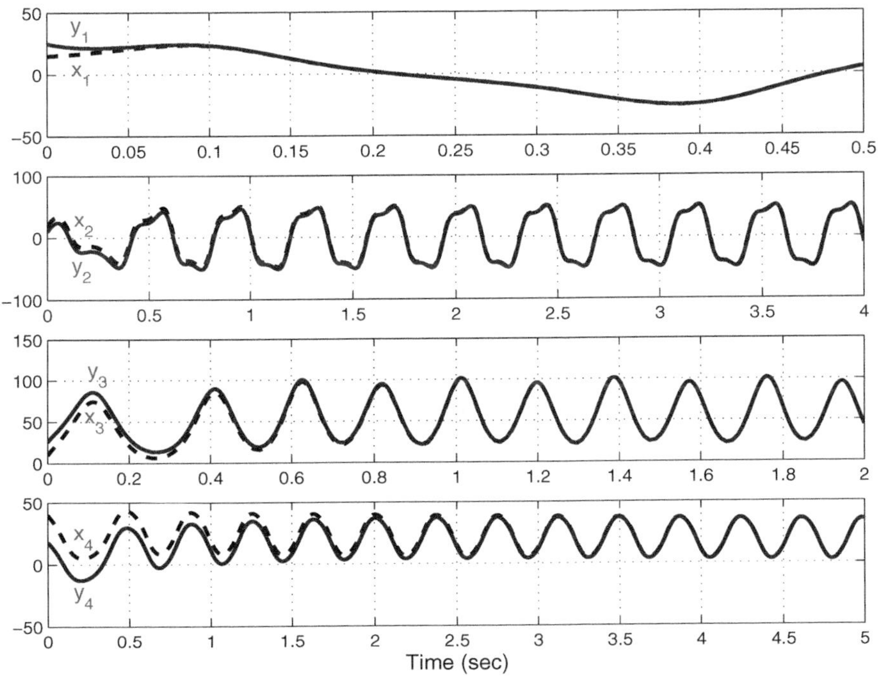


**Fig. 4.** Synchronization of the Hyperchaotic Wang and Pang Systems

## 5 Conclusions

In this paper, we have used nonlinear control method based on Lyapunov stability theory to achieve global chaos synchronization for the identical hyperchaotic Pang systems (2011) and non-identical hyperchaotic systems, *viz.* hyperchaotic Pang system (Pang and Liu, 2011) and hyperchaotic Wang system (Wang and Liu, 2006). Numerical simulations are also given to validate all the synchronization results derived in this paper. Since the Lyapunov exponents are not required for these calculations, the nonlinear control method is very effective and convenient to achieve global chaos synchronization for the hyperchaotic systems addressed in this paper.

## References

1. Alligood, K.T., Sauer, T., Yorke, J.A.: Chaos: An Introduction to Dynamical Systems. Springer, New York (1997)
2. Fujikasa, H., Yamada, T.: Stability theory of synchronized motion in coupled-oscillator systems. Progr. Theoret. Phys. 69, 32–47 (1983)
3. Pecora, L.M., Carroll, T.L.: Synchronization in chaotic systems. Phys. Rev. Lett. 64, 821–824 (1990)
4. Pecora, L.M., Carroll, T.L.: Synchronizing chaotic circuits. IEEE Trans. Circ. Sys. 38, 453–456 (1991)
5. Lakshmanan, M., Murali, K.: Chaos in Nonlinear Oscillators: Controlling and Synchronization. World Scientific, Singapore (1996)
6. Han, S.K., Kerrer, C., Kuramoto, Y.: Dephasing and burstling in coupled neural oscillators. Phys. Rev. Lett. 75, 3190–3193 (1995)
7. Blasius, B., Huppert, A., Stone, L.: Complex dynamics and phase synchronization in spatially extended ecological system. Nature 399, 354–359 (1999)
8. Kwok, H.S., Wallace, K., Tang, S., Man, K.F.: Online secure communication system using chaotic map. Internat. J. Bifurcat. Chaos. 14, 285–292 (2004)
9. Kocarev, L., Parlitz, U.: General approach for chaos synchronization with applications to communications. Phys. Rev. Lett. 74, 5028–5030 (1995)
10. Murali, K., Lakshmanan, M.: Secure communication using a compound signal using sampled-data feedback. Applied Math. Mech. 11, 1309–1315 (2003)
11. Yang, T., Chua, L.O.: Control of chaos using sampled-data feedback control. Internat. J. Bifurcat. Chaos. 9, 215–219 (1999)
12. Ott, E., Grebogi, C., Yorke, J.A.: Controlling chaos. Phys. Rev. Lett. 64, 1196–1199 (1990)
13. Park, J.H., Kwon, O.M.: A novel criterion for delayed feedback control of time-delay chaotic systems. Chaos, Solit. Fract. 17, 709–716 (2003)
14. Wu, X., Lü, J.: Parameter identification and backstepping control of uncertain Lü system. Chaos, Solit. Fract. 18, 721–729 (2003)
15. Samuel, B.: Adaptive synchronization between two different chaotic dynamical systems. Adaptive Commun. Nonlinear Sci. Num. Simul. 12, 976–985 (2007)
16. Yu, Y.G., Zhang, S.C.: Adaptive backstepping synchronization of uncertain chaotic systems. Chaos, Solit. Fract. 27, 1369–1375 (2006)
17. Park, J.H., Lee, S.M., Kwon, O.M.: Adaptive synchronization of Genesio-Tesi system via a novel feedback control. Physics Lett. A. 371, 263–270 (2007)
18. Park, J.H.: Adaptive control for modified projective synchronization of a four-dimensional chaotic sytem with uncertain parameters. J. Comput. Applied Math. 213, 288–293 (2008)

19. Yau, H.T.: Design of adaptive sliding mode controller for chaos synchronization with uncertainties. Chaos, Solit. Fract. 22, 341–347 (2004)
20. Chen, H.K.: Global chaos synchronization of new chaotic systems via nonlinear control. Chaos, Solit. Fract. 23, 1245–1251 (2005)
21. Sundarapandian, V., Suresh, R.: Global chaos synchronization for Rössler and Arneodo chaotic systems. Far East J. Math. Sci. 44, 137–148 (2010)
22. Pang, S., Liu, Y.: A new hyperchaotic system from the Lü system and its control. J. Comput. Applied Math. 235, 2775–2789 (2011)
23. Wang, F.Q., Liu, C.X.: Hyperchaos evolved from the Liu chaotic system. Chin. Phys. 15, 963–968 (2006)
24. Hahn, W.: The Stability of Motion. Springer, New York (1967)

# Mining Web Path Traversals Based on Generation of FP Tree with Utility

E. Poovammal and Pillai Cigith

Department of Computer Science, SRM University,
Kattankulathur, Chennai 603203, Tamil Nadu, India
epsrm@yahoo.com, cigith@rediffmail.com

**Abstract.** Web Mining is mining information from the Web. Web Usage Mining extracts user accessing patterns and gives the desired web page quickly. While accessing web, the user activities gets recorded in a web server log. Using data mining algorithm, if mined properly, the web log acts as the gateway to user's interests. But, merely considering presence or absence of web page in web log does not gauge its importance to user. Hence, in this work two parameters namely *frequency of the access pattern* and *utility value* are used to find better access patterns of the user and succeeded. This paper uses the method of Frequent Pattern (FP) Tree generation to find frequency of access pattern and the time spent by user on each web page is considered as the utility value.

**Keywords:** FP Tree, Utility, Usage patterns, Web Usage Mining, frequent user access patterns.

## 1 Introduction

The World Wide Web (WWW) is a huge galaxy of information which acts as a source for rich and dynamic collection of hyperlink data with usage characteristics. Data mining techniques are widely applied to obtain useful and understandable patterns from these massive sources of data. All the data mining techniques exploited on the Web comes under the category of Web mining. Web mining can be broadly classified into three parts, i.e. content mining, usage mining, and link structure mining. Web Content Mining describes the discovery of useful information from the web content/data/documents such as text, image, audio and video. Web Structure Mining is the process of discovering knowledge from the World Wide Web organization and links across the webs. Web Usage Mining is the process of extracting interesting patterns or knowledge from various web access log records.

A Web server usually registers a log entry for every access of a Web page in a web log. Web Usage Mining deals with performing data mining techniques on web log entries to find association patterns, sequential patterns, and trends of Web accessing. Analyzing and exploring regularities in web log entries can identify potential customers for electronic commerce, improve the quality and performance of Web services system, and optimize the site architecture to cater to the preference of end users.

D. Nagamalai, E. Renault, and M. Dhanushkodi (Eds.): CCSEIT 2011, CCIS 204, pp. 94–100, 2011.

When it comes to mining frequent patterns, most of the studies carried out previously adopted an Apriori [1]-like approach based on the anti-monotone property: if any length k pattern is not frequent in the database, its length (k + 1) super-pattern can never be frequent.

But an Apriori-like algorithm can suffer from two nontrivial costs:

- It is costly to handle a huge number of candidate sets.
- It is tedious to repeatedly scan the database and check a large set of candidates by pattern matching

Hence, we use another data mining algorithm known as FP Tree [2],[3] generation which have the following advantages:

- It does not involve creation of candidate sets
- It makes use of a compact data structure called as FP Tree which is an extended prefix tree structure storing important information about frequent patterns.
- It makes use of partitioning based divide- and- conquer approach instead of Apriori's level wise approach.

However, considering only the binary frequency for each website of a web traversal path is not a good measure for finding interesting web traversal paths because each user may spend different time period for each website.

This paper uses FP tree with utility value to give more relevant results to the user. This approach first pre-processes the web server log and with that data as input to the FP Tree algorithm mines for finding out frequent patterns. After mining the utility value is calculated which results in useful patterns to the user and helps in decision making.

The remainder of this paper is divided as follows. In section 2 we describe related work. In section 3 we describe the problem definition. In section 4 we present the experiments and their results. Finally conclusions are presented in section 5.

## 2 Related Work

Chen et al. [4] developed two efficient algorithms for determining large reference sequences for web traversal paths. The first one, called full-scan (FS) algorithm, solved discrepancy between traversal patterns and association rules. The second one, called selective-scan (SS) algorithm is able to avoid database scans in some passes so as to reduce the disk I/O cost involved. But both these algorithms had some limitations like they required large number of database scans, candidate generation for large databases where difficult.

Sequential pattern mining [5] discovered sequential patterns from customer sequence databases. But sequential pattern mining suffered from mining closed and maximal sequential patterns as well as mining approximate sequential patterns. The Two-Phase algorithm [6], [7] was developed, based on the definitions of [8] to find high utility itemsets using the downward closure property of Apriori. The authors [6], [7] have defined the transaction weighted utilization (twu) and proved that it is possible to maintain the downward closure property. This algorithm suffers from the same problem of the level-wise candidate generation-and-test methodology.

A novel approach of finding frequent patterns without generating candidate sets was proposed in the FP Tree generation algorithm. This method uses a data structure called FP Tree which stores the database in a compressed form. A brief comparison between Apriori algorithm and FP Tree algorithm with its advantages and disadvantages can be studied in [9].

Zhou et al. [10] introduced the concept of utility in web path traversal mining model. They adopted the definitions of utility from the high utility pattern mining model [6], [7], [8], [11]. They expressed the browsing time of a user as a utility of a website. Their algorithm is based on the Two-Phase [6], [7] high utility mining algorithm which suffers from the level-wise candidate generation-and-test methodology of the Apriori algorithm. Hence, they also suffered from generating too many candidate patterns and need several database scans to find out the frequent web traversal paths. Also combining both frequency and utility to find web path traversals was still not explored.

So, a more effective algorithm EUWPTM (Effective Utility based Web Path Traversal Mining) [12] was introduced which was a modification of the previous work and is based on pattern sequential approach.

## 3 Problem Definition

We start with the basics of all the important terms that lead to a formal definition of web path traversal mining using both FP Tree generation and Utility value.

A Web server log file contains requests made to the Web server, recorded in a chronological order. The most popular log file formats are the Common Log Format (CLF) and an extended version of the CLF, the Extended CLF. A line in the ECLF is as shown in Figure 1. Some of the fields it contains are as follows:

- The client's host name or its IP address
- The date and time of the request
- The operation type (GET, POST, HEAD, etc.)
- The status code of the request (200, 404 etc.)
- Referrer Link, type of OS and the type of browser.

A sample weblog data is as follows

```
192.168.0.1 - - [29/Sep/2004:17:10:31 +0200] \GET /axis/people.shtml
HTTP/1.1"200 8289 \http://www-sop.inria.fr/axis/table.html" \Mozilla/4.0
(compatible; MSIE6.0; Windows NT 5.2; .NET CLR 1.1.4322)"
```

**Fig. 1.** A web request from INRIA's Web Server Log in the ECLF Format

When a user starts his session he moves from one web page to a different web page. He moves around the Web and leaves back a trace of his activities in the web server log. Pre-processing the web server log is the most important step in using the log for mining. After pre-processing is done all the vital information needs to be stored in a database. This database can be a data structure also.

A web path traversal is the set of links accessed by the user in his one particular session. For e.g. if 'a',' b', 'c', 'd' are four websites and the user first goes to web page 'a' followed by 'c', 'b' and 'd' in one session then the path traversal for the user is a-c-b-d.

Let I = {i1, i2 ...im} be a set of links, and a Weblog W = {T1, T2... Tn} where Ti is a traversal path which contains a set of links in I.

The **support** (or occurrence frequency) of a pattern A, which is a set of links, is the number of traversals containing A in W. A, is a **frequent pattern** if A's support is no less than a predefined minimum support threshold S. Given a Weblog W and a minimum support threshold, S, the problem of finding the complete set of frequent path traversals is called the frequent path traversal mining problem. These web path traversals are provided as the input to the FP Tree algorithm [2].

In this algorithm, first a data structure called frequent-pattern tree is constructed which stores significant quantitative information about the frequent patterns. Secondly, an FP-tree based pattern-fragment growth mining is developed starting from a length-1 pattern (initial suffix pattern) examining only its conditional-pattern base (a sub database which consists of the set of frequent items co-occurring with the suffix pattern), constructs its (conditional) FP Tree and performs mining recursively with such a tree.

The pattern growth is obtained by concatenating the suffix pattern with the newly generated from the conditional FP Tree. And finally, the search techniques used is a partitioning-based, divide-and conquer method rather than Apriori-like level-wise generation of the combinations of frequent itemsets. It employs the least frequent items as suffix, which offers good selectivity. All these techniques contribute to substantial reduction of search costs.

Merely considering the presence or absence of a web page does not really judge the importance of a web page. Different web pages mean different to different users. Hence some parameter is chosen which acts as a measure of the importance to the particular user. Here the parameter taken is the time spent by each user on the web page known as the utility [10] value.

### 3.1 The Proposed Approach

Pre processing the web server log [13] results in the removal of useless data like searches made by bots, spiders, Page Not Found error(status code:404) etc. This cleaned data is the input to the FP Growth algorithm. The utility value is calculated using the following steps:

i. The identification of users is done based on their ip addresses. Here the assumption made is that each individual user uses the same machine.
ii. The utility value of a web page is calculated using the timestamp values of consecutive requests of the user. For e.g.

**Table 1.** A sample server log

| Users | Links Traversed |
|---|---|
| 174.129.130.230 | A(04:23:56) |
| 83.109.145.234 | P(04:24:07) |
| 174.129.130.230 | C(04:24:11) |
| 83.109.145.234 | S(04:24:13) |

**Table 2.** Transformed one

| Users | Links Traversed |
|---|---|
| 174.129.130.230 | A(04:23:56), C(04:24:11) |

Consider the sample of the pre-processed server log shown in Table 1. Table 1 is pre processed and the result is shown in Table 2.

And finally the utility value for page A is calculated as follows:

$$\text{Utility (A)} = \text{Timestamp (C)} - \text{Timestamp (A)} \quad (1)$$

So, in this example Utility (A) = 15 secs.

iii. At the end, those links are selected whose utility value satisfies a minimum utility threshold (Ω) value.

## 4 Experimental Results

To evaluate the performance of the proposed method, experiments were conducted on the log file *access.log* of web server of www.vinrcorp.com which had 389 entries. For description of the entries in detail refer section 3 of this paper. Out of these entries there were requests from spiders and robots which are not useful in mining. So we removed them and the data was cleaned. This reduced the number of entries on which the algorithm is applied. After cleaning the data 362 entries were obtained. These entries were logged for 81 different users who searched for 138 unique web pages. Figure 2 shows the snapshot of the output of the parsed server log. The programs were written in Java and the IDE used was NetBeans 6.9.1. The programs were run on Windows 7 operating system on an Intel Core i3 processor with 2.26 Ghz CPU and 4GB RAM.

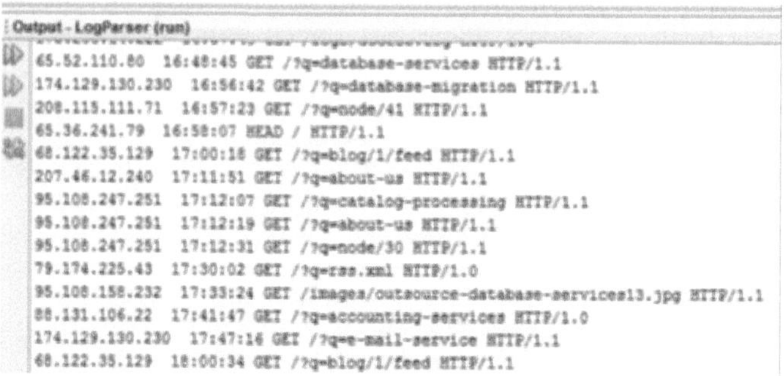


**Fig. 2.** shows a snapshot of the output of the parsed web server log. The values are separated by tab. To the extreme left is the ip address, in the middle is the time of access and at the end the link accessed.

The *BMS-WebView-1* contains several months' worth of click-stream data from two e- commerce web sites. Each transaction in these datasets is a web session

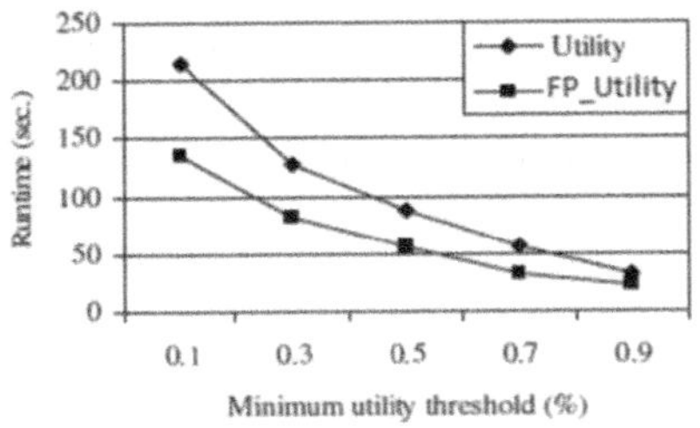


**Fig. 3.** shows the runtime performance curve for the *BMS-Webview-1* dataset. It is measured against minimum utility threshold on the x-axis and runtime (in secs) on y-axis.

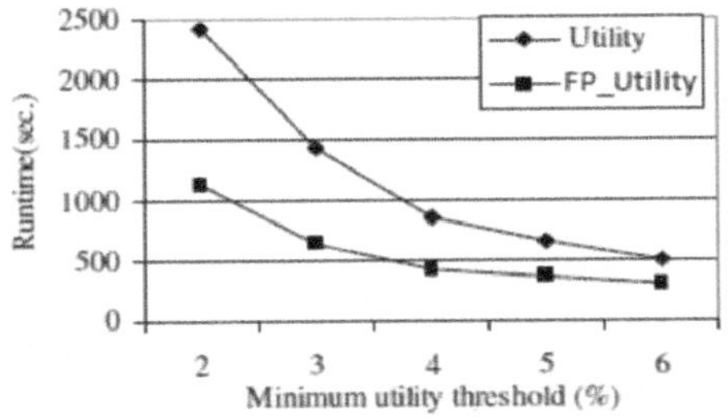


**Fig. 4.** shows the runtime performance curve for *kosarak* dataset. It is measured against minimum utility threshold on the x-axis and runtime (in secs) on y-axis.

consisting of all the product detail pages viewed in that session. The goal for the dataset is to find associations between products viewed by visitors in a single visit to the web site. This dataset contain 59,602 transactions (with 497 distinct items). Figure 3 shows the runtime performance curve for *BMS-WebView-1* dataset.

Another dataset, the *kosarak* dataset contains web click-stream data of a Hungarian on-line news portal. It is a big dataset containing almost one million transactions (990,002) and 41,270 distinct items. Figure 4 shows the runtime performance curve for the *kosarak* dataset.

## 5 Conclusion

Discovering frequent Web accessing sequences from Weblog databases is not only useful in improving the website design but also lead to better marketing decisions. But considering only the frequency of the web pages does not reflect the importance of the web page for the user. It is because the user might not visit an important web page as frequently as say the user visiting web pages for daily news or even checking his personal e-mails. So taking into account web pages only on the basis of frequency will easily neglect a web page which is not frequent but important to the user. The quality of the web pages thus obtained makes a big impact on the time taken for searching effectively.

In this paper, an attempt has been made to combine two methods namely frequency of the access pattern and the Utility value. FP Growth algorithm finds the frequently accessed patterns while Utility mining technique incorporates the 'utility value' which measures the significance of a web page for the user.

The proposed method efficiently reduces the performance time. For a particular user this method gives a set of links which satisfies a minimum support threshold and also arranges all the links traversed by the user in descending order of the utility value.

The resulting links obtained from the proposed method are checked for both 'frequency' and 'interestingness' and it is found that the results are significant for decision making process. The links which were previously discarded because of their low frequency but were important were also included in the final result.

## References


1. Agrawal, R., Srikant, R.: Fast Algorithms for Mining Association Rules in Large Databases. In: 20th International Conference on Very Large Data Bases, pp. 487–499 (1994)
2. Han, J., Pei, J., Yin, J.: Mining Frequent Patterns without Candidate Generation. In: ACM-SIGMOD International Conference Management of Data, Dallas, TX, pp. 1–12 (2000)
3. Aiman, M.S., Dr. Dominic, P.D.D., Dr. Azween, B.A.: A Comparative Study of FP-growth Variations. J. Computer Science and Network Security 09 (2009)
4. Chen, M., Park, J., Yu, P.: Efficient data mining for path traversal patterns. IEEE Transactions on Knowledge and Data Engineering, 209–221 (1998)
5. Yao, H., Hamilton, H.: Mining itemset utilities from transaction databases. Data & Knowledge Engineering 59, 603–626 (2006)
6. Liu, Y., Liao, W., Choudhary, A.: A fast high utility itemsets mining algorithm. In: 1st International Conference on Utility-Based Data Mining, pp. 90–99 (2005)
7. Liu, Y., Liao, W., Choudhary, A.: A Two Phase algorithm for fast discovery of High Utility of Itemsets. In: Ninth Pacific-Asia Conference on Knowledge Discovery and Data Mining, pp. 689–695 (2005)
8. Yao, H., Hamilton, H., Butz, C.: A Foundational Approach to Mining Itemset Utilities from Databases. In: Third SIAM International Conference on Data Mining, pp. 482–486 (2004)
9. Santosh Kumar, B., Rukmani, K.: Implementation of Web Usage Mining Using APRIORI and FP Growth Algorithms. J. Advanced Networking and Applications 400(01), 400–404 (2010)
10. Zhou, L., Liu, Y., Wang, J., Shi, Y.: Utility-based Web Path Traversal Pattern Mining. In: Seventh IEEE International conference on Data Mining Workshops, pp. 373–378 (2007)
11. Pei, J., Han, J., Asl, B., Wang, J., Pinto, H., Chen, Q., Dayal, U., Hsu, M.: Mining sequential patterns by pattern-growth: The PrefixSpan Approach. IEEE Transactions on Knowledge and Data Engineering 16, 1424–1440 (2004)
12. Chowdhury, F., Syed, T., Byeong-Soo, J., Young-Koo, L.: Efficient Mining of Utility Based Web Path Traversals. In: International Conference on Advanced Communication Technology 2009 (2009) ISBN 978-89-5519-139-4
13. Mohd, W., Mohd, H.M., Hafizul, F.H., Mohamad, F.M.M.: Data Pre-processing on Web Server Logs for Generalized Association Rules Mining Algorithm. World Academy of Science, Engineering and Technology 48 (2008)

# Evaluating Degree of Dependency from Domain Knowledge Using Fuzzy Inference System

Vibha Gaur and Anuja Soni

Department of Computer Science
University of Delhi
{3.vibha,30.anuja}@gmail.com

**Abstract.** Inter-agent communication is one of the main concerns of agent oriented requirements engineering that is delineated as managing inter-dependencies and interaction among various agents performing collaborative activities. To carry out cooperative activities, tasks are disseminated and delegated to other agents with a rationale of sharing mutual expertise and potential. So an agent may be dependent on other agent for accomplishing a goal or for want of a resource to achieve that goal. This requires an agent to quantify the dependency needs termed as Degree of Dependency (DoD) to realize whether it should delegate the task to another agent, if yes, then to whom so that overall quality of Multi-Agent System (MAS) is not compromised. To evaluate DoD, this work employs domain knowledge that is an inclusive knowledge of an environment containing the business rules, credentials and reports to understand the business needs precisely. As the domain knowledge may be fuzzy and uncertain, Fuzzy Inference system is utilized to evaluate DoD from domain knowledge. This will assist the developer to address inter-agent coordination issues without squandering resources and hence in building MAS of high quality.

**Keywords:** Fuzzy Inference System, Degree of Dependency, MAS, Domain Knowledge, PARETO's rule.

## 1 Introduction

In software engineering community, there is an ever increasing demand to amend and streamline software projects in terms of Multi-Agent Systems (MAS). MAS involve a large number of agents playing different roles, interacting with each other to achieve personal and common goals [1], [2]. To achieve common goals, tasks are distributed and entrusted to other agents with purpose to share proficiency and potential; to work in parallel or sequence on common problems. By depending on others, an agent may be able to accomplish goals that are difficult or infeasible to achieve otherwise. Requirements analysis necessitates supporting ways for identifying and analyzing dependencies in inter-agent coordination.

In literature various requirements frameworks such as Formal Tropos [1], [5], i*[6], REF [7] have been recommended for inter-agent coordination. Tropos [1] is well known; being one of the most cited methodologies and is suitable for the very

D. Nagamalai, E. Renault, M. Dhanushkodi (Eds.): CCSEIT 2011, CCIS 204, pp. 101–111, 2011.

early phases of requirements analysis of Agent-Oriented Systems. In it dependency requirements are modeled using UML. Tropos exhibits diverse kinds of dependencies viz. goal, task and resource by employing agent interaction diagarms and dependency UML diagrams. i* [6] recommends a modeling framework that focuses on strategic actor relationships. It is employed for obtaining a better understanding of the organizational relationships among the various organizational agents. REF [7] involves novel requirements engineering techniques integrated with software quality modeling approaches to provide an environment in which the agents can freely collaborate to obtain goals, tasks and resources.

However all these frameworks of early phase of requirements engineering focus more on capturing various kinds of dependencies instead of their degree and do not support the ways to examine dependency needs of an agent. This work facilitates to quantify dependency needs of an agent entitled as Degree of Dependency (DoD) that would assist the developer in taking decision of delegating a task to another agent in MAS so that its quality may not compromised. As an information system is implanted in the domain knowledge of a specific business or activities in an organization, the acquaintance with target business area is indispensable[3], [4].

As the domain knowledge may be fuzzy, incomplete and uncertain, this work employs the Fuzzy Inference System with a number of inference rules to fuzzify domain characteristics and result in a single crisp output value entitled as Degree of Dependency. In addition, PARETO's rule is applied to identify critical tasks of the goals with high degree of dependency to overcome the delay in pursuing these goals.

The organization of this paper is as follows. The Section 2 introduces Fuzzy Inference System and PARETO's rule. Section 3 presents Fuzzy Inference Model for DoD using Domain Knowledge. Section 4 illustrates the proposed methodology with an experimental study and analyses the results. Finally section 5 concludes the paper.

## 2 Fuzzy Inference System

Fuzzy logic is widely used due to its ability in representing the vagueness and imprecise information. Mamdani's Fuzzy Inference System (FIS) is widely known and used in developing fuzzy models. It consists of rules of the form "IF $x_1$ is very less AND $x_2$ is less THEN y is high,'' where very less, less and high are linguistic terms with functional forms like Gaussian etc., also known as membership functions[8],[9]. Implication is applied for each rule. Generally min operator representing the AND method is used for implication.Aggregation is used to unify the outputs of all the rules resulting into a single fuzzy set. The aggregated output function is defuzzified in a single crisp number using defuzzification method viz. Centroid method [8].The framework of deriving DoD from domain knowledge is shown in fig. 1 that is explained in detail in experimental study in section 4.

**Pareto's rule:** Dr. Juran's study of the "vital few and trivial many", the principle that 20 percent of something always are accountable for 80 percent of the outcome, was entitled as Pareto's principle or the 80/20 rule[10]. This Pareto's rule would be utilized in deciding DoD at task level as explained in section 4.

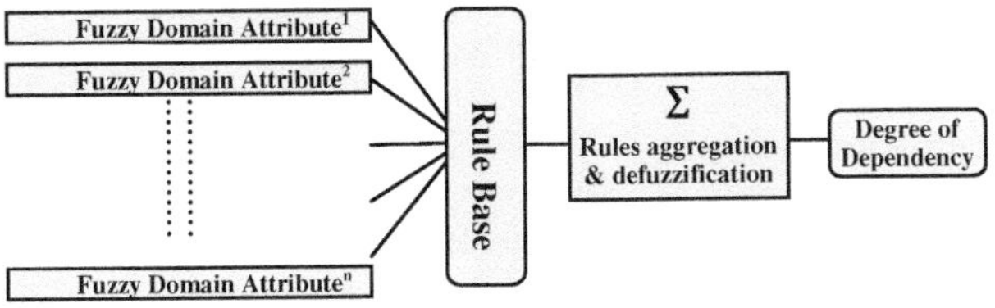


**Fig. 1.** Structure of Mamdani's Fuzzy Inference System

## 3 Fuzzy Inference Model for DoD Using Domain Knowledge

Advances in networking technology have revitalized agent technology for engineering complex distributed software systems. However, the success of agent-based system is subject to the adoption of effective agents' cooperation that would be required to share expertise for achieving goals of MAS [2].

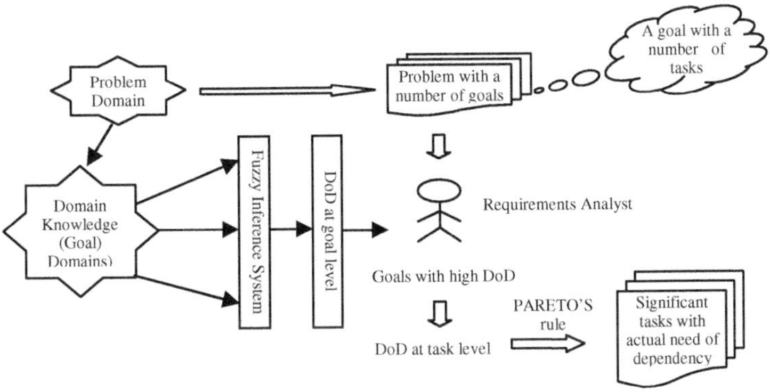


**Fig. 2.** Fuzzy Inference Model for DoD using Domain Knowledge

Degree of Dependency is defined as a parameter to quantify an agent's dependency needs for delegating a goal in a distributed environment. For designating a goal to other agents, an agent requires to exercise its own dependency requirements which may originate from domain knowledge. The focal business area is entitled as problem domain and supportive detailed information of problem is characterized as domain knowledge [3], [4]. This study is aimed to obtain DoD of an agent w.r.t. a goal exercised over domain knowledge of business area. The fig. 2 illustrates a frame work to derive DoD from domain knowledge of the problem domain which can be described by the following subsections.

### 3.1 Degree of Dependency

Degree of Dependency signifies the quantification of dependency needs of a depender agent for accomplishing a goal. A dependency describes an agreement between two actors, the depender and the dependee [5]. Depender agent necessitates identifying the critical goals with very high or high DoD from the list of goals. The following section describes the domain knowledge required to compute DoD.

### 3.2 Domain Knowledge

In software engineering, domain knowledge is knowledge about the environment in which the target system functions [3], [4]. Domain knowledge is essential to be learned from software users in the problem domain viz. domain experts. Domain Knowledge utilized in this work consists of a number of goal domains that specify the problem. Goal domain is an extensive term which incorporates all attributes along their domain that characterize the problem. It is an area of expertise of an application that needs to be investigated and converted in a set of rules in knowledge base, by knowledge engineers to achieve a goal. An example of goal domain identified from procurement data is shown in fig. 3 which further is described in detail in section 4.

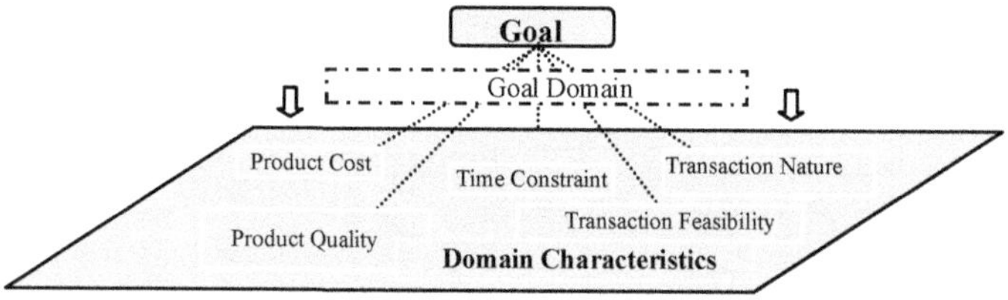


**Fig. 3.** Domain characteristics in contributing DoD

As the domain knowledge is fuzzy, uncertain and ambiguous, this work applies FIS on domain knowledge to evaluate DoD in the following sub-section.

### 3.3 Methodology to Evaluate Degree of Dependency from Domain Knowledge

In a MAS, to carry out collaborative activities, tasks are delegated to other agents with a rationale of contributing mutual potential and expertise[1], [5], [6], [7]. For quantifying the dependency requirements of a goal, this work defines DoD which is a parameter in [0, 1] exercised over domain knowledge. Following steps are performed to evaluate DoD of an agent at goal and task level from domain knowledge:

i. Obtain domain knowledge for a specific goal from domain expert or real data set of business environment.
ii. Derive domain characteristics of the goals from domain knowledge with the assistance of domain experts.
iii. Fuzzify all inputs formulated on domain characteristics in the form of fuzzy sets via Gaussian membership function.
iv. Formulate a fuzzy set represented by a membership function on output variable. In this work, DoD is treated as output variable.
v. Once fuzzy inputs and outputs parameters are obtained, observe varying fuzzy nature of outputs as per fuzzy inputs leading to a number of rules.
vi. Apply FIS to obtain final aggregated crisp value DoD at goal level.
vii. Decompose goal into a number of tasks.

viii. If DoD at goal level, is 0.5 or more, belonging to high or very high category, further analyze it at task level.
ix. Obtain importance of tasks with respect to a goal in human linguistic terms viz. less, high, very high from domain experts.
x. Linguistic terms obtained are converted into corresponding crisp values using defuzzification method mentioned in our previous work [13] using the following equations. The crisp interval $^{\alpha}A$ by α-cut operation against fuzzy number (l,m,u) can be obtained as[9],[13]:

$$^{\alpha}A = [\, I_{\alpha l} \,,\, I_{\alpha u}] = [(m - l)\alpha + l, -(u - m)\alpha + u] \quad (1)$$

where a α-cut is viewed as a bridge between fuzzy sets and crisp sets.
Crisp values against the crisp intervals representing the degree of optimism for domain experts, is computed as [9], [13]:

$$CW_{tsk} = \mu\, I_{\alpha u} + (1 - \mu)\, I_{\alpha l} \,, \text{where } \mu \in [0,1] \quad (2)$$

where $CW_{tsk}$ is crisp weight at task level, $I_{\alpha l}$ =lower bound of crisp interval, $I_{\alpha u}$=upper bound of crisp interval, μ is called as the index of optimism representing optimistic level of domain experts as optimistic, moderate or pessimistic. Higher value of μ represents the higher degree of optimism. Similarly value of α shows the level of confidence in most promising crisp value.

xi. $CW_{tsk}$ is converted to a normalized value in [0,1] i.e. $CW_{tsk[0,1]}$ by dividing it by summation of all task level weights of respective goal [13].
xii. To assess DoD at task level, distribute DoD at goal level to its corresponding tasks using the following equation:

$$DoD_{atTskLvl} = CW_{tsk[0,1]} * DoD_{atgoal} \quad (3)$$

where $DoD_{atgoal}$ refers to DoD at goal level and $D_{atTskLvl}$ to DoD at task level.

xiii. Apply Pareto's Principle-the 80-20 Rule in identifying most critical tasks for which depender agent needs to be dependent on other agents.

The value of DoD obtained above would assist the developer in deciding the number of optimized dependee agents during requirements analysis so as to obtain quality of MAS in limited resources.

## 4 Experimental Study

To see the application of the method, an experiment was carried out to compute Degree of Dependency of PurchaseHeadAgent in Agent-Mediated Materials Procurement Environment (AMMPE). AMMPE is composed of PurchaseHeadAgent, RawMaterialAgents, SparesAgents, PackagingAgents, ConsumableAgents and MiscelleneousItemsAgents involved with procuring various items viz. raw materials, spares, packaging, consumable and miscellaneous for various projects distributed over various locations. PurchaseHeadAgent is concerned with the actual assessment of dependency pertaining to various procurement activities.

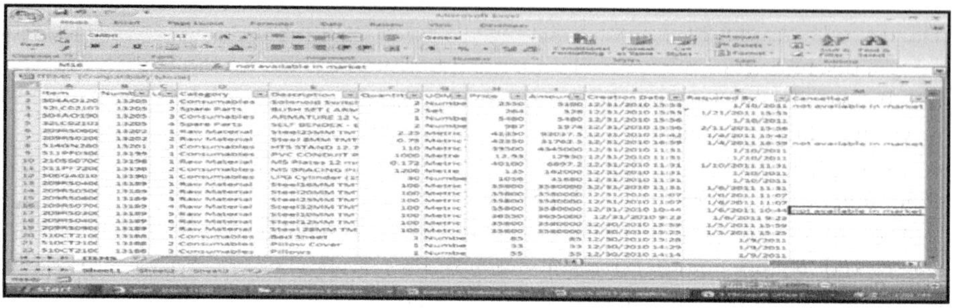

**Fig. 4.** Data set formulated on quarterly report of purchase order

| Lead time for Procurement of Items | | | |
|---|---|---|---|
| S no | Items | Item type | Lead time |
| 1 | PQC Material | raw material | 7 days |
| 2 | Plastic films, plastic drums, iron drums | packaging | 1 week |
| 3 | Crusher parts and Crusher parts | spares | 2 - 4 weeks |
| 4 | Automotive parts | spares | 10 days |
| 5 | GI sheets and steel | miscellaneous | 2 weeks |
| 6 | TIL parts | spares | 3-4 weeks |
| 7 | Stationery | consumable | 4 days |
| 8 | Oils and greese | consumable | 1 week |
| 10 | Generqal Hardware | miscellaneous | 7 days |
| 11 | Electrical items | spares | 2 weeks |
| Conditions | | | |
| 1) | Subject to payment of party. | | |
| 2) | Subject to availabilty of Road permit | | |

**Fig. 5.** Lead time of various procurement materials

The data set of procurement area of an Infrastructure industry was utilized to capture the values of fuzzy input variables characterizing the domain attributes explained in detail in section 4.1. The data set consists of quarterly report of 10 years record of data of procurement activity as shown in fig. 4 and lead time of procurement items illustrated in fig. 5. Lead time is the time elapsed between the initiation of procurement action and the receipt into the supply system of the material procured [12]. The purchase order of procurement activities consists of the quarterly description of various materials viz. raw material, consumable items, spares, packaging, consumable and miscellaneous items in the infrastructure industry.

The present experimental work in concern with the AMMP activity is classified under following three categories:

## 4.1 Fuzzification of Domain Characteristics and DoD, Preprocessing of Purchase Order Report

Fuzzification of domain characteristics along with Degree of Dependency and preprocessing of purchase order can be described in following subsections.

### 4.1.1 Identification of Domain Characteristics

Fifteen domain experts from materials management were involved to capture and formulate five domain characteristics namely Product Cost, Transaction Feasibility, Time Constraint, Product Quality and Transaction Nature for the procurement data set of infrastructure industry which are further defined.

**Product Cost** is the summation of acquiring all the components required to build a product. The product with very high cost requires significant attention and henceforth high DoD in obtaining that product within assigned time constraints. **The**

**Transaction Feasibility** is concerned with the availability of product in the market. Less feasible product requires high efforts in procuring it. **Time Constraint** is concerned with the time restriction imposed on a transaction. For example time constraint in obtaining raw material is always rigid as time delay may hamper the entire production work in the industry. Hence the activity with tight time constraint requires additional agents and efforts leading to high DoD. **Product Quality** is related to the collection of features and characteristics of a product that contribute to its ability to meet given requirements. The product with high quality acquires high cost and henceforth requires significant concern. **Transaction Nature** signifies the nature of transaction as highly critical, moderately critical or trivial. The items viz. raw materials, which are directly required in the production process, are considered as highly critical as their shortage would have a severe impact on the production. Highly critical items require momentous attention in comparison to other trivial items employed in industry.

#### 4.1.2 Fuzzification of Domain Characteristics Utilized as Input Variables

Aforesaid domain characteristics viz. **ProductCost, TimeConstraint** are fuzzified and various levels of Gaussian membership functions against the linguistic terms are defined with the assistance of domain experts as shown in fig. 6,7,8,9 and 10.

**Fig. 6.** Variation of Product Cost

**Fig. 7.** Variation of Transaction Feasibility

**Fig. 8.** Variation of Time Constraint

**Fig. 9.** Variation of Product Quality

**Fig. 10.** Variation of Transaction Nature

**Fig. 11.** Output Variable or DOD

#### 4.1.3 Fuzzification of DoD Employed as Output Variable

The variation of the **DoD** as output variable is shown in fig. 11.The output varies from very less to very high signifying the level of inter-agent dependency in this experiment.

#### 4.1.4 Preprocessing of Purchase Order to Assess Domain Characteristics

The data set of purchase order was worked out to assess the values of various domain characteristics in the following manner. The **cost** of various materials was obtained by the equation below:

$$\text{Product cost}_{\text{MtrlType[p]}} = \sum_{q=1}^{m} \text{Price}_{\text{itm[q]}_{\text{MtrlType[p]}}} * \text{Qty}_{\text{itm[q]}_{\text{MtrlType[p]}}} \quad (4)$$

where n refers to total number of categories of various materials and m number of items under category MtrlType[p], where1<=p<=n.

$\text{ProductCost}_{\text{MtrlType[q]}}$ is the total cost of $q^{th}$ material, $\text{Price}_{\text{itm[q]}_{\text{MtrlType[p]}}}$ refers to price of $q^{th}$ item of MtrlType[p] and $\text{Qty}_{\text{itm[q]}_{\text{MtrlType[p]}}}$ is the quantity of $q^{th}$ item of materialtype[p]. The cost of various materials computed using equation (4) over quarterly report of 10 years record of data is normalized in [0,1] using the equation (5) which further exhibits the behavior very less, less, high or very high.

$$\text{Product cost}_{\text{MtrlType[p]}_{[0\ 1]}} = \text{ProductCost}_{\text{MtrlType[p]}} / \text{HighestCost} \quad (5)$$

**Time Constraint** is derived from required date of items from purchase order in the following manner. If (required time is less than lead time of item) then Time Constraint is tight. If (required time is more than lead time of item) then Time Constraint is lenient and If (required time is equal to lead time of item) then Time Constraint is medium. **Transaction Nature** is directly related to Time Constraint. Tight Time Constraint signifies the Transaction Nature as critical, lenient Time Constraint indicates Transaction Nature as trivial and medium as moderately critical.

Various levels of **Product Quality** are mapped to the extent of quality standards followed for manufacturing process such as ISO,QS9000 etc. Less : signifies that no quality standards are adopted, Medium: refers to less adherence of quality standards, High : refers to strict adherence of standards. Levels of quality are directly related to the Transaction Nature. The critical Transaction Nature of product requires the high quality, moderately critical as medium quality and trivial as less quality. The status of **Transaction Feasibility** of the product as very less, less, high and very high would be captured from the frequency of cancelled items encapsulated in data set.

### 4.2 Obtaining Degree of Dependency at Goal Level

To show the application of aforesaid proposed methodology, an experiment was conducted using Mamdani's Fuzzy Inference System [8], [9] for evaluating DoD of PurchaseHeadAgent for obtaining raw material. Rules were formulated in the rule base of the Fuzzy Inference System with the help of domain experts. The team of 15 domain experts from the infrastructure industry was involved to provide the assistance in formulating the inference rules. Inference rules can be characterized by inferring various output values by blending several input values. The inference rules in rule base pertaining to raw material with a goal domain of less TransactionFeasibility, tight TimeConstraint, very high ProductCost, high ProductQuality and highly critical TransactionNature are represented as follows:

If (TransactionFeasibility is less) and (TimeConstraint is tight) and (ProductCost is high) and (ProductQuality is high) and (TransactionNature is highlycritical) then (DoD is high)

The other rules were formulated in the rule base in the similar way. Similarly, DoD for obtaining spares, packaging, consumable and miscellaneous items was obtained as shown in table 1.

### 4.3 Obtaining DoD at Task Level and Implication of PARETO'S Rule

Goals were further decomposed into a number of tasks. DoD for critical goals was distributed at task level using equation 3. The importance of tasks in human linguistic terms, captured from various domain experts was defuzzified into crisp values as explained in section 3.3. Thereafter PARETO's rule was applied to overlook trivial tasks and pick out critical tasks to be delegated to efficient procurement agents.

### 4.4 Result Analysis

The decomposition of various procurement goals into a number of tasks along with their DoD is illustrated table 1. Raw materials and spares are the critical goals with very high and high DoD and rest of other goals viz. obtaining packaging, consumable and miscellaneous items with DoD less than 0.5 are considered as trivial.

**Table 1.** Decomposition of goals into a number of tasks along with their DoD

| Procurement Goals | Tasks | DoD at goal level | Goal Status | Crisp Weights $CW_{tsk[0,1]}$ | DoD at task level | Resulted Task Status (using PARETO'S rule) |
|---|---|---|---|---|---|---|
| Raw Material | Planning | | | 0.4375 | 0.3067 | Critical |
| | Ordering | 0.7010 | Critical | 0.3437 | 0.2410 | Critical |
| | Delivery | | | 0.0937 | 0.0657 | Trivial |
| | Receiving | | | 0.0937 | 0.0657 | Trivial |
| | Issuing | | | 0.0312 | 0.0219 | Trivial |
| Spares | Planning | | | 0.4375 | 0.2678 | Critical |
| | Ordering | 0.6121 | Critical | 0.3437 | 0.2104 | Critical |
| | Delivery | | | 0.0937 | 0.0574 | Trivial |
| | Receiving | | | 0.0937 | 0.0574 | Trivial |
| | Issuing | | | 0.0312 | 0.0191 | Trivial |
| Packaging | Planning | 0.2913 | Trivial | | | |
| Consumable items | Ordering<br>Delivery<br>Receiving | 0.2900 | Trivial | | | |
| Miscellaneous items | Issuing | 0.2000 | Trivial | | | |

Fig.12 evidently segregates various goals with various levels of dependencies formulated on a number of rules in rule base.

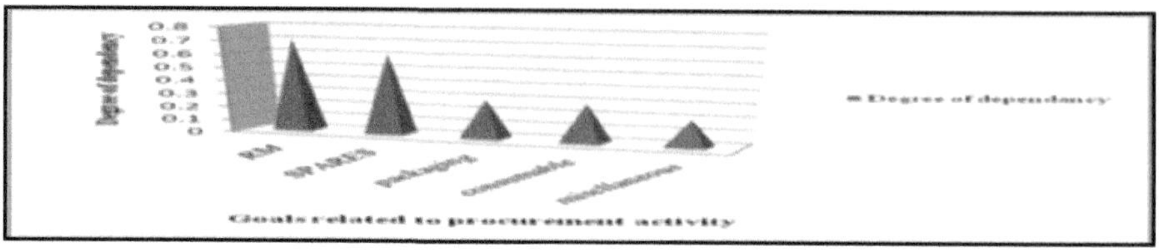

**Fig. 12.** DoD with respect to various goals

As evident from fig. 13, it is observed that in raw material activity by applying Pareto's rule, 70%- 80% of DoD is derived from two activities PLANNING and ORDERING.

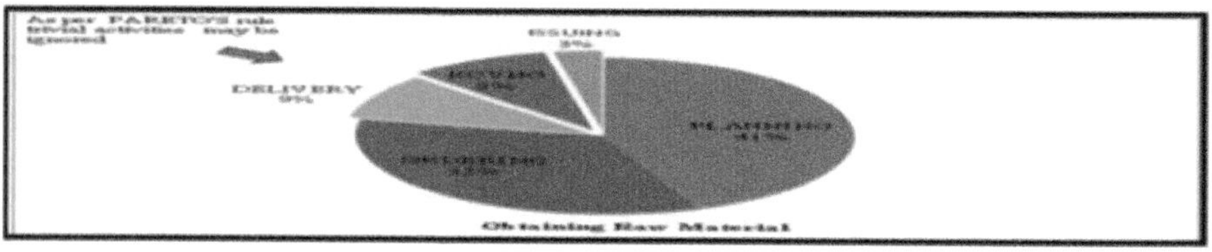

**Fig. 13.** DoD at task level

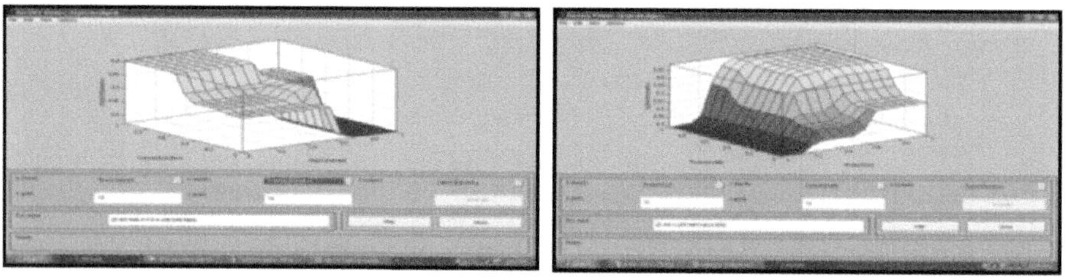

**Fig. 14.** DOD w.r.t. (a) Transaction Nature, Time Constraint (b) Product Cost, Product Quality

Hence dependent agent needs to be dependent for these two critical activities and rest of other activities viz. DELEVERY, RECEIVING, ISSUING less contribute in DoD , henceforth dependent agent may manage these activities of his own to avoid unnecessary communication overheads.

Fig. 14 shows the fuzzy surface plot obtained in the present experiment in order to depict the relationship between input and output variables. Fig. 14 (a) exhibits the relationship between the input variables Transaction Nature and Time Constraint on one hand and the output variable DoD on the other. DoD is inversely proportional to Time Constraint and directly proportional to Transaction Nature. It signifies that highly critical item viz. raw material urgently required for the production with tight time constraint requires high DoD for accomplishing it. Fig. 14 (b) exhibits that DoD is directly proportional to Product Cost as well as Product Quality signifying that higher product cost incurred in obtaining a goal and higher quality required necessitates higher dependency requirements in the form of DoD. The relationship thus obtained between input variables and output variable DoD would assist PurchaseHeadAgent in planning future strategies for the adjustments of input variables with inappropriate infrastructure and very high DoD.

## 5 Conclusion

The advent of ubiquitous computing has revolutionized distributed Multi-agent systems. However, its successful application is subject to the adoption of effective agents' communication that would be needed to share expertise for achieving goals of MAS. Tasks are disseminated and delegated to other agents with the intent of propagating proficiency and capability. Before indulging in the process of coordination, agents are required to assess their own dependency needs. To quantify dependency needs of an agent, this work provides a dependency metric entitled as DoD that would facilitate an agent to estimate its exertion load devised in the form of Degree of Dependency in a distributed environment. To evaluate DoD, this work employs Fuzzy Inference System exercised over domain knowledge that may be fuzzy, uncertain and ambiguous. PARETO's rule is applied to signify the inevitability of dependency of an agent for critical tasks and ignore trivial tasks. This study would assist developer to address dependency requirements of an agent during requirements engineering and provide him a sound underpinning in estimating number of agents to execute these dependency needs, which consequently may eschew unnecessary communication overheads.

## References


1. Bresciani, P., Perini, A.: Tropos: An Agent-Oriented Software Development Methodology. Autonomous Agents and Multi-Agent Sytems 8, 203–236 (2004)
2. Bedi, P., Gaur, V.: Optimal Partner Selection Model for Cooperation. In: 3rd Indian International Conference on Artificial Intelligence, pp. 1297–1307 (2007)
3. Osada, A.: The Role of Domain Knowledge Representation in Requirements Elicitation. In: 25th Conference on IASTED International Multi-Conference: Software Engineering (2007)
4. Rose, P.: A Method and Framework for Domain Knowledge Assisted Requirements Evolution. In: International Conference on Software Engineering (2009)
5. Castro, J., Kolp, M.: Towards requirements-driven information systems engineering: the Tropos project. Elsevier Information Systems 27, 365–389 (2002)
6. Singh, Y.: Evaluation of Agent Oriented Requirements Engineering Frameworks, vol. 2, pp. 33–38. IEEE, Los Alamitos, ISBN: 978-0-7695-3336-0,2008
7. Donzelli, P.: A goal-driven and agent-based requirements engineering framework*. Springer, Heidelberg (2004), doi:10.1007/s00766-003-0170-4
8. Zadeh, L.A.: Outline of a new Approach to the Analysis of Complex Systems and Decision Processes. IEEE SMC-3, 28–44 (1973)
9. Sivarao, P.B.: GUI Based Mamdani Fuzzy Inference System Modeling To Predict Surface Roughness in Laser Machining. IJECS 9(9)
10. Chen, Y.-S., Chong, P.P.: Mathematical and computer modelling of the Pareto principle. Mathematical and Computer Modelling 19(9), 61–80 (1994)
11. Klir, G.J.: Fuzzy Sets and Fuzzy Logic. PHI Publications (1995) ISBN:81-203-1136-1
12. Srinivasan, S.: Multi-Agent System Supply Chain Management in Steel Pipe Manufacturing. IJCSI 7(4) (2010) ISSN : 1694-0814
13. Gaur, V., Soni, A.: An Integrated Approach to Prioritize Requirements Using Fuzzy Decision Making. IACSIT IJET 2(4), 320–328 (2010) ISSN: 1793-8236

# Novel Methodologies to Avert Hacker

Paul Bharath Bhushan Petlu[1], N.L. Kumar Anantapalli[1],
Devaki Pendlimarri[2], and M. Muralidhara Rao[1]

[1] Swarnandhra College of Engineering and Technology
Narsapur – 534 280, India
{trishipaul,kumaranantapalli,drmmrao54}@gmail.com
[2] Swarnandhra Institute of Engineering and Technology
Narsapur – 534 280, India
pdevaki02@gmail.com

**Abstract.** With the advent of internet the world has become very small in the arena of communication. The information can be communicated from one place to another place through the internet very easily and with great speed. Before communicating with each other the user must be authenticated by the remote system to provide security to the information. The concept of username with password is an efficient method of maintaining shared secret information between a user and a remote system. The information that is communicated through the internet passes across an unsecured network. This gives security breaches to the information of the user authentication and is being hacked by the hacker. Even though the user credentials are encrypted into a cipher text, still with the constant monitoring, the hacker is able to guess and decrypt the user credentials. Here, we are presenting some novel methods to prevent the hacker in knowing the user credentials even though it passes through the unsecured networks.

**Keywords:** Attacks, hacker, authentication, network/information security, security, time-slot based algorithms, multiple databases, cipher text.

## 1 Introduction

With the advent of internet the world has become very small in the arena of communication. The use of computers has been increased due to the vast utilization of internet for e-mails, e-commerce, social networking etc. The information that is being communicated through internet passes through the unsecured networks, which gives chance to the security breaches [4] and [5]. There are several techniques to provide security to the information communicated between two users. The username along with password is the efficient method of maintaining a shared secret between the user and the remote system. The other way of authenticating the user is using biometrics. This technique uses the human parts to authenticate the user, such as, thumb impression, retinal of eye etc. To implement biometrics, we need additional hardware and its technology support. If the hardware fails then we cannot authenticate the user. The username along with password can be proved as an efficient technique for the user authentication by adding additional protection and security measures. As the

D. Nagamalai, E. Renault, M. Dhanushkodi (Eds.): CCSEIT 2011, CCIS 204, pp. 112–120, 2011.

information passes through the unsecured networks, the attacker/hacker will attack/hack the information and do harm to the system or individual information.

One of the main threats to the data transmitted is sniffing. This involves using a wired or wireless sniffer to listen to the network traffic. Sniffing may occur as a passive eavesdropping or active interception, such as man-in-the-middle attack with an attacker/hacker serving as an intermediary between the two systems through which message passes. If the sniffer knows the packet structure, most sniffers have an ability to decode and analyze the data gathered. Some sniffers can automatically filter out usernames along with passwords from other observed information, providing the attacker with an uncluttered view of captured account information and data.

Another threat against the password transmission is attacks. Classes of attacks might include passive monitoring of communications, active network attacks, close-in attacks, exploitation by insiders, and attacks through the service provider. Some of the attacks are [7]:

## 1.1 Replay Attacks

A breach of security in which information is stored without authorization and then retransmitted to trick the receiver into unauthorized operations such as false identification or authentication or a duplicate transaction. For example, messages from an authorized user who is logging into a network may be captured by an attacker and resent (replayed) the next day. The retransmission of valid logon messages is sufficient to gain access to the network even though the messages may be encrypted and the hacker does not know the actual keys and passwords.

## 1.2 Passive Attacks

A passive attack on a cryptosystem is one in which the cryptanalyst cannot interact with any of the parties involved, attempting to break the system solely based upon observed data i.e. the cipher text. This can also include known plaintext attacks where both the plain text and its corresponding cipher text are known. While most classical ciphers are vulnerable to this form of attack, most modern ciphers are designed to prevent this type of attack above all others.

## 1.3 Active Attacks

An intrusion into a computer network which attempts to delete or modify the data stored on the computers which form part of the network. This is one of the most serious forms of attack since the operations of many companies' critically depend on data.

## 1.4 Password Attacks

Password cracking is the process of recovering passwords from data that has been stored in or transmitted by a computer system. A common approach is to try repeatedly in guessing for the password. The purpose of password cracking might be to help a user to recover a forgotten password (though installing an entirely new password is less of a security risk, but involves system administration privileges), to gain unauthorized access to a system, or as a preventive measure by system administrators to check for easily crack-able passwords. On a file-by-file basis, password cracking is utilized to gain access to digital evidence but access to a particular file is restricted.

### 1.5 Phishing Attacks

Phishing is a way of attempting to acquire sensitive information such as usernames, passwords and credit card details by masquerading as a trustworthy entity in an electronic communication. Communications purporting to be from popular social web sites, auction sites, online payment processors or IT administrators are commonly used to lure the unsuspecting public. Phishing is typically carried out by e-mail spoofing or instant messaging and it often directs users to enter details at a fake website whose look and feel are almost identical to the legitimate one. Phishing is an example of social engineering techniques used to deceive users and exploit the poor usability of current web security technologies. Attempts to deal with the growing number of reported phishing incidents include legislation, user training, public awareness and technical security measures.

### 1.6 Hijack Attack

Session hijacking is the exploitation of a valid computer session — sometimes also called a *session key* — to gain unauthorized access to information or services in a computer system. In particular, it is used to refer to the theft of a magic cookie used to authenticate a user to a remote server. It has particular relevance to web developers, as the HTTP cookies used to maintain a session on many web sites can be easily stolen by an attacker using an intermediary computer or with access to the saved cookies on the victim's computer.

### 1.7 Spoof Attack

Spoofing attack is a situation in which one person or program successfully masquerades as another by falsifying data and thereby gaining an illegitimate advantage.

### 1.8 Buffer Overflow

In computer security and programming, a buffer overflow, or buffer overrun, is an anomaly where a program, while writing data to a buffer, overruns the buffer's boundary and overwrites adjacent memory. This is a special case of violation of memory safety.

### 1.9 Exploit Attack

The attacker knows of a security problem within an operating system or a piece of software and leverages that knowledge by exploiting the vulnerability.

Here, we suggest some novel set of methodologies to prevent the hacker or attacker from decrypting the user credentials. The hacker/attacker tries to know the encrypted user credentials by constant monitoring.

## 2 Background

The username along with the password is an efficient method of maintaining a shared secret between the user and the remote system. The user credentials are communicated through the unsecured network, which causes to security breach. In order to provide security, the user credentials are encrypted into a cipher text by using the Data Encryption Standard algorithm. But, still by constant monitoring the attacker/hacker is able to guess and decrypt the cipher text. The username along with the password can be proved as an efficient authentication technique by giving more protection and security measures to the user credentials that are being communicated through the network. Some novel methods are suggested by the authors to provide more protection and security to the user credentials. They are: *cipher text based on the time-stamp, multiple user passwords, and multiple cipher texts.* These methods can be applied to prevent the hacker/attacker in decrypting the user credentials that are being communicated through the unsecured network by constant monitoring.

## 3 Methodology

Here, we suggest some novel set of methodologies to prevent the hacker/attacker from capturing the username and the password via an unsecured network. Most of the services that are provided through the internet are using the username with password mechanism for the user authentication. This username and password will be given to the user at the time of creation of the new account. Later, whenever the user wants to communicate with the system provided through the internet, he/she is identified and authenticated using this username with password. As this username and password information passes to the remote system via unsecured networks, causes a breach to the hackers/attackers to decrypt the username and password. To prevent the hacker/attacker from capturing the username and the password, we suggest the below novel set of methodologies.

### 3.1 Cipher Text Based on the Time-Stamp

The round the clock i.e. 24 hours is divided into number of *equal* or *unequal* time-slots. Each time-slot is associated with an algorithm using which the user credentials are encrypted into cipher text.

*Equal time-slots:* The 24 hours are divided into number of time-slots. Each time-slot $t_1, t_2 \ldots t_n$ is of equal time period such as 1 hour or 1½ hour or 2 hours etc. depending upon the value of *n, total number of time-slots*. For example, if $n = 12$ then each time-slot is of two hour time period.

*Unequal time-slots:* The 24 hours are divided into number of time-slots of different time periods. For example, $t_1 = 1hr, t_2 = 2hrs, t_3 = 2hrs...t_n = 1\frac{1}{2}hr$ .

In either the case the below equation must be satisfied:

$$N = \sum_{a=1}^{n} t_a = 24hours \quad (1)$$

Each time-slot is represented as $t_1, t_2...t_n$ where $n$ represents the number of time-slots divided per day. Each time-slot $t_i$ is associated with an algorithm $algorithm_i$ , where $1 \le i \le n$ . There is an algorithm $standard\_algorithm$ , which is used to encrypt the user credentials i.e., username and password at the first time of user account creation.

There are many systems offering their services through the internet. Every user need user account to access the services of the system. This user account holds the username along with the password to authenticate the user uniquely. After creation of user account, whenever the user wants to access the services of the system, the user credentials are used to authenticate the user by the *sign up* page. This sign up page will encrypt the user credentials using the $algorithm_i$ associated with the time-slot $t_i$ when the user submits the credentials. This encrypted data will be sent to the server along with the time stamp as a packet. After receiving this packet the server will decrypt the encrypted data with the use of $algorithm_i$ associated with the same time-slot $t_i$ received along with the packet. This gives the original username and the password as entered by the user. Due to security reasons, as the username and the passwords shouldn't be stored as a plain text in the database, the server will again encrypt the data using $standard\_algorithm$ and store this into the database as shown in Fig 1. With this the user will have an account through which he/she can access the services of the organization through the internet at any time.

Whenever the user wants to access the services of the system, he/she is prompted to submit the user credentials for the user identification and authentication. After submitting these credentials the *sign in* page will encrypt the user credentials using $algorithm_i$ associated with the time-slot $t_i$ that the user has submitted the details. This encrypted data will be sent to the server along with the time stamp as a packet through the network. After receiving the packet, the server will decrypt the data using $algorithm_i$ associated with the same time-slot $t_i$ which is being received along with the packet. This will give the original username and password as submitted by the user. The server encrypts the decrypted user credentials using the algorithm $standard\_algorithm$ , and verifies in the database. Once the encrypted user credentials matches with the user credentials stored in the database, the server will create a session through which the user can communicate with the system for further access.

Each time-slot is associated with different algorithms; the details submitted by the user will be encrypted into different cipher texts, which will prevent the hacker/attacker in guessing the user credentials. More number of time-slots and/or data encryption algorithms will lead to more prevention of hacker/attacker. By applying the below strategies, even the hacker can be averted more. They are:

- *Changing encryption algorithms:* The data encryption algorithms associated with each time-slot can be changed in regular intervals.
- *Repeated algorithms:* The set of data encryption algorithms can be repeated to some time-slots.
- *Changing time periods:* The time period of the time-slots can be changed in regular intervals.
- *Hybrid approach:* The above strategies can be applied simultaneously.

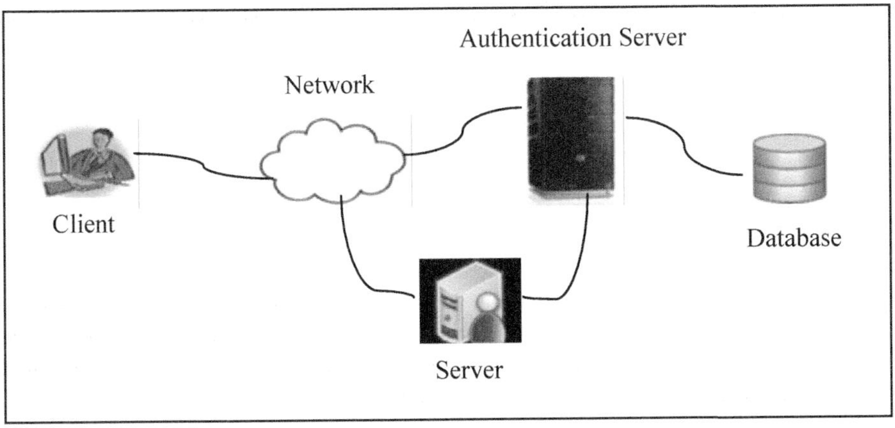


**Fig. 1.** Generation of cipher text based on time-stamp with two tier architecture

The problem in the above methodology is that, the number of decryption and encryption operations at the server side will increase the burden on the server. We can reduce this burden by using the authentication server. The user identification and authentication process will be carried out by the authentication server instead of the main server. Once the user is authenticated as a valid user, then the Authentication Server will communicate this to the Main Server to create a session which will further allow the user to communicate with the server. With the introduction of authentication server we have additional cost, space and technology.

### 3.2 Multiple User Passwords

The services of the system though the internet can be accessed by the users having an account. This account has the information of the user credentials, username along with the password. In this methodology, we propose that, at the time of account creation, the user may select the number of passwords he would like to have. This number may be restricted either to two or three. The user is required to choose two or three passwords instead of one at the time of account creation. The chosen username and the password(s) are stored in the database at the server as *username, password1, password2, and password3*. Certain password selection measures are to be taken while choosing the passwords, such as, complexity of the password in terms of size, combination of alphabets and digits, etc., which results the password to be strong. The user is able to access the services of the system by submitting the user credentials, username along with any one of the three passwords. The user credentials submitted by the user are verified with the stored information to authenticate the user. Once the user is authenticated as a valid user, the server will create a session through which the user is able to access the system for further communication.

### 3.3 Multiple Databases

The methodology is as shown in Fig. 2. The round the clock, i.e. 24 hours is divided into number of time-slots such as $t_1, t_2 ... t_n$, where $n$ represent the number of time-slots. Each time-slot $t_i$ is associated with an encryption algorithm $algorithm_i$, where $1 \leq i \leq n$. Each distinct data encryption algorithm is associated with a database. These databases are represented as $db_1, db_2 ... db_m$ where $m$ represents the number of distinct data encryption algorithms. The encryption algorithm $standardalgorithm$ is used to encrypt the user credentials into cipher text at the time of creation of a user account.

Every user can access the system with a user account. At the time of creation of user account, the submitted user credentials by the user are encrypted into a cipher text using the algorithm $standardalgorithm$. This encrypted cipher text will be sent to the server along with the time stamp as a packet. After receiving this packet, the server will decrypt the cipher text with the use of the algorithm $standardalgorithm$. This gives the original credentials as submitted by the user. The server will encrypt these user credentials into different cipher texts $cipher_1, cipher_2 ... cipher_m$ using the algorithms $algorithm_1, algorithm_2 ... algorithm_m$ respectively. These encrypted cipher texts are stored in the respective databases $db_1, db_2 ... db_m$.

After creation of a new account, whenever the user wants to communicate with the system, the user is prompted to submit the user credentials for the user identification and authentication. After submitting these details the *sign in* page will encrypt the user credentials into a cipher text using the algorithm $algorithm_i$, where $i$ represent the time-slot the user has submitted the details. This encrypted cipher text will be sent to the server as a packet through the network. The server will verify this cipher text with the databases $db_1, db_2 ... db_m$ simultaneously. If the user submitted credentials matches with the previously stored user credentials in any one of the database, then the user is authenticated as a valid user. Once the user is authenticated as a valid user, the server will create a session through which the user can communicate with the system for further access.

In every time-slot, as the data encryption algorithm is changing the details entered by the user will be encrypted into different cipher texts, which will prevent the hacker/attacker in guessing the user credentials. Hence, the more number of time-slots and/or data encryption algorithms will lead to more restriction to the hacker/attacker. To provide more protection and security to the user credentials from the man-in-the-middle, the below strategies can be applied:

- *Changing encryption algorithms:* The data encryption algorithms associated with each time-slot can be changed in regular intervals.
- *Repeated algorithms:* The set of data encryption algorithms can be repeated to some time-slots.

- *Random selection of encryption algorithms:* The data encryption algorithms to each time-slot can be selected randomly.
- *Changing time periods:* The time period of the time-slots can be changed in regular intervals.
- *Hybrid approach:* The above strategies can be applied simultaneously.

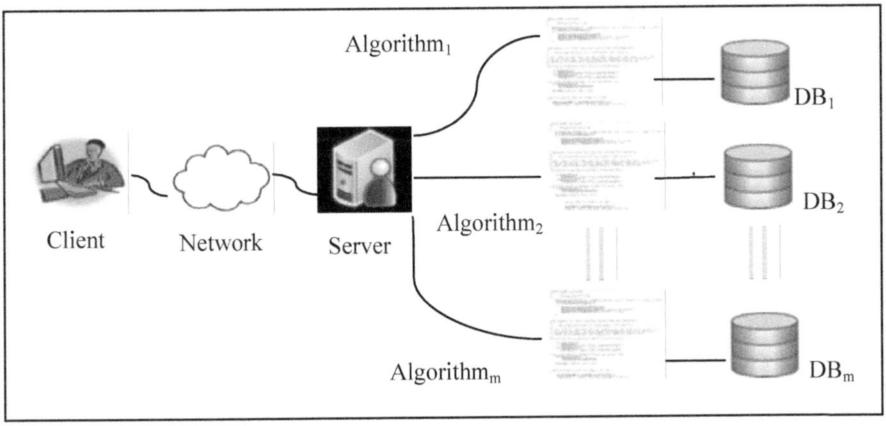


**Fig. 2.** Methodology for generating different forms of cipher texts for the user credentials

**Note:** The same set of data encryption algorithms must be used. Addition of a new algorithm leads to the addition of a database at the server side.

The problem here is that the server is maintaining the multiple databases to maintain the user credentials in different forms of encrypted cipher texts using different algorithms. However, the maintenance of multiple databases can lead to an advantage in the case of database recovery. Suppose, one database is crashed due to some reasons, we can recover the information from the remaining databases. Hence, more number of databases will lead to a more security in database recovery.

## 4 Conclusion and Future Work

Since the advent of internet there is a vast utilization of internet for e-mailing, e-commerce etc. by the numerous users being at different places. The concept of username and password is an efficient method of maintaining a shared secret between the server and a remote client system. The communication between the server and the client system is via an unsecured network. This gives breaches to the hacker/attacker to know the username and the password of the user by sniffing, hacking etc. being a man-in-the-middle. To restrict the man-in-the-middle from hacking/attacking, we proposed three methodologies. Each methodology has its own advantages as well as disadvantages. The selection of methodology may depend upon the available technology and the cost of establishment of the methodology. Further, with the innovative methods, techniques and algorithms the data that has been passed through the network can be made more secured. Hence, the man-in-the-middle can be prevented or nullified more from either hacking or attacking the packets.

## References

[1] Scarfone, K., Souppaya, M.: Guide to Enterprise Password Management (Draft). Recommendations of the National Institute of Standards and Technology, Special Publication 800-118, 38 pages (2007)

[2] Elminaam, D.S.A., Kader, H.M.A., Hadhoud, M.M.: Performance Evaluation of Symmetric Encryption Algorithms. IJCSNS International Journal of Computer Science and Network Security 8(12), 280–286 (2008)

[3] Identification and Authentication, NIST Computer Security Hand Book, http://www.sos.cs.ru.nl/applications/courses/security2008/nistiadraft.pdf

[4] Raj, A.K., Tewari, R.R., Upadhyay, S.K.: Different Types of Attacks on Integrated MANET- Internet Communication. International Journal of Computer Science and Security (IJCSS) 4(3), 265–274 (2010)

[5] Network security types of attack passive attack active attack, http://www.computernetworkingnotes.com/ccna_certifications/types_of_attacks.htm

[6] Petlu, P.B.B., Pendlimarri, D., Anantapalli, N.L.K., Naga Srinivas, P.V., Muralidhara Rao, M.: Proceedings of the IC2IT, International Conference on Computing and Information Technology, held on 11-12, 2011 at King Mongkut's University of Technology, North Bangkok, Bangkok, Thailand, pp: 116-121 (May 2011)

[7] http://en.wikipedia.org/wiki/Password_cracking

# Multi-Domain Meta Search Engine with an Intelligent Interface for Efficient Information Retrieval on the Web

D. Minnie[1] and S. Srinivasan[2]

[1] Madras Christian College, Chennai, India
minniearul@yahoo.com
[2] Anna University of Technology Madurai, Madurai, India
sriniss@yahoo.com

**Abstract.** A web search engine searches for information in the World Wide Web. The user needs to search for information in various domains such as financial and healthcare. The user is expected to remember various Vertical Search Engine names to extract appropriate information. A Vertical Search Engine is a domain specific search engine that provides the user with results for specific queries on that domain. Often the user is unable to get the exact information from the web due to the different page ranking techniques followed by individual search engines. Meta Search Engines solve this problem to a certain level by sending the user's search queries to various search engines and combining the search results. This paper proposes Multi Domain Meta Search Engines that facilitate efficient Information Retrieval on multiple domains. User interfaces for selecting domain, category and search engines is presented. An intelligent User Interface is also designed to get the user query and to send it to appropriate search engines. Few algorithms are designed to combine results from various search engines and also to display the results.

**Keywords:** Information Retrieval, Web Search Engine, Vertical Search Engine, Meta Search Engine.

## 1 Introduction and Related Work

The World Wide Web (WWW) has evolved into a huge repository of information. The web resources increase everyday and that in turn increases the complexity of accessing the web data. There is a need for efficient searching techniques to extract appropriate information from the web, as the users require correct and complex information from the web.

### 1.1 Web Search Engine

A Web Search Engine is a search engine designed to search WWW for information about given search query and returns links to various documents in which the search query's key words are found. The types of search engines [1 – 3] used in this paper are General Purpose Search Engines such as Google, Vertical Search Engines and Meta Search Engines.

D. Nagamalai, E. Renault, M. Dhanushkodi (Eds.): CCSEIT 2011, CCIS 204, pp. 121–129, 2011.

### 1.2 Vertical Search Engine

A Vertical Search Engine is a domain specific search engine that provides the user with results for specific queries on that domain. It indexes web pages that are related to a particular domain. The filter component of Vertical Search Engine classifies the web pages downloaded by the crawler into appropriate domains [4]. Medical search engines such as "MedSearch", facilitate the user to get information on medical domain [5 - 7]. Search engines assume that users are capable of creating appropriate search queries. Vertical Search Engines such as iMed creates search queries for the user by interacting with the user [8].

### 1.3 Meta Search Engine

Different Search Engines follow different page ranking techniques and hence approximately 20% of the results are same for a specific search query. The user may fail to receive the necessary information. This gives rise to the use of Meta Search Engines [9]. Meta Search Engine accepts a search query from the user and sends the search query to a limited set of Search Engines. The queries may be sent as it is or it may be modified to suit the requirements of the search engines. The results are retrieved from the various search engines and they are combined to produce a result set and are given to the user.

### 1.4 Information Retrieval

Information Retrieval is the science of searching and retrieving information. Web search is also a type of information retrieval as the user searches the web for information. The efficiency of a search facility is measured using two metrics Precision and Recall. Precision specifies whether the documents retrieved are relevant and Recall specifies whether all the relevant documents are retrieved.

$$\text{Precision} = \text{Relevant and Retrieved} / \text{Retrieved} \quad (1)$$

$$\text{Recall} = \text{Relevant and Retrieved} / \text{Relevant} \quad (2)$$

Precision is calculated as the ratio of the relevant web pages that are retrieved to the number of web pages that are retrieved. Recall is not calculated as it is not possible to find the number of relevant web pages out of the millions of web pages in the WWW.

## 2 Web Search Engine Architecture

A web search engine consists of a Web Crawler or Spider, Indexer, Query Parser and Ranker. The architecture of a web search engine is shown in Fig. 1.

Web search engines work by storing information about many web pages, which they retrieve from the WWW itself. These pages are retrieved by a Web crawler which follows every link it sees. The contents of each page are then analyzed to determine how it should be indexed.

Web pages are filtered using Term Frequency-Inverse Document Frequency (TFIDF) scores for that page. TFIDF is calculated as the product of Term Frequency (TF) of a term t in document d and Inverse Document Frequency (IDF) of a term t.

$$TFIDF_{td} = TF_{td} * IDF_t \quad (3)$$

Term Frequency ($TF_{td}$) of document d and term t is given as the ratio of the number of occurrence of a term in a web page to the total number of words in that page.

$$TF_{td} = \text{No. of term t in document d / Total no. of words in document d} \quad (4)$$

Inverse Document Frequency ($IDF_t$) of a term specifies the uniqueness of a document. It is given as a ratio of the total number of documents to the number of documents containing the term.

$$IDF_t = \log [\text{Total no. of documents (N) / No. of documents with term t}] \quad (5)$$

The higher value for $IDF_t$ specifies that the document is unique as the term is present in few documents. The lower value specifies that the term is present in many documents and hence the document is not unique even though the Term Frequency of the document could be high.

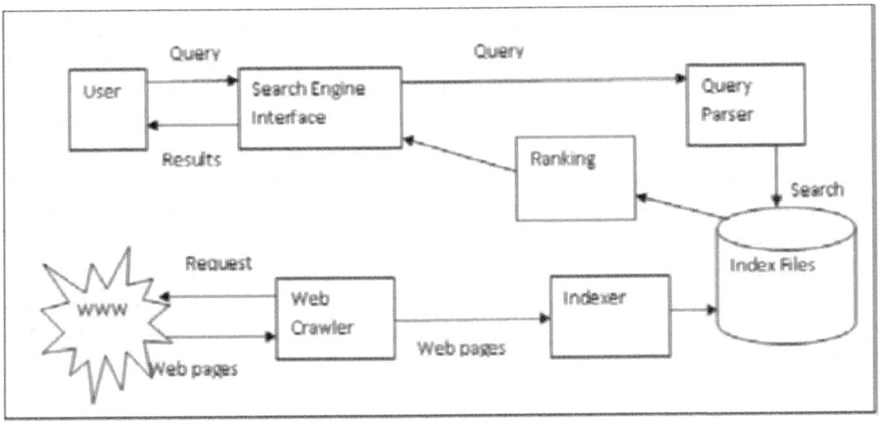


**Fig. 1.** Architecture of Web Search Engine

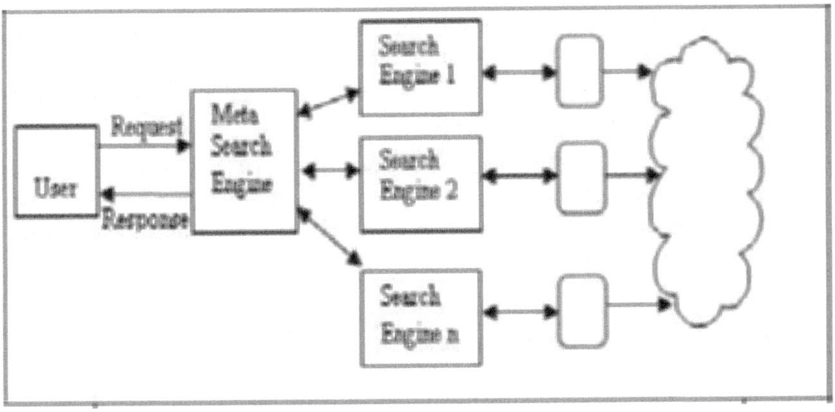


**Fig. 2.** Meta Search Engine Architecture

## 3 Meta Search Engine Architecture

Meta Search engine is not a repository of physical data. It receives request from the user and sends the request to various search engines. The search engines check their indices and extract a list of web pages as links and pass the result to the Meta Search Engine. The Meta Search Engine receives the links, applies few algorithms, ranks the results and finally displays the result. The Meta Search Engine architecture is shown in Fig.2.

## 4 Methodology

Few Web Search Engines[10] such as Google, Yahoo, MSN and Vertical Search Engines[11] such as Yahoo Health, webMD, MSNHealth and Meta Search Engines [12] such as DogPile, MetaCrawler, MonsterCrawler are used in this study. A Multiple Search Engine [13] is also analyzed in which a Search Engine can be selected from a list of search engines and queries can be sent to it.

Four topics Cancer, Diabetes, Paracetamol and Migraine are identified to be searched in the nine search engines. The search results are classified under the categories Relevant, Reliable, Redundant and Removed based on the retrieved web page contents. 20 results from each of the nine Search Engines are considered for each topic and are given in table 1. The interpretation of the results is shown in Fig.3.

The precision values are calculated using the ratio of relevant documents to the retrieved documents and is presented in Table 2. The precision details are also plotted in a graph and are shown in Fig. 4.

The user searches for financial information, healthcare information and so on in the web. The user is expected to remember various search engine names to extract necessary information. The search engines also have a biased view of presenting the documents and the user may miss the necessary information.

Hence the facility to get efficient domain related information on multiple domains is a need of the hour. Multi-Domain Meta Search Engine [14] sends queries to various search engines and consolidates the results from the search engines.

**Table 1.** Relevance, Reliability, Redundant and Removed details for Search Engines

| Search Engine | Relevant | Reliable | Redundan | Removed |
|---|---|---|---|---|
| DogPile | 18.75 | 16.25 | 1.75 | 0.5 |
| Google | 19.25 | 19.25 | 0.25 | 0.25 |
| Meta Crawler | 17.75 | 14.25 | 4 | 0 |
| Monster Crawler | 19 | 14 | 1 | 0 |
| MSN | 18.25 | 12.5 | 1.5 | 1.25 |
| MSN Health | 1 | 1 | 0 | 0 |
| WebMD | 17 | 17 | 0.25 | 0 |
| Yahoo | 19.25 | 17.25 | 2.75 | 0 |
| Yahoo Health | 11 | 11 | 0 | 0 |

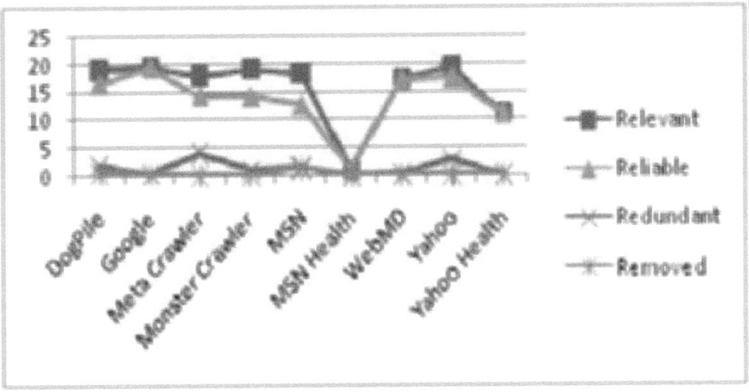


**Fig. 3.** Search Engine's Relevant, Reliable, Redundant and Removed details

**Table 2.** Precision value table

| Search Engine | Precision | Search Engine | Precision |
|---|---|---|---|
| DogPile | 0.94 | MSN Health | 0.25 |
| Google | 0.96 | WebMD | 0.85 |
| Meta Crawler | 0.89 | Yahoo | 0.96 |
| Monster Crawler | 0.95 | Yahoo Health | 0.55 |
| MSN | 0.91 | | |

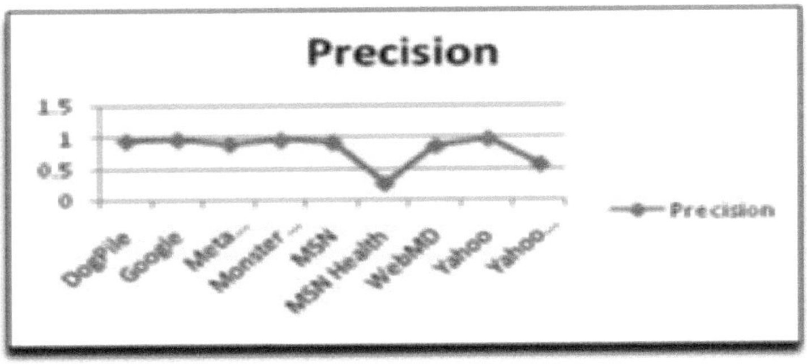


**Fig. 4.** Precision for various Search Engines for the given topics

## 5 Proposed Multi-Domain Meta Search Engine

We propose a user-friendly Multi-Domain Meta Search Engine that sends search queries to various search engines and to retrieve results from them. Various cases of selection of search engines are proposed in this paper.

Various cases of Meta Search Engines are designed and analyzed in this study. The basic architecture of the Multi-Domain Meta Search Engine is given in Fig. 5. In the first model the Meta Search Engine was designed to send queries to Search Engines such as Google, Yahoo, AltaVista and AskJeeves. The queries were formed for a specific domain by adding the domain name as part of the search query. An interface for the Meta Search Engine was formed consisting of push buttons to select the domain and text box to enter the query string. The query string to be sent to the search engines are formed with the + symbol applied on the string entered and the selected domain name.

The second model of Multi Domain Meta Search Engine aims at providing an efficient information retrieval on a particular domain for the user by accessing various Vertical Search Engines.

The result of the query can be shown as a list of results from each search engine in individual windows, or as a list of results in different frames of a same window or as a single ordered list of results.

## 6 Multi-Domain Meta Search Engine Operations

The Multi-Domain Meta Search Engine consists of a user interface for domain and category selection, a user interface for search engine selection, query formulator, dispatcher, results receiver, filter, ranker and display of results.

### 6.1 User Interface – Domain and Category Selection

The Domain selector presents a list of domains in this interface to the user. The domains are listed in the alphabetical order for easier access.

The user is allowed to enter the request as a query. The user can directly move to search or can select the domain from which the information is requested. User can also select the category of his request such as research papers, news articles, and homepages and so on so that the search can be performed on relevant pages.

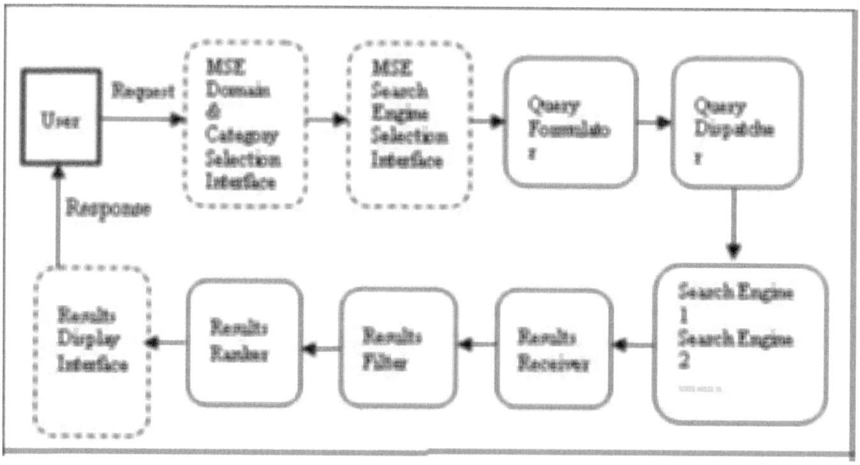


**Fig. 5.** Multi-Domain Meta Search Engine Architecture

### 6.2 User Interface – Search Engine Selection

The Search Engine Selection Interface presents to the user a list of Search Engines that are appropriate for the selected domain and category. Vertical Search Engines if available for a specific domain is also listed here. The user has the option to select one or more of the search engines. If the user has not selected the search engines then all the listed search engines are selected for further processing.

### 6.3 Query Formulator and Dispatcher

As different Search Engines follow different styles for the representation of the query search string, the search query strings are generated for a given user input. Different query strings are generated as per the requirement for the different search engines. The query strings are then sent to various search engines to extract desired results from the search engines.

### 6.4 Results Receiver

The user generally searches only the first and second pages of a search result and also the most relevant results are presented in those pages. The Search Engines provide the best results also in the top few pages. Hence 25 results from each Search Engine are selected as the results from the Search Engines. The results are stored for further processing.

### 6.5 Filtering and Ranking Results

The results from a single search engine are found to be having redundant links and removed links during some of the searches. These links are removed from the results set from each of the search engine results in the first step. All the results are populated and the redundant links across the different search engine results are eliminated. The ranking used by the different search engines is used to rank the results.

### 6.6 Results Generation in Different Windows

The results from the various search engines are displayed in different windows. The user has the freedom to look into the results given in various windows. The advantage of this method is that the user can select as many Search Engines as needed. There is no need for the ordering and ranking of results. The disadvantage of the method is that the user has to switch between windows for effective usage of the system.

### 6.7 Results Generation in Multiple Frames of Same Window

The results from various search engines are displayed in various frames of the same window. The output window is designed to display results from 4 search engines simultaneously. This method is faster to use but the user has to make the decision about the result links to be visited. The result for the searches on Cancer, Diabetes, Paracetamol and Migrane for this case is given in Fig.6 and the precision details are given in Fig. 7. In this method also the ordering and ranking of the results is not done as the result from the various search engines are used as it is.

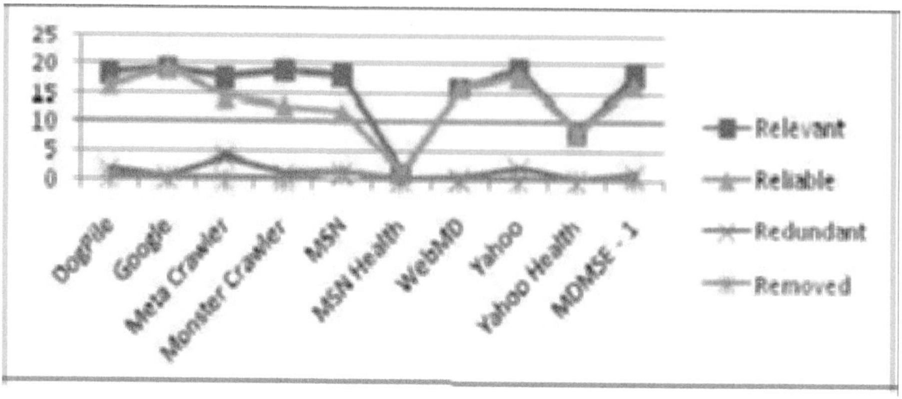


**Fig. 6.** Relevant, Reliable, Redundant and Removed data of Multi-Domain Meta Search Engine

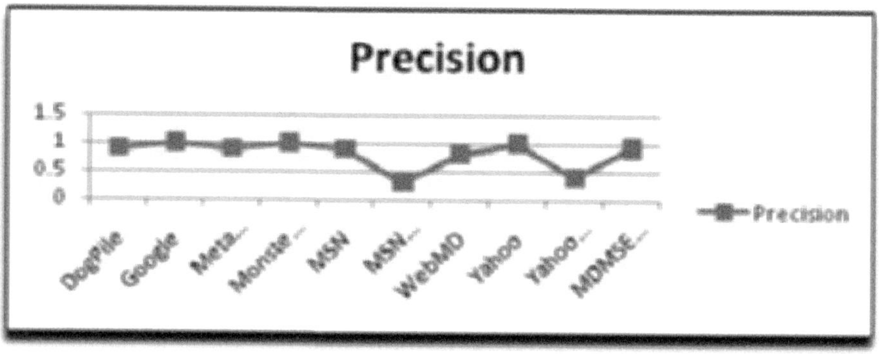


**Fig. 7.** Precision for MDMSE and various Search Engines for the given topics

It can be seen from the figures 6 and 7 that the Multi-Domain Meta Search Engine's performance is better than the performance of individual search engines.

### 6.8 Results Generation in Same Window

Alternatively the ordering process is to be performed and the ranked list is to be displayed along with the rank and the name of the search engine that has produced the link.

## 7 Conclusion and Future Work

This paper analyses two Multi Domain Meta Search Engines that provide an information retrieval of information on various domains for the user using various Search Engines that are already available. The Web Search Engine, Vertical Search Engine and Meta Search Engine features are also presented. Few Search Engines are tested for the relevancy, reliability, redundancy and availability of search results for few topics. A domain and category selection user interface is presented. A query interface is designed to send user's search query to the various search engines and the resultant links are consolidated to display the results to the user.

The system can be extended to process the content of the web pages in addition to processing of the links.

**Acknowledgement.** The authors wish to thank Ms.L.R.Rita, project student, Department of Computer Science, Madras Christian College, Chennai, India for her contribution towards development of the Multi-Domain Meta Search Engine.

## References

1. Dunham, M.H., Sridar, S.: Data Mining. Pearson Education, London (2007)
2. Kosla, R., Blockeel, H.: Web Mining Research: A Survey. SIGKDD Explorations 2(1), 1–15 (2000)
3. Jeyaveeran, N., Haja Abdul Khader, A., Balasubramaniyan, R.: E-Learning and Web Mining: An Evaluation. In: Proceedings of the 2nd International Conference on Semantic e-Business and Enterprise Computing (2009)
4. Shettar, R., Bhuptani, R.: A Vertical Search Engine – Based on Domain Classifier. International Journal of Computer Science and Security 2(4), 18–27
5. Luo, G., Tang, C.: On Iterative Intelligent Medical Search. In: SIGIR 2008, The 31st Annual International ACM SIGIR Conference, Singapore, July 20-24, pp. 3–10 (2008)
6. Luo, G.: Intelligent Output Interface for Intelligent Medical Search Engine, Association for the Advancement of Artificial Intelligence. In: Proceedings of the Twenty-Third AAAI Conference on Artificial Intelligence, pp. 1201–1206 (2008)
7. Ilic, D., Bessell, T.L., Silagy, C.A., Green, S.: Specialized Medical search-engines are no better than general search-engines in sourcing consumer information about androgen deficiency. Human Reproduction 18(3), 557–561 (2003)
8. Luo, G.: Design and Evaluation of the iMed Intelligent Medical Search Engine. In: Proceedings of the 2009 IEEE International Conference on Data Engineering, ICDE 2009, pp. 1379–1390 (2009)
9. Chan, K.-P.: Meta search engine. Theses and dissertations. Paper 232 (2007), `http://digitalcommons.ryerson.ca/dissertations/232`
10. General Search Engine web sites, `http://google.com`, `http://google.com`, `http://google.com`
11. Vertical Search Engine web sites, `http://health.yahoo.net`, `http://www.webmd.com`, `http://health.msn.com`
12. Meta Search Engine web sites, `http://dogpile.com/`, `http://metacrawler.com`, `http://monstercrawler.com`
13. White, R.W., Richardson, M., Bilenko, M.: Enhancing Web Search by Promoting Multiple Search Engine Use. In: Proceedings of the 31st Annual International ACM SIGIR Conference on Research and Development in Information Retrieval (SIGIR 2008), Singapore, July 20-24 (2008)
14. Minnie, D., Srinivasan, S.: Meta Search Engines for Information Retrieval on Multiple Domains. In: Proceedings of the International Joint Journal Conference on Engineering and Technology (IJJCET 2011), pp. 115–118. Gopalax Publications & TCET (2011)

# Improved Video Watermarking Using Nested Watermark, Wavelets and Geometric Warping

T. Jayamalar[1] and V. Radha[2]

[1] Research Scholar, Avinashilingam Deemed University for Women, Coimbatore, India
jayamalarskjin@gmail.com
[2] Associate Professor, Avinashilingam Deemed University for Women, Coimbatore, India
radharesearch@yahoo.com

**Abstract.** The advancements in communication software and hardware is increasing the amount of information in the form of text, image, audio and video being shared. This increase arise the issue of protecting the intellectual property. This paper studies the protection of video signals through watermarking. The system proposed uses a combination of visual cryptography and wavelets to create a nested watermark which is embedded into the raw uncompressed video signal using a geometric warping technology. To improve robustness against compression, the method embeds the watermark in relevant area using a block selection method. The experimental results prove that the proposed system is efficient in terms of capacity and robustness against various attacks.

**Keywords:** Video Watermarking, Nested Watermarking, Geometric Warping, Wavelets.

## 1 Introduction

The current era of information explosion is envisaging tremendous growth in communication medium and hardware growth. This in turn, is generating a huge volume of digital signals in the form of text, images, videos and audio content. The World Wide Web (WWW) technology is providing an easy-to-access medium, which is allowing a large number of people to share information in a cost effective manner. However, these advancements prove to be challenging while taking the intellectual content protection into consideration. To answer this challenge, techniques like encryption, stegnography, cryptography and watermarking are being used. Among these techniques, digital watermarking techniques have gained more attention while proving the integrity and authenticity of the owner ([25], [17], [18]).

The art of embedding hidden data into a host signal is termed as 'digital watermark'. The techniques can be used for copyright protection, fingerprinting, copy protection, broadcasting monitoring and data authentication. The techniques of watermarking can be text-based watermarking, image watermarking ([19], [7], [9], [14]), video watermarking ([5],[10]), audio watermarking and 3D watermarking [26]. Among them image and video watermarking are two important areas that has attracted several researchers ([33], [21]).

D. Nagamalai, E. Renault, M. Dhanushkodi (Eds.): CCSEIT 2011, CCIS 204, pp. 130–141, 2011.

All these techniques have the common objective of embedding a secret data to protect a signal. With video watermarking it works with the principle that they should ensure digital ownership, have the ability to track the source of the digital video, have quick and easy way to detect video manipulation, have no visible video degradation and should have easy implementation procedure. The basic characteristics are imperceptibility, reliability, low complexity of watermarking algorithm and security of the hiding place.

In general scenario, watermarks are embedded into video signal either during or after compression ([29], 28]). This has the advantage that watermarking process is not influenced by the compression algorithm. However, the watermarks are generally not robust to a transcoding of the video. Thus, embedding the watermark to the uncompressed video data will have the advantage of being robust to the transcoding of the video. The challenge here is that since both watermarking technique and compression algorithm both seek to use irrelevant data, the embedded watermark might be removed by the compression algorithm. It would be better if the watermark is embedded in the relevant data so that it is not removed by compression algorithm. A method based on geometric warping was suggested by [27]. This method was efficient in terms robustness to lossy compression and improved the quality by using a block linking approach. This method had the advantage of not requiring the original video while extracting the watermark. However, this technique reduced the watermark capacity. This method is referred as base system in this paper. Aggarwal *et al.* [1] proposed another efficient watermarking algorithm to improve the payload and robustness without affecting the image quality. They used a concept of nested watermarking where one watermark is embedding into another watermark and the resultant image is embedded into the cover watermark. The concept of nested watermark improved payload and was able to maintain the quality to a great extent.

Motivated by these two works, in the present work, a single copyright message is divided into two shares using Visual Cryptography (VC). The second share is then embedded into first share, thus creating a new watermark image. This image is then embedded into an uncompressed video signal using geometric warping method as suggested by [27].

The rest of the paper is organized as follows. Section 2 presents a brief study on the previous works related to the topic. In Section 3, the basic concepts of visual cryptography (VC) and geometric warping are described. Section 4 explains the proposed watermarking algorithm. Section 5 presents the experimental results and Section 6 concludes the work with future research directions.

## 2 Previous Studies

Solutions to video watermarking have been studied by various researchers who work with both raw uncompressed video data [8] and compressed video data [2]. Video watermarking techniques can be mainly grouped as spatial domain approaches and frequency domain approaches.

The spatial domain watermarking techniques embed the watermark by modifying the pixel values of the host image directly. Least Significant bit (LSB) technique is the most frequently used method [16]. Both [30] and [13] embedded the watermark in saturation on the HIS (Hue, Saturation, Intensity) color space. A variable block size based adaptive watermarking in spatial domain was proposed by [15]. In a later period, [32] proposed a probability block based watermarking method for color image with fixed block size. Recently, an algorithm based on chaotic maps was proposed by [34] to determine the pixel bit for embedding. The main advantages of pixel based methods are that they are conceptually simple and have very low computational complexities. However, they also exhibit some major limitations. The need for absolute spatial synchronization leads to high susceptibility to de-synchronization attacks; lack of consideration of the temporal axis results in vulnerability to video processing and multiple frame collusion; and watermark optimization is difficult using only spatial analysis techniques.

Frequency domain approaches use transformation techniques to embed the watermark into a video sequence. Discrete Cosine Transformation (DCT), Discrete Fourier Transformation (DFT) and Discrete Wavelet Transformation (DWT) are the three main methods of data transformation. A subsample based watermarking technique was proposed by [20], where the DCT coefficients of the subimages were utilized to store the watermark. The authors of [6], proposed an algorithm which was based on embedding the watermark image three times at three different frequency bands, namely, low, medium and high and the results proved that the watermark cannot be totally destroyed by either low pass, medium or high pass filter. In [29], two complementary watermarks were embedded into the host image in order to make it difficult for attackers to destroy both of them. The main benefit obtained from frequency domain techniques is that they can take advantage of properties of alternate domains to address the limitations of pixel-based methods. Besides, analysis of the host signal in a frequency domain is a prerequisite for applying more advanced masking properties of the HVS to enhance watermark robustness and imperceptibility. Generally, the main drawback of transform domain methods is the high computational requirement.

Another field where scientists focus is on embedding into a compressed domain which can be done during or after the encoding process ([27],[28]). Alternatively, techniques to work on raw format have also received attention. In general, invisible watermark is embedded into irrelevant information of video data which can not be perceived by human eyes. This results in problems because compression algorithms, which also try to remove irrelevant information, might remove watermark accidentally. In [27], this problem was solved by embedding watermark in the relevant information of videos in an imperceptible manner. This approach is based on geometric warping of blocks. This approach though very efficient produced visible artifacts and video quality after extraction was low. Method to prevent these artifacts and improve the video quality was proposed by [27]. They used geometric warping method to embed watermarks, which did not require the original video during watermark extraction. This method improved [23] by removing artifacts and video quality, but was slow for applications where time was significant. This paper proposes a solution to solve this problem.

# 3 Methods Used

The concepts behind the various methods used in the proposed watermarking scheme are explained in this section.

## 3.1 Visual Cryptography

Visual Cryptography (VC), a paradigm introduced by [22], was initially used to encrypt material like written text, printed text, pictures, in a secure manner. It is a Visual Secret Sharing Scheme (VSSS) which uses the Human Visual System (HVS) to decrypt a secret message without expensive and complicated decoding process [31]. The basic VC system starts with the encoding phase, where a secret image is divided into a collection of 'm' black and 'n' white pixels. Each collection of m x n pixels is referred to as a share, which will resemble a noisy image when separately. During decoding phase, these shares or subset of shares are stacked together which will allow the visual recovery of the secret message. It has been applied to many applications, including but not restricted to E-voting system [24], financial documents [11], information hiding [4], general access structures [3], visual authentication and identification [23]. More detailed information about visual cryptography can be found in [35].

Out of the many VC schemes proposed, the present work considers the (2,2) VSS scheme proposed by [22]. In this algorithm, each pixel of the copyright image is expanded into 2 * 2 pixels. To share a white pixel of the secret image, one row from the first 6 rows of table1 is chosen randomly. Similarly, the two shares of a black pixel are determined by a random selection from the 6 last rows of Fig. 1. As a result, an M*N pixels secret image is expanded into two 2M*2N pixels share-images.

Considering security of the method, presence of only one share image reveals nothing about the corresponding secret image, i.e., each 2*2 pixels block of one share-image may correspond to either a white pixel or a black pixel of the secret image. Stacking the shares of a black secret pixel results in 4 black subpixels, whereas only 2 black subpixels is gained by stacking shares of a white secret pixel [12]. So, secret image is revealed to human eyes by stacking the shares without performing any cryptographical computations.

The scheme explained here is used in the proposed system and has the following advantages.

- Simple to implement
- Encryption do not required any heavy computations
- Decryption algorithm not required (uses only Human Visual System).
- Even infinite computation power can not predict the secret message.

### 3.2 Watermarking Using Geometric Warping

The geometric warping method used in this paper follows the methodology proposed by [27] and the readers are referred to this paper for a detailed explanation. A consolidated explanation is presented in this section.

The geometric warping works by embedding the watermark in the relevant and visible part of the video in an imperceptible manner. This approach uses the Peak Signal to Noise Ratio (PSNR) to perform geometric warping and work on the principal that the visible-imperceptible approach can be realized if the watermark is visible according to PSNR definition but imperceptible if someone sees the video. This method changes the border position of the objects as they are untouched by the compression algorithms. To decide for each block, which of the object borders to use during embedding, a Normed Centre of Gravity (NCG) value is used. The NCG value is calculated as follows.

1. Compute the mean values of the rows and columns of the block and store it in $M_x$ and $M_y$.
2. Using $M_x$ and $M_y$, compute x- and y-coordinate of the NCV. (This arranges the two vectors in two circles).
3. Compute the 2-D vector $v_k$ (k=x or y) using Equation (1)

$$v_k = \begin{pmatrix} \sum_{i=1}^{n} m_k(i).\cos\left(\frac{\pi}{n} + \left((i-1).\left(\frac{2\pi}{n}\right)\right)\right) \\ \sum_{i=1}^{n} m_k(i).\sin\left(\frac{\pi}{n} + \left((i-1).\left(\frac{2\pi}{n}\right)\right)\right) \end{pmatrix} \quad (1)$$

4. For each vector, compute the vector angles (Equation 2) and the vector lengths Lx, Ly.

$$k = \frac{n\theta_k}{2\pi} \quad \text{for}\,(k = x \text{ or } y) \quad (2)$$

The main advantage of NCG is that it is robust again any compression algorithms. The NCG of blocks with distinct object borders are robust to compression algorithms while NCG blocks without distinct object borders are not robust. Distinct object borders result in high vector lengths, using which the robustness of the NCG x,y-coordinates can be predicted. The NCG x,y- coordinates distortion decreases with increase in vector length. Through experiments, it was proved that the distortion for vector lengths greater than 430 is lower than 0.25 pixels for compression rates upto QP less than equal to 40. Hence, a watermark can be embedded by quantizing the values of the x,y- coordinates of blocks with L > 430 b using a quantization step size of one. This watermark will be robust to compression rates upto QP less than equal to 40.

During embedding of watermark, the first step is to find the vector length $L_{min}$ which ensures the required robustness. After which, the blocks with $L > L_{min}$ are chosen. To embed on watermark bit, the values of the x,y- coordinates of the NCG are quantized. The quantization is done by geometric warping of the block. The direction and strength of warping depends on the original and the quantized NCG-coordinates. To prevent block artifacts to neighbor blocks, the warping process uses different weights for the strength of warping the single pixels. The weighting is computed by using a quadratic function.

This geometric warping process is used to change the NCG x,y- coordinates.

## 4 Proposed Watermarking Scheme

The proposed watermarking algorithm consists of two stages. The first stage is the embedding stage and the second is extraction of the copyright. The process is shown in Fig. 1 and Fig. 2.

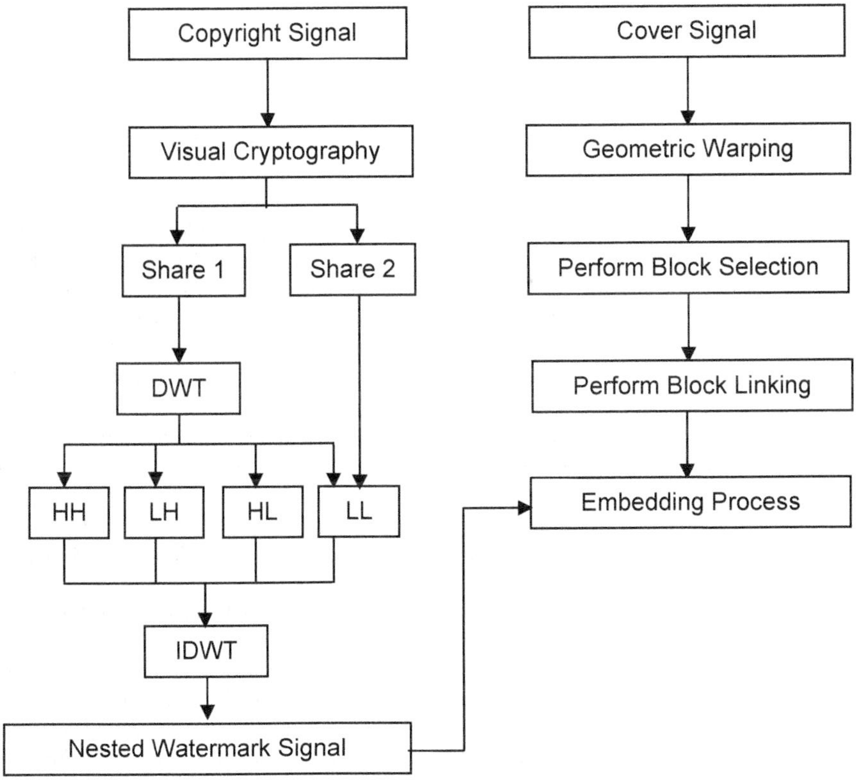


**Fig. 1.** Embedding Process

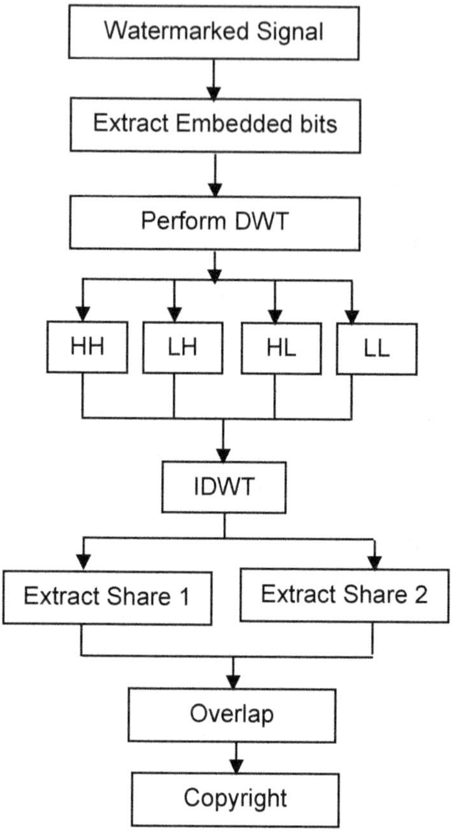


**Fig. 2.** Watermark Extraction Process

### 4.1 Embedding Process

The first step is to create nested watermark and the procedure is given in Fig. 3.

1. Perform Visual Cryptography on the original copyright image and obtain Share 1 and Share 2.
2. Read Share 1
3. Perform 2D DWT on Share 1 to obtain LL, LH, HL and HH subbands.
4. Read Share 2
5. Add Share 2 in LL subband
6. Perform IDWT to get nested watermark.
7. Calculate Peak Signal to Noise Ratio of nested watermark with original copyright image.

**Fig. 3.** Creating Nested Watermark

The second step is the watermarking process using geometric warping. To avoid blocking artifacts the geometric warping method uses a block selection procedure. In block selection technique, the selected blocks should maintain minimum distance d to each other in order to prevent artifacts and flickering effects. The minimum distance can be calculated using Equation (3)

$$[d_i = \lfloor \log(d.(i_{size} - 1) + 1) + 0,.5 \rfloor \quad (3)$$

where d = distance and i = dimensions (x, y or t).

The outline of the block selection algorithm is given as follows:

(i) The block group is chosen which is of arbitrary forms.
(ii) The minimum distance dx, dy and dt is computed for each dimension x, y and t of the block group.
(iii) The numbers of blocks inside an ellipsoid are counted for each block. The semi-axes of the ellipsoids are defined by dx, dy and dt. The current block defines the center of the current ellipsoid.
(iv) The blocks with the least neighbors are chosen for watermarking. If more than one chosen block is inside an ellipsoid, only one of them (the first) is used for watermarking. This process is unambiguous.
(v) The chosen blocks and all blocks within the ellipsoids around the chosen blocks will not be considered in the next steps.
(vi) The steps 3-4 are repeated until all blocks are chosen for watermarking or have a distance lower than the minimum to a chosen block.

The blocks chosen using the above algorithm are used for watermark embedding. The watermarking information is embedded by moving the x, y-coordinates on the quantization lattice to minima/maxima. This is done by geometric warping.

### 4.2 Extraction Process

To extract the watermark bits, the blocks with $L > L_{min}$ are chosen. From the NCG coordinates of the chosen blocks the embedded bit values can be computed. Thus, to extract the watermark, the original video is not needed. The NCG is used to detect the watermarked blocks and compute the embedded watermark bit. After successful extraction of the watermarked bits, the next step is to extract the nested watermark. The procedure used is given in Fig. 4.

1. Identify blocking containing watermark bits and extract them to get the nested watermark
2. Apply 2D-DWT to obtain LL, LH, HL and HH subbands
3. Extract Share1 and Share 2 parts
4. Overlap to obtain the copyright image.

**Fig. 4.** Extraction Process

## 5 Experimental Results

Several experiments were conducted to evaluate the performance of the proposed system. During experimentation several samples were used which included standard test videos and commercially-available movies. The results observed had similar pattern. The results projected in this paper are based on a video clip having 1526 frames (352 x 289 frame size). The experiments were conducted on a Pentium IV with 512 MB memory. The attacks taken into consideration are frame dropping, frame averaging and compression. After extracting and refining the watermarks, a similarity measure, called Normalized Correlation (NC) is used (Equation 4) which gives the cross-correlation, normalized by the reference watermark energy giving unity as the peak correlation.

$$NC = \frac{\Sigma_i \Sigma_j W_{ij} * RW_{ij}}{\Sigma_i \Sigma_j W_{i^2 j}} \tag{4}$$

where $W_{ij}$ is the original watermark and $RW_{ij}$ is the extracted watermark.

Number of frames dropped is a common attack encountered by video watermarking. The reason behind this is the large amount of redundancies that exists between the frames. The effect of this attack is projected in Fig. 5 when tested with different percentage of frames dropped. From the results, it could be seen that the proposed watermarking scheme is better with this type of attack.

Frame averaging and statistical analysis is another common attack to the video watermark. When attackers collect a number of watermarked frames, they can estimate the watermark by statistical averaging and remove it from the watermarked video. Fig. 6 shows the results obtained while taking the frame averaging attack into consideration. The results show that fact that the proposed scheme can resist statistical averaging quite well than the base model.

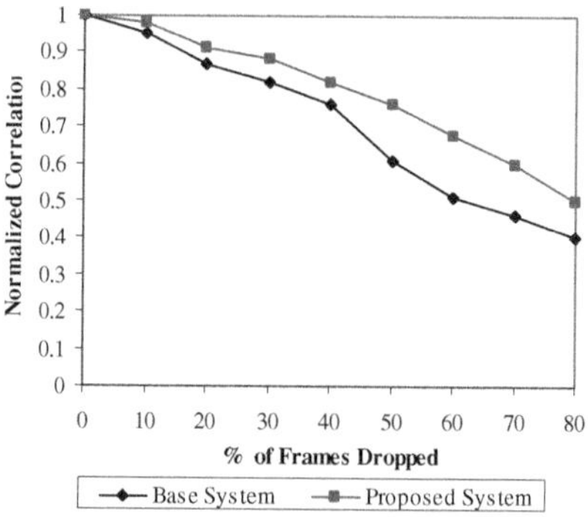


**Fig. 5.** Number of Frames Dropped

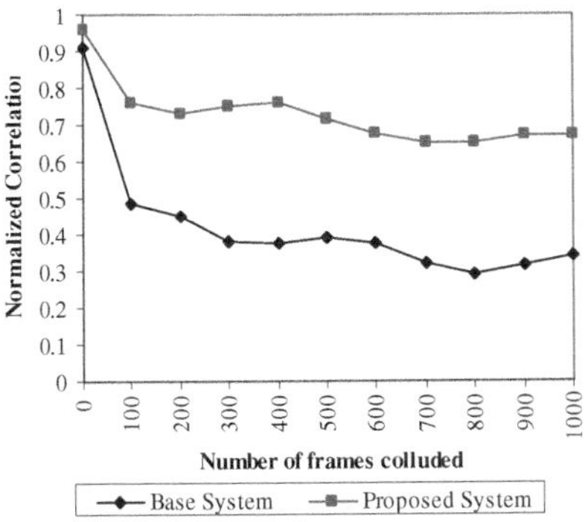


Fig. 6. Frame Averaging

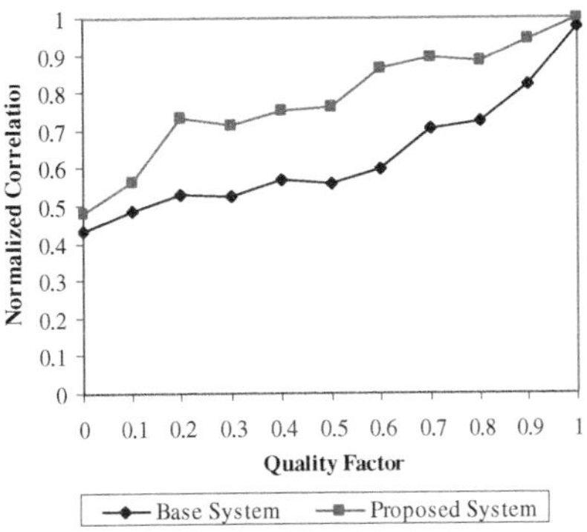


Fig. 7. Compression Attacks

The NC values of the extracted watermarks with different quality factors are shown in Fig. 7. From the experiment, it is noted that the proposed scheme improves the robustness for watermark protection. It can further be noted, with high quality factor, the algorithm is better in resistance.

From the various results obtained, it can be concluded that the proposed watermarking scheme is superior to the existing system.

## 6 Conclusion

This paper proposed a technique for watermarking a video signal with a secret copyright image. The proposed technique is an amalgamation of various techniques like Visual Cryptography, Wavelets and Geometric Warping. Visual cryptography is a concept that is normally used to for hiding multiple watermarks. In this paper, a novel idea of splitting the copyright image into two and creating a nested watermark by storing one split into another is proposed. This method increases the capacity of the watermark. Experimental results proved that the proposed technique is efficient in terms of capacity and robustness against attacks. The present work uses only black and white copyright images, which can be changed to color images and a color visual cryptography method can be combined with this. Moreover, applicability of wavelet packets can be considered for more efficiency.

## References

1. Aggarwal, D., Kaur, S., Anantdeep, E.: An Efficient Watermarking Algorithm to Improve Payload and Robustness without Affecting Image Preceptual Quality. Journal of Computing 2(4) (2010)
2. Arena, S., Caramma, M.: Digital watermarking applied to MPEG2 coded video sequence exploiting space and frequency masking. In: Proceedings International Conference on Image Processing (ICIP 2000), Vancouver, Canada, vol. 3, pp. 438–441 (2000)

3. Ateniese, G., Blundo, C., De Santis, A., Stinson, D.R.: Visual cryptography for general access structures. Information Computation 129, 86–106 (1996)
4. Bonnis, A., Santis, A.: Randomness in secret sharing and visual cryptography schemes. Theor. Comput. Sci. 314, 351–374 (2004)
5. Checcacci, N., Barni, M., Bartolini, F., Basagni, S.: Robust video watermarking for wireless multimedia communications. In: Proceedings IEEEE Wireless Communications and Networking Confernce, WCNC 2000, vol. 3, pp. 1530–1535 (2000)
6. Chun-Shien, L., Shih-Kun, H., Chwen-Jye, S., Mark, L.H.: Cocktail watermarking for digital image protection. IEEE Transactions on Multimedia 2, 209–224 (2004)
7. Craver, S., Wu, M., Liu, B.: What can we reasonably expect from watermarks? In: Proceedings IEEE Workshop on the Applications of Signal Processing to Audio and Acoustics, pp. 223–226 (2001)
8. Ejima, M., Miyazaki, A.: A wavelet-based watermarking for digital images and video. In: Proceedings International Conference on Image Processing (ICIP 2000), Vancouver, Canada, vol. 3, pp. 678–681 (2001)
9. Foo, S., Yeo, T., Huang, D.: An adaptive audio watermarking system. In: Proceedings IEEE Region 10 International Conference on Electrical and Electronic Technology, vol. 2, pp. 509–513 (2001)
10. Hartung, F., Girod, B.: Watermarking of Uncompressed and Compressed Video. IEEE Transaction Signal Processing 66(3), 283–301 (1998) (Special issue on Watermarking)
11. Hawkes, L., Yasinsac, A., Cline, C.: An Application of Visual Cryptography to Financial Documents; technical report TR001001, Florida State University (2000)
12. Houmansadr, A., Ghaemmaghami, S.: A Digital Image Watermarking Scheme Based on Visual Cryptography. In: International Symposium on Telecommunications, pp. 1–5 (2005)
13. Huang, P.S., Chiang, C.S., Chang, C.P., Tu, T.M.: Robust spatial watermarking technique for colour images via direct saturation adjustment. IEEE Proceedings Vision, Image and Signal Processing 152, 561–574 (2005)
14. Kaabneh, K., Youssef, A.: Muteness-based audio watermarking technique. In: Proceedings International Conference on Distributed Computing Systems Workshop, pp. 379–383 (2001)
15. Kimpan, S., Lasakul, A., Chitwong, S.: Variable block size based adaptive watermarking in spatial domain. In: IEEE International Symposium on Communications and Information Technology, ISCIT 2004, vol. 1, pp. 374–377 (2004)
16. Lee, Y.K., Chen, L.H.: High capacity image steganographic model. IEEE Proceedings Vision, Image and Signal Processing 147, 288–294 (2000)
17. Lin, E., Podilchuk, C., Kalker, T., Delp, E.: Streaming Video and Rate Scalable Compression: What Are the Challenges for Watermarking? In: Proceedings the SPIE International Conference on Security and Watermarking of Multimedia Contents III, San Jose, CA, vol. 4314, pp. 22–25 (2001)
18. Lu, C., Huang, S., Sze, C., Liao, H.Y.M.: Cocktail watermarking for digital image protection. IEEE Transactions on Multimedia 2, 209–224 (2000a)
19. Lu, C., Liao, M., Chen, L.: Multipurpose audio watermarking. In: Proceedings 15th International Conference on Pattern Recognition 2000, vol. 3, pp. 282–285 (2000b)
20. Lu, W., Lu, H., Chung, F.L.: Robust digital image watermarking based on subsampling. Applied Mathematics and Computation 181, 886–893 (2006)
21. Mohammed, A.A., Hussein, J.A.: Efficient Video Watermarking Using Motion Estimation Approach. In: 2009 Eight IEEE/ACIS International Conference on Computer and Information Science (ICIS 2009), pp. 593–598 (2009)

22. Naor, M., Shamir, A.: Visual cryptography. In: De Santis, A. (ed.) EUROCRYPT 1994. LNCS, vol. 950, pp. 1–12. Springer, Heidelberg (1995)
23. Naor, M., Pinkas, B.: Visual authentication and identification. In: Kaliski Jr., B.S. (ed.) CRYPTO 1997. LNCS, vol. 1294, pp. 322–336. Springer, Heidelberg (1997)
24. Paul, N., Evans, D., Rubin, A., Wallach, D.: Authentication for remote voting, workshop on human-computer interaction and security systems, Fort Lauderdale, Florida (April 2003)
25. Piva, A., Bartolini, F., Barni, M.: Managing copyright in open networks. IEEE Transactions on Internet Computing 6(3), 18–26 (2002)
26. Praun, E., Hoppe, H., Finkelstein, A.: Robust Mesh Watermarking. ACM SIGGRAPH (1999)
27. Pröfrock, D., Schlauweg, M., Müller, E.: Video Watermarking by Using Geometric Warping Without Visible Artifacts. In: Proc. of Information Hiding (8thIH 2006), Alexandria, USA (2006)
28. Pröfrock, D., Richter, H., Schlauweg, M., Müller, E.: H.264/AVC video authentication using skipped macroblocks for an erasable watermark. In: Proc. of Visual Communications and Image Processing - VCIP, vol. 5960 (2005)
29. Qui, G., Marziliano, P., Ho, A.T.S., He, D., Sun, Q.: A hybrid watermarking scheme for H.264/AVC video. Proc. of the ICPR 4, 865–868 (2004)
30. Ren-Junn, H., Chuan-Ho, K., Rong-Chi, C.: Watermark in color image. In: Proceedings of the first International Symposium on Cyber Worlds, pp. 225–229 (2002)
31. Tai, G.C., Chang, L.W.: Visual Cryptography for Digital Watermarking in Still Images. In: Aizawa, K., Nakamura, Y., Satoh, S. (eds.) PCM 2004. LNCS, vol. 3332, pp. 50–57. Springer, Heidelberg (2004)
32. Verma, B., Jain, S., Agarwal, D.P., Phadikar, A.: A New color image watermarking scheme, Infocomp. Journal of Computer Science 5(2), 37–42 (2006)
33. Wang, C., Nie, X., Wan, X., Wan, W.B., Chao, F.: A Blind Video Watermarking Scheme Based on DWT," iih-msp. In: 2009 Fifth International Conference on Intelligent Information Hiding and Multimedia Signal Processing, pp. 434–437 (2009)
34. Wu, X., Guan, Z.H.: A novel digital watermark algorithm based on chaotic maps. Physics Letters A 365, 403–406 (2007)
35. Yang, C.: A note on Efficient Color Visual Encryption, vol. 18, pp. 367–372 (2002)

# Robust Fault Tolerant AOMDV Routing Mechanism in MANET

Jayalakhsmi Vaithiyanathan and B. SenthilrajaManokar

Computer Applications Department, Sudharsan Engineering College, Sathyamangalam,Tamilnadu, India
jayasekar1996@yahoo.co.in

**Abstract.** Wireless mobile ad-hoc networks are characterized as networks without any physical connections. Multipath routing is a new extension to many traditional routing protocols in MANETs. Using multipath routing in MANETs can save energy, reduce frequent routing update, enhance data transmission rates, and increase wireless network bandwidth. There are numerous Multi-paths on-demand routing algorithms exist to discover several paths instead of one, once the routing is performed. In this paper a new Robust Fault Tolerant AOMDV (RFTA) algorithm is proposed to improve the fault tolerance and efficient routing in mobile adhoc network (MANET). The algorithm is compared with various multi path routing protocols like SMR, SMS and MDSR. The efficiency of the protocol has been evaluated on different scenarios using performance metrics such as packet delivery ratio and end-to-end delay.

**Keywords:** AODMV , end-to-end delay, fault tolerance, packet delivery ratio, SMR, SMS.

## 1 Introduction

MANET is the network completely self-organizing and self-configuring, requiring no existing network infrastructure or administration. Some routing policies [1] [2] [3] have been proposed in order to establish and maintain the routes in MANET, and DSR is a generally utilized one. There are many common routing algorithms used in Adhoc networks but AODV [4, 5] and DSR [6], which both are on-demand algorithms. In some multi-path algorithms once different paths are discovered, they are all stored but only one of them is used for transferring the data. The other stored paths will become useful once the current one is broken. There are also other multi-path algorithms that transfer data over all discovered paths concurrently which reduces end to end delay and increases end to end bandwidth.

Some key challenges Associated with an ad hoc environment, and some specific characteristics.

*Dynamic topology*: The routing between the different nodes Consists of multi hop, therefore routes may change Irregular. Due to the nodes Movement, the routing information will need to be updated more often than its wired counterpart. This results in more routing overhead information, which in turn increases the use of the

D. Nagamalai, E. Renault, M. Dhanushkodi (Eds.): CCSEIT 2011, CCIS 204, pp. 142–150, 2011.

radio medium resources and puts great emphasize in choosing routing protocol, when all the different protocols uses different strategies to establish and maintain routes [7].

*Multi-hop:* The data packets have to be forwarded between multiple intermediate nodes to reach the destination. In a MANET the topology may change constantly, due to possibility of node movement [8] Bandwidth-constrained.

*Variable capacity links*: As a contrast to hardwired networks, communication over wireless medium will continue to have less bandwidth capacity. In addition signal interference, noise and fading, will most resulting in even less useful bandwidth. Congestion is very common, will likely approach or exceeds the network capacity, as the mobile networks often are simply an extension of a fixed infrastructure [9].

The typical problems encountered in designing single path protocols for a mobile wireless environment also arise in designing multipath protocols. Mobility makes it difficult or impossible to maintain a global view of the network. Mobility also has implications in terms of caching policies. If cached information at intermediate nodes is frequently out of date, caching can degrade routing performance because detecting inaccurate routing. Information is not instantaneous. In addition to the dynamic topology, unreliable and range limited wireless transmission makes resiliency a requirement rather than an enhancement in a routing solution. Since mobile transmitters are likely to be battery powered, routing protocols need to minimize the communication for coordinating network nodes. At the protocol level, the design of multipath routing needs to consider failure models, characteristics of redundant routes, coordinating nodes to construct routes, mechanisms for locating mobile destination and intermediate forwarding nodes, and failure recovery.

The paper is organized in 5 sections. Section 2 includes some definitions that gives a useful background to this paper. Section 3 has a Review of Multipath Routing Protocols. Section 4 gives the proposed RFTAOMDV algorithm. Section 5 illustrates the performance evaluation with the other protocols. Section 6 presents conclusions at the end.

## 2 Related Work

Most of multi-path routing algorithms such as AOMDV [10] have their basis in a single-path algorithm due to their efficiency. One drawback with multi-path algorithms is that they send more routing request packets, comparing to single path algorithms and this makes them more complicated.

### 2.1 AOMDV

It is a simple, yet high efficient multi-path algorithm which tries to find Link-Disjoint paths. In this algorithm the source broadcasts routing request packet to all its neighbors. Once neighbors of the source node receive the RREQ packet, they add their own address to the packet; announcing themselves as the founder of a path, and broadcast it. Every intermediate node by receiving a request packet will check to see if it has already received a same request and if not, then it reconstructs and broadcasts

the request packet. If the same request was processed previously the following conditions will be investigated and if met, the packet would be accepted and will be inserted in the route table; otherwise the packet will be ignored.
These conditions are:

1. Whether the request packet is received from a new neighbor node.
2. Whether the request packet is received from a new path founder node.
3. Whether the number of hops in the request packet is less than the existing ones.

The destination node will send as many responses as number of request packets it has received and hereby informs the source node of the paths.

### 2.2 Dynamic Source Routing (DSR)

This routing protocol consists of two mechanisms: the *Route Discovery* and the *Route Maintenance* that permit to completely maintain and automatically determine routes: Route Discovery is in charge to find routes between the nodes into the network by broadcasting RREQ. DSR is a reactive protocol, a source node S starts searching routes only if it needs to send a packet to a destination D. Any intermediate node that receives a non-duplicate RREQ appends its address to the source route list in the RREQ packet and rebroadcast it and the destination node sends a RREP back to the source. If an intermediate node has some requested information, it may send an RREP back to the source. Finally, the source node obtains several routes to reach the destination.

## 3 Review of Multipath Routing

### 3.1 Split Multipath Routing (SMR)

SMR[11] protocol provides a way of determining maximally disjoints paths. Paths are maximally disjoint when they are node disjoint, but when there are no node-disjoint paths available, the protocol minimizes the number of common nodes. Multiple routes are discovered on demand, one of which is the path with the shortest delay. The routes established by the protocol are not necessarily equal in length. When a source needs a route to a destination but no information is available, it floods a RREQ message on the entire network. Because of this flooding, several duplicates that traversed through the network over different routes reach the destination. The destination then selects multiple disjoint paths, and sends RREP packets back via the chosen routes. Since the destination needs to select disjoint paths, source routing is used. The complete route information is in the RREQ packets. Furthermore, intermediate nodes are not allowed to send RREPs, even when they have route information to the destination. If nodes reply from their cache as in DSR, it is very difficult to determine maximally disjoint multiple routes since the destination does not receive enough RREQs and will not know the information of routes formed from intermediate nodes cache. In SMR a new packet-forwarding approach is introduced. Instead of dropping all duplicate RREQs, intermediate nodes only forward the packets, which used a different incoming link than the first received RREQ and whose hop count is not larger than that of the first received RREQ The SMR algorithm is optimized when the destination selects the two routes that are maximally). One of the two routes is the route with the shortest delay; the path taken by the first RREQ the destination receives. This path is used to minimize route acquisition latency required by on-demand protocols. When the first RREQ is received, the destination sends a RREP to the source via this path. The RREP includes the entire path so the intermediate nodes can determine where to forward the packet. After having sent the first RREP, the destination waits a certain amount of time to receive more RREQs and determine all possible routes. Because the destination then knows route information from all possible routes it can determine the maximally disjoint route to the already replied route.

### 3.2 Multi-Path DSR (MP-DSR)

M-DSR first extends DSR by labeling the initial route request received at a destination node the primary source route. additional route requests received are from node disjoint routes, given that route requests are forwarded at most once by any given node involved in any given route discovery. These additional routes are generally longer, as these route requests arrive after the primary source route's request arrives. Route replies for each route are sent to the source. The M-DSR route usage policy dictates secondary routes are only used after the primary route has failed. The use of previously discovered secondary routes reduces route discovery overhead. If only one RREQ is received by the destination, that means there is only one path from source to destination, M-DSR acts essentially the same as the single path DSR protocol. M-DSR offers two mutually exclusive multi-path maintenance heuristics. In M-DSR have been developed two versions of protocols. It prevents intermediate node route caching for alternate routes, which implies that only the source is capable of fixing a broken path.

### 3.3 Shortest Multipath Source (SMS)

SMS routing protocol is proposed based on DSR. It builds multiple partial-disjoint paths from source S to destination D in order to avoid the overhead of additional route discoveries and to quick recovery in case of route breaks. Improved performance in terms of Fraction of packets delivered, end-to-end delay and routing overheads is obtained. SMS earns multiple partial-disjoint paths that will bypass at least one

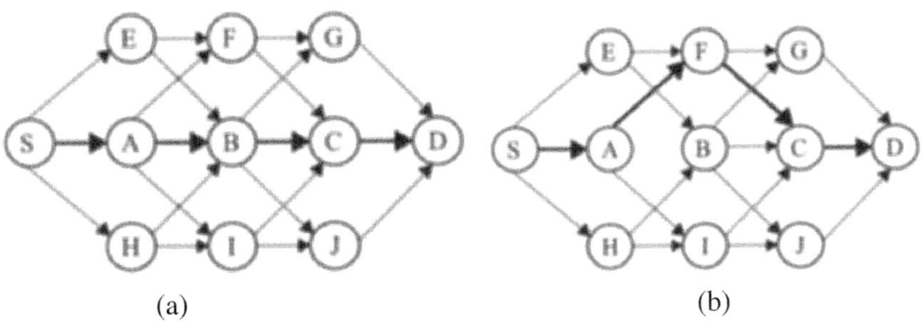


**Fig. 1.** (a) Partial Disjoint Multiple Paths (b) Select the alternate path

intermediate node on the primary path that is shown in Fig. 1-a.Consider the case of traffic flowing between nodes Sand D using link A-B as a segment of the path. In case of a link failure between A and B, the source node will search for an alternate route that does not contain node B that is shown in Fig. 1-b. An alternative route between source S and destination D is (S, A, F, C, D).

# 4 Proposed RFTA Algorithm

In the proposed RFTA( Robust Fault Tolerant AOMDV) algorithm , there is no need for the intermediate nodes to have Route Cache tables and as a result the destination will receive all the path-request messages from different paths. Also in the proposed algorithm, each node should save the RREQ messages it receives from other nodes in RREQ_Seen table, so that it can respond to the queries it receives from its neighbors. Also, there is an additional field in RREQ_Seen table of each node, called Neighbor_Count (N_C) which will be used to count the number of active neighbors identified after sending the RREQ message. In order to let the subsequent nodes to know the total number of active neighbors of the traversed nodes along a path, a new field called Neighbor Count is added to the headers of both RREQ and RREP messages. Also two new messages, RREQ_Query and RREQ_Query_Reply, are added to the route discovery process.

## 4.1 Pseudo Code of RFTA

The steps taken by source node, destination node and intermediate nodes are listed in the following tables.

**Table 1.** Algorithm for the Source node in RFTA

| Steps |
| --- |
| 1. If there is data to be sent to a certain destination and there is no valid path for that destination, broadcast the RREQ packet<br>2. Wait for first RREP packet to arrive.<br>3. After receiving the first RREP packet, wait for a certain amount of time, then from received paths choose those ones that have the least active neighbors and starts load balancing data on these paths. |

**Table 2.** Algorithm for the Destination node in RFTA

| Step |
| --- |
| 1. Send back a RREP packet to all the nodes from which a RREQ packet is received |

**Table 3.** Algorithm for the Intermediate node in RFTA

Steps

1. Once a RREQ message is received and is acceptable(based on three aforesaid conditions in the Related Works section) perform the following steps:
   i. Save this message in the RREQ_Seen table
   ii. Construct the RREQ_Query packet
   iii. Send the RREQ_Query packet to the neighbours asking them if they have already seen the same RREQ message before
   iv. Wait for a certain amount of time for RREQ_Query_Reply messages from neighbours
   v. Increase the ActiveNeighborCount field based on the positive responses that are received
   vi. Broadcast the RREQ message with the new ActiveNeighborCount value
2. Once a RREQ_Query message is received one of the following steps are performed:
   i. If based on the RREQ _Seen table this is a new RREQ message. Store the RREQ in the RREQ Seen table.
   ii. If based on RREQ_ Seen table the same RREQ has been received from the same node, ignore the query.
   iii. If the RREQ is an existing one in the RREQ _Seen table and it has been received from a new neighbor but the same RREQ had not been broadcasted before ,then only respond with sending back the positive RREQ_ Query_ Reply.
   iv. If the RREQ is an existing one in the RREQ_ Seen table and it has been received from a new neighbor and the same RREQ has already been broadcasted then respond with sending back the positive RREQ_ Query_ Reply and also increment the After A_N_C field of corresponding RREQ in the RREQ_ Seen table.
3. Once the RREP message is received add the content of the corresponding After N_C to the Active NeighborCount field of RREP and send the RREP.

# 5 Performance Evaluation Results

## 5.1 Simulation Environments

The simulation environment is ns2[12]. The network includes 50 mobile host nodes which are placed in 1000*500 meter flat area in random and the number of source nodes was 30 of 50 hosts. Each node has a radio propagation range of 250 m and channel capacity was 2Mbps [5]. The IEEE 802.11 Distribution Coordination Function (DCF) is used as Medium Access Control (MAC).A traffic generator was random constant bit rate (cbr) and Tep was established between nodes. The size of data payload was 512 bytes. The random waypoint model was used as a mobility model. Sending rate of packets was set to 3 packets per second and the simulation was run in 30 seconds as a pause time. The maximum and minimum speed was varied between 0 and 10 *m/s* [6].

Three important metrics are evaluated:

- *Throughput*: measured as the ratio of the number of data packets delivered to the number of data packets sent by the source.
- *End-to-end delay*: measured as the average end-to-end latency of data packets.
- *Routing loads*: measured in terms of the total number of routing packets transmitted per data packet delivered at the destination. Each broadcast transmissions was count as one single transmission.

## 5.2 Simulation Results

Throughput of the four protocols is shown in Fig. 2. From the results SMS outperform SMR especially when the mobility increases. In DSR only one route is used for each session and when that route is invalidated the source uses the cached route that is learned from overhearing packets. Intermediate nodes that have cached

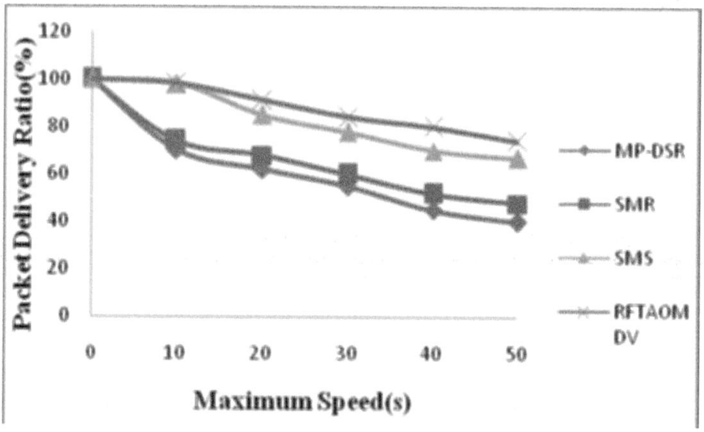


**Fig. 2.** Fraction of Packets Delivery vs. Speed

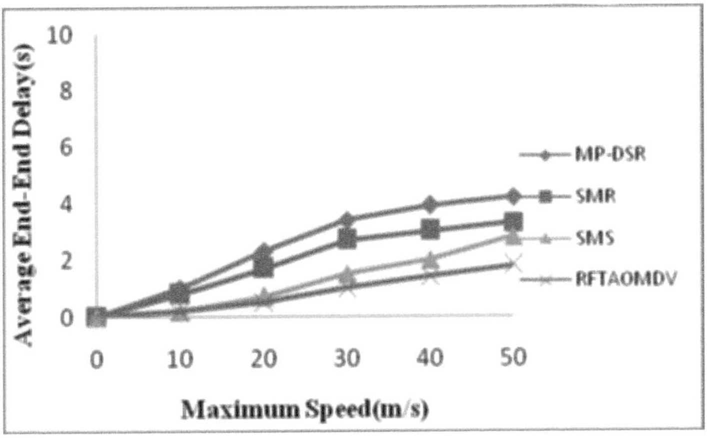


**Fig. 3.** End-to-End Delay vs. Speed

routes to the destination provide those routes to the source by sending RREPs. SMS outperforms DSR, SMR and MP-DSR due to lower overheads in the route discovery process since the destination node sends multiple RREP. In moderately loaded networks, the loss of RREP packets results in the failure of the route discovery process. In on demand protocols the source uses an exponential back-off algorithm to limit the rate at which new route discoveries are initiated. Since the back-off time is large compared to send buffer capacity, application packets within the back-off time may be dropped due to buffer overflow. But with this proposed RFTA algorithm which use multiple route replies show higher Throughput than the other protocols.

The end-to-end delay of the four protocols is shown in Fig 3. SMR has the longest delay in mobile scenarios because it delivers data packets on routes longer than those of SMR. In addition, MP-DSR yields longer delays in reconstructing routes and the period of time the data packets are buffered at the source node during route recovery results in larger end-to-end delays.

SMR uses the remaining valid route when one of the multiple routes is disconnected, and hence no route acquisition latency is required. The availability of shorter alternate routing paths in the former eliminates route discovery latency that contributes to the delay when an active route fails. When a congestion state occurs in a path, the source node can re-distribute incoming data packets to alternative routing paths, thereby reducing the queue waiting time of data packets.

The proposed algorithm routing outperforms the other protocols at lower mobility and node densities. SMS finds the shortest alternative routes in a single route discovery process, reducing delay and routing overheads incurred when recovering from route breaks. SMR overloads the network with control messages and data packets are dropped at full buffer of intermediate nodes. Routing load savings for M-DSR at higher nobilities and node densities came from a large saving on route requests.

## 6 Conclusion

The proposed Robust Fault Tolerant AOMDV algorithm explores alternate disjoint routes that can be useful in case the primary route breaks. Two variations are explored. In the first, only the source gets multiple alternate routes. In the second, each intermediate node on the primary route gets an alternate route. The key advantage is the reduction in the frequency of route discovery flood, which is recognized as major overhead in on demand protocols. In any case, any form of multipath technique always performs substantially better than single path routing. The simulation study also demonstrates that even though the multipath routing brings down routing load, it also increases end-to-end Delay, as alternate routes tend to be longer than primary routes. This presents subtle tradeoffs between delay and routing load. Simulation results indicate that the proposed RFTA outperforms SMS, SMR and MP-DSR on consideration of a number of metrics and is a candidate that, potentially, offers improved routing performance. And there has been a noticeable improvement in the packet delivery ratio and also in the reduction of end-to-end delay comparing to other protocols.

## References

[1] Johnson, D.B., Maltz, D.A., Hu, Y.C.: The Dynamic source routing protocol for mobile ad hoc networks (DSR). IETF draft-ietf-manet-dsr- 10.txt (July 2004)

[2] Perkins, C.E., Belding-Royer, E., Das, S.: Ad hoc on demand distance vector (AODV) routing. IETF RFC3561 (2003)

[3] Perkins, C.E., Watson, T.J.: Highly dynamic destination sequenced distance vector routing (DSDV) for mobile computers. In: Proc. of ACM SIGCOMM, London, UK (1994)

[4] Perkins, C.E., Belding-Royer, E.M., Das, S.: Ad hoc On-Demand Distance Vector (AODV) Routing. RFC 3561 (July 2003)

[5] Perkins, C.E., Royer, E.M.: The Ad hoc On-Demand Distance Vector Protocol. In: Perkins, C.E. (ed.) Adhoc Networking, pp. 173–219. Addison-Wesley, Reading (2000)

[6] Johnson, B., Maltz, D.A.: Dynamic Source Routing in Ad-Hoc Wireless Networks. Mobile Computing 353, 153–181 (1996)

[7] Li, X., Cuthbert, L.: A Reliable Node-Disjoint Multipath Routing with Low Overhead in Wireless Adhoc Networks. In: Proceedings of the Seventh ACM/IEEE International Symposium on Modeling, Analysis and Simulation of Wireless and Mobile Systems, Venezia, Italy, pp. 230–233 (2004)

[8] Mueller, S., Ghosal, D.: Multipath Routing in Mobile Ad-hoc Networks: Issues and Challenges. Invited paper in Lecture Notes in Computer Science (2004)

[9] Nasipuri, A., Castaneda, R., Das, S.: Performance of Multipath Routing for On-demand Protocols in Mobile Ad-hoc Networks. ACMlKluwer Mobile Networks and Applications Journal 6, 339–349 (2001)

[10] Marina, M.K., Das, S.R.: On-demand Multipath Distance Vector Routing for Ad Hoc Networks. In: Proc. of 9 th IEEE Int. Conf. on Network Protocols, pp. 14–23 (2001)

[11] Lee, S.j., Gerla, M.: Split Multipath Routing with Maximally Disjoint Paths in Ad-hoc Networks. In: Proceedings of IEEE International Conference on Communication (ICC), Elsinki, Finland, pp. 3201–3205 (2001)

[12] Network Simulator-2, `http://www.isi.edu/nsnam/ns`

# ZKIP and Formal System

M. Thiyagarajan[1] and S. Samundeeswari[2]

[1] Professor,
School of Computing, SASTRA
m_thiyagarajan@yahoo.com
[2] Assistant Professor,
School of Computing, SASTRA
samu_sivi@yahoo.com

**Abstract.** Every crypto system is an interactive system, every interactive system is a formal system, the results of the formal system can be proved through the Godel's incompleteness Theorem. Proving techniques in the solution of problems designed on formal systems employ logical connectivity. This logical programming is possible from the proof of Godel's incompleteness theorem in the formal systems. Any intractable problems can be solved by heuristic approach. One such approach is the Godel's attempt, we realize the automated steps translating the proof of Godel's theorem as our zero knowledge interactive steps. The implementation of this approach is being illustrated.

## 1 Introduction

The implementation validation of protocols as ZKIP is our contribution in this area of cryptography. On classification of complexity classes which are driven from the study of formal language and automated theory. Such hypothesis helps us to decide on the particular complexity class. Some of the problems studied as NP complete problems adapt themselves to a reduction to polynomial time problem. This type of conversion is perceived as a function which converts Non Interactive Zero-Knowledge Protocol [2] to Zero-knowledge Interactive Protocol. Godel's Incompleteness theorem and the proof by the earlier researchers in the foundation of mathematics pave the way in the conversion of Non Interactive Protocol to Interactive protocol.

In the most recent techniques of proving general results on formal systems is the Zero knowledge Interactive Proof. Here the Godel's theorem and proof are represented in the formal system set up between corporate concepts such as consistency and $\omega$_consistency[6]. The system is developed through Godel number, Godel sequence and primitive formulas. The connectivity of these concepts with propositional connectives and quantifiers is correct because they are contained in ordinary logical connectives. The matching of expression for the Godel formula through consistency and $\omega$_consistency are used in developing the interaction between the prover and verifier. The input is a goal which may contain a constraint such as a time limit and a partial proof, which is a list of proof steps. A proof step contains a proposition or an assumption or an explanation which gives as arguments, the proceeding proof steps are used to derive the propositions. A look up table is

D. Nagamalai, E. Renault, M. Dhanushkodi (Eds.): CCSEIT 2011, CCIS 204, pp. 151–161, 2011.

supplied by the prover at the request of the verifier, in proving the two main implications of the Godel's incompleteness theorem.

ZKIP[3] contains steps which are related to reasoning process of mathematics and guide the search for the proof reductio ad absurdum is used to prove the negation of a proposition. We generate formulas and sets which form the central ideas of significant proof. The architecture and model of operation of ZKIP are illustrated in detail by our proof of Godel's incompleteness theorem, which says, every formal number theory contains an un decidable formula, i.e., neither the formula nor its negation are provable in the theory.

Here construction in the proof of Godel's namely Godel number, Godel sequence and Godel formula together with formal system developed with these put two different main streams consistency and $\omega$-consistency. This leads to a learning procedure of a verifier to get the steps from the prover who gives directions and paves the way for subsequent calculations and decisions leading to the ultimate proof. Godel's theorem and the proof are represented in the formal languages which incorporate concepts such as concepts of substitutions with special relational procedure which models the ordinary representation of proofs in soft computing.

# 2 Preliminaries

## 2.1 Definition 1

A formal theory is consistent if for no formula F both F and –F are provable. A formalization is : [ 5]

$$allT(fnt(T) \to (consistent(T) \leftrightarrow allF(provable(F,T) \to not(provable(\neg F,T)))))$$

## 2.2 Definition 2

A formal theory is $\omega - consistenet$ if every formula $F(x)$ has the property: If $\vdash F(n), i.e., F(n)$ is provable, for all natural numbers $n$, then $\neg\forall xF(x)$ is not provable. A formulization is : [5]

$$\begin{aligned}
&allT(fnt(T) \to \\
&(\omega - consistenet(T) \leftrightarrow \\
&allF, x(al\ln(number(n) \to \\
&provable(sub(F, x, num(n)), T) \to \\
&not(provable(\neg\forall xF, T)))))
\end{aligned} \tag{1}$$

There is a closed formula $F$ of formal number theory such that

- If the theory is consistent, $F$ is not provable, and
- If the theory is $\omega - consistent$, $\neg F$ is not provable.

A formalization is:

$$\begin{aligned}&allT\,(fnt(T) \rightarrow \\ &(existsF\,(closed - formula(F,T)\,\& \\ &(consistent(T) \rightarrow not(provable(F,T)))\,\& \\ &(\omega - consistenet(T) \rightarrow not(provable(\neg F,T)))))\end{aligned} \tag{2}$$

## 2.3 Definition 3: Formal System [6]

A formal system is $\omega$-inconsistent if it contains a formula $\varphi(x)$ such that both

$$(\exists x)\,\varphi(x)\,(\exists\, x)\;\varphi\,(x) \tag{3}$$

and

$$\iota\varphi(0),\quad \iota\varphi(\bar{1}),\quad \iota\varphi(\bar{2}),\ldots \tag{4}$$

are provable in the system. If no such formula exists, then the system is called $\omega$-consistent. Evidently $\omega$-consistency implies ordinary consistency (since inconsistency implies all formulas provable), but the converse does not necessarily hold.

## 2.4 Proposition 1 [6]

If the formal system S is $\omega$-consistent, then the formula $(\exists y)\;\Gamma(\bar{i}, y)$ is not provable in $S$.

## 2.5 Proof 1

Suppose that it is provable. Then $B(j,k)$ cannot hold any $k$, since otherwise (3.10) is provable in $S$ and $S$ is inconsistent. Thus, for no $k$ does $B(j,k)$ hold. That is, $B(f(i,i),k)$ holds for no $k$, implying that $G(i,k)$ holds for no $k$. Thus, for each natural number $k$. $\sim\Gamma(i,\bar{k})$ is provable in $S$. In particular then all formulas

$$\sim\Gamma(\bar{i},0),\qquad \sim\Gamma(\bar{i},\bar{1}),\qquad \sim\Gamma(\bar{i},\bar{2}), \tag{5}$$

are provable in $S$. But, since $(\exists y)\;\Gamma(\bar{i}, y)$ is also provable, this implies that $S$ is $\omega$-consistent. We must conclude, then, that $(\exists y)\;\Gamma(\bar{i}, y)$ is not provable in $S$.

## 2.6 Theorem 1 [6]

If the system $S$ is $\omega$-consistent, then it cannot be completeness.

The main problem in the proof of Godel is a construction of an un decidable problem. In order to construct such a formula a mathematician may apply elementary rule for the formulation of formula to a known formula and resulting attempt to prove one of the resulting formulas is un decidable.

### 2.7 Proposition 2 [6]

If a formal system, $S$, such as that described above, is consistent, the formula (3.10) is not provable in $S$.

### 2.8 Proof 2

Suppose that it is provable in $S$, and let $k$ be the sequence number of such a proof. Then $B(i,k)$ holds and, by virtue of $j = f(i,i)$, $G(\bar{i},\bar{k})$ holds. But then $\Gamma(\bar{i},\bar{k})$, is that provable in $S$, for it may be shown that, whenever particular numbers are in some recursive relationship, the formula of the system which expresses this fact is formally provable. It then follows easily that $(\exists y)\ \Gamma(\bar{i}, y)$ is provable. Since our starting assumption was that $(\exists y)\Gamma(z, y)$. is provable in $S$, it follows that $S$ is not consistent $S$. However, $S$ was assumed consistent, so we conclude that $(\exists y)\Gamma(z, y)$. is not provable in $S$.

To show that the formula

$$(\exists y)\ \Gamma(\bar{i}, y) \tag{6}$$

Is also not provable, Godel used the assumption of "$\omega$ - consistency";

### 2.9 Properties of Zero Knowledge Proof

Zero-knowledge proofs (ZKPs) are interactive proofs (protocols) in which the prover proves the knowledge (or the possession) of a secret without revealing any information about the secret itself. ZKPs can be used whenever there is critical data to exchange while only proving the possession of such data is needed, without a real need for exchanging the critical data. [7]

#### 2.9.1 Completeness Property

For any $c > 0$ and sufficiently long $x \notin L$, Probability ($V$ accepts $x$) $> 1 - |x|^{-c}$. In other words, an interactive proof (protocol) is complete if, given an honest prover and an honest verifier, the protocol succeeds with overwhelming probability.[1]

### 2.9.2 Soundness Property

For any $c > 0$ and sufficiently long $x \notin L$, Probability ($V$ accepts $x$) $< |x|^{-c}$, i.e. negligible, even if the prover deviates from the prescribed protocol. In other words, if a dishonest prover $P'$ can successfully execute the protocol with non negligible probability, then $P'$ has knowledge essentially equivalent to the actual secret [1].

### 2.9.3 Zero Knowledge Proof

An interactive proof $< P,V >$ is called zero-knowledge if for every probabilistic polynomial-time $V^*$, there exists a probabilistic expected polynomial-time simulator (algorithm) $M_{v*}$ that on inputs $x \in L$ produces probability distributions $M_{v*}(x)$ polynomially indistinguishable from the distributions $< P,V^* > (x)$. [1]

### 2.9.4 Interactive Proof (IP)

An interactive proof $< P,V >$ for language $L$ is a two-party protocol in which a computationally unrestricted prover, $P$ , interacts with a probabilistic polynomial-time verifier, $V$ by exchanging message. Both parties share a common input $x$. At the end, $V$ either accepts or rejects and both completeness and soundness properties hold. [1]

## 3 Objective

In this section, we first describe the model of our ZKIP protocol for Godel. Then, we present the pseudo code representation of the Godel incompleteness theorem.

### 3.1 Our Model

Following sequence of steps leading to a goal

1. The proof attempts to have a proposition. If the proposition is not contained in the proof, we go to universal proposition as implication, an existential proposition or a negation
2. Existential proposition $\exists xp(n)$
3. Component – domain and ranges of elementary functions which form the elementary building stars of formulas and prediction.
4. Basic of negation reduction , contradiction and formulas
5. Data exchanged --basic parameters can reaches through Godel building blocks.
6. If the theory is consistency F is not provable.
7. If theory is $\omega$-consistent negation F is not provable.

   The problem is un decidable if there is no algorithm that takes input and incidences of the problem and determines whether that answer to that incidence is yes or no. The theory of un decidability is concerned with the existence and non existence of algorithm for solving problems with an infinity of incidence.

The main problem in the proof of Godel's theorem is the construction of an undecidable formula, for which elementary rules are applied for the formation of formulas to a known formula and attempt to prove that one of the resulting formulas is undecidable. An example of a rule for the formation is that $\neg F$ is a formula if $F$ is a formula. Another rule is that $\forall x F$ is a formula if $x$ is a variable and $F$ a formula. The application of the two rules to the variable $x$ and the formula $A(x)$ yields the formulas $\neg A(x)$ and $\forall x A(x)$.

### 3.2 Pseudo Code

**Step 1:** Generation of Godel number and Godel sequence

**Step 2:** The statement of the formal system and the composition taken in the form of Godel number and Godel sequence.

**Step3:** This formalization is translated in to a formal system in which either a statement A or its negation is true.

**Step 4:** The process consists of considering statements on the premises as logical statements using implication and contradiction.

**Step 5:** These steps are developed in the existence of an undecidable formula. This formula if it is true for all possible combination of earlier implication, contradiction and truthful nature makes the statement universe. That is true of all combinations of truth value for the component statements.

Careful connectivity of following statements yields the proof of the incompleteness theorem.

**Composition**

$$closed - formula\ (\forall y \neg A(k,y), T) \tag{7}$$

**Contradiction**

$$\begin{aligned}
&allF\,(provable(F,T) \rightarrow not(provable(\neg F,T))) \\
&assume(provable(\forall y \neg A(k,y),T)) \\
&consistenet(T) \\
&not(provable(\forall y \neg A(k,y),T)) \\
&assume(A(k,n)) \\
&not(A(k,n))
\end{aligned} \tag{8}$$

**Forward**

$$
\begin{aligned}
&existsn\ (gnp\ (n, \forall y \neg A(k, y), T\ )) \\
&assumen\ \ (gnp\ (n, \forall y \neg A(k, y), T\ )) \\
&A(k, n) \\
&provable\ \ (A(k, n), T \\
&provable\ \ (\exists y A\ (k, y).T\ ) \\
&provable\ \ (\neg \forall\ y \neg A(k, y), T\ )
\end{aligned} \tag{9}
$$

$$
\begin{aligned}
&not(provable(\neg \forall y A(k, y), T) \\
&gnp(n, \forall y \neg A(k, y), T) \\
&existsn(gnp(n, \forall y \neg A(k, y), T)) \\
&provable(\forall y \neg A(k, y), T)
\end{aligned} \tag{10}
$$

# 4 Data and Analysis

## 4.1 Derivation of Godel Number Formula

We introduce the formal system in the foundation of mathematics through the derivation of Godel number and description of Godel sequence. The crucial concept $w$-consistency is used to state and prove the Godel's incompleteness theorem.

Consider a formula system whose primitive symbols $F, G, \ldots,$ and further symbols $S, 0, = \iota, \supset$, (formulas,). In the interpreting formulas constructed from these symbols we shall regard the individual variables as ranging over the natural numbers, while the functional symbols stand for properties and relations of numbers, $S$ will denote the successor function (so that $S_x$ is the number $x+1$), while $0, =,$ and the logical connectives have their usual meaning. Thus the formula

$$(x)F(x). \supset .(\exists y)(G(y) = SSO)) \tag{3.7}$$

Thus the formula would be read "If ever number $x$ has the property $F$, then there exists a number $y$ such that either $y$ has the property $G$, or $y$ is 2". In the formal system, incidentally, we shall denote SSO by $\bar{2}$, SSSO by $\bar{3}$, etc.

We precede to "arithmetize" the formal system as follows. We first assign to each primitive symbol a coordinate in the following system as follows. We first assign to each primitive symbol a coordinate in the following manner (in the groupings below, symbols in the first line are within the system, and the numbers in the second line are the coordinates of the symbols directly above):

For the basic logical symbols we make the correspondence:

| | | | | | | |
|---|---|---|---|---|---|---|
| 0 | S | ~ | ⊃ | ( | ) | = |
| 2 | 3 | 4 | 5 | 6 | 7 | 8 |

For the individual variables (which can be denumerable infinite in number):

| | | | |
|---|---|---|---|
| $x$ | $y$ | $z$ | ... |
| 11 | $11^2$ | $11^3$ | ... |

For functions of the first order (functions of individual variables):

| | | |
|---|---|---|
| $F_1$ | $G_1$ | $H_1$ |
| 13 | $13^2$ | $13^3$ |

and so on, using powers of the $(n+5)^{th}$ prime number for function symbols of the nth order.

To a formula in the system will correspond, then, a sequence of numbers, namely, those corresponding to the successive symbols of the formula.

Thus, to

$$(x)F1(x) \tag{11}$$

Will correspond to the sequence

6 11 7 13 6 11 7

We can make a unique number correspond to the formula by taking the product of the successive prime numbers (in their natural order) with powers equal to the numbers of the symbols (in the order in which they occur). Thus to formula above will correspond the number

$$2^6 . 3^{11} . 5^7 . 7^{13} . 11^6 . 13^{11} . 17^7 \tag{12}$$

We shall call this the Godel number of the formula.

## 4.2 Description of Godel Sequence [41]

For any sequence of formulas we may make a correspond unique number

$$2^{n(i)} . 3^{n(2)} ... p_k^{n(k)} \tag{13}$$

Where $n(1), n(2), ...n(k)$ are the successive Godel numbers of the formulas of sequence. We call (3.8) the sequence number of the sequence of formulas. In particular every formal proof (that is proof with in the system) corresponds a sequence number, since a formal proof of a formula $F$ consists of a finite sequence of formulas each of which is an axiom, or derived from axioms and preceding formulas by the allowable rules of inference of the system, the last formula of the sequence being $F$ itself.

Let $B(x, y)$ denote that $y$ is the sequence number of a formal proof in which the last formula has Godel number $x$. And let $f(x, y)$ be the Godel number of the formula obtained by replacing, in the formula having Godel number $x$, each occurrence of the particular individual variable z by $\bar{y}$. Godel shows that both $B$ and $f$ are recursive.

Now form the relation $B(f(x,x), y)$. This is also recursive: denote it by $G(x, y)$. Being recursively definable, it has a counterpart in the system which denote by $\Gamma(x, y)$. Then by virtue of the rules for constructing formulas, the system contains the formula.

$$\iota(\exists y)\Gamma(z, y). \tag{14}$$

Let Godel number of this formula be $i$.

A new formula is obtained by replacing $z$ in $2^{n(i)}.3^{n(2)}\ldots p_k^{n(k)}$ by $\bar{i}$

$$j = f(i,i) \tag{15}$$

### 4.3 Validation

From the above discussion we can decide on the completeness of a formal system. As many NP complete problem can be reduced to polynomial time problem, the $\omega$-consistency helps the reduction of formal system to one of polynomial formal system.

Here the Proof of the Godel's Incompleteness theorem is used as a Zero Knowledge Interactive Proof. The formal system's complexity is an undecidable problem, We do not have any proof to validate a proposition in the formal system or its negation. The system is converted to an interactive proof system where there is an efficient proof verification algorithm called a verifier following the conditions of completeness and soundness. This is an attribute of the Godel's assertion that every consistent formal system is not provable and $\omega$_consistency formal system is not also provable in number theory. This is whole analysis of ZKI protocol developed by Golwasser, Micali, and Rockoff[]

## 5 Conclusion

Formal system addresses the problem based on logic programming in almost all branches of science, engineering and technology. This system is based on semantic and other aspects of programming techniques and their certainty alone with due consideration on acceptable valid conclusions from the major premises. On the other hand ZKIP (zero knowledge interactive protocol) can be applied to situations under deterministic interactive and non interactive type of proving techniques as well as uncertainties dealt with probabilistic approach. The step involved in the automated proof of Godel's incompleteness theorem regarded as interactive steps. This analytical approach leads to understand crypto system protocol design as ZKIP. This work being general in nature can be adapted to any type of protocols designing a crypto system as a formal system. As the logical programming can be implemented in standard packages, this approach translates the code for implementing in an easy way. We feel strongly ZKIP attains better status in the proving techniques than formal systems.

## References


[1] Almuhammadi, S.: One-Round Zero Knowledge Proofs And Their Application In Cryptographic Systems. University of California, Berkeley (2005)

[2] Blum, M., Feldman, P., Micali, S.: Non interactive zero-knowledge and its application. In: Proc. Twentieth Annual ACM Symposium on Theory of Computing, Chicago, pp. 103–112 (1988)

[3] Diffie, W., Hellman, M.E.: New directions in cryptography. IEEE Trans. Inform. Theory IT-22(6), 644–654 (1976)

[4] Goldwasser, S., Milcli, S., Rackoff, C.: The knowledge complexity of interactive proof-systems. SIAM Journal on Computing 18, 186–208 (1989)

[5] Amman, K.: An Automatic proof of Godel's Incompleteness theorem. Arttificial Intelligence, Vol 61, 291–306 (1993)

[6] Ray, M., Wilder, L.: Introduction to the Foundation of Mathematics. John Wiley and Sons, Inc., New York (1965)

[7] Rhee, M.Y.: Cryptography and Secure Communications. McGraw-Hill, Korea (1994)

# Appendix

## Problem for Structured proof of Godel's Incompleteness theorem and its analysis

```
#include<iostream.h>
#include<conio.h>
#include<math.h>
void main()
{
clrscr();
int seq_no,i,seq[20],p[20];
double ans=1;
cout<<"GODEL SERIES IMPLIMENTATION\n";
cout<<"enter the number of sequence to used\n";
cin>>seq_no;
cout<<"\nenter the sequence\n";
for(i=1;i<=seq_no;i++)
cin>>seq[i];//entering the sequence...
//prime number sequence...
p[1]=2;p[2]=3;p[3]=5;p[4]=7;p[5]=11;p[6]=13;p[7]=17;
p[8]=19;p[9]=23;p[10]=29;p[11]=31;p[12]=37;p[13]=41;p[14]=43;
cout<<"\nsequence is\n";
for(i=1;i<=seq_no;i++)
cout<<" "<<seq[i];
cout<<"\n";
for(i=1;i<=seq_no;i++)
cout<<p[i]<<".";
for(i=1;i<=seq_no;i++)
ans*=pow(p[i],seq[i]);
cout<<"\nthe number is:"<<ans;
getch();
}
```

The input is as follows:
GODEL SERIES IMPLIMENTATION
enter the number of sequence to used
3
enter the sequence
6
11
6
The output is given by:
sequence is
$2^6\ 3^{11}\ 5^6$
the number is:1.77147e+11

# Personalized Web Based Collaborative Learning in Web 3.0: A Technical Analysis

S. Padma[1] and Ananthi Seshasaayee[2]

[1] Research Scholar, Bharathiar University, Coimbatore
padmanivasan@gmail.com
[2] Quaid-e-Millath Govt. College for women, Chennai
ananthiseshu@gmail.com

**Abstract.** Web 3.0 is an evolving extension of the current web environment. Information in web 3.0 can be collaborated and communicated when queried. Web 3.0 architecture provides an excellent learning experience to the students. Web 3.0 is 3 D, media centric and semantic. Web based learning has been on high in recent days. Web 3.0 has intelligent agents as tutors to collect and disseminate the answers to the queries by the students. Completely Interactive learner's query determine the customization of the intelligent tutor. This paper analyses the Web 3.0 learning environment attributes .

**Keywords:** E learning, Web 3.0, Semantic ,Collaborative learning.

## 1 Introduction

Web 3.0 is an evolving extension of www, in which the information can be shared and interpreted by other software agent to find and integrate applications to different domains.

Web 3.0 provides integrated real time application environment to the user. The applications are majorly involved in searching using semantic web, 3D web and are media centric. Web 3.0 supports pervasive components. Each component and its relations are represented below.

In web 3.0, web is transformed into database or Data Web wherein the data which are published in the web is reusable and can be queried. This enables a new level of data integration and application interoperability between platforms. It also makes the data openly accessible from anywhere and linkable as web pages do with hyperlinks. Data web phase is to make available structured data using RDF. The scope of both structured and unstructured content would be covered in the full semantic web stage. Attempts will be to make it widely available in RDF and OWL semantic formats.

The driving force for web 3.0 will be artificial intelligence. Web 3.0 will be intelligent systems or will depend on emergence of intelligence in a more organic fashion and how people will cope with it. It will make applications perform logical reasoning operations through using sets of rules expressing logical relationships between concepts and data on the web. With the realization of the semantic web and its concepts web 3.0 will move into Service Oriented Architecture.

The evolution of 3D technology is also being connected to web 3.0 as web 3.0 may be used on massive scale due to its characteristics.

D. Nagamalai, E. Renault, M. Dhanushkodi (Eds.): CCSEIT 2011, CCIS 204, pp. 162–170, 2011.

Web 3.0 is media centric where users can locate the searched media in similar graphics and sound of other media formats.

The pervasive nature of web 3.0 makes the users of web in wide range of area be reached not only in computers and cell phones but also through clothing, appliances, and automobiles.

Learning process in Web 3.0 is a highly sophisticated for the learners. The learners have access the unbelievable knowledge source. The tutors are intelligent agents who are customized for the learners.

### 1.1 Review of Related Work

Claudio Baccigalupo and Enric Plaza discussed in the paper poolcasting : a social web radio architecture for Group Customization about Pool casting a social web radio architecture in which groups of listeners influence in real time the music played on each channel. Pool casting users contribute to the radio with songs they own, create radio channels and evaluate the proposed music, while an automatic intelligent technique schedules each channel with a group customized sequence of musically associated songs [13].

M.T.Carrasco Benitez discussed in the paper Open architecture for multilingual social networking about an open architecture for all the multilingual aspects of social networking. This architecture should be comprehensive and address well-trodden fields such as localization, and more advanced multilingual techniques to facilitate the communication among users [14].

Autona Gerber, Alta van der Merwe, and Andries Barnard discussed in the paper A functional Semantic web architecture about the CFL architecture which depicts a simplification of the original architecture versions proposed by Bernes-Lee as a result of the abstraction of required functionality of language layers. Gerber argues that an abstracted layered architecture for the semantic web with well defined functionalities will assist with the resolution of several of the current semantic web research debates such as the layering of language technologies [15].

## 2 Web Based Learning -- The Current Scenario

Web based learning is the electronically supported learning and teaching which are procedural in character and aim to effect the construction of knowledge with reference to individual experience, practice and knowledge of the learner. Information and communication systems, whether networked or not, serve as specific media to implement the learning process[1].

The current web environment is on the transition state between web 2.0 and web 3.0. The Earlier Web based learning process was instructional. Using the internet technologies the students acquired knowledge. In Web 2.0 the learner is powered by the social softwares like blogs, wiki's , podcasts and virtual worlds. The value addition of social software has made the learner to be participatory in the learning process. Social networking software's are participatory knowledge acquisition phenomenon which shares the knowledge of the society.

A typical environment of learning in Web 2.0 is read and write .The primary participants are learner and a tutor. The learner can be an active participant in the social networking sites and acquire knowledge. When the learner opts for undergoing classes in a particular topic, he learns from the static knowledge of the tutor which are programs. It includes static contents and the learner can ask questions and answers if supported by the programs will be displayed. The tutors look and appearance is 2D.

The main disadvantage of Personalized learning in Web 2.0 is although the environment is participatory, it is static and predefined.

It does not includes the collective knowledge of current advancements in the topic. The interface is simple to use. The Learner can have only limited queries. The dissemination of knowledge by the tutor is determined by the developer.

## 3 Technical Background of Web 3.0

Web 3.0 architecture makes the learner sophisticated. Resource Definition Frame work (RDF), SPARQL, WebDAV, site Specific API's ,FOAF and SSL are the components of Web 3.0.

Resource Definition Frame work is a meta data data model . The resources are transformed into statements of the form subject-predicate-object triple expressions. The subject denotes the resource, predicate denotes traits or aspects of the resource and express a relationship between the subject and the object. For example to represent the notion "The sky has the color blue" in RDF is as the triple: a subject denoting "the sky", a predicate denoting "has the color", and an object denoting "blue". A collection of RDF statements intrinsically represents a labeled, directed multi-graph. As such, an RDF-based data model is more naturally suited to certain kinds of knowledge representation'2'. In web 3.0 Personalized learning is accomplished by associating the queries by learners and the knowledge of the tutor together with the available web resources in RDF format so that the query can be associated with the knowledge.

SPARQL protocol for RDF query language process query which consist of triple patterns, conjunctions , disjunctions and optional patterns. SPARQL facilitates personalized web based learning in web 3.0 by accepting patterns, multiple queries and complicated queries by the learner [3]. It also empowers the intelligent agent tutor to collaborate multiple web resources into meaningful and understandable text and visuals of the learner.

Web-based Distributed Authoring and Versioning (WebDAV) is a set of methods based on the Hypertext Transfer Protocol (HTTP) for collaboration between users in editing and managing documents and files stored on World Wide Web servers. The most important features of the WebDAV protocol include:locking (overwrite prevention ),properties (creation, removal, and querying of information about author, modified date),name space management ( ability to copy and move Web pages within a server's namespace )and collections (creation, removal, and listing of resources)[4].In Personalized collaborative environment webDAV assists the intelligent agent tutor to interact with the web servers to collect and produce cognizant knowledge to to the learner.

Site specific API's interacts with the RDF and produces required output of the learner. Site specific API's provides the functionality to the learner by making the query understandable to the tutor in precise form. It supports the intelligent agent tutor to interact with the web data resources in world wide web servers to collaborate the data[5]. Site Specific API's are useful while modifying the existing personalized learning web sites of web 3.0 so that the functionality can be reused and remixed again.

FOAF+SSL is a decentralized secure authentication protocol using the FOAF profile information and SSL security layer. It makes ease the role of the intelligent agent tutor in web 3.0 by maintaining FOAF cloud of the persons , activities their relations to other people and objects[5]. The intelligent agent tutor of web 3.0 need not maintain a centralized machine readable ontology . The security issues are maintained by SSL.

The aggregation of these components is utilized in the personalized collaborative learning of web 3.0.

Technology plays a vital role in all aspects of personalized learning. The technologies which are utilized in personalized collaborative learning includes Artificial intelligent, Automated reasoning, Cognitive architecture, Composite applications, Distributed computing, Knowledge representation, Ontology, cloud computing, Grid computing, Scalable vector graphics, Semantic web, Semantic WiKi and Software agents.

## 4 Personalized Web Based Learning in Web3.0

In web 3.0 the two major components are the learner and the 3D tutor. The learner is the human who intends to acquire knowledge about a specific subject. The tutor is an intelligent agent which delivers the collective knowledge to the learner.

The learner initially starts to take up the tutorial. The learner specify the tutor's look. He can specify the gender of the tutor. The look or avatar of tutor can also be specified by the learner. During the learning session, the learner must be completely interactive. His interactive queries, interpretations and examples defines the intelligence of the tutor. The learner can request more examples, working 3D models , clarifications, justifications, applications at any point of time. The output can be audio, video, 3D, or text as opted by the learner. It can be a combination of the output formats also depending on the preferences of the learner. The learner can demand aggregations and mash up of websites based on the topic. He can have choice of collaboration of best web resources and technologies available till date.

The tutor's look is specified by the user. The tutor intuitively asses the intelligence of the learner. Based on the intelligence and personal preferences of learner the tutor must deliver the knowledge. The tutor must collaborate knowledge from various web resources , filter the irrelevant knowledge and share it. The tutor must give outputs according to the personal preferences of the learner and also prefer the alternative output form if one type is not sufficient. It must understand the language of the learner and interact accordingly. The agent must aggregate the web resources required by the learner which is updated till date. The technologies and tools used must be the suitable by analyzing all the till date available technologies and tools.

## 5 Attribute Analysis of Personalized Web Based Learning in Web 3.0

Attributes in web 3.0 are the characteristics which determine the behaviour of the applications in web 3.0 architecture. Attribute analysis determines the key factors in developing web 3.0 applications. The attributes considered are 3D,Mash ups, Preferences regarding e-tutor and Speech recognition.

## 6 Data Collection

Data collection is a process of preparing and collecting data[10]. The method of Data collection adopted is Questionnaire. A Questionnaire is designed with a closed type question. Responses of the questionnaire is got from 100 Post Graduate Compute stream Students at a private university in Chennai.

The questionnaire sample is given below.

Questionnaire:

| s.no | Question | Response type |
|---|---|---|
| 1. | Do you require a 3D tutor ? | Yes/no |
| 2. | Can the Demos for your learning be in 3D? | Yes/no |
| 3. | Do you wish to ask a knowledge based random query to the etutor ? | Yes/no |
| 4. | Will you prefer a collective text output ? | Yes/no |
| 5. | Do you want a collective knowledge of the web sites of your choice ? | Yes/no |
| 6. | Do you want the tutor to teach according to your knowledge level ? | Yes/no |
| 7. | Will you like to opt the gender of your tutor ? | Yes/no |
| 8. | Should your tutor know about your preferences ? | Yes/no |
| 9. | Should the tutor have a collective knowledge of the best web sites available ? | Yes/no |
| 10. | Do you wish your speech to be recognized by your tutor ? | Yes/no |

| 3D tutor | 3D demos | Random question | Mash up of knowledge | Custom Mash up | eTutor teaches according to the level | Choose look of etutor | intutive_reasoning | concise_knowledge | Speech recognitio |
|---|---|---|---|---|---|---|---|---|---|
| 0 | 1 | 1 | 0 | 1 | 1 | 1 | 1 | 1 | 1 |
| 0 | 0 | 1 | 1 | 1 | 1 | 1 | 1 | 1 | 0 |
| 1 | 0 | 1 | 1 | 1 | 1 | 1 | 1 | 1 | 0 |
| 0 | 0 | 1 | 1 | 1 | 1 | 1 | 1 | 1 | 0 |
| 1 | 1 | 1 | 1 | 1 | 1 | 1 | 1 | 1 | 1 |
| 0 | 0 | 0 | 0 | 1 | 1 | 0 | 0 | 1 | 0 |
| 1 | 1 | 0 | 0 | 1 | 1 | 0 | 0 | 1 | 1 |
| 1 | 1 | 1 | 1 | 1 | 1 | 1 | 1 | 1 | 1 |
| 0 | 0 | 1 | 1 | 1 | 1 | 1 | 1 | 1 | 0 |
| 1 | 0 | 1 | 1 | 1 | 1 | 1 | 1 | 1 | 0 |
| 0 | 1 | 1 | 1 | 1 | 1 | 1 | 1 | 1 | 1 |
| 0 | 0 | 1 | 1 | 1 | 0 | 1 | 1 | 0 | 0 |
| 1 | 1 | 1 | 1 | 1 | 0 | 1 | 1 | 0 | 1 |
| 0 | 0 | 1 | 1 | 1 | 1 | 1 | 1 | 1 | 0 |
| 1 | 1 | 1 | 1 | 1 | 1 | 1 | 1 | 1 | 1 |
| 1 | 1 | 1 | 1 | 1 | 1 | 1 | 1 | 1 | 1 |
| 1 | 1 | 1 | 0 | 1 | 1 | 1 | 0 | 1 | 1 |
| 1 | 1 | 1 | 0 | 1 | 1 | 1 | 0 | 1 | 1 |

**Fig. 1.** Sample Data

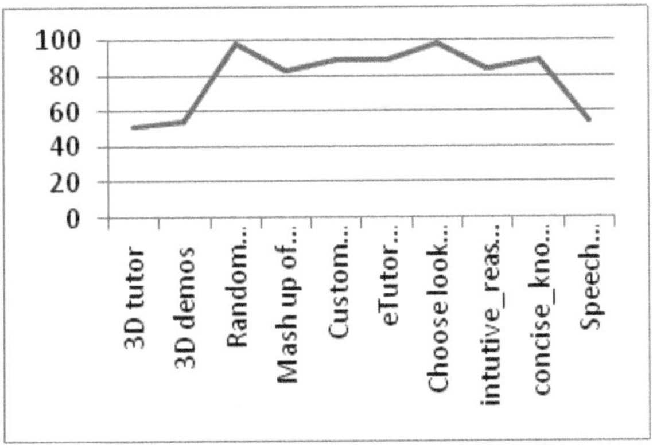


**Fig. 2.** Comparison of requirements

**The Top requirements are** Random question, Custom Mash up, Etutor teaches according to the level Choose look of etutor, concise_knowledge.

## 7 Finding the Frequent Item Sets in the Attributes Considered Using Apriori Algorithm

One of the most popular data mining approaches is to find frequent itemsets from a transaction dataset and derive association rules. Finding frequent itemsets (itemsets with frequency larger than or equal to a user specified minimum support) is not trivial because of its combinatorial explosion. Once frequent itemsets are obtained, it is straightforward to generate association rules with confidence larger than or equal to a user specified minimum confidence.

Apriori is a seminal algorithm for finding frequent itemsets using candidate generation [12]. It is characterized as a level-wise complete search algorithm using anti-monotonicity of itemsets, "if an itemset is not frequent, any of its superset is never frequent". By convention, Apriori assumes that items within a transaction or itemset are sorted in lexicographic order. Let the set of frequent itemsets of size $k$ be $Fk$ and their candidates be $Ck$ . Apriori first scans the database and searches for frequent itemsets of size 1 by accumulating the count for each item and collecting those that satisfy the minimum support requirement. It then iterates on the following three steps and extracts all the frequent itemsets[11].

Apriori algorithm was adopted to the data set. Let R1 represent Random question, R2 represent Custom Mash up, R3 represent Etutor teaches according to the level ,R4 represent Choose look of etutor, R5 represent concise_knowledge.

# 8 Results

| A-Priori parameters | |
|---|---|
| Support min | 0.50 |
| Confidence min | 0.75 |
| Max rule length | 2 |
| Lift filtering | 1.00 |

**Requirement matrix:**

A requirement matrix is constructed based on the analysis done using apriori algorithm.

| | R1 | R2 | R3 | R4 | R5 |
|---|---|---|---|---|---|
| R1 | 0 | 0 | 0 | 1 | 0 |
| R2 | 0 | 0 | 1 | 0 | 1 |
| R3 | 0 | 1 | 0 | 0 | 1 |
| R4 | 1 | 0 | 0 | 0 | 0 |
| R5 | 0 | 1 | 1 | 0 | 0 |

From the matrix representation it is evident that,

{R1 => R4}

{R2 => R3 ,R5}

{R3=>R2,R5}

{R4=>R1}

{R5=>R2,R3}

R2,R3,R5 are strongly inter related requirements

R1 ,R4 are strongly inter related requirements

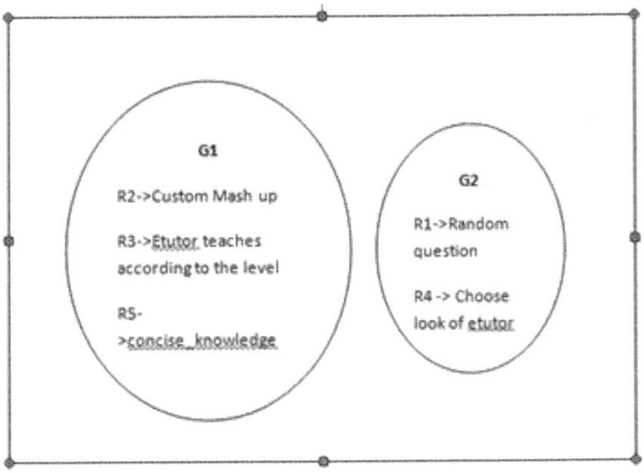


**Fig. 3.** Defining groups

# 9 Conclusion

From the analysis ,it is concluded that Custom Mash up, Etutor teaching according to the level of the learner , and concise_knowledge of the tutor are the interrelated requirements by the students for designing a tutorial web3.0 site. Learner asking Random question and learner choosing the look of etutor are the interrelated requirements by the students for designing a tutorial web 3.0 site. If any one of the interrelated requirement are required by the specific clients , then the developer can include all the inter related requirements in the a tutorial web 3.0 product for better customer satisfaction.

# 10 Future Work

By identifying the other attributes of web 3.0, the interdependencies of attributes can be identified to ease the design and development of web developer fraternity.

# References

[1] Tavangarian, D., Leypold, M., Nölting, K., Röser, M.: Is e-learning the Solution for Individual Learning? Journal of e-learning (2004)
[2] http://en.wikipedia.org/wiki/Resource_Description_Framework
[3] http://www.w3.org/TR/rdf-sparql-protocol/
[4] http://en.wikipedia.org/wiki/WebDAV
[5] http://richard.cyganiak.de/2008/papers/rwwod-tr-
[6] http://en.wikipedia.org/wiki/FOAF_(software)
[7] Bahrami, A.: Object Oriented Systems Development –using the unified modeling language. McGRaw-Hill International, New York
[8] http://www.sequencediagrameditor.com/diagrams/AutomatedRegistrationSystem.htm

[9] Carrasco Benitez, M.T.: Open architecture for multilingual social networking. Position paper for the W3CWorkshop on the Future of Social Networking (FSN), January 15-16 (2009)
[10] http://en.wikipedia.org/wiki/Data_collection
[11] http://www.hkws.org/seminar/sem433-2006-2007-no69.pdf
[12] Agrawal, R., Srikant, R.: Fast algorithms for mining association rules. In: Proceedings of the 20th VLDB Conference, pp. 487–499 (1994)
[13] Baccigalupo, Plaza, C., Poolcasting, E.: A Social Web Radio Architecture for Group customization Automated Production of Cross Media Content for Multi-Channel Distribution. In: AXMEDIS apos; 2007 Third International Conference, November 28-30, pp. 115–122 (2007)
[14] Carrasco Benitez, M.T.: Open architecture for multilingual social networking. Position paper for the W3C workshop on the Future of Social Networking (FSN), January 15-16 (2009)
[15] Gerber, A., van der Merwe, A., Barnard, A.: A functional semantic architecture, http://www.eswc2008.org/final-pdfs-for-web-site/fisr-1.pdf

# Rule Acquisition in Data Mining Using a Self Adaptive Genetic Algorithm

K. Indira[1], S. Kanmani[2], D. Gaurav Sethia[2], S. Kumaran[2], and J. Prabhakar[2]

[1] Department of Computer Science,
[2] Department of Information Technology,
Pondicherry Engineering College,
Puducherry, India
{induharini,kanmani.n,gaurav.sethia7,kumarane.90}@gmail.com,
prabhakar_pec64@yahoo.co.in

**Abstract.** Rule acquisition is a technique of data mining that is used to deduce inferences from large databases. These inferences cannot be noticed easily without data mining. *Genetic algorithms* (GAs) are considered as a global search approach for optimization problems. Through the proper evaluation strategy, the best "chromosome" can be found from the numerous genetic combinations. In the self-adaptive genetic algorithm, its main thought is to let control parameter (crossover rate, mutation rate) adjusted adaptively within the proper range, thus achieve a more optimum solution. It is proved that the self-adaptive genetic algorithm is with excellent convergence and higher precision than the traditional genetic algorithm.

**Keywords:** Association rule mining, Genetic algorithm, Crossover, Mutation, Fitness, Support, Confidence.

## 1 Introduction

Mining is used to refer to the process of searching through a large volume of data, stored into a database, to discover useful and interesting information previously unknown. Association rule mining is a type of data mining. It is the method of finding the relations between entities in databases. Association rule mining is mainly used in market analysis, transaction data analysis or in the medical field. For example, in a medical database, diagnosis is possible provided the symptoms or in case of supermarket, the relation between the purchase of different commodities can be obtained. Such inferences are drawn using association rule mining and can be used for making decisions.

There are some well known techniques for association rule mining. Some of the well known algorithms are Apriori, constraint based mining, Frequency Pattern Growth Approach, genetic algorithm. There have been several attempts for mining association rules using Genetic Algorithm.

The main reason for choosing a genetic algorithm for data mining is that a GA performs global search and copes better with attribute interaction when compared with the traditional greedy methods, based on induction.

D. Nagamalai, E. Renault, M. Dhanushkodi (Eds.): CCSEIT 2011, CCIS 204, pp. 171–178, 2011.

Genetic algorithm is evolved from Charles Darwin's 'Survival of the fittest' theory. It is based on individual's fitness and genetic similarity between the individuals. Breeding occurs in every generation and eventually it leads to better and optimal group in the later generations. [1] Analyses the mining of Association Rules by applying Genetic Algorithms.

[2] introduces CBGA approach that hybridizes constraint-based reasoning within a genetic algorithm for rule induction. The CBGA approach uses Apriori algorithm to improve its efficiency.

[3], [4] discuss some variations of the traditional Genetic algorithms in the field of data mining. [3] is based on a evolutionary strategy and [4] adopts a self adaptive approach. The self adaptive modification on a GA has never been attempted on association rule mining before, but as this promises to be very promising in improving the efficiency, it has been taken up.

The main modules in data mining process are

i. Data cleaning: also so known as data cleansing, is a phase in which noise data and irrelevant data are removed from the collection.
ii. Data selection: at this step, the data relevant for the analysis is decided on and retrieved from the large data collection.
iii. Data mining: it is the crucial step in which clever techniques are applied to extract patterns potentially useful.

A brief introduction about Association Rule Mining and GA is given in Section 2, followed by the proposed system in section 3. In section 4 the parameters used in association rule mining using SAGA are defined. Section 5 presents the experimental results followed by conclusion in the last section.

## 2 Association Rules and Genetic Algorithm

### 2.1 Association Rules

An important type of knowledge acquired by many data mining systems takes the form of if-then rules [5]. Such rules state that the presence of one or more items implies or predicts the presence of other items. A typical rule has the form

If A, B, C…n then Y

The two parameters with respect to if-then rules are described below.

The *confidence* [6] for a given rule is a measure of how often the consequent is true, given that the antecedent is true. If the consequent is false while the antecedent is true, then the rule is also false. If the antecedent is not matched by a given data item, then this item does not contribute to the determination of the confidence of the rule.

The *support* indicates how often the rule holds in a set of data. This is a relative measure determined by dividing the number of data that the rule covers, i.e., that support the rule, by the total number of data in the set.

## 2.2 Genetic Algorithm

Genetic algorithm [7] is simulated in the natural environment of biological evolution and genetics and the formation of an adaptive search algorithm for global optimization probability. Genetic algorithm is suitable for solving problems characterized by large space, multi-peak, non-linear, global optimization problem of high complexity display. The parameters of the problem to be resolved into a binary code or the decimal code (also into other hexadecimal code) that the gene, a number of genes form a chromosome (individual), a number of chromosomes is similar to natural selection, crossover and mutation matching algorithms, after repeated iteration (that is, hereditary from generation to generation) until the final results of the optimization. The use of genetic algorithms to solve the problem involved the following seven key factors: encoding, fitness function, selection operator, the crossover operator, mutation operator, control parameters.

In the traditional genetic algorithm, the crossover rate and mutation rate are fixed values which are selected based on experience. Generally we believe that when the crossover rate is too low, the evolutionary process can easily fall into local optimum to result in groups of premature convergence due to population size and the lack of diversity. When the crossover rate is too high, the process is optimized to the vicinity of optimal point and the individual is difficult to reach optimal point which can slow the speed of convergence significantly though groups can ensure the diversity.

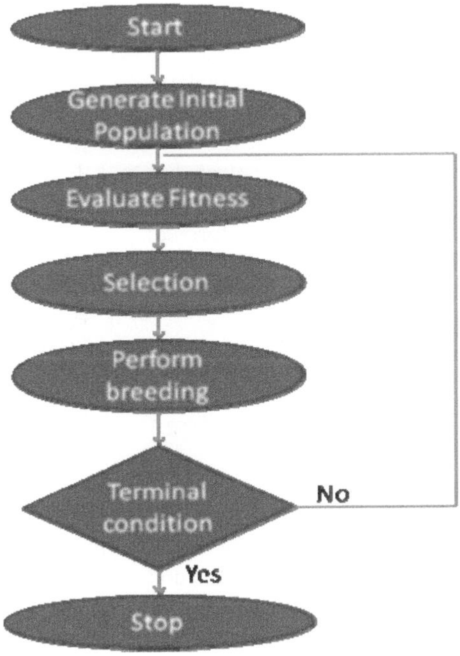


**Fig. 1.** Flow Chart Traditional GA

# 3 Proposed System

To overcome the drawbacks of traditional genetic algorithm, SAGA is proposed. SAGA involves changing the crossover and mutation rates adaptively [8]. The main purpose of setting mutation operator is to maintain the diversity of population and avoid stagnation of evolution. In the traditional genetic algorithm the mutation rate is fixed and after several iterations, the group's quality will gradually to converge and to form inbreeding. The organized adaptive genetic algorithm has a higher robustness, global optimality and efficiency.

```
Procedure SAGA

Begin
        Initialize population p(k);
        Define the crossover and mutation rate;
Do
{
        Do
        {

        Calculate support of all k rules;
        Calculate confidence of all k rules;
        Obtain fitness;
        Select individuals for crossover / mutation;
           Calculate the average fitness of the n and (n-1) the generation;
           Calculate the maximum fitness of the n and (n-1) the generation;
           Based on the fitness of the selected item, calculate the new crossover and
        mutation rate;
           Choose the operation to be performed;

           } k times;
}
```

# 4 Parameters in Genetic Algorithm

## 4.1 Selection of Individuals

Chromosomes are selected from the population for breeding. According to Darwin's evolution theory the best ones should survive and create new offspring. According to *Roulette Wheel Selection,* Parents are selected according to their fitness. The better the chromosomes are, the more chances to be selected they have, because the selection depends on fitness.

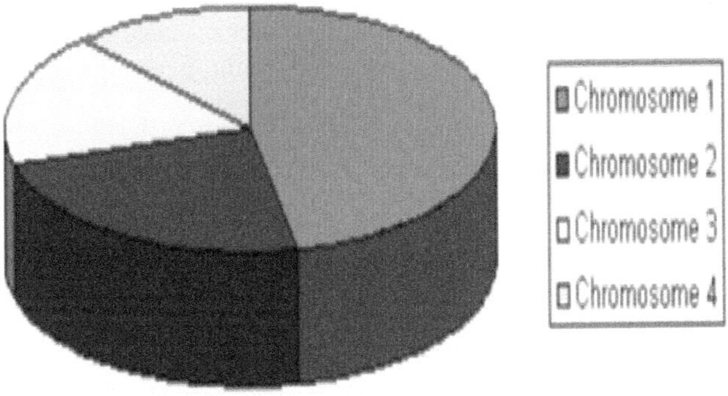


**Fig. 2.** Roulette Wheel Selection

### 4.2 Fitness Function

Given a particular chromosome, the fitness function returns a single numerical "fitness," or "figure of merit". This value is proportional to the "utility" or "ability" of the individual which that chromosome represents.

This paper adopts minimum support and minimum confidence for filtering rules. Then correlative degree is confirmed in rules which satisfy minimum support-degree and minimum confidence degree. After support-degree and confidence-degree are synthetically taken into account, fit degree function is defined as follows.

$$Fitness(X) = R_s \cdot \frac{Supp(X)}{Supp_{min}} + R_c \cdot \frac{Conf(X)}{Conf_{min}} \quad (1)$$

In the above formula, $R_s + R_c = 1$ ($Rs \geq 0$_$Rc \geq 0$) and $Supp_{min}$, $Conf_{min}$ are respective values of minimum support and minimum confidence. By all appearances_ if the $Supp_{min}$ and $Conf_{min}$ are set to higher values, then the value of fitness function is also found to be high.

### 4.3 Crossover Operator

Crossover selects genes from parent chromosomes and creates a new offspring. The most common form of crossover is single point crossover in which a crossover point on both parents is selected and child 1 is head of chromosome of parent 1 with tail of chromosome of parent 2 and child 2 is head of 2 with tail of 1.

### 4.4 Mutation Operator

Mutation changes randomly the new offspring. For binary encoding we can switch a few randomly chosen bits from 1 to 0 or from 0 to 1. Mutation provides a small amount of random search, and helps ensure that no point in the search has a zero probability of being examined.

### 4.5 Number of Generations

The generational process of mining association rules by Genetic algorithm is repeated until a termination condition has been reached. Common terminating conditions are:

- A solution is found that satisfies minimum criteria.
- Fixed number of generations reached.
- The highest ranking solution's fitness is reaching or has reached a plateau such that successive iterations no longer produce better results.
- Manual inspection.
- Combinations of the above.

### 4.6 Self Adaptive

The use of a fixed mutation probability Pm, when Pm value is small, does not have an impact on the mutation operator. When Pm value is great, it could undermine the group's excellent genes, the algorithm does not even slow down the convergence. Here, a method of adaptive mutation rate is used as follows:

$$p_m^{(n+1)} = \lambda p_m^{n} \sqrt{\frac{\sum_{i=1}^{m} \left(f_{max}^{(n+1)} - f_i^{(n+1)}\right)^2}{\sum_{i=1}^{m} \left(f_{max}^{(n)} - f_i^{(n)}\right)^2}} \tag{2}$$

$p_m^{n}$ is the nth generation mutation rate, $p_m^{(n+1)}$ is the $(n+1)^{th}$ generation mutation rate. The first generation mutation rate is $p_m^{0}$. $f_i^{(m)}$ is the fitness of the nth individual stocks i. $f_{max}^{(n+1)}$ is the highest fitness of the $(n+1)^{th}$ individual stocks. $f_i^{(n)}$ is the fitness of the nth individual i. m is the number of individual stocks. λ is the adjustment factor.

## 5 Experimental Studies

The objective of this study is to compare the accuracy achieved on different datasets using a traditional GA and a SAGA. The encoding of chromosome is binary encoding with fixed length. The fitness function adopted is as given.

Three datasets namely Lenses, Haberman and Car evaluation from UCI Machine Learning Repository have been taken up for experimentation. Lenses dataset has 4 attributes with 24 instances. The second dataset is Haberman which has 4 attributes and 306 instances. The final one is the car evaluation dataset, which has 6 attributes and 1728 instances. The Algorithm is implemented using Java.

The accuracy and the convergence rate are recorded in the table below. Accuracy is the count of rules matching between the original dataset and resulting population divided by the number of instances in dataset. The convergence rate is the generation at which the fitness value becomes fixed.

The parameters for the fitness function and the other parameters are chosen for best performance of a traditional GA. The parameters are set such that the convergence is fastest and the number of matches is maximum and SAGA is run with the same parameters and the results obtained are tabulated below. Accuracy is in terms of percentage with respect to number of matches.

**Table 1.** Default GA parameters

| **Parameter** | **Value** |
|---|---|
| Population Size | Varies as per the dataset |
| Initial Crossover rate | 0.9 |
| Initial Mutation rate | 0.1 |
| Selection Method | Roulette wheel selection |
| Minimum Support | 0.2 |
| Minimum Confidence | 0.8 |

**Table 2.** Accuracy comparison between GA and SAGA when parameters are ideal for traditional GA

| **Dataset** | **Traditional GA** | | **Self Adaptive GA** | |
|---|---|---|---|---|
| | **Accuracy** | **No. of Generations** | **Accuracy** | **No. of Generations** |
| Lenses | 75 | 38 | 87.5 | 35 |
| Haberman | 52 | 36 | 68 | 28 |
| Car Evaluation | 85 | 29 | 96 | 21 |

**Table 3.** Accuracy comparison between GA and saga when parameters are according to termination of SAGA

| **Dataset** | **Traditional GA** | | **Self Adaptive GA** | |
|---|---|---|---|---|
| | **Accuracy** | **No. of Generations** | **Accuracy** | **No. of Generations** |
| Lenses | 50 | 35 | 87.5 | 35 |
| Haberman | 36 | 38 | 68 | 28 |
| Car Evaluation | 74 | 36 | 96 | 21 |

From the above table we can conclude that the Self adaptive GA performs better than a traditional GA in both aspects, i.e. the convergence rate and the accuracy. The accuracy in case of the Haberman dataset is low because one of the parameter in the data is age. As age can take a wide range of values and only perfect matches are considered, the accuracy comes down.

When the algorithm comes to an end, the parameter values for mutation rate and crossover rate are changed because of self adaptivity. If the new values are set as the original values for a GA, then the performance of the GA is as below.

The table shows that the accuracy of the traditional GA goes down if the parameters are set in accordance with the termination Condition mutation rate of SAGA, this is because, when the SAGA ends, the mutation rate might take a high value, which when applied to a GA, will bring down the accuracy. The fitness threshold plays a major role in deciding the efficiency of the rules mined and convergence of the system.

## 6 Conclusion

Genetic Algorithms have been used to solve difficult optimization problems in a number of fields and have proved to produce optimum results in mining Association rules. When Genetic algorithm is used for mining association rules the GA parameters decides the efficiency of the system. Once the optimum values are fixed for individual parameters, then making the algorithm self adaptive increases the efficiency because it changes the mutation and crossover rate adaptively thus making the algorithm more intelligent. When the mutation rate is varied with respect to the result from the previous generation the accuracy increases. The efficiency of the methodology could be further explored on more datasets with varying attribute sizes.

## References

1. Collard, M., Francisi, D.: Evolutionary Data Mining: An Overview of Genetic-Based Algorithms. In: 8th IEEE International Conference on Emerging Technologies and Factory Automation, vol. 1, pp. 3–9 (2001)
2. Chiu, C., Hsu, P.-l.: A Constraint Based Genetic algorithm approach for Mining Classification Rules. IEEE Transactions on Systems, Man and Cybernetics 35, 305–320 (2005)
3. Saggar, M., Agarwal, A.K., Lad, A.: Optimization of Association Rule Mining using Improved Genetic Algorithms. IEEE, Transaction on System, Man and Cybernetics 4, 3725–3729 (2004)
4. Zhu, X., Yu, Y., Guo, X.: Genetic Algorithm Based on Evolution Strategy and the Application in Data Mining. In: First International Workshop on Education Technology and Computer Science, ETCS 2009, vol. 1, pp. 848–852 (2009)
5. Cattral, R., Oppacher, F., Dwego, D.: Rule Acquisition with Genetic Algorithm. In: Congress on Evolutionary Computation, CEC 1999, vol. 1 (1999)
6. Dai, S., Gao, L., Zhu, Q., Zhu, C.: A Novel Genetic Algorithm Based on Image Databases for Mining Association Rules. In: IEEE Conference on Computer and Information Science, pp. 977–980 (2007)
7. Wu, Y.-T., An, Y.J., Geller, J., Wu, Y.T.: A Data Mining Based Genetic Algorithm. In: IEEE Workshop on Software Technologies for Future Embedded and Ubiquitous Systems (2006)
8. Li, J., Feng, H.R.: A Self-Adaptive Genetic Algorithm Based on Real Code, pp. 1–4. Capital Normal University, CNU (2010)

# Dynamic Multi Dimensional Matchmaking Model for Resource Allocation in Grid Environment

Japhynth Jacob[1], Elijah Blessing Rajsingh[2], and Isaac Balasingh Jesudasan[3]

[1] Assistant Professor, St. Mother Theresa Engineering College
jafijacob@yahoo.co.in
[2] Director, KSCST, Karunya University
elijahblessing@karunya.edu
[3] Director, St. Mother Theresa Engineering College
isaacbalasinghj@yahoo.co.in

**Abstract.** Resource allocation and management in the Grid environment is a complex undertaking as resources are distributed and heterogeneous in nature. Optimally assigning jobs to resources is one of the key issues in Grid environment. This work introduces a new Multi Dimensional Matchmaking Framework for resource allocation on computational grids. Making the environment intelligent and monitoring the dynamic behavior of resources gives accurate prediction of resource allocation. The dynamic dimensional based resource allocation enhances the overall performance of services.

**Keywords:** Matchmaking, Intelligence, Guaranteed time.

## 1 Introduction

Grids are distributed computing models that enable decentralized collaboration by integrating computers into networks. In the Grid environment, shared resources and users typically span different organizations. Before resources can be allocated to run an application, a user or agent must select resources appropriate to the requirements of the application. This process of selecting resources based on application requirements is called “resource matching”. Nowadays many matchmaking based technologies addresses the issue of Grid resource management. The grid matchmaking process involves three types of agents: consumers (requesters), producers (providers), and a matchmaking service. A matchmaking service uses a matching algorithm to evaluate a matching function which computes the matching degree. This paper introduces a new multi dimensional matchmaking framework. The multi dimensional matchmaking process considers static, dynamic and behavioral parameters. The correlation between the request and the resource is ranked and optimal resource is allocated. Intelligence is added in this model to speed up the allocation process. User satisfaction level is retrieved for every job.

## 2 Related Work

In condor G matchmaking mechanism [1] users describe their requirements in class ads. Classads describes attributes of request (i.e. storage space, library files required

D. Nagamalai, E. Renault, M. Dhanushkodi (Eds.): CCSEIT 2011, CCIS 204, pp. 179–185, 2011.

etc…). Resources also describes class ads describing their attributes. Condor matchmaker matches the request with the resources. This system lacks support for parallel jobs, lacks integration with GIS, and is not normalized. A semantic matchmaker service on grid [10] uses separate ontology for request and service (ontology as set-of-concept-definitions or specification of a conceptualization).Ontology translator translates various formats into one. Matchmaking is done by using terms defined in ontologies. It is based on static information. Semantic based matchmaking uses ontology based matchmaking and uses grid service technology to access it. But in this system users should express their preferences manually by calculating the ranking function and there are no application level descriptions in request ontology. In matching service for an intelligent grid [2], each request is characterized by set of properties. Each property is a tuple (name, value).Matching service uses matching algorithm to evaluate a matching function which computes matching degree. It considers only static information. Online resource matching for heterogeneous grid environments [4] takes into account the dynamically changing conditions on resources and request. Resource State Repository (RSR) stores the real time status of all resources. Each job requires resources from various grid sites. It uses integer programming to solve the problem of heterogeneity. Each job has one dependency on each resource type. This system lacks fairness and preemption. Semantic matchmaking of grid services using parameter matchmaking algorithm matches request with resources based upon three parameters; input, output and functionalities. Matching degrees are computed by applying different filters. The matchmaking algorithm semantically compares IOF of request, service and computes various degrees of matches. Here the least rank is taken as the final rank. However other higher ranks have lost its significance. There may be a chance for that parameter to be a significant parameter to compare. In Interoperating grids through delegated matchmaking, delegation chain has been formed. But it has worst resource selection. In Grid resource discovery approach based on matchmaking engine overlay [3] centralized matchmaker becomes a bottleneck when requests and resources increase. It also leads to single failure problem.

## 3 Problem Statement

Condor G matchmaking mechanism lacks support for parallel jobs, lacks integration with GIS and ranks are not normalized. In a semantic matchmaker service on grid, users should express their preferences manually by calculating the ranking function and there are no application level descriptions in request ontology. Resource matchmaking service for intelligent grid considers only static information. Online resource matching for heterogeneous grid environments has no fairness in allocation and has no preemption. Delegated matchmaking has worst resource selection. Problems in matchmaking are summarized as: Ranks of the matched resources are not normalized, even after allocating best resource, job's successful execution till the end is not assured, Needs dynamic updation about the status of jobs, Speed of matching process needs to be improved.

## 4 Multi Dimensional Matchmaking Framework

Multi dimensional matchmaking framework assigns weight value for each parameter to normalize the ranks. It includes behavioral parameters to avoid unpredictable failures, dynamic parameters to dynamically update the status. The intelligent part will speed up the matchmaking process. First filtering stage considers Static parameters like resource type, resource name, IP address, node memory size, node CPU speed, node hard disk capacity, no of nodes, total memory size, total CPU rates, total disk capacity, total CPU speed. The next stage of filtering considers dynamic parameters like Process deadline or guaranteed time, economic efficiency, security or dedication rate. The final filtering stage considers behavioral parameters like hit ratio, abandon rate, occupation rate .Parameters in each stage will be given weight value according to their importance in each stage. This weight value is multiplied with matching degree. The highest value is taken for each stage. The average of 3 stages is the final matching value. The process repeats until it searches all possible resources. The resource with highest value is chosen as the best resource.

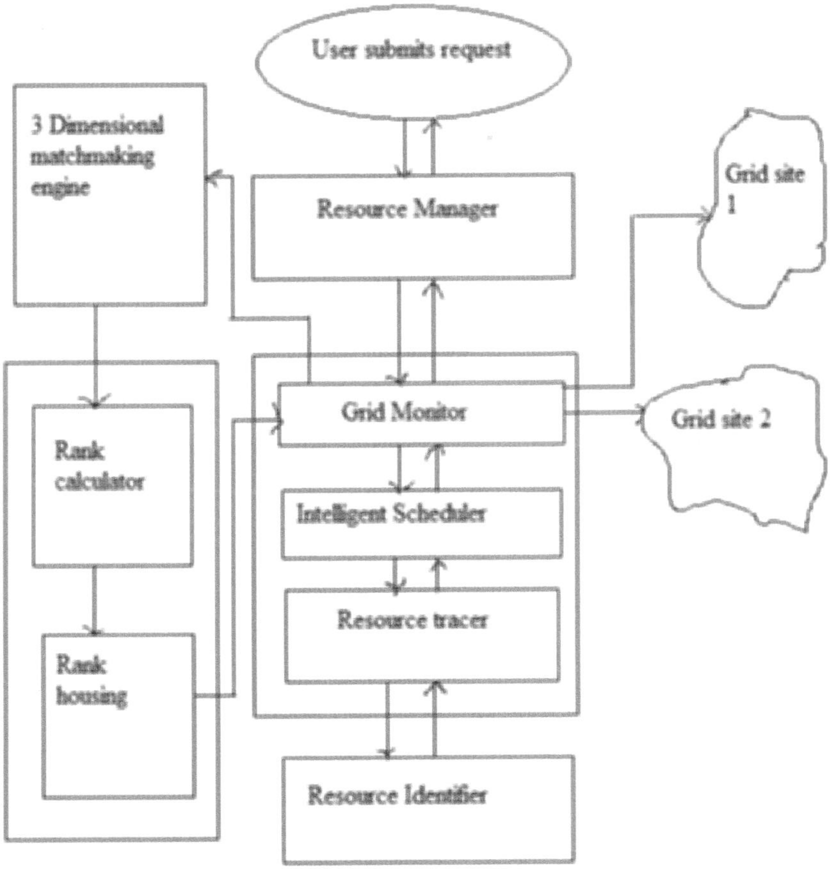


**Fig. 1.** Matchmaking architecture

Intelligent scheduler gets feedback from users after they have successfully completed their jobs. It stores their level of satisfaction. If their level of satisfaction is nearly equal to the threshold level, it stores the location of resources they have used in temporary hashing. Next time if same job arrives, the temporary hashing gives the location of resources that have successfully completed that job previously. Every similar type of resources in grid site will have a separate capacity value 'C'. If the newly arrived resource capacity value is greater than the resource the old resources' reference will be replaced by the newly joined resource location. Matchmaking process is analyzed by considering four cases

**Case 1:** Resource ingress Low and customer ingress Low
**Case 2:** Resource ingress Low and customer ingress High
**Case 3:** Resource ingress High and customer ingress Low
**Case 4:** Resource ingress High and customer ingress High.

## 4.1 Matchmaking Algorithm

User submitting the request Ri to resource manager RM
RM checks whether Ri = {any one of the old request}
If Ri = {any one of the old request}
{Ri is directed to GM
//GM retrieves all the resource that is already used by the earlier old request (since it is similar to the newly arrived request now)
//GM then directs the request along with the list of used resources to HMM
Ri {$r_i$ i=1-n}→HMM
//HMM calculates the transition probability of the retrieved request
//Transition probability of resource is high if
*any new resource which is of the same type but with advanced features/capacity
*any resources left the grid site or any resource newly added into the grid site.
//The state transition probability is low if
*no new resources are being added up with advanced features into the grid site
*no resources left the grid site or no resources newly added into the grid site
If {State transition probability is high
// recalculate the rank with the newly joined resources.
Ri →Pi i=1-n
//store the result in rank house
//use the old rank calculated for the request
Ri→X}}
If Ri = new request
{ //Direct the request to the matchmaking engine.
//Calculate new rank for the resources retrieved.
//Allocate highly ranked resource to the user and stores the rank in rank housing.
}

# 5 Performance Evaluation

## 5.1 Experiment Setup

The objective of this research is to assign optimal resource to job, to speed up the allocation process, to make the system more reliable. It is assumed that all service providers are available at all time provided they are free with their resources. The number of request served per second increases as the system receives more number of jobs. The optimal resource is selected by considering assignment model, correlation model, testing of samples and measures of dispersion.

Let $\boldsymbol{X}$ represents job, then $\boldsymbol{X} = \{X_1, X_2, \ldots, Xn\}$
Where $n$ is number of parameters.
Let $R_1, R_2, R_3 \ldots Rz$ are resources, where $z$ is number of resources.
Let $R_1 = \{Y_{11}, Y_{12}, Y_{13}\}$
$R_2 = (Y_{21}, Y_{22}, Y_{23}\}$
.
.
$Rz = \{Y_{z1}, Y_{z2}, Y_{z3}\}$

The correlation between the request and the resource is calculated using

$$Rz = \frac{n.\sum_{n=1}^{3} XnYzn - \sum_{i=1}^{3} Xn.Yzn}{\sqrt{n.\sum_{n=1}^{3}(Xn)^2 - (\sum_{n=1}^{3} Xn)^2} * \sqrt{n.\sum_{n=1}^{3}(Yzn)^2 - (\sum_{n=1}^{3} Yzn)^2}} \quad (1)$$

The behavior of service is calculated by testing of samples

$$Bs = \frac{x - \mu}{\sigma \sqrt{n}} \quad (2)$$

$x$ represents old data output
$\mu$ represents expected output
$\sigma$ represents deviation from previous behavior
$n$ represents number of previous samples
$Bs$ represents behavior of service.

# 6 Conclusion

In this paper we have proposed a multi dimensional matchmaking model to select optimal resource to job. The intelligent environment prevents the starvation of low priority jobs in the remote site. By analyzing the job and resource in multi dimensions, best resource can be allocated to jobs. The intelligent scheduler reduces the search time irrespective of number of resources. Job entry and exit dynamicity is considered while allocating resources for jobs. Measuring user satisfaction level and threshold value gives accurate prediction in allocation. Our proposed system ensures that higher priority jobs are always treated better than low priority jobs leading to an overall betterment of the QoS.

## References


1. Imamagic, E., Radic, B., Dobrenic, D.: An Approach to Grid Scheduling By Using Condor-G Matchmaking Mechanism. Journal of Computing and Information Technology - CIT 14(4), 329–336 (2006), doi:10.2498/cit.2006.04.09.
2. Bai, X., Yu, H., Ji, Y., Marinescu, D.C.: Resource Matching and a Matchmaking Service for an Intelligent Grid. World Academy of Science, Engineering and Technology 1 (2005)
3. Liu, Y., He, H.: Grid Resource Discovery Approach Based on Matchmaking Engine Overlay. In: Third International Conference on Semantics, Knowledge and Grid
4. Vijay, C., Naik, L.K., Liu, C., Wagner, J.: On-line ResourceMatching for Heterogeneous Grid Environments. In: IEEE International Symposium on Cluster Computing and the Grid (2005)
5. Li, F., Qi, D.: Research on Grid Resource Allocation Algorithm Based on Fuzzy Clustering. In: Second International Conference on Future Generation Communication and Networking (2008)
6. Clematis, A., Corana, A., D'Agostino, D., Galizia, A., Quarati, A.: Job_resource matchmaking on Grid through two-level benchmarking. Future Generation Computer Systems (June 2010)
7. Shu, G., Rana, O.F., Avis, N.J., Chen, D.: Ontology-based semantic matchmaking approach. Advances in Engineering Software 38, 59–67 (2007)
8. Han_, W., Shi, X., Chen, R.: Process-context aware matchmaking or web service composition. Journal of Network and Computer Applications 31, 559–576 (2008)
9. Wanga, C.-M., Chenb, H.-M., Hsuc, C.C., Lee, J.: Dynamic resource selection heuristics for a non-reserved bidding-based Grid Environment. Future Generation Computer Systems 26, 183–197 (2010)
10. Grigori, D., Corrales, J.C., Bouzeghoub, M.: Behavioral matchmaking for service retrieval: Application to conversation protocols. Information Systems 33, 681–698 (2008)
11. Asaduzzaman, S., Maheswara, M.: Utilizing unreliable public resources for higher profit and better SLA Compliance in computing utilities. J. Parallel Distrib. Computing 66, 796–806 (2006)
12. de Assunção, M.D., Buyya, R.: Performance analysis of allocation policies for intergrid resource provisioning. Information and Software Technology 51, 42–55 (2009)
13. Yu, H., Bai, X., Marinescu, D.C.: Workflow management and resource Discovery for an intelligent grid. Parallel Computing 31, 797–811 (2005)
14. Foster, I., Kesselman, C.: The GRID 2: Blueprint For a New Computer Infrastructure. Elsevier, Amsterdam (2005)
15. Liu, S.: User-Centric Resource Allocation Hierarchy in Grid Computing. In: The Sixth International Conference on Grid and Cooperative Computing, GCC 2007 (2007)
16. Altunay, M., Brown, D., Byrd, G., Dean, R.A.: Evaluation of Mutual Trust during Matchmaking. In: Proceedings of the Fifth IEEE International Conference on Peer-to-Peer Computing, P2P 2005 (2005)
17. Li, H., Buyya, R.: Model-based simulation and performance evaluation of grid scheduling strategies. Future Generation Computer Systems 25, 460–465 (2009)

18. Comito, C., Gounaris, A., Sakellariou, R., Talia, D.: A service-oriented system for distributed data querying and integration on Grids. Future Generation Computer Systems, doi:10.1016/j.future.2008.11.009.
19. Kosar, T., Balman, M.: A new paradigm: Data-aware scheduling in grid computing. Future Generation Computer Systems 25, 406–413 (2009)
20. Korkhova, V.V., Moscicki, J.T., Krzhizhanovskaya, V.V.: Dynamic workload balancing of parallel applications with user-level scheduling on the Grid. Future Generation Computer Systems 25, 28–34 (2009)

# A Comparative Study of CSO and PSO Trained Artificial Neural Network for Stock Market Prediction

Suresh Chittineni[1], Vabbilisetty Mounica[1], Kaligotla Abhilash[1], Suresh Chandra Satapathy[1], and P.V.G.D. Prasad Reddy[2]

[1] Anil Neerukonda Institute of Technology and Sciences
Sangivalasa-531162, Visakhapatnam, Andhra Pradesh, India
{sureshchittineni,mounica.vabbilisetty,abhilash.kaligotla, sureshsatapathy}@gmail.com
[2] Department of CS&SE, Andhra University Engineering College
Visakhapatnam, Andhra Pradesh, India
prasadreddy.vizag@gmail.com

**Abstract.** Stock market prediction is the act of trying to determine the future value of a company stock of other financial Instrument traded on a financial exchange. This paper presents a comparison between, PSO and CSO trained Neural Network to predict the stock rates by preparing data which acts as input. The data is prepared in such a way that the external factors like traditional issues can be mitigated. Earlier Neural Network was trained using Back Propagation algorithm but it converges to local optima and cannot be applied to discrete functions. Sow we have chosen PSO and CSO optimization algorithm to train the Neural Network. The results show that training neural network with such data gives a better performance.

**Keywords:** PSO Neural Network, and CSO Based Neural Network, Pre Processed Data.

## 1 Introduction

The importance of wealth in life is necessary for everything and it brings comfort and luxury. Wealth doesn't mean a recurring earning/salary but it does refer to a fixed asset, an investment or savings so it is a challenging and daunting task to find out which is more effective and accurate method for stock rate prediction. The successful prediction of a stock's future price could yield significant profit. [2] Prediction of financial markets has long been an attraction in the minds of Equity investors. Technical Analysis [11] provides a framework for studying investor behavior, and generally focuses only on price and volume data. Technical Analysis using this approach has short-term investment horizons, and access to only price and exchange data. With the advent of powerful computers much attention has been focused on this field. Equity market prices depend on many influences. Key factors that influence future equity prices can be broadly divided into quantitative and qualitative types.

D. Nagamalai, E. Renault, M. Dhanushkodi (Eds.): CCSEIT 2011, CCIS 204, pp. 186–195, 2011.

Primary quantitative factors include open rate, high rate, low rate, close rate and volume for individual equities. Qualitative factors include socio-economic, political, international, regional and performance factors to name but a few. The aim of this paper is to compare the pso and cso based neural network so as to find a technique that can predict stock market more accurately.

Preliminary research performed on Indian National Stock Exchange market has suggested that the inputs to the system may be taken as: previous day's closing rate and volume of last trading day for PSO[3] and CSO based neural network. After the inputs have been determined, the data have been gathered for NTPC stock for the period of Jan-2007 to DEC-2009 for training PSO and CSO based neural network. For testing purpose we have used testing data of NTPC Stock for the period of 1 year i.e. JAN-2010 to DEC-2011. Training and testing is performed using two network architectures.

1) One Hidden Layer PSO-Neural Network.
2) One Hidden Layer CSO-Neural Network.

The results are compared between PSO-NN (Particle Swarm Optimization Based Neural Network) and CSO-NN (Cat Swarm Optimization Based Neural Network).

## 2 PSO Based Neural Network

### A. Overview

Particle swarm optimization (PSO) is a population based stochastic optimization technique developed by Dr. Eberhart and Dr. Kennedy in 1995, inspired by social behavior of bird flocking or fish schooling.

PSO shares many similarities with evolutionary computation techniques such as Genetic Algorithms (GA). The system is initialized with a population of random solutions and searches for optima by updating generations. However, unlike GA, PSO has no evolution operators such as crossover and mutation. In PSO, the potential solutions, called particles, fly through the problem space by following the current optimum particles.

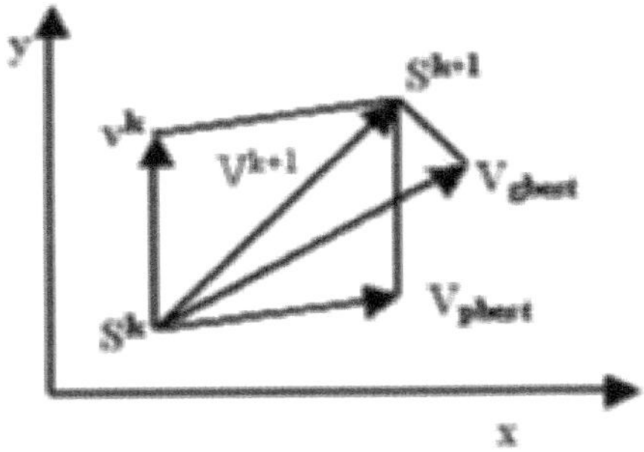


**Fig. 1.** Conceptual diagram explaining PSO

$S^k$: current searching point
$S^{k+1}$: modified searching point
$V^k$: current velocity
$V^{k+1}$: modified velocity
$V_{Pbest}$: velocity based on pbest.
$V_{gbest}$: velocity based on gbest.

Each particle keeps track of its coordinates in the problem space that is associated with the best solution (fitness) it has achieved so far. (The fitness value is also stored.) This value is called *pbest.* When a particle takes all the population as its topological neighbors, the best value is a global best and is called *gbest.*

The particle swarm optimization concept consists of, at each time step, changing the velocity of (accelerating) each particle toward its *pbest* and *gbest* locations. Acceleration is weighted by a random term, with separate random numbers being generated for acceleration toward *pbest* and *gbest* locations.

## B. Training Neural Network with PSO

The algorithm for PSO training is as follows:

Step 1: randomly generate the initial positions and velocities of particles of, say; p strings each of length d, which depends on the number of weights required.

Step 2: compute the fitness of each particle.

Step 3: get the values of gbest and pbest, where gbest is the global best among all the particles and pbest is the personal best till this iteration.

Step 4: update the velocities of the particles using the equation (1) below:

$$\text{Newpvel}=C1*\text{pvel} + C2*\text{rand}*(\text{ppos-gbest})+C3*\text{rand}*(\text{ppos-pbest}) \quad (1)$$

Where,

Pvel: current velocity of the particle.
C1, C2, C3: random values.
Ppos: current position of the particle.
Newpvel: updated velocity of the particle.

Step 5: Update the position of the particle using the equation (2) below:

$$\text{Newppos}=\text{ppos}+\text{newpvel} \quad (2)$$

Where,

Newppos: updated position of the particle.
Ppos: current position of the position of the particle.

Step 6: continue the above-mentioned steps till some stopping criterion is satisfied.
Step 7: Now with the best particle the stock rates are predicted.

### C. PSO – NN Organization

Hybrid approach offer strong advantage over either rule based or unaided neural network approaches [8] PSO based neural network offer good generalization [5] abilities although it may be difficult to determine the optimal network configuration and network parameters [4][6]. The architecture of PSO based neural network used is as follows.

1) Input layer with 2 nodes
2) One hidden layer with 2 nodes [7]
3) Output layer with one node.

## 3 CSO-Based Neural Network

### A. Overview

CSO is generated by observing the behavior of cats, and composed of two sub-models by simulating the behavior of cats[9]. Two of the major behavioral traits of cats are modeled for the proposed Cat Swarm Optimization (CSO). These are termed "seeking mode" and "tracking mode". Combination of these two modes allows CSO a better performance. Details of the two modes are given below.

**Rest and alert – seeking mode:** This sub mode is used to model the cat during a period of resting but being alert – looking around its environment for its next move. Seeking mode has four essential factors, which are designed as follows: seeking memory pool (SMP), seeking range of the selected dimension (SRD), counts of dimension to change (CDC) and self position consideration (SPC). SMP is used to define the size of seeking memory of each cat, indication any points sort by the cat. The cat would pick a point from the memory pool, according to rules described later. SRD declares the mutative ration for the selected dimensions. While in seeking mode, if a dimension is selected for mutution, the difference between the old and new values may not be out of range, the range defines by the SRD. CDC tells how many of the dimensions will be varied. All these factors play important roles in seeking mode. SPC is a Boolean valued variable, and indicates whether the point at which the cat is already standing will be one of the candidate points to move to. SPC cannot influence the value of SMP.

**Movement – tracing mode:** Tracing mode is the sub-model for modeling the case of the cat in tracing targets.

Once a cat goes into tracing mode, it moves according to its' own velocities for each dimention.

### B. Training Neural Network with CSO

As already discussed, CSO has two sub modes, namely seeking mode and tracing mode. We combine these two modes into the algorithm.

Seeking mode is described below:

Step 1: Make j copies of the present position of $cat_k$, where j=SMP. If the value of SPC is true, let j=(SMP=1), then retain the present position as one of the candidates.

Step 2: For each copy, according to CDC, randomly plus or minus SRD percents the present values and replace the old ones.

Step 3: Calculate the fitness values (FS) of all candidate points.

Step 4: Randomly pick the point to move to from the candidate points, and replace the position of $cat_k$.

The action of tracing mode can be described as follows:

Step 1: Update the velocities for every dimension ($v_k$, d) according to equation (1).

Step 2: Check if the velocities are in the range of maximum velocity. In case the new velocity is over range, it is set equal to the limit.

Step 3: Update the position of $cat_k$ according to equation (2).

$$V_{k,d} = V_{k,d} + r1.c1.(xbest, d - x_{k,d}),\quad d\text{-}1, 2, 3, \ldots.M \tag{1}$$

where $x_{best,d}$ is the position of the cat, who has the best fitness value; $x_{k,d}$ is the position of $cat_k$, $c_1$ is a constant and $r_1$ is a random value in the range of [0, 1].

$$X_{k,d} = X_{k,d} + V_{k,d} \tag{2}$$

**C. CSO-NN Organizations**

Hybrid approach offer strong advantage over either rule based or unaided neural network approaches [8]. CSO based neural network offer good generalization abilities although it may be difficult to determine the optimal network configuration and network parameters [4][6]. The architecture of CSO based neural network used is as follows:

1) Input layer with 2 nodes
2) One hidden layer with 2 nodes[7]
3) Output layer with one node.

## 4 Experiment Results

The sytem has been developed and tested on Windows XP operating system. We have used MATLAB for programming. Normalization is a key part of the input for neural networks and should enable more accurate predictable rates. Normalized data is used for training PSO based neural network and CSO based neural network. We normalize inputs so that input values lied between 0 and 1. Simulation data was sourced from Indian National Stock Exchange (NSE).

## A. Preparation of the Data Sets

Stock market depends upon many Qualitative factors like socio-economic, political, international, regional and performance factors, so the system which we tune should be in such a way that it is able to predict the stock rates accurately at least to some extent and quantify some of the external factors. Traders invest their money in shares, whose minds are closely tied up with sentiments. They sometimes choose a particularly day and start investing money on that particular day only.

Now the data set required to train such a Neural Network System is prepared by taking a particular day's stock values from the entire data of 3 years. This method is used to aim at a better result than normal data. [1]. Thus in this paper the traditional issues are taken into account to quantify the external factors and produce good results.

Input attributes should be carefully selected to keep the dimensionality of input vectors relatively small [10]. As we know close rate and volume are primary quantitative factors for individual equities and from quantitative factors the key qualitative factor of the market sentiment can be derived. So we used close rate and volume of stocks as our input in PSO and CSO based neural network and next stock rate as our target for training networks. PSO-NN and CSO-NN is trained on data set of NTPC for the years of Jan 2007 to Dec 2009 after training, testing is done on data set of 1 year from JAN 2010 to DEC 2010 of NTPC stock. PSO-NN and CSO-NN performance is compared on stock rates of NTPC for the specified period.

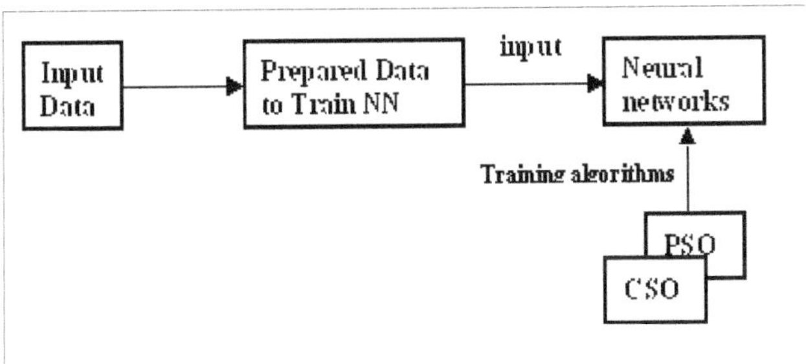


**Fig. 2.** Block Diagram explaining the working with the prepared data

In Fig3 and Fig4, we have taken the close rate and volume of Friday and predicted the open rate of Monday, thus 45 days of prediction is done in this NN. The CSO-NN system demonstrated better performance in comparison to PSO-NN on stock rates of NTPC for the specified period. On comparing it has been found that, CSO-CNN system is able to predict stock price movement of NTPC correctly with MSE 0.0012 as shown in Fig 3, while performance of PSO Based NN system with MSE 0.0017 shown in Fig 4.

In Fig5 and Fig6, we have taken the close rate and volume of Wednesday and predicted the open rate of Thursday, thus 49 days of prediction is done in this NN. The PSO-NN system demonstrated better performance in comparison to CSO-NN on stock rates of NTPC for the specified period. On comparing it has been found that, PSO-NN system is able to predict stock price movement of NTPC correctly with MSE 0.0013 as shown in Fig 5, while performance of CSO Based NN system with MSE 0.0017 shown in Fig 6.

The following graphs show the prediction of stock market for 45 week days (Monday). The red lines in the graph indicates the expected trend and the blue lines indicate the calculated trend obtained by training the neural network system for over a period 3 years.

The graph in Fig 3 given below represents the output of CSO trained neural network system that is expected to predict stock rates for 45 days. In this graph the 0 to 10 days period the trend is followed. From 1 0 to 15 days the graph is almost the same. From 15 to 40 days period the trend remained same. From 40 to 45 days again the trend of the graph remained same with small deviations. On the whole the trend is depicted every time.

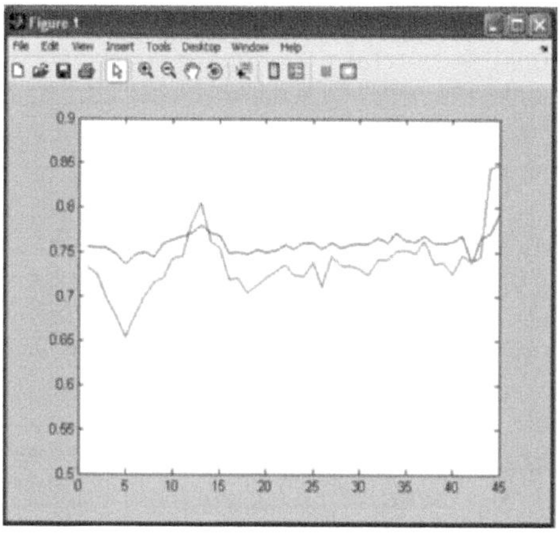


**Fig. 3.** Output of CSO trained neural network system for Mondays

This graph is Fig 4 below represents the output of PSO trained neural network system that is expected to predict stock rates for 45 days. Over here the trend of the calculated values remained the same all through with the expected values with a few deviations occurring.

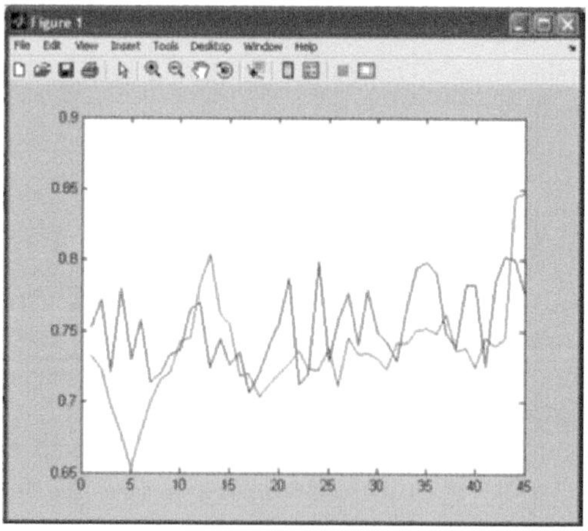


**Fig. 4.** Output of PSO trained neural network system for Mondays

The following graphs show the prediction of stock market for 49-week days (Thursday). The red line in the graph indicates the expected trend and the blue lines indicate the calculated trend obtained by training the neural network system for over a period 3 years.

The graph in Fig 5 given below represents the output of PSO trained neural network system that is expected to predict stock rates for 49 days. Over here the trend of the calculated values remained the same all through with the expected values with a few deviations occurring.

This graph in Fig 4 below represents the output of CSO trained neural network system that is expected to predict stock rates for 49 days. In this graph the 0 to 10 days period the trend is followed. From 1 0 to 17 days the graph is almost the same. From 18 to 46 days period the trend remained same. From 46 to 49 days, small deviations have occurred. On the whole the trend is depicted every time.

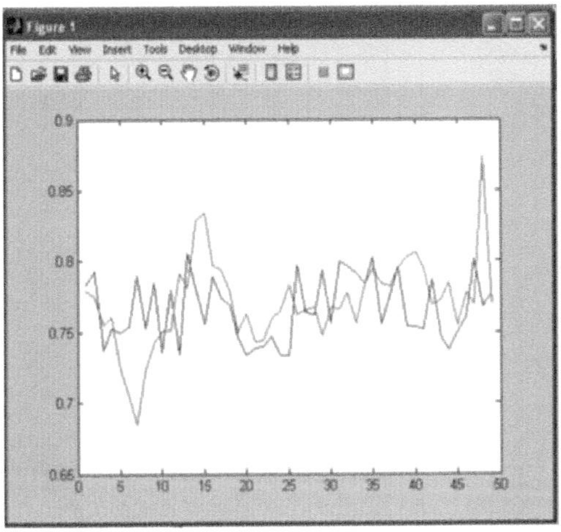


**Fig. 5.** Output of PSO trained neural network system for Thursdays

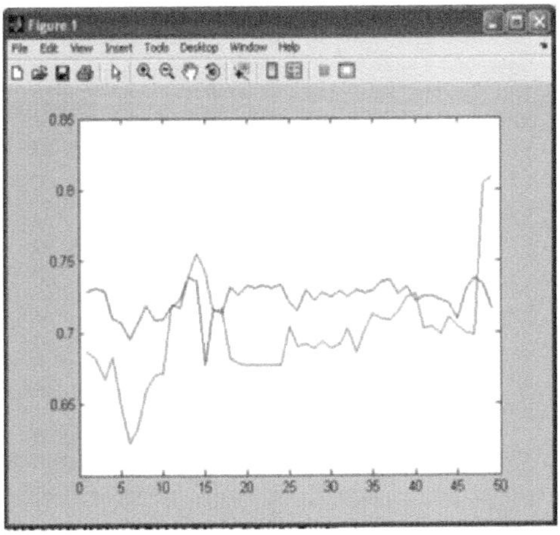


**Fig. 6.** Output of CSO trained neural network system for Thursdays

The deviations in both the above shown graph highlights the fact that the system that we used is a trained neural network system but in reality due to some external aspects like social, political and economical factors influence in day-to-day stock market prediction. This makes the prediction of the stock market movement very difficult. This variation in the graph can be mitigated if the external factors are quantified. Quantifying the external factors can be a further research.

**Table 1.** Comparative details between the two algorithms

| S.No | Training algorithm | Data set | No of days trained | No of days predicted | MSE value |
|---|---|---|---|---|---|
| 1 | PSO-Monday | NTPC | 3 YRS | 45 | 0.0017 |
| 2 | CSO-Monday | NTPC | 3 YRS | 45 | 0.0012 |
| 3 | PSO-Thursday | NTPC | 3 YRS | 49 | 0.0013 |
| 4 | CSO-Thursday | NTPC | 3 YRS | 49 | 0.0017 |

## 5 Conclusion

This paper has compared the forecasting accuracies of PSO based neural network and CSO based neural network. Results showed that under certain circumstances that prevailed on Monday, CSO based neural network has outperformed in predicting this stock movement in comparison to PSO based neural network. While under circumstances that prevailed on Thursday. PSO based neural network has outperformed in predicting this stock movement in comparison to CSO based neural network. By this approach of preparing data we are able to predict stock market rates better than ordinary data, this result is also not very accurate because stock rates not only depend on traditional sentiment factors but also on other factors like political issues, and all other issues related to the company which are difficult to be quantized. A further research can be done in quantifying the external factors.

## References

[1] Suresh, Ch., Abhilash, K., Mounica, V., Satapathy, S.C.: Cat Swarm Optimization Based Neural Network and Particle swarm optimization Based Neural Network in Stock Rates Prediction

[2] Khan, A.U., et al.: Genetic Algorithm Based Back propagation Neural Network Performs better than Back propagation Neural Network in Stock Rates Prediction, Glow Gift, Bhopal (2005)

[3] Zhang, X., Chen, Y.: Stock Index Forecasting Using PSO Based Selective Neural Network Ensemble
[4] Hamid, S.A.: Primer on using neural networks for forecasting market variables. In: Proceedings of the a Conference at School of Business, Southern New Hampshire University (2004)
[5] Majhi, R., Panda, G.: Robust Prediction of Stock Indices using PSO based Adaptive Linear Combiner (2009)
[6] Adya, M., Collopy, F.: How Effective are Neural Networks at Forecasting and Prediction? A Review and Evaluation. Journal of Forecasting (1998)
[7] Khan, A.U., et al.: Stock Rate Prediction Using Back Propagation Algorithm: Analyzing the prediction accuracy with different number of hidden layers, Glow Gift, Bhopal (2005)
[8] Bergerson, K., Wunsch, D.: A commodity trading model based on a neural network-expert system hybrid. In: IJCNN 1991, Seattle International Joint Conference, July 8-14, vol. 1, pp. 289–293 (1991)
[9] Chu, S.-C.: Computational Intelligence Based on the Behaviour of Cats-And it's Applications, Taiwan (2006)
[10] White, H.: Economic prediction using neural networks: The case of IBM daily stock returs. In: Neural Networks in Finance and Investing, ch. 18, pp. 315–328. Probus Publishing Company (1993)
[11] Mizuno, H., Kosaka, M., Yajima, H., Komoda, N.: Application of Neural Network to Technical Analysis of Stock Market Prediction. Studies in Informatic and Control 7(3), 111–120 (1998)

# Scaled Conjugate Gradient Algorithm in Neural Network Based Approach for Handwritten Text Recognition

Haradhan Chel[1], Aurpan Majumder[2], and Debashis Nandi[3]

[1] Bharat Sanchar Nigam Limited, India
[2] NIT Durgapur, Dept. of Electronics and Communication, India
[3]NIT Durgapur, Dept. of Information Technology, India

**Abstract.** Handwritten text and character recognition is a challenging task compared to recognition of handwritten numeral and computer printed text due to its large variety in nature. Neural Network based approach provides most reliable performance in handwritten character and text recognition but recognition performance depends upon some important factors like no of training samples, reliable features and no of features per character, training time, variety of handwriting etc. Important features from different types of handwriting are collected and are fed to the neural network for training. More no of features increases testing efficiency but it take longer time to converge the error curve. To reduce this training time effectively proper algorithm should be chosen so that the system provides best train and test efficiency in least possible time that is to provide the system fastest intelligence. In this paper we have used *Scaled Conjugate Gradient Algorithm*, a second order training algorithm for training of neural network. It provides faster training with excellent test efficiency. A scanned handwritten text is taken as input and character level segmentation is done. Some important and reliable features from each character are extracted and used as input to a neural network for training. When the error level reaches into a satisfactory level ($10^{-12}$) weights are accepted for testing a test script. Finally a lexicon matching algorithm solves the minor misclassification problems.

**Keywords:** Transition Feature, Sliding Window Amplitude Feature, Contour feature, scaled conjugate gradient.

## 1 Introduction

The endeavor of researchers for recognition of handwritten text through computer learning started in the sixtieth decade. Along with the up gradation of computational techniques it was always tried to implement the human perception into mathematical logic which is used to train the computer system. Different researchers took different approaches [7],[ 9], [20] to recognize handwritten text. Almost all the researchers have chosen some common logical steps for recognition of handwritten text such as Segmentation, Feature Extraction and pattern classification technique. Most of the approaches were on Neural Classifier [6], [14], [23], Hidden Markov Model [9], Fuzzy Classifier, or hybridized technique like Neuro-fuzzy, Neuro-GA, or some other

D. Nagamalai, E. Renault, M. Dhanushkodi (Eds.): CCSEIT 2011, CCIS 204, pp. 196–210, 2011.

statistical methods [25]. Neural Classifier has high discriminative power [1],[11],[12] for different patterns. Handwritten text contains wide varieties of styles in nature. It is really difficult to get real character level segmentation. A lot of works has been done on segmentation of hand written text in last two decades. Improper segmentation results inaccurate feature extraction and poor recognition performance. Similarly reliable and only discriminative features give better recognition performance. If number of features is increased recognition performance increases but it increases computational complexities and leads to much longer time of training of the neural network. The most essential thing is to choose a proper training algorithm which train faster the network with large no of features and provide best recognition performance. Many of these algorithms are based on Gradient Descent Algorithm such as Back Propagation Algorithm [15],[18]. But all the algorithms are practically inapplicable in large scale systems because of its slow nature and it's performance also depends on some user dependent parameters like learning rate and momentum constant. Some second order training algorithm such as Fleture Reeves [5], Polak Riebier[16],[19] or Powell-Beale Restarts algorithm[24] may be applied in appropriate applications. However *Scaled Conjugate Gradient* algorithm [22] results in much better recognition performance. By virtue of Scaled Conjugate Algorithm fastest training is obtained with almost 98 percent recognition efficiency. This paper not only shows the comparison of different second order training algorithm but also the reliable feature extraction and efficient lexicon matching technique. A small relevant description of Scaled Conjugate Algorithm is also given in later section. Different feature extraction schemes are also described in brief and finally the result of all experiments.

## 2 Image Preparation and Segmentation

A handwritten text is scanned through optical scanner and stored in bitmap image format. The image is converted in to a binary image. One important point may be mentioned regarding image preparation is choosing proper threshold so that image does not get any discontinuity in any continuous part. Image is segmented [4],[8],[ 3] text level to word level and word level to character level segmentation all relevant information such as no of word, no of characters, no of lines are stored. The algorithm for segmentation is used in image separation may be described below [20].

Algorithm 1:

Step1: The base lines such as upper, lower and middle base lines are calculated.
Step 2: Word is over segmented.
Step 3: The word segmentation points are modified using various rules.
Step 4: Undesired and incorrect segmentation points are removed.
Step 5: The correct segmentation points are used for final segmentation of words.
Step 6: Position of characters and words in the script are stored.

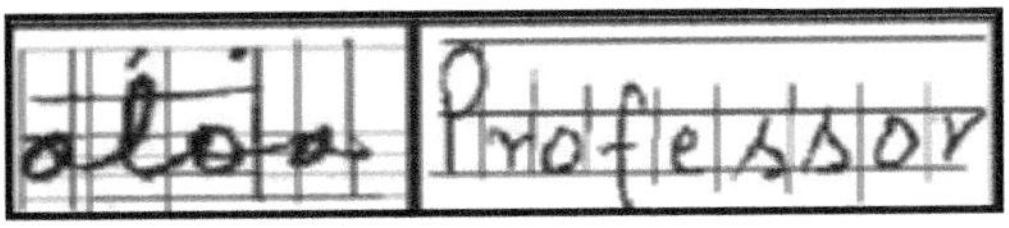

# 3 Feature Extraction

As it was mentioned earlier in introduction that reliable and meaningful feature increases the recognition performance. Reliable feature means those features which produce almost same value irrespective of slight variations in different sample images of same character. In this experiment three kinds of features extracted and a total 88 nos. of features were collected per character. Before feature extraction all the segmented character is resized [17] in a fixed size (60 x40). One runtime separated unbounded character image, bounded image and resized image is shown in figure 1. A short description of all four kinds of features may be mentioned below.

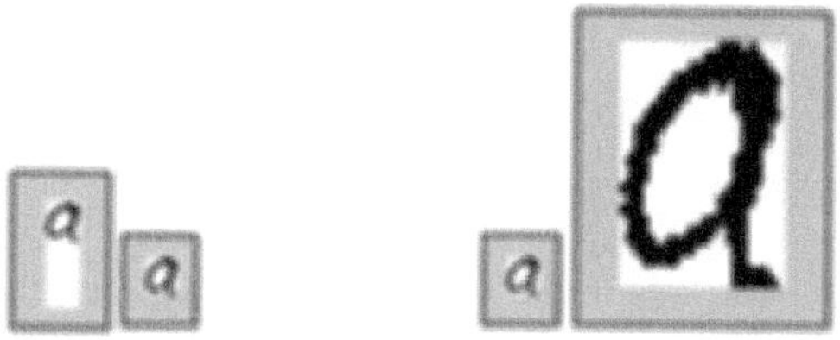

**Fig. 1.** Unbounded image, Bounded image and Resized Image

## 3.1 Transition Feature

Suppose the character image has a fixed dimension of X rows and Y columns. An M x N grid is superimposed over the image as shown in figure 2. The grid is prepared in such a way that all start and end values of each row and column got an integer value. The start values of row m and column n may be written as under

$$x_1 = y_1 = 1 \tag{1}$$

$$x_m = Int\left[\frac{(m-1)X}{M-1}\right] \qquad m = 2,3 \ldots M \tag{2}$$

$$y_n = Int\left[\frac{(n-1)Y}{N-1}\right] \qquad n = 2,3 \ldots N \tag{3}$$

$Int\ [x]$ refers to nearest integer value to x. Assume the intensity of coordinate(x,y) of the X x Y image is A(x,y). The original image is scanned along every row and column and the gradient information along each row and columns are collected.

$$\Delta(x_m, y) = \ A(x_m, y+1) - A(x_m, y) \tag{4}$$

$$where\ \ y = 1,2 \ldots\ldots.\ \ Y - 1$$

And

$$\Delta(x, y_n) = A(x+1, y_n) - A(x, y_n) \tag{5}$$

$$where\ x = 1,2 \ldots\ldots.\ \ X - 1$$

As the image is a binary image both $\Delta(x_m, y)$ $\Delta(x, y_n)$ return one of the three values 0, 1 or -1. 0 indicate no transition 1 means white to black transition and -1 indicate transition from black to white. A 5x5 grid over character image is shown on figure 2.a and transitions are also shown in figure 2.b. In this case we would consider only -1 value and in each row and columns of the M x N grid total number of transitions in every row and columns are important features of the image. As it was mentioned earlier that the approach of the experiment is neural network based and the immediate operation after feature extraction is training, it is found experimentally that neural network works reliably and faster if the input feature values are within 0 to 1 limit. As in this case all values are more than 1, they are normalized between 0 and 1 by a nonlinear transfer function as mentioned below.

$$f(x) = \frac{1}{1+e^{-kx}} \quad \dots\dots\dots\dots\dots 0 \le x \le 5 \tag{6}$$

k is a constant and values may be chosen between .3 to 1 for better result. The upper limit of x is taken as 5 because maximum possible white to black transition in handwritten characters found 5. In this way we found (M+N) nos. feature for each handwritten character. At the time of scanning the original image through all M rows and N columns using M x N grid and the coordinates of first and last transitions in each rows and columns are stored and a new kind of transition features were collected. Suppose in $m^{th}$ row first and last transition occurs at $(x_m, y_f)$ and $(x_m, y_l)$ then two features can be collected from that row as mentioned below

$$F_1 = \frac{y_f}{Y} \tag{7}$$

$$F_2 = \frac{Y - y_l}{Y} \tag{8}$$

Similarly if in $n^{th}$ column of the M x N grid first and last transition occurs at $(x_f, y_n)$ and $(x_l, y_n)$ then two features can be collected from that column as mentioned below

$$F_3 = \frac{x_f}{X} \tag{9}$$

$$F_4 = \frac{X - x_l}{X} \tag{10}$$

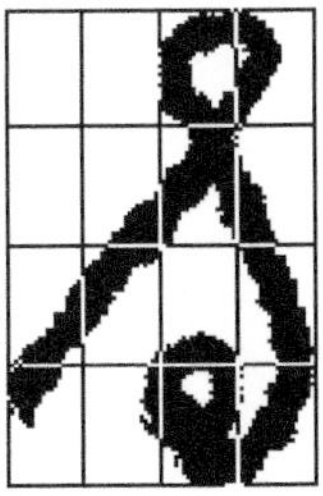

**Fig.2. a.** 5 x 5 grid over original image

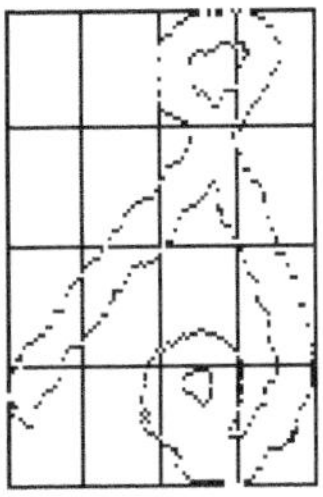

**Fig. 2. b.** 5 x 5 grid over grid image

### 3.2 Sliding Window Amplitude Feature

In this feature extraction scheme the image is subdivided into some overlapping windows such that half of the window breadth and width are overlapped with previous and next window along row and column wise except the windows at boundaries and corners. The windows at left boundary have no windows left at left side that it will overlap. Similarly top, bottom and right boundary side windows do not overlap with any windows beyond their boundary. The corner block has only two blocks to overlap for example the top left corner has only right and down windows to overlap. Black pixel density in each block or window is calculated. The original image of handwritten character has a fixed size of X rows and Y columns. As shown in figure (2.a) the image is superimposed with M x N grid. The four corners of the rectangular window R(m,n) are [($x_m$,$y_n$) ($x_m$, $y_{n+2}$) ($x_{m+2}$, $y_n$) ($x_{m+2}$, $y_{n+2}$)] .Values of $x_m$ and $y_n$ is defined in eqn. 1-3. Each block produces one feature and the feature is normalized between 0 and 1 and a total (M-2) (N-2) features are collected with normalized feature values [13]. In our experiment we have taken both M and N as 8 and a total 36 features were collected per character.

$$f = \frac{no\ of\ black\ pixel\ in\ the\ block}{total\ no\ of\ pixel\ in\ the\ block}$$

### 3.3 Contour Feature

The most important feature extraction scheme is contour feature [2][ 3]. Image is scanned and filtered. The filtered image is superimposed with an 11 X 1 grid as shown in figure (3.a) to find the external boundary intersection points of the character and total 22 such points are collected. All the 22 points are numbered in a circular fashion that from the top to bottom as 1 to 11 and from bottom to top as 12 to 22. All points are connected with the next point to draw an exterior boundary contour as shown in figure (3.b). For example the $n_{th}$ line is bound between two coordinate ($x_n$,$y_n$) and ($x_{n+1}$, $x_{n+1}$).. The alignment of all the boundary lines is unique property of that image and it differs from character to character. The two unique parameters of a straight line that is the gradient and a constant value which solely depend on the coordinates of the two points are taken as feature normalizing by a nonlinear hyperbolic function. Equation of a straight line and its representation by two points may be mentioned by equation 12-14. Suppose a straight line in equation is bounded between two points ($x_1$, $y_1$) and ($x_2$, $y_2$). The m is the gradient and c is a constant parameter of the straight line. Both m and c are two unique parameters which contain the boundary patterns of the characters.

$$y = mx + c \tag{12}$$

$$m = \frac{y_1 - y_2}{x_1 - x_2} \tag{13}$$

$$c = \frac{y_1 x_2 - y_2 x_1}{x_2 - x_1} \tag{14}$$

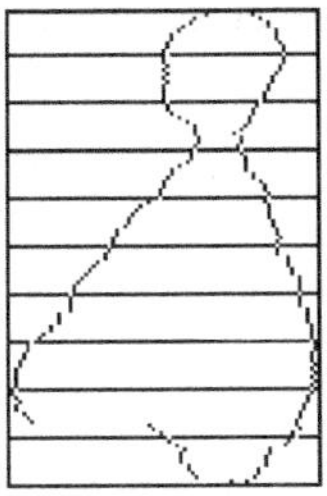

**Fig. 3.a.** 11 X 1 grid superimposed over boundary image

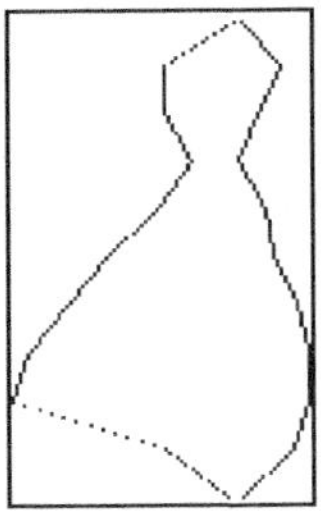

**Fig. 3.b.** Contour of image represented with straight line

Both the m and c may have any positive and negative values between negative infinity and positive infinity. As the immediate process after feature extraction is training of neural network with feature matrix all the feature values should be normalized between 0 and 1 as it was shown by experiment that neural network shows better response if the inputs are in range within 0 to 1 limit. For normalizing all the values of gradient (m) and constant (c) a hyperbolic nonlinear function is used as transfer function which is shown in equation 15 and a graphical representation is also shown in figure (4).In equation 15 any value of x maps the output between 0 and 1.k is a constant and value of it depends upon the no of features extracted. For 22 nos. of contour feature a value range from k=.3 to .5 is selected.

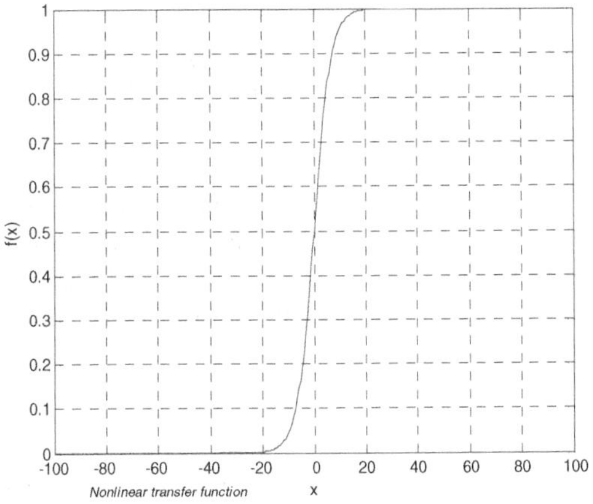


**Fig. 4.** Nonlinear Hyperbolic Transfer Function

$$f(x) = \frac{1}{1 + e^{-kx}} \tag{15}$$

If no of feature is increased its value should be decreased and similarly if no of features are decreased its value should be increased. In figure 4 plots for x versus f(x) is shown for an x range from -100 to 100. In this way 22 nos. features are collected and finally put in to the feature matrix. In our experiment we have taken three kinds of features extraction schemes and we have taken 30 transition features, 36 sliding window amplitude feature and 22 contour features and a total 88 nos. features are collected. For training ten sets of capital letters (A to Z) and ten sets of Small letters (a to z) written over two A4 size pages are taken and processed as described in the previous sections. Finally the feature matrix having dimension 520x 88 is fed to the neural network for training. The description of neural network and algorithm used for training are described in the next section.

## 4 Training of Neural Network

The immediate operation after feature extraction is training of a neural network. In this experiment we have used both single and multilayer feed forward network. It is not surprising that increase in no of feature improves recognition efficiency. But in opposite side it takes a longer time to train the neural network. In our experiment we have taken 88 nos. different kinds of features and 520 training character set as a result total 45760 nos. of features were collected. But theses large nos. of features takes very longer time to train the network if a proper training algorithm is not chosen. We have used different conjugate gradient methods for training and the speed of training improved significantly. All the algorithms are used for experiment are second order conjugate direction method such as Fleture-Reeves [15], [21], Powell Beale [24], Polak Ribiere [16],[ 19] and Scaled Conjugate Gradient algorithm by Moller [22] .A comparative study also been made among the above algorithms and a comparison of converging performance of all the second order conjugate gradient methods is presented in result section. It has also been found experimentally that among the entire training algorithms Powell Beale method and Scaled conjugate methods have shown better performance in respect of quick convergence of the error curve. Using *Scaled Conjugate Gradient* algorithm [22] the Hessian matrix of error equation is always positive over all iterations. But in all other algorithms mentioned above this thing is not guaranteed. This property of SCG algorithm increases learning speed reliably in successive iteration. Let us we define an error function Taylor Series expansion.

$$E(\widetilde{w}+\Delta\widetilde{w}) = \mathrm{E}(\widetilde{\mathrm{w}}) + \mathrm{E}'(\widetilde{\mathrm{w}})^{\mathrm{T}}\Delta\widetilde{\mathrm{w}} + \Delta\widetilde{\mathrm{w}}^{\mathrm{T}}\mathrm{E}''(\widetilde{\mathrm{w}})\Delta\widetilde{\mathrm{w}} + \qquad (16)$$

Suppose the notations are used as $A = E'(\widetilde{w})$ and $H = E''(\widetilde{w})$. Weight vector in $n^{th}$ iteration may be mentioned as $\widetilde{w}_n$. H is the Hessian matrix and A is the local gradient vector.

We will use a second order approximated error equation for further calculations and the same equation may be as follows

$$E(\widetilde{w}+\Delta\widetilde{w}) = \mathrm{E}(\widetilde{\mathrm{w}}) + \mathrm{E}'(\widetilde{\mathrm{w}})^{\mathrm{T}}\Delta\widetilde{\mathrm{w}} + \Delta\mathrm{w}^{\mathrm{T}}\mathrm{E}''(\widetilde{\mathrm{w}})\Delta\widetilde{\mathrm{w}} \qquad (17)$$

The solution of the quadratic difference equation may be found as follows [15].

$$\Delta\widetilde{\mathrm{w}} = \mathrm{H}^{-1}\mathrm{A} \qquad (18)$$

The above solution can be achieved subject to condition that H is positive definite matrix. The above equation states that how much amount of shifts is required for all the weights so that the error curve converges significantly. The above equation (18) is the essence of Newtown's Method [15] and from the equation it is found that the equation can be converged in one step if the inverse of the Hessian matrix is calculated. But in practical situation this phenomenon does not occur because of the some constraints as mentioned below.

a) When number of Weights is more calculation of inverse of the Hessian Matrix is computationally expansive and highly time consuming matter.
b) There is no guarantee that the inverse of the Hessian Matrix will be positive definite.
c) The system converges in one step if and only if the error equation is perfectly quadratic in nature. But generally all error equation has some higher order terms.

To avoid the above limitations the only way was found to achieve the solution through an iterative process. Suppose a set of nonzero vectors $p_1, p_2, p_3 \ldots\ldots\ldots p_N$ are basis vectors in $R^N$ and are H conjugate. If H is a positive definite matrix then the following conditioned must be satisfied [10].

$$\begin{aligned} \tilde{p}_k^T H \, \tilde{p}_i &= 0 \qquad for\ all\ k\ and\ i\ except\ k = i \\ \tilde{p}_k^T H \, \tilde{p}_i &> 0 \qquad for\ k = i \end{aligned} \tag{19}$$

Let $x_n = \tilde{p}_n^T H \, \tilde{p}_i$ . When k= i= n then $x_n$ is nonzero quantity.

$$\begin{aligned} x_n &= \tilde{p}_n^T H \, \tilde{p}_n \\ &= \tilde{p}_n^T \tilde{s}_n \end{aligned} \tag{20}$$

Here $\tilde{s}_n = H \, \tilde{p}_n = E''(\tilde{w}_n) \, \tilde{p}_n$. The idea to estimate the term $s_n$ with a non symmetric approximation [22] it may be written as

$$\tilde{s}_n = \frac{E'(\widetilde{w_n} + \sigma_n \tilde{\tilde{p}}_n) - E'(\widetilde{w_n})}{\sigma_n} \qquad \text{Where } 0 < \sigma_n \ll 1 \tag{21}$$

The above equation is the essence of second order error estimation used by Hestene [10] in conjugate gradient method. Scaled Conjugate Gradient method is a slightly different concept applied over it. In eqn.( 20) the sign of $x_n$ may be negative or positive but by definition if H is a positive definite matrix then $x_n$ must be a greater than zero. Hestene [10] combines the concept of introducing a dumping scalar value and modifies the equation as in eqn. 22. The other steps are same to the conjugate direction method. The function of the scalar parameter $\lambda$ is to compensate

$$s_n = \frac{E'(\widetilde{w_n} + \sigma_n \tilde{p}_n) - E'(\widetilde{w_n})}{\sigma_n} + \lambda_n \tilde{p}_n \ \text{ Where } 0 < \sigma_n \ll 1 \tag{22}$$

the indefiniteness of value of $E''(w_k)$ when it is negative. In every iteration sign of $x_n$ is checked i.e. whether $x_n > 0$ or $x_n < 0$. When $x_n$ is less than zero the value of $\lambda_n$ is increased and similarly when $x_n$ is greater than zero the value of $\lambda_n$ is decreased. All other steps are similar to the conjugate direction method.

The whole process is governed by basically two up gradation equation. Firstly the weight up gradation equation (eqn. 23) and secondly the basis vector up gradation equation (eqn. 24). The target is to find such a solution set of weight vector that H become positive definite. As the iterations go forward the first order error gradient $E'(w_n)$ decreases and at last it reaches very near to zero. The equations are as follows

$$\tilde{w}_{n+1} = \tilde{w}_n + \alpha_n \tilde{p}_n \tag{23}$$

$$\tilde{p}_{n+1} = \tilde{r}_n + \beta_n \tilde{p}_n \tag{24}$$

Where $r_n = -E'(w_n)$, $\alpha_n$ and $\beta_n$ may be calculated in each iteration by the following equations.

$$\alpha_n = \frac{\tilde{p}_n^T \tilde{r}_n}{\tilde{p}_n^T H \tilde{p}_n} = \frac{\tilde{p}_n^T \tilde{r}_n}{x_n} \tag{25}$$

And

$$\beta_n = \frac{\langle \tilde{r}_{n+1}, \tilde{r}_{n+1}, \rangle - \langle \tilde{r}_{n+1}, \tilde{r}_n \rangle}{\tilde{p}_n^T H \tilde{p}_n} \tag{26}$$

The most important matter is to discuss that how the value of $\lambda$ in eqn. 22 is chosen. As described earlier that, the function of $\lambda_n$ is to check the sign of $x_n$ in each iteration and to set a proper value of $\lambda_n$ so that $\tilde{x}_n$ get a positive value.

Let it is found that $x_n \leq 0$ and $\lambda_n$ is raised by $\overline{\lambda_n}$ so that $s_n$ get a new value

$$\tilde{\tilde{s}}_n = \tilde{s}_n + (\overline{\lambda_n} - \lambda_n)\tilde{p}_n \tag{27}$$

$$\tilde{x}_n = \tilde{p}_n^T \tilde{\tilde{s}}_n \tag{28}$$

Putting the value of $\tilde{\tilde{s}}_n$ from eqn. no 27 in eqn. 28

$$\tilde{x}_n = x_n + (\overline{\lambda_n} - \lambda_n)|\tilde{p}_n|^2 > 0 \tag{29}$$

$$\Rightarrow \overline{\lambda_n} > \lambda_n - \frac{x_n}{|\tilde{p}_n|^2} \tag{30}$$

From eqn. 29 some guide line is found that the new $\lambda_n$ should be greater than by -$\frac{x_n}{|\tilde{p}_n|^2}$ but no such particular value can be obtained that so that the optimal solution can be obtained. However we have used

$$\overline{\lambda_n} = 3\left(\lambda_n - \frac{x_n}{|\tilde{p}_n|^2}\right) \tag{31}$$

The above assumption is put into eqn. 29 we get

$$\tilde{x}_n = -2x_n + 3\lambda_n|\tilde{p}_n|^2 \tag{32}$$

Combining eqn. 32 and 25 we found the modified value of $\alpha_n$

$$\alpha_n = \frac{\tilde{p}_n^T \tilde{r}_n}{-2x_n + 3\lambda_n|\tilde{p}_n|^2} \tag{33}$$

the above equation clears that with the increase of the value of $\lambda_n$ decrease the height of step size and decrease in $\lambda_n$ increases the height of step size which agrees to all the assumptions made earlier.

In the above steps it was realized that how the Hessian Matrix remain positive in all iterations using a scalar parameter $\lambda$. But the second order approximation of error equation which was used during entire calculation cannot assure for the best performance though we get positive definite Hessian matrix over all iterations. The above calculation steps assure that the error curve will be converging towards minima but in some situation choosing a proper value of λ is required otherwise the rate of convergence become slow. So a proper mechanism for scaling λ is adopted. A parameter ρ is defined which measure in what extent the second approximation of the error curve matches the original error curve. The following equation defines ρ as

$$\rho_n = \frac{E(\widetilde{w}_n) - E(\widetilde{w}_n + \alpha_n \tilde{p}_n)}{E(\widetilde{w}_n) - E_{2q}(\alpha_n \tilde{p}_n)} \tag{34}$$

Here $\rho_n$ measures how finely the second order approximated error equation $E_{2q}(\alpha_n \tilde{p}_n)$ matches to the $E(\widetilde{w}_n + \alpha_n \tilde{p}_n)$ . The above equation tells than the value of λ nearer to 1 means better the approximation. For faster convergence λ is scaled by the following equations.

$$\begin{aligned} \lambda_n &= \frac{1}{3}\lambda_n \qquad if\ \rho_n > .75 \\ \lambda_n &= \lambda_n + \frac{x_n(1-\rho_n)}{|p_n{}^2|} \qquad if\ \rho_n < .25 \end{aligned} \tag{35}$$

In our experiments the initial values was chosen as $\lambda_1 = 10^{-3}$ and $\sigma = 10^{-5}$ was taken and in later steps $\sigma_n = \frac{\sigma}{|\tilde{p}_n|^2}$ was assumed. Initially weights are chosen as random nos. between 0 and 1. At starting of training in first iteration $p_1 = r_1 = -E'(\widetilde{w}_1)$ is assumed. The algorithm may be summarized as below.

a) Initialization of parameters like σ,p,r in first iteration.
b) Calculation of second order parameters like s,x and σ
c) Check whether Hessian matrix H is positive definite or not
d) If false adjust the value of s by increasing λ and recalculate s again.
e) Calculate $\rho$ and readjust the value of λ
f) Calculate step size
g) If error > minimum error limit go to next iteration.
h) Accept the weight vector for test.

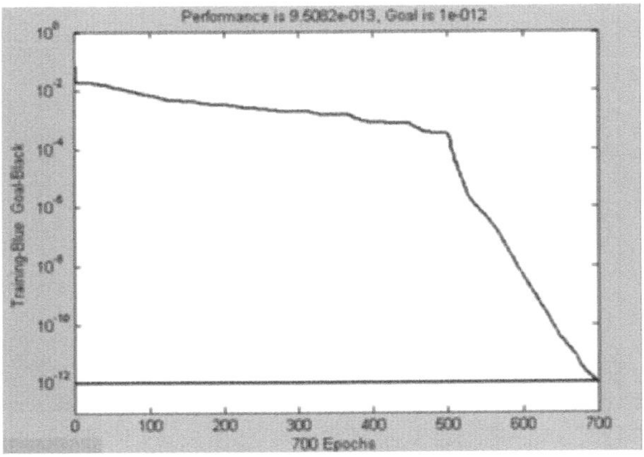


**Fig. 5.** epoch versus error curve

During the training operation the epoch versus error data is collected and the nature of the convergence is also noted. A run time error curve is plotted in figure 5 which shows that the nature of convergence is very fast and it takes just one or two min for completion of training and error limit reaches in $10^{-12}$ range within only 700 epoch.

## 5 Result and Discussion

The experiment for recognizing handwritten text was conducted in MATLAB environment where two types of images are used as input i.e. train image and test image. The purpose of the experiment was to recognize individual's handwriting by a neural network which is trained to identify the patterns of handwriting of the same person. Here both the train and test script was written by same person. For this reason this experiment was not conducted and compared by any standard data base like CEDER or IAM data base of handwritten text. Two major features are highlighted in this experiment which differs it from all other researches [6],[13],[23] done earlier may be mentioned that is i) the superfast training speed ii) high recognition performance. Ten sets of handwritten capital and small letters of English alphabets are taken as training script and various handwritten texts written by same person are used as test script. A sample train and test scripts are shown in figure 6 and figure 7. All the characters are written in natural way over A4 sheet. In both training and testing stage common character level segmentation and feature extraction is done over the train and test script. Training characters are English alphabets so there is no concept of forming words but test script is a text or may be called as group of words without any special symbol like ;, : and ?. In case of test script the segmentation is done from text level to word level and later from word level to character level. The start and end locations of each word in the text are stored. After recognition of all the characters in test script the computer printed words are regrouped by the NN output characters using the start and end location information of each words. Character level and Word level efficiency may be defined as under.

*Character Level Efficiency = (No of characters recognized correctly)/(Total no of characters in the script)*

*Word Level Efficiency = (No of words recognized correctly)/(Total no of words in the script)*

One character misclassification in a word decreases the word level recognition performance. But it gets easily be improved if the same word is checked through a

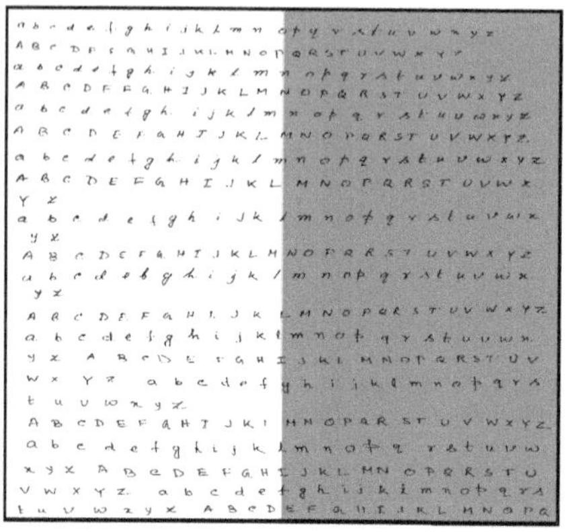

**Fig. 6.** Train script

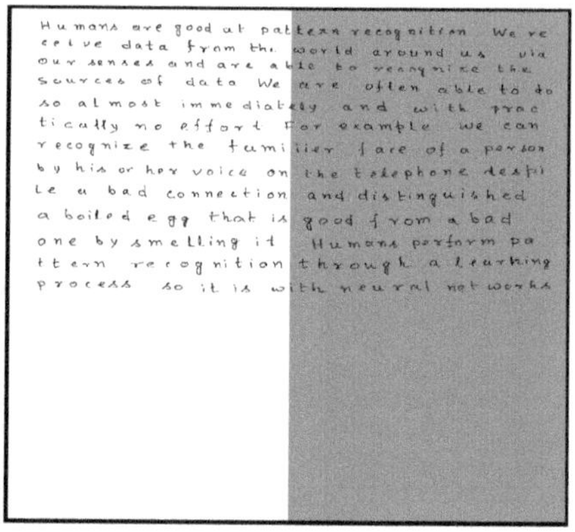

**Fig. 7.** Test script

**Table 1.** Comparison between different training algorithms

| No of Features/ character | NN structure | Training algorithm | No of epochs | Train efficiency | Error limit |
|---|---|---|---|---|---|
| 88 | 88,120, 52 | SCG | 700 | 100% | $10^{-12}$ |
| 88 | 88,120, 52 | Powell-Beale | 1480 | 100% | $10^{-12}$ |
| 88 | 88,120, 52 | Fleture Reeves | 5658 | 100% | $10^{-12}$ |

dictionary for validity of the word. If the word exists in the dictionary it returns the same word otherwise nearest word of same length is returned. For implementing this one dictionary is formed in MATLAB as an m-file which contains more than two thousand common words taken from a standard oxford dictionary. In this paper various algorithms were used for training such as a) Scaled Conjugate direction Method b) Powell Beale method and c) Fleture-Reeves method among the three algorithm Scaled conjugate algorithm shown the best training speed. A comparative statement is described in table.1.the above table shows that total 88 nos. of features were taken and NN structure indicates 52 nos. of output neurons for 52 characters (a-z and A-Z) and the result was obtained for 120 nos. of hidden neurons. Both the three algorithms provide good convergence of the error curve but Scaled Conjugate Gradient algorithm provides fastest training speed and the training completes within 700 epoch which takes less than one minute for reaching 100 percent train efficiency. The neural network trained with ten sets of handwritten alphabets that is total 520 nos. of characters when asked to make response to test feature matrix its response is really satisfactory only a few misclassification arises for very similar type of characters and for very bad handwritings. A character level recognition performance is shown in table 2 which shows that using double hidden layer with 150 and 100 hidden neurons we get maximum 83.60 percent case sensitive character level recognition and a

**Table 2.** Character level classification performance

| NN structure | MSE obtained | Train efficiency % | Test efficiency (Character Level) % | |
|---|---|---|---|---|
| | | | Case sensitive | Case insensitive |
| 88,160,52 | $9.63x10^{-13}$ | 100 | 81.43 | 89.17 |
| 88,200,52 | $9.75x10^{-13}$ | 100 | 82.64 | 89.42 |
| 88,250,52 | $9.32x10^{-13}$ | 100 | 79.25 | 88.70 |
| 88,120,80,52 | $9.44x10^{-13}$ | 100 | 74.17 | 87.24 |
| 88,150,100,52 | $9.69x10^{-13}$ | 100 | 83.60 | 92.32 |
| 88,200,100,52 | $9.98x10^{-13}$ | 100 | 81.67 | 87.72 |

**Table 3.** Performance before and after lexicon matching

| NN Structure | Without Lexicon Matching | | After Lexicon Matching | |
|---|---|---|---|---|
| | Character Level (%) | Word Level (%) | Character Level (%) | Word Level (%) |
| 88,160,52 | 89.17 | 62.87 | 98.43 | 96.74 |
| 88,200,52 | 89.42 | 48.65 | 99.02 | 97.83 |
| 88,250,52 | 88.70 | 47.56 | 99.02 | 97.83 |
| 88,120,80,52 | 87.34 | 38.96 | 98.43 | 95.65 |
| 88,150,100,52 | 92.32 | 66.13 | 98.43 | 95.65 |
| 88,200,100,52 | 89.72 | 47.56 | 99.02 | 98.91 |

maximum 92.32 percent case insensitive character level recognition performance is achieved. Though the degree of recognition is good but it may further be improved by lexicon matching technique which was discussed earlier. The word label performance after lexicon matching is shown in table 3. This shows that the case insensitive performance shown in table 3 improves by maximum 99.02 percent and because some of the word which contain lesser misclassified characters get corrected by the lexicon matching technique which increases word level recognition and when word get corrected the misclassified characters are also changes to its correct version. As a whole both the word and character level performance increases. In this experiment maximum word level and character level recognition performance achieved as 98.91 and 99.02 percent respectively. Though the result of the experiment was very well and most of the words and characters were recognized correctly except a few misclassifications that were found during experiment which may be shown in table 4.

**Table 4.** General misclassifications

| Original | Recognized | Original | Recognized |
|---|---|---|---|
| H | M ,n | n | m |
| r | v, Y | Y | r,y |
| e | c,l,C | y | r,Y |
| a | Q, u, l ,G | c | C |
| f | t | C | c |
| s | a,b, | O | o |
| l | x | o | O,b,D |
| I | j | j | i |
| b | o | t | f,F,q |

## 6 Conclusion

An efficient approach using a second order training algorithm **is** presented in this paper for recognition of handwritten alphabet based text. When no. of features are higher and no. of output classes are 52 all first order training algorithms basically fails or generates very poor result in training. Basic reason is the slower convergence of error curve and proper second order training algorithm become suitable replacement of those algorithms. Using this algorithm Hessian matrix of error equation always remain positive definite and in every iteration the error curve converges in a faster way. In this paper comparison between several second order training algorithms has been shown and it was found experimentally that Scaled Conjugate Gradient algorithms works with fastest speed and recognition performance is also excellent. The experiment was done using a Core 2 Duo (1.73 GHz) processor, 1 GB RAM, windows XP platform and MATLAB software. Training part was very fast and most of the time it completed within one or two minutes. But regarding the complexity of the test script, scripts characters are simple, easy to understand by human eye. Test script contains some inter-line horizontal space and some inter-word and inter-character space also. This pattern is not always available in natural handwriting. So it needs more experimental effort for faithful conversion from difficult handwriting to text conversion. However, this approach may be a true guideline for future research for giving computer an intelligence which a human being applies everyday and at every moment.

## References


1. Cho, S.-B.: Neural-Network Classifiers for Recognizing Totally Unconstrained Handwritten Numerals. IEEE Trans. on Neural Networks 8, 43–53 (1997)
2. Verma, B.: A Contour Code Feature Based Segmentation For Handwriting Recognition. In: 7th IAPR International conference on Document Analysis and Recognition, ICDAR 2003, pp. 1203–1207 (2003)
3. Strathy, N.W., Suen, C.Y., Krzyzak, A.: Segmentation of Handwritten Digits using Contour Features. In: ICDAR 1993, pp. 577–580 (2003)
4. Casey, R.G., Lecolinet, E.: A Survey of Methods and Strategies in Character Segmentation. IEEE Trans. Pattern Analysis and Machine Intelligence 18, 690–706 (1996)
5. Fletcher, R., Reeves, C.M.: Function minimization by conjugate gradients. The Computer Journal 7, 149–153 (1964)
6. Gorgevik, D., Cakmakov, D.: An Efficient Three-Stage Classifier for Handwritten Digit Recognition. In: ICPR, vol. 4, pp. 507–510 (2004)
7. Fink, G.A., Plotz, T.: On Appearance-Based feature Extraction Methods for Writer-Independent Handwritten Text Recognition. In: Proceedings of the 2005 Eight International Conference on Document Analysis and Recognition (ICDAR 2005), pp. 1520–5263 (2005)
8. Dunn, C.E., Wang, P.S.P.: Character Segmentation Techniques for Handwritten Text A Survey. In: Proceedings of the IInd International Conference on Pattern Recognition, The Hague, The Netherlands, pp. 577–580 (1992)
9. Zimmermann, M., Bunke, H.: Optimizing the Integration of a Statistical Language Model in HMM based Offline Handwritten Text Recognition. In: Proceedings of the 17th International Conference on Pattern Recognition (ICPR 2004), pp. 1051–4651 (2004)
10. Hestenes, M.: Conjugate Direction Methods In Optimization. Springer, New York (1980)
11. Lee, S.-W.: Off-Line Recognition of Totally Unconstrained Handwritten Numerals Using Multilayer Cluster Neural Network. IEEE Trans. on Pattern Analysis and Machine Intelligence 18, 648–652 (1996)
12. Gader, P.D., Mohamed, M., Chiang, J.-H.: Handwritten Word Recognition with Character and Inter-Character Neural Networks. IEEE Transactions on Systems, Man, and Cybernetics-Part B:Cybernetics 27, 158–164 (1997)
13. Blumenstein, Verma, B.: Neural–based solutions for the segmentation and recognition of difficult handwritten words from a benchmark database. In: Proc. 5th International Conference on Document Analysis and Recognition, Bangalore, India, pp. 281–284 (1999)
14. Chiang, J.-H.: A Hybrid Neural Model in Handwritten Word Recognition. Neural Networks 1I, 337–346 (1998)
15. Haykin, S.: Neural Networks A comprehensive Foundation, 2nd edn.
16. Duda, R.O., Hart, P.E., Stock, D.G.: Pattern classification, 2nd edn.
17. Pratt, W.K.: Digital Image Processing, 3rd edn.
18. Rajeshkaran, S., Vijayalakshmi Pai, G.A.: Neural Networks, Fuzzy Logic, and Genetic Algorithms Synthesis and Applications. Eastern Economy
19. Khoda, K.M., Liu, Y., Storey, C.: Generalized Polak –Ribiere Algorithm. Journal of Optimization Theory and Application 75(2) (November 1992)

20. Verma, B., Lee, H.: A Segmentation based Adaptive Approach for Cursive Hand written Text Recognition Neural Networks. In: International Joint Conference on IJCNN 2007, August 12-17, pp. 2212–2216 (2007), doi:10.1109/IJCNN.2007.4371301
21. Fletcher, R.: Practical Methods of Optimization. John Wiley & Sons, New York (1975)
22. Moller, M.F.: A Scaled Conjugate Gradient Algorithm For Fast Supervised Learning. Neural Networks 6, 525–533 (1993)
23. Blumenstein, M., Verma, B.: Neural-based Solutions for the Segmentation and Recognition of Difficult Handwritten Words from a Benchmark Database. In: Proc. 5th International Conference on Document Analysis and Recognition, Bangalore, India, pp. 281–284 (1999)
24. Dai, Y.H., Yuan, Y.: Convergence properties of the Beale-Powell restart algorithm. Sci. China Ser. A 41, 1142–1150 (1998)
25. Pal, U., Sharma, N., Wakabayashi, T., Kimura, F.: Off-line handwritten character recognition of Devanagari script. In: Proceedings of 9th International Conference on Document Analysis and Recognition, vol. 1, pp. 496–500 (2007)

# A Linear Framework for Identification of Ship and Its Velocity Estimation Using SAR Images

P. Subashini and M. Krishnaveni

Department of Computer Science,
Avinashilingam Deemed University for Women, Coimbatore
TamilNadu, India
{mail.p.subashini,krishnaveni.rd}@gmail.com

**Abstract.** Satellite-based Synthetic Aperture Radar (SAR) provides a powerful vessel surveillance capability in front of time consuming traditional reconnaissance methods. Nevertheless, due to the presence of speckle and to the reduced dimensions of the targets compared to the sensor resolution, the automatic interpretation of SAR images is often complicated even though vessels undetected are sometimes visible by eye. Therefore, the main difficulties of automatic ship detection will be reviewed and the main drawbacks will be identified here. As a result of this preliminary analysis, the application of transformations is found to be a useful mathematical object for this purpose. Hence based on the capabilities of the image analysis, the objective is achieved in such a way that it provides better understanding and further exploitation of satellite high resolution images. More specifically, a novel method for ship detection based on multiscale tools is proposed, in which SAR images are over here.

**Keywords:** Ship detection, velocity, transformation, segmentation, SAR images.

## 1 Introduction

In navy radar applications the presentation of the radar image traditionally has been the way for the radar operator to interpret the radar information manually. The large increase in calculation capacity of the image processing in modern radar systems had the effect that the detection and extraction of targets has been automated[9]. With powerful image processing techniques and algorithms, modern radar systems has the possibility to extract targets and their velocity from the surrounding background. A condition for this automatic detection is that the radar image should be relatively free from undesired signals. Such undesired signals can be rain clutter, sea clutter, measuring noise, landmasses, birds etc. Doppler filtering is often used to remove these undesired signals and extract the interesting part of the radar signal. By literature given by I-I Lin and Victor Khoo [2], revelation is that with modern image processing techniques be able to improve the radar image and investigate an automatic classification and presentation of the objects. The analysis of ship wakes in SAR imagery with specialized algorithms can provide significant information about a wake's associated vessel, including the approximate size and heading of the ship. The

D. Nagamalai, E. Renault, M. Dhanushkodi (Eds.): CCSEIT 2011, CCIS 204, pp. 211–224, 2011.

velocity of the vessel can be estimated by measuring the disarticulation of the ship relative to the height of the wake. Image processing algorithms such as the Fourier Transform and the Radon Transform allow the user to manipulate SAR images in a way that dramatically increases the chance of detecting ship wakes[14][11]. The paper is organized as follows: Section 2 deals with the structure of the framework for target detection. Section 3 deals with the preprocessing the image before identification. Section 4 converses the proposed algorithm for detecting the ship with high probability of detection. Section 5 and 6 converse the determination of target centre and wake representation through transformation Section 7 explains the experimental results and findings of the target detection and estimation of vessel velocity. The paper concludes with the possible improvements and the future work of the system.

## 2 Structure of the Framework for Target Detection

A digital image is built up with a number of elements, called pixels, usually in matrix form. The matrix size corresponds to the image size. Other forms of image representation are chains and graphs. In radar terminology, the pixels in a radar image are named radar bins or just bins. The pixel value corresponds to its brightness or color in the picture. The brightness can be used to interpret different pictures in different ways, depending of what the brightness mirrors.

In surface radar images, the brightness reflects the received signal strength. Images can be of different kinds. One way to sort images is to divide them into intensity and binary images. The binary image has only two levels, 0 and 1. Multispectral images, or intensity images, have more levels. The proposed work is given in figure 1. This proposed framework combines the existing and recent techniques of image processing techniques for detecting target. The output of the system delivers a Matlab-based application development platform intended for detection system. The design let the user to examine, devise, and evaluate algorithms and applications using SAR images. It offers standardization for the mentioned purpose that the user does not require comprehensive knowledge of the target hardware.

## 3 Preprocessing SAR Image

This is the first and lowest level operation to be done on images. The input and the output are both intensity images[1]. The main idea with the preprocessing is to suppress information in the image that is not relevant for its purpose or the following analysis of the image. The pre-processing techniques use the fact that neighboring pixels have essentially the same brightness. There are many different pre-processing methods developed and for different purposes. Interesting areas of pre-processing for this work is image conversion, noise suppression and edge detection. Image smoothing is a local pre-processing technique, which mainly is used for noise suppression. The method calculates new pixel values in a local neighborhood. In image enhancement, restoration is handled over the SAR image which otherwise called restoration.

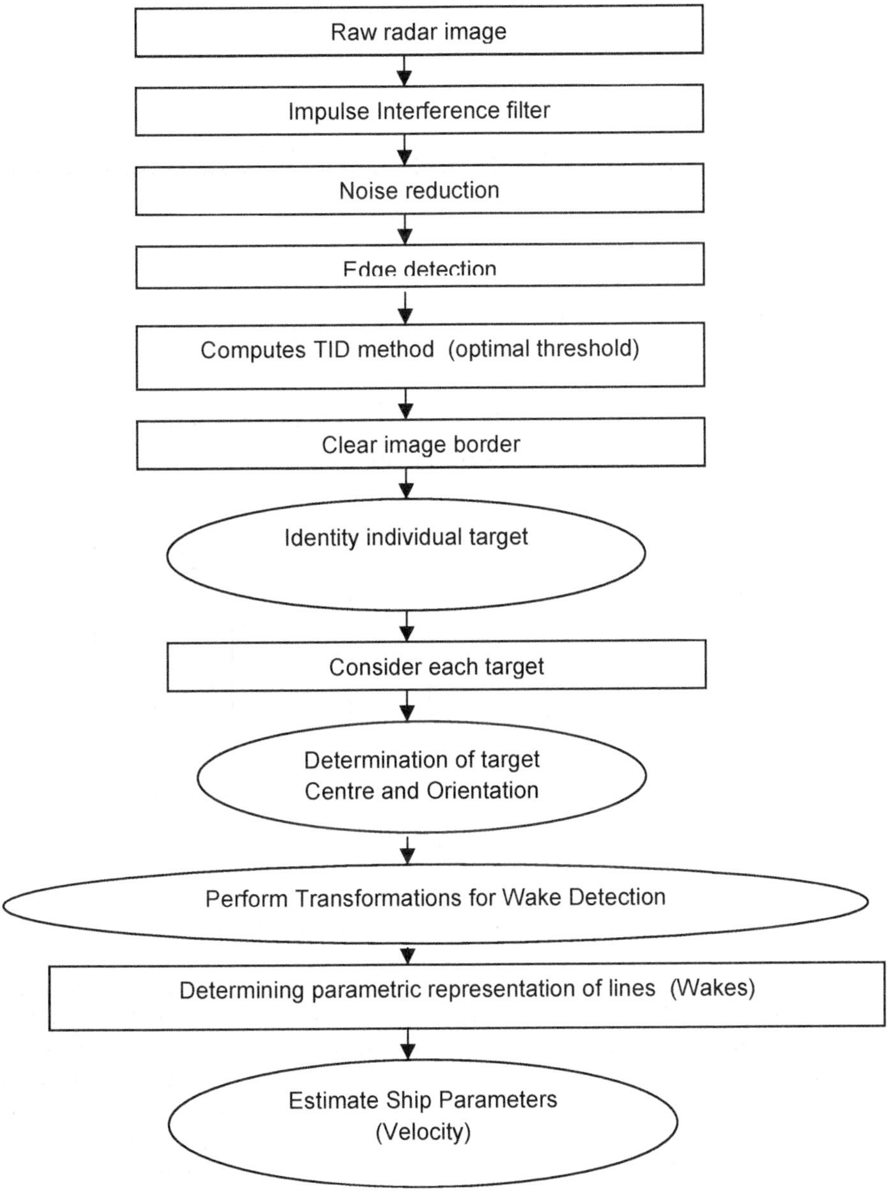


**Fig. 1.** A proposed Linear framework for ship and wake detection

### 3.1 Noise Removal

According to the characteristics of SAR images it contains speckle noise [7]which henceforth reduced using three filters namely (i) Inverse filter (ii) Wiener filter (iii) Lucy Richardson filter .On the experimental verification, the result obtained articulates that wiener filter gives better outcome for noise suppression.

**Inverse filter -**Inverse filter is a form of high pass filer. Inverse filtering responds very badly to any noise that is present in the image because noise tends to be high frequency. Works better for a noiseless case[5].

**Wiener filter** - The Wiener filtering is a linear estimation of the original image.It is also called regularized inverse filter.It removes the additive noise and inverts the blurring simultaneously.

**Lucy Richardson filter -** Lucy-Richardson denoising accounts on photon counting noise with a Poisson distribution. The Lucy-Richardson algorithm generates a restored image through an iterative method. Lucy-Richardson maximum likelihood algorithm addresses complex image restoration tasks.

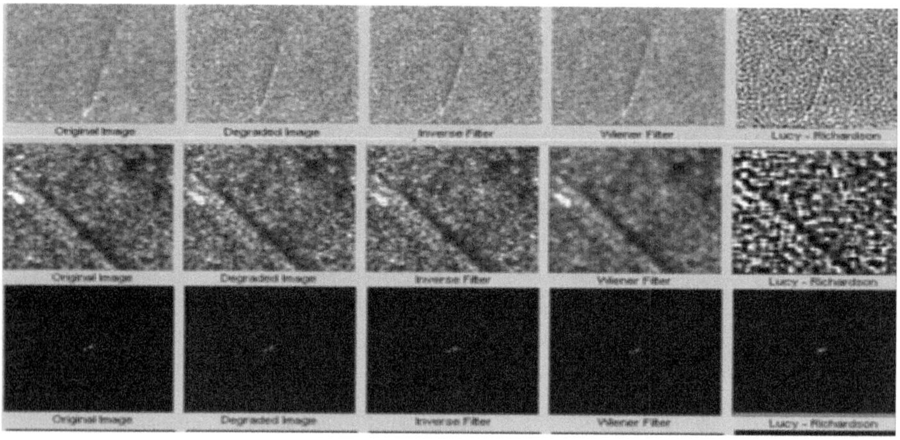


**Fig. 2.** Subjective evaluation for Noise filters

### 3.2 Edge Detection

**Roberts operator**

The Roberts operator performs a simple, quick to compute, 2-D spatial gradient measurement on an image[7]. The gradient magnitude is given by

$$|G| = \sqrt{Gx^2 + Gy^2} \tag{1}$$

And the angle of orientation of the edge giving rise to the spatial gradient (relative to the pixel grid orientation) is given by:

$$\theta = \arctan(Gy / Gx) - 3\pi / 4 \tag{2}$$

**Prewitt operator**

The result of using this edge operator is given by.

$$|G| = \max\left(|G_i| : i = 1 \quad to \quad n\right) \tag{3}$$

For each pixel the local edge gradient *magnitude* is estimated with the maximum response of all *kernels* at this pixel location.

**Laplacian of Gaussian (LoG)**

The Laplacian *L(x, y)* of an image with pixel intensity values *I(x, y)* is given by:

$$L(x.y) = \frac{\partial^2 I}{\partial x^2} + \frac{\partial^2 I}{\partial y^2} \tag{4}$$

Since the input image is represented as a set of discrete pixels, a discrete convolution kernel can approximate the second derivatives in the definition of the Laplacian.

**Table 1.** Advantages and disadvantages of Edge Detection Techniques

| Operator | Advantages | Disadvantages |
|---|---|---|
| **Classical (Prewitt, Roberts..)** | ➢ Simplicity,<br>➢ Detection of edges and their orientations | ➢ Sensitivity to noise,<br>➢ Inaccurate |
| **Canny** | ➢ Using probability for finding error rate<br>➢ Localization and response<br>➢ Improving signal to Noise ratio<br>➢ Better detection specially in noise conditions | ➢ Complex Computations<br>➢ False Zero crossing<br>➢ Time Consuming |
| **Sobel** | ➢ Higher weights are assigned to the pixels close to the candidate pixels.<br>➢ It masks for horizontal and vertical edge detection | ➢ Coupling between image pixels on a single scale |

**Canny operator**

The Canny method finds edges by looking for local maxima of the gradient of image. The method uses two thresholds, to detect strong and weak edges, and includes the weak edges in the output only if they are connected to strong edges. The edge gradient and direction can be determined using

$$G=\sqrt{G_x^{\ 2}+G_y^{\ 2}}$$

$$\theta=\arctan(\frac{G_y}{G_x}) \tag{5}$$

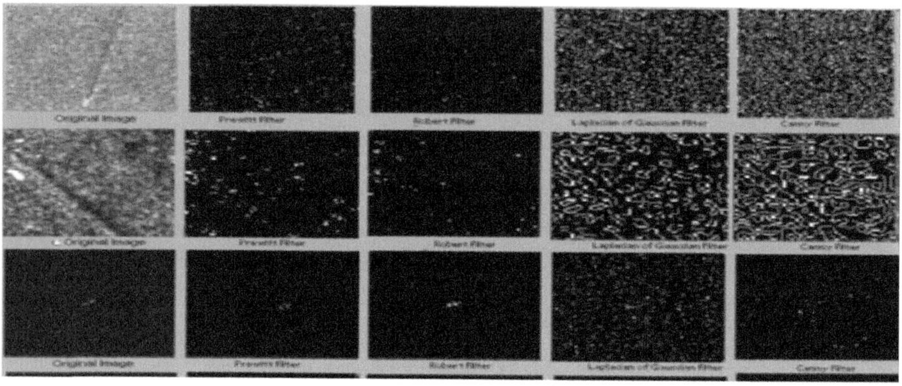

**Fig. 3.** Subjective evaluation for edge operators

## 4 Identifying Individual Target

Segmentation of target in SAR images is an important step in any ship detection system and many different approaches to the problem have been proposed so far[13]. The approaches of the paper are Adaptive thresholding (manual and automatic) and Distribution thresholding[3]. Two methods for thresholding are taken. The hard part of the thresholding is (1) Selecting suitable threshold for any input image (2) Deciding the proper class hierarchy for modular implementation (3) Implementing path traversal and nonmaximal suppression algorithms. In simple implementations, the segmentation is determined by a parameter *T* known as a threshold which is a result of analysis of the attribute *Y(x,y)*. In the simplest algorithm implementation, each pixel of the image is compared with the threshold *T*. If the pixel's value of the attribute *Y(x,y)* is lower than the threshold, the pixel is qualified to the background, otherwise the pixel is marked as object's pixel.

**Manual –**A binary output image one containing only 0's (background) and 1's (object),may be created by applying a threshold:

$$V_{out}(i,j)=\left\{\begin{matrix}0\\ \overline{1}\end{matrix}\right. if_{otherwise} v_{in}(i,j)<threshold \tag{6}$$

This can be extended to as many different classes as necessary by defining thresholds $t_1,t_2$ such that :

$$\begin{aligned} &0 \text{ if } 0 \le v_{in}(i, j) < t_1 \\ &1 \text{ if } t_1 \le v_{in}(i, j) < t_2 \\ &2 \text{ if } t_2 \le v_{in}(i, j) < t_3 \end{aligned} \tag{7}$$

The most common and reliable method of finding the thresholds is manual selection.

### 4.1 Iterative Threshold Selection Method

The methodology of the threshold value selection starts from segmenting the grayscale of the neighborhood into two parts using an initial threshold $T_0$ which is set to the neighborhood average lightness value $L_N$. Then the means of the gray values associated with the object's pixels $L_o$, and the background pixels $L_B$, are computed. The new threshold value $T_1$ is now set to the average of these two means as shown in the following equation:

$$T_1 = \frac{L_B + L_O}{2} \tag{8}$$

The process is repeated based upon a new threshold, until a threshold averages to a constant value. This situation corresponds to the following equation.

$$T_i = T_{i+1} \tag{9}$$

The size of the neighborhood is determined by a 3x3 size mask.

### 4.2 Threshold Intensity Distribution Method

The segmentation of the reference image is designed to group pixels of similar gray-levels[4]. Since the sea surface and the ship targets are usually poorly textured, the gray-level segmentation is likely to group pixels belonging to objects or background. If the objects and background occupy different ranges of gray levels, then it can mark the object pixels by a process called thresholding[6].

In intensity distribution B (i,j) is a binary image (pixel are either 0 or 1) created by thresholding F(i,j)

$$\begin{aligned} &B(i,j) = 1 \text{ if } F(i,j) < t \\ &B(i,j) = 0 \text{ if } F(I,j) >= t \end{aligned} \tag{10}$$

It is assumed that the 1's are the object pixels and the 0's are the background pixels.

The Histogram (h) - gray level frequency distribution of the gray level image F.

$h_F(g)$ = number of pixels in F whose gray level is g
$H_F(g)$ =number of pixels in F whose gray level is <=g

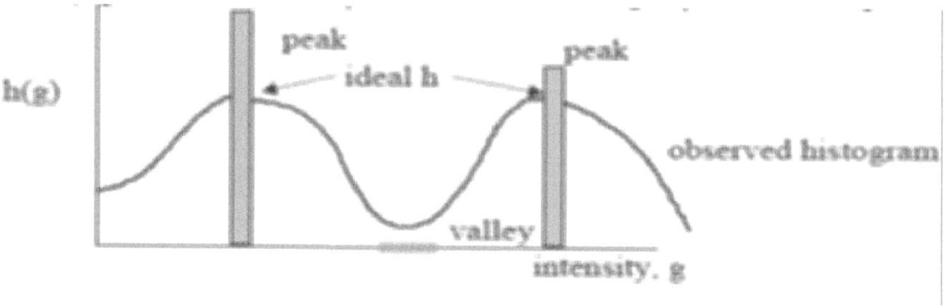


**Fig. 4.** Intensity distribution

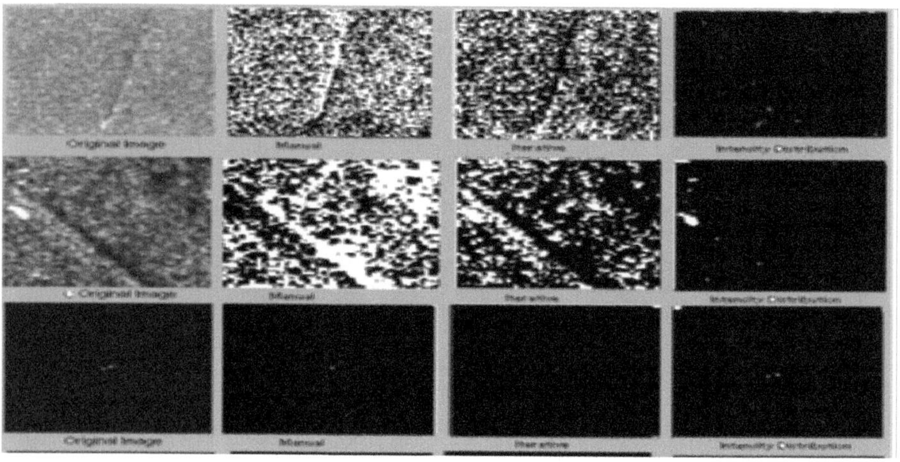

**Fig. 5.** Subjective evaluation of Segmentation methods (a) Original image (b) iterative method (c) intensity distribution method

This method is a probabilistic method that makes parametric assumptions about object and background intensity distributions and then derives "optimal" thresholds.

# 5 Determine Target Centre and Finding Velocity Using Wake Representation

*(i) Two features [10] of the target is obtained using property method*

Centroid'-- 1-by-n dimensions (L) vector; the center of mass of the region. The first element of Centroid is the horizontal coordinate (or x-coordinate) of the center of mass, and the second element is the vertical coordinate (or y-coordinate). All other elements of Centroid are in order of dimension.

Orientation' -- Scalar; the angle (in degrees) between the x-axis and the major axis of the ellipse that has the same second-moments as the region. This property is supported only for 2-D input label matrices.

*(ii) Determine parametric representation of wake*

The angle f is calculated as $\phi = \tan^{-1}(m)$, and a = f +90° . The Radon transform is performed to all ranges of a and r . However, in the case of wake detection, it is not necessary to apply the Radon transform to all possible a since turbulent wake is parallel to the ship[2][12]. Nevertheless, to account for possible errors, a range of a is used. This range is chosen as ± 20° from a with interval every 5° , i.e. the range of a used is í a -20° , a -15° , a -10° , a -5° , a , a +5° , a +10° ,a +15° ,a +20° ý . Similarly, it is not necessary to use the full range of r since ship can only travel at some maximum speed and thus both the azimuth shift D and r are limited. After the transformation, the maximum and minimum of all the g(a ,r ) are calculated. The maximum will correspond to a bright turbulent wake while the minimum will be associated with a dark turbulent wake and the wake position in terms of (a ,r ) can be identified.

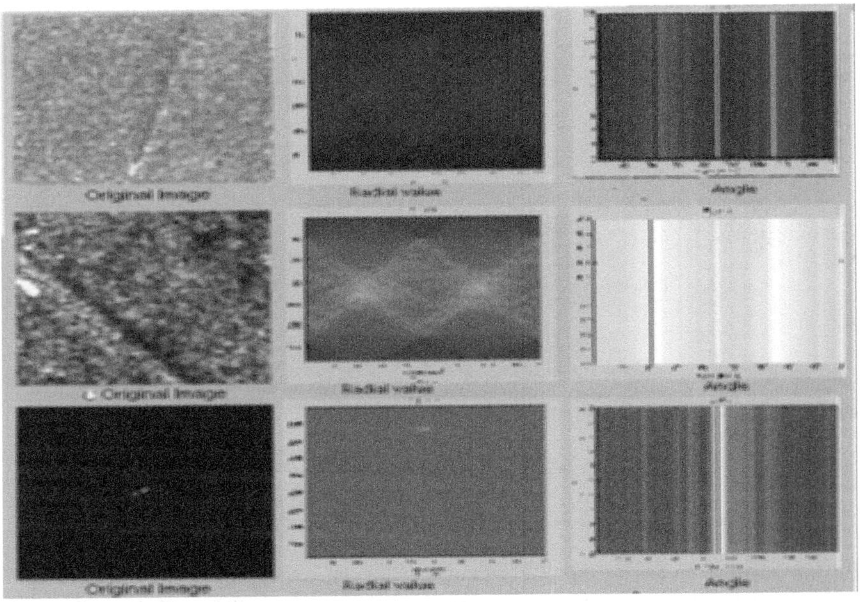


**Fig. 6.** Wake representation using radon transform

## 6 Performance Evaluation

The measurement of the image quality is divided into two main categories, subjective and objective. The subjective category is a measurement of how a human observe and interpret the image. The objective category is a technical measurement and the image is often compared with a known reference image[8]. The image used for image processing is in the format of .PNG and the size of the image is 256 x 256. It is necessary to find out the best method that is suitable of estimation. The performance evaluation of the adopted operation is quantified by following assessment.

### 6.1 Noise Removal

**Peak to signal noise ratio (PSNR)**

$$PSNR = 10\log_{10}\left(\frac{255^2}{MSE}\right) \tag{11}$$

Larger the PSNR value [40] the performance is good and smaller the MSE value the error rate is low.

**Mean Square Error (MSE)**

MSE is calculated using formula (1):

$$MSE = \frac{1}{MN}\sum_{i=1}^{M}\sum_{i=1}^{N}\left[g(i,j) - f(i,j)\right]^2 \tag{12}$$

where, $M$ and $N$ are the total number of pixels in the horizontal and the vertical dimensions of image. $g$ denotes the Noise image and $f$ denotes the filtered image.

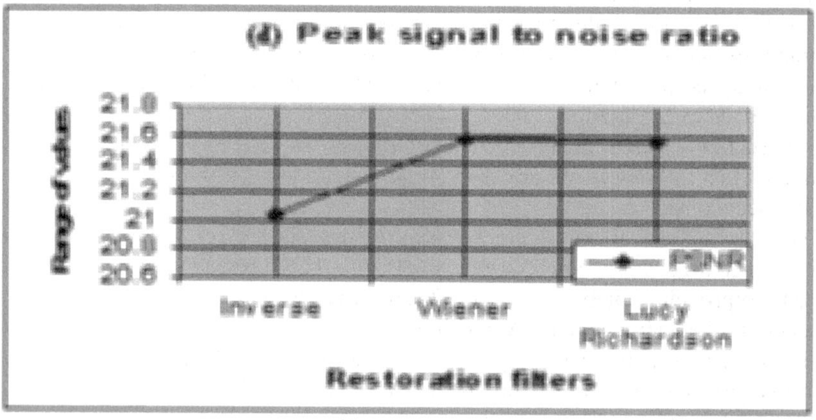


**Fig. 7.** Comparison based on PSNR

### 6.2 Edge Detection

The overall image sharpness is measured by computing the average difference between pixel pairs at all locations within the input image. The pixel subtraction operator takes two images as input and produces as output a third image whose pixel values are simply those of the first image minus the corresponding pixel values from the second image. If the operator computes absolute differences between the two input images then the formula is given in the Eq:

$$Q = \left|P_1(i,j) - P_2(i,j)\right| \tag{13}$$

where Q is the absolute difference and the $P_1$ and $P_2$ are the pixel of original image and filtered image. The absolute difference is calculated between the image in memory and the current image.

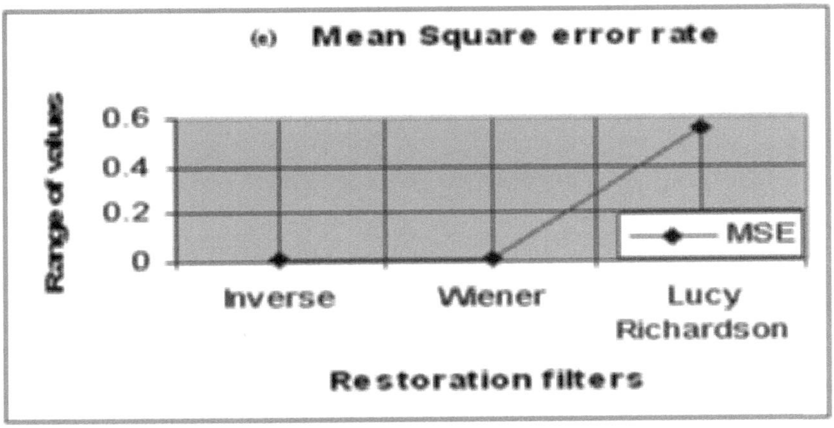


**Fig. 8.** Comparison based on MSE

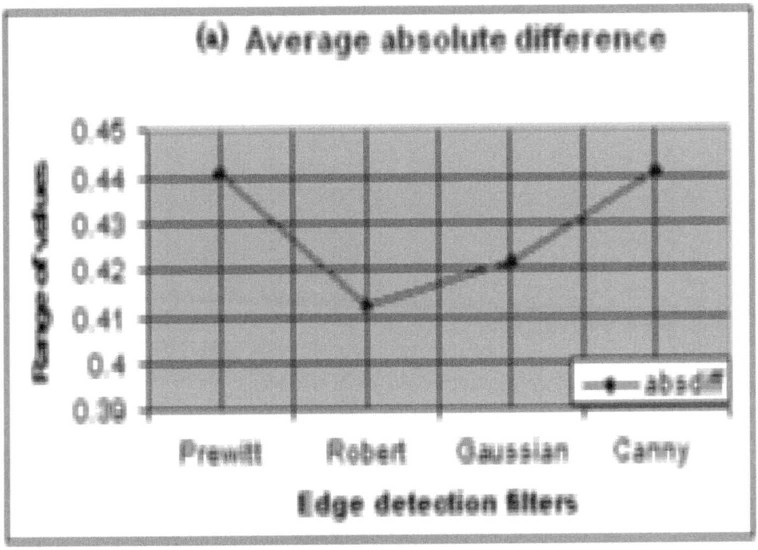


**Fig. 9.** Comparison based on AAD

### 6.3 Segmentation

**Structural content.** A structural content is used to find mask structural defects and noise, but are prone to periodic artifacts. This examines to classify image regions hierarchically based on the level of structural content.

$$SC = \sum_{j=1}^{M}\sum_{k=1}^{N} x_{j,k^2} 2/ \sum_{j=1}^{M}\sum_{k=1}^{N} x'_{j,k^2} 2/ \tag{14}$$

Structural content of a segmented image is calculated by summation of the original image by the division of the segmented image.

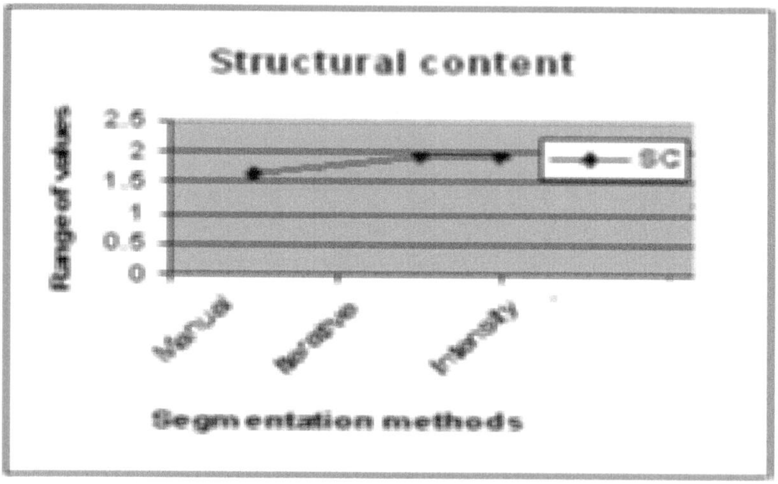


**Fig. 10.** Comparison based on PSNR

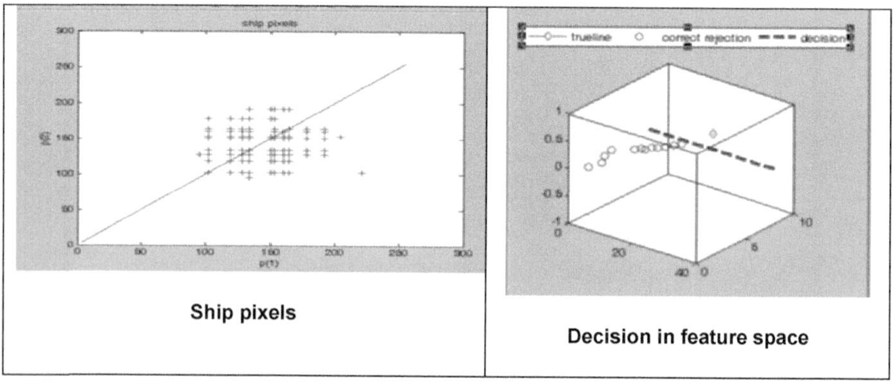


**Fig. 11.** performance of target detection

### 6.4 Estimating the Velocity of the Target

The velocity is estimated with the target centre and the wakes angle values and the calculation in given in figure 12. In figure 12, the SAR image taken indicates the speed of the vessel as 12.5 m/s. As per the image analysis the estimated value from the experiment is equal to the insitu value of the SAR image.

| | **Image for Analysis** |
|---|---|
| <br>Ship displaced from wake indicate speed (left with a speed of 6 m/s, and the **right with 12.5 m/s** ERS-2 03 APR-1996 03:29:28 Hr descending mode,incident angle 23 lat:01.64 long 102.69 | 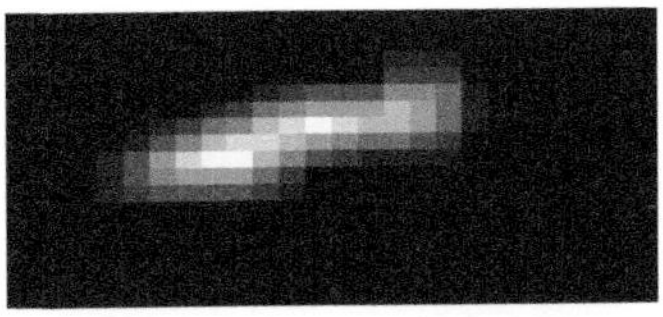<br>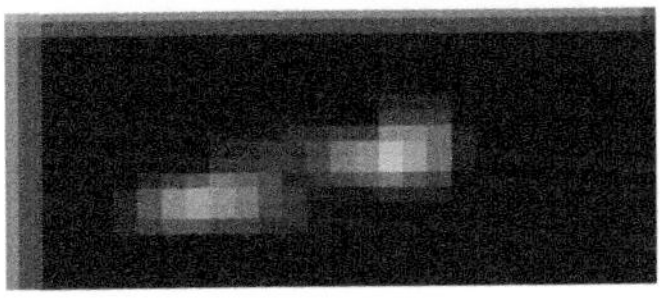 |
| 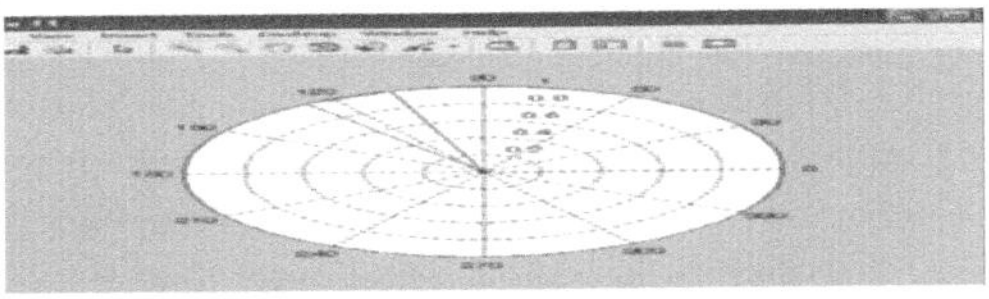<br>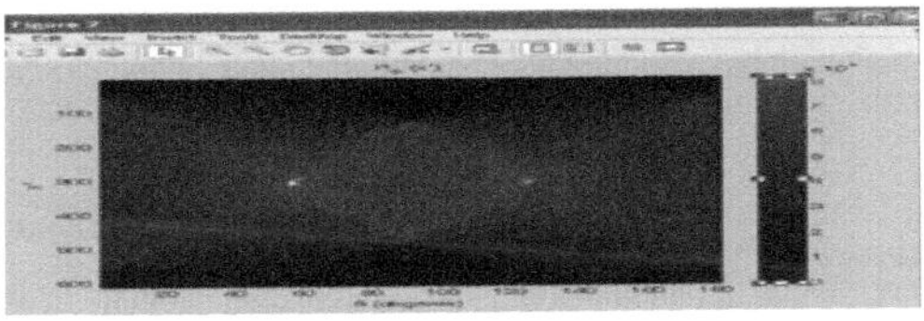 | **Calculated velocity**<br>**Centroid: [12.1389 10.6667]**<br>**Orientation: 21.0129**<br>$\upsilon = \frac{-1}{44}\left(\frac{\rho}{\lvert\cos\phi\rvert\sin\alpha}\right)$<br>**velocity = 12.31527 m/s** |

**Fig. 12.** Estimation of ship velocity

## 7 Conclusion

In this paper, individual object detection problem is analyzed for different SAR images. At the first glance, the problem is separated into two classes: identifying the target and determining the wakes for finding velocity using SAR images. The solutions proposed in preprocessing are: enhancement, noise removal, edge detection. The noise elimination method gives promising results in the controlled environments. The reliability of the edge detection process turns out to be the most important step in this method. The best results can be obtained, when "canny method" is used for background extraction. Thresholding and morphological operations steps are also very important for binarization and localization of the detection results. Automatic threshold selection is also used for segmentation. The experiments show that adaptive selection of the threshold for each image pair gives more promising results than a global threshold selection. Wake detection is handled with all the conventional transform method and the radon transform gives better solution for finding the velocity of the ship. The results of this detection scheme are reliable and satisfactory while the computation time is quite high compared to background elimination method. In the validation sequences, the detection problem is much more integrated with new algorithms to bring out optimal output. However, the segmentation process is very critical in this approach and it may require manual interaction for optimization. Therefore, more consistent unification approaches, which include temporal analysis, is required and is left as a future work.

## References

[1] Tekalp, A.M., Parbolic, G.: Digital Restoration of Images scanned from photographic paper. Journal of Electronic Imaging 2(c), 19–27 (1993)

[2] Lin, I.-I., Khoo, V.: Computer-based algorithm for ship detection from ERS SAR imagery. ESA.Earthonline.net

[3] Kittler, J., Illingworth, J.: Threshold selection based on a simple image statistic. CVGIP (30), 125–147 (1985)

[4] Otsu, N.: A thresholding selection method from graylevel histogram. IEEE Transactions on Systems, Man, and Cybernetics (9), 62–66 (1979)

[5] Hu, J., Yan, H., Hu, X.: Removal of impulse noise in highly corrupted digital images using a relaxation algorithm. Optical Engineering 36(03), 849–856 (1997); Proc SPIE

[6] Tsai, D.M., Chen, Y.H.: A fast histogram-clustering approach for multilevel thresholding. Pattern Recognition Letters 13(4), 245–252 (1992)

[7] Robinson, G.: Edge Detection by Compass Gradient Masks. Computer Graphic Image Processing 6, 492–501 (1997)

[8] Wang, Z., Bovik, A.C.: A universal image quality index. IEEE Signal Processing Letters 9(3), 81–84 (2002)

[9] Lopes, A., Nezry, E., Touzi, R.: Structure detection and statistical adaptive speckle filtering in SAR images. International Journal of Remote Sensing 14, 1735–1758 (1993)

[10] Mukherjee, A., Parui, S.K., Chaudhuri, B.B., Krishnan, R., Rao, K.K.: Detection of linear features in satellite imagery using robust estimation. In: Proceedings of the 12th International Conference on Pattern Recognition, Jerusalem, pp. 514–516 (1994)

[11] Krogagar, E., Czyz, Z.H.: Properties of the sphere, diplane, helix (target scattering matrix decomposition). In: Saillard, J., et al. (eds.) Proc. JIPR-3, Nantes, France, March 21-23, pp. 106–114 (1995)

[12] Cameron, W.L., Youssef, N.N., Leung, L.K.: Simulated polarimetric signatures of primitive geometrical shapes. IEEE Trans. on GRS 34(3), 793–803 (1996)

[13] Touzi, R., Charbonneau, F.: Characterization of target symmetric scattering using polarimetric SARs. IEEE Trans. on GRS 40(11), 2507–2516 (2002)

[14] Freeman, A., Durden, S.L.: A three-component scattering model for polarimetric SARdata. IEEE Trans. on GRS 36(3), 963–973 (1996)

# Creational Patterns to Create Decorator Pattern Objects in Web Application

Vijay K. Kerji

Department of Computer Science, PDA College of Engineering, Gulbarga, India
vkkerji@gmail.com

**Abstract.** Creational Design Patterns play an important role in making the software applications most reusable and adaptable to future changes in software requirements. Creational patterns abstract (encapsulate) how objects are created in the software system. They enable client code independent of object creation, object composition and how the objects are represented. Abstract Factory creational design pattern is one of the object creational patterns which help software system independent of how objects are created. In this paper, we proposed and applied creational patterns (Abstract Factory and Singleton) to achieve the above mentioned objectives. Creation of Decorator Pattern objects in DMS application is accomplished by Abstract Factory and Singleton patterns. This change enabled the application more adaptable to the changes (e.g. any addition of new user) without changing the client code and enhanced the reusability of the system with increased ease of system maintenance. Proposed changes are applied to the DMS application using .NET framework, C# and ASP.NET.

**Keywords:** Design Patterns, Abstract Factory, Singleton, Reusability, Web Application.

## 1 Introduction

*Design Patterns* refer to reusable or repeatable solutions that aim to solve similar design problems during development process. Various design patterns are available to support development process. Each pattern provides the main solution for particular problem. From developer's perspectives, design patterns produce a more maintainable design. While from user's perspectives, the solutions provided by design patterns will enhance the usability of web applications [1] [2].

Normally, developers use their own design for a given software requirement, which may be new each time they design for a similar requirement. Such method is time consuming and makes software maintenance more difficult. Adapting the software application to future requirements changes will be difficult if adequate care is not taken during design [3].

Adopting Design Pattern while designing web applications can promote reusability and consistency of the web application. Without the adoption of the design patterns or wrong selection of them will make the development of web application more complex and hard to be maintained. Thus knowledge in design patterns is essential in selecting

D. Nagamalai, E. Renault, M. Dhanushkodi (Eds.): CCSEIT 2011, CCIS 204, pp. 225–232, 2011.

the most appropriate patterns in the web application. Unfortunately, the ability of the developers in applying the design pattern is undetermined. The good practice of using design patterns is encouraged and efforts have to be taken to enhance the knowledge on design patterns [1].

Creational design patterns abstract the instantiation process. They help make a system independent of how its objects are created, composed, and represented. A class creational pattern uses inheritance to vary the class that's instantiated, where as an object creational pattern will delegate instantiation to other objects. Creational patterns become important as systems evolve to depend more on object composition than class inheritance. As that happens, emphasis shifts away from hard coding a fixed set of behaviors toward defining a smaller set of fundamental behaviors that can be composed into any number of more complex ones. Thus creating objects with particular behaviors requires more than simply instantiating class [4].

In case of web applications, it is sometimes necessary to add additional responsibilities to the web page dynamically and to cater to the changing requirements. Such additions should be transparent without affecting other objects. There can be possibility where in these responsibilities can be withdrawn. In such scenario, Decorator Pattern can be used to add the additional responsibilities to the default web page (whose some of the contents are always constant). For Example if the application has different users who can log into it (such as Admin, Program Manager and Stakeholder).Admin will have different user interface when compared to other users but some of the user interface elements may be similar across all users. In such scenario, we can use Decorator Pattern to decorate the changing user interface elements depending on the user type, keeping the remaining elements constant across all type of users [3].

In such applications, Decorator Pattern objects can be created using Abstract Factory creational pattern. In this work, concrete decorator objects are created with such Abstract Factory pattern without specifying their concrete classes. This encapsulates the object instantiation process from the client thus making the system more flexible to future changes in the requirements. Additional new users can be added to the system without applying changes to the client side code. Client side code is hidden from the type of current user instantiated by the Abstract Factory pattern thus making the system more reusable and easier to maintain. Since the application typically needs only one instance of a concrete factory in abstract factory pattern, we used Singleton design pattern to implement the concrete factory object.

We applied the required changes to the DMS application [3] to achieve the above mentioned objectives. The resulting application showed considerable improvement in performance when multiple users are added. It abstracts the client code from changes in concrete decorator object instantiation.

The rest of the paper is organized as follows. Section 2 describes Abstract Factory and Singleton Patterns and their applicability. Section 3 briefs about DMS web based application. Implementation and sample code is listed in section 4. Finally conclusion is presented in section 5.

## 2 Abstract Factory and Singleton Patterns

Abstract factory provide an interface for creating families of related or dependent objects without specifying their concrete classes. It helps to control the classes of objects that an application creates. Because a factory encapsulates the responsibility and the process of creating product objects, it isolates clients from implementation classes. Clients manipulate instances though their abstract interfaces. Product class names are isolated in the implementation of the concrete factory; they do not appear in client code [4].

Abstract Factory declares an interface for operations that create abstract product objects. Concrete Factory implements the operations to create concrete product objects. Abstract Product declares an interface for a type of product object. Concrete Product defines a product object to be created by the corresponding concrete factory. It implements the Abstract Product interface. Client uses only interfaces declared by Abstract factory and Abstract Product classes.

The factory completely abstracts the creation and initialization of the product from the client. This indirection enables the client to focus on its discrete role in the application without concerning itself with the details of how the product is created. Thus, as the product implementation changes over time, the client remains unchanged. The most important aspect of this pattern is the fact that the client is abstracted from both the type of product and the type of factory used to create the product. Furthermore, the entire factory along with the associated products it creates can be replaced in a wholesale fashion. Modifications can occur without any changes to the client[5].

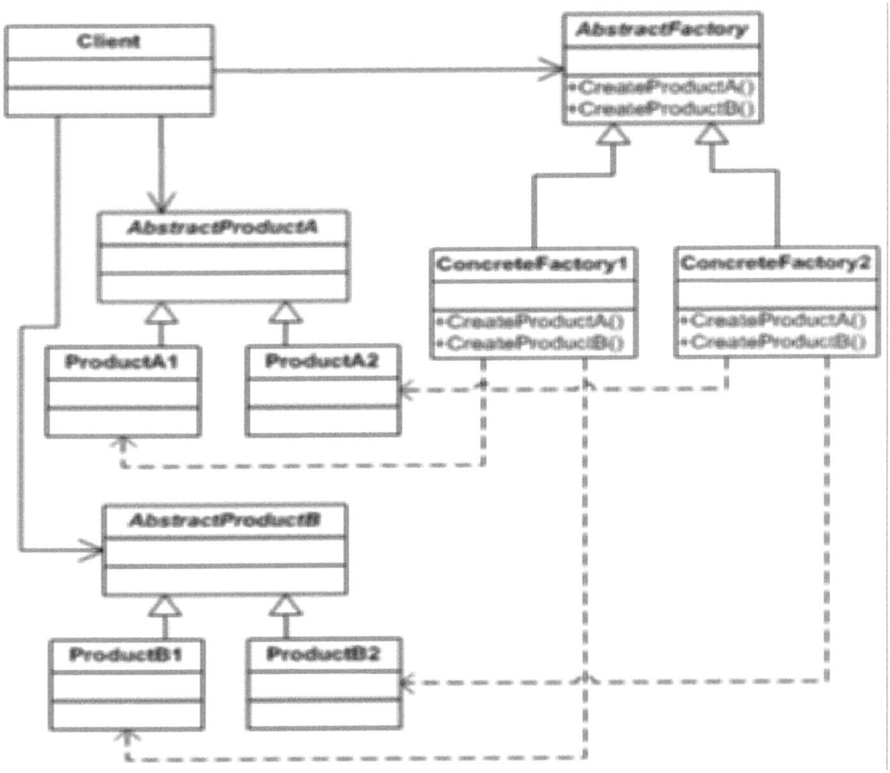


**Fig. 1.** Structure of Abstract Factory Pattern

In our work, Abstract Product is that of Decorator object of the DMS application. Concrete products are different types of users (Admin, Program Manager and Stake Holder). Essentially, Decorator Pattern objects are made as that of Abstract Product and its Concrete Product objects. Individual Concrete Factory objects are created for each of the user types (Concrete Factory1 for user type Admin, Concrete Factory2 for user type Program Manager and Concrete Factory3 for user type Stake Holder). In this design, Client is unaware of which concrete product and concrete factory classes are created since it access the required objects embedded in abstract class pointer. Hence the client code is detached from underlying concrete object creation and making the system more adaptable to future requirement changes and easier to maintain it.

Singleton Pattern ensures a class has only one instance and provides global point of access to it. Since there is only one instance of the Concrete Factory class, it is implemented using Singleton Pattern so that client can have access to it from well known access point.

## 3 DMS Application

Development Management System is a web based application which allows different types of user (Admin, Stake Holder and Program Manager) to login and operate the DMS application. User page has certain functionality similar across all type of users. Changed functionality is implemented using corresponding Decorator objects, using XML input data queried through XPath [3].

With reference to Decorator Pattern, DefaultUserView class is of the type ConcreteComponent which is sub classed from Component base class. Decorator class is sub classed from Component class which is a base class for Concrete Decorators. Concrete decorators can be Admin, Stakeholder or Program Manager decorator classes. Dynamic user page is decorated with any of the concrete decorators depending upon the type of user logged into the system. Also, the contents of the decorator are read from XML using X-path query [3].

In our work, changes are incorporated in Decorator Pattern object instantiation process. It is achieved through Abstract Factory pattern which abstract clients from how the objects are represented, composed and instantiated. Admin, Stake Holder and Program Manager Objects of Decorator design pattern are now become part of the Abstract Factory pattern. Since there can be only one instance of the Concrete Factory object for a particular logged in user, Singleton design pattern is used so that client can access this one and only instance for further usage.

## 4 Implementation and Sample Code

In this work, we applied changes to a web based application called Development Management System. This application allows different types of user (Admin, Stake Holder and Program Manager) to login and operate the DMS application. User page has certain functionality similar across all type of users. Changed functionality is implemented using corresponding Decorator objects, using XML input data queried through XPath [3].

AbstractFactory class(Listing1) is a Singleton Design Pattern class which creates the required Concrete Factory object depending the type of user logged into the system. Instance is a static function which returns the single instance of the concrete factory object. getInstance is a helper function which returns the appropriate concrete factory object depending the user type. User type can be read from either configuration file or registry. Abstract factory also has an important abstract function, CreateUser which will be implemented by the concrete factory objects to create the respective user type object.

```
abstract class AbstractFactory
{
  //one and only singleton instance
  private static AbstractFactory instance = null;

      //function to access the single instance object
      public static AbstractFactory Instance()
      {
            if(instance == null)
            {
                  instance = getInstance();
            }
            return instance;
      }
      //helper function to create proper factory object
      Private static AbstractFactory getInstance()
      {
            string userType; //get the user type from
            //.ini file or registry
            //Create factory object depending on user
            //type
            Switch(userType)
            {
                  Case "Admin":
                        return  new  ConcreteFactory1();
                  Case "StakeHolder":
                        return new ConcreteFactory2();
                  Case "ProgramManager":
                        return new ConcreteFactory3();
            }

      }
      //abstract class & function to create decorator
      //object
            Public abstract Decorator CreateUser();
}
```

**Listing 1.** Abstract Factory Singleton class

ConcreteFactory1 class (Listing 2 ) is a sub class of AbstractFactory class which is responsible for creating AdminDecorator object corresponding to Admin user type. AdminDecorator takes DefaultUserView as an input argument to its constructor and AdminDecorator object is returned to the client which has called CreateUser function.

```
class ConcreteFactory1 : AbstractFactory
{
      Public override Decorator CreateUser()
      {
            //Default user view as an input to decorator
            DefaultUserView duv = new DefaultUserView();
            //create&return Admin Decorator object
            return new AdminDecorator(duv);
      }
}
```

**Listing 2.** ConcreteFactory1 class for Admin Decorator

ConcreteFactory2 class (Listing 3) is responsible for creating stake holder decorator object. ConcreteFactory is a subclass of AbstractFactory class and implements CreateUser function declared in AbstractFactory base class. Again the StakeHolderDecorator object is passed to the client as Decorator object pointer, without revealing the underlying object in it.

```
class ConcreteFactory2 : AbstractFactory
{
      Public override Decorator CreateUser()
      {
            DefaultUserView duv = new DefaultUserView();
            //create&return Stake Holder Decorator object
            return new StakeHolderDecorator(duv);
      }
}
```

**Listing 3.** ConcreteFactory2 class for Stake Holder Decorator

ConcreteFactory3 class (Listing 4) is responsible for creating ProgramManager Decorator object. ConcreteFactory is a subclass of AbstractFactory class and implements CreateUser function declared in AbstractFactory base class. Again the ProgramManagerDecorator object is passed to the client as Decorator object pointer, without revealing the underlying object in it.

AdminDecorator is a ConcreteDecorator (Listing 5) which adds the responsibility of rendering the Admin related XHTML to the web browser. AdminDecorator will copy the visual component object to its base class visual component member variable. It implements the Render function which in turn reads the XML file to read its required contents and develops the XHTML as per the XML data [3].

```
class ConcreteFactory3 : AbstractFactory
{
      Public override Decorator CreateUser()
      {
            DefaultUserView duv = new DefaultUserView();
            //create&return Program Mgr Decorator object
            return new ProgramManagerDecorator(duv);
      }
}
```

**Listing 4.** ConcreteFactory3 class for Program Manager Decorator

```
public class AdminDecorator : Decorator
{
      //constructor for Admin Decorator
      Public AdminDecorator(VisualComponent vc)
      {
            base.vc = vc;
      }
            ...
}
```

**Listing 5.** Admin Decorator class derived from Decorator

```
      AbstractFactory  af = AbstractFactory.Instance();
      Decorator dc = af.CreateUser();

      //Client unaware of the type of user
      string totalXHTML = dc.Render();
      //write the XHTML as string to the browser
      Response.Write(totalXHTML);
```

**Listing 6.** Client code to render the XHTML

Client Code (Listing 6) shows how the final XHTML contents are retrieved from Decorator object. Client code has been simplified and is now unaware of the user type logged into the system. It gets single instance of concrete AbstractFactory object and in turn created the proper Decorator Object which will render the XHTML code to the browser.

Above code snippet reveals us that the client code is hidden from the instantiation process of concrete decorator objects. AbstractFactory object creation is implemented with the Singleton design pattern to ensure that only one instance of the concrete factory object is created in the application to provide the global point of access. Hence the above method of implementation makes the maintenance of software easier and any changes in the user requirements can be handled without changing the client part of the code. It also improved the performance of the application when more number of users is added to the system.

## 5 Conclusion

Software Design Patterns are proven design methods which when used carefully in the implementation of software development will lead to several advantages. Reusability, Encapsulation and Ease of system maintenance are some of the benefits one can incorporate into software systems if these proven design patterns are used in the system. Creational patterns belong to that category of design pattern which will help in abstraction of object instantiation process. They help make a system independent of how objects are created, composed, and represented. Abstract Factory and Singleton patterns belong to creational pattern category, which are used in this work to encapsulate the Decorator Pattern object instantiation process. Concrete Decorator objects are created by respective Concrete Factory objects depending on the type of user logged into the system. This resulted in improved application performance when additional user types are added to the system and easier to maintain the application code for future requirement changes.

## References


1. Thung, P.L., Ng, C.J., Thung, S.J., Sulaiman, S.: Improving a Web Application Using Design Patterns: A Case Study
2. Pawan, V.: Web Application Design Patterns. Morgan Kaufmann, CA (2009)
3. Kerji, V.K.: Decorator Pattern with XML in web based application. In: ICNCS 2011, Kanyakumari, India (2011)
4. Gamma, E., Helm, R., Johnson, R., Vlissides, J.: Design Patterns Elements of Reusable Object Oriented Software. Addison Wesley, Reading
5. Matic, D., Butorac, D., Kegalj, H.: Data Access Architecture in object oriented application using Design Patterns

# Scalable Geographic Based Multicasting (SGBM) Protocol

H. Venkateswaran and A. John Prakash

Department of Computer Science, Anna University,
Chennai-600 025
venkatphd@gmail.com

**Abstract.** The Scalable Geographic Based Multicast (SGBM) protocol is designed for the Mobile Adhoc Networks (MANET), which overcomes the difficulties in group membership management, multicast packet delivery and maintenance of multicast structure over the dynamic network. Furthermore a zone is divided based on the threshold value, so that the packet delivery is maintained easily and the time spent on the data forwarding will be less. The performance of the protocol was studied by extensive simulations. The result depicts that the protocol we designed is more reliable, efficient and scalable to a large group size or a network size.

**Keywords:** Multicast routing, geographic multicast, mobile ad hoc network, scalable.

## 1 Introduction

Mobile ad hoc network (MANET) consists of many mobile hosts connected by wireless links. Each node operates not only as an end-system, but also as a router to forward packets. In MANET, the network topology may change frequently due to the nodes' movements [1]. A routing protocol is always needed to forward packets quickly to the destination and adapt to the topology changes. Multicasting allows one entity to communicate efficiently with multiple entities residing in a subset of the nodes in a network. Although multicasting can be simulated using multiple point-to-point messages, mechanisms that enable multidestination delivery using a single group address can provide transparency, greater efficiency, and concurrency [2].

Many multicasting protocols have been designed, but due the dynamic network topology changes the group membership management is not properly maintained, so that there is a delay in the multicast packet delivery.

To have an efficient multicast packet delivery a tree structure is designed. However it is difficult to maintain a tree structure in mobile adhoc network, and the tree connection is easy to break. So then came the mesh based protocol which is an alternative for tree based, where we have redundant path. The challenge in the mesh based protocol is to support reliable and scalable multicast in a MANET with these based schemes, as it is difficult to manage group membership, find and maintain multiple paths in a network topology changes [3].

D. Nagamalai, E. Renault, M. Dhanushkodi (Eds.): CCSEIT 2011, CCIS 204, pp. 233–243, 2011.

To overcome these problems, we have designed a protocol called SGBM (Scalable Geographic Based Multicasting) protocol which can scale to a large network size. Here we introduce a virtual tree structure which is different from other tree structure. The virtual tree structure can be formed during the multicast packet forwarding with the help of node positioning. It also introduces a zone based scheme to efficiently handle the group membership management. The zone is also formed virtually and the zone where the node is located is calculated based on the node position.

We discuss rest of the paper as follows: In section 2 we see the related work on MANET multicast protocol. Section 3 follows as the motivation, where the limitations in the related work are discussed. In section 4, the features of SGBM are discussed. In section 5, the SGBM architecture and group membership management is explained. In section 7, the preference analysis and finally in section 8, we conclude the paper.

## 2 Related Works

In earlier days the Adhoc multicast protocol was a tree based protocol (e.g. AMRoute). It creates a bidirectional shared multicast tree using unicast tunnels to provide connections between multicast group members. Each group has at least one logical core that is responsible for member and tree maintenance [4]. The mesh-based protocols (e.g., FGMP [5], Core-Assisted Mesh protocol [6], ODMRP [7]) are proposed to enhance the robustness by providing redundant paths between the source and destination pairs at the cost of higher forwarding overhead. The conventional tree based protocols consist of three components [8]:

1. Group membership management. The group membership changes frequently as each node may join or leave a multicast group randomly, and the management becomes harder for a large group.

2. Creation and maintenance of a tree- or mesh-based multicast structure. In these protocols, the structures are based on some non-geographic mechanisms, which make the tree-based structure not so robust, while the mesh-based ones achieve the robustness at the cost of inefficiently utilizing network resource. Also the nongeographic routing mechanisms prohibit these protocols from scaling to a large network size.

3. Multicast packet forwarding. The multicast packets are forwarded along the pre-built tree or mesh structure, but the pre-built paths are vulnerable to be broken over the dynamic topology, especially in a large network with potentially longer paths.

Beside these components there is a need for the geographic based multicast protocol to scale for the large network size and to maintain the group membership management. The geographic multicast protocols presented in [9], [1] and [10] need to put the information of all the group members into the packet header, which creates a lot of overhead for a large group, so they are only applicable for the small group case [8].

## 3 Motivation

In this we point out the limitations of the related works.

1. Difficult to maintain the tree structure.
2. Difficult to manage group membership.
3. Maintenance of multicast structure over the dynamic network topology for a large group size or network size.
4. Problem in periodic flooding.

In particular, AMRoute performed well under no mobility, but it suffered from loops and inefficient trees even for low mobility [4]. AMRIS was effective in a light traffic environment with no mobility. But its performance was susceptible to traffic load and mobility [11]. CAMP showed better performance when compared to tree protocols but with mobility, excessive control overhead caused congestion and collisions that resulted in performance degradation [6]. ODMRP was very effective and efficient in most of our simulation scenarios. The protocol showed a trend of rapidly increasing overhead as the number of senders increased [7]. These limitations motivated us to develop the protocol called the SGBM (Scalable Geographic Based Multicast) protocol.

## 4 Contribution

SGBM features are discussed in this section, The protocol can scale to a large network size, where the packet transmission takes place in a dynamic mobile environment. For the packet transmission, it uses several virtual architecture, where there is no need to maintain the state information. This protocol is different from other tree based protocol, where there is no need to maintain the tree structure explicitly. Here it creates the virtual tree structure and transmits the data's more efficiently. This protocol also handles the group membership management by a zone based scheme. The zone structure is also formed virtually. It also avoids the need of network periodic flooding of source information.

## 5 SGBM Architecture

The above figure gives the architecture of the SGBM, which is also similar to the RSGM. SGBM is made up of two-tier architecture. In the lower tier, the zone is built based on position information and a leader is elected on demand in the zone of the group members. The leader manages the group membership. At the upper tier, the leaders of the member zone report the source directly via a virtual reverse tree based path [3]. SGBM differs from RSGM only in the multicast packet delivery. It is described in detailed in section 7.

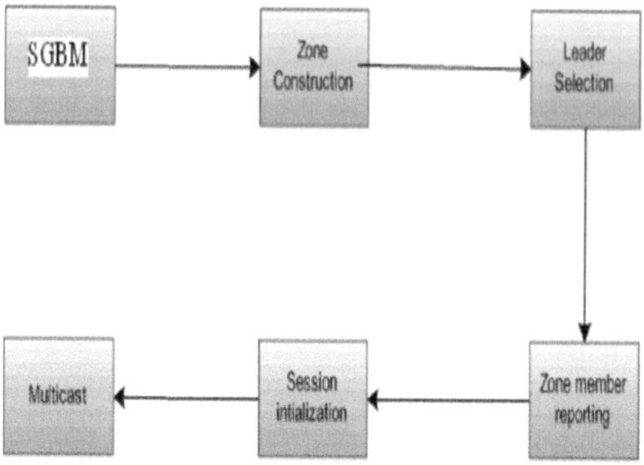


**Fig. 1.** SGBM Architecture

## 6 Group Management

The group membership management is managed at two tiers. The zone is built on the lower tier and a leader is elected for each zone. At the upper tier the leader, of the zone report the source directly via a virtual reverse tree path. A node joining or leaving the zone should be reported to the zone leader for the periodic membership change. When a node joins a zone it should be updated to the corresponding zone leader, if there is no leader in the zone, the node announces it as the leader and distributes the data.

## 7 Multicast Packet Delivery

The source takes care of the multicast packet delivery to each zone. Using optimization technique a threshold value is assigned to each zone, and number of nodes in the zone is based on the threshold value. If the nodes exceed the threshold value, then we split the zone into two halves, and checking for the condition that exceeds the threshold value. If so, again split the zone. Here the time taken to forward the data packets will be less. The zone leader distributes the data's within the zone. Since we split the zone, the packet delivery takes place quickly and more efficiently. Then the next hop is based on the geographic broadcast forwarding through a virtual tree based path where it reduces the time delay and forwarding bandwidth.

## 8 Performance Analysis

In this section, we study the performance of SGBM by simulations [3]; SGBM is implemented in the Glomosim [15]. The performance of the protocol is impacted on Packet Delivery Ratio, Control overhead, Average End-to-End delay, joining delay.

Packet Delivery Ratio (PDR) – The ratio of the number of packets received at the destination and number of packets sent by the source.

Control overhead – The ratio of packet delivery time and number of packets delivered. Average

End-to-End Delay – The ratio of end-to-end delay and number of packets received, measured in milliseconds. It includes the processing time and queuing time of packet in each node in the route.

Joining delay – The average time interval between a member joining a group and its first receiving of the data packet from that group. To obtain the joining delay, the simulations were rerun with the same settings except that all the members joined groups after the sources began sending data packets.

The performance of the protocol is impacted on zone size, multicast packet delivery ratio, time delay, forwarding bandwidth. The following graphs give the end to end path length delay, control overhead and packet delivery ratio in SGBM (i.e;RSGM IM).

**Table 1.** The parameters used in network simulation are listed below

| | |
|---|---|
| Channel Type | Channel/Wireless Channel |
| Radio-Propagation Model | Propogation/TwoRayGround |
| Network Interface Type | Phy/WirelessPhy |
| MAC Protocol Type | MAC/802.11 |
| Interface Queue Type | Queue/DropTail/PriQueue |
| Link Layer Type | LL |
| Radio Link Control | RLC |
| Routing Protocol | SGBM |
| No. of Nodes | 200 |
| Simulation Time | 120 Seconds |
| Terrain Range | 1000 M x 1000 M |
| Packet Size | 512 bytes |

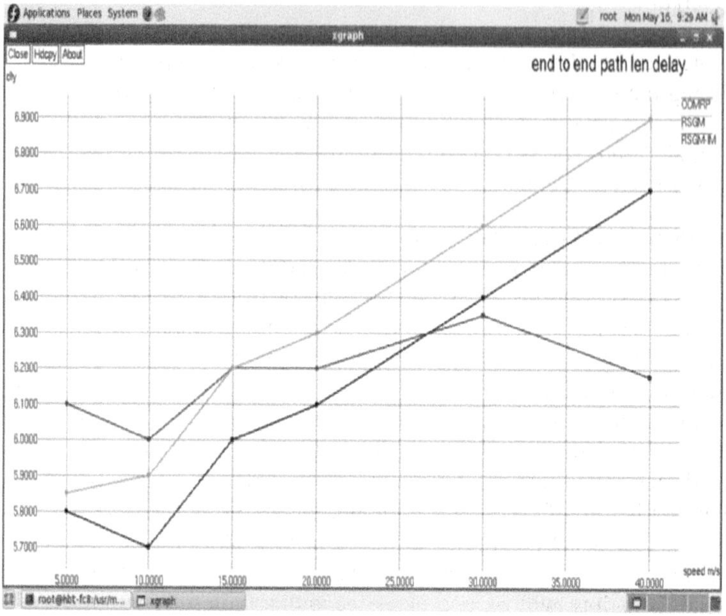


**Fig. 2.** SGBM End-to-End Path Length Delay

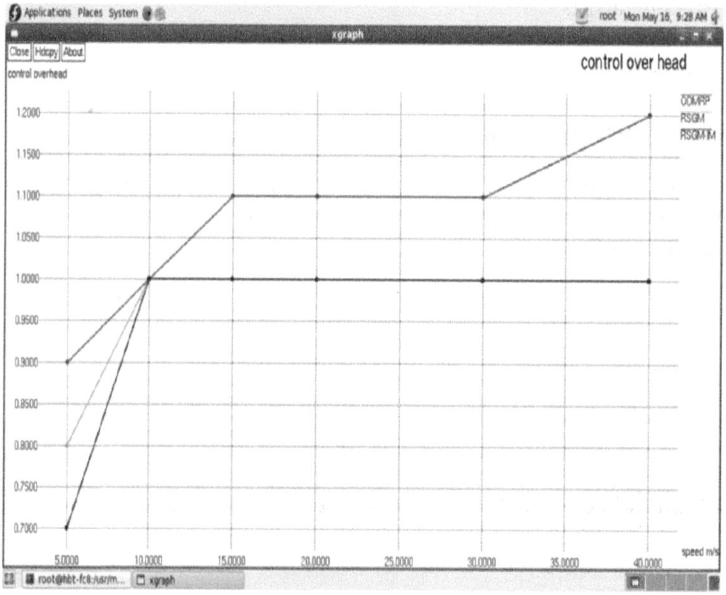


**Fig. 3.** SGBM Control Overhead

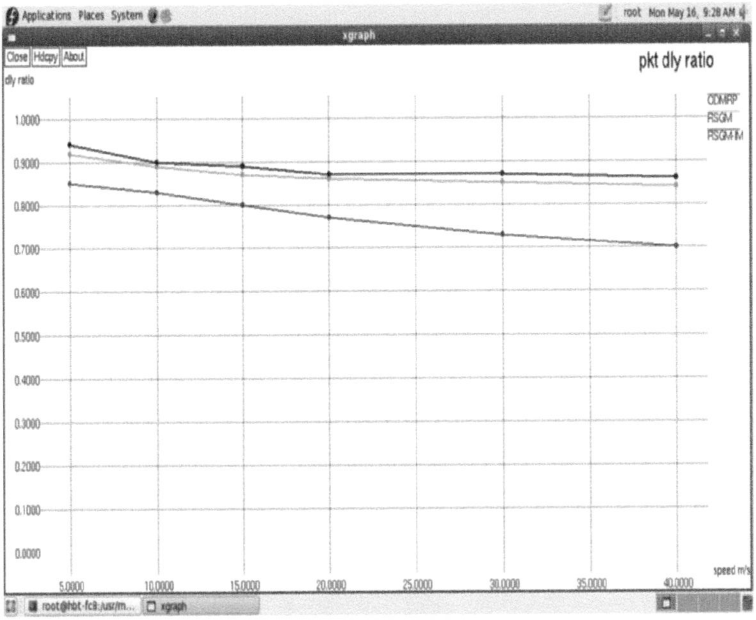


**Fig. 4.** SGBM Packet Delivery Ratio

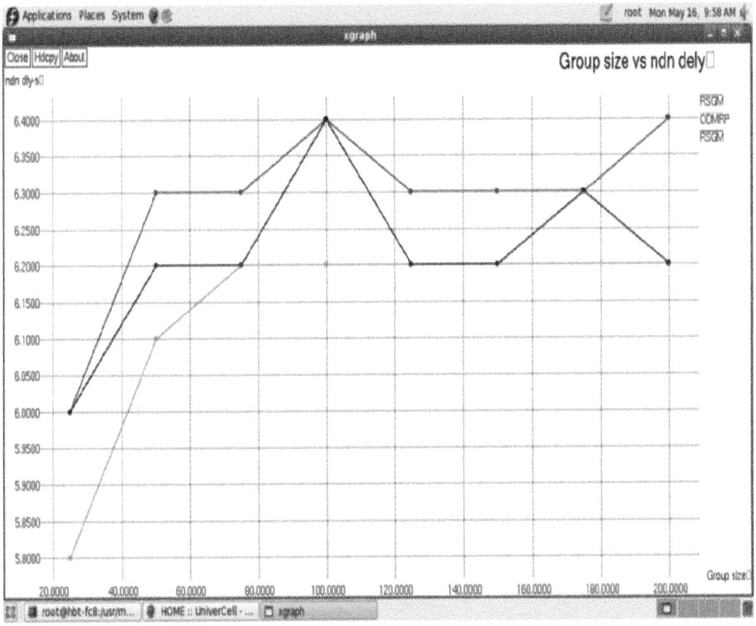


**Fig. 5.** SGBM Group Size vs. End to End delay

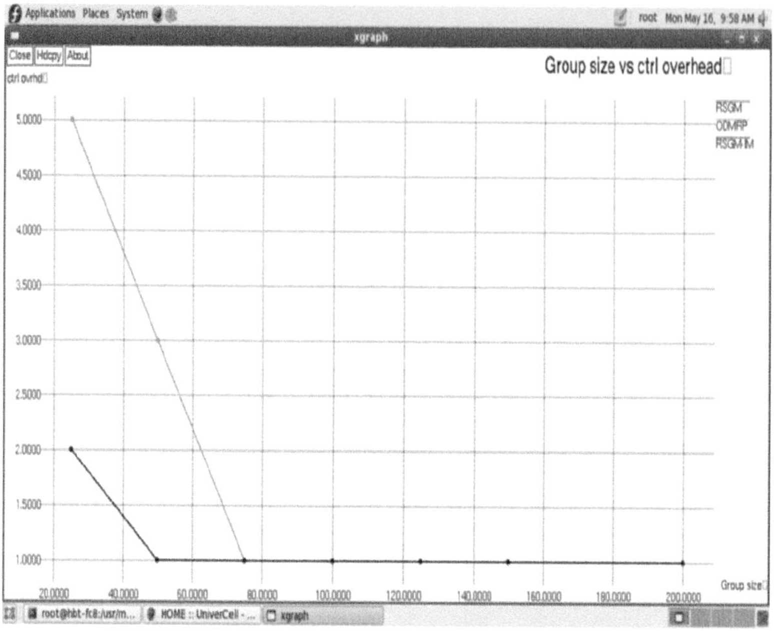


**Fig. 6.** SGBM Group Size vs. Control Overhead

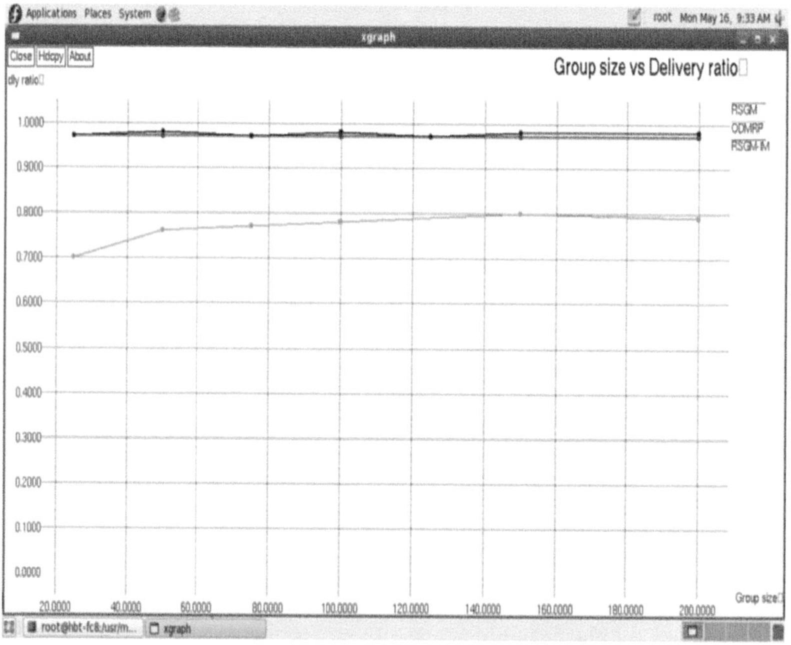


**Fig. 7.** SGBM Group Size vs. Packet Delivery Ratio

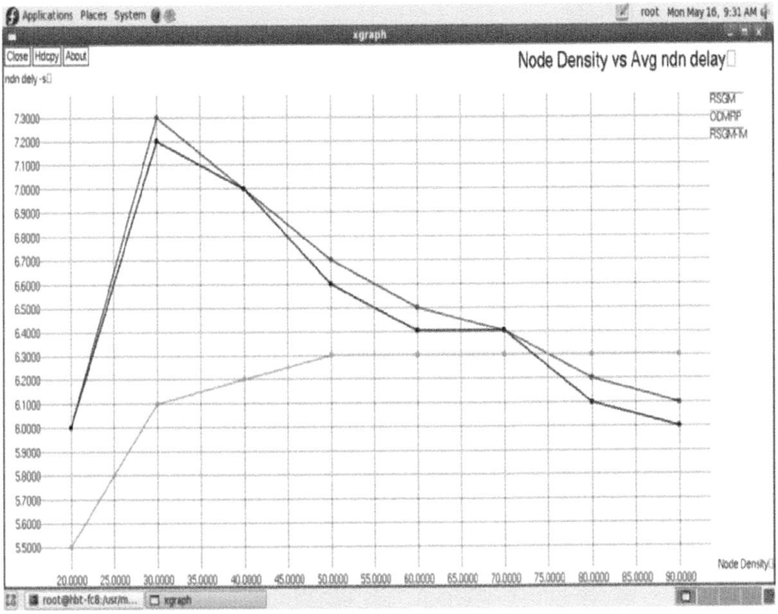


**Fig. 8.** SGBM Node Density vs. Average End to End delay

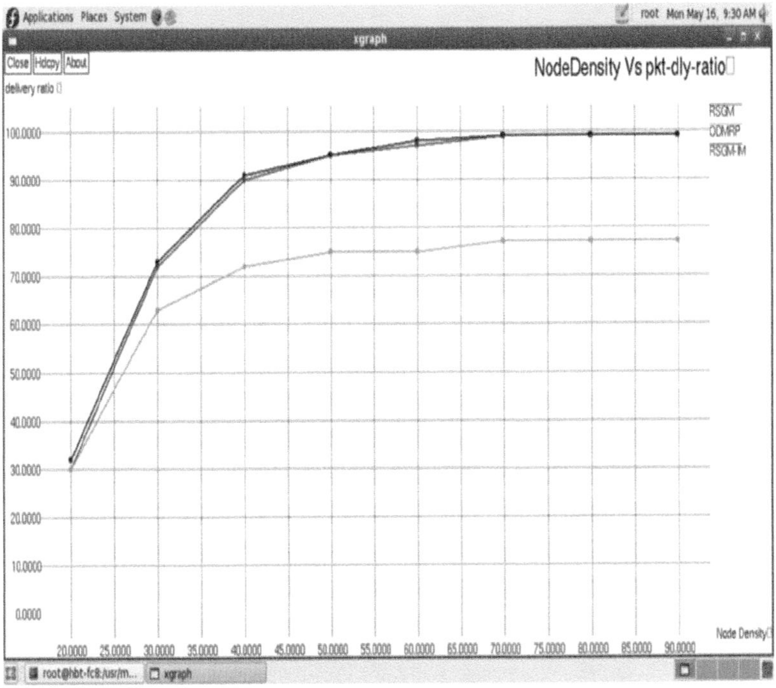


**Fig. 9.** SGBM Node Density vs. Packet Delivery Ratio

## 9 Conclusion

In this paper, we designed the Scalable Geographic Based Multicasting Protocol (SGBM) for Mobile Adhoc Network. In SGBM, the data's are forwarded from source to destination in a virtual tree based path without the need of explicitly creating and maintaining a tree structure. Scalable membership management is achieved through a zone structure.A home zone is defined to provide location and address service for the sources to avoid the periodic network-range flooding of source information, and the location service for group members is combined with the membership management to avoid the overhead of searching for addresses and positions of all group members through an outside location server. The position information is used in SGBM to guide the zone structure building, membership management and packet forwarding, which reduces the maintenance overhead and leads to more robust multicast forwarding upon the topology changes. We also handle the empty zone problem which is challenging for the zone-based protocols. Our studies show that our protocol not only would outperform the existing geographic multicast protocol and conventional multicast protocol but would also scale to a large group size and large network size. Specifically, our protocol would be more robust to network dynamics.Finally we conclude the paper as the protocol SGBM is more better than ODMRP and RSGM protocol.Hence it is given as RSGMIM(RSGM improved)in the graph.

## References

1. Chen, K., Nahrstedt, K.: Effective Location-Guided Tree Construction Algorithms for Small Group Multicast in MANET. In: Proc. IEEE INFOCOM, pp. 1180–1189 (2002)
2. Auerbach, J., Gopal, M., Kaplan, M., Kutten, S.: Multicast Group Membership Management. IEEE/ACM Transactions on Networking 11(1), 166–175 (2003)
3. Xiang, X., Wang, X., Yang, Y.: Stateless Multicasting in Mobile Adhoc Networks. IEEE Transactions on Computers 59(8), 1076–1090 (2010)
4. Lee, S., Su, W., Hsu, J., Gerla, M., Bagrodia, R.: A Performance Comparison Study of Ad Hoc Wireless Multicast Protocols. In: Proc.IEEE INFOCOM (2000)
5. Chiang, C.-C., Gerla, M., Zhang, L.: Forwarding Group Multicast Protocol (FGMP) for Multihop. Mobile Wireless Networks, Cluster Computing 1(2), 187–196 (1998); Special Issue on Mobile Computing
6. Garcia-Luna-Aceves, J.J., Madruga, E.: The Core-Assisted Mesh Protocol. IEEE J. on Selected Areas in Comm. 17(8), 1380–1394 (1999)
7. Gerla, M., Lee, S.J., Su, W.: On-Demand Multicast Routing Protocol (ODMRP) for Ad Hoc Networks, Internet Draft, draftietf-manet-odmrp-02.txt (2000)
8. Xiang, X., Wang, X., Zhou, Z.: An Efficient Geographic Multicast Protocol for Mobile Ad Hoc Networks, pp. 1–12 (1999)
9. Basagni, S., Chlamtac, I., Syrotiuk, V.R.: Location Aware, Dependable Multicast for Mobile Ad Hoc Networks. Computer Networks 36(5/6), 659–670 (2001)
10. Mauve, M., Fubler, H., Widmer, J., Lang, T.: Position-Based Multicast Routing for Mobile Ad-Hoc Networks. In: Proc. ACM MOBIHOC, Poster Section (June 2003)

11. Wu, C., Tay, Y., Toh, C.K.: Ad Hoc Multicast Routing Protocol Utilizing Increasing Id-Numbers (AMRIS) Functional Specification, Internet Draft (November 1998)
12. Paliwoda, K.: Transactions involving Multicast. Computer Communication Rev. 11(6), 313–318 (1988)
13. Deering, S.E.: Multicast Routing in internetworks and Extended LANs. Computer Communication Rev. 18(4), 55–64 (1988)
14. Deering, S.E.: Host extensions for IP Multicasting, Networking Group, RFC 988 (1986)
15. UCLA Parallel Computing Laboratory, GloMoSim (2010), `http://pcl.cs.ucla.edu/projects/glomosim/`

# Median Adjusted Constrained PDF Based Histogram Equalization for Image Contrast Enhancement

P. Shanmugavadivu[1], K. Balasubramanian[2], and K. Somasundaram[3]

[1]Associate Professor, Department of Computer Science and Applications, Gandhigram Rural Institute – Deemed University, Gandhigram, India
psvadivu67@gmail.com
[2] Associate Professor, Department of Computer Applications, PSNA College of Engineering and Technology, Dindigul, India
ksbala75@gmail.com
[3] Professor, Department of Computer Science and Applications, Gandhigram Rural Institute – Deemed University, Gandhigram, India
somasundaramk@yahoo.com

**Abstract.** A novel Median adjusted Constrained PDF based Histogram Equalization (MCPHE) technique for contrast enhancement is proposed in this paper. In this method, the probability density function of an image is modified by introducing constraints prior to the process of histogram equalization (HE). This technique of contrast enhancement takes control over the effect of HE so that it enhances the image without causing any loss to its details. A median adjustment factor is then added to the result, which normalizes the change in the luminance level after enhancement. This factor suppresses the effect of luminance change due to the presence of outlier pixels. The outlier pixels of highly deviated intensities have greater impact in changing the contrast of an image. Experimental results show that the proposed method gives better results in terms of PSNR and SSIM values when compared to the existing histogram based equalization methods.

**Keywords:** Contrast Enhancement, Histogram, Histogram Equalization (HE), Probability Density Function (PDF), Cumulative Density Function (CDF).

## 1 Introduction

Contrast enhancement plays a vital role in image processing for both human and computer vision. It is used as a preprocessing step in medical image processing, texture synthesis, speech recognition and many other image/video processing applications [3], [7], [9]. Different techniques have already been developed for this purpose. Some of these methods make use of simple linear or nonlinear gray level transformation functions [1] while the rest use complex analysis of different image features such as edge, connected component information and so on.

Histogram is a statistical probability distribution of each gray level in a digital image. Histogram Equalization (HE) is a very popular technique for contrast enhancement of images[1]. It is the most commonly used method due to its simplicity

D. Nagamalai, E. Renault, M. Dhanushkodi (Eds.): CCSEIT 2011, CCIS 204, pp. 244–253, 2011.

and comparatively gives better performance on almost all types of images. HE transforms the gray levels of the image based on the probability distribution of the input gray levels.

The histogram-based equalization techniques are classified into two principal categories as global and local histogram equalization. Global Histogram Equalization (GHE) uses the histogram information of the entire input image in its transformation function. Though this global approach is suitable for overall enhancement, it fails to preserve the local brightness features of the input image. Normally in an image, the high frequency gray levels dominate the low frequency gray levels. In this situation, GHE remaps the gray levels in such a way that the contrast stretching is restricted to some dominating gray levels having larger image histogram components and causes significant contrast loss for other smaller ones.

Local histogram equalization (LHE) [1] tries to eliminate such problems. It uses a small window that slides through every pixel of the image sequentially. The block of pixels that are masked by the window are considered for HE. Then, the gray level mapping for enhancement is done only for the center pixel of that window. Thus, it makes use of the local information remarkably. However, LHE requires high computational cost and sometimes causes over-enhancement in some portions of the image. Moreover, this technique has the problem of enhancing the noises in the input image along with the image features. The high computational cost of LHE can be minimized using non-overlapping block based HE. Nonetheless, these methods produce an undesirable checkerboard effects on enhanced images.

Histogram Specification (HS) [1] is another method in which the expected output of image histogram can be controlled by specifying the desired output histogram. However, specifying the output histogram pattern is not a simple task as it varies from image to image. A method called Dynamic Histogram Specification (DHS) [10] generates the specified histogram dynamically from the input image. Though this method preserves the original input image histogram characteristics, the degree of enhancement is not significant.

Brightness preserving Bi-Histogram Equalization (BBHE) [2], Dualistic Sub-Image Histogram Equalization (DSIHE) [4] and Minimum Mean Brightness Error Bi-Histogram Equalization (MMBEBHE) [6] are the variants of HE based contrast enhancement. BBHE divides the input image histogram into two parts based on the mean brightness of the image and then each part is equalized independently. This method tries to overcome the problem of brightness preservation. DSIHE method uses entropy value for histogram separation. MMBEBHE is an extension of BBHE method that provides maximal brightness preservation. Though these methods can perform good contrast enhancement, they also cause more annoying side effects depending on the variation of gray level distribution in the histogram. Recursive Mean-Separate Histogram Equalization (RMSHE) [5], [8] and Recursive Sub-image Histogram Equalization (RSIHE) [11] are the improved versions of BBHE. However, they are also not free from side effects. Recursively Separated and Weighted Histogram Equalization (RSWHE) is presented in [15]. To achieve sharpness in images, a method called Sub-Regions Histogram Equalization (SRHE) is recently proposed [16] in which the image is partitioned based on the smoothed intensity values, obtained by convolving the input image with a Gaussian filter. In this paper, we propose a Median adjusted Constrained PDF based Histogram Equalization (MCPHE) method by extending Weighted Thresholded HE (WTHE) [12] method for contrast enhancement.

Section 2 discusses the popular HE techniques. In section 3, the principle of the proposed method is presented. Section 4 presents the information about two widely used statistical parameters to assess the image quality. The results and discussions are given in section 5. In section 6, the conclusion is given.

## 2 HE Techniques

In this section, the existing HE approaches such as GHE, LHE, various partitioning approaches and WTHE are reviewed briefly.

### 2.1 Global Histogram Equalization (GHE)

For an input image F(x, y) composed of discrete gray levels in the dynamic range of [0, L-1], the transformation function $C(r_k)$ is defined as:

$$S_k = C(r_k) = \sum_{i=0}^{k} P(r_i) = \sum_{i=0}^{k} \frac{n_i}{n} \quad (1)$$

where $0 \le S_k \le 1$ and k = 0, 1, 2, …, L-1, ni represents the number of pixels having gray level $r_i$ , n is the total number of pixels in the input image and $P(r_i)$ represents the Probability Density Function of the input gray level ri. Based on the PDF, the Cumulative Density Function is defined as $C(r_k)$. The mapping given in equation (1) is called Global Histogram Equalization or Histogram Linearization. Here, $S_k$ is mapped to the dynamic range of [0, L-1] by multiplying it with (L-1).

Using the CDF values, histogram equalization maps an input level k into an output level $H_k$ using the level-mapping equation:

$$H_k = (L-1) \times C(r_k) \quad (2)$$

For the traditional GHE described above, the increment in the output level $H_k$ is given by:

$$\Delta H_k = H_k - H_{k-1} = (L-1) \times P(r_k) \quad (3)$$

The increment of level $H_k$ is proportional to the probability of its corresponding level $k$ in the original image. For images with continuous intensity levels and PDFs, such a mapping scheme would perfectly equalize the histogram in theory. But, the intensity levels and PDF of a digital image are discrete in practice. In such a case, the traditional HE mapping is not ideal and it results in undesirable effects where the intensity levels with high probabilities often become over-enhanced and the levels with low probabilities get less enhanced and their frequencies get either reduced or even eliminated in the resultant image.

### 2.2 Local Histogram Equalization (LHE)

GHE takes the global information into account and can not adapt to local light conditions. Local Histogram Equalization (LHE) performs block-overlapped histogram equalization. LHE defines a sub-block and retrieves its histogram information. Then, histogram equalization is applied at the center pixel using the CDF of that sub-block. Next, the window is moved by one pixel and sub-block histogram equalization is repeated until the end of the input image is reached. Though LHE cannot adapt to partial light information, it still over-enhances certain portions depending on its window size. However, selection of an optimal block size that enhances all parts of an image is not an easy task to perform.

### 2.3 Histogram Partitioning Approaches

BBHE tries to preserve the average brightness of the image by separating the input image histogram into two parts based on the input mean and then equalizing each of the parts independently. RMSHE, RSIHE and RSWHE partition the histogram recursively. Here, some portions of histogram among the partitions cannot be expanded much, while the outside region expands significantly that creates unwanted artifacts. This is a common drawback of most of the existing histogram partitioning techniques since they keep the partitioning point fixed throughout the entire process of equalization. In all the recursive partitioning techniques, it is not an easy job of fixing the optimal recursion level. Moreover, as the recursion level increases, recursive histogram equalization techniques produce the results same as GHE and it leads to computational complexity.

### 2.4 Weighted Thresholded HE (WTHE)

WTHE is a fast and efficient method for image contrast enhancement [12]. This technique provides a novel mechanism to control the enhancement process, while being adaptive to various types of images. WTHE method provides a good trade-off between the two features, adaptivity to different images and ease of control, which are difficult to achieve in the GHE-based enhancement methods. In this method, the probability density function of an image is modified by weighting and thresholding prior to HE. A mean adjustment factor is then added with the expectation to normalize the luminance changes. But, while taking the mean of the input and reconstructed images, the highly deviated intensity valued pixels known as outlier pixels are also taken into account. This will not effectively control the luminance change in the output image, which is a major drawback of this method. The outlier pixels are the pixels which are usually less in number and are having distant intensity values than the other pixels.

## 3 Median adjusted Constrained PDF Based Histogram Equalization (MCPHE)

The proposed method, MCPHE is an extension of WTHE which performs histogram equalization based on a modified histogram. Each original probability density value $P(r_k)$ is replaced by a Constrained PDF value $P_c(r_k)$ yielding:

$$\Delta H_k = (L-1) \times P_c(r_k) \tag{4}$$

The proposed level mapping algorithm is given below.

i. Input the image, $F(i, j)$ with a total number of 'n' pixels in the gray level range [0, L-1]
ii. Compute the Probability Density Function (PDF), $P(r_k)$ for the gray levels
iii. Find the median value of the PDFs, $M_1$
iv. Compute the upper constraint '$P_u$' :
$P_u = v * max(PDF)$ where $0.1 < v < 1.0$
v. Set power factor '$r$' where $0.1 < r < 1.0$
vi. Set lower constraint factor $P_l$ be as low as 0.0001
vii. Compute the constrained PDF, $P_c(r_k)$

$$P_c(r_k) = \mathrm{T}(P(r_k)) = \begin{cases} P_u & \text{if } P(r_k) > P_u \\ \left(\dfrac{P(r_k) - P_l}{P_u - P_l}\right)^r * P_u & \text{if } P_l < P(r_k) \le P_u \\ 0 & \text{if } P(r_k) \le P_l \end{cases} \tag{5}$$

viii. Find the median value of the constrained PDFs, $M_2$
ix. Compute the median adjustment factor as:
$M_f = M_2 - M_1$
x. Add $M_f$ to the constrained PDFs
xi. Compute cumulative probability density function, $C_c(F(i,j))$
xii. Apply the HE procedure (level mapping) as:

$$F'(i, j) = (L-1) \times C_c(F(i, j)) \tag{6}$$

where $F'(i, j)$ is the enhanced image.

The transformation function T(.) in equation (5) transforms all the original PDF values between the upper constraint $P_u$ and the lower constraint $P_l$ using a normalized power law function with index $r > 0$.

In this level-mapping algorithm, the increment for each intensity level is decided by the transformed histogram given in equation (5). The increment is controlled by adjusting the index $r$ of the power law transformation function. For example, when $r<1$, the power law function gives a higher weightage to the low probabilities. Therefore, the lower probability levels are preserved and the possibility of over-enhancement is less. The effect of the proposed method approaches that of the GHE, when $r \to 1$. When $r > 1$, more weight is shifted to the high-probability levels and MCPHE would yield even stronger effect than the traditional HE.

The upper constraint $P_u$ is used to avoid the dominance of the levels with high probabilities when allocating the output dynamic range. The lower constraint $P_l$ is used to find the levels whose probabilities are too low. The $P_l$ value is set to be as low as 0.0001. Any pixel having its probability less than the lower constraint $P_l$ is having very low impact in the process of contrast enhancement. It can be observed from equation (5) that when $r = 1$, $P_u$=1 and $P_l$=0, the proposed MCPHE reduces to the GHE.

After the constrained PDF is obtained from equation (5), the median adjustment factor ($M_f$) is calculated by finding the difference between the median value of the constrained PDFs and the median value of the original PDFs. Then, the $M_f$ is added to the constrained PDFs which will effectively control the luminance change, since the outlier pixels are ignored while computing the medians.

The cumulative density function (CDF) is obtained as:

$$C_c(k) = \sum_{m=0}^{k} P_c(m), \qquad for\ k = 0, 1, ..., L-1 \tag{7}$$

Then, the HE procedure is applied as given in equation (6). This will not cause serious level saturation (clipping), to the resulting contrast enhanced image.

## 4 Measurement Tools to Assess Image Quality

There are two well known statistical parameters namely Peak Signal to Noise Ratio (PSNR) and Structural Similarity Index Matrix (SSIM) which are usually employed to assess the after-enhancement image quality [11].

### 4.1 Peak Signal-to-Noise Ratio

The peak signal-to-noise ratio (PSNR) is computed using the source image *X(i, j)* of size M by N pixels and the enhanced image *Y(i, j)*. Errors are computed only on the luminance signal so that the pixel values of *X(i, j)* ranges between black (0) and white (255). First the mean square error (MSE) of the reconstructed image is computed as follows:

$$MSE = \frac{\sum_{i=1}^{M}\sum_{j=1}^{N}[X(i,j)-Y(i,j)]^2}{M \times N} \tag{8}$$

The root mean squared error (RMSE) is the square root of MSE. The PSNR in decibels (dB) is computed as:

$$PSNR = 20\ \log_{10}\left(\frac{\max(Y(i,j))}{RMSE}\right) \tag{9}$$

### 4.2 Structural Similarity Index Matrix (SSIM)

The Structural Similarity Index Matrix (SSIM) is defined as:

$$SSIM(X,Y) = \frac{(2\mu_X\mu_Y + C_1)(2\sigma_{XY} + C_2)}{(\mu_X^2 + \mu_Y^2 + C_1)(\sigma_X^2 + \sigma_Y^2 + C_2)} \tag{10}$$

where X and Y are the reference and the output images respectively; $\mu_X$ is the mean of X, $\mu_Y$ is the mean of Y, $\sigma_X$ is the standard deviation of X, $\sigma_Y$ is the standard deviation of Y, $\sigma_{XY}$ is the square root of covariance of X and Y, whereas $C_1$ and $C_2$ are constants. The SSIM value between two images X and Y is generally in the range 0 to 1. If X=Y, then the SSIM is equal to 1 which implies that when the SSIM value is nearing 1, the degree of structural similarity between the two images are more.

## 5 Results and Discussion

The performance of the newly developed method, MCPHE is tested on various standard images such as Einstein, Village, Bottle, House and Peppers out of which Einstein is given in Fig. 1(a). To compare the performance of MCPHE, the image is enhanced with the contemporary enhancement techniques like GHE, LHE, BBHE, DSIHE, HS, RMSHE and WTHE. For all these methods, the performance is measured qualitatively in terms of human visual perception and quantitatively by computing the Peak Signal-to-Noise Ratio (PSNR) values and Structural Similarity Index Matrix (SSIM) values which are given in Table 1 and Table 2 respectively.

The original Einstein image is given in Fig. 1(a). The enhanced images of the same by GHE, LHE, BBHE, DSIHE, HS, RMSHE, WTHE and are shown in Fig. 1(b) - 1(h) respectively. Fig. 1(i) is the result of MCPHE which clearly shows the improvement in image quality than the other HE techniques. Similarly, MCPHE is found to produce better results for the other images such as village, Bottle, House, Peppers etc in terms of visual perception, compared to other methods. The PSNR values obtained for various images including the above said images are furnished in Table 1. It can be noted that for all the images, the PSNR values of MCPHE is higher than all other methods. The SSIM values are calculated for various methods including MCPHE and are presented in Table 2. MCPHE is found to generate higher SSIM value than the contemporary methods. Hence, the results of the study clearly reveal that MCPHE produces better image enhancement compared to the contemporary methods in terms of PSNR and SSIM.

**Table 1.** Comparison of PSNR values of various methods and MCPHE

| Method / Image | GHE | LHE | BBHE | DSIHE | HS | RMSHE (r=2) | WTHE | MCPHE |
|---|---|---|---|---|---|---|---|---|
| **Einstein** | 14.9772 | 6.3961 | 16.1678 | 16.5590 | 16.1721 | 24.3070 | 22.2114 | 26.6039 |
| **Village** | 17.4181 | 5.7367 | 22.2929 | 21.4134 | 19.0391 | 26.8144 | 34.5230 | 39.5615 |
| **Bottle** | 12.6761 | 8.7862 | 18.4965 | 17.6775 | 12.9664 | 28.4096 | 30.6490 | 34.3248 |
| **House** | 10.7989 | 4.3828 | 14.4538 | 12.9997 | 11.0700 | 24.1902 | 24.8067 | 33.0999 |
| **Peppers** | 22.6708 | 4.9869 | 26.6785 | 26.6894 | 24.8050 | 27.1802 | 32.5050 | 37.0061 |

**Table 2.** Comparison of SSIM values of various methods and MCPHE

| Method / Image | GHE | LHE | BBHE | DSIHE | HS | RMSHE (r=2) | WTHE | MCPHE |
|---|---|---|---|---|---|---|---|---|
| **Einstein** | 0.6672 | 0.0046 | 0.7099 | 0.6950 | 0.7430 | 0.8033 | 0.8770 | 0.9699 |
| **Village** | 0.7814 | 0.0068 | 0.8924 | 0.8843 | 0.8596 | 0.8868 | 0.9950 | 0.9982 |
| **Bottle** | 0.7517 | 0.0067 | 0.8847 | 0.8700 | 0.7685 | 0.9433 | 0.9796 | 0.9919 |
| **House** | 0.6582 | 0.0042 | 0.7475 | 0.7098 | 0.6727 | 0.8936 | 0.9475 | 0.9942 |
| **Peppers** | 0.9286 | 0.0052 | 0.9542 | 0.9553 | 0.9577 | 0.9365 | 0.9918 | 0.9974 |

(a) Original (b) GHE (c) LHE

(d) BBHE (e) DSIHE (f) HS

(g) RMSHE (r=2) (h) WTHE (i) MCPHE

**Fig. 1.** Einstein image (a) Original; results of (b) GHE (c) LHE with window size = 5 (d) BBHE (e) DSIHE (f) HS (g) RMSHE (r=2) (h) WTHE (i) MCPHE

## 6 Conclusions

The proposed MCPHE contrast enhancement method is proved to be a better approach for low contrast images. Experimental results on standard images show that the degree of enhancement measured in terms of PSNR and SSIM values for the proposed method is higher than those of the existing histogram based equalization techniques. Moreover, this method preserves the details of the input images during enhancement. This method can effectively be employed in video processing, medical image processing etc.

## References


1. Gonzalez, R.C., Woods, R.E.: Digital Image Processing, 2nd edn. Prentice Hall, Englewood Cliffs (2002)
2. Kim, Y.: Contrast enhancement using brightness preserving bihistogram equalization. IEEE Transactions on Consumer Electronics 43(1), 1–8 (1997)
3. Wahab, A., Chin, S.H., Tan, E.C.: Novel approach to automated fingerprint recognition. IEEE Proceedings on Vision, Image and Signal Processing 145(3), 160–166 (1998)
4. Wan, Y., Chen, Q., Zhang, B.: Image enhancement based on equal area dualistic sub-image histogram equalization method. IEEE Transactions on Consumer Electronics 45(1), 68–75 (1999)
5. Chen, S., Ramli, A.R.: Contrast Enhancement using Recursive Mean-Separate Histogram Equalization for Scalable Brightness Preservation. IEEE Transactions on Consumer Electronics 49(4), 1301–1309 (2003)
6. Chen, S., Ramli, A.R.: Minimum mean brightness error bi-histogram equalization in contrast enhancement. IEEE Transactions on Consumer Electronics 49(4), 1310–1319 (2003)
7. Pei, S.C., Zeng, Y.C., Chang, C.H.: Virtual restoration of ancient Chinese paintings using color contrast enhancement and lacuna texture synthesis. IEEE Transactions on Image Processing 13(3), 416–429 (2004)
8. Chen, S., Ramli, A.R.: Preserving brightness in histogram equalization based contrast enhancement techniques. Digital Signal Processing 14, 413–428 (2004)
9. Torre, A., Peinado, A.M., Segura, J.C., Perez-Cordoba, J.L., Benitez, M.C., Rubio, A.J.: Histogram equalization of speech representation for robust speech recognition. IEEE Transactions on Speech Audio Processing 13(3), 355–366 (2005)
10. Sun, C.C., Ruan, S.J., Shie, M.C., Pai, T.W.: Dynamic contrast enhancement based on histogram specification. IEEE Transactions on Consumer Electronics 51(4), 1300–1305 (2005)
11. Sim, K.S., Tso, C.P., Tan, Y.Y.: Recursive sub-image histogram equalization applied to gray-scale images. Pattern Recognition Letters 28, 1209–1221 (2007)
12. Wang, Q., Ward, R.K.: Fast image/video contrast enhancement based on weighted thresholded histogram equalization. IEEE Transactions on Consumer Electronics 53(2), 757–764 (2007)
13. Menotti, D., Najman, L., Facon, J., Araujo, A.A.: Multi-histogram equalization methods for contrast enhancement and brightness preserving. IEEE Transactions on Consumer Electronics 53(3), 1186–1194 (2007)

14. Sengee, N., Choi, H.K.: Brightness Preserving Weight Clustering Histogram Equalization. IEEE Transactions on Consumer Electronics 54(3), 1329–1337 (2008)
15. Kim, M., Chung, M.G.: Recursively Separated and Weighted Histogram Equalization for Brightness Preservation and Contrast Enhancement. IEEE Transactions on Consumer Electronics 54(3), 1389–1397 (2008)
16. Ibrahim, H., Kong, N.S.P.: Image Sharpening Using Sub-Regions Histogram Equalization. IEEE Transactions on Consumer Electronics 55(2), 891–895 (2009)

# Isolated Word Recognition System Using Back Propagation Network for Tamil Spoken Language

V. Radha, C. Vimala, and M. Krishnaveni

Department of Computer Science,
Avinashilingam Deemed University for Women,
Coimbatore, Tamilnadu, India
radharesearch@yahoo.com
{vimalac.au,krishnaveni.rd}@gmail.com

**Abstract.** Recently with the wide development of computers, various forms of information exchange between human and computer are discovered. At present, interacting with the computer using speech is one of the active scientific research fields especially for people with disabilities who face variety of barriers to computer use. Such research in Automatic Speech Recognition (ASR) is investigated for different languages because each language has its specific features. This paper presents a speech recognition system for individually spoken word in Tamil language using multilayer feed forward network which falls under the category, "Networks for Classification and Prediction" and has widespread interesting applications and functions related to speech processing. To implement the above system, initially preprocessing is done with the input signal and the speech features being the main part of speech recognition system, are analyzed and extracted via Mel Frequency Cepstral Coefficients (MFCC). These feature vectors are given as the input to the Feed-Forward Neural Network for classifying and recognizing Tamil spoken word. Experiments are done with sample Tamil speech signals and its performance are measured based on Mean square error (MSE). The results indicate that the adopted network with specified parameters have produced the best MSE.

**Keywords:** Elliptic Filter, Mel Frequency Cepstral Coefficients, Feed-forward neural networks, Tamil Speech Recognition, Mean square error, and Peak signal to Noise Ratio.

## 1 Introduction

Speech is one of the oldest and most natural means of information exchange between human beings. As the new generation of computing technology, it comes as the next major innovation in man-machine interaction. Obviously such interface would yield great benefits which can be accomplished with the help of Automatic Speech Recognition (ASR) system. It is a process by which a machine identifies speech. It takes a human utterance as an input and returns a string of words or phrases as output. Recently, Neural Network (NN) [1] is considered as one of the most successful information processing tool that has been widely used in speech recognition [2][3].

D. Nagamalai, E. Renault, M. Dhanushkodi (Eds.): CCSEIT 2011, CCIS 204, pp. 254–264, 2011.

The conventional neural networks Multi-Layer Perceptron (MLP) have been increasingly used for word recognition and also for other speech processing applications. The main objective of this paper is to implement classification and recognition system for isolated Tamil [6][7]spoken words. To carry out this task preprocessing [4][5] is done before feature extraction[2][10] which includes filtering, [4] and windowing. The best filtered speech signal is chosen and fed as an input. These filtered outcomes are evaluated based on MSE and PSNR values. For extracting specific features from the speech the popular feature extraction [2][10] technique MFCC is used. Using this 12 feature vectors are extracted and they are given as the input to the network. Finally, the speech recognition [2][3] is developed using feed-forward neural networks [1]and their performances are evaluated based on its default measure MSE.

The paper is organized as follows. Section 2 explains the system overview. Section 3 explains the preprocessing [5] steps involved in this system. Section 4 gives details about feature extraction technique based on MFCC. Section 5 deals with Feed forward neural network techniques and its performance in Tamil [6][7] speech signal. Section 6 explores the performance evaluation of the adopted method. Finally, the conclusion is summarized in section 7 with future work.

## 2 System Overview

The performance of the speech recognition [2][3] system depends on the classification techniques such as Neural Networks, Hidden Markov Models, Bayesian networks and Dynamic Time Warping etc. Among these approaches Neural networks (NNs) [1]

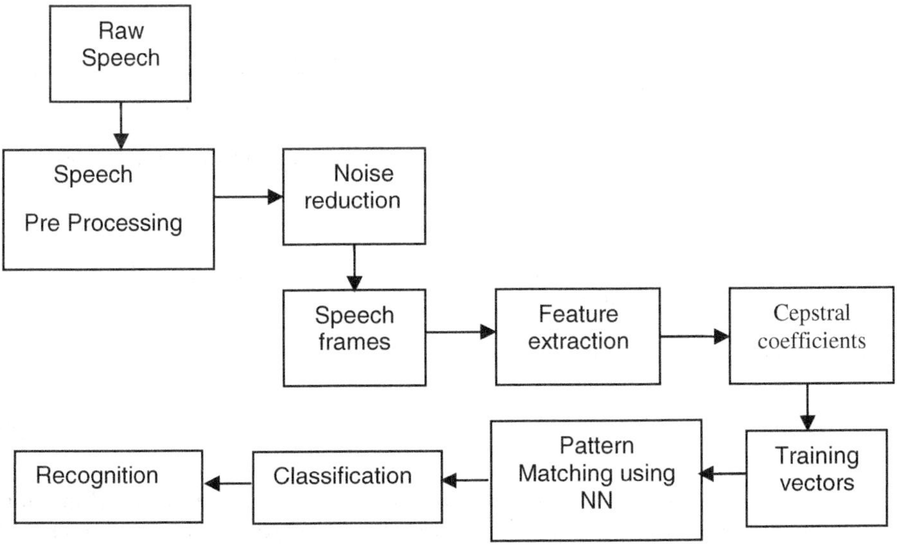


**Fig. 1.** System overview of speech recognition based on NN

have proven to be a powerful tool for solving problems of prediction, classification and pattern recognition. Rather than being used in general-purpose speech recognition applications it can handle low quality, noisy data and speaker independence applications. Such systems can achieve greater accuracy than HMM based systems, as long as there is training data and the vocabulary is limited.

One of the most commonly used networks based on supervised learning algorithm is multilayer feed forward network is implemented in this paper for classifying and recognizing Tamil speech [6][7]. In Tamil language, the pronunciation of independent letters and group of letters forming words are not different. Tamil speech recognizing system [7][8] does not require the support of a dictionary. Thus the recognizing process in Tamil speech [9] is fairly simple compared to English.

To implement the system, initially the speech data is preprocessed using filtering, framing [4] and windowing techniques. Subsequent to that, the enhanced signal is given as the input to the MFCC algorithm to extract features. These feature vectors also called cepstral coeffients given as the input to the network. After that the network is trained with these input vectors and the target vectors. Finally classification and recognition is done based on pattern matching. The figure 1 demonstrates the overall structure of the system.

## 3 Preprocessing

To enhance the accuracy and efficiency of the speech signals, they are generally pre-processed before further analysis. Since the selected dataset usually needs to be pre-processed prior to being fed into a NN this is done here. Speech signal pre-processing includes digital filtering, [4] and windowing.

| No of samples | Original Signal | Output signal |
|---|---|---|
| 1 | 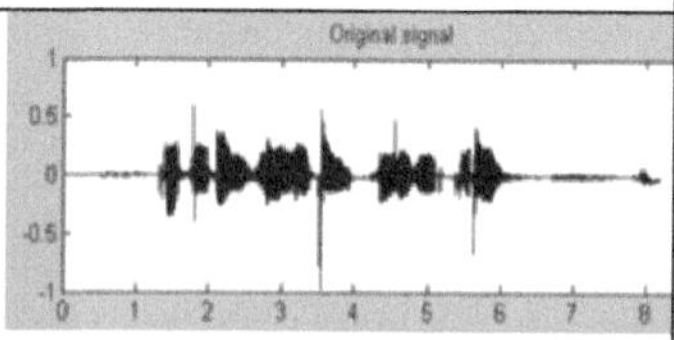  | 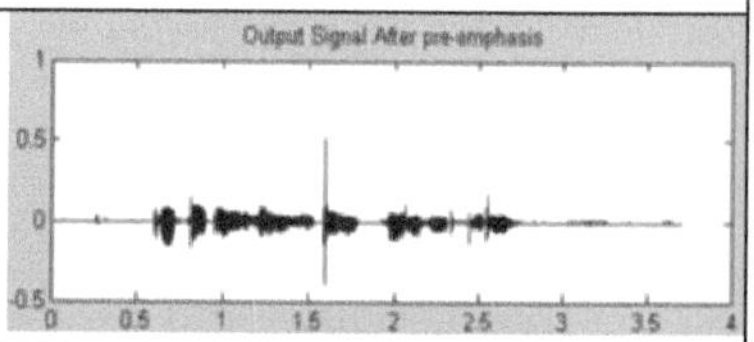  |
| 2 | 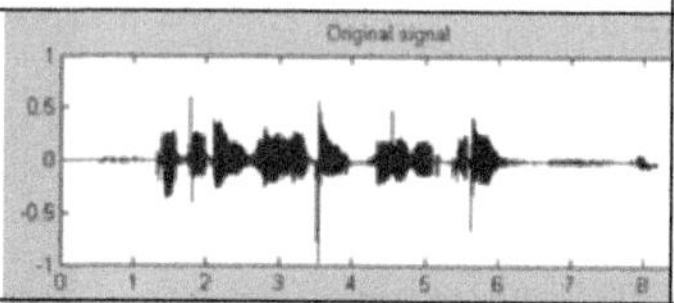  | 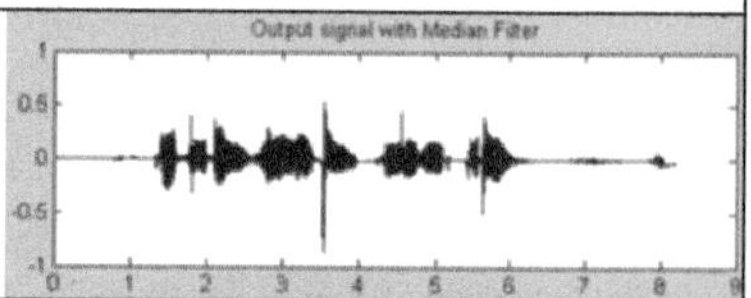  |
| 3 | 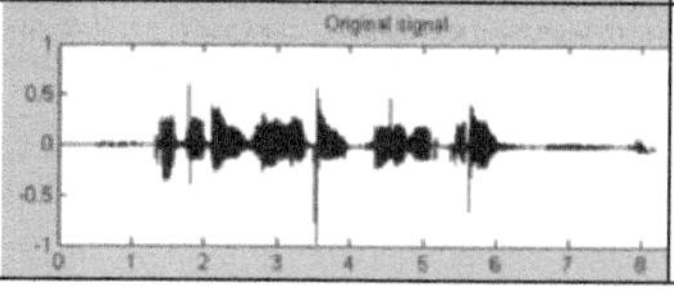  | 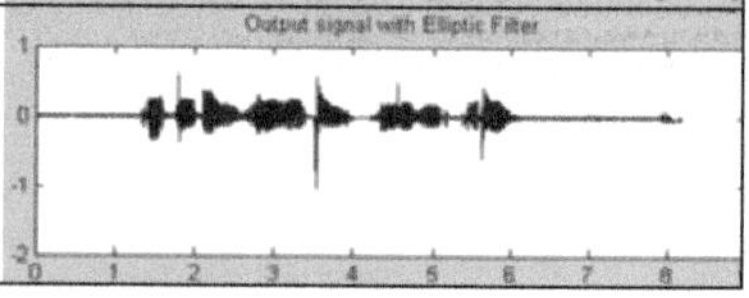  |

**Fig. 2.** Original signals and the resulted signals after applying filters

### 3.1 Filtering

One of the foremost processing steps in speech processing is to apply filters to the signal. It is essentially used for speech enhancement by reducing the background noise and to improve the quality aspects of speech. To reduce noise in the signal three types of filters are used namely pre emphasis, median and elliptic. All these filters are implemented and based on the experimental results, it was found that the elliptic filter performs well among the other filters. Elliptic filter provides the lowest order implementation of the classical filters for the same frequency and rejection requirements. Elliptic filters are equiripple in the pass band and the stop band. The finite zeros of transmission, which allow the filter to have a narrower transition band, determine the ripple response in the stop band. The above taken filter results are evaluated both subjectively and objectively. This resulted signal is used for further processing of a system. The figure 2 shows the subjective evaluation of the above three filters.

### 3.2 Framing and Windowing

The further preprocessing [4] step includes framing [4] and windowing. In this step the continuous speech signal is blocked into frames of N samples, with adjacent frames being separated by M (M < N). The first frame consists of the first N samples. The second frame begins M samples after the first frame, and overlaps it by N - M samples and so on. This process continues until all the speech is accounted for within one or more frames. Typical values for N and M are 256.

The next step in the preprocessing [5] is to applying window for each individual frame so as to minimize the signal discontinuities at the beginning and end of each frame. It is used for minimizing the spectral distortion by using the window to taper the signal to zero at the beginning and the end of each frame. It is defined as $w(n), 0 \le n \le N-1$, where N is the number of samples in each frame and then the result of windowing is the desired signal. Windowing is done by using the equation(1).

$$y_l(n) = x_l(n)w(n), \quad 0 \le n \le N-1 \tag{1}$$

Window choice is an important task in any signal processing applications. Among the different types of windows like triangular, Blackman etc, the hamming window best suites for speech signal processing. Typically the Hamming window is defined as in the equation (2).

$$w(n) = 0.54 - 0.46\cos\left(\frac{2\pi n}{N-1}\right), \quad 0 \le n \le N-1 \tag{2}$$

## 4 Feature Extraction Using MFCC

The motivation of feature extraction [2] [10] is to convert speech waveform to a parametric representation at a lower information rate for further analysis. Speech signals have non-stationary characteristics. As a result, the speech waveforms are commonly small frames (typically 5 ms to 40 ms) in which the signal characteristics are considered quasi-stationary to allow for short-term spectral analysis and feature extraction. A wide range of feature extraction techniques are available such as Linear Predictive Coefficients (LPC), Linear Predictive Cepstral Coefficients (LPCC) and Mel Frequency Cepstral Coefficients (MFCC). In this research work the most frequently used MFCC parameters are considered to determine the best feature set for the Tamil[6][7] speech database. The steps involved in MFCC can be seen in the following figure 3 which illustrates the functioning process of deriving the Mel Frequency Cepstral Coefficients.

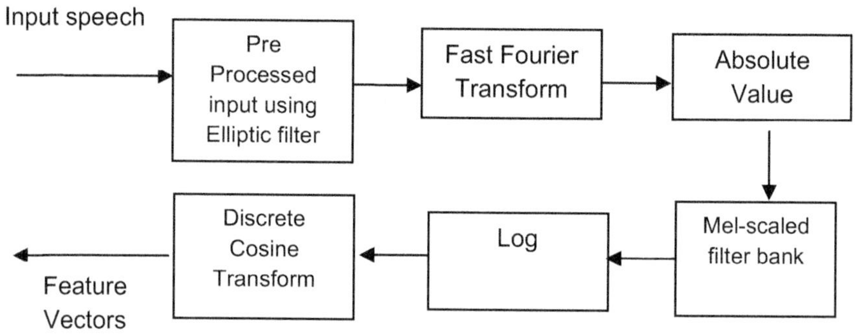


**Fig. 3.** Steps of MFCC

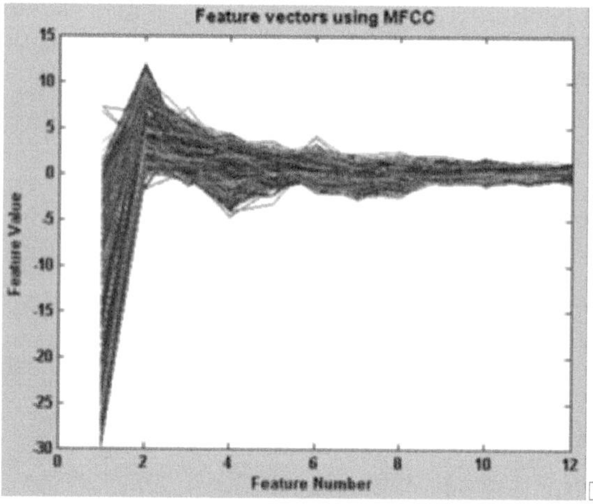


**Fig. 4.** MFCC coefficients

Initially, the preprocessed speech sample is taken as an input and FFT is applied which provides the absolute value. After that, the bank filter is used for warping the Mel frequency. According to this, the width of the triangular filters varies and so the log total energy in a critical band around the center frequency is included. After warping the numbers of coefficients are obtained which is done by the equation(3).

The extracted 12 feature vectors are as follows:

**Table 1.** Sample Feature vectors of MFCC

| No of Samples | 12 coefficients values of MFCC | | |
|---|---|---|---|
| 1 | 3.262109483760267 | 2.183415580698431 | -3.211822134287218 |
| | 7.959173232010254 | -1.489845448833876 | 0.787602786178414 |
| | 5.759973017258244 | -2.252154484623268 | 3.439043146191077 |
| | 1.25409424132334 | -0.129311243184615 | 0.300936550868309 |
| 2 | 3.798478622210788 | 1.74797281436433 | -2.489657282715821 |
| | 8.522089004065837 | -3.47052936681966 | 0.302290925428428 |
| | 6.570803874923982 | -2.833757855234683 | 4.19327339637328 |
| | 1.515215876790111 | -0.349448852656072 | 0.614526124621857 |
| 3 | 3.262109483760267 | 2.183415580698431 | -3.211822134287218 |
| | 7.959173232010254 | -1.489845448833876 | 0.787602786178414 |
| | 5.759973017258244 | -2.252154484623268 | 3.439043146191077 |
| | 1.25409424132334 | -0.129311243184615 | 0.300936550868309 |
| 4 | 8.418557880841314 | 5.409326742576696 | 0.103873482360843 |
| | 10.423558371669165 | -1.562692756940896 | 1.840185046791586 |
| | 3.560342622423488 | -6.684312284440906 | 1.217165610703996 |
| | -0.449879780800645 | -0.702780799553466 | 0.272368699987399 |
| 5 | 3.556871227862671 | 1.880082703099456 | -2.677798789621644 |
| | 9.584605229643586 | -1.913588118828719 | 1.378513202846451 |
| | 5.8273459608221 | -6.567998823813931 | 2.083529875349709 |
| | 0.019901182114678 | -1.264406283818704 | 0.614171471695149 |

$$Y(k)=\sum_{j=1}^{N/2} S(j)H_k(j) \tag{3}$$

Finally the Inverse Discrete Fourier Transformer is used for the cepstral coefficients calculation. It transforms the log of the quefrench domain coefficients to the frequency domain where N is the length of the DFT which is defined in equation (4).

$$c(n)=\frac{1}{N'}\sum_{k=0}^{N'-1} Y(k)e^{j\frac{k2\pi}{N'}n} \tag{4}$$

For this work, the MFCC is implemented with 24 filters and 12 dimensions. It uses 256 size of NFFT. The following figure 4 and table 1 shows the 12 feature vectors derived from MFCC for the sample Tamil [7][8]speech signals. These coeffients are taken as the input vector for the network.

## 5 Feed Forward Neural Network for Tamil Spoken Word Recognition

A feed forward neural network [10] is a biologically inspired classification algorithm. It consists of a (possibly large) number of simple neuron-like processing units, organized in layers. Every unit in a layer is connected with all the units in the previous layer. These connections are not all equal and may have a different strength or weight. The weights on these connections encode the knowledge of a network. They usually consist of three to four layers in which the neurons are logically arranged. The first and last layers are the input and output layers respectively and there are usually one or more hidden layers in between the other layers. The term feed-forward means that the information is only allowed to "travel" in one direction. This means that the output of one layer becomes the input of the next layer, and so forward. Feed-forward networks are advantageous as they have the fastest models to execute.

In the Network, the input layer does have a part in calculation, rather than just receiving the inputs. The raw data is computed, and activation functions are applied in all layers. This process occurs until the data reaches the output neurons, where it is classified into a certain category. The operation of this network can be divided into two phases:

1. The learning phase
2. The classification phase

During the learning phase the weights in the FFNet will be modified. All weights are modified in such a way that when a pattern is presented, the output unit with the correct category, hopefully, will have the largest output value. In the classification phase the weights of the network are fixed. A pattern, presented as the input will be transformed from layer to layer until it reaches the output layer. Now classification can occur by selecting the category associated with the output unit that has the largest output value. In contrast to the learning phase classification is very fast. The experimental results are presented in the following section.

## 6 Performance Evaluation

Performance evaluation is done for both filtering techniques and the NN separately. The best filtering signal for feature extraction is evaluated based on Mean Square Error Rate (MSE) and Peak Signal to Noise Ratio (PSNR) values. The MSE value is calculated by the equation(5)

$$MSE = \frac{\Sigma\left( y_i - \hat{y}_i \right)}{n - p} \tag{5}$$

and PSNR value is calculated by the equation (6)

$$PSNR = 10\log_{10}\left[\frac{R^2}{MSE}\right] \tag{6}$$

The average values of 10 speech samples were taken for the performance evaluation. The figure 5 (a) and (b) presents the performance evaluation of three filters based on MSE and PSNR values. Among these filters Elliptic filter offer least MSE value and high peak signal value.

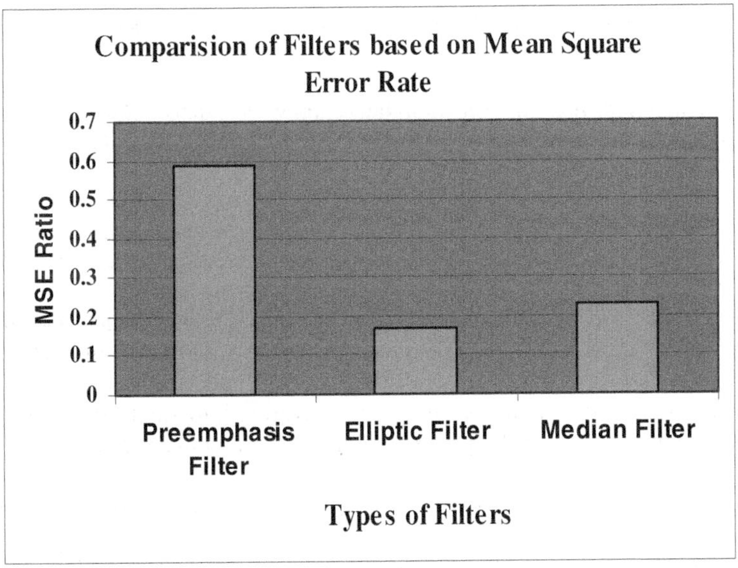


**Fig. 5(a).** Performance evaluation of filters based on MSE

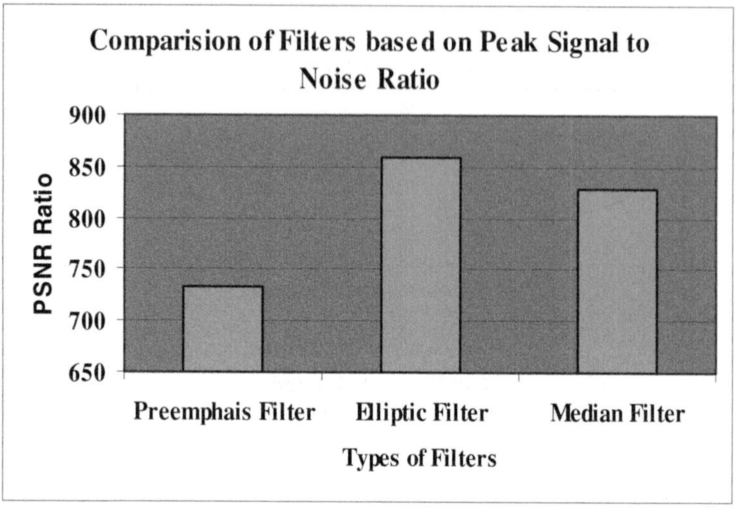


**Fig. 5(b).** Performance evaluation of filters based on PSNR

In this research work, the input speech signal is preprocessed with elliptic filter. This filtered signal is divided into short frames and hamming window is applied to reduce the discontinuities between these frames. For further process of creating and training a network, the values given for the input vector are 12 coefficients of MFCC. There are 12 neurons in the first layer and 2 neurons in the second (output) layer. The transfer function used in the first layer is tan-sigmoid, and the output layer transfer function is linear. With standard back-propagation, the learning rate is held constant throughout training. The performance of the algorithm is very sensitive to the proper setting of

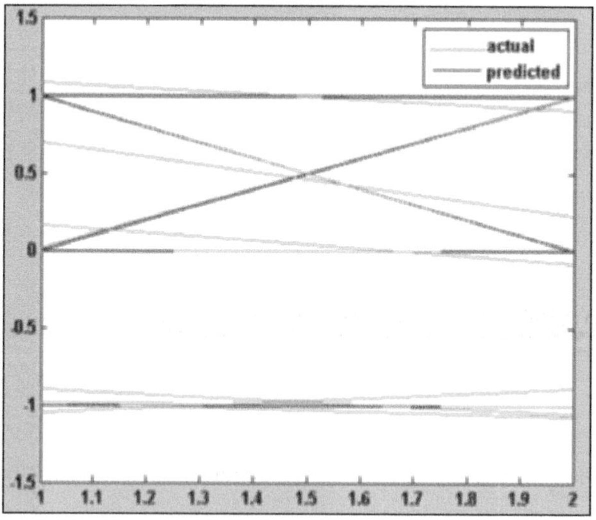


**Fig. 6(a).** Actual and predicted value

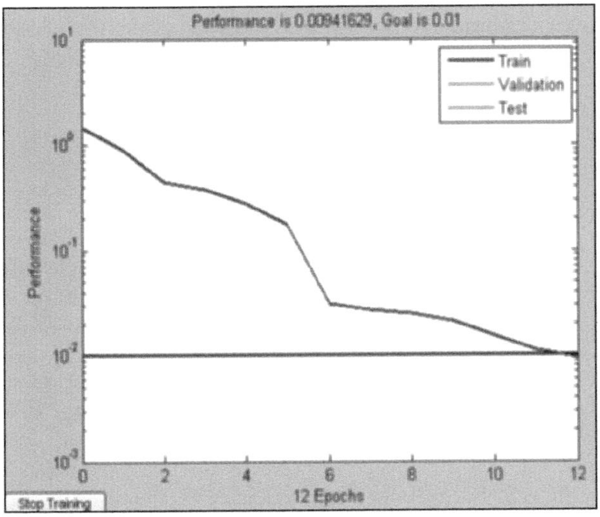


**Fig. 6(b).** Performance of the Network

the learning rate. If the learning rate is set too high, the algorithm may oscillate and become unstable. If the learning rate is too small, the algorithm will take too long to converge. However this sensitivity can be improved if it allows the learning rate to change during the training process. For this research work the learning rate of 0.03 is used and epoch count was limited to 300. The following figure 6 (a) and (b) shows the experimental results of actual and predicted value of a sample Tamil speech[7][8] signal. The performance goal is achieved with the parameters specified for the network.

The mean square error rate is the default performance evaluation measure for the adopted network. It is computed by "summing the squared differences between the predicted values vs. actual value, and then divide the sum by the number of components. A threshold of 0.01 mean square error is good. Our research work achieved minimum MSE for the Tamil speech [9] signal i.e 0.00571769/0.01.

## 7 Conclusion

The latest developments in the research of neural networks are providing new and improved methods of tackling complex problems and tedious tasks such as speech recognition. For human beings, speech is a natural and simple task. However, it is an extremely complex and difficult job to make a computer respond to spoken commands. In this paper such an important task is carried out for recognizing Tamil spoken words. The important feature vectors are extracted from the enhanced speech signal using MFCC and they are given as the input to the feed forward neural network. The adopted network is trained with these input and target vectors. In the successful implementation the results were found to be satisfactory considering less number of training data. In future more number of isolated and continuous words to be trained and tested with this network which may provide functional alternatives to the standard input operations using keyboard and mouse. This preliminary experiment will helps to develop Automatic Speech Recognition system for Tamil language using different approaches like Hidden Markov Models or with other hybrid techniques.

## References

[1] Ahmad, A.M., Ismail, S., Samaon, D.F.: Recurrent Neural Network with Backpropagation through Time for Speech Recognition. In: Intemational Symposium on Communications and Information Technologies (ISCIT 2004), Sapporo, Japan, October 26-29 (2004)

[2] Saheli, A.A., Abdali, G.A., Suratgar, A.A.: Speech Recognition from PSD using Neural Network. In: Proceedings of the International MultiConference of Engineers and Computer Scientists, IMECS 2009, Hong Kong, March 18-20, vol. I (2009)

[3] Hasnain, S.K., Beg, A.: A Speech Recognition System for Urdu Language. In: International Multi-Topic Conference (IMTIC 2008), Jamshoro, Pakistan, pp. 74–78 (2008)

[4] Hasnain, S.K.: Recognizing Spoken Urdu Numbers Using Fourier Descriptor and Neural Networks with Matlab. In: Second International Conference on Electrical Engineering, March 25-26, University of Engineering and Technology, Lahore (2008)

[5] Melin, P., Urias, J., Solano, D., Soto, M., Lopez, M., Castillo, O.: Voice Recognition with Neural Networks, Type-2 Fuzzy Logic and Genetic Algorithms. Engineering Letters, 13(2), EL_13_2_9 (2006); Advance online publication: 4 August

[6] Murugavel, M.: Speech Recognition Model for Tamil Stops. In: Proceedings of the World Congress on Engineering, WCE 2007, London, U.K, July 2-4, vol. I (2007)

[7] Saraswathi, S., Geetha, T.V.: Design of language models at various phases of Tamil speech recognition system. International Journal of Engineering, Science and Technology 2(5), 244–257 (2010)

[8] Saraswathi, S., Geetha, T.V.: Implementation of Tamil Speech Recognition System Using Neural Networks. In: Manandhar, S., Austin, J., Desai, U., Oyanagi, Y., Talukder, A.K. (eds.) AACC 2004. LNCS, vol. 3285, pp. 169–176. Springer, Heidelberg (2004), doi:10.1007/978-3-540-30176-9_22(2004).

[9] Thangarajan, R.: Word and Triphone Based Approaches in Continuous Speech Recognition for Tamil Language. WSEAS Transactions on Signal Processing 4(3) (March 2008)

[10] Uma Maheswari, N., Kabilan, A.P., Venkatesh, R.: A Hybrid model of Neural Network Approach for Speaker independent Word Recognition. International Journal of Computer Theory and Engineering 2(6), 1793–8201 (2010)

# Enhanced Hybrid Compression Models for Compound Images

D. Maheswari and V. Radha

Department of Computer Science,
Avinashilingam Deemed University for Women,
Coimbatore, India
mahileni@gmail.com
radharesearch@yahoo.com

**Abstract.** This paper presents an efficient compound image compression method based on block and layer–based segmentation techniques. Two hybrid models are proposed for segmenting compound images. The first model combines layer-based and block-based techniques for segmentation and compression of compound images. Several experiments were conducted to evaluate both models in terms of compression ratio, PSNR and time.

**Keywords:** Compound Image, Layer–based, Block-based, Segmentation, Compression.

## 1 Introduction

Compound images are different kind of images that contain both palletized regions [1], which has text or graphics, and continuous tone regions. The example of typical compound images are hybrid documents, contents captured from screen, slides, and web-pages. In this fast transmission world, compound images need to be compressed and transmitted in a secure way. Compressing compound images with a single algorithm that simultaneously meets the requirements for text, image and graphics has been elusive and thus requires new algorithms that can competently reduce the file size without degrading the quality [2]. The compound image compression performance basically depends on the segmentation result. A segmentation process is used where regions of similar data types are grouped together. After successful segmentation existing techniques that best suits each data type can be used to achieve best compression results[3].Most of the segmentation algorithms proposed in the literature belong to three categories, namely object-based, block-based and layer-based segmentation [4]. Many researchers follow either layer or block-based methods [5][6][7] some commercial compressors like DjVu[8.] ,Digipaper [9] , JPEG 2000[10] also exists. This paper proposes yet another hybrid method combining both block and layer-based segmentation techniques for compressing compound image compression. The main goal is to develop a hybrid model to handle the overlapping regions of compound images. For this purpose two hybrid models were proposed. Model 1

D. Nagamalai, E. Renault, M. Dhanushkodi (Eds.): CCSEIT 2011, CCIS 204, pp. 265–274, 2011.

combines block-based segmentation with wavelet packets for analyzing compression of overlapping regions. Model 2 combines both the layer -based and block -based techniques for efficient handling of overlapping images.

The paper is organized as follows: Section 1 gave a brief introduction to the topic under discussion. Section 2 discusses the compound image segmentation. Section 3 explains proposed hybrid models. Section 4 presents the results and evaluates the proposed system. Section 5 concludes the paper.

## 2 Compound Image Segmentation Techniques

The compound image segmentation has three categories. Normally researchers follow two types of compound image segmentation techniques, They are

1. Layer- based segmentation and
2. Block-based segmentation.

This paper combines both of these techniques. In layer-based segmentation, a page is divided into rectangular layers. In block-based segmentation page is divided into rectangular blocks where each block follows exact object boundaries.

### 2.1 Layer-Based Segmentation

The steps followed during layer–based segmentation and compression, are,

- Segmentation of image into layers
- Production of a binary mask image
- Data filling of sparse foreground image
- Data filling of sparse background image
- Compression of each layer with an appropriate encoder
- Package of the whole representation into one meaningful stream for transmission or storage.

Most layered coding algorithms use the standard three layers Mixed Raster Content (MRC) representation.

### 2.2 Block-Based Segmentation

In this technique the compound image is divided in blocks of a certain size (e.g. 8x8) and a classification method is applied to the block to segment the image into blocks, which can later be compressed using different techniques. The classification information is also coded before coding the block pixels and usually it depends only on the values of the pixels inside a block and possibly the decision on the neighboring blocks.

# 3 Hybrid Models

The block and layer-based techniques separately gives better performance. Combining the two segmentation methods gives much better results than the individual. For this purpose hybrid model 1 and model 2 are proposed.

## 3.1 Model 1

Model 1 uses block classification method to identify various blocks in the compound image. As the first step, the image pixels are grouped into three classes, namely Low-Gradient, Mid-Gradient and High-Gradient pixels. In this model the compound image is first divided into 16x16 blocks. Using the gradient values and a combination of decision rules, the block is classified into five different regions, namely, background block, text block, picture block, graphics block and overlapping blocks. Out of all the five blocks, the main challenge was faced with the overlapping block, which had mixed contents. For this purpose a wavelet packet-based compression technique is used. The overall process is given in Figure 1.

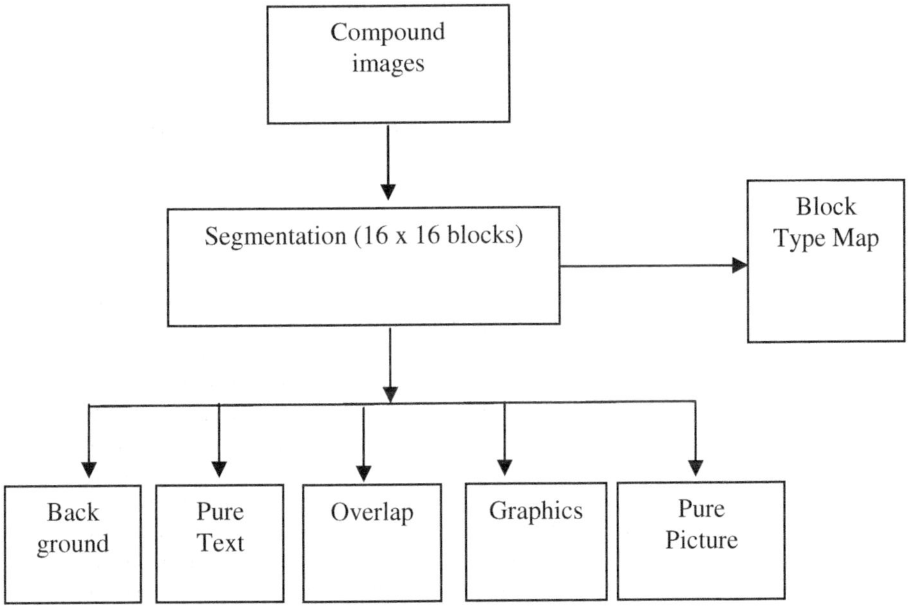


**Fig. 1.** The Proposed Model 1 Segmentation Technique

During segmentation, two features, namely, histogram and gradient of the block is used. Using the gradient values, the histogram distribution for each pixel group is computed. For this purpose the following decision rules are used, that are shown in Figure 2.

| | | |
|---|---|---|
| **Rule 1** : *If HG + LG < $T_1$* | → | *Picture block* |
| **Rule 2** : *If HG < T2 and LG > T3 and Gray level Range > T4* | → | *Background block* |
| *Else* | → | *Picture block* |
| **Rule 3** : *If Major colors Number ≤ 2 and HGP>T6* | → | *Text block* |
| *Else if colors ≥ 2* | → | *Graphics block* |
| *Else* | → | *Picture bock* |
| **Rule 4** : *If HG ≥ T7* | → | *Overlap block* |
| *Else* | → | *Picture block* |

**Fig. 2.** Decision Rules of Model 1

After segmenting a compound image using wavelet packet tree the image is divided into background, pure text, overlap, graphics and pure picture regions. The background block is compressed using arithmetic coder, the pure text block, graphics, pure picture blocks are compressed using XML coder [3], Palette based coder and modified JPEG coder respectively. Then the overlap block is compressed using the proposed wavelet packet tree coder. The overlapping block, which cannot be compactly represented both in spatial and frequency domain contain strong high frequency signals due to text edges. In order to fully utilise these during compression a wavelet packet tree based compression is used.

The tree construction procedure is as follows: Information content of decomposed component (approximation and details) may be greater than the information content of components, which has been decomposed. This method reduces the time complexity of wavelet packets decomposition and selects the sub-bands that have vital information. The decision of further decomposition and cost calculation is based

1. *Let level counter = 1*
2. *Let the current node = input image.*
3. *Decompose current node using wavelet packet tree.*
4. *Find the entropy of current node.*
5. *Find the entropy of decomposed components.*
6. *Compare the entropy of parent node with the sum of the entropy of child node. If the sum of the entropy of child nodes is less than that of parent node, then child node will be considered as a leaf node of a tree and repeat the steps 3-6 for each child nodes considering it as current node. Otherwise parent node acts as a leaf node of a tree.*
7. *Stop.*

**Fig. 3.** Algorithm for Overlapping Block

on a strategy, which decides during run time whether to retain or prune the decomposed components. After the best basis has been selected based on cost function, the image is represented by a set of wavelet packets coefficients, which is further compressed using run length encoder to give better overall compression. The algorithm is given in Figure 3.

### 3.2 Model 2

In model 2, the two techniques layer-based and block-based are combined to form a hybrid compound image compression technique. The method proposed uses model 1 for segmentation, but differs in the handling of the overlapping block. The overlapping block is dealt with separately and uses the layer-based approach to segment the block into three layers text, picture and mask. Then the various compression schemes suiting the need of block or layer are applied. The text layer is compressed using the XML coder , The mask layer is compressed using JBIG coder and the picture layer is compressed using the modified JPEG coder. This model combines the advantages of both layer-based and block-based approaches and hence can be considered as an improvement of both the approaches.

The segmentation technique for overlapped block in model 2 is given in Figure 4.

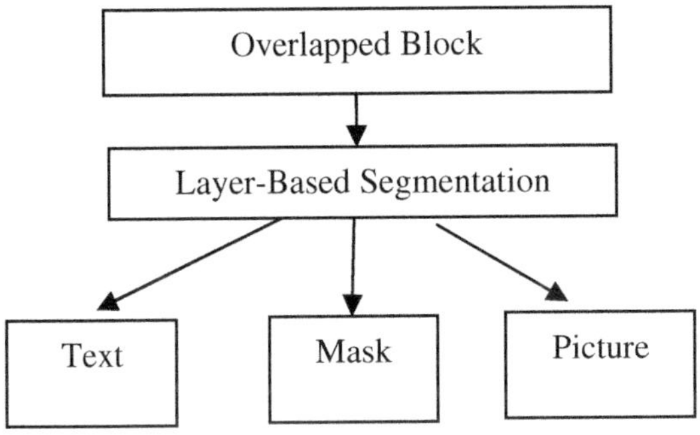


**Fig. 4.** Overlapped block Model 2

## 4 Experimental Results

The main objective of the tests was to find which of the proposed algorithms is best suited for compound color image compression.

### 4.1 Test Images

In order to verify the performance of the proposed schemes, experimental evaluation is performed using a series of different benchmark images. The experimentation is done in three types of compound images, which are scanned document images, computer generated images and photographic images with compound nature. The following Figure 5 is the test images.

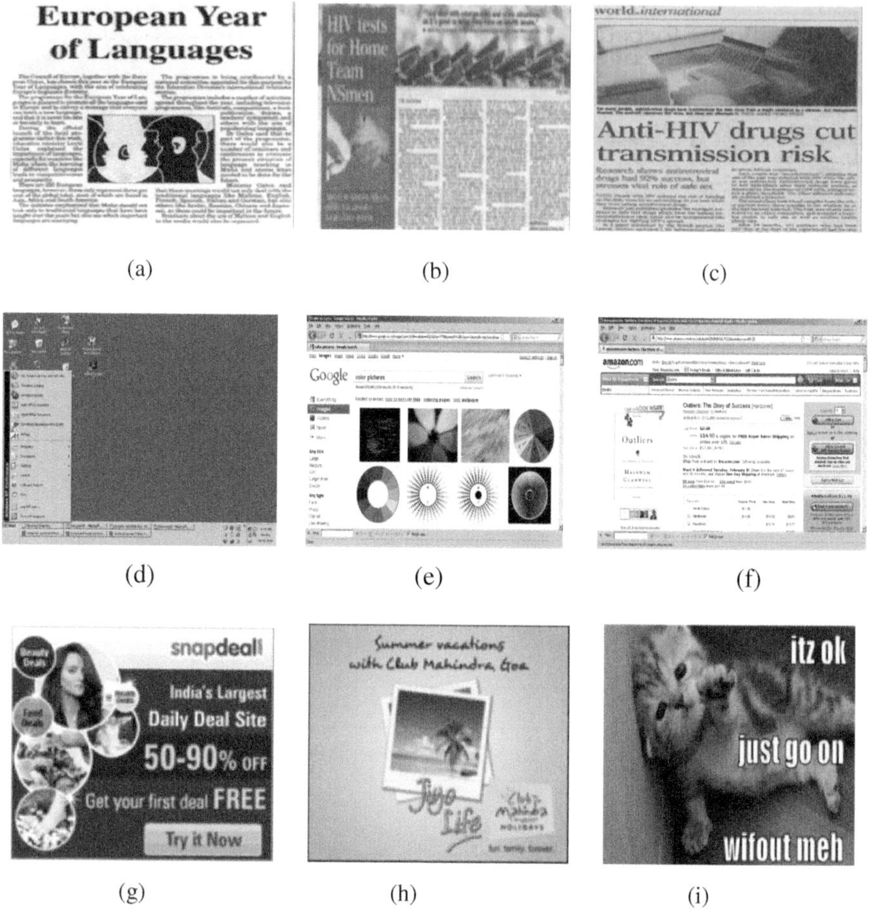


**Fig. 5.** Test Images

For experimentation, 9 images named (a to i) were selected and the selection was done in such a way that they were combination of all classes of compound images.

## 4.2 Performance Metrics

The performance of the proposed models was evaluated using different parameters like compression ratio, compression and decompression time. One of the objectives of the proposed compression models is to maintain the visual quality of the decompressed image. The quality of the decompressed image was ascertained by using the quality metric Peak Signal to Noise Ratio (PSNR).

**Compression Ratio.** The compression ratio of the test images were presented in Table 1. The proposed models are compared with traditional compressors JPEG and DjVu.

**Table 1.** Compression Ratio(%)

| Images | Model 1 | Model 2 | JPEG | DjVu |
|---|---|---|---|---|
| a | 45.02 | 47.91 | 37.43 | 39.89 |
| b | 39.56 | 43.02 | 35.00 | 36.51 |
| c | 39.12 | 42.51 | 34.78 | 36.24 |
| d | 42.77 | 45.91 | 36.39 | 37.67 |
| e | 39.12 | 40.32 | 36.47 | 37.12 |
| f | 39.25 | 41.67 | 35.92 | 36.23 |
| g | 42.26 | 44.32 | 35.64 | 37.62 |
| h | 43.38 | 45.66 | 36.60 | 38.84 |
| i | 40.67 | 43.95 | 36.67 | 38.74 |

From the above Table 1, it can be seen that both model 1 and model 2 are enhanced versions of compression schemes. However model 2 produces better compression ratio for all the test images than model 1.

**Compression Time and Decompression Time.** The following Table 2 shows the result of compression and decompression time for all the models. The compression and decompression time are the time taken for encoding and decoding an image.

**Table 2.** Compression and Decompression Time (Seconds)

| Images | Model 1 | | Model 2 | | JPEG | | DjVu | |
|---|---|---|---|---|---|---|---|---|
| | CT | DT | CT | DT | CT | DT | CT | DT |
| a | 0.77 | 0.65 | 0.75 | 0.62 | 1.02 | 0.79 | 0.89 | 0.73 |
| b | 0.72 | 0.66 | 0.72 | 0.64 | 1.02 | 0.82 | 0.89 | 0.75 |
| c | 0.71 | 0.65 | 0.71 | 0.63 | 0.98 | 0.83 | 0.84 | 0.77 |
| d | 0.73 | 0.67 | 0.72 | 0.66 | 0.87 | 0.83 | 0.83 | 0.72 |
| e | 0.82 | 0.74 | 0.81 | 0.72 | 0.89 | 0.83 | 0.85 | 0.75 |
| f | 0.84 | 0.75 | 0.83 | 0.73 | 0.96 | 0.89 | 0.93 | 0.75 |
| g | 0.72 | 0.65 | 0.71 | 0.62 | 0.92 | 0.84 | 0.82 | 0.76 |
| h | 0.73 | 0.64 | 0.71 | 0. 63 | 0. 93 | 0.84 | 0.84 | 0.76 |
| i | 0.73 | 0.63 | 0.72 | 0.62 | 0.94 | 0.75 | 0.89 | 0.71 |

From above Table 2, again model 1 and model 2 achieves compression in a minimum period of time compared with other traditional compressors.

**Peak Signal to Noise Ratio.** The PSNR is used to judge the quality of decompressed image with that of the original image. The Figure 6 compares the PSNR values of proposed algorithms.

db
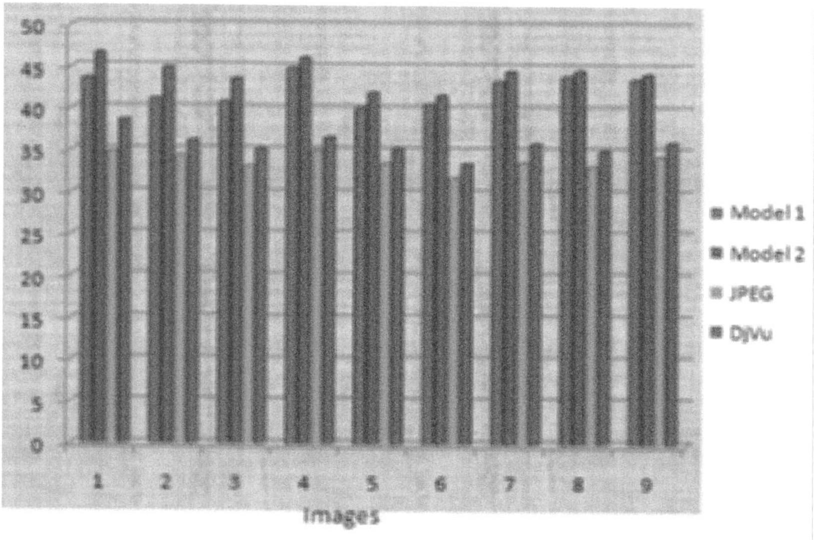


**Fig. 6.** PSNR Value of Four Models

The high PSNR value is obtained by model 2 when compared with other models. To show the improvement of the proposed algorithms, a Visual comparison of the image (h), before and after compression is shown in Figure 7.

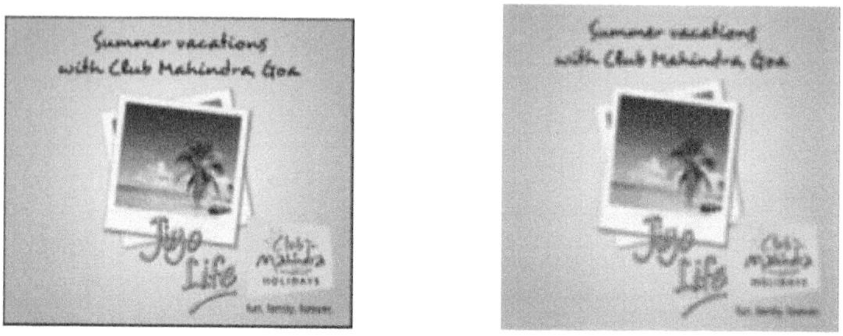

Original Image Model 1

**Fig. 7.** Visual Comparison of Test Image

Model 2

JPEG

DjVu

**Fig. 7.** *(Continued)*

From the above results, Model 1 and Model 2 achieves better results compared with the traditional models.

## 5 Conclusion

This paper proposed two hybrid compound image compression models. The first model divides the image into five blocks and applies block-based segmentation. The second model uses block-based segmentation which is used in the first model and for overlapping block it again applies layer-based segmentation. The proposed models are compared with traditional JPEG and DjVu. During experimentation, it was found that the proposed Model 2 produces good results compared with other models.

## References


1. Liu, D., Ding, W., He, Y., Wu, F.: Quality-biased Rate Allocation for Compound Image Coding with Block Classification, pp. 4947–49500 (2006), 0- 7803-9390-2/06/$20.00 ©2006 IEEE
2. Maheswari, D., Radha, V.: Enhanced Layer Based Compound Image Compression. In: Proceedings of the First Amrita ACM-W Celebration of Women in Computing, pp. 209–216 (2010) ISBN:978-1-4503-0194-7

3. Maheswari, D., Radha, V.: Secure Layer Based Compound Image Compression using XML Compression. In: 2010 IEEE International Conference on Computational Intelligence and Computing Research, pp. 494–498 (2010) ISBN:978-1-4244-5966-7
4. Ding, W., Lu, Y., Wu, F.: Enable efficient compound image compression in H.264/AVC intra coding. In: IEEE International Conference on Image Processing, vol. 2, pp. II-337–II-340 (2007)
5. Radha, V., Aparna, R., Maheswari, D.: Performance evaluation of H.264/AVC intra compound image compression system. International Journal of Computer Applications, Foundation of Computer Science 1(10), 48–54 (2010)
6. Xi, Q., Wu, X., Zhang, S.: Compound Image Compression with Multi-step Text Extraction Method. In: Fifth International Conference on Intelligent Information Hiding and Multimedia Signal Processing, pp. 1270–1273 (2009)
7. Zaghetto, A., de Queiroz, R.L.: Iterative pre-and postprocessing for MRC layers of scanned documents. In: 15th IEEE International Conference on Image Processing, ICIP 2008, pp. 1009–1012 (2008)
8. Bottou, L., Haffner, P., Howard, P., Simard, P., Bengio, Y., LeCun, Y.: High quality document image compression using DjVu. Journal of Electronic Imaging 7(3), 410–425 (1998)
9. Huttenlocher, D., Felzenszwalb, P., Rucklidge, W.: DigiPaper: A versatile color document image representation. In: Proc. ICIP, vol. I, pp. 219–223 (1999)
10. Sharpe, L.H., Buckley, R.: JPEG 2000 .jpm file format: a layered imaging architecture for document imaging and basic animation on the web. In: Proceedings of SPIE, vol. 4115, pp. 464–475 (2000)

# Index Based Ordered Broadcast with Status (IOBS) Algorithm for Consistent Data Broadcast

K. Chitra[1], K. Senthamarai Kannan[2], and S. Abirami[3]

[1] Asst.Prof.,Computer Science,
Govt. Arts & Science college for women,
Pudukottai, Tamilnadu
[2] Professor & Head of Statistics,
Manonmaniam Sundaranar University,
Tirunelveli-12, Tamilnadu
[3] Lecturer in BCA , Dept. of Mathematics,
Thiagarajar college (Autonomous),
Madurai-9, Tamilnadu

**Abstract.** Mobile computing system is more popular as people need data on move. Many emerging mobile database applications demand broadcast based data dissemination, ie, the data broadcasted to more number of clients without any request. In this paper we address the problem of preserving data consistency and currency among mobile transactions, while the broadcast and update transactions executed concurrently on server without any control. We propose an efficient serializability based algorithm (IOBS) for ensuring the consistency of entire database. This algorithm has the following properties.

- The order of update transactions seen by the different mobile clients will be same.
- Consistency and currency are ensured without contacting the server.
- The size of control information for conflict checking is smaller.

**Keywords:** Data Dissemination, Consistency , Serializability, Currency.

## 1 Introduction

Data broadcasting method is very effective technique for disseminating information to massive number of clients when the data base size is small. There are two types[13] [16] of information delivery for mobile data applications .First is push based data delivery(Data broadcast method) broadcast the data items in continuous cycles to the mobile clients, who were all tuned in to the channel. Second is on demand data delivery (pull based delivery) broadcast the data to specific clients who issue request to the server. For example, in E-commerce application and road traffic information systems the database server broadcast data continuously to mobile clients. Mobile clients can get the broadcasted data without the communicating the server.

In data broadcast[3] [15]applications, the asymmetric data flow from servers to clients is more than data flow from clients to servers. Data items stored on database server are highly sensitive. The database server in mobile computing system has to

D. Nagamalai, E. Renault, M. Dhanushkodi (Eds.): CCSEIT 2011, CCIS 204, pp. 275–286, 2011.

broadcast the fresh data items while the update processes executed concurrently with broadcast processes. If there is no control over broadcast transaction and update transaction then the mobile client may read outdated data items (inconsistent data).Concurrency control [14] is one of the important components of transaction management. Traditional concurrency control algorithm (2PL) is not suitable for this kind of system due to the overhead in setting locks. Since with the limited bandwidth available for clients, broadcasting consistent data is challenging research issue. In this paper we guarantee the broadcast of consistent data items (correct data items) and current data item (fresh data items) to the mobile clients. We propose a new algorithm(IOBS)for consistency and currency of data items using serialization criterion. Serializability is a commonly accepted criteria for correctness of set of transactions which are executed concurrently[1].The transactions has serializable property ,if the effect of concurrent transactions execution should be as though all the transactions executed in some serial order. This algorithm ensures data consistency for read only mobile transactions[2].

The following summarizes the contributions of the paper.

1. This algorithm is suitable for large size of updates.
2. The size of control information is very smaller. Since it is suitable for low bandwidth network.
3. This algorithm broadcast the same ordered set of updated data items to all mobile clients.
4. IOBS algorithm broadcast, rebroadcast updated data items with New status for consistency and currency of data items.
5. IOBS ensures consistency over multiple broadcast cycles.

The organization of remaining part of the paper is as follows, Section 2 reviews related works on consistent delivery, Section 3 describes our system model, Section 4 discuss the data inconsistency problem with examples. Section 5 states our new algorithm. finally conclusion and future work of the paper is in Section 6.

## 2 Related Work

Mobile computing system becomes a very active research area. Concurrency control is the most important component of broadcast based data delivery. However traditional concurrency control protocols (2PL) are not suitable for mobile computing system. Due to their over head in detecting conflicts and overhead for locking conflicts [4].

In [5], two approaches were proposed. One is based on broadcasting multiple versions of data items and another one is based on conflict serialization graph. Multi version broadcast approach broadcast the list of different versions of updated data. In invalidation based consistency scheme, server broadcast the control information, invalidation report and serialization graph for each cycle. Upon the receipt of graph, client update its local copy of graph using invalidation report and control information. Each client maintains its local copy of the graph to ensure the serializability. But this approach suffers from the overhead in broadcasting serilization graph and also each client has to listen the channel to maintain its updates.

In [6] F-Matrix and R-Matrix algorithms were proposed for data concurrency. This algorithm broadcast the matrix (m x n) for each cycle. ie. C(i,j) represents the latest cycle number in which some transactions affects the value of Obj and writes to Obi. Then Mobile client performs consistency checking using matrix before reading the broadcast data item. Even though this algorithm broadcast fresh data items but it needs high bandwidth to broadcast the matrix control information and the server has to do expensive computation for control matrix.

In [7], [9], [10], they suggest serializability based mechanisms to maintain the freshness of broadcast data. These approaches abort the update transactions if any one of the following condition meet. If a transaction updates the data items of Not Read Set and Already Read Set.If a transaction updates data items of Not Read Set and read the data items of NUS(set of data .items written by the update transaction after the broadcast transaction). It shows simple but it suffers from overhead of aborting update transactions. Since, these protocol are not suitable for broadcasting sensitive data items.

In multi version data broadcast approach [11],the server broadcast multiple version of each data item. To reduce the latency of client transactions, caching mechanism ,broadcast disk organization approach were suggested. But these approaches require high bandwidth for broadcasting multiple versions.

In [12], they suggested hybrid rebroadcast/ invalidation report mechanism. The server process is same as that of UFO algorithm[4]. The client maintains the schedule between update transaction and mobile transaction to be serializable by combining the rebroadcast and IR method. This algorithm broadcast the ordered read set to mobile clients, But the client hast to contact with the server even for read only transactions. it's performance suffer from restart process of mobile transactions.

To our knowledge, UFO algorithm in [4] has been shown to be more efficient algorithm.In this algorithm serializability is used as consistency criterion. During the execution phase the set of data items of broadcast and update transactions are compared. If there is any overlap (ie,updated data item is in already broadcast list) this algorithm rebroadcast them. Otherwise the serialization order is U→BT. This approach broadcast the fresh data item without any control information and it has minimal impact on server. But the different mobile transactions are allowed to perceive different serialization orders of update transactions[8] (ie, the order of read operation of mobile clients is not taken in to account) It is not suitable if the data conflict size is larger and there is a overhead due to the division of execution of update transaction in to two phases.

## 3 Transaction Organization

A mobile computing system provides data, voice and video services between clients from anywhere, any time and in any form. Mobile computing system consists of database servers and a number of clients. In such environment, the server has high broadcast bandwidth while the band width of client is low. The database server maintains the dynamic information of external environment. Such information are highly sensitive. Such as stock quotes information system, location of moving object, Ecommerce applications. To maintain the freshness, newly updated data items of database server have to be committed and broadcasted instantly while the status of external environment has changed. Otherwise the values of updated data items may become useless. The mobile clients may get outdated data values. For example, in stock information system if a client tune in to the mobile network to read the current value of a stock. The updated data item(fresh data item) is very important to his business decision.

Our system model based on a broadcast environment in which database server periodically broadcast the data items to clients. The period for which broadcasting data items is called a broadcast cycle. The length of the broadcast cycle may be fixed or variable. [4]. No cache is in mobile clients. Mobile clients can fetch the data items from the broadcast channel without contacting data base server. In further, we assumed that 2PL (Two phase lock in method) [1] is used for concurrency control among update transaction at data base server.

### 3.1 Mobile Client's Read Only Transactions and Server's Update Transactions

Mobile clients do not need to continuously listen broadcast. They can tune in to read specific item. To do so, clients must have the prior information about the structure of broadcast. The mobile clients issue the request (Read Only), called mobile transactions(MBT). We have modeled the broadcast process of database server as broadcast transaction (BRT) and update process generated by external environment as update transaction (UPT). They can execute concurrently on data base server.

### 3.2 Assumptions

- The data base server maintains each data item in a row with unique row index.
- All the update transactions (UPT) are executed only on server.
- All the mobile transactions are read only
- All the mobile clients can get the same ordered set of updated data.
- The number of update transactions on server is lower than the number of broadcast transactions.
- For each update transaction , Broadcast transaction broadcast latest version of data item, updated status information.

## 4 Inconsistency Problem in Data Dissemination

If data base server allows the interleaved execution of update and broadcast transaction without any control, the mobile transactions may read inconsistent data item. In this paper, consistency of data item is ensured using serializability criterion[1]. In the following three examples [4], we discuss the data inconsistency problem of stock information system . we consider D1 as stock value of Reliance, D3 as stock value of Ranbaxy, D5 as stock value of Tata.

**Example 1:** The data conflicts between update transaction and mobile transaction.

Suppose the update transaction (UPT) update the stock values of D1 and D3, a mobile transaction (MBT) needs to read the stock value of D1 and D3. Then the schedule is,

i).BRT broadcast D3.
ii).MBT reads D3.
iii).UPT updates D1.
iv).UPT updates D3.
v).BRT broadcast D1.
vi).MBT reads D1.

The mobile transactions (MBT) may read inconsistent data items . The serialization graph is cyclic, such as MBT→UPT→MBT it is non serializable[4]. The reason is that BRT broadcast D1 before to the execution of update transaction. MBT couldn't read updated value of D1.

**Example 2:** The data conflicts with more update transactions (UPT).

Suppose two update transaction (UPT1, UPT2), such as UPT 1 updates the value of D3, D1 and UPT 2 updates the values of D1, D5. Then the schedule is,

i).BRT broadcast D1.
ii).MBT reads D1.
iii).UPT1 updates D3.
iv).UPT1 updates D1.
v).UPT2 updates D1.
vi).UPT2 updates D5.
vii).BRT broadcast D5.
viii).MBT reads D5.

The serialization graph is cyclic, such as UPT2→MBT→UPT1→UPT2.

**Example 3:** Data conflicts with multiple broadcast cycles.

Suppose a mobile transaction may not able to find all his data items in a single broadcast cycle. In consistency may occur over more than one broadcast transactions. For example, if the schedule is,

i).BRT1 broadcast D1.
ii).MBT reads D1.
iii).End of BRT1
iv).UPT1 Updates D1
v).UPT1 Updates D3
iv).BRT2 broadcast D3.
vii).MBT reads D3.

The serialization graph is cyclic, such as MBT→UPT1→ MBT.

The data inconsistency problems in the three examples can be resolved using our IOBS algorithm.

## 5 Index Based Ordered Broadcast with Status Algorithm (IOBS)

When the broadcast transaction is being processed, there can be two types of update transactions. One is update transaction executed before the broadcast transaction, another is update transaction executed after broadcast transaction. Since instead of aborting update transaction, the IOBS algorithm detects the data conflicts among update and broadcast transaction to maintain the data consistency and currency.

The basic process of IOBS algorithm is to preserve the data consistency for the update transactions serialized before broadcast transaction and the update transaction serialized after broadcast transaction. The implementation of IOBS algorithm will be discussed in detail in section **5.4.**

## 5.1 Details of IOBS Algorithm

The IOBS algorithm will be executed for each broadcast cycle indefinitely at data base server. This algorithm has three phases,

- Update phase
- Commit phase
- Broadcast phase.

### 1. Update phase

During update phase, it is assumed that a well formed concurrency control protocol 2PL [1] is adopted for preserving data consistency among update transactions. In this phase the broadcast transaction is blocked. Since we have assumed the number update transactions is smaller than broadcast transaction. This phase maintains ordered updated data items and the ordered (based on the order of updations) row indices of updated data items.

### 2. Commit Phase

After the completion of update phase, commit phase will be executed. During this phase each row index from ordered list of row indices of updated data items (UPI) is compared with the index of data items to be broadcasted (BRI). If there is a match, the server broadcast the corresponding data item from UPD and with the status value "New". Then the newly broadcasted data is removed from BRD to avoid duplicated data broadcast. Since the serialization order is UPT→ BRT. Otherwise serialization order is BRT→ UPT. (ie. The index of updated data item may be in the index list of already broadcasted list) This algorithm rebroadcast newly updated data item from UPD and with the status value "New". At the end of this phase UPI, UPD are reinitialized for the next update transaction.

### 3. Broadcast Phase

During this phase IOBS algorithm broadcast the data items alone from BRD .Since this algorithm resolved the data inconsistency problem by rebroadcasting newly updated values with the status of "New". Then all the mobile transactions can read the fresh data items in the order of update transactions executed at server. The IOBS algorithm detect the data conflict across the broadcast transactions also.

## 5.2 The IOBS Algorithm

In this section, we will state the IOBS algorithm. First we defined the following quantities.

UPD=The set of updated data items of an update transaction on current broadcast cycle.

UPI= Ordered indices set of updated data items of an update transaction on current broadcast cycle.

BRD=The set of data items to be read by the broadcast transactions.

BRI= The set of indices of data items to be read by the broadcast transaction .

The data base server execute the following algorithm for each cycle indefinitely.

```
Ser()
{
 While (true){
 If (UPT is in execution)
{
    Block BRT.
    UPI+= Ordered Index of updated data item.
   UPD+= Ordered Updated data item.
}
else if (UPT is committed)
{
      for each ( X in UPI)
         {
      If (X ∈  BRI)
      {//serialization order is UPT →BRT.
      Remove data item from BRD where  row index=X
      Broadcast data item T, status =New from UPD where
row index = X.
         }
         else
         {//serialization order is BRT →UPT.
          Rebroadcast data item T, status = New from UPD where
          row index = X.
        }
      }
       UPI={ }, UPD={ }
   }
else
{//no dependency between UPT and BRT.
Broadcast each data item T from BRD.}
}// End while}
```

### 5.3 Examples

Referring to the examples in section 4, we will discuss the schedule under IOBS algorithm.

**Example 1**

i).BRT broadcast D3.

ii).MBT read D3.

iii).UPT1 updates D1.
iv).UPT1 updates D3.
v).UPI={index of D1, D3}.
UPD={updated values of D1, D3.}
BRD={D1 }
BRI={index of D1}
Compare the two sets, UPI and BRI.
vi).BRT broadcast D1 with status=New
and remove it from BRD.
vii).MBT reads D1 with the updated New.
viii).BRT rebroadcast D3 with status =New.
ix).MBT reads D3 with status =New.

**Example 2**
i).BRT broadcast the stock value of D1.
ii).MBT reads D1 .
iii).UPT1 updates D3.
iv).UPT1 updates D1.
v).UPI={index of D3, D1}
UPD={new value of D3, D1}.
BRD={D5}.
BRI={ index of D5}.
Compare the two set of UPI and BRI.
vi).BRT rebroadcast D3 with status=New
vii).MBT reads D3 with status=New
viii).BRT rebroadcast D1 with status=New
ix).MBT reads D1 with status=New
x).UPT2 updates D1.
xi).UPT2 updates D5.
xii)UPI={index of D1, D5}
UPD={new value of D1, D5}
BRD={D5 }
BRI={ index of D5}
Compare UPI and BRI.
xiii).BRT rebroadcast D1 with status=New
xiv).MBT reads D1 with status=New
xv).BRT broadcast D5 with status= New.
and remove it from BRD.
xvi).MBT reads D5 with status=New

**Example 3**
i).BRT1 broadcast D1
ii).MBT reads the D1
iii).End of BRT1
iv).UPT1 updates D1
v).UPT1 updates D3
vi).UPI={index of D1, D3}
UPD={new values of D1, D3}

BRD={ D3}
BRI={ index of D3}
Compare UPI and BRI.
vii).BRT2 Rebroadcast D1 with status= New
ix).MBT reads D1 with status= New
xi).BRT2 broadcast D3 with status=New and remove it from BRD
x).BRT2 reads D3.

## 5.4 Implementation of IOBS Algorithm

### 5.4.1 Properties

The important properties of IOBS algorithm are

i).The size of the control information used to retain the data consistency is minimum.
ii).It ensures the data consistency and ordered read set for mobile clients.
iii).The impact of this algorithm on server is minimum.

The implementation of IOBS algorithm is simple and does not require any algorithm for mobile clients. Mobile transaction can access the broadcast data without contacting server. All the conflict checks will be executed on server. The overheads of the algorithm are:

1.comparison between index set of updated data items and index set of broadcasted data items to maintain the serializability.
2. The broadcast transaction is blocked while the update transaction is in execution.
3.rebroadcast of conflicting data items.
4.broadcast of updated data items with status information.

The waiting time of broadcast transaction is short since number of update transactions are smaller than broadcast transactions. The IOBS algorithm broadcast data items along with status only for updated data items .Hence overhead for broadcasting status should be minimum.

The advantage of rebroadcast is that every time most updated data items are broadcasted. Index set comparison speed up the conflict checking process and the usage of index set simplifies the large size of updates.

### 5.4.2 Correctness

We have seen how IOBS algorithm has low over head and minimal impact on server. Four tasks of IOBS algorithm summarized in the following fig.1.

**Task 1**

It corresponds to the first if part in 5.2. In this task broadcast transaction is stopped, ordered index set of updated data items(UPI) and UPD framed.

**Task 2**

It corresponds to the second if part in 5.2. In this task, the two sets UPI and BRI were compared. For each index of UPI, if the index is in BRI then this phase broadcast the fresh value of updated data item with New status and it removes the corresponding data item, index from BRD and BRI to avoid the duplicated data broadcast. ie, this task will be executed if the update transactions executed before broadcast transactions.

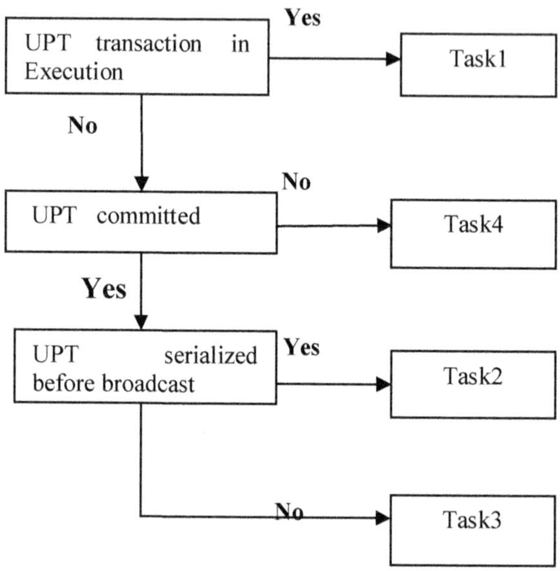


**Fig. 1.** Three phases of IOBS Algorithm

**Task 3**
It corresponds to the third if part in 5.2.In this task , the index of UPI may be in index set of already broadcasted list then the algorithm rebroadcast data item with New status. ie, this task would be executed if the update transactions executed after broadcast transactions.

**Task 4**
It corresponds to the last if part in 5.2.If there is no update transactions ,this task broadcast the data items without any status.

## 6 Conclusion and Future Work

The IOBS algorithm is an efficient algorithm to control the concurrent execution of update and broadcast transactions. This algorithm broadcast s the same ordered set of updated data items to all mobile clients. The ordered broadcast is very useful for dynamic databases. Mobile clients can take instant decisions by using the dynamic data bases . IOBS algorithm is suitable for large size of updations like video, Image ,etc., since row index is used for data conflict checks.

In addition, IOBS algorithm broadcasts the fresh data along with the New status information. Broadcasting status information is use full for the mobile clients to get current and consistent data items even on network disconnections. The followings are threads for future work,

1)The mobile client can issue only read-only query to database server.

2) Detection of data conflict is complicated.

3)Overhead due to rebroadcast of updated data items while the update transactions serialized after broadcast transactions.

## References

1. Date, C.J.: An Introduction to Database systems, 7th edn. Pearson Education, London (2000)
2. Varadarajan, R., Chithra, M.: Scheduling data reads and writes using feedback on a weakly connected environment. International Journal of Computer Applications in Technology 34(3) (2009)
3. Varadarajan, R., Chithra, M.: Reducing latency through efficient channel allocation methods for multiversion data dissemination in mobile computing environment. International Journal of Internet Technology and Secured Transactions 1(3/4) (2009)
4. Lam, K.Y., Chan, E., Au, M.-W.: Broadcast of Consistent Data to Read-Only Transactions from Mobile Clients. In: Proceedings of 2nd IEEE Workshop on Mobile Computing Systems and Applications, New Orleans, Louisiana, USA (February 1999) (3)
5. Pitoura, E.: Supporting Read-Only Transactions in Wireless Broadcastin. In: Quirchmayr, G., Bench-Capon, T.J.M., Schweighofer, E. (eds.) DEXA 1998. LNCS, vol. 1460. Springer, Heidelberg (1998)
6. Shanmugasundaram, J., Nithrakashyap, A., Sivasankaran, R., Ramamritham, K.: Efficient Concurrency Control for Broadcast Environments. In: Proceedings of ACM SIGMOD International Conference on Management of Data, Philadelphia, June 1-3 (1999) (8)
7. Lee, V.C.S., Lam, K.-W., Wu, S., Chan, E.: Broadcasting Consistent Data in Mobile Computing Environments. In: Proceedings of the seventh Real Time Technology and Applications Symposium, RTAS 2001. IEEE Computer Society, Washington, DC (2001)
8. Young, C.-R., Chiu, G.-M.: An Efficient Protocol for Disseminating Consistent Data in Broadcast Environments. In: International Workshop on Ubiquitous Data Management (2005)
9. Poon, B.H.C., Lam, K.-W., Lee, V.C.S.: Broadcasting Consistent Data in Mobile Environments. In: 2004 IEEE International Conference on Mobile Data Management (MDM 2004), p. 169 (2004)
10. Hin, B., Poon, C., Lam, K.-W., Lee, V.C.S.: Server-side Broadcast Transaction in Mobile Computing Environment. In: International Conference on Information Technology: Coding and Computing (ITCC 2004), vol. 1, p. 511 (2004)
11. Pitoura, E., Chrysanthis, P.K.: Multiversion Data Broadcast. IEEE Transactions on Computers 51(10), 1224–1230 (2002)
12. Lam, K.-Y., Chan, E., Leung, H.-W., Au, M.-W.: Concurrency Control Strategies for Ordered Data Broadcast in Mobile Computing Systems. Inf. Syst. 29 (2004)
13. Krishna Mohan Rao, S., Venugopal Reddy, A.: Data Dissemination in Mobile Computing Environment. BVICAM'S International Journal of Information Technology 1(1) (January-July 2009)

14. Moiz, S.A., Rajamani, L., Pal, S.N.: Commit Protocols in Mobile Environments:Design& Implementation. International Journal of Database Management Systems (IJDMS) 2(3) (August 2010)
15. Acharya, S., Alonso, R., Franklin, M., Zdonik, S.: Broadcast Disks: Data Management for Asymmetric Communication Environments. The Kluwer International Series in Engineering and Computer Science 353, 331–361 (1996)
16. Acharya, S., Franklin, M., Zdonik, S.: Balancing Push and Pull for Data Broadcast. In: Proceedings of the 1997 ACM SIGMOD International Conference on Management of Data, vol. 26(2) (1997)

# Capturing High-Level Semantics of Images in Web Documents Using Strength Matrix

P. Shanmugavadivu[1], P. Sumathy[1], and A. Vadivel[2]

[1] Department of Computer Science and Applications
Gandhigram Rural Institute, Dindigul, Tamil Nadu, India
psvadivu@yahoo.com, sumathy_bdu@yahoo.co.in
[2] National Institute of Technology, Trichy, Tamil Nadu, India
vadi@nitt.edu

**Abstract.** The multimedia information retrieval from distributed environment specifically from World Wide Web is considered as one of the challenging issues to be addressed by the researchers. It has been perceived that the description of images using low-level features apparently increases the semantic gap. The information present in a HTML document in the form of textual keywords can be extracted for capturing semantic information with the view to narrow the semantic gap. The high-level textual information of images can be extracted and associated with the textual keywords that narrow down the search space and improve the precision of retrieval. In this paper, a strength matrix, is being proposed, which is based on the frequency of occurrence of keywords and the textual information pertaining to image URLs. The strength of these textual keywords are estimated and used for associating these keywords with the images present in the documents. The high-level semantics of the image is described in the HTML documents in the form of image name, ALT tag, optional description, etc., is used for estimating the strength. The effectiveness of information retrieval of the proposed technique is found to be comparatively better than many of the recently proposed retrieval techniques. The experimental results of the proposed method endorse the fact that image retrieval using image information and textual keywords is better than those of the text based and the content-based approaches.

**Keywords:** Multimedia Information Retrieval, Web Image Retrieval, High-level Features, Textual Keywords.

## 1 Introduction

The revolutionized advent of internet and the ever-growing demand for information sprawled in the World Wide Web has escalated the need for cost-effective and high-speed information retrieval tools. Many attempts have been made to use image contents as a basis for indexing and images retrieval. In early 1990, researchers have built many image retrieval systems such as QBIC [1], Photobook [2], Virage [3], Netra [4] and SIMPLIcity [5] etc., which are considered to be different from the conventional image retrieval systems. These systems use image features such as color,

D. Nagamalai, E. Renault, M. Dhanushkodi (Eds.): CCSEIT 2011, CCIS 204, pp. 287–296, 2011.

texture, shape of objects and so on whereas the recently devised image retrieval systems use text as well as image features. In the text based approach, the images are manually annotated by text descriptors and indexed suitably to perform image retrieval. However, these types of systems have two major drawbacks in annotating the keywords. The first drawback is that a considerable level of human intervention is required for manual annotation. The second one is the inaccuracy annotation due to the subjectivity of human perception.

To overcome the aforesaid drawbacks in text-based retrieval system, content based image retrieval (CBIR) has been introduced [6]. However, these systems are suitable only for domain-specific applications. The low-level features such as color, texture and shape are efficiently used for performing relevant image retrieval [7] - [8]. Color histograms such as Human Color Perception Histogram [9] - [10] as well as color-texture features like Integrated Color and Intensity Co-occurrence Matrix (ICICM) [11] - [12] show high precision of retrieval in such applications. However, the inevitable semantic gap that exists between low-level image features and the user semantics drift the performance of CBIR still far from user's expectations [13]. In addition, the representation and storage of voluminous low-level features of images may be a costly affair in terms of processing time and storage space. This situation can be effectively handled by using keywords along with the most relevant textual information of images to narrow down the search space.

For this purpose, initially it is essential to explore the techniques to extract appropriate textual information from associated HTML pages of images. Many research methods have been proposed on the content structure of HTML document, including image title, ALT tag, link structure, anchor text and some form of surrounding text [20]. The main problem of these approaches is that the precision of retrieval is lower. This disadvantage has triggered the task of developing adaptive content representation schemes, which can be applied to a wide range of image classification and retrieval tasks. Further, the design techniques are also required to combine the evidences extracted from text and visual contents with appropriate mechanisms to handle large number of images on the web. Many recent techniques classify the images into one or more number of categories by employing learning based approach to associate the visual contents extracted from images with the semantic concept provided by the associated text. The principal challenge is to devise an effective technique to retrieve web-based images that combine semantic information from visual content and their associated HTML text.

This paper proposes a faster image retrieval mechanism, which is tested on a large number of HTML documents. For this purpose, the HTML documents are fetched, using a web crawler. The content of the HTML documents is segregated into text and images and HTML tags. From the text, keywords are extracted and these keywords are considered to be the relevant keywords to represent the high level semantics of the images contained in the same HTML document.

In the following sections of the paper, the related works are presented. In section 3, the proposed method is elaborated along with a suitable example. The experimental result is presented in section 4. The conclusion is given in the last section of the paper.

## 2 Related Works

The retrieval of images from the Web has received a lot of attention recently. Most of the early systems have employed text based approach, which exploits how images are structured in web documents. Sanderson and Dunlop [16] were among the first to model image contents using a combination of texts from associated HTML pages. The content is modeled as a bag of words without any structure and this approach is found to be ineffective for indexing. Shen *et al* [14] have built a chain related terms and used more information from the Web documents. The proposed scheme unifies the keywords with the low-level visual features. The assumption made in this method is that some of the images in the database have been already annotated in terms of short phrases or keywords. These annotations are assigned either using surrounding texts of the images in HTML pages or by speech recognition or manual annotations. During retrieval, user's feedback is obtained for semantically grouping keywords with images. Color moments, color histograms, Tamura's features, Co-occurrence matrix features and wavelet moments are extracted for representing low-level features. Keywords in the document title of HTML page and image features have been combined for improving the retrieved documents of news category [17]. In this technique, from a collection of 20 documents chosen from one of the news site has been used and 43 keywords along with HSV based color histogram are constructed. While constructing histogram, saturation and hue axes are quantized into 10 levels to obtain H×S histogram with 100 bins. However, this technique is found to perform well for a small number of web pages and images only. In general, image search results returned by an image search engine contain multiple topics and organizing the results into different semantic clusters may help users. Another method has been proposed for analyzing the retrieval results from a web search engine [20]. This method has used Vision-based Page Segmentation (VIPS) to extract semantic structure of a web page based on visual presentation [18]. The semantic structure is represented as a tree with nodes, where every node represents the degree of coherence to estimate the visual perception.

Recently, a bootstrapping approach has been proposed by Huamin Feng *et al* (2004), to automatically annotate a large number of images from the Web [20]. It is demonstrated that the co-training approach, combines the information extracted from image contents and associated HTML text. Microsoft Research Asia [21] is developing a promising system for Web Image Retrieval. The purpose is to cluster the search results of conventional Web, so that users can find the desired images quickly. Initially, an intelligent vision based segmentation algorithm is designed to segment a web page into blocks. From the block containing image, the textual and link information images are extracted. Later, image graph is constructed by using block-level link analysis techniques. For each image, three types of representations are obtained such as visual feature based representation, textual feature based representation and graph based representation. For each category, several images are selected as non representative images, so that the user can quickly understand the main topics of the search results. However, due to index based retrieval, the time for processing is found to be on the higher side. Rough set based model has proposed for decompositions in information retrieval [22]. The model consists of three different knowledge granules in incomplete information system. However, while WWW documents are presented along with images as input, the semantic of images are exactly captured and thus retrieval performance is found to be lower. Hence, it is important to narrow down the semantic gap between the images and keywords present in WWW.

The textual information of WWW documents, which is the high-level semantics, can be effectively used for defining the semantics of the images without really using the low-level features. This kind of approach simplifies the semantic representation for fast retrieval of relevant images from huge voluminous data. This paper proposes a scheme for extracting semantic information of images present in WWW documents using only the textual information. The relationship between the text and images present in WWW documents estimated with frequency of occurrence of keywords and other important textual information present in image link URLs. Based on the experimental results, it is observed that the performance of the system is better than that of Google (www.google.com).

## 3 Strength Matrix between Keyword and Image High-Level Features for Capturing Image Semantics

Let H be the number of HTML pages, I be the number of images and K be the set of keywords. Thus,

$$H = \{h_1, h_2, h_3 \cdots h_n\},\ I = \{i_1, i_2, i_3 \cdots i_m\} \text{ and } K = \{k_1, k_2, k_3 \cdots k_l\}$$

where $n$, $m$ and $l$ denote the total number of HTML pages, images and keywords respectively.

In order to effectively capture the semantics of $I$ present in $H$, $K$ can be used. The association between $K$ and $H$ can be written as:

$$stg(K \leftrightarrow H) = \left( \frac{stg(K)}{Max(stg(K))} \right) \tag{1}$$

This denotes the association between keywords to HTML pages and can be estimated using the frequency of occurrence of $K$ in $H$. Since, $K$ is the total number of keywords in a single HTML page '$h_p$' may contain only '$k_q$'of keywords. Now, the relation between each keyword '$k_j$' where $(j = 1,2,3,\cdots,q)$ with a single HTML document can be written as

$$stg(k_j \leftrightarrow h_p) = \left( \frac{stg(k_j)}{Max(stg(k_j))} \right)_{j=1,2,\ldots,q} \tag{2}$$

The above equation denotes the association between each keyword $K_j$ in a single HTML document '$h_p$'. Similarly, the relationship between a page and image can be derived. From Eq. 2, we get the importance i.e. strength of each keyword '$k_j$' in a document '$h_p$'. The strength is the measure of the frequency of occurrence and a keyword with a maximum frequency of occurrence is assigned higher strength value, which is 1. Similarly, all the keywords present in a particular document is assigned a strength value by which the importance of that particular keyword is estimated. The example depicted in Table 1 elucidates the estimation of strength using frequency of occurrence of a keyword in a HTML document.

Let the number of keywords in a HTML document is 5 and maximum frequency of occurrence (FOC) is 30.

**Table 1.** Strength Matrix using Frequency of occurrence

| HTML Page | FOC | Strength |
|---|---|---|
| $K_1$ | 10 | 0 |
| $K_2$ | 3 | 1 |
| $K_3$ | 30 | 1 |
| $K_4$ | 8 | 0 |
| $K_5$ | 5 | 1 |

From the above example, we can observe that not all keywords are equally important. It is sufficient to consider only a set of keywords $k_{stg}$ such that the strength of these keywords is greater than a threshold $t_{stg}$. In our approach, we have fixed this threshold as 25% of the maximum strength value. Now, it is important to estimate the strength of the keywords with the images. We have used Image Title, ALT tag, link structure and anchor text as high-level textual feature and feature vector is constructed. Thus, the relation

$$stg\left(k_j \leftrightarrow I_{hp}\right) = stg\left(k_j \leftrightarrow h_p\right) + m\left(TAG, k_j\right) + m\left(IN, k_j\right) + m\left(LS, k_j\right) + m\left(AT, k_j\right) \quad (3)$$

in above equation j = 1, 2, q and $m$ is a string matching function with either 0 or 1 as the output. The output value of each component of above equation is in the rage of [0-1]. This relation is effective in capturing the importance of a keyword in a document and that of images. Both the strength value as well as image related textual information are combined to narrow down the semantic gap of image Sample strength matrix with all features has been depicted in table 2.

**Table 2.** Strength Matrix using both frequency of occurrence and image high-level features

| Keyword | FOC | $stg(k_j \leftrightarrow h_p)$ | $m(TAG, k_j)$ | $m(IN, k_j)$ | $m(LS, k_j)$ | $m(AT, k_j)$ |
|---|---|---|---|---|---|---|
| $K_1$ | 1 | 0.033 | 1 | 0 | 0 | 0 |
| $K_2$ | 30 | 1 | 1 | 1 | 1 | 0 |
| $K_3$ | 20 | 0.66 | 1 | 0 | 0 | 1 |
| $K_4$ | 8 | 0.26 | 1 | 1 | 1 | 0 |
| $K_5$ | 5 | 0.16 | 1 | 0 | 0 | 1 |

In the above table, $K_i$ is a keyword extracted from HTML documents. While extracting the keywords, the stop words are removed and the stemming operation is carried out. The high-level semantic information of the images can be extracted from the above table. Say for example, frequency of $K_2$ is high and also it is matching with most of the image related textual string and thus $K_2$ has more association with the HTML page and captures the semantics of the images.

$m(TAG, k_j), m(IN, k_j), m(LS, k_j)$ and $m(AT, k_j)$ are the match functions to match the similarity between images TAG and keyword, image name and keyword, link structures and keyword and Anchor Text with keyword respectively.

## 4 Experimental Results

For the purpose of crawling, the HTML documents along with the images, an internet crawler was also developed. The various stages of experimental setup is shown in Fig. 1

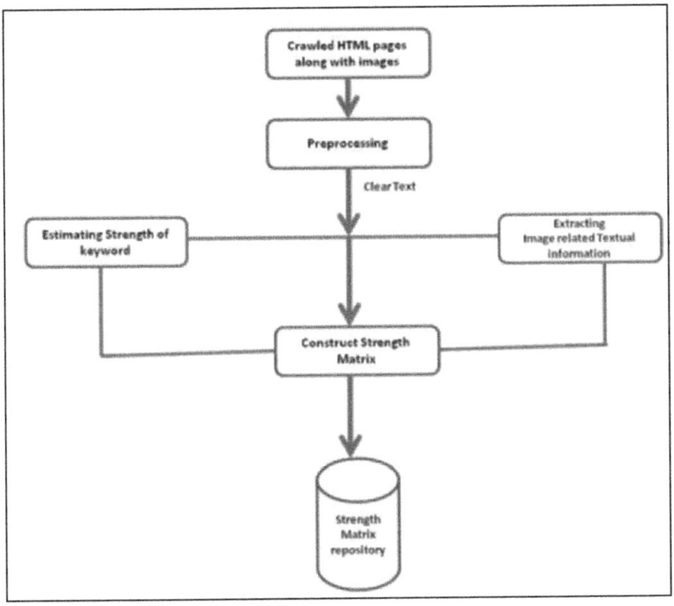


**Fig. 1.** Stages of Experimental Setup

The performance of the proposed technique is measured based on the crawled HTML pages along with the images from WWW. The textual keywords are extracted and the frequency of occurrence of all the keywords is calculated. The text information from URL link of images is also extracted. The strength matrix is constructed using this information stored in a repository. During querying, the query keyword is found in the repository and based on the strength value, the result is ranked. In the experiment, many web documents from various domains such as sports, news, cinema, etc have been crawled. Approximately, 10,000 HTML documents with 30,000 images have been extracted and preprocessed for retrieval. The web crawler provides HTML pages along with the image content. The text in each page is categorized into normal text and image related text. While extracting the keywords, the stop words are removed and the stemming operation is carried out. Once the keywords and image related text are extracted, the weighted matrix is constructed and it is stored in the repository. For each page, this matrix is constructed and stored in the repository along with a relation between the document and image present. While making a query based on textual keywords, search will be carried out in the repository and the final ranking of retrieval is based on the weighted value. The images with higher weights are ranked first and will tip the retrieval result.

Fig.2 illustrates the obtained sample result for a keyword "A Vadivel" in www.google.com. The fourth, fifth and nineteenth images are relevant to the query and however, these images are ranked lower. The outcome of the proposed retriever for the same keyword as query, is shown below in Fig. 3 It is observed from Fig. 3 that the proposed system has retrieved the relevant image from WWW and ranked the relevant images higher. This is due to the fact that the strength matrix constructed from each page effectively captures the association between images and keywords. In addition, manually looked into the textual content of each HTML page and estimated the strength and are given for each image as the percentage of the estimated strength value was also computed manually. For evaluating the performance of the proposed approach in our system, the precision of retrieval is used as the measure. Moreover, the obtained results are compared with some of the recently proposed similar approach and are presented in Fig. 4. The average precision (P) in percentage for 10, 20, 50 and 100 nearest neighbors is given. We have compared performance with Hierarchical clustering (HC) [21]; Rough Set (RS) based approach [22] and Bootstrap Framework (BF) [20]. From the results shown in Fig. 4, it is observed that the performance of the proposed approach is quite encouraging. The precision of retrieval using the strength matrix is found to be high compared to others. The reason for this performance enhancement is due to the effectiveness of strength matrix in capturing the high level semantics of the images.

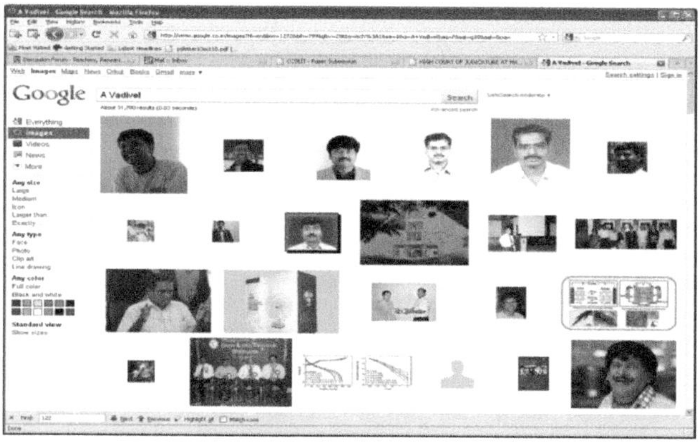


**Fig. 2.** Sample query for retrieving images from Google.com

**Fig. 3.** Sample retrieval result from the proposed system for query keyword "A Vadivel"

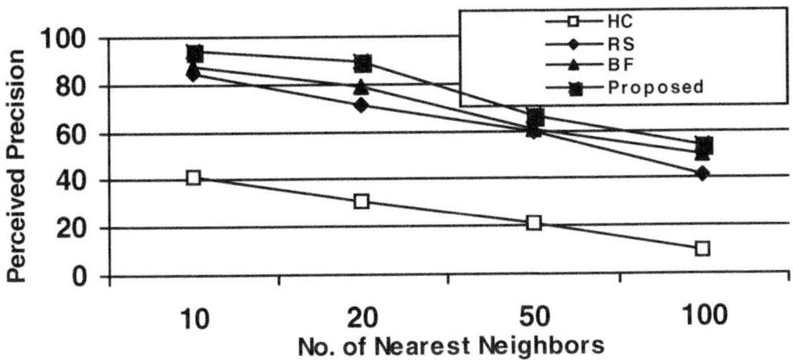


**Fig. 4.** Comparison of Precision of Retrieval using strength matrix

## 5 Conclusion

The role of textual keywords for capturing high-level semantics of an image in HTML document is studied. It is felt that the keywords present in HTML documents can be effectively used for describing the high-level semantics of the images present in the same document. Additionally, a web crawler was developed to fetch the HTML document along with the images from WWW. Keywords are extracted from the HTML documents after removing stop words and performing stemming operation. The strength of each keyword is estimated and associated with HTML documents for constructing strength matrix. In addition, textual information presents in image URL is also extracted and combined with the strength matrix. It is observed from the experimental result that both textual keywords and keywords from image URL achieves high average precision of retrieval.

## References


1. Faloutsos, C., Barber, R., Flickner, M., Hafner, J., Niblack, W., Petkovic, D., Equitz, W.: Efficient and Effective Querying by Image Content. J. Intell. Inf. Syst. 3(3-4), 202–231 (1994)
2. Pentland, A., Picard, R.W., Scaroff, S.: Photobook: Content-based manipulation for image databases. Int. J. Comput. Vision 18(3), 233–254 (1996)
3. Gupta, R.J.: Visual Information Retrieval. Commun. ACM 40(5), 70–79 (1997)
4. Ma, W.Y., Manjunath, B., Netra: A toolbox for navigating large image databases. In: Proceedings of International Conference on Image Processing, pp. 568–571 (1997)
5. Wang, J.Z., Li, J., Wiederhold, G.: Simplicity: semantics-sensitive integrated matching for picture libraries. IEEE Trans. Pattern Mach. Intell. 23(9), 947–963 (2001)
6. Liu, Y., Zhang, D.S., Lu, G., Ma, W.-Y.: A survey of content-based image retrieval with high-level semantics. Pattern Recognition 40(1), 262–282 (2007)
7. Long, F., Zhang, H.J., Feng, D.D.: Fundamentals of content-based image retrieval Multimedia Information Retrieval and Management. Springer, Berlin (2003)
8. Rui, Y., Huang, T.S., Chang, S.-F.: Image retrieval.: Current techniques, promising directions, and open issues. J. Visual Commun. Image Representation 10(4), 39–62 (1999)

9. Vadivel, A., Shamik, S., Majumdar, A.K.: Robust Histogram Generation from the HSV Color Space based on Visual Perception. International Journal on Signals and Imaging Systems Engineering 1(3/4), 245–254 (2008)
10. Gevers, T., Stokman, H.M.G.: Robust Histogram Construction from Color Invariants for Object Recognition. IEEE Transactions on Pattern Analysis and Machine Intelligence 26(1), 113–118 (2004)
11. Vadivel, A., Shamik, S., Majumdar, A.K.: An Integrated Color and Intensity Co-Occurrence Matrix. Pattern Recognition Letters 28(8), 974–983 (2007)
12. Palm, C.: Color Texture Classification by Integrative Co-Occurrence Matrices. Pattern Recognition 37(5), 965–976 (2004)
13. Liu, Y., Zhang, D., Lu, G.: Region-based image retrieval with high-level semantics using decision tree learning. Pattern Recognition 41, 2554–2570 (2008)
14. Shen, H.-T., Ooi, B.-C., Tan, K.-L.: Giving meaning to WWW images. ACM Multimedia, 39–47 (2000)
15. Yanai, K.: Generic image classification using visual knowledge on the web. ACM Multimedia, 167–176 (2003)
16. Sanderson, H.M., Dunlop, M.D.: Image retrieval by hypertext links. ACM SIGIR, 296–303 (1997)
17. Zhao, R., Grosky, W.I.: Narrowing the Semantic Gap—Improved Text-Based Web Document Retrieval using Visual Features. IEEE Transactions on Multimedia 4(2), 189–200 (2002)
18. Cai, D., He, X., Ma, W.-Y., Wen, J.-R., Zhang, H.: Organizing WWW Images based on the Analysis of Page Layout and Web Link Structure. In: Proc. Int. Conf. on Multimedia Expo, pp. 113–116 (2004)
19. Cai, D., Yu, S., Wen, L.R., Ma, W.Y.: VIPs: a vision based page segmentation algorithm. Microsoft Technical Report, MSR-TR-2003-79 (2003)
20. Feng, H., Shi, R., Chua, T.-S.: A bootstrapping framework for annotating and retrieving WWW images. In: Proceedings of the ACM International Conference on Multimedia (2004)
21. Cai, D., He, X., Li, Z., Ma, W.-Y., Wen, J.-R.: Hierarchical clustering of WWW image search results using visual, textual and link information. In: Proceedings of the ACM International Conference on Multimedia (2004)
22. Wu, C., Hu, X.: Applications of Rough set decompositions in Information Retrieval. International Journals of Electrical and Electronics Engineering 4(4) (2010)

# Multiple Access Scheme Using MSE-OFDM with Pseudo Random Cyclic Prefix

Biswajit Datta[1], Anupam Karmakar[2], and Mrinal K. Naskar[3]

[1] Department of Electronics and Communication Engineering,
College of Engineering and Management,
Kolaghat, West Bengal,
Pin – 721171, India
data_biswajit@yahoo.com

[2] Department of Electronic Science,
University of Calcutta, Kolkata,
West Bengal,
Pin – 700 009, India
anupamkarmakar@yahoo.co.in

[3] Department of Electronics and Telecommunication Engineering,
Jadavpur University,
West Bengal,
Pin – 700 032, India
mrinalnaskar@yahoo.co.in

**Abstract.** A new multiple access scheme using Multi-Symbol Encapsulated Orthogonal Frequency Division Multiplexing (MSE-OFDM) is proposed. This is basically a time division multiple access (TDMA) technique utilizing the MSE-OFDM. The Bit Error Rate (BER) performance of the TDMA system using conventional OFDM is slightly better than that of the proposed system. This weakness is compensated as the proposed technique exhibits an improvement in bandwidth efficiency compared to the TDMA using conventional OFDM.

## 1 Introduction

Orthogonal Frequency Division Multiplexing (OFDM) is a digital multicarrier modulation technique. The technique of multicarrier modulation was first proposed in 1966 by R.W. Chang. But at that time the technique did not become popular because of complexity in its implementation. The concept of OFDM was first proposed in mid-1980s [1]. But the use of OFDM for wireless communications became possible only in early 1990s with the advances in hardware for digital signal processing. The main advantage of OFDM is its robustness to frequency selective fading and multipath propagation which has made it a natural choice for high speed data communication in mobile environment. In the recent years OFDM has found its applications in Digital Audio Broadcasting (DAB), Digital Video Broadcasting (DVB) and High Speed Digital Subscriber Line (HDSL). OFDM has been set

D. Nagamalai, E. Renault, M. Dhanushkodi (Eds.): CCSEIT 2011, CCIS 204, pp. 297–306, 2011.

as the standard of air interference for Wireless LANs (802.11a and 802.11g). In spite of its growing applications OFDM suffers from some inherent drawbacks like lower bandwidth efficiency due to the Cyclic Prefix (CP) overhead, high Peak-to-Average-Power-Ratio (PAPR) and stringent requirements for time and frequency synchronisation. The Multiple-Symbol Encapsulated OFDM (MSE-OFDM) technique [2,3] overcomes some of the above mentioned drawbacks to a great extent.

In MSE-OFDM technique [2,3] a number of OFDM symbols are grouped together and are protected by a single CP. CP in this case comprises of last few samples of the last OFDM symbol. The length of CP is at least equal to the length of the Channel Impulse Response (CIR). There are two proposed variations of MSE-OFDM [2], viz., (i) CP reduced MSE-OFDM and (ii) FFT-size reduced MSE-OFDM. In CP reduced MSE-OFDM, shown in Figure 1, a number of conventional OFDM symbols are put together into a frame with a single CP. The fame is known as an MSE-OFDM frame. As there is a single CP for a group of OFDM symbols the CP-reduced MSE-OFDM exhibits an improvement of bandwidth efficiency compared to the conventional OFDM. In case of FFT-size reduced OFDM, shown in Figure 2, the size of the MSE-OFDM frame is made equal to that of a conventional OFDM symbol. Thus the size of individual OFDM symbol is much less than that of the conventional one. This implies a less number of subcarriers and hence a lower PAPR. The smaller number of subcarriers also implies higher spacing between subcarriers. The frequency synchronisation error depends on the relative frequency offset, i.e., the ratio of the actual frequency offset to the subcarrier spacing. Hence more subcarrier spacing means less frequency synchronisation error. The time synchronisation error and bandwidth efficiency in this case, however, remain unchanged compared with the conventional OFDM.

Considering the advantages of MSE-OFDM, in this paper we propose a multiple access scheme using MSE-OFDM for mobile environment. Here we consider a typical mobile environment consisting of a single Base Station (BS) or Access Point (AP) and multiple Mobile Stations (MS) which communicate to an external network via the BS. In this case we propose a multiple access scheme for data transmission from the BS to different MS's. Two multiple access schemes using MSE-OFDM have been proposed in [4] and [5]. In the first case [4] a complete MSE-OFDM frame is necessary at the receiver for frequency equalisation. In the second case [5] also all OFDM symbols of a complete frame are required at the receiver for the removal of Inter Symbol Interferences (ISI) and Inter Carrier Interferences (ICI). Hence the data destined for different MS's cannot be put together into one MSE-OFDM frame as proposed in [4,5]. Also both the schemes are not suitable for packet transmission because in both cases continuous transmission from all sources have been proposed. In that case there will be reduction in battery life of the MS's.

The rest of the paper is organised as follows. The proposed multiple access schemes using MSE-OFDM has been described in Section 2. This section discusses the transmitter, receiver and the equalisation technique for the proposed multiple access scheme. The system performance has been discussed in Section 3 using the simulation results. Section 4 concludes the paper.

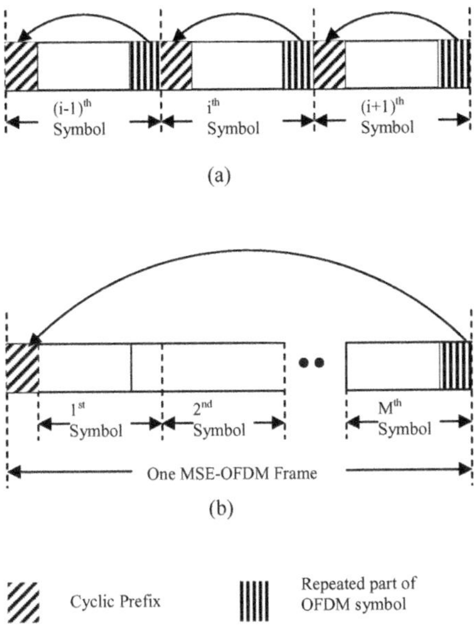


**Fig. 1.** Structure of CP-reduced MSE OFDM. (a) Conventional OFDM symbols. (b) An MSE-OFDM frame.

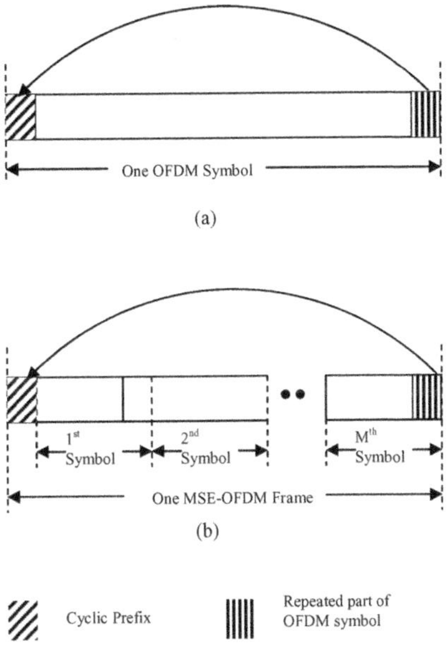


**Fig. 2.** Structure of FFT-size reduced MSE OFDM. (a) Conventional OFDM symbols. (b) An MSE-OFDM frame.

## 2 Proposed Multiple Access Scheme

In this proposed system the input data packets for different users are stored into a buffer at the BS. From the buffer the packets are transmitted to the destination MS's in sequence of their arrivals. Each data packet forms a complete MSE-OFDM frame so that the complete frame is demodulated at the destination MS only. Several MSE-OFDM frames will comprise a one multi-access frame. The packet for each MS will get transmission opportunity in its allocated slot in the multi-access frame. In our case we proposed an MSE-OFDM system using pseudo random cyclic prefix [6]. Here each MSE-OFDM frame uses one pseudo random sequence as the CP and different pseudo random sequences will be used for different destination MS's. At each MS the received CP will be correlated with its own pseudo random sequence code. The MS that results the highest value of correlation will be able to receive the rest of the packet. Also the same PN sequence will be used at the end of each frame so that the received signal has a cyclic look.

### 2.1 Transmitter

MSE-OFDM transmitter [6] is shown in Figure 3(a). In this case each OFDM symbol consists of N complex exponentials or subcarriers which are modulated by the complex data symbol X. Thus an OFDM symbol may be represented by

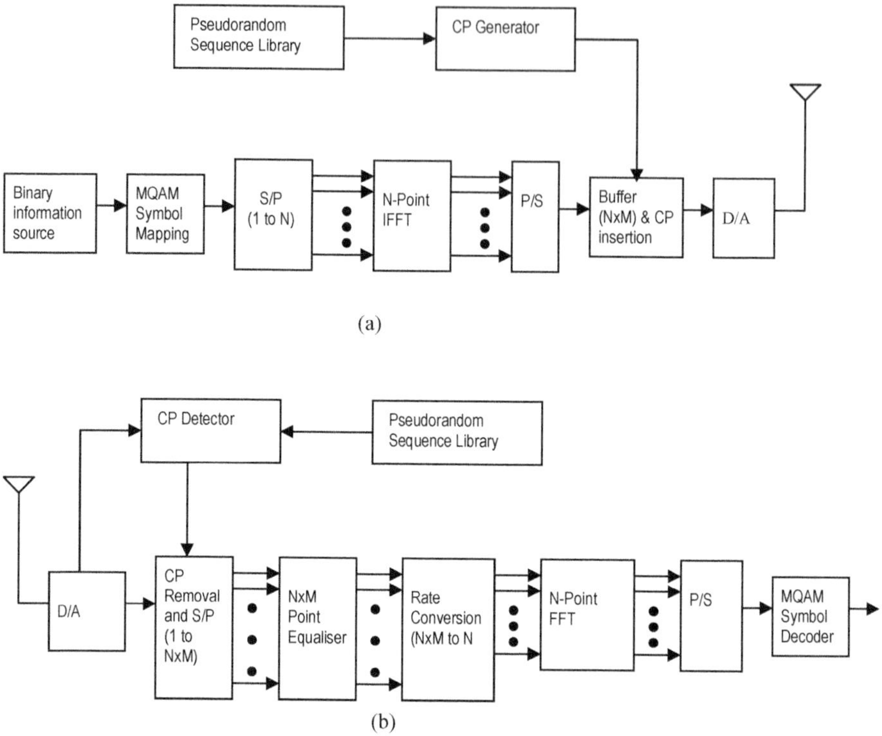


**Fig. 3.** Block Diagrams (a) MSE-OFDM Transmitter, (b) MSE-OFDM Receiver

$$x[n] = \frac{1}{\sqrt{N}} \sum_{k=0}^{N-1} X(k) e^{j2\pi \frac{nk}{N}}, n = 0,1,2,...,N-1 \tag{1}$$

Each MSE-OFDM frame consists of M such N-Point OFDM symbols so that the $l^{th}$ frame of the MSE-OFDM signal can be written as

$$s_l = \sum_{k=0}^{N-1} C_{l,p}(k)\psi_1(n,k) + \sum_{i=0}^{M-1}\sum_{k=0}^{N-1} X_{l,i}(k)\psi_2(n-iN-P,k) \tag{2}$$

where $C_{l,p}$ is the cyclic prefix using a pseudo random sequence. The symbol $i$ indicates the $i^{th}$ OFDM symbol of the $l^{th}$ frame. $\psi_1(n,k)$ and $\psi_2(n,k)$ are two multiplexing window functions corresponding to the cyclic prefix and the M information carrying OFDM symbols defined as follows

$$\psi_1(n,k) = \begin{cases} 0 & ,0 \le n \le P-1 \\ 1 & ,elsewhere \end{cases} \tag{3}$$

and

$$\psi_2(n-iN-P,k)=\begin{cases}\frac{1}{\sqrt{N}}e^{\frac{j2\pi k(n-P-iN)}{N}} & ,P\le n\le MN+P-1\\ 0 & ,elsewhere\end{cases} \tag{4}$$

At the transmitter side of the base station data packets from various sources will be stored in the buffer from where the packets will be transmitted either in sequence of their arrival or according to some pre-defined priorities. Each data packet will comprise one complete frame of MSE-OFDM signal. The CP will be appended at the transmitter as per the destination MS address. Different CP's will be used for different mobile stations.

## 2.2 Receiver

In this case the transmission medium is broadcast in nature. So at the receiver of each MS it will be necessary to perform time domain correlation analysis between the received CP and a predefined pseudo random code generated within each MS. The receiver that generates the highest correlation will be able to receive the complete MSE-OFDM frame. Block diagram of the MSE-OFDM receiver is shown in Figure 3(b).

## 2.3 Equalisation

The equalisation technique used here is different from that of the conventional OFDM technique because of the unique frame structure of the MSE-OFDM. In this case an MN-Point equalisation is employed. Hence in this case an additional MN-Point DFT and IDFT will be required to be done before and after the frequency domain equalisation respectively. As a pseudo random sequence has been used as CP, the same sequence must also be added to the end of each MSE-OFDM frame to give the signal a cyclic look so that the received signal will be the cyclic convolution between the transmitted signal and the channel impulse response. Thus the transmitted signal in (2) can be written as a vector

$$s_l=[C_{l,p}(0),\ldots,C_{l,p}(P-1),x_{l,0}(0),\ldots,x_{l,M-1}(0),\ldots,x_{l,M-1}(N-1),C_{l,p}(0),\ldots C_{l,p}(P-1)]^T \tag{5}$$

The MSE-OFDM signal has now been modified according to accommodate the pseudo random sequence postfix. The postfix is stripped off at the receiver along with the CP.

The received signal vector $r_l$ corresponding to the transmitted signal vector (6) can be expressed as

$$r_l = C.s_l + w_l \tag{6}$$

where $w_l$ is an additive white Gaussian noise (AWGN) vector with the size of $s_l$ and $C$ is the channel matrix of size $[MN+3P-1,MN+2P]$ as given by

$$C = \begin{vmatrix} h_0 & 0 & & & & 0 \\ h_1 & h_0 & & & & \\ & & & & & \\ h_{L-1} & & h_1 & h_0 & & \\ 0 & & & & & 0 \\ & & h_{L-1} & & h_1 & h_0 \\ & & & h_{L-1} & & h_1 \\ & & & & & \\ 0 & & & & 0 & h_{L-1} \end{vmatrix} \quad (7)$$

A cyclic convolution occurs between the CP removed received signal $\tilde{s}_l$ and $h$ so that the following transform pair holds.

$$\tilde{s}_l \otimes h + \tilde{w}_l \Leftrightarrow DFT(\tilde{s}_l).H + \tilde{W}_l = DFT(\tilde{r}_l) \quad (8)$$

where $H$ and $\tilde{W}_l$ are the Fourier Transform of $h$ and $\tilde{w}_l$. The size of the DFT here is $MN$ points. It is assumed that the channel matrix is known from the channel estimation. The effect of the channel on the received signal can now be removed by a single tap frequency domain equalisation and the signal is converted back to time domain for FFT demodulation according to the following relation

$$\tilde{r}_l^{FEQ} = IDFT\{ \frac{DFT(\tilde{r}_l)}{H} \} + w_l^{FEQ} \quad (9)$$

where $w_l^{FEQ}$ has the same statistical property as that of the AWGN. The signal $\tilde{r}_l^{FEQ}$ is broken down into M-OFDM symbols in sequence for demodulation using N-Point FFT.

## 3 Performance Analysis and Discussions

As already discussed, there are two different implementations of MSE-OFDM. The CP-reduced MSE-OFDM configuration shows improvement of bandwidth efficiency as there is reduction in the number of CP insertion. This system has reduced robustness in time-selective channel and hence it is suitable for static or slowly varying channels. However the FFT-size reduced MSE-OFDM configuration does not show any improvement in terms of bandwidth efficiency as the MSE-OFDM frame duration is kept the same as the symbol duration of the conventional OFDM. It is a better option in time varying channels. Our multiple access system is proposed for the time invariant channels; therefore we use the CP-reduced case. We thus get a definite improvement in terms of bandwidth efficiency of the system.

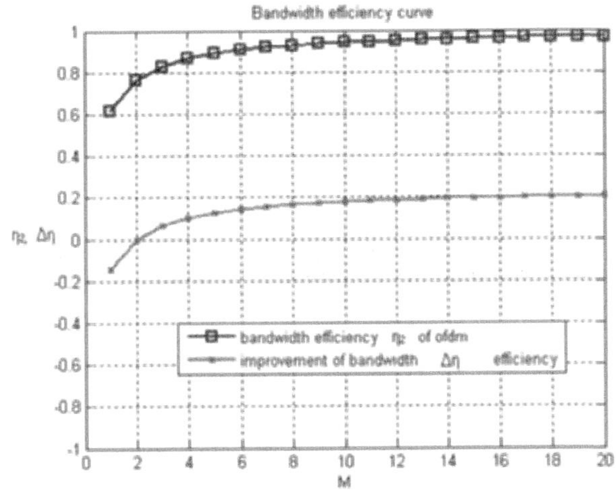


**Fig. 4.** Bandwidth Efficiency Curve

### 3.1 Bandwidth Efficiency

We consider that each subcarrier conveys a discrete symbol taken from a two-dimensional signal space containing $2^v$ signal points. Then for the N-point conventional OFDM with a CP of length P the bandwidth efficiency $\eta_1$ is given by [7]

$$\eta_1 = v \frac{N}{N+P} \text{ b/s/H} \tag{10}$$

In the case of MSE-OFDM, the OFDM frame consists of MN signal points, the CP and a postfix by CP in order to introduce cyclic convolution. Hence the bandwidth efficiency is given by

$$\eta_2 = v \frac{MN}{(MN+2xP)} \text{ b/s/H} \tag{11}$$

The improvement in bandwidth efficiency is given by:

$$\Delta\eta = \eta_2 - \eta_1 = \frac{NP(M-2)}{(N+P)(MN+2xP)} \tag{12}$$

Figure 4 shows both the bandwidth efficiency of MSE-OFDM symbols and also the trend in improvement of bandwidth efficiency for different values of M. The analysis has been done taking N=64 and P=16; modulation type being BPSK. It shows that there will be improvement in bandwidth efficiency of the system with increasing the value of M. However, it can also be observed that the improvement in bandwidth efficiency approaches the saturation value of P/(N+P) = 0.2.

### 3.2 BER Performance of Multiple Access MSE-OFDM System

We consider that $\Delta k$ represents the relative frequency offset (ratio between actual frequency offset and subcarrier spacing) and $\Delta n$ is the relative timing offset (ratio between timing offset and sampling interval). Then the bit error rate for the BPSK system for the $k^{th}$ subcarrier in one OFDM symbol can be written as [4]

$$P_{k,e} = erfc\sqrt{\frac{E_b}{N_o + \sigma^2{}_{ICI} + \sigma^2{}_{ISI}}} \tag{13}$$

where the variance of ICI is given by

$$\sigma^2{}_{ICI} = \sum_{l=0, l\neq k}^{N-1} X^2(l)H^2(l)\left\{\frac{\sin^2(\pi\Delta k)}{N^2 \sin^2(\pi(l-k+\Delta k)/N}\right\} \tag{14}$$

H(k) is the channel transfer function at the $k^{th}$ subcarrier and the variance of ISI is given by

$$\sigma^2{}_{ISI} = X^2(k)\left[\sum_{n=1}^{\infty} \frac{[(j\pi\Delta k(N-1) + j2\pi k\Delta n + j2\pi\Delta k\Delta n)/N]^n}{n!}\right]^2 \tag{15}$$

The average bit error rate can be given as

$$P_e = \frac{1}{N}\sum_{k=0}^{N-1} P_{k,e} \tag{16}$$

One aim of this study is to analyse the BER performance of the proposed multiple access system and also to compare it with the same using conventional OFDM system. Therefore, we conducted a set of simulations for both the cases of conventional OFDM and CP-reduced MSE-OFDM. The parameters used in simulations are N=64, P=16 and the channel impulse response (CIR) for three different users:

h1 = [1 0 0 0 0 0]
h2 =[0.9285 0 0.3714 0 0 0]
h3 =[-0.6154 0.7692 0 0.1538 0.0769]

One simulation we have conducted for the conventional OFDM system. Here we have used a set of random bit sequences for three different users and inserted the appropriate CPs. Above three different frequency selective paths have been selected for the three users. At the receivers the frequency equalization has been performed using the appropriate transfer functions. Another set of simulation have been conducted on the same data for the proposed MSE-OFDM system. Here we have used the same set of CIRs, but the frequency equalisation has been performed after MN-point FFT on the received MSE-OFDM frame.

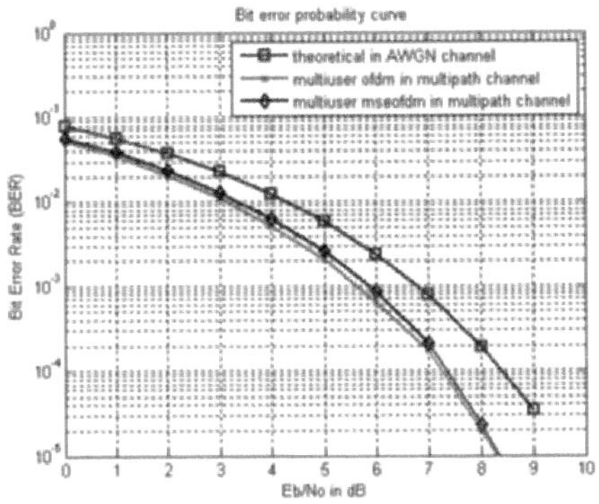


**Fig. 5.** BER performance curve for user 1

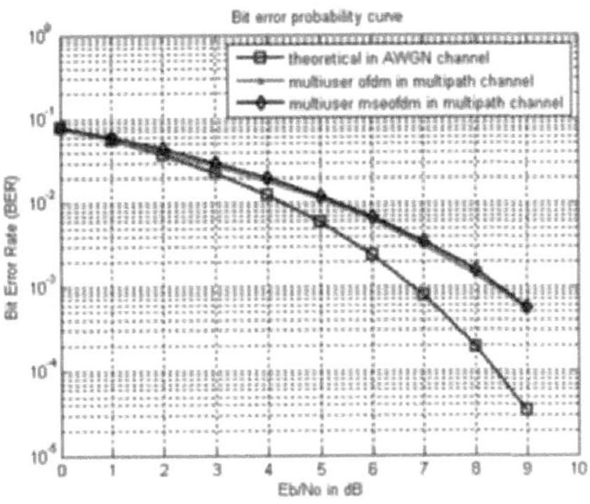


**Fig. 6.** BER performance curve for user 2

The results of simulation are shown in Figures 5, 6 and 7. From the Figures it can be observed that there are little differences in the BER performance of MSE-OFDM compared to that of the conventional OFDM. The reason behind this is that in case of MSE-OFDM the individual OFDM symbols are not separated by guard time whereas in case OFDM individual symbols are separated by the guard intervals. Thus the symbols in MSE-OFDM suffer from more ISI.

It can further be said that the better performances by both the OFDM and MSE-OFDM compared to the theoretical values are due to the channel inversion that takes place at the receiver equalizer.

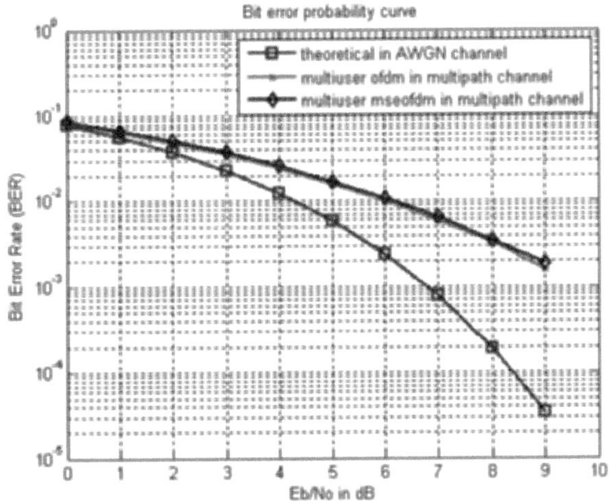


**Fig. 7.** BER performance curve for user 3

## 4 Conclusions

The proposed multiple access method is basically a time division multiple access technique. This method is obviously better than the other multi-access techniques

[4,5] because in this proposed technique all users receive their individual MSE-OFDM frames completely for frequency equalization and detection during its assigned time slot. There can be improvement in the BER performance in case of multiple access MSE-OFDM if the receiver structure with reduced complexity [6] using hybrid equalization technique is adopted. But in that case the receiver becomes complex.

## References


1. Cimini Jr., L.J.: Analysis and Simulation of a Digital Mobile Channel Using Orthogonal Frequency Division Multiplexing. IEEE Transactions on Communications 33, 665–675 (1985)
2. Wang, X., Wu, Y., Chouinard, J.-Y.: On the comparison between conventional OFDM and MSE-OFDM systems. In: IEEE Global Telecommunications Conference, GLOBECOM 2003, December 1-5, vol. 1, pp. 35–39 (2003)
3. Chouinard, J.-Y., Wang, X., Wu, Y.: MSE-OFDM: A new OFDM transmission technique with improved system performance. In: Proceedings of IEEE International Conference on Acoustics, Speech, and Signal Processing (ICASSP 2005), March 18-23, vol. 3, pp. iii/865–iii/868 (2005)
4. Singh, P., Chakrabarti, S., Raja Kumar, R.V.: Multiple Access Using MSE-OFDM. In: 5th IEEE Consumer Communications and Networking Conference, CCNC 2008, January 10-12, pp. 403–406 (2008)
5. Singh, P., Chakrabarti, S., Kumar, R.V.R.: A multiple access scheme using MSE-OFDM with reduced complexity. In: IEEE Region 10 Conference TENCON 2008, November 19-21, pp. 1–4 (2008)
6. Wang, X., Wu, Y., Wu, H.C., Gagnon, G.: An MSE-OFDM System with Reduced implementation Complexity Using Pseudo Random Prefix. In: IEEE Global Telecomm. Conference, GLOBECOM 2007, vol. 1, pp. 2836–2840 (December 2007)
7. Wang, X., Wu, Y., Chouinard, J.-Y., Wu, H.-C.: On the design and performance analysis of multisymbol encapsulated OFDM systems. IEEE Transactions on Vehicular Technology 55, 990–1002 (2006)

# Elimination of Irrelevancy during Semantic Service Discovery Using Clustering Approach

Chellammal Surianarayanan and Gopinath Ganapathy

Department of Computer Science, Bharathidasan University, Khajamalai Campus, Tiruchirappalli, 620 023, Tamil Nadu, India
chelsrsd@rediffmail.com
gganapathy@gmail.com

**Abstract.** Semantic web service discovery process consumes a significant amount of time even to perform a typical service match of limited functional capabilities. Modern business integration requires services need to be dynamically discovered and composed quickly. In this work, a similarity based clustering approach is suggested for quick elimination of irrelevant services during discovery. In this method, all the available published services are clustered by service description similarity in prior to semantic matching. This yields clusters of similar services. After services are clustered, when a query is submitted, firstly, a particular cluster to which the query belongs to is found out. Then semantic matching of capabilities will be performed only to that particular cluster ignoring all other service clusters as irrelevant to the query. The proposed method is tested for its performance with a small test collection and results are presented.

**Keywords:** semantic service discovery, semantic reasoner, service clustering, candidate service, semantic matching.

## 1 Introduction

In automatic service discovery, a programme will autonomously match a given query with the available published services and discover a list of matched services without human intervention. Traditional Universal Description Discovery and Integration (UDDI) based discovery supports only manual keyword based searching and category based searching facilities that limits the capability and efficiency of queries [1]. Though few of the limitations of UDDI described in [2-3] are met with Web Service Description Language (WSDL), lack of explicit semantics in WSDL prevents clients from automatically discovering services without human assistance [4]. Semantic service description languages such as [5-6] enable service providers to express their intended semantics and bring maximal automation and dynamism in service discovery and composition [7].

Generally, semantic service discovery frameworks employ semantic reasoners such as Pellet and RACER for finding semantic relations between different concepts during service matching. Despite its accuracy, semantic reasoning is heavy in time [8]. Generally business transactions involve multiple services from different domains need to be composed dynamically in a complex chain [9].

D. Nagamalai, E. Renault, M. Dhanushkodi (Eds.): CCSEIT 2011, CCIS 204, pp. 307–315, 2011.

Optimization techniques viz., optimization at semantic reasoner and optimization at matcher [10] are used to achieve the two critical needs, viz., 'performance' and 'scalability' during service discovery. As the first method depends on the internal architecture of the reasoner, the second method is used predominantly. The second method attains improved performance by reducing the number of services that really need semantic matching of capabilities. Generally, for any given query, there may be a huge number of obviously irrelevant services available. For example, services that serve financial functionalities are irrelevant to a query which asks for temperature of a location. Optimization at matcher removes such irrelevant services quickly and produces a reduced list of 'candidate services' that really need semantic matching.

In this work, a simple agglomerative clustering based approach is suggested for identifying candidate services for semantic matching. In this approach, all the available published services are grouped into clusters by similarity in prior to discovery. When a query is submitted, firstly the specific cluster to which the query belongs to is found out. Then the semantic matching of capabilities is employed only to that specific cluster, considering all other clusters as irrelevant to the query.

## 2 Related Work

Various techniques such as clustering, inverted indexing, caching, WordNet based similarity metric, etc. are used to achieve optimization at matching component. An extended Jacquard similarity metric based star clustering is used in [11] for efficient web service retrieval. In [12], K-Means clustering is used to filter out irrelevant services along with Probabilistic Latent Semantic Analysis method to capture hidden semantics of the query. In [13], Semantic Web Services Clustering method is used as a layer between UDDI registry and Internet to group services into domains. In [14], a single linkage based clustering with Jaccard similarity is used to reduce the overhead significantly and locate the suitable services quickly. A method proposed in [15] uses WordNet similarity for finding similar services and clusters them into various categories. An inverted indexing based approach presented in [16] accelerates the removal of irrelevant services and facilitates automatic service composition.

The present work is a complementary approach in using a simple text based similarity to find the 'service description similarity' among services. It finds the service description similarity between all pair-wise combinations of services. A dissimilarity matrix is constructed. Then the dissimilarity matrix is given as input for hierarchical agglomerative clustering. This yields clusters of similar services. The method quickly identifies the candidate services as it simply uses text similarity approach.

## 3 Proposed Method

A three-step clustering approach is proposed to find candidate services for semantic matching. The steps include computation of pair-wise similarity score between all possible service pairs, construction of dissimilarity matrix and clustering of services by similarity. The methods of similarity computation, construction of dissimilarity matrix and clustering of services are explained in the following sub sections. The overview of the proposed approach is shown in Fig. 1.

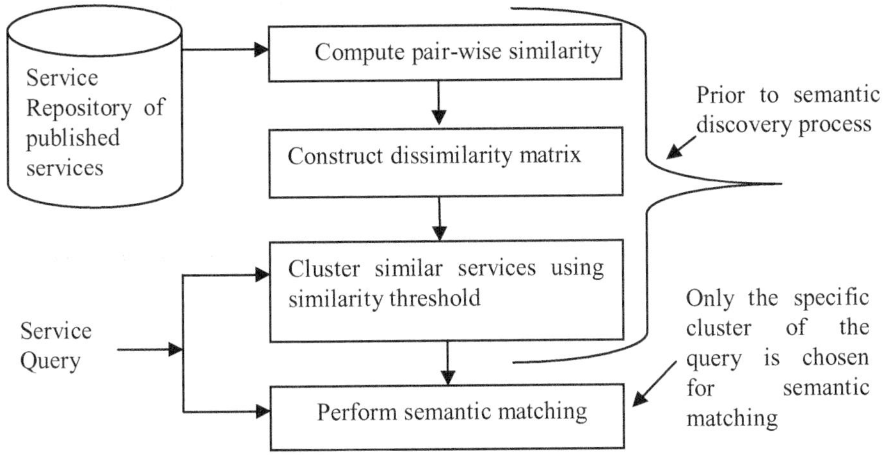


**Fig. 1.** Proposed clustering approach

### 3.1 Similarity Computation

Irrespective of various kinds of semantic service description languages which are used to describe the interface of services, any service description contains details about service name, service description and a set of operations. This first level service profile is used in computing the service description similarity among services.

A service (S) is expressed as S= (SN, SD, $OP_i$) where SN represents the service name, SD represents the textual description of the service and $OP_i$ represents names of all operations associated with the service.

The service description similarity between any two services is computed using the following steps.

Step 1: Extract service name, service description and names of all operations from service description as individual terms for the pair of services for which the service description similarity score is to be found out

Step 2: Perform the following pre-processing steps on the contents extracted from the pair of service descriptions under similarity computation. The pre-processing steps include removal of common words (stop words) such as ‘this’, ‘the’, ‘returns’, ‘and’, etc., removal of insignificant service terms and conversion of words in different Parts-Of-Speech (POS) into words of single POS.

For example, let us consider a textual service description as “this finds temperature of a location”. This description contains the common words, ‘this’, ‘of’ and ‘a’. From an investigation with service description files, it is found that, most of the service descriptions contain insignificant terms such as ‘request’, ‘response’, ‘get’, ‘soapIn’, ‘soapOut’, ‘HttpRequest’, ‘’HttpResponse’, ‘portType’, ‘Result’, ‘GetIn’, ‘GetOut’, ‘interface’, etc. Operation names may also contain their service names as prefixes. For example, the operation name, ‘CurrencyExchange_getRate’ contains its service name, ‘CurrencyExchange’ as its prefix. During pre-processing common and insignificant terms are removed. Further the service terms which occur in plural forms are converted into singular forms.

Step 3: Construct 'representative strings' using tokens obtained in Step 2 with duplicate terms eliminated.

Step 4: Find the normalized cosine similarity between the representative strings of pair of services. The above similarity represents the service description similarity score between the services.

The computation of service description similarity is illustrated as follows. Let us consider two services 'PhoneValidator' and 'PhoneVerification' as S1 and S2.

Step 1: Extraction of service name, service description, and operation names for S1 and S2 as individual terms.

**Contents extracted for S1.** Service Name: PhoneValidator. Service Description: Validates area codes and prefixes Operation names: PhoneValidatorSoap, GetRemainingHits.

**Contents extracted for S2.** Service Name: Phone3T Service Description: Validates a phone number and provides additional information Operation names: PhoneVerify

Step 2: Pre-processing of the contents extracted from S1 and S2

The common words 'and' in service description of S1 and 'a' and 'and' in S2 are removed. Then, the service insignificant terms, 'Get', 'Soap' in S1 and '3T' in S2 are removed. While converting words in different POS into words of single POS, the words, 'validator' and 'validates' in S1 are converted into single POS term in verb form as 'validate'. The words 'prefixes' and 'hits' are converted into their singular forms.

Step 3: Let T1 and T2 represent all the individual terms obtained after preprocessing the descriptions of S1 and S2 respectively.

T1 = {phone, validate, validates, area, codes, prefix, phone, validate, prefix, remaining, hits}

T2 = {phone, validate, phone, number, provides, additional, information, phone, verify}.

With removal of duplicate words, the representative strings for S1 and S2 are formed as

RS1 = {phone, validate, area, codes, prefix, remaining, hit}

RS2 = {phone, validate, number, provides, additional, information, verify}

Step 4: Normalized Similarity (NS) between RS1 and RS2 is computed using cosine similarity metric.

### 3.2 Construction of Dissimilarity Matrix

The construction of dissimilarity matrix is illustrated with 4 services as follows. Let the services are denoted as S1, S2, S3 and S4. The possible combinations of service-pairs are (S1-S1), (S1-S2), (S1-S3), (S1-S4), (S2-S2), (S2-S3), (S2-S4), (S3-S3), (S3-S4) and (S4-S4). The normalized dissimilarity score between a pair of services is given as Dissimilarity, D=(1-NS) where NS is the normalized cosine similarity score of the pair of services After computing dissimilarity values, a dissimilarity matrix is constructed. For the above 4 services, the dissimilarity matrix is constructed as shown in Fig. 2.

$$\begin{array}{c|cccc} & S1 & S2 & S3 & S4 \\ \hline S1 & 0 & 0.1 & 0.3 & 0.2 \\ S2 & 0.1 & 0 & 0.4 & 0.5 \\ S3 & 0.3 & 0.4 & 0 & 0.1 \\ S4 & 0.2 & 0.5 & 0.1 & 0 \end{array}$$

**Fig. 2.** Dissimilarity matrix for four services

### 3.3 Clustering of Services

After the construction of dissimilarity matrix, the services are clustered using hierarchical agglomerative clustering with group average scheme as clustering criterion. During clustering a similarity threshold of 30% is specified as stopping criterion.

## 4 Experimental Setup

The following steps are used to test the performance of the proposed approach.

Step: 1 Prepare test data which should contain groups of related/similar services. For example, in a test data of 10 services, 5 are 'zip-code' related services, 2 are 'sms' related services and remaining 3 are 'email' related services.

Step 2: Compute all pair-wise cosine similarity scores. Derive dissimilarity scores and construct dissimilarity matrix

Step 3: Employ hierarchical agglomerative clustering method

Step 4: Compare the clusters returned in Step 3 with the original groups of related services to evaluate the accuracy of the proposed approach.

An experimental setup as shown in Fig 3 is constructed. After the construction of dissimilarity matrix the services are clustered using Matlab clustering tool.

A test data of 10 test services with 3 service groups is constructed using services collected from "www.xmethods.com" and "www.remotemethods.com. The related service groups along with the number of services in each group is given in Table 1

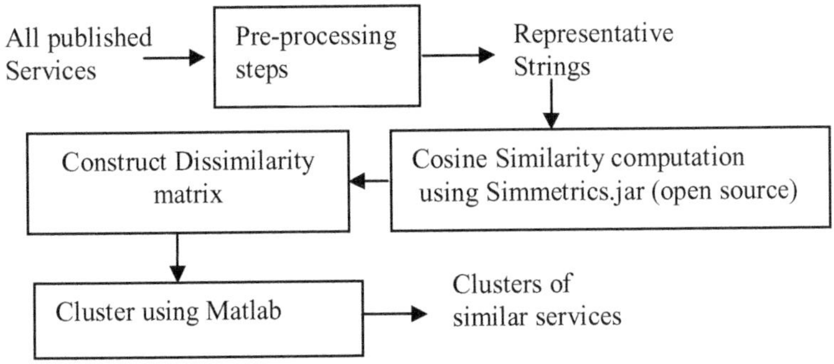


**Fig. 3.** Experimental setup for the proposed method

**Table 1.** Related service groups in test data

| # | Service group | Number of similar services in each group |
|---|---|---|
| 1 | Zip code service group | 5 |
| 2 | Email | 3 |
| 3 | SMS | 2 |

**Table 2.** URLs of WSDL files of Test Services

| Service id | URL of WSDL |
|---|---|
| S1 | http://ws.strikeiron.com/zipdistance?wsdl |
| S2 | http://www.info-me-sms.it/ws.php?wsdl |
| S3 | http://www.ripedevelopment.com/webservices/ZipCode.asmx?wsdl |
| S4 | http://ws.xwebservices.com/XWebEmailValidation/V2/XWebEmail Validation.wsdl |
| S5 | http://www.codebump.com/services/zipcodelookup.asmx?wsdl |
| S6 | http://www.webservicemart.com/uszip.asmx?wsdl |
| S7 | http://ws2.fraudlabs.com/mailboxvalidator.asmx?wsdl |
| S8 | http://www.barnaland.is/dev/sms.asmx?wsdl |
| S9 | http://ws.strikeiron.com/zipinfo3?wsdl |
| S10 | http://ws.strikeiron.com/EmailVerify4?wsdl |

**Table 3.** Representative Strings of Test Services

| Service id | Representative Strings |
|---|---|
| S1 | {zip, distance, city, remaining, hit} |
| S2 | {send, sms, check} |
| S3 | {zip, code, city, state, area, time} |
| S4 | {validate, email} |
| S5 | {zip, code, three, digit, distance, user, place} |
| S6 | {validate, zip} |
| S7 | {mail, validate} |
| S8 | {send, sms} |
| S9 | {zip, city, state, remaining, hit, status} |
| S10 | {email, verify, status, remaining,, hit} |

The test services are given service ids as S1, S2, S3, S4, S5, S6, S7, S8, S9 and S10. The URLs of the WSDL files and the representative strings of the test services are given in Table 2 and Table 3 respectively.

The normalized service description similarity score is calculated using the experimental setup. From the similarity score, the dissimilarity score is computed for each pair. For each service, the dissimilarity score with every other possible service pair is computed. For illustration, dissimilarity score of all pairs of S1 and S2 are shown in Table 4.

**Table 4.** Dissimilarity Score of all Service Pairs for S1

| Service Pairs for S1 | Dissimilarity score | Service Pairs for S2 | Dissimilarity score |
|---|---|---|---|
| S1-S2 | 1 | S2-S1 | 1 |
| S1-S3 | 0.635 | S2-S3 | 1 |
| S1-S4 | 1 | S2-S4 | 1 |
| S1-S5 | 0.662 | S2-S5 | 1 |
| S1-S6 | 0.684 | S2-S6 | 1 |
| S1-S7 | 1 | S2-S7 | 1 |
| S1-S8 | 1 | S2-S8 | 0.184 |
| S1-S9 | 0.27 | S2-S9 | 1 |
| S1-S10 | 0.8 | S2-S10 | 1 |

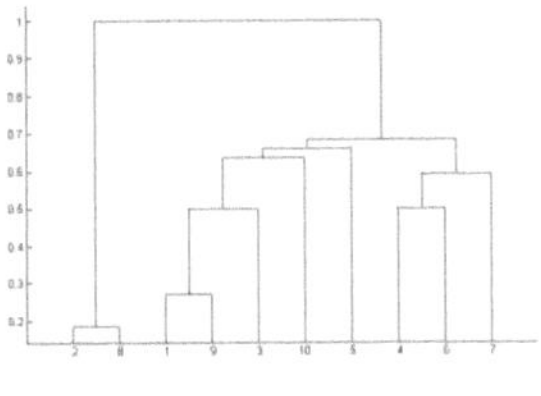

**Fig. 4.** Dendrogram obtained for test data

The test data results in a 10 × 10 dissimilarity matrix with all zeros in the diagonal. The dissimilarity matrix is symmetric. The matrix is given as input to Matlab clustering tool in vector format. The dendrogram obtained is shown in Fig. 4.

## 5 Results and Conclusion

From dendrogram, it is found with the dissimilarity of 70% (i.e. with 30% similarity threshold) three clusters are formed. Let the clusters be cluster-1, cluster-2 and cluster-3. The clusters obtained are given in Table 5.

**Table 5.** Clusters obtained using proposed approach

| # | Cluster | Services in the cluster |
|---|---|---|
| 1 | Cluster -1 (sms group) | S2 and S8 |
| 2 | Cluster -2 (Zip group) | S1, S9,S3,S10 in and S5 |
| 3 | Cluster-3 (email group) | S4,S6,S7 |

From the above results, it is noted that out of 10 services, two services namely, S6 and S10 are grouped in different groups. i.e. S6 is expected to be grouped in Cluster-2. But the approach grouped it in Cluster-3. Similarly, the service S10 is expected to be grouped in Cluster-3. But the approach grouped it in Cluster-2. When the results are analyzed, it is found that the cause of wrong grouping is because of the presence of terms 'validate' and 'remaining' and 'hit'. These terms are commonly used both in email and zip group of services. The proposed method results in a clustering accuracy of 80% with this simple data set. It is planned to apply the proposed method to huge data set and compare it with other standard procedures for further evaluation.

## References

[1] Sohali, A., Zamanifar, K.: Matching Model for Semantic Web Services Discovery. Journal of Theoretical and Applied Information Technology 7(2), 139–144 (2009)

[2] Lee, K.-H., Lee, M.-Y., Hwang, Y.-Y., Lee, K.-C.: A Framework for XML Web Services Retrieval with Ranking. In: IEEE International Conference on Multimedia and Ubiquitous Engineering, pp. 773–778 (2007)

[3] Plebani, P., Pernici, B.: URBE: Web Service Retrieval based on Similarity Evaluation. IEEE Transactions on Knowledge and Data Engineering 21(11), 1629–1642 (2009)

[4] Guo, R., Le, J., Xia, X.L.: Capability Matching of Web Service Based on OWL-S. In: Proceedings of the 16th International Workshop on Database and Expert Systems Applications, pp. 653–657. IEEE Computer Society, Los Alamitos (2005)

[5] Kopecky, J., Vitvar, T., Bournez, C., Farrell, J.: SAWSDL: Semantic Annotations for WSDL and XML Schema. IEEE Internet Computing 11(6), 60–67 (2007)

[6] Klusch, M., Kaufer, F.: WSMO-MX: A hybrid Semantic Web service matchmaker. Web Intelligence and Agent Systems 7(1), 23–42 (2009)

[7] Sycara, K., Paolucci, M., Ankolekar, A., Srinivasan, N.: Automated discovery, interaction and composition of Semantic Web Services. Journal of Web Semantics 1(1), 27–46 (2003)

[8] Mokhtar, S.B., Kaul, A., Georgantas, N., Issarny, V.: Towards Efficient Matching of Semantic Web Service Capabilities. In: International Workshop on Web Services Modeling and Testing, pp. 137–152 (2006)

[9] Milanovic, N., Malek, M.: Current Solutions for Web Service Composition. IEEE Internet Computing 8(6), 51–59 (2004)

[10] Lv, Q., Zhou, J., Cao, Q.: Service Matching Mechanisms in Pervasive Computing Environments. In: International Workshop on Intelligent Systems and Applications, ISA, pp. 1–4 (2009)

[11] Shou, D., Chi, C.-H.: A Clustering-Based Approach for Assisting Semantic Web Service Retrieval. In: IEEE International Conference on Web Services, Beijing, pp. 838–839 (2008)

[12] Ma, J., Zhang, Y., He, J.: Efficiently finding web services using a clustering semantic approach. In: ACM International Conference Proceeding Series, Beijing, vol. 292, Article No.5 (2008)

[13] Nayak, R., Lee, B.: Web Service Discovery with additional Semantics and Clustering. In: Proceedings of the IEEE/WIC/ACM International Conference on Web Intelligence, pp. 555–558. IEEE Computer Society, USA (2007)

[14] Liu, P., Zhang, J., Yu, X.: Clustering-Based Semantic Web Service Matchmaking with Automated Knowledge Acquisition. In: Liu, W., Luo, X., Wang, F.L., Lei, J. (eds.) WISM 2009. LNCS, vol. 5854, pp. 261–270. Springer, Heidelberg (2009)

[15] Qu, X., Sun, H., Li, X., Liu, X., Lin, W.: WSSM: A WordNet-Based Web Services Similarity Mining Mechanism. In: Computation World: Future Computing, Service Computation, Cognitive, Adaptive, Content, Patterns, pp. 339–345. IEEE Computer Society, Greece (2009)

[16] Gao, T., Wang, H., Zheng, N., Li, F.: An Improved Way to Facilitate Composition-Oriented Semantic Service Discovery. In: International Conference on Computer Engineering and Technology, Singapore, pp. 156–160 (2009)

# Leaf and Flower Recognition Using Preferential Image Segmentation Algorithm

N. Valliammal and S.N. Geethalakshmi

Avinashilingam Deemed University for Women, Coimbatore-641043
Valli.p.2008@gmail.com
sngeethalakshmi@yahoo.com

**Abstract.** Automatic plant classification systems are essential for a wide range of applications including environment protection, plant resource survey, as well as for education. With the aid of advanced information technology, image processing and machine learning techniques, automatic plant identification and classification will enhance such systems with more functionality, such as automatic labeling and flexible searching. Image segmentation and object recognition are two aspects of digital image processing which are being increasingly used in many applications including leaf recognition. In this paper, the Preferential Image Segmentation (PIS) method is used to segment an object of interest from the original image. A probabilistic curve evolution method with particle filters is used to measure the similarity between shapes during matching process. The experimental results prove that the preferential image segmentation can be successfully applied in leaf recognition and segmentation from a plant image.

**Keywords:** Preferential image segmentation, Tree of shapes, Filters, Segmentation.

## 1 Introduction

Digital image processing is the use of the algorithms and procedures for operations such as image enhancement, image compression, image analysis, mapping, geo-referencing, etc. The influence and impact of digital images on modern society is tremendous and is considered as a critical component in variety of application areas including pattern recognition, computer vision, industrial automation and healthcare industries. The topics in image processing can be roughly categorized as follows: imaging, image denoising, image restoration, image coding, image segmentation, image transformation, object representation and recognition, and content-based image retrieval. Typical tasks for computer vision are scene interpretation, object recognition, optical character recognition, registration, feature extraction and video tracking.

There are many kinds of plants that exist on the earth. Plants play an important role in both human life and other lives that exist on the earth. Due to various serious issues like global warming, lack of awareness of plant knowledge, etc., the categories of plant is becoming smaller and smaller. Fortunately, botanists are realizing the importance of protecting plants and are discovering methods to protect them. In a

D. Nagamalai, E. Renault, M. Dhanushkodi (Eds.): CCSEIT 2011, CCIS 204, pp. 316–325, 2011.

similar fashion, there is also a need for recognizing a plant by its category, to help agriculturists / farmers during reproduction of extinct plants. The main challenge faced by them is the identification of the category to which a particular plant belongs to. To help them in this challenging venture, several researches (Man *et al.*, 2008; Lee and Chen, 2006) are conducted to automatically classify a plant into a category, when its leaf sample image is given as input.

Plant has plenty use in foodstuff, medicine and industry. And it is also vitally important for environmental protection. However, it is an important and difficult task to recognize plant species on earth. Designing a convenient and automatic recognition system of plants is necessary and useful since it can facilitate fast classifying plants, and understanding and managing them (Tzionas *et al.*, 2005). The classification of plant leaves is a vital process in botany and in tea, cotton and other industries. Moreover, it can also be used for early diagnosis of certain plant diseases.

The paper is organized as follows. Section 2 describes the general architecture used by plant classifiers. Section 3 explains the preferential segmentation, while section 4 summarizes the results and section 5 shows the conclusion and references.

## 2 General Architecture

The general process for plant classification through leaf recognition is given in Figure 1.

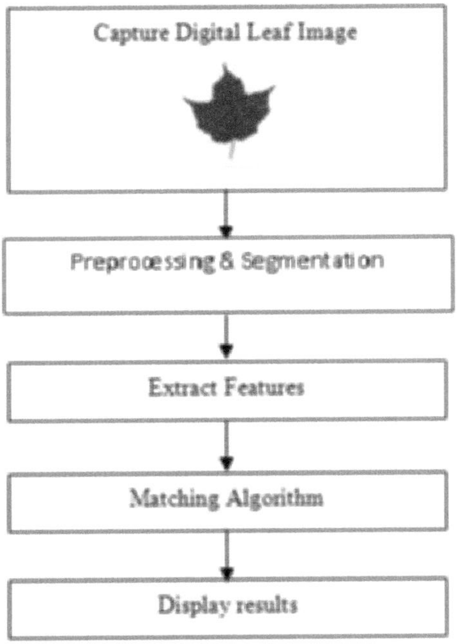


**Fig. 1.** General architecture of Automatic Plant Classification system

The performance of object recognition depends on the performance of preprocessing and segmentation. Segmentation is defined as the process of partitioning a digital image into multiple segments, where the pixels inside each segment share certain visual characteristics such as color, intensity or texture. Adjacent regions are significantly different with respect to the same characteristics. The goal of segmentation is to simplify and/or change the representation of an image into something that is more meaningful and easier to analyze (Shapiro and Stockman, 2001). Image segmentation locates objects and boundaries (lines, curves, etc.) in images. After segmentation, the image is represented by a set of segments, which collectively represents the entire image. The process of plant identification through leaf recognition is explained below.

## 2.1 Capturing Leaf Image

The leaf image is acquired by scanners or digital cameras. Since classification is a complex operation that needs high memory usage [13], normally, lossy compression format like JPEG / BMP / GIF is used. For convenience sake all the leaf images will be of the same size and resolution. Plants are basically classified according to shapes, colors and structures of their leaves and flowers [14]. However, to recognize the plant based on 2D images, it is difficult to analyze shapes and structures of leaves since they have complex 3D structures. In addition, the colors of leaves are always green; shades and the variety of changes in atmosphere and season cause the color feature having low reliability. Therefore, to recognize various plants using their leaves, the obtained leaf image in RGB format will be converted to gray scale before preprocessing. The formula used for converting the RGB pixel value to its grayscale counterpart is given in Equation 1.

$$\text{Gray} = 0.2989 * R + 0.5870 * G + 0.1140 * B \quad (1)$$

where R, G, B correspond to the color of the pixel, respectively.

## 2.2 Preprocessing

The leaf image preprocessing refers to the initial processing of input leaf image to correct the geometric distortions, calibrate the data radiometrically and eliminate the noise and clouds that present in the data. These operations are called preprocessing because they normally carried out before the real analysis and manipulations of the image data occur in order to extract any specific information. The aim is to correct the distorted or degraded image data to create a more faithful representation of the real leaf.

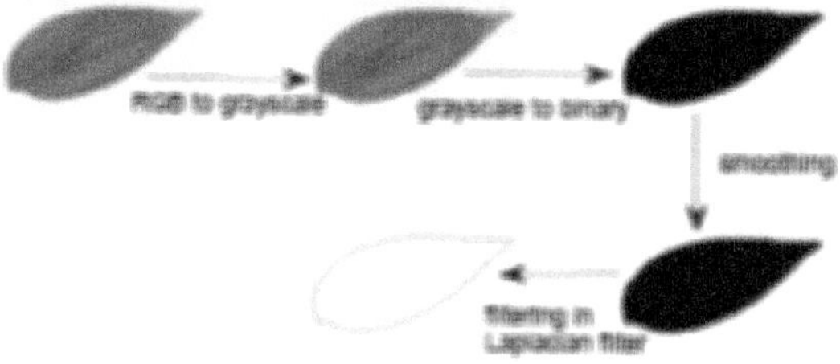


Preprocessing Example

Various preprocessing techniques are then used to enhance the leaf image obtained. Several techniques like boundary enhancement, smoothening, filtering, noise removal, etc. can be applied to improve the quality of the leaf image.

Several approaches to segmentation have been proposed which can be categorized into cluster based techniques (Pauwels and Frederix, 1999), histogram based techniques (Zwiggelaar, 2010), compression based techniques (Ballester *et al.*, 2007), edge detection, region growing methods (Chen and Georganas, 2008), Graph partitioning methods (Sumengen and Manjunath, 2006), watershed transformation (Wenzhong and Xiaohui, 2010), etc. Another popular method used is the curve evolution methods (Farzinfar *et al.*, 2010). These methods evolve the initialized curve(s) to the boundaries of objects in an image. The evolution of the curves may be driven by image gradient information (Kichenassamy *et al.*, 1995; Caselles *et al.*, 1997), region information (Chan and Vese, 2001; Jehan-Besson *et* al., 2003), or their combination (Tsai *et al.*, 2001). The image segments generated are enclosed by closed contours and are unsupervised which make it an ideal candidate for image post processing. These methods are more suitable for simple images and their performance degrades both in terms of segmentation and complexity, with complicated images (Paragios and Deriche, 2000). In such cases, the utilization of prior information is necessary for curve evolution methods in complicated image segmentation applications.

Several methods have been proposed that utilize prior information for supervised image segmentation. These methods usually propose new variational energy functions which integrate both the prior information and the gradient/region information in the image to be segmented. The minimizations of these functions can lead to segmentation results. Shape priors are utilized in (Cremers and Funka-Lea, 2005; Cremers *et al.*, 2002; Chen, 2002) and (Leventon *et* al., 2000). Both intensity priors and shape priors are applied in Leventon *et al.*, (2000). Natural image statistics are employed for natural image segmentation in Heiler and Schnorr (2005). These methods usually work better than unsupervised methods. However, shape priors are primarily incorporated in supervised segmentation methods, and the information contained in the intensities is not always fully utilized. For example, the methods in Freedman *et al.* (2005) and Freedman and Zhang (2004) make use of the intensity histograms only, and the spatial information is ignored. Furthermore, these methods usually have initialization problems because the energy functions have multiple local minimums. The methods are sensitive to the initial locations of the evolving curves.

In some cases, the user may be interested only in finding the location of an object within the image. In such cases the background and other non-interested objects need not be clustered. To perform such a task, the process of segmentation is combined with object recognition. This concept is the key idea behind "Preferential Image Segmentation" (PIS). PIS method is described as a method to preferentially segment objects of interests from an image and ignore the remaining portions of the image for this application. An example is shown in Figure 2.

PIS can be considered as a method similar to object detection, where rough location of the object of interest is identified (Agarwal *et al.*, 2004). The drawback behind this method is that it often neglects to provide the exact boundaries which are very important for segmentation. Similarly, classification methods (Aujol *et al.*, 2003), image parsing methods (Tu *et al.*, 2005), wavelets (Aijol *et al.*, 2003) and variational methods (Samson *et al.*, 2000) can be considered similar to PIS. However, these methods classify the image as a whole instead of separating it into meaningful objects and additionally they are computationally intense.

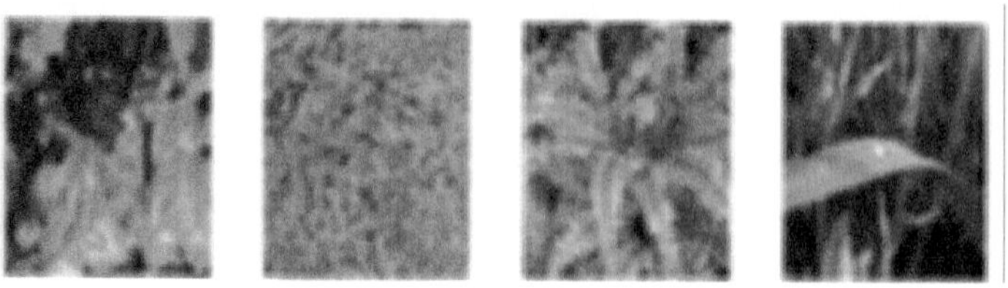

(a) (b) (c) (d)

**Fig. 2.** Sample Leaf images

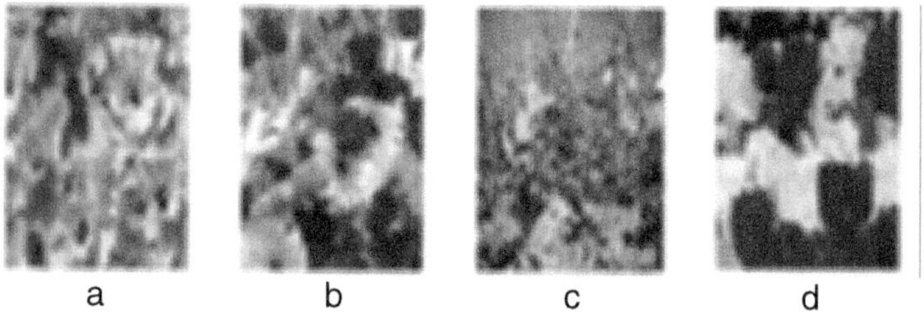


**Fig. 3.** Sample Flower images

This paper analysis the usage of PIS to segment leaves from a plant. Segmenting leaves from plant / tree is considered important in plant identification system. Automatic plant identification by computers is a very important task for agriculture, forestry, pharmacological science and etc. The general plant identification system is composed of four steps, namely, (i) preprocessing (ii) segmentation (iii) feature extraction and (iv) matching process. In this paper a method for leaf segmentation as proposed by Pan *et al.* (2009) is analyzed. The paper is organized as below. This section provided a brief introduction to the topic under discussion. Section 2 gives an overview of the PIS technique used. Section 3 tabulates and discusses the results obtained during experimentation and Section 4 presents a conclusion.

## 3 Preferential Image Segmentation (PIS)

The PIS is explained using an example image. Consider a situation where the image in Figure 2a has to be identified in a clutter image as in Figure 2b.

Tree of shapes provides a natural way to represent the spatial relationships of the shapes in an image. It represents images based on the techniques of contrast-invariant mathematical morphologies. This method is based on the theory of image representation using connected components of set of finite perimeters in the space of functions with Weakly Bounded Variations (WBV). It shows that an image, if taken as a function of weakly bounded variation is guaranteed to be decomposed into connected components with closed boundary this is an extension to classical methods where an image is taken as a piecewise- smooth function. The representation of an image using a tree of shapes utilizes the inferior or the superior of a level line to represent an object, and considers the boundary of the inferior area as the shape of the object. Therefore, only closed shapes are generated. This representation also provides a tree structure to represent the spatial relationship for the objects in an image. Concept of "shape" is introduced to generate a unique inclusion tree for an image. A shape is defined as the connected components of a level set and the holes inside them. The whole image acts as the root of the tree located at the top level. The shapes in the same level are spatially disjoint in the image. The shapes in the lower level are spatially included in the shapes in the next higher level. The tree of shapes, therefore, provides a natural way to represent the spatial relationships between the shapes in the image.

The tree of shapes is used as a candidate to obtain the intensity information in the priori image which is used in preferential segmentation. A common method used to construct tree of shapes to both images (Figures 3a and 3b), whose features are then compared to identify the ratio of similarity between the images. However, the main disadvantage of this approach is the amount of shapes (features) generated, which is very huge. For examples, the example picture produces 3832 shapes (Figure 2c). The closed curves in Figure 3c corresponds to the boundaries of the shape in the tree of shapes. These discovered boundaries provide valuable information which can be utilized during shape similarity measure.

As the number of shapes increases, the computational complexity also increases, thus necessitating to reduce the number of shapes used during segmentation and shape comparison. This is performed using the intensity information that is contained inside the shape. The intensity information by means of the tree of shapes includes the following features.

1) The number of objects (Nd) contained directly in the shape, which corresponds to the number of direct children of the shape in the tree, Nd=2 for the indicated shape in Figure 2a.
2) The total number of objects (Nt) contained in the shape, which corresponds to the total number of children below the shape in the tree Nt=3 for the indicated shape in Figure 2a.
3) The relative area change between the shape and its direct children. Suppose the area of the shape is S, and the areas of its direct children are $S_i$, where $1 \leq i \leq Nd$, the relative area change is then defined as

$$A = \prod_{i=1}^{Nd} \frac{S_i}{S}$$

4) The rough similarities of the boundaries of the shapes, which can be represented as the ratio of the circumferences (C) squared to the area (S) of the boundaries of the shapes (Gonzalez and Woods, 2002), i.e., $R=C^2/S$.

These features for the two shapes should be very close to match. Exact matching may not be achieved because of the differences between shapes, the effects of noise, lighting changes and cluttered background. Thresholds should be set for coarse matching to affect a rapid exclusion of most candidate shapes and can be adjusted according to the application.

The number of candidate shapes decreases substantially by means of the intensity features extracted from the tree of shapes. In the case of Figure 3b, the feature decreases the number of candidate shapes from 3832 to 977; the feature decreases the number from 977 to 83; the feature decreases the number from 83 to 1; the feature retains this candidate. The candidate shape left over matches the prior shape, as shown in Figure 2d. The process takes 1.70 s.

In most cases, however, the candidate shapes will decrease from thousands to tens. Curve matching is then performed on the boundaries of the resulting candidate shapes and the prior shapes. The candidate shape which best matches the prior's curve is taken as the preferential segmentation result.

The curve matching steps are outlined below:

1) Extraction of the level lines for each image. The level set representations (1), (2) are utilized here for the extraction. The level lines are defined as the boundaries of the connected components.
2) Affine filtering of the extracted level lines at several scales. This step is applied to smooth the curves using affine curvature deformation to reduce the effects of noise.
3) Local encoding of pieces of level lines after affine normalization. Both local encoding and affine normalization are designed for local shape recognition methods. This step will help to deal with occlusions in real applications.
4) Comparison of the vectors of features of the images. Euclidean distance is utilized to compare the feature vectors.

The performance of curve matching between two curves is calculated after affine filtering, curve normalization and local encoding. Suppose $C_1$ and $C_2$ and are two curves for matching, and $S_1$ and $S_2$ are pieces from and $C_1$ and $C_2$, respectively, then the performance to take $S_1$ and $S_2$ as the matching between $C_1$ and $C_2$ is calculated using the formula as in (1).

$$\text{Score} = \frac{l_1 x l_2}{L_1 x L_2} \tag{1}$$

where $l_1$=arclength($S_1$), $l_2$=arclength($S_2$), $L_1$= arclength($C_1$) and $L_2$ = arclength($C_2$). The maximum score over all possible matching pieces is taken as the matching between the curves $C_1$ and $C_2$. The implementation and description is provided in Lisani *et al.* (2003).

This paper introduced the usage of PIS segmentation for plant and flower recognition. The PIS method integrated image segmentation and object recognition using the tree of shapes, which is an approach to the mathematical description of morphology. The system was tested with several images and the results are discussed in the next section.

## 4 Results

The system was evaluated using a series of benchmark images belonging to two categories. The first is the leaf images along with plant images. The second is flower images and their corresponding images. A total of 500 plant and flower images were collected (Set I). All these images were RGB color images of 256 x 256 pixels size. A total of 100 leaf and single flower images were stored separately (Set II). These pictures also belonged to RGB color space and care was taken to collect different sized images ranging from roughly 0.8 to 2 times the size of plants in the first set. Sample images of plants from Set I are shown in Figure 4. In a similar fashion, Figure 5 shows sample images of flowers in Set I. Example leaf and flower images in Set II are shown in Figure 4 and Figure 5 respectively.

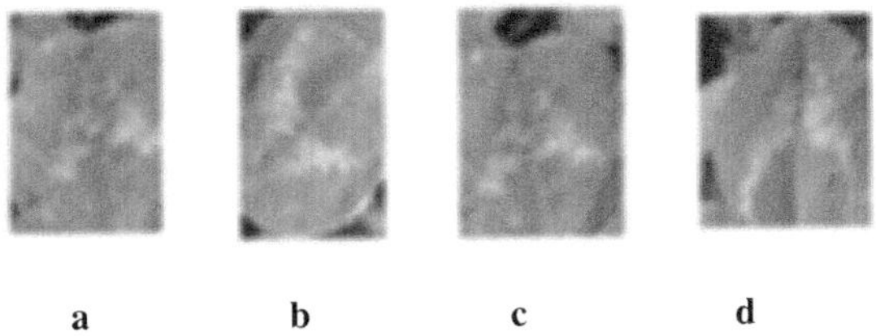

**Fig. 4.**

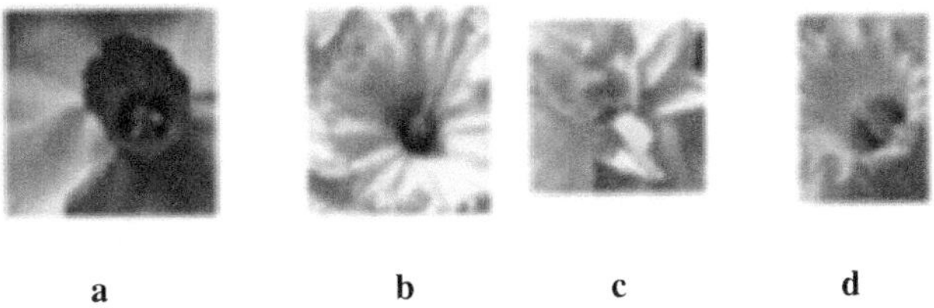

**Fig. 5.**

**Fig. 6.** Leaf Recognition

The visual results obtained while using preference image segmentation with plants and flowers are given in this section. Figure 6 and 7 shows the image after PIS.

**Fig. 7.** Flower Recognition

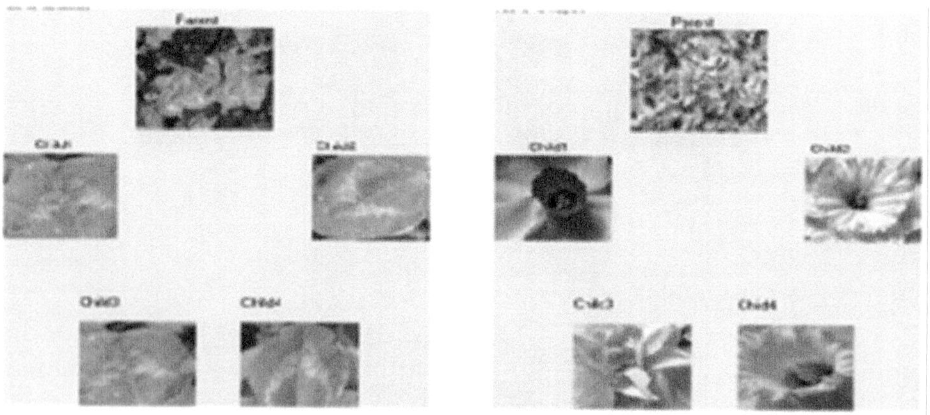

**Fig. 8.** Tree of Shapes for leaf and flower

The tree of shapes obtained for Figure 6 and Figure 7 are shown in Figure 8. From the results, it is evident that the PIS algorithm is a good candidate for leaf and flower segmentation and object recognition from plants and set of flowers.

## 5 Conclusion

A probabilistic curve evolution method with particle filters is used to measure the similarity between shapes during matching process. The experimental results prove that the preferential image segmentation can be successfully applied in leaf recognition and segmentation from a plant image. In summary, the proposed method models an image with a function of bounded variation and utilizes a tree of shapes for image representation. This method encodes the prior information for preferential segmentation as a tree of shapes. The method is invariant to translation, rotation and scale transformations because both feature extraction and boundary matching are invariant to these transformations. All these features and the positive results obtained prove that the Preferential Image Segmentation method can be successfully exploited in leaf identification for plant recognition systems. Future research directions include testing the image with more flowers and leaf images and incorporating the PIS method into the leaf recognition system.

## References


[1] Agarwal, S., Awan, A., Roth, D.: Learning to detect objects in images via a sparse, part-based representation. IEEE Trans. Pattern Anal. Mach. Intell. 26(11), 1475–1490 (2004)

[2] Aujol, J., Aubert, G., Blanc-Féraud, L.: Wavelet-based level set evolution for classification of textured images. IEEE Trans. Image Process. 12(12), 1634–1641 (2003)

[3] Caselles, V., Kimmel, R., Sapiro, G.: Geodesic active contours. Int. J. Comput. Vis. 22(1), 61–79 (1997)

[4] Chan, T.F., Vese, L.A.: Active contours without edges. IEEE Trans. Image Process. 10(2), 266–277 (2001)

[5] Chen, L., Georganas, N.D.: Region-based 3D Mesh Compression Using an Efficient Neighborhood-based Segmentation. Simulation, Society for Computer Simulation International 84(5), 185–195 (2008)

[6] Chen, Y.: Using prior shapes in geometric active contours in a variational framework. Int. J. Comput. Vis. 50(3), 315–328 (2002)

[7] Cremers, D., Funka-Lea, G.: Dynamical statistical shape priors for level set based sequence segmentation. Variational and Level Set Methods Comput. Vis., 210–221 (2005)

[8] Cremers, D., Tischhauser, F., Weickert, J., Schnorr, C.: Diffusion snakes: Introducing statistical shape knowledge into the Mumford–Shah functional. Int. J. Comput. Vis. 50(3), 295–313 (2002)

[9] Farzinfar, M., Xue, Z., Teoh, E.K.: A novel approach for curve evolution in segmentation of medical images. Comput. Med. Imaging Graph 34(5), 354–361 (2010)

[10] Freedman, D., Zhang, T.: Active contours for tracking distributions. IEEE Trans. Image Process. 13(4), 518–526 (2004)

[11] Cotton Incorporated USA (2009), The classification of Cotton, `http://www.cottoninc.com/ClassificationofCotton`

[12] Kumar, N., Pandey, S., Bhattacharya, A., Ahuja, P.S.: Do leaf surface characteristics affect agrobacterium infection in tea [camellia sinensis (l.) o kuntze]? J. Biosci. 29(3), 309–317 (2004)

[13] National Institute for Agricultural Botany. Chrysanthemum Leaf Classification, Cambridge (2005)

[14] Tzionas, P., Papadakis, S., Manolakis, D.: Plant leaves classification based on morphological features and a fuzzy surface selection technique. In: Fifth International Conference on Technology and Automation, Thessaloniki, Greece, pp. 365–370 (2005)

[15] Wu, S.G., Bao, F.S., Xu, E.Y., Wang, Y., Chang, Y., Xiang, Q.: A Leaf Recognition Algorithm for Plant Classification Using Probabilistic Neural Network. In: IEEE International Symposium on Signal Processing and Information Technology, pp. 11–16 (2007)

# Analysis of Next Hop Selection for Geocasting in VANET

Sanjoy Das and D.K. lobiyal

School of Computer and Systems Sciences
Jawaharlal Nehru University,
New Delhi, India
{sdas.jnu,lobiyal}@gmail.com

**Abstract.** Geocasting is a special variant of multicasting, where data packet or message is transmitted to a predefined geographical location i.e., known as geocast region. The applications of geocasting in VANET are to disseminate information like, collision warning, advertising, alerts message, etc. The topology of VANET is changes frequently and rapidly due to the fast movement of vehicles. Therefore, the link time between two neighboring vehicles exist for a short span of time. This makes delivery of messages to a geocast region most challenging issue in VANET. In this paper, we have proposed an analytical model based on connectivity and selection of next node. For the better network connectivity in multi hop vehicular ad hoc network each cell should have at least one node. The numerical results show how connectivity between nodes depends on the transmission range of nodes.

**Keywords:** VANET, Geocast, neighbour node, probability, Cell.

## 1 Introduction

VANET is a special class of Mobile Ad hoc Network (MANET).A Mobile Ad hoc network is a dynamically reconfigurable wireless network with no fixed infrastructure. Every node in this network behaves like a router to relay a message from one node to another. In MANET; nodes are laptops, PDAs, palmtops, and other mobile devices whereas in VANET [1] nodes are vehicles. In addition, the other characteristics which differentiate VANET from MANET are mobility of nodes; structure of the geographical areas, delay constraint, privacy, etc.., node movement depends on the structure of road or structure of city or terrain etc.

While delivering message from source to destination node, if destination node is not within the transmission range of source node then the source node send the message to the destination node with the help of intermediate nodes. The ad hoc network is multi-hop in nature and message delivery in depends on the connectivity among the intermediate nodes. The message delivery from one location to another location is done with the help of intermediate nodes. The aim of Geocast routing protocols is to deliver a message from one location (i.e. sender) to a predefined location known as Geocast region with optimal number of node and time period. It is

D. Nagamalai, E. Renault, M. Dhanushkodi (Eds.): CCSEIT 2011, CCIS 204, pp. 326–335, 2011.

desirable that protocols should maintain the low end to end delay and high accuracy in delivering message i.e., high delivery success ratio, low overhead and total numbers of participating hops should be minimum.

The rest of paper is organized as follows. Section 2 presents work related to the geocast protocols. In section 3 overview of our proposed mathematical model is presented. In section 4 probabilistic analysis of one hop neighbor and effect of various network parameters is discussed. Finally, the work presented in this paper is concluded in section 5.

## 2 Related Work

Extensive works have been carried out by researchers, academicians and industries for successfully dissemination of messages from source to destination. There are several projects [2], [3], [4], [5] on VANET i.e. [CarTalk, FleetNet–Internet on the Road, NoW (Network on Wheel)] are going on for its deployment in the real world. The main focus of all these projects is to provide safety, and timely dissemination of message from one location to another location. One of the message delivery protocols proposed for VANET tries to deliver a message to a geographic region rather than to a node called geocast routing.Many protocols have been developed for geocast routing such as LAR [6], LBM [7], GeoTORA [8] is modified TORA, GRID protocol is modified to GeoGRID [9], DREAM [10], GRUV [11], etc. A comprehensive survey of geocasting protocol is presented in [12]. A Voronoi diagram based geocast protocol has been proposed in [13]. None of these protocols considered the road structure since they have been primarily designed for MANET. In[22], the authors proposed different modified LAR [6] algorithms. They have modified the request zone. Through simulation, the authors have established that their proposed algorithms reduces route request overhead as compared to original LAR. The performance analysis shows that their method outperforms original LAR especially, in a dense and highly dynamic ad hoc network. The mathematical analysis of VANET and ad hoc network characteristics and network connectivity are proposed in [14], [15], [16], [17], [18],[19].

The geocasting is the variant of multicasting, in geocasting source node will send message to a particular geographical area. We divided the geocast method in two phases. In the Phase-I, the source node will send a message to a node inside the geocast region and in the phase-II, the receiving node of the geocast region will deliver the message to the entire member in the geocast region. The node which is move to the geocast region will automatically become the member of the geocast region. In Phase-II, the message delivery inside the geocast region is done by simple flooding techniques. Here, in this paper we are only focused our work on the phase –I. In phase I, we have described the probabilistic analysis of nodes and effect of various network parameters i.e., transmission range of node, node density.

## 3 Proposed Model

We have considered the multi-hop environment, because it's very rare that source and destination node fall in each other transmission range. As there is no direct connectivity between source and destination node, to route the message intermediate nodes plays a vital role. The intermediate nodes are act as relay node. We have considered highway scenario to deliver message from source to geocast region shown in fig-1.

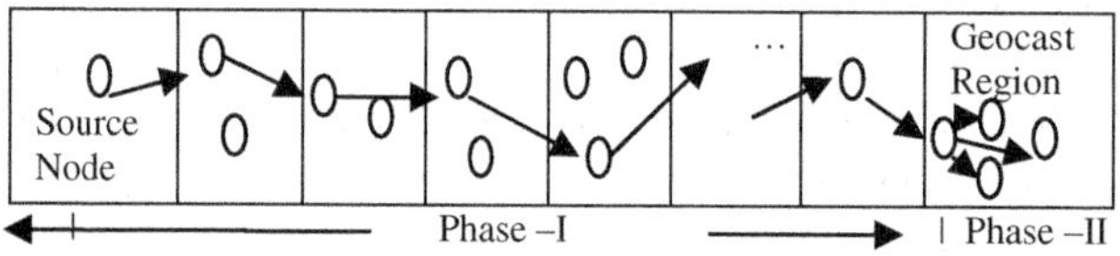


**Fig. 1.** Simple Scenario of Geocast on Highway

The highway is divided into small size cells [17] shown in fig.2. The cell size is considered in this paper as a rectangular structure. The cell length is equal to the transmission range of any nodes i.e. R and width is W. The transmission range for all nodes is fixed. The highway is divided into n number of equal sized cells. The advantage of cell is that it is used to identify the location of the vehicles.

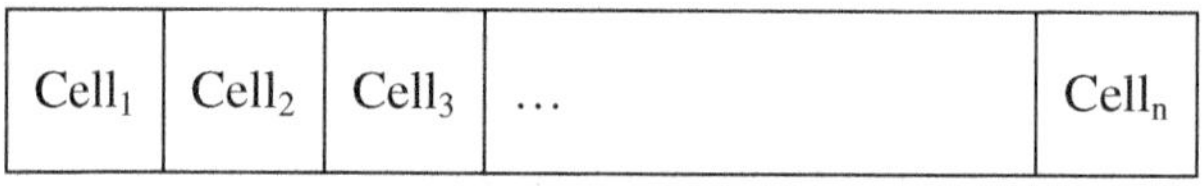


**Fig. 2.** Single lane highway scenarios

The minimum number of nodes required to send a message from source to destination is depend on node's transmission range and distance between them. To compute the distance between two nodes we have to know the coordinate of both the nodes. Let's take (x1, y1) and (x2, y2) are the coordinates of two nodes. The Euclidian distance D between two nodes can be calculated by using equation-1

$$D = \sqrt{(x2 - x1)^2 + (y2 - y1)^2} \, . \tag{1}$$

The two nodes are able to communicate to each other if the Euclidian distance between them is equal to or less than the transmission range.

## 4 Probabilistic Analysis of Neighbour Nodes

For our analysis, we assume that all nodes in the network have the same transmission range. The others parameters used in the analysis are given in the table-1

### 4.1 Analysis of One Hop Neighbour

We have considered that a highway is divided into number of cell of rectangular shape. The size we considered here for highway is more realistic. The length of cell is L and width is w. The area of each cell will be $C_i$= L*W. The total area of network will be A= $C_i$*$N_c$. The probability of a node having m numbers of neighbours depends on the size of a cell and transmission range of nodes. The probability of a node having m numbers of neighbour nodes can be calculated according to theorem-1 proposed in [20] with the help of the following equation.

$$P\,(N_{neigh}=m)=\binom{N-1}{m}\left(\frac{W*R}{C_i*N_c}\right)^{m}\left(1-\frac{W*R}{C_i*N_c}\right)^{N-1-m};\,0\leq m\leq N-1\ . \quad (2)$$

and the average number of neighbours per node can be calculated as follows in equation -3.

$$\overline{m} = \left(E_{neigh}\right) = (N-1)*\left(\frac{W*R}{C_i*N_c}\right). \quad (3)$$

**Table 1.** Symbol notation

| Symbols | Description |
|---|---|
| N | Total number of nodes in network |
| A | Area of network |
| R | radio range |
| m | Average number of one hop neighbours |
| $N_{neigh}$ | Number of one hop neighbours |
| $N_c$ | Total number of cell. |
| $C_i$ | Area of Cell. |

#### 4.1.1 Numerical Results

Fig 3 shows the one hop neighbour probability with varying the number of cells for 300 nodes in the network. We have considered value of radio transmission range 250, number of nodes fixed to 300 and number of cells 20, 30 and 50 given in Table 2. The result shows that when the number of cells is 20, the probability of having maximum number of one hop neighbours is 0.1. When the number of cells increased to 30 and 50 the probability of having maximum number of one hop neighbours is 0.13 and 0.17. The probability of having maximum number of one neighbour node is high when the no of cell is 20. This probability decreases as the numbers of cell increases.

**Table 2.** Simulation parameters

| Parameter | Values |
|---|---|
| $N_c$ | 20,30,50 |

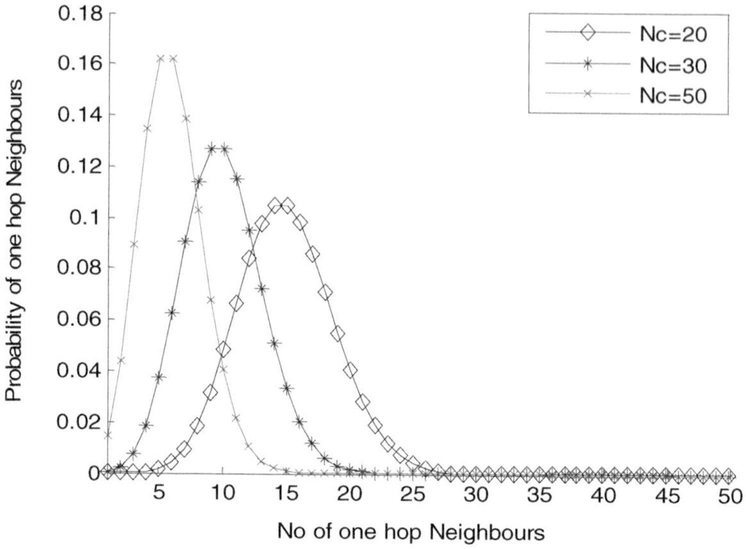


**Fig. 3.** Probability of one hop neighbours with varying number of cells

**Table 3.** Simulation parameters

| Parameter | Values |
|---|---|
| N | 100,200,300 |

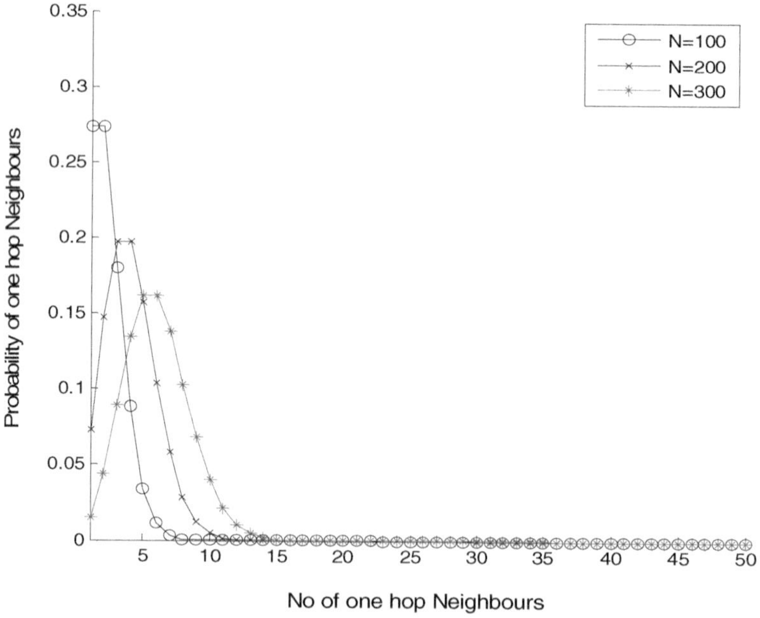


**Fig. 4.** Probability of one hop neighbours with varying number of nodes

Fig 4 shows the one hop neighbour probability with varying the number of nodes for 50cells in the network given in Table 3. For a network size of100 nodes, the probability of having maximum one hop neighbour is nearly 0.27. For a network size of 200 and 300 nodes, the probability of having maximum number of one hop neighbours is 0.2 and 0.17, respectively. The probability of having maximum numbers of one hop neighbour is high when the number of node is high.

### 4.2 Probabilistic Analysis of Nodes Present in a Cell

We assume that the highway is a two dimensional structure. We have divided the entire road structure in to number of cells of size w*L. The total number of nodes N is uniformly distributed over $L_0$ to $L_{max}$ area on the Highway. Here, we have analyzed the probability of number of nodes, $n_o$ present in each cell. In case each cell has $n_0$ numbers of node, the network will be fully connected. It is very essential in multi hop ad hoc environment that connectivity among nodes always exists for uninterrupted successful data transmission.

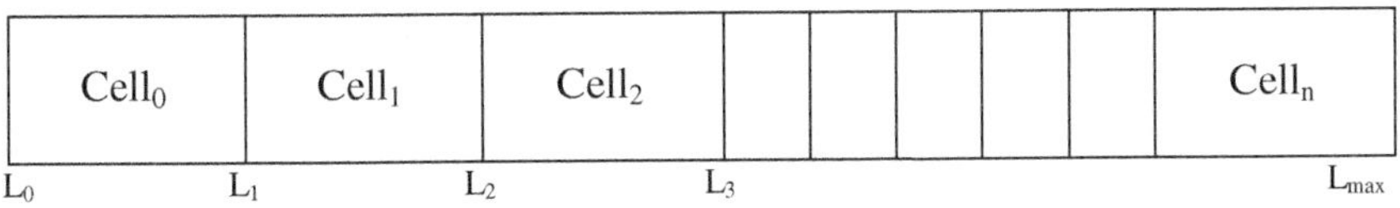


**Fig. 5.** Cells are divided in to subarea

According to [21] we computed the probability of number of nodes, $n_0$ present in a cell. We assume that $n_0$out of N nodes are located in the interval $[L_i, L_j]$, where $0 \leq L_i \leq L_j \leq L_{max}$ and i<0 and j<0.The total area of the network is denoted by A and each cell area is denoted by $C_i$ where 0<i<n. The expected number of nodes present in a unit area is denoted by $\rho$ =N/A.

A random variable d denotes the number of nodes within the given interval. The probability that a node is placed within the interval $[L_i, L_j]$ is $p = (L_j-L_i)/L_{max}$. The probability that $n_0$ of N nodes are placed in a cell area $C_i$ is

$$P\,(d=n_0) = \frac{\left(\frac{Ci*N}{A}\right)^{n_0}}{n_0!} * e^{-\frac{Ci*N}{A}}$$

$$= \frac{(Ci*\rho)^{n_0}}{n_0!} * e^{-Ci*\rho} \; . \tag{4}$$

A node can transmit data within $\Pi * R^2$ area. But in our proposed model the area is not circular, as we divided the road into cells of size w*L. Therefore, the probability that a randomly chosen node has $n_0$neighbours is

$$P\,(d=n_0) = \frac{(\rho*W*R)^{n_0}}{n_0!} * e^{-\rho*W*R} . \tag{5}$$

For large value of N and A, we can approximate this solution with a Poisson distribution, i.e.

$$P\ (d{=}n_0) = \frac{(Np)^{n_0}}{n_0!} * e^{-Np}. \tag{6}$$

The node density $\rho$ = N/A is the expected number of nodes present per unit area. For given density $\rho$ we can calculate the probability that $n_0$ nodes are present in an interval of length $L_{21} = L_2 - L_1$ as

$$P\ (d{=}n_0) = \frac{(\rho L_{21})^{n_0}}{n_0!} * e^{-\rho L_{21}}. \tag{7}$$

### 4.2.1 Numerical Results

In fig. 6 shows the probabilitics analysis of numbers of nodes to be present in each cell area.The different parameters considered for the analysis is given in table-4. We

**Table 4.** Simulation parameters

| Parameter | Values |
|---|---|
| N | 50,100,150,200,250,300,350…2000 |

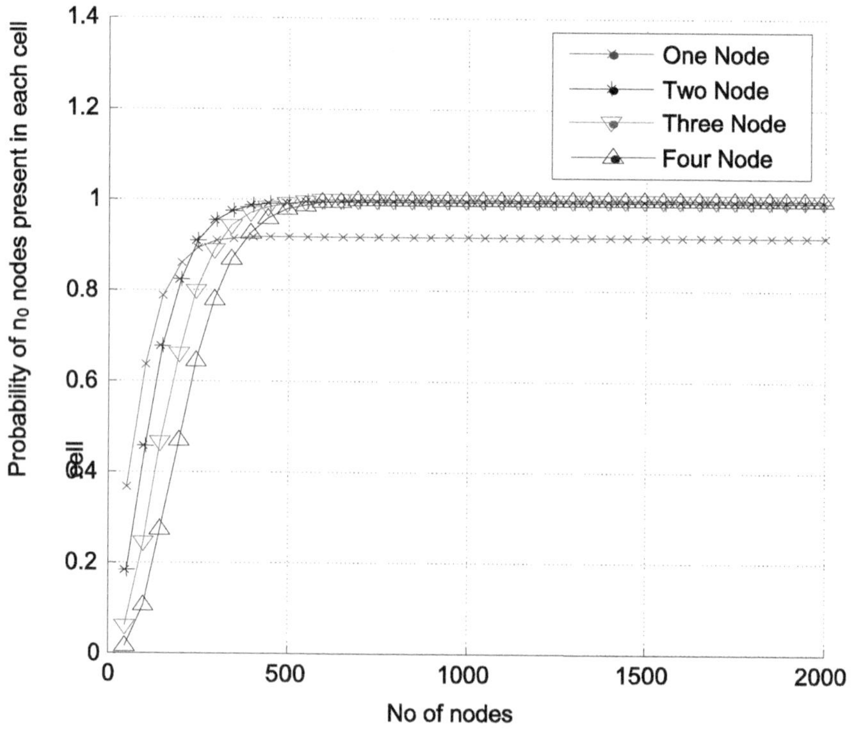


**Fig. 6.** Probability of having $n_0$ numbers of node present in each cell Vs Node Density

have fixed the number of cells and the transmission range of node. The node density is vary here. We have shown the probability of 0,1,2,3 and 4 nodes present in each cell.The initial value of node is 50 and incremented by 50 maximum is 2000. When the number of node is 50 probability having zero node present in a cell is greater than 0.2. The probability of 3 and 4 nodes present in each cell is near to 0. When the number of nodes are increases upto 500 the probability of having 2,3 and 4 nodes in each cell is 1. As the number of nodes exceeds 500, it shows that 3 to 4 nodes are always present in each cell. As the number of nodes increases, presence of nodes in each cell is also increases and the network becomes fully connected.

In fig. 7 shows the probabilitics analysis for numbers of nodes to be present in each cell area. The parameters considered for this analysis is given in Table-5. The node density and the transmission range vary in this analysis. When the transmission range

**Table 5.** Simulation parameters

| Parameter | Values |
|---|---|
| R | 100,110,120,...300 |

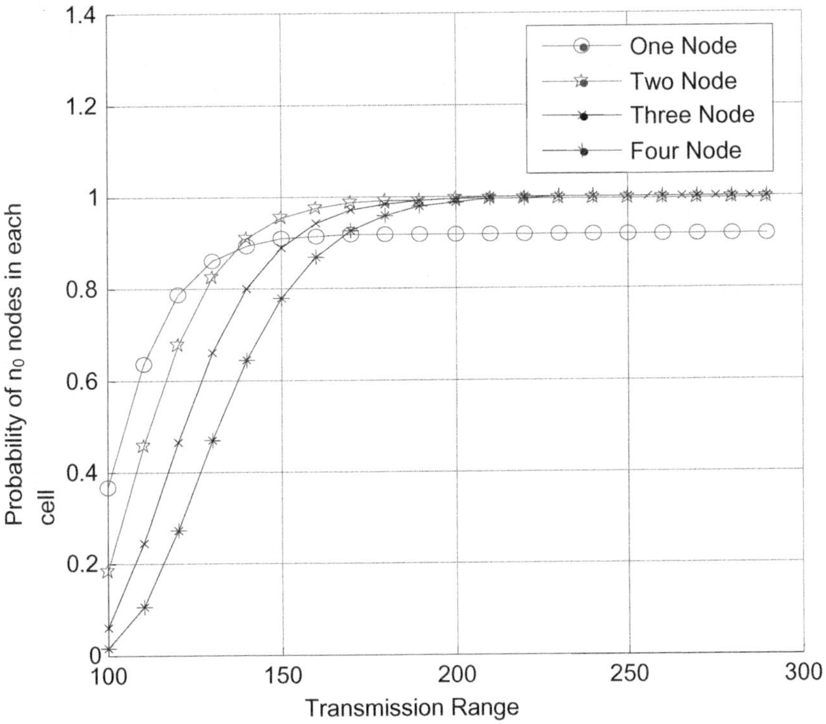


**Fig. 7.** Probability of having $n_0$ numbers of node present in each cell Vs Transmission Range

is 100 and the node density is low , the probability of having 0 or 1 node in a cell is 0.3. When the node density and the transmission range is increasing probability of having nodes in cells is also increasing. The probability that at least one node is available in any cell becomes constant at 0.9 after transmission range reach to 150 with increasing number of nodes. Further, any cell having atleast 2,3 or 4 node its probability become constant to 1 after transmission range exceed 200. If we transmit data with high transmission range, the nodes available in the network will be always connected and successfully deliver the data to the destination.

## 5 Conclusion

In this paper we have presented a geocast protocol for Vehicular Ad hoc network. In this protocol, we have proposed a model for highway scenario in which the highway is divided in to number of cells. The probabilistic analysis of nodes present in each cell is shown in our analysis. For the better network connectivity in multi hop vehicular ad hoc network each cell should have at least one node. We have analyzed the impact of various network parameters like node density, transmission range and network size and their impact in network connectivity. The simulation work is implemented using Matlab. The results shows that when the number of node is 50 probability having zero node in a cell is greater than 0.2. As the number of nodes exceeds 500, it shows that 3 to 4 nodes are always present in each cell. Further, with respect to transmission range the probability of at least one node is present in any cell becomes constant at 0.9 after transmission range reach to 150 with increasing number of nodes. The probability of having more than 1 node become constant to 1 after transmission range exceed 200.

## References

1. Schmitz, R.: Analysis of Path Characteristics and Transport Protocol Design in Vehicular Ad Hoc Networks. In: IEEE 63rd International Conference Vehicular Technology VTC, pp. 528–532 (2006)
2. `http://vanet.info/projects`
3. The NoW: Network on wheels Project, `http://www.network-on-wheels.de/about.html`
4. CarTalk, `http://www.cartalk2000.net/`
5. FleetNet–Internet on the Road, `http://www.fleetnet.de/`
6. Ko, Y.B., Vaidya, N.: Location-aided routing (LAR) in mobile ad hoc networks. In: Proceedings of the ACM/IEEE International Conference on Mobile Computing and Networking (MOBICOM 1998), pp. 66–75 (1998)
7. Ko, Y.B., Vaidya, N.: Geocasting in Mobile Ad-hoc Networks: Location-Based Multicast Algorithms. In: 2nd IEEE Workshop on Mobile Computing Systems and Applications, New Orleans, Louisiana (February 1999)
8. Ko, Y.B., Vaidya, N.: GeoTORA: A Protocol for Geocasting in Mobile Ad Hoc Networks. In: IEEE International Conference on Network Protocols, Osaka, Japan (2000)

9. Liao, W.H., Tseng, Y.C., Sheu, J.P.: GeoGRID: A Geocasting Protocol for Mobile Ad Hoc Networks Based on GRID. J. Internet Tech. 1(2), 23–32 (2000)
10. Basagni, S., Chlamtac, I., Syrotiuk, V.R., Woodward, B.A.: A distance routing effect algorithm for mobility (DREAM). In: Proc. MOBICOM, pp. 76–84 (1998)
11. Zhang, G., Chen, W., Xu, Z., Liang, H., Dejun, M., Li, G.: Geocast Routing in Urban Vehicular Ad Hoc Networks. In: Lee, R., Hu, G., Miao, H. (eds.) Computer and Information Science 2009, vol. 208, pp. 23–31. Springer, Heidelberg (2009)
12. Maihofer, C.: A Survey of Geocast Routing Protocols. IEEE Communications Surveys & Tutorials 6(2), 32–42 (2004)
13. Stojmenovic, I.: Voronoi Diagram and Convex Hull Based Geocasting and Routing in Wireless Networks, University of Ottawa, TR-99-11 (1999)
14. Antunes, N., Jacinto, G., Pacheco, A.: On the minimum hop count and connectivity in one-dimensional ad hoc wireless networks. Telecommunication Systems 39(2), 137–143 (2008)
15. Seada, K., Helmy, A., Govindan, R.: On the Effect of Localization Errors on Geographic Face Routing in Sensor Networks. In: IEEE/ ACM 3rd International Symposium on Information Processing in Sensor Networks IPSN, Berkeley, California, USA, pp. 71–80 (2004)
16. Seada, K., Helmy, A.: Efficient Geocasting with Perfect Delivery in Wireless Networks. In: IEEE Wireless Communications and Networking Conference WCNC, pp. 2551–2556 (2004)
17. Stojmenovic, I.: Geocasting with Guaranteed Delivery in Sensor Networks. IEEE Wireless Communications 11(6), 29–37 (2004)
18. Khabazian, M., Ali, M., Mehmet, K.: A Performance Modelling of Connectivity in Vehicular Ad Hoc Networks. IEEE Transactions on Vehicular Technology 57(4), 2440–2450 (2008)
19. Harb, S.M., McNair, J.: Analytical Study of the Expected Number of Hops in Wireless Ad Hoc Network. In: Li, Y., Huynh, D.T., Das, S.K., Du, D.-Z. (eds.) WASA 2008. LNCS, vol. 5258, pp. 63–71. Springer, Heidelberg (2008)
20. Chen, Q.J., Kanhere, S.S., Hassan, M.: Modeling path length in wireless ad hoc network. UNSW-CSE-TR-0617 (August 2006), `ftp://ftp.cse.unsw.edu.au/pub/doc/papers/UNSW0617.pdf`
21. Bettstetter, C.: On the minimum node degree and connectivity of a wireless multi-hop network. In: Proceedings of the ACM MobiHoc, pp. 80–91 (2002)
22. DeRango., F., Ha, A., Molinaro, A., Marano, S.: A Modified Location Aided Routing Protocol for the Reduction of Control Overhead in Ad-hoc Wireless Networks. In: ICT 2003, vol. 2, pp. 1033–1037 (2003)

# Multi Objective Genetic Approach for Solving Vehicle Routing Problem with Time Window

Padmabati Chand and J.R. Mohanty

School of Computer Engineering,
KIIT University, Bhubaneswar,
Odisha, India
{padma024,jnyana1}@gmail.com

**Abstract.** Vehicle routing problem with time window (VRPTW) is a NP-Complete and a multi-objective problem. The problem involves optimizing a fleet of vehicles that are to serve a number of customers from a central depot. Each vehicle has limited capacity and each customer has a certain demand. Genetic Algorithms maintain a population of solutions by means of a crossover and mutation operators. For crossover and mutation, best cost route crossover techniques and exchange mutation procedure is used respectively. In this paper, we focus on three objectives of VRPTW i.e. number of vehicles, total cost (distance), and time window violation (routing time). The proposed Multi Objective Genetic Algorithm (MOGA) finds optimum solutions effectively.

**Keywords:** Vehicle Routing Problem (VRP), Genetic Algorithm, Multi-Objective Optimization, Best Cost Route Crossover, Exchange Mutation.

## 1 Introduction

Routing and scheduling problems are important elements of many logistics systems. A lot of researches are devoted to finding effective methods to tackle these complex combinatorial problems [1]. One of the major research topics has been the Vehicle Routing Problem with Time Windows (VRPTW).The Vehicle Routing is a complex combinatorial optimization problem. Vehicle routing problem as the efficient use of a fleet of vehicles must make a number of stops to pick up and deliver passengers or products. The term customer is used to denote the stops to pick up and deliver the product. Every customer has to be assigned to exactly one vehicle in a specific order, which is to be done with respect to the capacity in order to minimize the total cost. The problem can be considered as a combination of the two well-know optimization problems i.e. the Bin Packing Problem (BPP) and the Travelling Salesman Problem (TSP). Relating this to the VRP, customers can be assigned to vehicles by solving BPP and the order in which they are visited can be found by solving TSP. The rest of the paper is organized as follows - section 2 gives a back ground study of the VRP, section 3 gives the multi objective genetic search of VRP, section 4 describes on experimental results.

D. Nagamalai, E. Renault, M. Dhanushkodi (Eds.): CCSEIT 2011, CCIS 204, pp. 336–343, 2011.

## 2 Background

The VRPTW can be defined as follows. Let G = (V, E) be a connected digraph consisting of a set of n + 1 nodes, each of which can be reached only within a specified time interval or time window, and a set E of arcs with non-negative weights representing travel distances and associated travel times.Let one of the nodes be designated as the depot.Each node i, except the depot, requests a service of size $q_i$. In the VRPTW all customers request the same type of service: pickups or deliveries. Moreover, each customer must be serviced once, thus prohibiting split service and multiple visits. The primary objective of the VRPTW is to find the minimum number of tours; such that each node is serviced within its time window and that the accumulated service up to any node does not violate the vehicle capacity [2]. The tours correspond to feasible routes starting and ending at the depot. Often a secondary objective is imposed to minimize the total distance traveled or to minimize total schedule time. All problem parameters, such as customer demands, travel times and time windows, are assumed to be known with certainty. The VRPTW is a basic distribution management problem that can be used to model many real-world problems [3]. Some of the most useful applications of the VRPTW include bank deliveries, postal deliveries, industrial refuse collection, national franchise restaurant services, school bus routing, security patrol services and JIT (just in time) manufacturing [4]. The VRPTW has been the subject of intensive research efforts for both heuristic and exact optimization approaches. The main objectives of VRPTW are:

- Minimizing the total distances traveled by the vehicles.
- Minimizing the number of vehicles used.
- Minimizing the time window violation (routing time).

## 3 Multi-Objective Genetic Search for the VRPTW

The ability of genetic algorithm (GA) is to simultaneously search different regions of a solution space which makes it possible to find a diverse set of solutions for difficult problems with non-convex, discontinuous and multi-modal solutions spaces. Multi-objective optimization, also known as multi-criteria optimization, is the process of simultaneously optimizing two or more conflicting objectives subject to certain constraints [5]. If a multi-objective problem is well formed, there should not be a single solution that simultaneously minimizes each objective to its fullest. In each case we are looking for a solution (for which each objective has been optimized) to the extent that if we try to optimize it any further, then the other objective(s) will suffer as a result. Finding such a solution and quantifying how much better this solution (compared to other such solutions) is the goal of setting up and solving a multi objective optimization problem [6].

In GA, each chromosome in the population pool is transformed into a cluster of routes. The chromosomes are then subjected to an iterative evolutionary process until a minimum possible number of route clusters is attained or the termination condition is met [7]. The evolutionary part is carried out as in simple GA using selection, crossover, and mutation operations on chromosomes. Tournament selection is used to perform fitness-based selection of individuals for reproduction. The crossover operator of GA may exploit structures of good solutions with respect to different objectives to create new non-dominated solutions in unexplored parts of the Pareto front. A crossover operator that ensures solutions generated through genetic evolution are all feasible is also proposed. Hence, both checking of the constraints and repair mechanism can be avoided, thus resulting in increased efficiency. In addition, most Multi-Objective Genetic Algorithm (MOGA) do not require the user to prioritize, scale, or weight objectives. Therefore, GA has been the most popular heuristic approach to multi-objective design and optimization problems [8].

### 3.1 Chromosome Representation and Initial Population Creation

In our approach, a chromosome representing route of length N, where N is the number of customers in a particular problem instance. A gene in a given chromosome indicates the original node number assigned to a customer, while the sequence of genes in the chromosome indicates the order of visitation of customers, '-' indicates a node representing a group of clustered customers that have already been committed to a given vehicle [9]. Thus, the chromosome consists of integers, where new customers are directly represented on a chromosome with their corresponding positive index number and each committed customer is indirectly represented within one of the groups (shown by a '–' mark ) representing a given deployed vehicle [10].

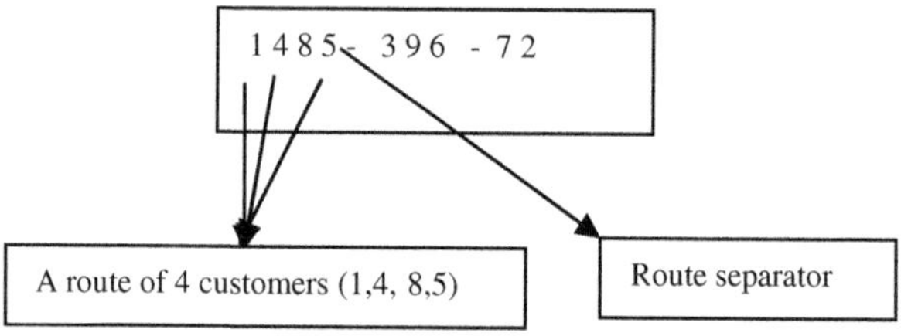


**Fig. 1.** Chromosome Representation. This figure represents a route of three sub routes (1 4 8 5), (3 9 6), (7 2). Each sub route is separated by '-'symbol.

### 3.2 Fitness Evaluation

In this approach, we use dominance-information of the individuals of the population by calculating for each individual i the number of alternatives $X_i$ from which this individual is dominated. For a population consisting of $n_p$ (no. of population) alternatives we get values of:

$$0 \le X_i \le n_p - 1 \qquad (1)$$

Individuals that are not being dominated by others should receive a higher fitness value than individuals that are being dominated.

We calculated the fitness value for each individual i by a linear normalization. Individuals with the lowest values of $X_i$ ($X_i = 0$) receive the highest corresponding value of $F_i = F_{max}$.

$$F_i = F_{max} - X_i \qquad (2)$$

### 3.3 Cross Over

Initial experiments using standard crossover operators such as Partially-Mapped-Crossover (PMX) and uniform order crossover (UOC) yielded non-competitive solutions. Hence, we utilized a problem-specific crossover operator that generates feasible route schedules. An example of the procedure utilized by the proposed crossover (Best Cost Route Crossover, BCRC) is given in Fig.2. According to Fig.2, two parents A and B are selected from the population. A route from each parent chromosome is randomly selected and the customer orders present in each route are removed from the other parent. Since '–' marks represent existing vehicles, their customers are left untouched. This means only integers which represent uncommitted customers are reinserted into the current chromosome. Then the customers that have been removed are reinserted at the location which minimizes the overall cost of the entire tour. This requires computing the cost of inserting each of the remaining customers at each location in the chromosome without constraint violation. If no insertion location for a particular customer is found, a new route is created.

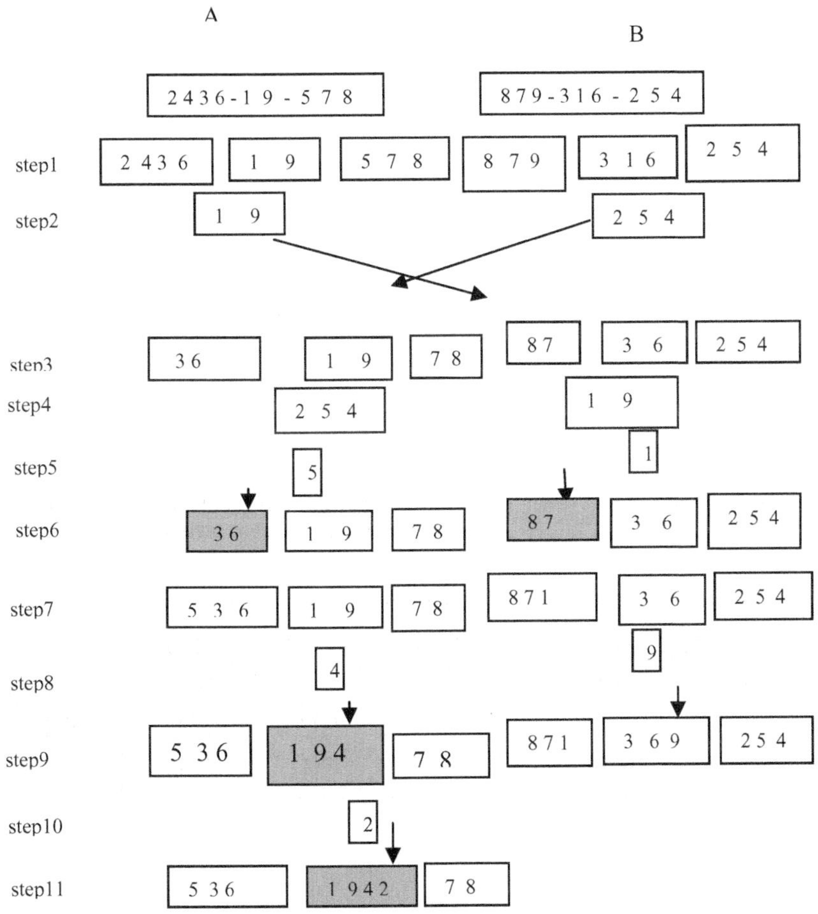


**Fig. 2.** Best Cost Route Cross Over

In the Fig. 2 A and B are two parents. In step 1 for parent A there is three routes (r1:2 4 3 6 r2:1 9 r3:5 7 8) similarly in B (r1:8 7 9 r2:3 1 6 r3:2 5 4). In step 2 we selected randomly a route from step 1 of parent A (r2:1 9) and in B (r3:2 5 4). In step 3 the route we selected from step 2 of parent A (1 9) that we removed from parent B of the given routes in step 1. Similarly the route we selected from step 2 of parent B (2 5 4) that we remove from parent A of the given routes in step 1. In step 4 the route we are deleted from parent A (2 5 4) and B (1 9) that we have again inserted. From the route (2 5 4) we have randomly selected a route suppose 5 that we have inserted again to the route (the route we got in step 3) by satisfying all the constraints i.e. vehicle capacity and also after inserting there should be optimum solution means distance should be minimum and minimum number of vehicles. The suitable location is darkening in figure and also there is an arrow mark in step 6. After inserting, the new route is (r1:5 3 6 r2:1 9 r3:7 8). Similarly we inserted 4 and 2. If any removed node is not satisfying constraints then make a new route. Similarly in parent B from 1 and 9 we have randomly selected a customer and that we have inserted by satisfying all the constraints. If not satisfying make a new route. So finally the optimum solution we got from parent A (r1:5 3 6 r2:1 9 4 2 r3:7 8) and B (r1:8 7 1 r2:3 6 9 r3:2 5 4) [11].

### 3.4 Mutation

Mutation is done by doing exchange mutation. Randomly we selected two genes from two sub routes and exchange it. But after exchanging the new route should satisfy the constraints [12]. Mutation is shown in the Fig. 3.

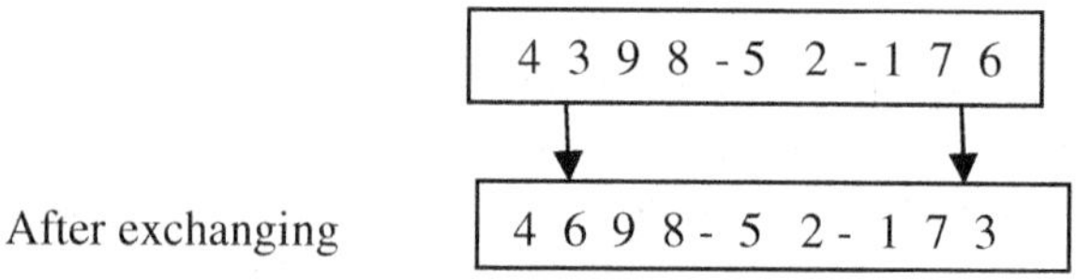


**Fig. 3.** Mutation. In this figure we randomly selected two genes from two different sub routes (gene 3 from sub route (4 3 9 8) and gene 6 from sub route (1 7 6)). Then two genes exchanged. After exchanging the new sub routes (4 6 9 8) and (1 7 3).

## 4 Computational Results

This section describes computational experiments carried out to investigate the performance of the proposed GA. By our given fitness function, it minimizes both number of vehicles and travel costs without bias. The algorithm was coded in mat lab and run on an Intel Pentium IV 1.6 MHz PC with 512 MB memory and it gives optimum result. We have applied MOGA to some instances from Solomon's benchmark set, it gives better result. Table 1 represents genetic algorithm (GA) parameters. This table shows number of parameters used to implement genetic algorithm. Number of times the program runs is 30. Cross over type is best cost route cross over (BCRC) and crossover rate 80. Mutation type is exchange mutation and mutation rate is 0.10. Table 2 represent experimental results of Solomon's data set. This table represents experimental results from some instance of Solomon's data set. Like C201_100 requires total no. of vehicles 3, total distance 897.24 and routing time 8456.29.

**Table 1.** Represents Genetic Algorithm Parameters

| Parameter type | Values |
|---|---|
| Initial Population | 100 |
| Number of runs | 30 |
| Crossover rate | 80 |
| Mutation rate | 0.10 |
| Selection type | Tournament |
| Crossover type | BCRC |
| Mutation type | Exchange |

**Table 2.** Represents Experimental Results of Solomon's data set

| Solomon's data set | Number of vehicles(NV) | Distance(D) | Routing Time |
|---|---|---|---|
| C201_100 | 3 | 897.24 | 8456.29 |
| R101_50 | 10 | 1324.57 | 2032.79 |
| R201_50 | 2 | 876.34 | 1234.63 |
| R201_100 | 8 | 1289 | 3068.76 |

The following figures represent experimental results according to the above table (table 2).Fig 4 represents initial population of 100 solutions. Fig 5 represents pareto optimal front (all the non-dominated solutions in a curve).Fig 6 and 7 based upon Solomon's data set. Both the figures represents route networks, where Fig.6 is the instance of R201_100 (number of vehicle=8, distance=1289, routing time=3068.76), Fig.7 is the instance of C201_100 (number of vehicle=3, distance=897.24, routing time=12107.76).

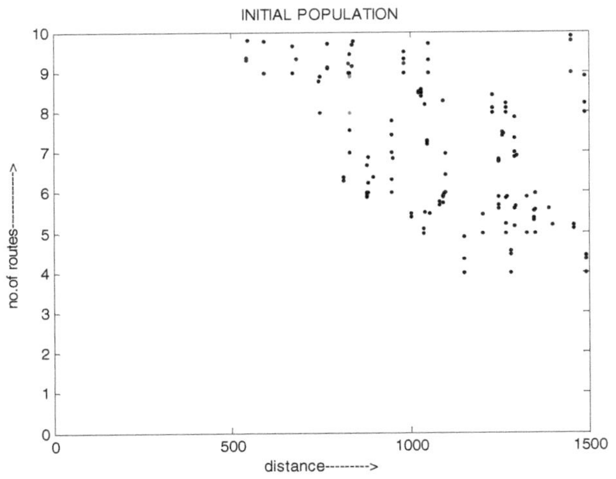


**Fig. 4.** Represents initial population (total no. of solutions). This figure shows x-axis represents total distance and y-axis represents total no. of routes per solution.

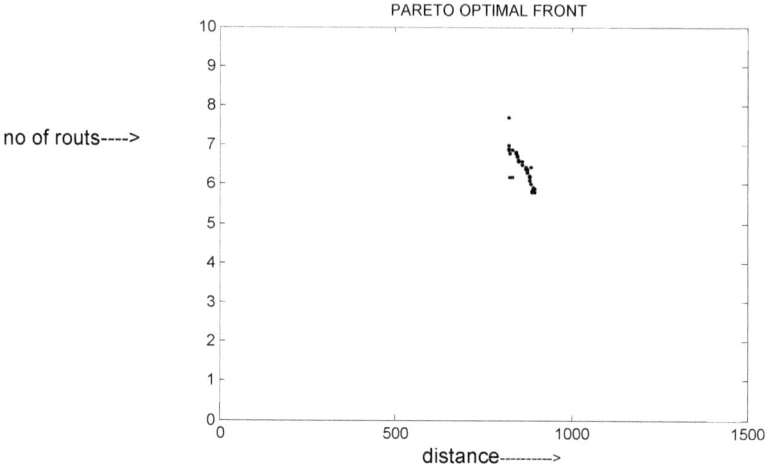


**Fig. 5.** Pareto Optimal Front. This figure represents number of non-dominated solutions. The above Figures Fig.4 represents initial population of 100 solutions. Fig.5 represents Pareto optimal front (represents all non-dominated solutions).Fig.6 and Fig.7, represents route networks of instance of R201_100 and C201_100 (Solomon's data set).

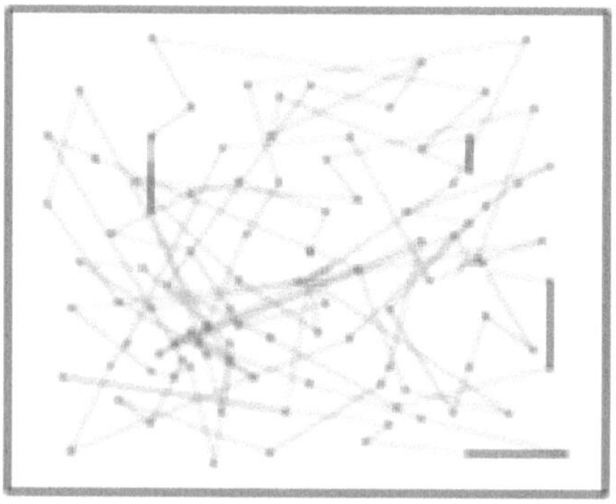

**Fig. 6.** Route network for Instance R201_100

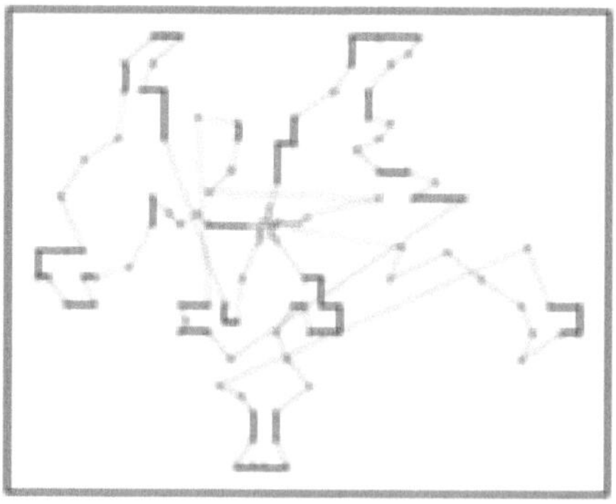

**Fig. 7.** Route network for instance C201_100

## 5 Conclusion

In this paper we presented a GA based approach for the static VRPTW. The approach was tested using problem instances reported in the literature, derived from publicly available Solomon's benchmark data for VRP. The experimental results showed that the GA approach was able to find high quality solutions. Our future goal is to generate larger problem instances, and further evaluate the GA's performance on these problems by considering other objectives like number of time window and speed of the vehicle.

## References


1. Desrosier, J., Dumas, Y., Solomon, M.M., Soumis, F.: Time constraint routing and scheduling. In: Ball, M.O., Magnanti, T.L., Monma, C.L., Nemhauser, G.L. (eds.) Handbooks in Operations Research and Management Science, vol. 8, pp. 35–139. Elsevier Science Publishers, Amsterdam (1995)
2. Garey, M.R., Johnson, D.S.: Computers and Intractability: A Guide to the Theory of NP-Completeness. W. H. Freeman and Company, New York (1979)
3. Braysy, O., Gendreau, M.: Vehicle routing problem with time windows, Part 1: Route construction and local search algorithms, SINTEF Applied Mathematics Report, Department of Optimization, Norway (2001)
4. Cordeau, J.F., Desaulniers, G., Desrosiers, J., Solomon, M.M., Soumis, F.: The VRP with time windows. In: Toth, P., Vigo, D. (eds.) The Vehicle Routing Problem. SIAM Monographs on Discrete Mathematics and Applications, ch. 7 (2001) (to appear)
5. Lenstra, J.K., Rinnooy Kan, A.H.G.: Complexity of vehicle routing problem with time windows. Networks 11, 221–227 (1981)
6. Braysy, O., Gendreau, M.: Vehicle routing problem with time windows, Part II: Metaheuristics, SINTEF Applied Mathematics Report, Department of Optimization, Norway (2001)
7. Kohl, N.: Exact methods for time constrained routing and related scheduling problems, PhD Thesis, Department of Mathematical Modeling, Technical University of Denmark (1995)
8. Zhu, K.Q.: A diversity-controlling adaptive genetic algorithm for the vehicle routing problem with time windows. In: Proceedings of the 15th IEEE International Conference on Tools for Artificial Intelligence (ICTAI 2003), pp. 176–183 (November 2003)
9. Goldberg, D.E.: Genetic Algorithms in Search, Optimization, and Machine Learning. Addison Wesley, Reading (1989)
10. Liu, B., Zhao, R.: Random programming and fuzzy programming, pp. 15–57. Tsing Hua University Publishing House, Beijing (1988)
11. Ombuki, B., Ross, J., Brian, H.F.: Multi Objective Genetic Algorithms for Vehicle Routing Problems with Time Windows. Applied Intelligence 24, 17–30 (2006)
12. Haupt, L., Randy, H., Ellen, S.: Practical Genetic Algorithms, 2nd edn., pp. 43–44. A John Wiley & Sons, Inc., Pennsylvania (2004)

# Application of Modified NSGA-II Algorithm to Reactive Power Optimization

S. Ramesh[1], S. Kannan[2], and S. Baskar[3]

[1] Arulmigu Kalasalingam College of Engineering, Krishnankoil, 626 126, India
[2] Kalasalingam University, Anand Nagar, Krishnankoil, 626 126, India
[3] Thiagarajar College of Engineering, Madurai, 625 015, India

**Abstract.** The optimal Reactive Power Dispatch (RPD) problem is a nonlinear constrained optimization problem. This paper presents true multi-objective solution set for multi-objective RPD problem. The objective functions are real power loss minimization and control variable adjustment costs. In this paper, a Modified Non-Dominated Sorting Genetic Algorithm version II (MNSGA-II) is proposed for solving RPD problem. For maintaining good diversity in the performance of NSGA-II, the concept of Dynamic Crowding Distance (DCD) is implemented in NSGA-II algorithm and given name as MNSGA-II. The standard IEEE 30-bus and IEEE 118 bus test systems are used. The results obtained using MNSGA-II are compared with NSGA-II and validated with reference Pareto front generated by conventional weighted sum method using Covariance Matrix Adapted Evolution Strategy (CMA-ES). The performance of NSGA-II and MNSGA-II are compared with respect to best, mean, worst and standard deviation of multi-objective performance measures.The results show the effectiveness of MNSGA-II and confirm its potential to solve the multi-objective RPD problem.

**Keywords:** Covariance Matrix Adapted Evolution Strategy, dynamic crowding distance, Multi-objective, Non-dominated sorting genetic algorithm, optimization, Reactive Power Dispatch.

## 1 Introduction

In general, RPD is a complex combinatorial optimization problem for a large scale power system involving nonlinear functions having multiple local minima and non-linear and discontinuous constraints. The aim of RPD problem is to minimize the network real power loss and to reduce the cost of control variable adjustments by regulating generator bus voltages, switching and changing transformer tap-settings [20]. With the increased loading of existing power transmission systems, the problem of voltage stability and voltage collapse, has become a major concern in power system planning and operation [22,14,3].

To solve the RPD problem, a number of conventional optimization techniques have been proposed. They include Gradient-based method, Non-linear programming (NLP), Quadratic programming (QP), Linear programming (LP) and Interior point methods [7,18,9,4]. Several evolutionary computation techniques like

D. Nagamalai, E. Renault, M. Dhanushkodi (Eds.): CCSEIT 2011, CCIS 204, pp. 344–354, 2011.

Genetic algorithm (GA) [12], Evolutionary programming (EP) [23] and Swarm intelligence [17] have been applied to solve the optimal RPD problems. However, this approach only results in a single optimal solution where trade-off between different components of the objective function must be fixed in advance of the solution. In order to provide a means to assess trade-off between two conflicting objectives, one may formulate the RPD problem as a multi-objective problem. Recently, the RPD dispatch problem is formulated as a multi-objective optimization problem. However, the problem is not treated as a true multi-objective problem [8,11]. It was converted to a single objective problem by linear combination of different objectives as a weighted sum [8]. Unfortunately, this requires multiple runs as many times as the number of desired Pareto-optimal solutions. In contrast, Evolutionary Algorithms (EAs) can find multiple optimal solutions in single simulation run due to their population approach [10,27]. Recently, some successful application of EAs to optimal RPD have been reported in [1,27], where minimizing voltage differences have been considered as an objective in addition to loss minimization. A similar multi objective RPD problem can be found in [25], where a fuzzy system is employed to adaptively adjust the parameters of PSO, such as the inertia weight and learning factors, during the evolutionary process. The NSGA-II algorithm has been demonstrated to be the most efficient algorithm for multi-objective optimization on a number of benchmark problems [6,5,27]. In addition, it has been shown that NSGA-II outperforms two other contemporary multi-objective evolutionary algorithms: Pareto-archived evolution strategy (PAES) [16] and Strength-Pareto Evolutionary Algorithm (SPEA) [26] in terms of finding a diverse set of solutions and in converging near the true Pareto-optimal set. Recently the NSGA-II algorithm is used in various power system problems such as economic dispatch [15] and Generation Expansion Planning [13]. But still NSGA-II algorithm suffers in maintaining diversity among the solutions in the Pareto front. This paper makes use of diversity maintenance strategy which is based on Dynamic Crowding Distance (DCD) introduced in [19].

The objective of this paper is to solve the RPD problem as a multi-objective optimization problem and solve it by using a modified NSGA-II (MNSGA-II) algorithm by means of incorporating DCD. To validate the performance of MNSGA-II and NSGA-II, conventional weighted sum method using Covariance Matrix Adapted Evolution Strategy (CMA-ES) is used.

The paper is organized as follows: Section 2 describes the multi-objective RPD problem formulations. Section 3 deals with MNSGA-II implementation to the RPD problem. Section 4 provides test results, and Section 5 sums up.

## 2 Problem Formulation

The optimal RPD problem is to optimize the steady state performance of a power system in terms of one or more objective functions while satisfying several equality and inequality constraints. Generally the problem can be represented as follows:

### 2.1 Objective Functions

**Real Power Loss ($P_L$).** This objective is to minimize the real power loss in transmission lines that can be expressed through the following equation

$$f_1 = P_L = \sum_{k=1}^{nl} g_k[V_i^2 + V_j^2 - 2V_iV_j\cos(\delta_i - \delta_j)] \tag{1}$$

where,
$nl$ is the number of transmission lines;
$g_k$ is the conductance of the kth line;
$V_i\angle\delta_i$ and $V_j\angle\delta_j$ are the voltages at end buses $i$ and $j$ of the $k^{th}$ line respectively.

**Cost of Control Variable Adjustment.** The proposed objective is to minimise the control variable adjustments that can be expressed as

$$f_2 = \sum_{k=1}^{N_C} w_k \left| \frac{u_k - u_k^0}{u_k^{max} - u_k^{min}} \right| \tag{2}$$

where $N_C$ is the number of control variables; $u_k$ and $u_k^0$ are the current and initial settings of the $k^{th}$ control variable respectively; and $w_k$ is a weighting factor to reflect the relative cost of the $k^{th}$ control variable. $u_k^{max}$ and $u_k^{min}$ are the maximum and minimum limits of the $k^{th}$ control variable.

The objective functions are subjected to following constraints

*Equality Constraints:*

$$\begin{aligned} 0 &= P_i - V_i \sum_{j\in N_i} V_j(G_{ij}\cos\theta_{ij} + B_{ij}\sin\theta_{ij}) i \in N_{B-1} \\ 0 &= Q_i - V_i \sum_{j\in N_i} V_j(G_{ij}\sin\theta_{ij} + B_{ij}\cos\theta_{ij}) i \in N_{PQ} \end{aligned} \tag{3}$$

where,
$V_i$ voltage magnitude at bus $i$.
$G_{ij}, B_{ij}$ mutual conductance and susceptance between bus $i$ and $j$
$\theta_{ij}$ voltage angle difference between bus $i$ and $j$
$N_{B-1}$ total number of buses excluding slack bus
$N_{PQ}$ set of $PQ$ buses
$N_i$ number of buses

*Inequality Constraints:* In the control variables, the generator bus voltages i.e. Automatic Voltage Regulator operating values ($V_G$) are taken as continuous variable, where as the transformer tap settings ($T_k$) is taken as discrete values. The load bus voltages ($V_{PQ}$), reactive power generation ($Q_G$) and Transmission line loading ($S_l$) are taken as state variables.
*Continuous Control Variables:*

$$V_{G_i}^{min} \leq V_{G_i} \leq V_{G_i}^{max}; i \in N_G \tag{4}$$

*Discrete Control variables:*

$$T_k^{min} \leq T_k \leq T_k^{max}; k \in N_T \tag{5}$$

*State Variables:*

$$\begin{array}{c} V_{PQ_i}^{min} \leq V_{PQ_i} \leq V_{PQ_i}^{max}; i \in N_{PQ} \\ Q_G^{min} \leq Q_G \leq Q_G^{max}; i \in N_G \\ S_l \leq S_l^{max} \end{array} \tag{6}$$

The objective function is calculated by using Newton-Raphson load flow calculation with the above mentioned equality and inequality constraints.

## 3 Implementation of MNSGA-II

The NSGA-II algorithm uses non-dominated sorting [6] method for ranking the individuals. The simulated binary crossover and polynomial mutation generates new offspring, and tournament selection is then used to select the population for the next generation. Elitism addresses the problem of losing good solutions during the optimization process due to random effects. One way to deal with this problem is to combine the old population and the offspring, i.e., the mating pool after variation, and to apply a deterministic selection procedure instead of replacing the old population by the modified mating pool. In this paper, NSGA-II algorithm with DCD (MNSGA-II) is used to solve the RPD problem.

### 3.1 Importance of MNSGA-II

It is of great importance for Pareto front with good convergence and diversity characteristics with respect to true Pareto front. A good diversity can give decision-makers more reasonable and efficient selections. Two aspects must be considered in the diversity, one is the span of Pareto front, and the other is the uniformity. The diversity maintenance strategy (DMS) is often realized in the process of population maintenance, which use truncation operator to wipe off individuals when the number of non-dominated solutions exceeds population size. In this paper, a DMS which is based on dynamic crowding distance (DCD) is discussed in [19].

In MOEAs, the horizontal diversity of Pareto front is very important. The horizontal diversity is often realized by removing excess individuals in the non-dominated set (NDS) when the number of non-dominated solutions exceeds population size. NSGA-II uses crowding distance (CD) measure as given in equation (7) to remove excess individuals. The individuals having lower value of CD are preferred to individuals with higher value of CD in removal process.

$$CD_i = \frac{1}{r}\sum_{k=1}^{r}\left|f_{i+1}^k - f_{i-1}^k\right| \tag{7}$$

where, $r$ is the number of objectives, $f_{i+1}^{k}$ is the $k^{th}$ objective of the $i+1^{th}$ individual and $f_{i-1}^{k}$ is the $k^{th}$ objective of the $(i-1)^{th}$ individual after sorting the population according to the CD. The major drawback of CD is lack of uniform diversity in the obtained non-dominated solutions. To overcome this problem, dynamic crowding distance (DCD) method is recently suggested in [19].

In this approach, one individual with the lowest DCD value is removed every time and recalculates DCD for the remaining individuals. The individuals DCD are calculated as follows:

$$DCD = \frac{CD}{\log\left(\frac{1}{V_i}\right)} \tag{8}$$

where $CD_i$ is calculated by equation (7) and $V_i$ is based on equation (9),

$$V_i = \frac{1}{r}\sum_{k=1}^{r}\left(\left|f_{i+1}^{k} - f_{i-1}^{k}\right| - CD_i\right)^2 \tag{9}$$

$V_i$ is the variance of CDs of individuals which are neighbors of the $i^{th}$ individual. $V_i$ can give information about the difference variations of CD in different objectives. Suppose the population size is $N$, the non-dominated set at $t^{th}$ generation is $Q(t)$ and its size is $M$. If $M > N$, DCD based strategy is used to remove $M - N$ individuals from non-dominated set. The summary of DCD algorithm is given below:

Step 1: **If** $|Q(t)| \leq N$ **then** go to Step 5, **else** go to Step 2.
Step 2: Calculate individuals DCD in the $Q(t)$ based on equation (8).
Step 3: Sort the non-dominated set, $Q(t)$ based on DCD.
Step 4: Remove the individual which has the lowest DCD value in the $Q(t)$.
Step 5: **If** $|Q(t)| \leq N$, stop the population, otherwise go to Step 2 and continue.

### 3.2 MNSGA-II Algorithm

The following steps can be adopted for the implementation of MNSGA-II algorithm.

Step 1: Identify the control variables like generator voltage limits, transformer tap setting limits for RPD problem
Step 2: Select the parameters like number of population, maximum number of generation, crossover and mutation probabilities.
Step 3: Generate initial population and evaluate the objective functions (i.e., $f_1, f_2$) for initial population.
Step 4: Set the generation count.
Step 5: Perform Simulated Binary crossover and polynomial mutation [5] for the set of individuals.
Step 6: Perform non-dominated sorting. (i.e., sorting the population according to each objective function value in ascending order of magnitude).
Step 7: Calculate DCD between the solutions based on the algorithm discussed in Section 3.1.

Step 8: Perform selection based on tournament selection [5] thereby a higher fitness is assigned to individuals located on a sparsely populated part of the front.

Step 9: Increment the generation count and repeat the steps from 6 to 9 until the count reaches the maximum number of generation.

## 4 Test Results

The implementation was done using MATLAB version 7.4, on a IBM PC with Intel core 2 quad processor having 1.86GHz speed and 3 GB RAM.

### 4.1 IEEE 30-Bus System

The representation of the IEEE 30 bus test system and the detailed data are given in [18]. The system has 6 generators, 4 transformers; therefore, the number of the optimized variables is 10 in this problem. For generator buses 1, 2, 5, 8, 11, and 13 the lower voltage magnitude limits are 0.95 p.u. and the upper limits are 1.1 p.u., and for the remaining buses are 1.05 p.u.. The transformer tappings are varied between 0.9 and 1.1 p.u. with the step size of 0.0001. The following control parameters are considered in the RPD problem.

1. Automatic Voltage Regulator (AVR) operating values (continuous variable).
2. On-Load Tap Changer (OLTC) (discrete variable).

These control parameters are treated in load flow calculation as follows: AVR operating values are treated as voltage specification values and OLTC tap positions are treated as tap ratio to each tap position.

**Generation of Reference Pareto-front.** To compare the performance of MNSGA-II and NSGA-II multiple run generated reference Pareto front is used. One of the classical methods for getting Pareto-optimal set is Weighted Sum method [[5], pp.50]. Using this method only one solution may be found along the Pareto front in single simulation run. Recently, CMA-ES has been applied to solve a non-linear multi-modal complex optimization problem successfully. One of the major advantages of CMA-ES algorithm is its consistency [21]. Hence in this paper CMA-ES is employed to obtain the Pareto front by using weighted sum of objectives. The problem was treated as single objective optimization problem by linear combination of objectives as follows.

$$Minimize\ C = w_1 \times f_1 + w_2 \times f_2 \tag{10}$$

where $w_1$ and $w_2$ are weighting factors and the sum of weighting factors must be 1. To get 30 non-dominated solutions the algorithm is applied 30 times with varying weighting factors as a random number varying between 0 and 1. The weight $w_1$ is calculated by selecting random number between 0 and 1. The weight $w_2$ is calculated by using the expression $w_2 = 1 - w_1$, so that the sum of weighting factors is always 1. For each run, 50000 function evaluations are performed for objective function evaluation.

In CMAES technique, the weights are randomly varied from 0 to 1 for each trial. A population size of 100 is used. About 30 points are obtained to construct the Pareto front through CMAES.

**Parameter Selection.** The results are sensitive to algorithm parameters. Hence, it is required to perform repeated simulations to find suitable values for the parameters. Optimal parameter combinations for different methods are experimentally determined by conducting a series of experiments with different parameter settings before conducting actual runs to collect the results. The population size is selected as 50 and generation size is selected as 500. The crossover probability ($P_c$) is selected between 0.8 and 0.95, in steps of 0.01 and for each $P_c$ performance is analyzed. It is found that $P_c = 0.89$, produces the best results. Other parameters such as mutation probability ($P_m$) crossover index ($\eta_c$) and mutation index ($\eta_m$) are selected as 0.01, 5 and 20 respectively.

**MNSGA-II Implementation.** The NSGA-II and MNSGA-II algorithm are applied to solve the RPD problem with 10 control variables. For analyzing the performance of MNSGA-II algorithm, same parameters are used. The best Pareto front obtained among 10 trial runs is represented in the Figure 1. The representation of Pareto front has been compared with NSGA-II, MNSGA-II and multiple run Pareto front in a two dimensional space.

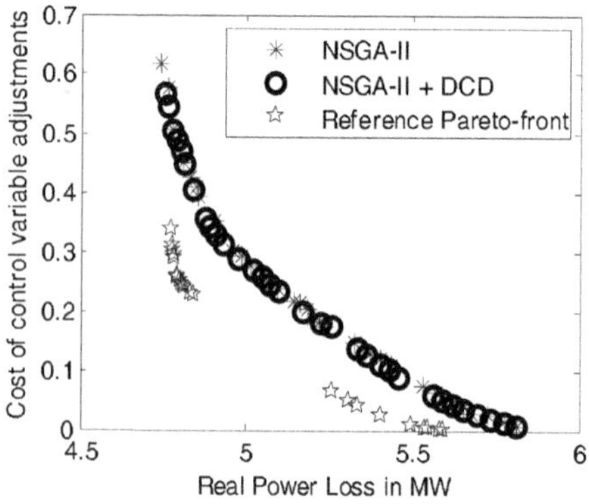


**Fig. 1.** Comparison of MNSGA-II, NSGA-II and Reference Pareto-front

In this paper, the four metrics namely gamma, spread [5], minimum spacing [2], and Inverted Generational Distance (IGD) [19] are considered. These metrics were helpful for evaluating closeness to the reference pareto-optimal front and also for evaluating diversity among Non-dominated solutions. The metrics are calculated with 50 numbers of non-dominated solutions. For the IEEE 30 bus system, 10 trial runs were conducted for NSGA-II and MNSGA-II. The extreme solutions of CMA-ES, NSGA-II and MNSGA-II can be seen from Table 1. Table 2 represents the statistical comparison of performance metrics for the test system and the bold numbers indicate the best values. It can be seen from the Table 1 that almost all the mean values of performance metrics obtained by NSGA-II with DCD is smaller than NSGA-II which means that NSGA-II with DCD can improve convergence at the same time of the improvement of diversity.

### 4.2 IEEE 118-Bus System

The implementation of MNSGA-II for standard IEEE 118-bus system is considered to study the convergence and diversity characteristics. The system has 54 generators, 186 branches, nine transformers (all are treated as OLTCs). The voltage settings and tap settings are taken from [24]. Here, the total number of variables to be optimized is 63. For the initial loading condition the real power loss is observed to be 132.86 MW and the corresponding cost of control variable adjustment is zero. Since the number of variables involved in 118-bus system is more than that of 30-bus system, the numbers of iterations require more to achieve better convergence characteristics of a Pareto front. Here, number of iterations used is 1800. The best results obtained for 118 bus-system is shown in Table 3.

**Table 1.** Best Results for IEEE 30-bus system

| Objective Function | CMA-ES | NSGA-II | MNSGA-II |
|---|---|---|---|
| $f_1$ in MW | 4.7740 | 4.7450 | 4.7560 |
| $f_2$ | 0.0000 | 0.0068 | 0.0056 |

**Table 2.** Statistical comparison of performance measures for IEEE 30-bus system

| Performance Measures | | NSGA-II | NSGA-II+DCD |
|---|---|---|---|
| Gamma ($\gamma$) | Mean | 0.1077 | **0.1073** |
| | $\sigma$ | 0.0014 | 0.0004 |
| Delta($\Delta$) | Mean | **1.2973** | 1.3141 |
| | $\sigma$ | 0.0011 | 0.0014 |
| Minimum | Mean | 0.0780 | **0.0749** |
| Spacing | $\sigma$ | 0.0038 | 0.0002 |
| Inverted Generational | Mean | 0.1794 | **0.1789** |
| Distance(IGD) | $\sigma$ | 0.0362 | 0.0138 |

**Table 3.** Best Results for IEEE 118-bus system

| Objective Function | CMA-ES | NSGA-II | MNSGA-II |
|---|---|---|---|
| $f_1$ in MW | 106.0000 | 122.5000 | 120.0000 |
| $f_2$ | 0.0000 | 0.0713 | 0.0669 |

MNSGA-II (NSGA-II with DCD) algorithm is performing better for RPD problem. By observing the Pareto fronts obtained from IEEE 30-bus and IEEE 118-bus system and from performance metrics it is concluded that the DCD has convergence to the reference Pareto front and also better lateral diversity characteristics among the set of solutions. The diversity characteristic of obtained Pareto front from NSGA-II with DCD can help the decision maker to have more choices in the selection of solution in a different operating condition of the system. This justifies the importance of multi-objective formulation of the RPD problem.

## 5 Conclusion

The Multi-objective RPD problem with conflicting objectives such as real power loss minimization and cost of control variable adjustments minimization is solved by MNSGA-II algorithm. For the first time, a true multi-objective solution set is given in this paper. The concept of DCD was employed in NSGA-II algorithm to improve convergence and maintain diversity among the set of solutions in the Pareto front. The NSGA-II, NSGA-II with DCD generated Pareto fronts are compared with reference Pareto front. With the help of statistical multi-objective performance measures, it can be concluded that the performance of NSGA-II with DCD is better than NSGA-II. The diversity characteristic of obtained Pareto front from NSGA-II with DCD can help the decision maker to have more choices in the selection of solution. The approach is quite flexible so that other formulations using different objectives and/or a larger number of objectives are possible.

**Acknowledgement.** The authors are thankful to the Department of Science and Technology, New Delhi for its support through the project SR/S4/MS/350/06 to the second author.

## References

1. Abido, M.A., Bakhashwain, J.M.: A Novel Multi-objective Evolutionary Algorithm for Reactive Power Dispatch Problem. In: ICECS 2003, pp. 1054–1057 (2003)
2. Bandyopadhyay, S., Pal, S.K., Aruna, B.: Multi-objective GAs, quantitative indices and pattern classification. IEEE Trans. Syst., Man, Cybern.-B 34(5), 2088–2099 (2004)
3. Belhadj, C., Mohamedi, R.: Voltage Stability Modelling and Real-Time Monitoring Using Expert System for Operation Assistance. IEEE Trans. Power Syst. 11, 1037–1045 (1996)
4. Da Costa, G.R.M.: Optimal reactive dispatch through primal-Dual method. IEEE Trans. Power Syst. 12, 669–674 (1997)
5. Deb, K.: Multi-Objective Optimization using Evolutionary Algorithms. John Wiley & Sons, Ltd., Singapore (2002)
6. Deb, K., Pratap, A., Agarwal, S., Meyarivan, T.: A Fast and Elitist Multi-objective Genetic Algorithm: NSGA-II. IEEE Trans. Evol. Comp. 6, 182–197 (2002)

7. Dommel, H.W., Tinny, W.F.: Optimal Power Flow Solutions. IEEE Trans. Power App. Syst. PAS-87, 1866–1876 (1968)
8. Durairaj, S., Kannan, P.S., Devaraj, D.: Multi-Objective VAR dispatch using Particle Swarm Optimization. Int. Journ. Emerg. Elect. Power Syst. 4(1) (2005)
9. Granville, S.: Optimal reactive power dispatch through interior point methods. IEEE Trans. Power Syst. 9(1), 136–146 (1994)
10. Gujarathi, A.M., Babu, B.V.: Multi-Objective Optimization of Industrial Styrene Reactor: Adiabatic and Pseudo-Isothermal Operation. Chemical Engineering Science 65(6), 2009–2026 (2010)
11. Hsaio, Y.T., Chaing, H.D., Liu, C.C., Chen, Y.L.: A Computer Package for Optimal Multi-objective VAR Planning in Large Scale Power Systems. IEEE Trans. Power Syst. 9, 668–676 (1994)
12. Iba, K.: Reactive power optimization by genetic algorithms. IEEE Trans. Power Syst. 9(2), 685–692 (1994)
13. Kannan, S., Baskar, S., McCalley, J.D., Murugan, P.: Application of NSGA-II Algorithm to Generation Expansion Planning. IEEE Trans. Power Syst. 24, 454–461 (2009)
14. Kessel, P., Glavitsch, H.: Estimating the voltage stability of power systems. IEEE Trans. Power Syst. 1(3), 346–354 (1986)
15. King, R.T.F.A., Rughooputh, H.C.S.: Elitist multiobjective evolutionary algorithm for environmental/economic dispatch. In: The 2003 Cong. Evol. Comp., vol. 2, pp. 1108–1114 (2003)
16. Knowles, J., Corne, D.: The Pareto archived evolution strategy: A new baseline algorithm for multiobjective optimization. In: Proc. of the 1999 Congress on Evolutionary Computation, pp. 98–105 (1999)
17. Lai, L.L., Nieh, T.Y., Vujatovic, D., Ma, Y.N., Lu, Y.P., Yang, Y.W., Braun, H.: Swarm Intelligence for Optimal Reactive Power Dispatch. In: IEEE / PES Trans. Distrib. Conf. and Exhibition, Asia and Pacific, Dalian, China (2005)
18. Lee, K.Y., Park, Y.M., Ortiz, J.L.: A United Approach to Optimal Real and Reactive Power Dispatch. IEEE Trans. Power App. Syst. PAS-104, 1147–1153 (1985)
19. Luo, B., Zheng, J., Xie, J., Wu, J.: Dynamic Crowding Distance - A New Diversity Maintenance Strategy for MOEAs. In: ICNC 2008, 4th Int. Conf. Natural Computation, vol. 1, pp. 580–585 (2008)
20. Mamandur, K.R.C., Chenoweth, R.D.: Optimal Control of Reactive Power Flow for improvements in voltage profiles and for real power loss minimization. IEEE Trans. Power App. Syst. PAS-100, 3185–3194 (1981)
21. Manoharan, P.S., Kannan, P.S., Baskar, S., Willjuice Iruthayarajan, M., Dhananjeyan, V.: Covariance matrix adapted evolution strategy algorithm-based solution to dynamic economic dispatch problems. Taylor & Francis - Engg. Optim. 41(7), 635–657 (2009)
22. Reis, C., Barbosa, F.P.M.: A Comparison of voltage stability indices. In: IEEE Mediterranean Electrotechnical Conference, MELECON 2006, pp. 1007–1010 (2006)
23. Wu, Q.H., Ma, J.T.: Power System Optimal Reactive Power Dispatch Using Evolutionary Programming. IEEE Trans. Power Syst. 10, 1243–1248 (1995)
24. Yan, W., Lu, S., Yu, D.C.: A Novel Optimal Reactive Power Dispatch Method Based on an Improved Hybrid Evolutionary Programming Technique. IEEE Trans. Power Syst. 19, 913–918 (2004)

25. Zhang, W., Yutian Liu, Y.: Multi-objective reactive power and voltage control based on fuzzy optimization strategy and fuzzy adaptive particle swarm. Elect. Power and Energy Syst. 30, 525–532 (2008)
26. Zitzler, E.: Evolutionary algorithms for Multi-objective optimization: Methods and applications. Doctoral dissertation ETH 13398, Swiss Federal Institute of Technology (ETH), Zurich, Switzerland (1999)
27. Zhihuan, L., Yinhong, L., Xianzhong, D.: Non-dominated sorting genetic algorithm-II for robust multi-objective optimal reactive power dispatch. IET Gen. Trans. and Dist. 4(9), 1000–1008 (2010)

# A Revised TESOR Algorithm for Perfect Routing

P.M. Joe Prathap[1], G.R. Brindha[2], and W. Vinil Dani[3]

[1] IT Dept., RMD Engineering College,
R.S.M. Nagar, Kavaraipettai, Tiruvallur District, India
joeprathappm@rediffmail.com
[2] ICT Dept., School of Computing,
SASTRA University, Thanjavur, India
brindha.gr@ict.sastra.edu
[3] EEE Dept., St.Xavier's Catholic College of Engineering,
Chunkankadai, Nagercoil
vinil_dani@yahoo.com

**Abstract.** In the modern world everything is faster. People are expecting optimality in each and every instance. In this today's era, shopping is an inseparable & essential activity. But in reality it creates lot of stress and utilizes not only our time but also energy. Recently cities are equipped with shopping malls, but still confusion and unnecessary wandering are unavoidable. Already we proposed a novel idea called TESOR (Time and Energy Saving Optimal Routing) algorithm using shortest path algorithm as a base one. Now in this article we further extended the TESOR based on number of adult customer, which leads great time and energy saver one. With that Revised TESOR almost 50% reduction can be achieved in terms of distance travel and spending time. The simulation analysis shows the effectiveness of the algorithm with proper analysis and comparative chart. The target domain of this paper is that of providing 100% end user satisfaction.

**Keywords:** routing, shortest path, optimization, energy efficient, TESOR.

## 1 Introduction

After focusing on the development, now it is the time for optimization. So many developments came to do efficient and faster work [1], [2], [3], but in some areas like shopping in large malls, the delay is unavoidable one. The shopping mall phenomenon came to existence by 1980's. Earlier days we have shop in different places, i.e. for each category we have to go different areas. We have to wander, to find the shops and then the products. Now shops are grouped together and available in a single shopping mall. Even though we have E-shopping we can't get real satisfying shopping experience. So real time shopping is an ever existing, inseparable concept in human life.

The malls try to attract customer with so many offers and services. But yet people are facing some unnecessary delays and confusion which will further burden their hectic schedule. The customer has to go here and there in order to do their shopping; for same category of products, we may have several shops inside the malls and there is a chance of non availability of shop for a particular category or some shops may

D. Nagamalai, E. Renault, M. Dhanushkodi (Eds.): CCSEIT 2011, CCIS 204, pp. 355–366, 2011.

declare holiday for a particular day. Already we proposed an idea in our innovative TESOR (Time and Energy Saving Optimal Routing) [4] algorithm, in which the optimized path and shop selection can be done and the customer will get a printout with the optimal routing graph. With this guidelines customer can shop all the items with good focus. Especially customer can avoid closed shops as well as he/she can get the assurance that the product is available in the shop in advance. Now, with revised TESOR even shopping become easiest one, because here we are checking the number of adult customers and with that value we are dividing the shopping area and allowing parallel shopping.

The reason behind this application is the motivation by different researches and their analysis. Some of them analyzed with dynamic programming algorithm [5] to solve the problem which is time dependant. Another algorithm determining the fastest travel time between two nodes based on a given starting time [6], [7], [8]. Subsequently, some authors proved that applying Dijkstra's naming method to the time-expanded network is guaranteed to return an optimal solution [9, [10], [11], [12], [13], [14]. More number of authors have tackled different variation of the least cost shortest path. These includes, multiple shortest paths in least cost [15] with single source [16] and turn constraints [17]. And also point to point connection problem, multicast trees with minimum cost are analyzed with TSP [18], [19], [20], [21], [22].

The rest of this paper is organized as follows. Section II analyzes the requirements by taking an ideal shopping mall as an example. Section III describes the revised TESOR algorithm and explains it with examples. Section IV demonstrates the soundness of the proposed scheme with system architecture and the management using revised TESOR. Section V depicts the simulation analysis with proper charts. Conclusions and areas of future research are presented in Section VI.

## 2 Requirement Analysis

Fig. 1 shows an idle shopping mall with 40 shops. Each shop has its own category of products. For example, sports shop may sell fitness equipments and sports accessories. Especially some categories have more than one shop in the mall. For

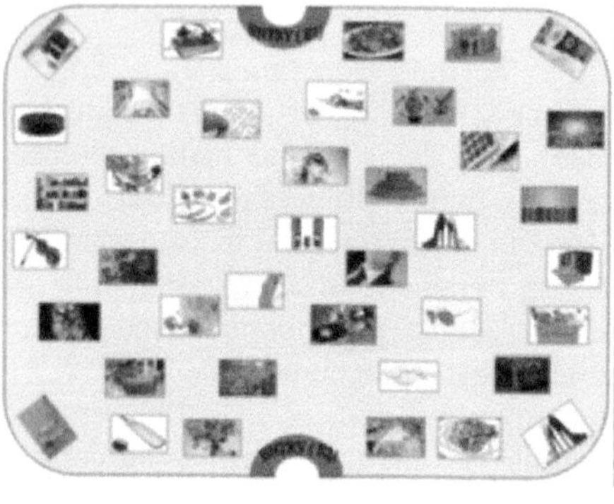

**Fig. 1.** Idle shopping mall

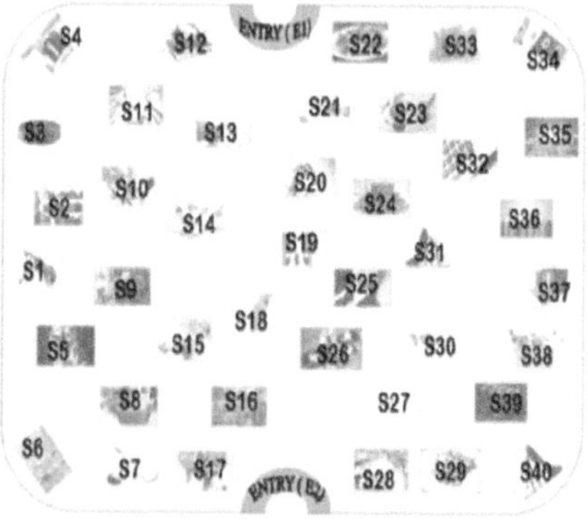


**Fig. 2.** Node Representation

example, for jewellary we have 3 distinct shops in this mall itself. Moreover this idle mall has two entries E1 and E2. Customers can choose any one of the entry based on their willingness. To analyze further this idle shopping mall is converted into mesh circuit as shown in Fig. 2. Here a unique shop id is assigned for each shop range from **S1** to **S40**. This id is used as an individual node name for our graph formation in future. While formulating graph care should be taken about similar shops. To do optimal shopping, end user need to choose only one shop from similar category.

## 3 Revised TESOR Algorithm

The existing TESOR algorithm has the following five major sections. Preliminary section explores the initial setup required like entry selection, distance between nodes, distance with entry, unique id assignment, etc. Second part tells about the customer interaction like category entry. Third section concentrates on multiple graph construction. Fourth part covers finding the shortest path in the graph. Fifth part focus on displaying optimal path graph and distance measures to customer.

The revised TESOR include all the above sections along with some extensions. Here from users we are getting the adult customer count. For example, a family contains totally 5 members with 3 adults and also they want to do parallel shopping, then they can enter the total number of adult as 2. So, one member can take care of children. Other two can do shopping without any tension. Obviously they can reduce the shopping time considerable amount.

```
REVISED TESOR ALGORITHM:
Begin
Initialize Total No. of Adult Customer → AC
If AC= 1
    TESOR();
Else
    OverallShopArea ← CONSTANT
    Division of shops based on AC
    No. of Dividend area = OverallShopArea / AC;
//Entry pt based division
    For i= 1 to No.of Dividend area
       TESOR();
Display graph:
No. of display graphs →  ∏_{i=1}^{N} Gi
    If No. of display graphs >1
          For i=1 to No. of display graphs
           Compare the shop category for duplicity
           If duplicity = TRUE
           Construct totally  ∏_{i=1}^{N} Di  possible graphs
          End for
          For i= 1 to No. of Dividend area
            For i=1 to nodes in G
              For k := 1 to n in G
                For j := 1 to n in G
                  path[i][j]   =   min   (   path[i][j],
```

```
            path[i][k]+path[k][j] )
            OD[i]=∑_{i=1}^{n} part i/parts rep. total dividend area
            at an instance.
            From OD[i] sort the values (Ascending)
            According to the minimum dist. Display the
            shortest cycle graph
TESOR()
Begin
PRELIMINARIES
For i = 1 to No.of shops
      ASSIGN
            Unique_ID= Si;
            Distance with Entry_Counters;
            Intershop distance;
            Shop Holiday details;
End  for
    If Entry_Counter =E1
            Origin_Node = E1;
    Else
            Origin_Node = E2;
    End If.
Continue ;
SHOP SELECTION PROCESS:
GET list of categories from Customers
List of Categories → List_of_Cat.
For i=1 to List_of_Cat
 CHECK availability of shops
 If Shop_Availability = TRUE and Shop_Holiday = FALSE
            Shop_ID  →   G (Vi,E)
Else
    Pass status information to Customers
    STATUS →FALSE
End for
GRAPH CONSTRUCTION:
Group the nodes which are same category using Si's and
name them as C1, C2,..
Mark shops with single category as permanent node
P1,P2…
    REPEAT
    Construct Graph Gi=(Vi,Ei)
    Include V's all P1, P2,…
          Include V's from C1,C2,….. one at a time
                  Until TotalGraphs → ∏_{i=1}^{N} Gi
      End Repeat
SHORTEST CYCLE:
Initial node → entry Counter
  For i=1 to nodes in G
```

```
      For k := 1 to n in G
        For j := 1 to n in G
          path[i][j] = min ( path[i][j], path[i][k]+path
[k][j] )
         Return the values
  OPTIMAL PATH DISPLAY:
  According to the minimum distance, display the shortest
cycle graph G (V,E)
  END TESOR
```

The algorithm was explained further by taking one real time scenario. Let us assume a family wants to visit thirteen different shops listed below in Table 1. The family has two adult members, and they want to do parallel shopping. Assume that preliminary setup already done as per our algorithm with AC=2. Assume that the entry point is selected as E1.

**Table 1.** Shopping item category list

| Sl. No | Category | Customer-1 | Customer-2 |
|---|---|---|---|
| 1 | Leather | S2,S4 | S39 |
| 2 | Musicals | S1,S12 | NA |
| 3 | Bakery | S3 | S3 |
| 4 | Furniture | NA | S33 |
| 5 | Sports | S7 | NA |
| 6 | Laptops | NA | S35 |
| 7 | Plastics | NA | S38 |
| 8 | Cosmetics | S17 | NA |
| 9 | Shoes | NA | S31, S40 |
| 10 | Men's Wear | NA | S25 |
| 11 | Mobiles | S19 | NA |
| 12 | Glass vessels | NA | S27 |
| 13 | Pet Shop | NA | NA |

Since AC value is 2, the area of the shopping mall should be divided into approximately two half's. The division is always based on entry point only. (i.e. From E1 division takes place. This is applicable for all kind AC values.) Now the divided section renamed as customer 1 and customer 2. The required shops for shopping also depicted in Table 1. From this table we got a clear picture about duplication. See for example, we have three distinct shops for leather (S1, S4 & S39). Our ultimate aim is the customer should avoid duplication and anyone of the customer can buy a single category. That is we need to avoid intra group duplication as well as inter group duplication. Note that some category may not be available in the shopping mall. In our example pet shop is not available. In this case this information should be passed to the customer. Moreover the holiday details about the shops also taken into account. So that if a shop is on holiday for that particular day, the node is automatically removed from the proposed graph formation.

With the revised TESOR algorithm, after preliminary setup, the TESOR function is called separately for customer 1 and customer 2. The details are shown in Table 2 and Table 3 respectively.

**Table 2.** Customer 1 possible paths

| Graph | Possible Path cycle | Total distance |
|---|---|---|
| G1 | E1→S3→S2→S1→S7→S17→S19→E1 | 57 |
| G2 | E1→S12→S3→S2→S7→S17→S19→E1 | 53 |
| G3 | E1→S4→S3→S1→S7→S17→S19→E1 | 59 |
| G4 | E1→S12→S4→S3→S7→S17→S19→E1 | <u>52</u> |

**Table 3.** Customer 2 possible paths

| Graph | Possible Path cycle | Total distance |
|---|---|---|
| G5 | E1→S33→S24→S25→S39→S38→S40→S35→E1 | 70 |
| G6 | E1→S33→S24→*S31*→S25→S39→S38→S35→E1 | 58 |
| G7 | E1→S33→S24→*S25*→S31→S38→S39→S35→E1 | 58 |

From Table 2 **G4** graph is taken as optimal one which has lowest distance value, whereas in customer 2 we have G6 and G7 with equal value 58. Arbitrarily anyone of the graph will be selected if we have equal values. In this example we pick **G6**.

Now, we have two optimal graphs G4 and G6. But here again we have to check for duplication among inter graphs. That is, G4 and G6 may have common category. For example (S4, S39) & (S3,S24) are belongs to same category. Now the next step in the algorithm is to find all possible combinations by including anyone of the shop at any one side of the customer without any repetition. After finding the combinations we have to sort the values by adding the lengths of various instances. These details are shown in Table 4.

**Table 4.** Possible instance details with total distance

| C1 | | | C2 | | |
|---|---|---|---|---|---|
| Instance | Path cycle | Length | Path cycle | Length | Total |
| A | E1-S12-S19-S17-S7-E1 | 46 | E1-S33-S24-S31-S25-S39-S38-S35-E1 | 63 | 109 |
| B | E1-S12-S4-S7-S17-S19-E1 | 50 | E1-S33-S24-S25-S31-S38-S35-E1 | 53 | <u>103</u> |
| | | | E1-S33-S24-S31-S25-S38-S35-E1 | 60 | 110 |
| C | E1-S12-S3-S7-S17-S19-E1 | 51 | E1-S33-S35-S31-S25-S39-S38-E1 | 68 | 119 |
| | | | E1-S33-S35-S25-S31-S38-S39-E1 | 63 | 114 |
| D | E1-S12-S4-S3-S7-S17-S19-E1 | 52 | E1-S33-S35-S31-S25-S38-E1 | 59 | 111 |
| | | | E1-S33-S38-S25-S31-S38-E1 | 53 | 105 |

From the Table 4, instance B become the best optimal path for both customer 1 and customer 2. Clearly by doing parallel shopping, overall distance travelled by an individual customer is almost reduced half. And the overall time spent in the shopping mall too reduced considerable amount of time.

Fig. 3 represents the graphical representation of the two final optimized distinct graphs. Suppose if we use ordinary TESOR for the same example (i.e. AC=1), we will get the following solution. These shops are needed to be visited. {[39,S4,S2], [S1,S12], [S3,S24], [S40,S31], S7, S17, S19, S25, S33, S35, S38} the same category of shops indicated within brackets. As per our TESOR formula $\prod_{i=1}^{N} Di$ we will get (3x2x2x2) = 24 combinations [21]. Further, in some combinations, the distance from one node to other nodes equal. In that case again we have to check inner combinations also. With that we got overall 40 distinct combinations. The optimal path cycle is given below. E1-S12-S4-S3-S7-S17-S19-S25-S31-S38-S33-S35-E1.

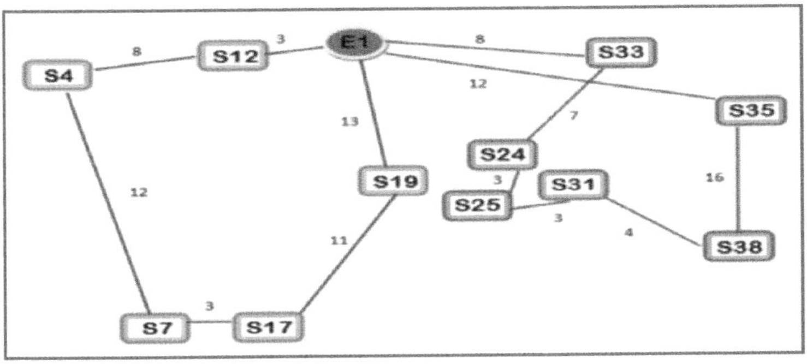


**Fig. 3.** Optimized graph for two customers

# 4 System Architecture and Management

Fig.4 depicts the overall system architecture of the TESOR system. All the shops in the malls are connected with the database content server. GUI for end user is

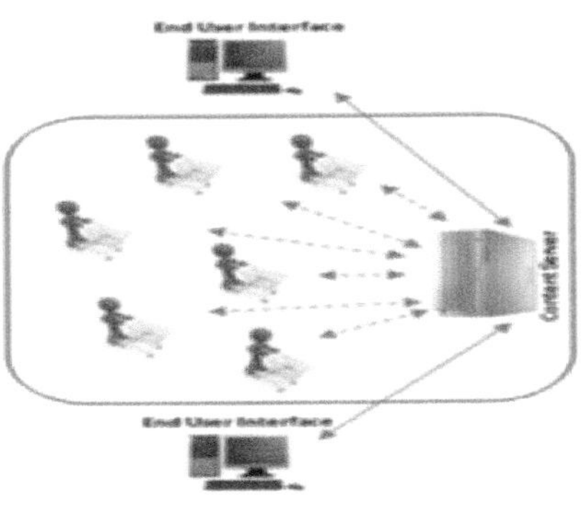


**Fig. 4.** The system architecture of TESOR

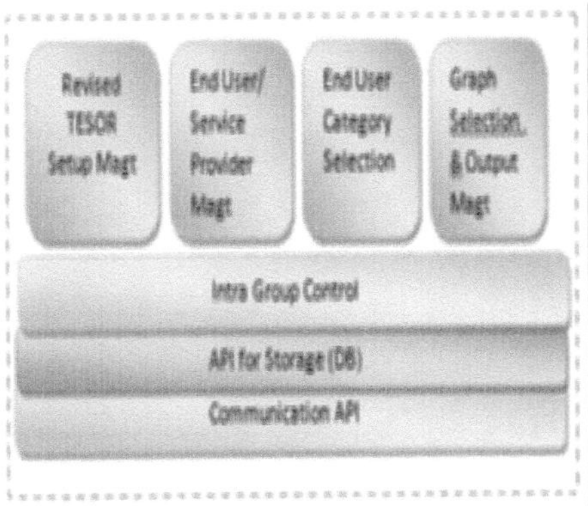


**Fig. 5.** Functional structure of TESOR

available in the entry area. This end user interface has the direct contact with the content server. So all the shop related updates are easily accessible by the end user. Information like holidays, non availability of certain product, discount details can be delivered to the end user easily.

Fig. 5 explores the functional structure of our TESOR software. The revised TESOR state transition diagram was depicted in Fig.6.

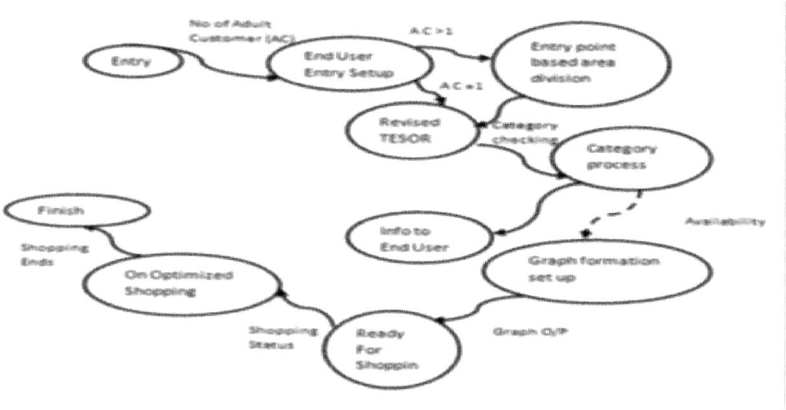


**Fig. 6.** Revised TESOR state transition diagram

## 5 Simulation Analysis

A simulation environment was developed with the following setup. The algorithm was trained up to 50 customers with shopping category range from 5 to 30. For every trial different values for AC taken ranges from 1 to 4. With each scenario various graph combinations and shortest length were analyzed. Especially for each case with TESOR and without TESOR path selection was performed and the performances were plotted further.

**Fig. 7.** Revised TESOR GUI

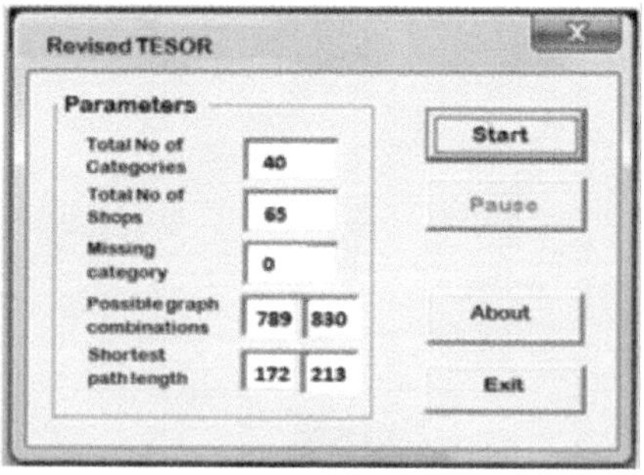


**Fig. 8.** Revised TESOR Simulation

Fig. 7 depicts the GUI for revised TESOR algorithm. Here the total number of adult customer will be entered. Based on that, the page is redirected to the simulation setup.

Fig. 8 shows the GUI for the simulation environment. Based on the values given by the user the optimal graphs will be generated as Fig. 9 by running revised TESOR.

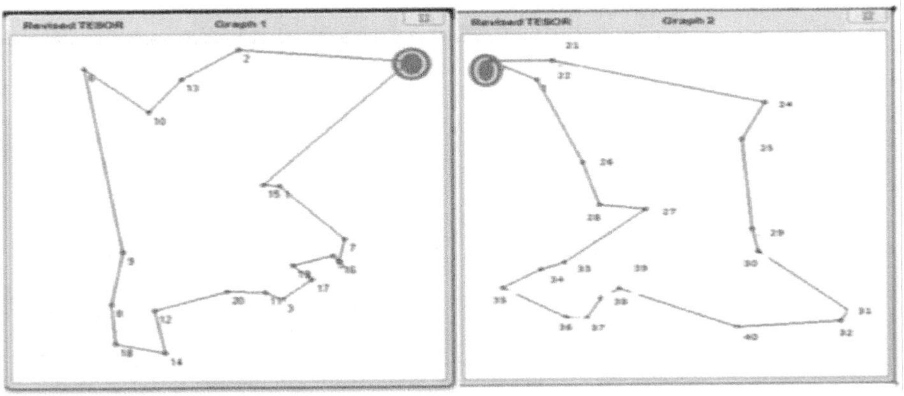

**Fig. 9.** Revised TESOR sample route sheet

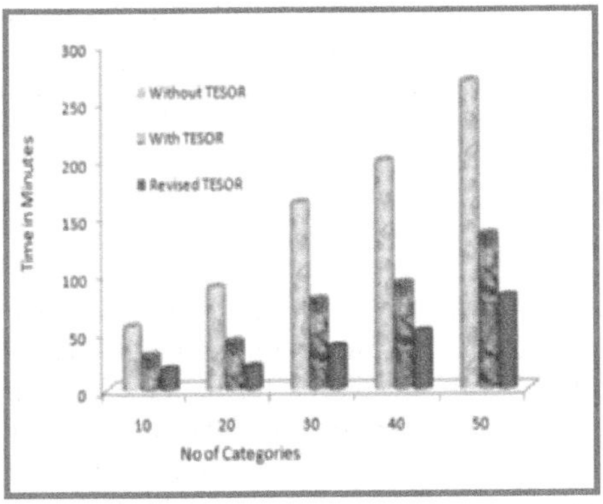


**Fig. 10.** Overall time utilization.

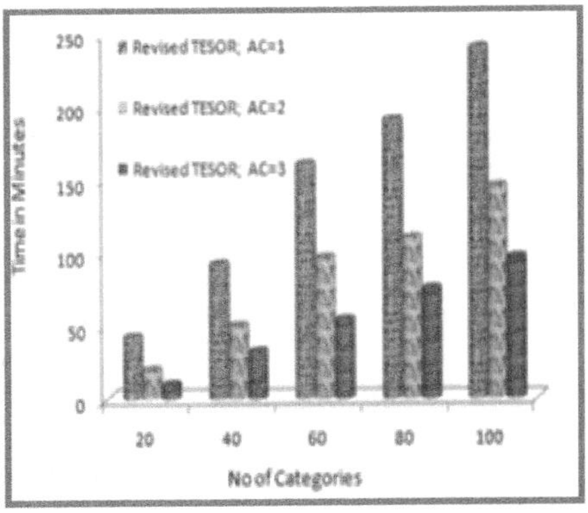


**Fig. 11.** Time utilization for different AC

Fig. 10 shows the comparative analysis of overall time utilized by the customer through TESOR algorithm, revised TESOR algorithm and with the regular shopping for the same scenario. From the below graph we can clearly understand the importance and advantages of our revised TESOR algorithm.

Fig. 11 compare the revised TESOR time utilization for various values of AC. If AC values increases automatically number of parallel shopping also increased. This will lead a great time minimization as clearly depicted in Fig.11.

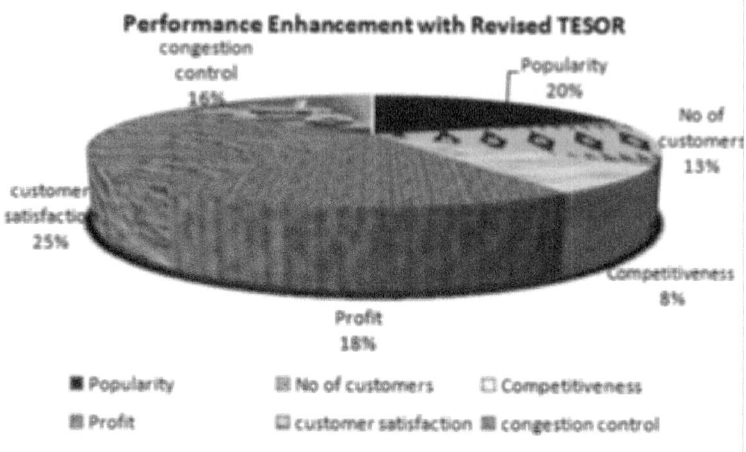


**Fig. 12.** Performance of Revised TESOR

Simulation work suggests that with our algorithm, the time spent inside the malls by a customer reduce approximately half way. This method helpful for people who are in hurry to do all the stuff( that is for fast moving people and modern world).

Fig. 12 shows the enhancment achieved in the performance of the shooping mall after adopting the proposed revised TESOR algorithm. With that the number of customer and popularity are increased drastically. Overall profit and customer satisfaction achiedved thus made a great impact in the society.

It also explores the various future enhancements that can be applied along with our proposed work.

## 6 Conclusion and Future Enhancement

In this paper, we have projected a innovative algorithm for finding an optimal solution based on shortest path problem with a specially-structured, for shop selection, graph construction and optimal path display function. We have shown that our previous algorithm TESOR is for AC with the value 1, but in revised TESOR further it si improved with AC values 2 or 3. We have also compared the difference and performance of both the algorithm.

Shopping malls also faced lot of competitions nowadays. To sustain its name and their customer lot of efforts were taken by them. Our proposed work is major break breakthrough for them if they applied it successfully. Our product not only provides better solution but also open further enhancements. This will play a small but better part for developing faster world. The shops can provide the offer pamphlets or information during the selection of their shop. So that that the customer can select the shops based on offers. This will create a competition environment between the service providers, from that the customer is benefited and service provider also increase their profit. The Graph can also include information about toilets, play area, water car parking, ATM and first aid area indication.

## References


1. Leng, J.F., Zeng, W.: An Improved Shortest Path Algorithm for Computing One-to-One Shortest Paths on Road Networks. In: Fac. of Inf. Eng., China Univ. of Geosci., April 26. IEEE, Wuhan (2010)
2. Kaplan, H., Lewenstein, H., Shafrir, N., Sviridenko, M.: Approximation Algorithms for Asymmetric TSP by Decomposing Directed Regular Multigraphs. In: Proc. 44th on Foundations of Comput. Sci., pp. 56–65. IEEE, Los Alamitos (2004)
3. Retvari, G., Biro, J., Cinkler, T.: On Shortest Path Representation. IEEE/ACM Transactions on Networking 15(6), 1293 (2007)
4. Joe Prathap, P.M., Brindha, G.R.: An Innovative TESOR Algorithm for Perfect Routing. In: International Conference on Signal Processing, Communication, Computing and Network Technologies, NCACNSP (2011) (accepted for publication)
5. Cooke, K.L., Halsey, E.: The shortest route through a network with time-dependent internodal transit times. Journal of Mathematical Analysis and Applications 14(3), 493–498 (1966)
6. Zhu, S., Huang, G.M.: A new parallel and distributed shortest path algorithm for hierarchically clustered data networks. IEEE Transactions Parallel and Distributed Systems 9(8) (September 1998)
7. Huang, G., Zhu, S.: A new distributed shortest path algorithm for hierarchically clustered data networks. In: Proceedings American Control Conference, 1995, June 21-23, vol. 3. Dept. of Electr. Eng., Texas A&M Univ., College Station, TX, IEEE (1995)
8. Orda, A., Rom, R.: Shortest-path and minimum-delay algorithms in networks with time-dependent edge-length. Journal of the Association for Computing Machinery 37(3), 607–625 (1990)
9. Soltani, A.R., Tawfik, H., Goulermas, J.Y., Fernando, T.: Path planning in construction sites: performance evaluation of the Dijkstra, Ap, and GA search algorithms. Advanced Engineering Informatics, 291–303 (May 29, 2003)
10. J.M.: A walk over the shortest path: Dijkstra's Algorithm viewed as fixed-point computation. Information Processing Letters 77, 197–200 (2001)
11. Jeffrey Solka, L., James Perry, C., Brian Poellinger, R., George Rogers, W.: Fast computation of optimal paths using a parallel Dijkstra algorithm with embedded constraints. Neurocomputing, 195–212 (February 10, 1994)
12. Cantone, D., Faro, S.: Two-Levels-Greedy: a generalization of Dijkstra's shortest path algorithm. Electronic Notes in Discrete Mathematics 17, 81–86 (2004)
13. Sedgewick, R., Vitter, J.: Shortest paths in Euclidean graphs. Algorithmica 1(1), 31–48 (1986)
14. Sherali, H., Ozbay, K., Subramanian, S.: The time-dependent shortest pair of disjoint paths problem: complexity, models, and algorithms. Networks 31, 259–272 (1998)
15. Rees, W.G.: Least-cost paths in mountainous terrain, Scott Polar Research Institute, University of Cambridge, Lensfield Road, Cambridge CB2 1ER, UK. Computers & Geosciences 30, 203–209 (2004)
16. Trtiff, J.L.: An experimental comparison of two distributed single-source shortest path algorithms. Parallel Computing, 1505–1532 (April 1995); Elsevier Science
17. Boroujerdi, A., Uhlmann, J.: An efficient algorithm for computing least cost paths with turn constraints. Information Processing Letters 67, 317–321 (1998); Elsevier Science (November 1997)

18. Natu, M., Fang, S.-C.: The point-to-point connection problem - analysis and Algorithms, Operations Research und industrial Engineering. Discrete Applied Mathematics 78, 207–226 (1997); Elsevier Science B.V. (December 1996)
19. Li, S., Melhem, R., Znati, T.: An efficient algorithm for constructing delay bounded minimum cost multicast trees. J. Parallel Distrib. Comput. 64, 1399–1413 (2004)
20. Kosaraju, S.R., Park, J.K., Stein, C.: Long tours and short superstrings. In: Proc. 35th Ann. Symp. on Foundations of Comput. Sci., pp. 166–177. IEEE Computer Society, Los Alamitos (1994)
21. Bellman, R.: Combinatorial Processes and Dynamic Programming. In: Bellman, R., Hall Jr, M. (eds.) Proceedings of Symposia in Applied Mathematics Combinatorial Analysis, pp. 217–249. American Mathematical Society, Providence (2004)
22. Kohn, S., Gottlieb, A., Kohn, M.: A Generating Function Approach to the Traveling Salesman Problem. In: ACM Annual Conference, pp. 294–300. ACM Press, New York (1997)

# An Accomplishment of New Level Register File Management Architecture for Montgomery Algorithm in Elliptic Curve

M. Prabu[1] and R. Shanmugalakshmi[2]

[1] Anna University Coimbatore,
Tamil Nadu,
India
[2] CSE,
Government College of Technology,
Tamil Nadu, India
{prabu_pdas,shanmuga_lakshmi}@yahoo.co.in

**Abstract.** In this paper, we present a background on Elliptic Curve Cryptography, along with the Montgomery operation, which plays major role on Elliptic Curve Cryptosystem and side channel attacks on EC-Montgomery operations. We have also provided a brief background on File Register architecture and proposed a new design level of EC-Montgomery addition algorithm. These design level architecture are designed in an efficient manner. Realizing the proposed design level of FPGA technology. Through this FPGA based technology very advantageous results were found when compared against other existing designs. Even though our design bears an extra computation time, it reduces the area space or register gate area space due to reassigning the design and power analysis attack.

**Keywords:** Montgomery Algorithm, Elliptic Curve Cryptosystem, Computation time, FPGA technology.

## 1 Introduction

In this article, a hardware implementation on register file architecture has been efficiently compared with two more algorithms namely Montgomery algorithm and modified Montgomery algorithm. And we have focused on Side Channel Attacks and its countermeasures on EC-Montgomery algorithm. Side Channel Attacks consists of a lot of attacks, here we mainly concentrated and worked on Power Analysis and Fault Analysis attacks[1,2]. Both the Power analysis and Fault Analysis can be worked based on logical designs. Many proposed implementations of EC- system are based on Montgomery algorithms.

Here we took a design level architecture as a major part and implemented both there algorithms (Montgomery and modified Montgomery). With the help of this

D. Nagamalai, E. Renault, M. Dhanushkodi (Eds.): CCSEIT 2011, CCIS 204, pp. 367–375, 2011.

design we got the quality rich design for the Montgomery algorithm, which is based on the countermeasures of Power Analysis and Fault Analysis.

### 1.1 Organization of This Paper

The rest of this paper is structured as follows. In Section 2, we recall the basic of the Elliptic curve cryptosystems, Montgomery Algorithm and Side Channel Attack. In Section 3, we introduce a Montgomery algorithm. In Section 4, the model approach is presented to performance of operation, and then we discuss its security and implementation properties. In Section 5 & 6, we turn our attention on File Register Architecture and design level. In Section 7, we make concluding remarks.

## 2 Background

### 2.1 Elliptic Curve Cryptography

ECC is a newly emerged public key cryptographic algorithm. It is younger than RSA and other cryptographic algorithms. Although, as Hankerson, Menezes and Vanstone (2004) state, Elliptic curve have been focused by researchers and algebraists in the middle of nineteenth century. In 1985, the first elliptic curve cryptography application has been discovered by Neal Koblitz and Victor Miller. The main advantage of ECC over RSA is its key size. For instance, it is widely accepted that 160-bit ECC offers equivalent security as 1024-bit RSA. This significant difference makes ECC especially attractive for applications on constrained environments as shorter key sizes are translated to less power and storage requirements, and reduced computing times.

The level of security provided by ECC increases faster with key size. ECC can also provide a faster implementation than RSA and DL systems, and utilize power and reduced bandwidth. In the last few years, confidence in the security of ECC had also arised, to the point where they had now been included in internationally recognized standards (especially IEEE std 1363, ANSI X9.62).

### 2.2 Side Channel Attacks

In recent years, cryptosystems have come under harass from various forms of side channel attacks. Kocher et al[4] discovered that cryptosystems implementations leak information, which can help an attacker to access secret data. One such technique for retrieving secret information is SPA. SPA involves monitoring the power consumption of cryptographic algorithms in single execution.

When a cryptographic algorithm is actually implemented, its weakness can be induced with some unanticipated ways. The attackers can track these weaknesses to circumvent the security of the underlying algorithms. These kind of attacks are referred and reviewed by Kelsey[9], as so called side channel attacks.

In fact, hardware encryption devices are ideal targets for side channel attacks. There attacks are divided into passive attacks (power consumption analysis, electromagnetic leaking analysis) and active attacks.

### 2.2.1. Power Analysis

The Power analysis can be classified into two parts, that is Simple Power Analysis(SPA) and Differential Power Analysis(DPA).

**Simple Power Analysis**

Simple Power Analysis is generally based on looking at the visual representation of the power consumption unit while an encryption operation is being performed. SPA is a technique that involves direct interpretation of power consumption measurements collected during various cryptographic operations in[9,10]. SPA can yield information about a device's operation as well as key material.

**Differential Power Analysis**

The goal of the Differential Power Analysis is to identify the secret key, or part of the secret key, for ciphering the information. The basic idea is to correlate the encrypted data and the power consumed by the device.

### 2.3 EC-Montgomery Algorithm

Elliptic curve is one of the strongest public-key cryptosystem, which is generally used for authentication The performance of such cryptosystems are primarily determined by effectively implementing modular multiplication and exponentiation. For performance as well as for physical security reasons, it is often advantageous to recognize them by using hardware. Hardware implementations of Elliptic curve cryptosystems are widely studied.

Montgomery algorithm developed a very efficient technique to compute in the group associated to an elliptic curve over a non-binary finite field $F(q)$ in which prime field is$F(p)$ The Montgomery algorithm proves useful for point compression in ECC. More precisely, instead of sending a point as part of some cryptographic protocol, there are variety of ways to perform the Elliptic Curve based Montgomery operations. Our aim, in this thesis is to give a crystal clear view of its operations and functions. The operations may be classified as Montgomery addition and Montgomery multiplication. A number of Montgomery multiplier, have been suggested [12, 13] and their hardware architectures are designed to deal with a maximum number of registers. However, when the computer processing power increases, it provides the same level security. Then the cryptosystem will verify the level of register management to be equalized.

## 3 Previous Work

In this section we introduce Yong Ki.L's and Ingrid.V's Montgomery compact scalar multiplication algorithm, which uses a common projective coordinate system [2] .The algorithm, is shown in Fig 1. A non-super singular elliptic curve $E$ over $GF(2m)$is a set of coordinate points $(x, y)$ satisfying $y2 + xy = x3 + ax2 + b$ with the point at infinity $O$, where $a, b, x, y \in GF(2m)$and $b \neq 0$.

Input : A point $P = (x, y) \in E$ and a positive integer $k = 2^{l-1} + \Sigma\ ki\ 2i$
Output: $Q = kP$
If ($k = 0$ *or* $x = 0$) then output(0,0) and stop
$X_1 \leftarrow x, Z_1 \leftarrow 1, X_2 \leftarrow x^4 + b, Z_2 \leftarrow X_2$
,
For $i = l - 2$ *to* 0 *do*
If $K_i = 1$ then

$$(X_1, Z_1) \leftarrow Madd\ (X_1, Z_1, X_2, Z_2), \tag{1}$$

$$(X_2, Z_2) \leftarrow Madd(X_2, Z_2, X_1, Z_1), \tag{2}$$

return $Q \leftarrow M_{XY}\ (X_1, Z_1, X_2, Z_2)$

**Fig. 1.** Montgomery Addition with Yong –Ingrid algorithm

The adding formula of $(X_{Add}, Z_{Add}) \leftarrow M_{add}(X_1, Z_1, X_2, Z_2)$ is defined in Eq.1

$$Z_{Add} = (X_1 + X_2)^2 \tag{3}$$

$$X_{Add} = x \times Z_{Add} + (X_1 \times X_2). \tag{4}$$

$Q \leftarrow M_{XY}(X_1, Z_1, X_2, Z_2)$ is the conversion of projective coordinate to affine coordinate. Yong-Ingrid adding algorithm is described in Fig 2. Where $c^2 = b$ .

The total number of registers in Fig 2. is six , i.e[6], the registers for $X_1, Z_1, X_2$, $Z_2, T_1$ and $T_2$. The total number of operations of Adding algorithms are 2 multiplications, 1 square, and 2 additions.

Adding Algorithm:

$$(X_1, Z_1) \leftarrow Madd(X_1, Z_1, X_2, Z_2)$$
$$T_2 \leftarrow X_1 + X_2$$
$$T_2 \leftarrow T_2 \cdot (T_1)$$
$$T_1 \leftarrow X_1 + X_2$$
$$X_1 \leftarrow x$$
$$X_1 \leftarrow T_2 \times X_1$$
$$X_1 \leftarrow X_1 + T_1$$

**Fig. 2.** Yong Ki L- Ingrid. V Adding Algorithm

Our proposed design of Montgomery Addition algorithm's implementation table shows six level register used for making two operations. This 6 level register concept is already done in[4]. But in our approach same level of registers are used in calculation part but there is a reduction in the implementation steps shown in Fig 7.

**Table 1.** Register File Management for Adding Algorithm

| Steps | Field | Operation | Reg 1 | Reg 2 | Reg 3 | Reg 4 | Reg 5 | Reg6 |
|---|---|---|---|---|---|---|---|---|
| 1 | Initial | | $X_1$ | $X_2$ | Z | | | |
| 2 | Shift | | | $X_1$ | $X_2$ | Z | | |
| 3 | Copy | | $X_1$ | $X_1$ | $X_2$ | Z | | |
| 4 | Add | $T_2 \leftarrow x_1+x_2$ | $T_2$ | $X_1$ | $X_2$ | Z | | |
| 5 | Shift | | | $T_2$ | $X_1$ | $X_2$ | Z | |
| 6 | Copy | | $T_2$ | $T_2$ | $X_1$ | $X_2$ | Z | |
| 7 | Shift | | | $T_2$ | $T_2$ | $X_1$ | $X_2$ | Z |
| 8 | Multiply | $T_2 \leftarrow T_2 2$ | $T_2$ | | | $X_1$ | $X_2$ | Z |
| 9 | Shift*4 | | | $X_1$ | $X_2$ | Z | $T_2$ | |
| 10 | Multiply | $T_1 \leftarrow X_1 x X_2$ | T1 | | $X_2$ | Z | $T_2$ | |
| 11 | Switch | | | $T_1$ | $X_2$ | Z | $T_2$ | |
| 12 | Load x | $X_1 \leftarrow x$ | $X_1$ | $T_1$ | $X_2$ | Z | $T_2$ | |
| 13 | Shift*2 | | $T_2$ | | $X_1$ | $T_1$ | $X_2$ | Z |
| 14 | Switch | | | $T_2$ | $X_1$ | $T_1$ | $X_2$ | Z |
| 15 | Multiply | $X_1 \leftarrow T_2 x X_1$ | $X_1$ | $T_2$ | | $T_1$ | $X_2$ | Z |
| 16 | Switch | | $T_2$ | $X_1$ | | $T_1$ | X2 | Z |
| 17 | Shift*5 | | $X_1$ | | $T_1$ | $X_2$ | Z | $T_2$ |
| 18 | Add | $X_1 \leftarrow X_1+T_1$ | $X_1$ | | $T_1$ | $X_2$ | Z | $T_2$ |

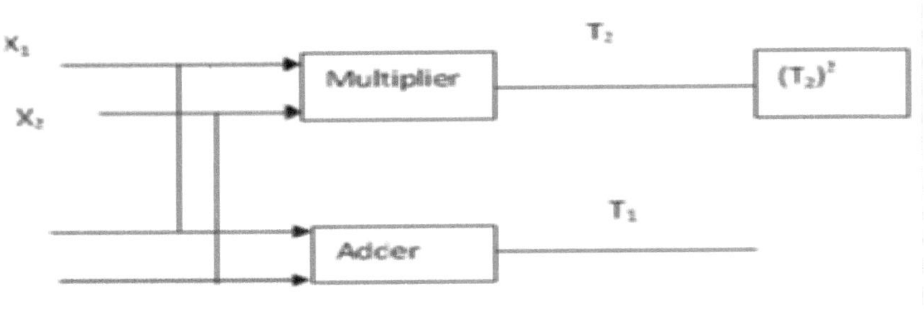


**Fig. 3.** Initial level of Design

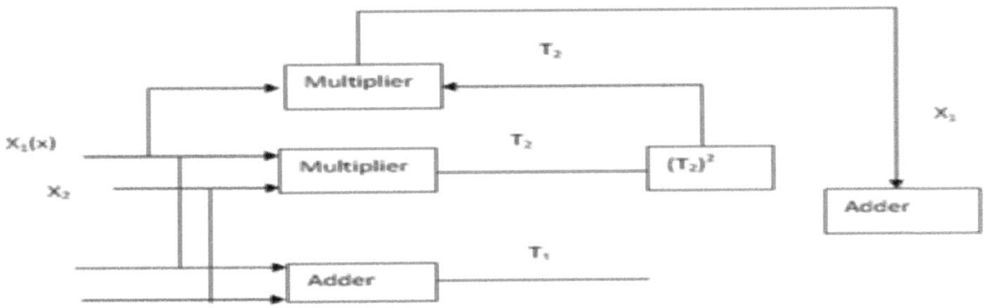


**Fig. 4.** Second Level of Design

The following two steps can be explained through the hardware design[3]. The inputs are X1 & X2 for the both multiplier and adder. After Multiplier the output (T2) can be extracted by the Square concept.

After completing the level 1 design, the second level is extracted by the level 1 design's output that is $(T_2)^2$ .After getting the result from the level one, it interacted to the input again X1(x), which is given by the user[7,8]. After that process, the resulting value will be stored in X1.

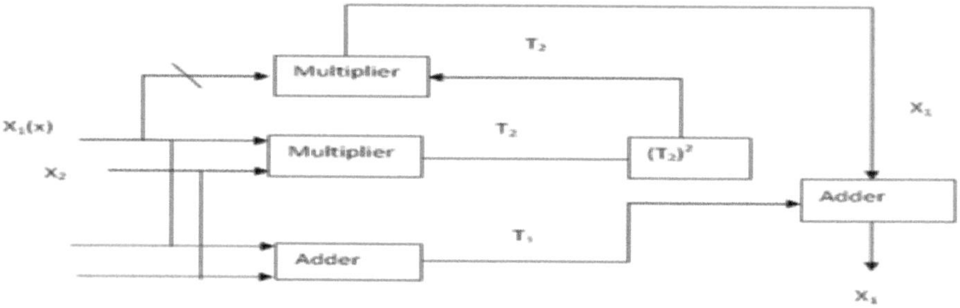


**Fig. 5.** Final Designing Section for Montgomery Algorithm

The final level. The latest X1 value is interacted with the adder with T1.

## 4 Performance Operation

In this article, we present a , modified version of file management algorithm, which is suitable for both hardware and software implementations. The shift operation shifts the register values in circular and the switch operation switches the values of Reg1

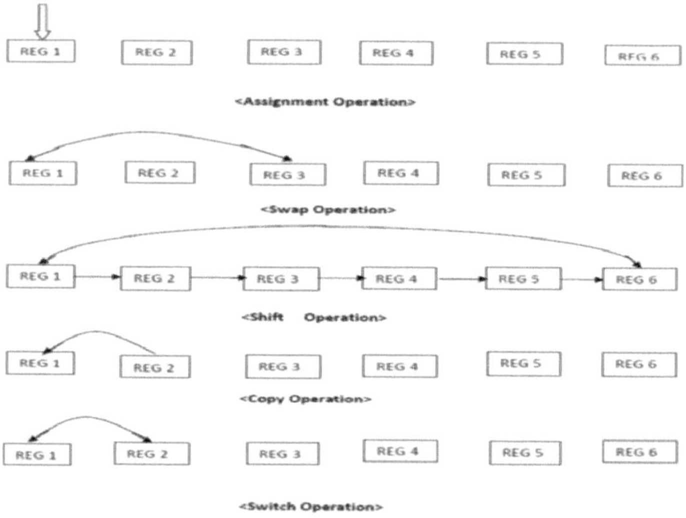


**Fig. 6.** Performance of operations

and Reg2. The copy operation replaces the value of Reg1 with Reg2. Note that the copy operation is required for field square process which is implemented as the field multiplication with two operands of the same value.

# 5 Algorithm for File Register Architecture

## 5.1 Proposed Algorithm

Initially declare $T_1 \leftarrow X_1.Z_2, T_2 \leftarrow Z_1.X_2$

Step 1: $M \leftarrow T_1 + T_2$

Step 2: $Z_1 \leftarrow M^2$

Step 3: $N \leftarrow T_1.T_2$

Step 4: $M \leftarrow x.Z_2$

Step 5: $X_1 \leftarrow M+N$

In our algorithm, we have to increase the temporary registers such as T1, T2, M and N. But our main role is to reduce the number of registers. For this problem,in our proposed work we proposed a new operation swap. It will help to swap the values from one register to another register while it will become empty. With the help of swap operation[5] , we had easily reduce the no registers as well as steps of execution.

## 5.2 Our File Register Architecture

**Table 2.** Register File Management for Our Modified Adding Algorithm

| Steps | Field | Operation | Reg 1 | Reg 2 | Reg 3 | Reg 4 | Reg 5 | Reg6 |
|---|---|---|---|---|---|---|---|---|
| 1 | Initial | | $X_1$ | $X_2$ | Z | | | |
| 2 | Shift | | | $X_1$ | $X_2$ | Z | | |
| 3 | copy | | $X_1$ | $X_1$ | $X_2$ | Z | | |
| 4 | Add | $T_2 \leftarrow X_1 + X_2$ | $T_2$ | $X_1$ | $X_2$ | Z | | |
| 5 | shift | | | $T_2$ | $X_1$ | $X_2$ | Z | |
| 6 | copy | | $T_2$ | $T_2$ | $X_1$ | $X_2$ | Z | |
| 7 | Shift | | | $T_2$ | $T_2$ | $X_1$ | $X_2$ | Z |
| 8 | | $T_2 \leftarrow T_2 2$ | $T_2$ | | | $X_1$ | $X_2$ | Z |
| 9 | Shift*4 | | | $X_1$ | $X_2$ | Z | $T_2$ | |
| 10 | | $T_1 \leftarrow X_1 x X_2$ | $T_1$ | | $X_2$ | Z | $T_2$ | |
| 11 | Shift | | $T_2$ | | $T_1$ | | $X_2$ | Z |
| 12 | swap | | | $T_2$ | $T_1$ | | $X_2$ | Z |
| 13 | | $X_1 \leftarrow x.\ T_2$ | $X_2$ | $T_2$ | $T_1$ | | $X_2$ | Z |
| 14 | Shift | | | $X_1$ | $T_2$ | $T_1$ | $X_2$ | Z |
| 15 | swap | | $X_1$ | | $T_2$ | $T_1$ | $X_2$ | z |
| 16 | | $X_1 \leftarrow X + T_2$ | $X_1$ | | $T_1$ | $T_2$ | $X_2$ | z |

# 6 Design Level Architecture

## 6.1 Proposed Design Level

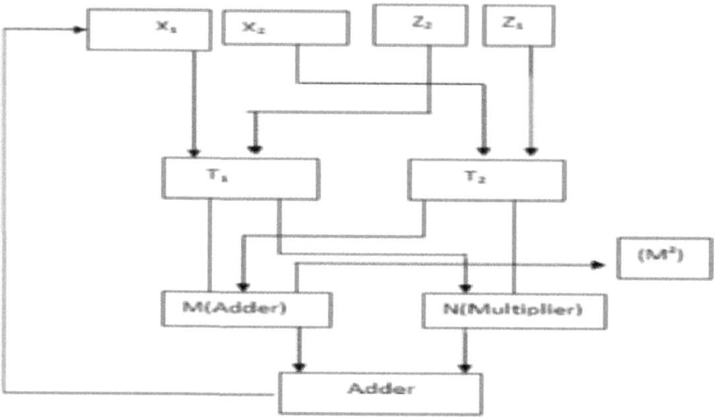


**Fig. 7.** Modified Design Level

The above mentioned design level, consist of eight registers in that 4 registers(T1,T2,M and N) were temporary. The modified process can be explained through the hardware design. The inputs are X1, X2, Z1,Z2 for the multiplier and adder.

# 7 Discussion and Conclusion

**Table 3.** Comparison of Computational workload

| Field Operation | Yong Ki.L and Ingrid.V algorithm | Our algorithm |
|---|---|---|
| Addition | 3 | 2 |
| Multiplication | 2 | 2 |
| Square | 1 | 1 |

In order to verify the design and the algorithm, we synthesized the proposed architecture and design using FPGA based Xilinx, Virtex-III platform. Through the Xilinx we designed our design architecture Fig.7[11]. After verifying the nature of algorithm, the computational load can be calculated and measured. In comparison with the known results of the algorithm more suitable for real-time applications. Reasons are optimistic about the results of our comparative study for all researchers on this important topic.

## References


1. Cilardo, A., Mazzeo, A., Mazzocca, N.: Representation of elements in F2m enabling unified field arithmetic for elliptic curve cryptography. Electronics Letters 41(4) (July 7, 2005)
2. Montgomery, P.: Speeding the Pollard and elliptic curve methods of factorization. Mathematics of Computation 48, 243–264 (1987)
3. López, J., Dahab, R.: Fast multiplication on elliptic curves over GF(2m) without precomputation. In: Koç, Ç.K., Paar, C. (eds.) CHES 1999. LNCS, vol. 1717, pp. 316–327. Springer, Heidelberg (1999)
4. Meloni, N.: Fast and Secure elliptic Curve Scalar Multiplication Over Prime Fields Using Special Addition Chains. Cryptology ePrint Archive: listing for 2006 (2006/216) (2006)
5. Paar, C.: Light-Weight Cryptography for Ubiquitous Computing. Invited talk at the University of California, Los Angeles (UCLA). Institute for Pure and Applied Mathematics (December 4, 2006)
6. Sakiyama, K., Batina, L., Mentens, N., Preneel, B., Verbauwhede, I.: Smallfootprint ALU for public-key processors for pervasive security. In: Workshop on RFID Security, 12 pages (2006)
7. Batina, L., Mentens, N., Sakiyama, K., Preneel, B., Verbauwhede, I.: Low-costElliptic Curve Cryptography for wireless sensor networks. In: Buttyán, L., Gligor, V.D., Westhoff, D. (eds.) ESAS 2006. LNCS, vol. 4357, pp. 6–17. Springer, Heidelberg (2006)
8. OztAurk, E.A., Sunar, B., Savas, E.: Low-power elliptic curve cryptography using scaled modular arithmetic. In: Joye, M., Quisquater, J.-J. (eds.) CHES 2004. LNCS, vol. 3156, pp. 92–106. Springer, Heidelberg (2004)
9. Satoh, A., Takano, K.: A Scalable Dual-Field Elliptic Curve Cryptographic Processor. IEEE Transactions on Computers 52(4), 449–460 (2003)
10. Menezes, J., van Oorschot, P.C., Vanstone, S.A.: Handbook of Applied Cryptography. CRC Press, Boca Raton (1997)
11. Walter, C.: Systolic Modular Multiplication. IEEE Trans. Computers 42(3), 376–378 (1993)
12. Kornerup, P.: A Systolic, Linear-Array Multiplier for a Class of Right-Shift Algorithms. IEEE Trans. Computers 43(8), 892–898 (1994)
13. Tsai, W.C., Shung, C.B., Wang, S.J.: Two Systolic Architecture for Modular Multiplication. IEEE Trans. VLSI 8(1), 103–107 (2000)

# Modified Partial Differential Equations Based Adaptive Two-Stage Median Filter for Images Corrupted with High Density Fixed-Value Impulse Noise

P. Shanmugavadivu and P.S. Eliahim Jeevaraj

Department of Computer Science and Applications,
Gandhigram Rural Institute, Deemed University,
Gandhigram-624 302, Tamil Nadu, India
psvadivu@yahoo.com, eliahimps@gmail.com

**Abstract.** This paper presents a newly devised noise filter viz., Modified Partial Differential Equation based Adaptive Two-Stage Median Filter (MPATS) to denoise the images corrupted by fixed-value impulse noise. The performance of the proposed filter is proved to be better in terms of Peak Signal - to - Noise Ratio and human visual perception. This filter is effective in denoising the highly corrupted images with the noise probability of even 90%.

**Keywords:** Fixed-value impulse noise, salt and pepper noise, standard median filter, PDE, highly corrupted image.

## 1 Introduction

Image restoration is a vital component of digital image processing which targets at the elimination or reduction of noise in corrupted images. The unwanted and irrelevant information present in any image is referred to as noise that degrades the quality of image and makes it unsuitable for high level processing [1]. Digital images are commonly corrupted due to faulty sensors in the capturing devices, atmospheric turbulence etc., during acquisition and during communication due to the limitations of the transmission media. If an image is corrupted with the maximum and minimum intensity of the dynamic ranges, it is termed as fixed-value impulse noise, whereas its variant namely random – value impulse noise corrupts the images by randomly assuming any value between the intensity intervals [2].

Review of literature reveals the fact that denoising potentials of non-linear filters are greater than that of linear filters like mean filters. Further, the non-linear filters are deemed meritorious over linear filters for their edge and detail preserving potentials. However, these filters have a major drawback of modifying the intensities of uncorrupted pixels too, during denoising, as they do not adopt any mechanism to discriminate between the corrupted and uncorrupted pixels. The next generations of filters known as adaptive filters invariably employ the strategy of noise detection prior to noise correction [3]-[4].

The median based filters such as Progressive Switching Median Filter (PSM), Iterative Median Filter (IMF), Signal Dependent Rank Order Median Filter (SDROM),

D. Nagamalai, E. Renault, M. Dhanushkodi (Eds.): CCSEIT 2011, CCIS 204, pp. 376–383, 2011.

Two-state Recursive Signal Dependent Rank Order Median Filter (SDROMR), Non-Recursive Adaptive-Center Weighted Median Filter (ACWM), Recursive Adaptive-Center Weighted Median Filter (ACWMR) , Russo's Fuzzy Filter (RUSSO), Zhang's filter (ZHANG) , Sun and Neuvo's Switching Based Filter (SUN) , Impulse Rejecting Filter (IRF) and Tri-State Median Filter (TSM) are devised with two fold benefit of noise detection and correction[5]-[6].

The proposed filter, "Modified Partial Differential Equation based Adaptive Two-Stage Median Filter for Images Corrupted with High Density Fixed-Value Impulse Noise (MPATS)", performs noise correction on the corrupted images using the principle of median and Partial Differential Equation (PDE). The Peak Signal-to-Noise Ratios (PSNR) of the proposed filter are observed to be higher than those of aforementioned filters. Moreover, the visual perception of the filtered image substantiates the inferences of the obtained PSNR values.

The noise model of fixed-value impulse noise is given in section 2. The principle of standard median filter is explained in section 3. The principle of PDE and the proposed algorithm is detailed in section 4 and 5 respectively. The results and discussion is presented in section 6 and the conclusions in 7.

## 2 Noise Model

The probability distribution of fixed value impulse noise in an image is uniform with respect to the minimum and maximum values of the intensity interval [7].

For a noise probability of p, the distribution of noise is described as:

$$N_{ij} = \begin{cases} I_{\max} \text{ with probability } p_1 \\ I_{\min} \text{ with probability } p_2 \end{cases} \tag{1}$$

where $N_{ij}$ are the corrupted pixels and $p_1=p_2=p/2$. The noisy image $X'_{ij}$ is represented as

$$X'_{ij} = \begin{cases} N_{ij} \text{ with probability p} \\ \text{I with probability } 1 - p \end{cases} \tag{2}$$

where, I represent the uncorrupted pixels in the image.

## 3 Standard Median Filter (SM)

For an input and output image, $X(\cdot,\cdot)$ and $Y(\cdot,\cdot)$ respectively, the SM filter uniformly substitutes the central pixel of each window of size n × n by the median of pixels covered by W. The output of standard median filter is given as:

$$Y_{i,j} = \text{median}\{X_{i-s,j-t} \mid (s,t) \in W\} \tag{3}$$

The coordinates of the pixels in W are given as:

$$W = \{(s,t) \mid -k \le s \le k, -k \le t \le k\}, \tag{4}$$

where k = (n-1)/2.

This process is successively repeated over the entire input image. Though SM is a standard image restoration technique, it removes the details such as thin lines and edges while denoising [8].

## 4 Partial Differential Equations (PDE)

In a networked mesh that is split into small grids, if the central point of the grid is unknown, then it is found from its neighboring points, using the principle of PDE [9]. The central point is estimated using Laplace equation as given in equation (5).

$$\frac{\partial^2 u}{\partial x^2} + \frac{\partial^2 u}{\partial y^2} = 0 \tag{5}$$

In a 3×3 grid (Fig. 1), if the derivatives are replaced by their difference approximations, then equation (5) transforms into equation (6) as:

$$M_{i,j} = \frac{1}{4}\left[M_{i-1,j-1} + M_{i+1,j-1} + M_{i-1,j} + 1 + M_{i+1,j+1}\right] \tag{6}$$

| | | |
|---|---|---|
| $M_{i-1,j-1}$ | | $M_{i-1,j+1}$ |
| | $M_{ij}$ | |
| $M_{i+1,j-1}$ | | $M_{i+1,j+1}$ |

**Fig. 1.** Illustration of Diagonal 5-point Formula

| | | |
|---|---|---|
| X | X | X |
| X | $M_{ij}$ | X |
| X | X | X |

**Fig. 2.** Illustration of 3×3 window

It is otherwise known as Diagonal 5 point formula. This second order PDE, finds application in the edge detection and noise removal [10]-[12].

## 5 Proposed Technique

The proposed filter Modified Partial Differential Equation based Adaptive Two-Stage Median Filter for Image Corrupted with High Density Fixed Value Impulse Noise (MPATS), works in two distinct phases as noise detection and noise correction.

In the noise detection phase, each pixel of the image is processed to identify the corrupted pixels. If the dynamic grayscale of the image is $[I_{min}, I_{max}]$ then, the pixel of intensity $I_{min}$ and $I_{max}$ are graded as corrupted pixels.

The proposed techniques MPATS, performs noise correction in two levels. The entire image is divided into a discrete set of subimages in the order of window W of size n x n. The layout of pixels aggregated by W is portrayed in Fig 2. The central pixel is the window is labeled as $M_{ij}$ and the other neighbouring pixels are labeled as X. The algorithm checks whether the central pixel of the window ($M_{ij}$) is corrupted or not. If found uncorrupted, the algorithm progresses to next subimage. Otherwise, the algorithm checks if any of the Diagonal 5 point formula pixels ($CD_P$) of $M_{ij}$ is corrupted or not. It groups the uncorrupted pixels of W as $U_P$ and the corrupted pixels of Diagonal 5 –point formula as $CD_P$. If all the pixels of Diagonal 5-point formula are uncorrupted then the median of those pixels is computed as $D_{MED}$ and $M_{ij}$ is replaced by $D_{MED}$. If those pixels are corrupted, the algorithm computes the median of Up in W as $U_{MED}$ and replaces each $CD_P$ in the Diagonal 5-point formula pixels by $U_{MED}$. According to equation (6), the mean is computed for those pixels. But, this algorithm computes median of those four pixels as $D_{MED}$, using which the corrupted central pixel $M_{ij}$ is replaced. The computation of median has an edge over mean when the distribution of data is uneven, as evidenced in images.

## 5.1 Algorithm of MPATS

The algorithmic description of the proposed filter is given below:

Step 1: Read the corrupted image X′.

Step 2: Divide X′ into subimages using overlapping sliding window W of sixe n × n.

Step 3: For each window of X′, check whether the central pixel $M_{ij}$ is corrupted or not.

Step 4: If $M_{ij}$ is uncorrupted, goto Step 3.

Step 5: If $M_{ij}$ is corrupted, then read the pixels ($CD_P$) of Diagonal 5-point formula and check if those pixels are corrupted or not.

a. If corrupted
   i. Label those pixels as $C_P$
   ii. Compute: $_{UMED} \leftarrow$ Median ( $U_P$), where $U_P$ represents the uncorrupted pixels in W
   iii. $C_P \leftarrow U_{MED}$
   iv. $D_{MED} \leftarrow$ Median [ $M_{i-1,j-1}$, $M_{i+1,j-1}$, $M_{i-1,j+1}$, $M_{i+1,j+1}$]
   v. $M_{ij} \leftarrow D_{MED}$
   vi. Goto Step 3.
b. If uncorrupted,
   i. $D_{MED} \leftarrow$ Median [ $M_{i-1,j-1}$, $M_{i+1,j-1}$, $M_{i-1,j+1}$, $M_{i+1,j+1}$]
   ii. $M_{ij} \leftarrow D_{MED}$
   iii. Goto Step 3.

Step 6: Stop.

## 5.2 Flowchart

The Pictorial representation of MPATS is depicted in Fig. 3.

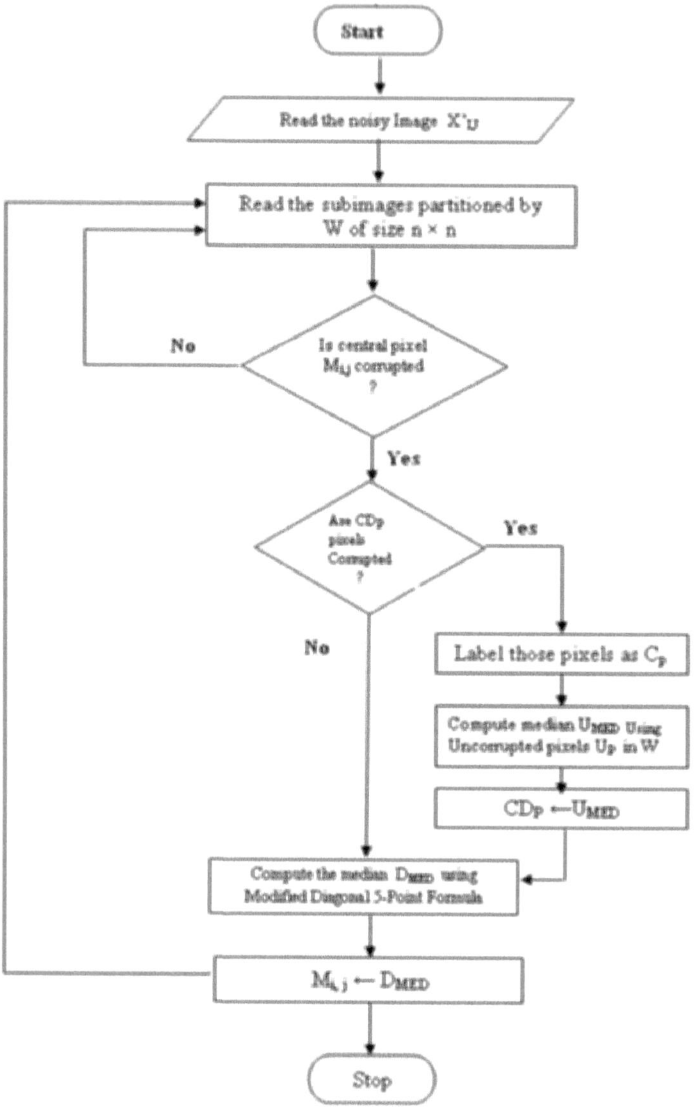


**Fig. 3.** Flowchart of MPATS

# 6 Results and Discussion

The proposed noise filter is developed in Matlab 6.5. The proposed filter was tested on many standard images like lena, cameraman, rice, saturn and mandrill. The window size was chosen as 3 × 3 and the noise probability was varied between 10% and 90%. The performance of the proposed filter was recorded in terms of PSNR and human visual perception.

**Table 1.** Comparison of PSNR values for Lena image

| Method | Noise Density (%) | | | | | | | | |
|---|---|---|---|---|---|---|---|---|---|
| | 10 | 20 | 30 | 40 | 50 | 60 | 70 | 80 | 90 |
| Corrupted | 15.5 | 12.4 | 10.7 | 9.4 | 8.5 | 7.7 | 7.0 | 6.4 | 5.9 |
| SM | 28.7 | 26.4 | 22.6 | 18.3 | 15.0 | 12.2 | 9.8 | 8.1 | 6.5 |
| CWM | 29.7 | 24.1 | 19.5 | 15.7 | 13.0 | 10.8 | 8.9 | 7.6 | 6.3 |
| PSM | 30.7 | 28.7 | 26.9 | 23.7 | 20.0 | 15.2 | 11.1 | 8.3 | 6.4 |
| IMF | 27.2 | 26.7 | 26.1 | 25.1 | 23.9 | 21.2 | 16.6 | 12.1 | 8.0 |
| SDROM | 30.3 | 26.7 | 22.0 | 17.6 | 14.4 | 11.7 | 9.4 | 7.8 | 6.4 |
| SDROMR | 30.5 | 28.5 | 25.7 | 23.5 | 20.7 | 17.8 | 14.2 | 10.6 | 6.9 |
| ACWM | 30.9 | 27.2 | 22.4 | 18.1 | 14.8 | 12.1 | 9.7 | 8.1 | 6.5 |
| ACWMR | 31.4 | 28.8 | 25.8 | 23.3 | 20.8 | 18.1 | 15.0 | 11.8 | 8.1 |
| RUSSO | 31.0 | 27.6 | 24.9 | 22.7 | 20.3 | 17.6 | 14.7 | 11.7 | 8.6 |
| ZHANG | 32.8 | 28.2 | 23.3 | 18.6 | 15.3 | 12.5 | 10.0 | 8.3 | 6.7 |
| SUN | 31.0 | 27.5 | 23.0 | 18.4 | 15.0 | 12.2 | 9.8 | 8.1 | 6.5 |
| IRF | 30.2 | 27.0 | 22.5 | 18.2 | 14.9 | 12.2 | 9.7 | 8.1 | 6.5 |
| TSM | 30.3 | 24.4 | 19.6 | 15.5 | 12.7 | 10.4 | 8.4 | 7.1 | 6.0 |
| MPATS (Proposed Filter) | 32.3 | 28.9 | 26.9 | 24.8 | 23.5 | 21.6 | 19.6 | 17.9 | 15.0 |

**Table 2.** Comparison of PSNR values for Mandrill image

| Method | Noise Density | | | | | | | | |
|---|---|---|---|---|---|---|---|---|---|
| | 10 | 20 | 30 | 40 | 50 | 60 | 70 | 80 | 90 |
| Corrupted | 15.7 | 12.6 | 10.9 | 9.6 | 8.6 | 7.9 | 7.2 | 6.6 | 6.1 |
| SM | 23.8 | 23.0 | 20.7 | 17.7 | 14.7 | 12.3 | 10.9 | 8.3 | 6.8 |
| CWM | 25.5 | 22.7 | 18.9 | 15.6 | 12.9 | 10.9 | 9.1 | 7.7 | 6.6 |
| PSM | 27.8 | 26.4 | 25.0 | 22.9 | 19.5 | 15.3 | 11.0 | 8.9 | 6.5 |
| IMF | 23.0 | 22.8 | 22.6 | 22.3 | 21.5 | 20.0 | 16.0 | 12.0 | 8.3 |
| SDROM | 26.0 | 24.0 | 20.8 | 17.4 | 14.3 | 11.8 | 9.6 | 7.9 | 6.6 |
| SDROMR | 25.8 | 24.6 | 23.2 | 21.8 | 19.7 | 17.6 | 14.2 | 10.8 | 7.1 |
| ACWM | 27.1 | 24.8 | 21.4 | 17.9 | 14.8 | 12.3 | 9.9 | 8.2 | 6.8 |
| ACWMR | 27.5 | 25.7 | 23.9 | 22.3 | 20.2 | 18.2 | 15.4 | 12.4 | 8.8 |
| RUSSO | 29.5 | 26.7 | 24.3 | 22.4 | 20.2 | 17.7 | 15.0 | 11.9 | 8.9 |
| ZHANG | 28.8 | 26.1 | 22.1 | 18.5 | 15.2 | 12.7 | 10.2 | 8.4 | 6.9 |
| SUN | 26.1 | 24.4 | 21.3 | 18.0 | 14.8 | 12.3 | 9.9 | 8.2 | 6.7 |
| IRF | 25.8 | 24.1 | 21.2 | 17.8 | 14.7 | 12.3 | 9.9 | 8.2 | 6.8 |
| TSM | 25.4 | 22.9 | 18.9 | 15.4 | 12.6 | 10.5 | 8.6 | 7.2 | 6.2 |
| MPATS (Proposed Filter) | 29.6 | 27.0 | 24.8 | 23.8 | 22.4 | 21.1 | 19.9 | 18.2 | 17.1 |

To ascertain the performance of MPATS, the PSNR values of lena and mandrill are combined with few high performing filters reported in the literature [2]. The PSNR values of lena and mandrill images are given in Table I and II respectively. It is apparent from these tables the PSNR values of MPATS for the lena image is higher than all the filters considered for comparison, except ZHANG for 10% of noise and IMF for 40% and 50% of noise, whereas the PSNR values of MPATS for mandrill are higher for all the noise probabilities. Moreover, the merits of the filter as reflected by the PSNR values are duly confirmed by the visual perceptions of the filtered images shown in Fig. 4.

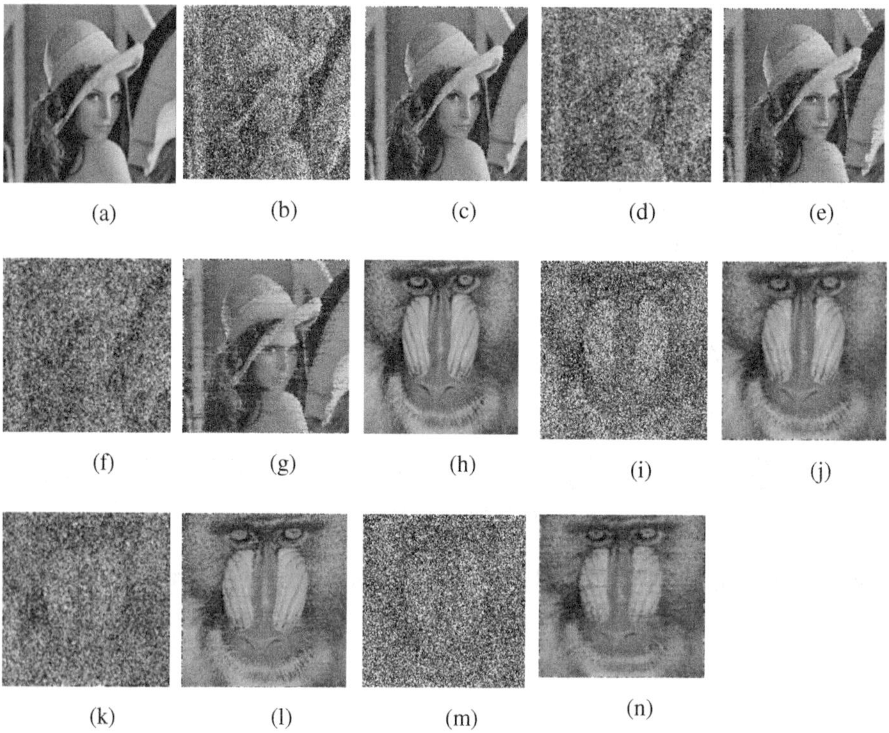

**Fig. 4.** (a) Original image of Lena (b) Lena image with 50% noise; (c) Filtered Image of (b); (d) Lena image with 60% noise; (e) Filtered Image of (d); (f) Lena Image with 70% noise; (g) Filtered Image of (f); (h) Original Image of Mandrill (i) Mandrill Image with 50% noise; (j) Filtered Image of (i); (k) Mandrill Image with 60 % noise; (l) Filtered Image of (k); (m) Mandrill Image with 70% noise; (n) Filtered Image of (m);

## 7 Conclusions

The proposed filter MPATS eliminates the fixed–value impulse noise remarkably from the corrupted images and ensures higher degree of restoration which is recorded in terms of PSNR and visual perceptions. Hence, this filter can be widely used in document imaging, medical imaging, bulk digitization of black and white films and image scanning. This filter also finds application in eliminating noise from Scanning Electron Microscope (SEM) images, used for the study of surface morphology, because SEM images are invariably degraded by noise.

## References

1. Gonzalez, R.C., Woods, R.E.: Digital Image Processing, 3rd edn. Pearson Prentice Hall, India (2009)
2. Besdok, E., Emin Yuskel, M.: Impulsive Noise Suppression for Images with Jarque – Bera Teat based Median Filter. International Journal of Electronics and Communications 59, 105–110 (2005)
3. Somasundaram, K., Shanmugavadivu, P.: Adaptive Iterative Order Statistics Filters. Journal of ICGST – GVIP 09, 23–32 (2009)
4. Ghandeharian, B., Yazdi, H.S., Homayouni, F.: Modified Adaptive Center Weighted Median Filter for Suppressing Impulsive Noise in Images. International Journal of Research and Reviews in Applied Sciences 01(03) (2009)
5. Satpathy, S.K., Panda, S., Nagwanshi, K.K., Ardill, C.: Image Restoration in Non – Linear Domain using MDB approach. IJICE 06(01) (2010)
6. Mansoor, M., Roomi, S., Lakshmi, I.M., Abhai Kumar, V.: A Recursive Gaussian Weighted Filter for Impulse Noise Removal. Journal of ICGST – GVIP 06(03), 33–37 (2006)
7. Somasundaram, K., Shanmugavadivu, P.: Impulsive Noise Detection by Second Order Differential Image and Noise Removal using Nearest Nieghbourhood Filter. International Journal of Electronics and Communications, 472–477 (2007)
8. Windya, P.S.: Fast Impulse Noise Removal. IEEE Trans. on Image Processing 10(01), 173–179 (2001)
9. You, Y.L., Kaveh, M.: Fourth order PDE for noise removal. IEEE Trans. on Image Processing 09(10), 1723–1730 (2000)
10. Shanmugavadivu, P., Eliahim Jeevaraj, P.S.: Fixed – Value Impulse Noise Suppression for Images using PDE based Adaptive Two -Stage Median Filter. In: ICCCET 2011 (IEEE Explore), pp. 290–295 (2011)
11. Shanmugavadivu, P., Eliahim Jeevaraj, P.S.: Selective 8-Neighbourhood based Medan Filter for Images Corrupted with High Density Fixed – Value Impulse Noise. In: 3rd IIST (2011)
12. Grewal, B.S.: Higher Engineering Mathematics, 2nd edn., pp. 1129–1130. Khanna Publishers, New Delhi (2007)
13. Bertozzi, A.L., Greer, J.B.: Travellling Wave Solutions of Fourth Order PDEs for Image Processing. SIAM, Journal of Mathematical Analysis 36(01), 36–68 (2004)

# Enhanced Classification Performance Using Computational Intelligence

Indrajit Mandal and N. Sairam

School of Computing
Sastra University, Tamil Nadu, India
{Indrajit,sairam}@cse.sastra.edu

**Abstract.** This paper presents a computational intelligence technique for enhancing the performance of classifier using a proposed algorithm called Modified Genetic Search Algorithms (MGSA) that avoids local bad search space with merit and scaled fitness variables, detecting and deleting bad candidate chromosomes, thereby reducing the number of individual chromosomes from search space and subsequent iterations in next generations. It addresses the strength of Modified Genetic Search algorithm combined with the Artificial Neural Network (ANN). In this work dynamic Backpropagation Neural Network is used. For training purpose, dynamic learning rate is used that causes the learning rate to decrease in subsequent epoch.

The combined MGSA-ANN is used for the classification of diabetes patients to identify positive and negative cases. It also discusses the main findings and concludes with promising result of the proposed model. The experimental results obtained by synergistic combination of Modified Genetic Search Algorithm with ANN surpass the performance of ANN by 1.4322%.

**Keywords:** Computational intelligence, Modified Genetic Search Algorithm, Backpropagation Neural Network, dynamic learning rate.

## 1 Introduction

The improvement in the performance of Artificial Neural Network has been discussed by Jiansheng and Mingzhe, 2005 [8] in different perspective. Here Back-propagation neural networks use multiple layer perceptron as system framework with MGSA is used and Back-propagation Algorithm as training rule. It helps to avoid network from diverging away from the targeted output and improves general performance. The experimental setup exploits the use of MSGA-ANN to classify patients as having diabetes using java programming. The real world data Pima Indian Diabetes database available and is adapted from UCI Machine Learning Repository that contains 768 instances. Besides, it contain eight numeric inputs and a target attribute class which takes on values 0 (tested-negetive) and 1 (tested-positive). The list of attribute (all numeric-valued) are listed below

1. 'preg'- Number of times pregnant
2. 'plas' -Plasma glucose concentration a 2 hours in an oral glucose tolerance test

D. Nagamalai, E. Renault, M. Dhanushkodi (Eds.): CCSEIT 2011, CCIS 204, pp. 384–391, 2011.

3. 'pres'-Diastolic blood pressure (mm Hg)
4. 'skin'-Triceps skin fold thickness (mm)
5. 'insu'-2-Hour serum insulin (mu U/ml)
6. 'mass'-Body mass index (weight in kg/(height in m)^2)
7. 'pedi'-Diabetes pedigree function
8. 'age'-Age of patients (years)
9. Class variable- tested_negative, tested_positive.

## 2 Artificial Neural Network

In this work, dynamic Backpropagation Neural Network with single hidden layer is used trained via sigmoid function in which the network parameters is possible to be monitored and modified while training time. For training purpose, dynamic learning rate is used that will cause the learning rate to decrease in subsequent epoch. The starting learning rate is divided by the epoch number, to determine the current learning rate. It helps avoiding the network from diverging away from the target output, as well as improves general performance. All the training data is normalized to enhance the network performance between 1 and -1. Brief summary of backpropagation algorithm as discussed by I. Mandal, 2010 [1].

The activation function used in the neural network training is sigmoid function mentioned in (1) where k is integer constant and x is the input value.

$$f(x) = \quad 1 / \sqrt{(1 + e^{(-kx)})} \tag{1}$$

Other network parameters
Number of neurons in hidden layer= (attribute+ classes)/2= (9+2)/2=6
Learning rate = 0.3
Momentum = 0.2
Number of epochs = 500.

## 3 Modified Genetic Simple Algorithm

As GA finds solution through evolution process, it is not inclined to good solution but move away from bad solutions. There is a chance that it may lead solution to a dead end. Also the number of iterations or generations required is very high. In the proposed MGSA, the idea is to identify and eliminate the bad chromosomes using merit and scaled variables so that the search space is reduced. Once the search space is minimized, containing prospective chromosomes then it leads to better optimization in search process. Here the merit and scaled variables refer to the classification error rates of chromosomes. Bad chromosomes indicate those individuals which lead to a dead end. Here is a brief summary of MSGA:

1) **Initial population generation.** The parameters of the model to be optimized are considered as chromosomes. A randomly chosen set of population of individuals is a binary string with a fixed length.

2) **Fitness function:** In each generation for which GA is run, the fitness of each individual is determined which is close to optimality. Here in the attribute selection problem, we have defined linear function **f `= af + b, where f `, f** are scaled and raw fitness values of chromosomes and **a,b** are constants.

3) **Detection and Elimination:** Those chromosomes having greater classification error rate determined by merit and scaled variables are determined and eliminated so that it is not allowed to be a candidate chromosome in the next generation and thereby filtering out the bad chromosomes and hence the number of iterations in the subsequent generations gets reduced.

4) **Selection:** A pair of best fit chromosomes or individuals which has least classification error rate is selected from the residual population.

5) **Crossover:** It is the reproductive stage where two new individuals are crossed with probability Pc to generate new pair of offspring.

6) **Mutation:** A single point alteration of bit string from zero to one and vice-versa with probability Pm is applied to selected individuals to generate a new pair of offspring and avoids premature convergence.

## 4 Experimental Design and Results

### 4.1 Case 1

The BPNN as said in Haykins, 1999 [9] is applied to the dataset to classify with 10-fold cross validation as mentioned by Picard and Cook,1984[7] and shows that it achieves a very impressive 75.7813 % classification accuracy. Here the data is divided into equally ten parts and nine are used for classification and one is used for testing. This is repeated for ten times and average accuracy is considered. Details are shown in table1.

A confusion matrix as discussed by Marom, Rokach, Shmilovici, 2010[5] is shown in table2 that contains the analytical details of classifications where eight attributes are input to it. Its performance is evaluated based on the data in the matrix for two class classifier. Accuracy is measured by Received Operator Characteristics (ROC) area as discussed by Kohavi and Provost, 1998[4], area under graph with TP as Y-axis and FP as X-axis and ranges from zero to one. For class tested-negative ROC plot =0.793 is shown in fig1 and Cost/benefit analysis is shown in fig 2 where gain is 141.21. The following tables 1, 2 show the confusion matrix and accuracy parameters respectively.

### 4.2 Case 2

In case 1 all the eight attributes are considered as input but in real world data irrelevant, redundant or noisy attributes are common phenomena, which impairs the classifier's result. The learning scheme Wrapper subset evaluation as discussed by Mallik, 2007[7] with ANN is now integrated with Modified Genetic Search algorithm. After filtration process only five attributes are selected as relevant. Attributes preg, skin, insu are eliminated.

The attributes of MGSA that includes

Population size: 30
Number of generations: 30
Probability of crossover: 0.6
Probability of mutation: 0.033

Now reclassifying the records ANN with 10-fold cross validation; however, this time only five attributes are input to the classifier. Selected attributes are plas, pres, mass, pedi, age.

In table3, every subset is a chromosome and merit is the fitness score reported by MGSA, which is equal to the corresponding classification error rate. Also, each chromosome's scaled fitness is shown in the scaled column where we use linear scaling technique to scale the values. By definition, the raw fitness and scaled fitness values have the linear relationship

$$f^{`} = a.f + b \quad (2)$$

where $f^{`}$ and f are the scaled and raw fitness values respectively. The constants **a** and **b** are chosen where

$$f^{`}_{avg} = f_{avg} \text{ and } f^{`}_{max} = K\, f^{`}_{avg} \quad (3)$$

The constant K represents the expected number of copies of the fittest individual in the population. Thus, by computing the average fitness values from table5, we obtain $f^{`}_{avg}$=0.055755 and $f_{avg}$=0.055753.

Solving equations (2) we get **a= 5.012392** and **b= -0.883779**. We use equation (3) to determine **K** i.e. **K=** $\mathbf{f^{`}_{max} / f^{`}_{avg}}$ **=0.27273/0.238443= 1.143795**

Observe in fifth row in table3 $\mathbf{f^{`}=0}$**.** The raw fitness value of **0.28516** corresponds to the largest classification error in the population produced by chromosome {3 6 7 8}, and as a result, $f^{`}$ is mapped to zero to avoid the possibility of producing negatively scaled fitness.

The improved classifier has accuracy level of 77.2135 %, which indicates that the second model outperforms the first model by 1.4322 % where the input to the later model is only five attributes indicated in ROC area. For class tested-negative ROC plot =0.809 is shown in fig3 and Cost/benefit analysis is shown in fig 4 where gain is 159.76 and observe the smoothness achieved in the graph. Moreover reducing the number of attributes in second model has increased the gain in cost benefit analysis. That is, the classification accuracy has improved positively where only five of eight attributes are specified in the input. Though the increased accuracy is not a dramatic improvement, it shows the strength of Modified Genetic Search algorithm is an appropriate algorithm for attribute selection process.

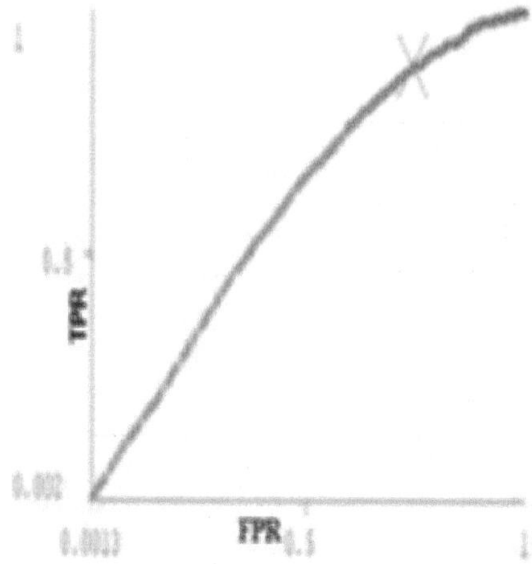


**Fig. 1.** ROC area=0.793 for case1

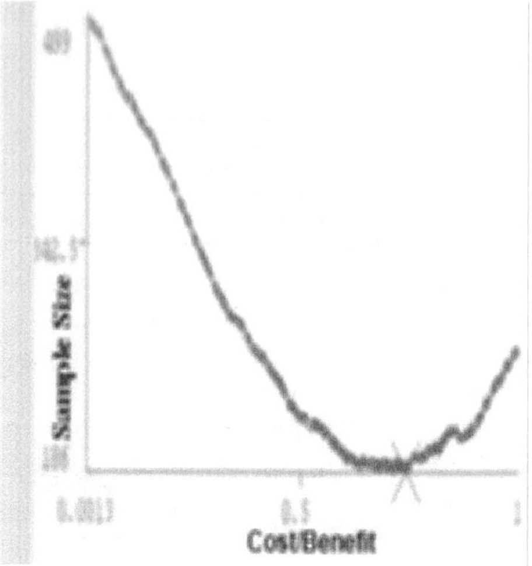


**Fig. 2.** Cost/benefit analysis for gain=141.21 in case 1 with 9 attributes

**Table 1.** Classification Accuracy by neural Network

| | TP Rate | FP Rate | Precision | Recall | F-Measure | ROC Area | Class |
|---|---|---|---|---|---|---|---|
| Accuracy by Class | 0.832 | 0.392 | 0.798 | 0.832 | 0.815 | 0.793 | tested_negative |
| | 0.608 | 0.168 | 0.66 | 0.608 | 0.633 | 0.793 | tested_positive |
| Weighted Average | 0.754 | 0.314 | 0.75 | 0.754 | 0.751 | 0.793 | |

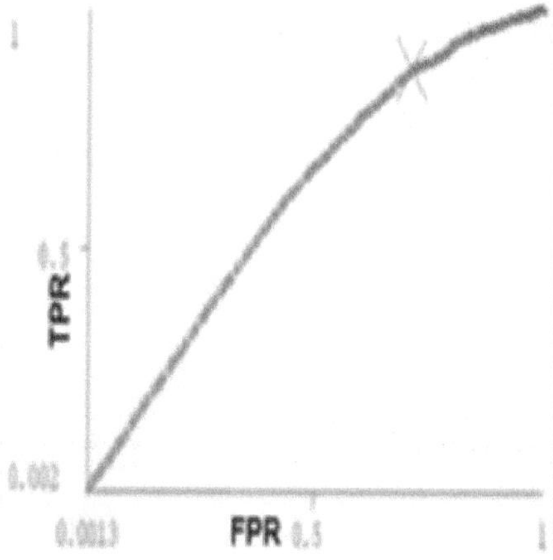


**Fig. 3.** ROC area=0.809 for case2

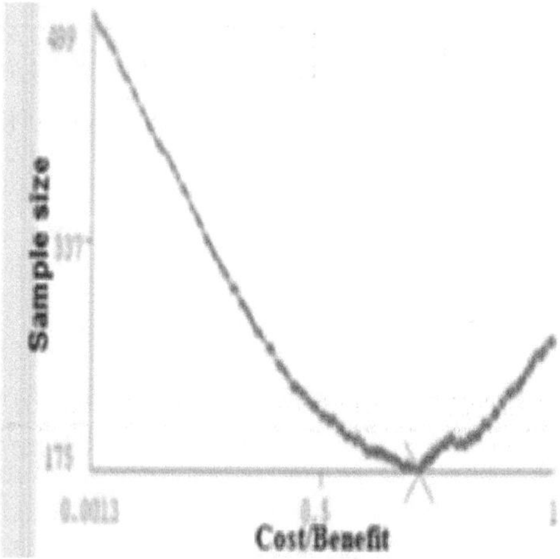


**Fig. 4.** Cost/benefit analysis for gain=159.76 in case2 with 5 attributes

**Table 2.** Confusion Matrix for neural network

| Predicted | | | |
|---|---|---|---|
| tested_negative | tested_positive | | |
| 416 | 84 | tested_negative | Actual |
| 105 | 163 | tested_positive | |

**Table 3.** Initial population characteristics for the 30 chromosomes

| merit | scaled | subset |
|---|---|---|
| 0.23073 | 0.27273 | 2 3 6 7 8 |
| 0.23073 | 0.27273 | 2 3 6 7 8 |
| 0.23073 | 0.27273 | 2 3 6 7 8 |
| 0.2349 | 0.25185 | 2 3 6 8 |
| 0.23073 | 0.27273 | 2 3 6 7 8 |
| 0.23984 | 0.22706 | 2 3 6 7 |
| 0.2388 | 0.23228 | 2 3 5 6 7 8 |
| 0.2349 | 0.25185 | 2 3 6 8 |
| 0.2474 | 0.18922 | 1 2 5 7 8 |
| 0.23073 | 0.27273 | 2 3 6 7 8 |
| 0.2388 | 0.23228 | 2 3 5 6 7 8 |
| 0.23073 | 0.27273 | 2 3 6 7 8 |
| 0.23073 | 0.27273 | 2 3 6 7 8 |
| 0.23073 | 0.27273 | 2 3 6 7 8 |
| 0.2388 | 0.23228 | 2 3 5 6 7 8 |
| 0.24193 | 0.21662 | 2 3 7 8 |
| 0.23073 | 0.27273 | 2 3 6 7 8 |
| 0.23073 | 0.27273 | 2 3 6 7 8 |
| 0.23073 | 0.27273 | 2 3 6 7 8 |
| 0.24193 | 0.21662 | 2 3 7 8 |
| 0.25755 | 0.13832 | 2 3 4 |
| 0.2388 | 0.23228 | 2 3 5 6 7 8 |
| 0.2388 | 0.23228 | 2 3 5 6 7 8 |
| 0.23073 | 0.27273 | 2 3 6 7 8 |
| 0.24323 | 0.2101 | 2 6 7 8 |
| 0.23073 | 0.27273 | 2 3 6 7 8 |
| 0.2349 | 0.25185 | 2 3 6 8 |
| **0.28516** | **0** | **3 6 7 8** |
| 0.23438 | 0.25446 | 2 5 6 7 8 |

**Table 4.** Classification Accuracy for neural network and genetic algorithm

| | TP Rate | FP Rate | Precision | Recall | F-Measure | ROC Area | Class |
|---|---|---|---|---|---|---|---|
| Accuracy by Class | 0.846 | 0.377 | 0.807 | 0.846 | 0.826 | 0.809 | tested_negative |
| | 0.623 | 0.154 | 0.684 | 0.623 | 0.652 | 0.809 | tested_positive |
| Weighted Average | 0.768 | 0.299 | 0.764 | 0.768 | 0.766 | 0.809 | |

**Table 5.** Confusion Matrix for neural network and genetic algorithm

| Predicted | | | |
|---|---|---|---|
| tested_negative | tested_positive | | |
| 423 | 77 | tested_negative | Actual |
| 101 | 167 | tested_positive | |

## 5 Conclusion

MGSA can perform global search rather than local search efficiently. It addresses the strength of proposed Modified Genetic Search algorithm combined with the Artificial Neural Network (ANN). For training purpose, dynamic BPNN trained with sigmoid function with learning rate to decrease in subsequent epoch and improve general performance. The proposed model MGSA addresses the local optima problem using merit and scaled variables in which the bad individual chromosomes are detected and deleted, reducing the search space and further iterations thereby improving efficiency. We have shown that MSGA is used for reducing the number of input parameters without sacrificing the performance of the classifier.

It shows synergistic performance of MGSA-ANN in the classification problem with 1.4322 % demonstrated with Pima Indian Diabetes dataset. The computational intelligence obtained from the MGSA helps to eliminate the irrelevant variables from the real world data set so it acts as subset selection evaluator.

Although in this paper there are only eight attributes, there are still 255 possible attribute subsets that can be given as input. If the numbers of attributes increases say hundred then there are 1.27x $10^{30}$ possible attribute subsets to choose. In such a hard situation like this MGSA may prove helpful in determining the optimal subset. From the experimental results, it is concluded that GA is an effective algorithm that can be exploited in complex situations. It can be used in various applications such as optimization of weights of links in neural networks. Further the performance of GA can be improved by parallel implementation using MPI and other techniques and can be applied to vast field of emerging applications.

## References


1. Mandal, I.: Software reliability assessment using artificial neural network. In: International Conference and Workshop on Emerging Trends in Technology, pp. 698–699. ACM, New York (2010)
2. Pima Indians Diabetes dataset, National Institute of Diabetes and Digestive and Kidney Diseases, UCI Machine Learning Repository, http://archive.ics.uci.edu/ml/datasets.html
3. Goldberg, D.E.: Genetic algorithms in search, optimization and machine learning. Addison-Wesley, Reading (1989)
4. Kohavi, Provost (1998), http://www2.cs.uregina.ca/~hamilton/courses/831/notes/confusion_matrix/confusion_matrix.html
5. Marom, N.D., Rokach, L., Shmilovici, A.: Using the Confusion Matrix for Improving Ensemble Classifiers. In: 2010 IEEE 26th Convention of Electrical and Electronics Engineers in Israel (2010)
6. Mallik, R.K.: The Uniform Correlation Matrix and its Application to Diversity. IEEE Transactions on Wireless Communications 6(5), 1619–1625 (2007)
7. Picard, R.R., Cook, R.D.: Cross-Validation of Regression Models. Journal of the American Statistical Association 79(387), 575–583 (1984), http://www.jstor.org/stable/2288403
8. Wu, J., Liu, M.: Improving Generalization Performance of Artificial Neural Networks with Genetic Algorithm. In: 2005 IEEE International Conference on Granular Computing, vol. 1, pp. 288–291 (2005)
9. Haykin, S.: Neural Networks-A Comprehensive Foundation, 2nd edn. Prentice Hall, Inc., New Jersey (1999)
10. Shi, H., Zhang, S.: Improving Artificial Neural Networks Based on Hybrid Genetic Algorithms. In: 5th International Conference on Wireless Communications, Networking and Mobile Computing, WiCom 2009, pp. 1–4 (2009)

# Feature Modeling of the Evolving Access Control Requirements

K. Shantha Kumari[1] and T. Chithralekha[2]

[1] Research Scholar,
Department of Banking Technology,
Pondicherry University
shanthajayakumar@gmail.com
[2] Reader,
Department of Computer Science,
Pondicherry University

**Abstract.** Access control mechanisms are an integral part of most modern software systems, but it is still not considered as an explicit part in the development process. Access control mechanisms and policies are generally added to existing systems as an afterthought, with all the problems of unsatisfied security requirements, integration difficulties, and mismatches between design models. In order to increase the overall system security, access control requirements should be taken into account early in the software development process. Due to the integration of Access control requirements early in the development process, potential security breaches can be detected and removed earlier. But, on the other hand, this integration results in spreading of access control functionalities across the software systems and making them tightly cohesive. When there is a need arises for updating the existing access control requirement, the changes would impact on several functional requirements. Moreover the access control requirements tend to evolve continually. Accordingly, the design phase should support both the static and dynamic changes of the access control requirements without making much impact on the functional requirements. To address this issue at the design level, we need an approach that could support to model the access control requirements along with the constraints and also capture the changes, without affecting the functional design. In this paper, we propose a Feature Conclave Model based on feature modeling, for analyzing and designing the evolving access control requirements along with the system design. The Feature Conclave Model views the access control requirements as "Features" and also provides step-by-step integration with the functional domain. Moreover, the new updates of the access control requirements can be represented as the variants and can easily be adapted within the system design.

**Keywords:** Access control requirements, Role based access control, Feature Modeling, Assembling, Verification.

## 1 Introduction

The need for Access control mechanisms in large scale software systems has become a major priority for all organizations. The increasing number of threats and

D. Nagamalai, E. Renault, M. Dhanushkodi (Eds.): CCSEIT 2011, CCIS 204, pp. 392–403, 2011.

vulnerabilities has repeatedly shown that the design and engineering of these systems from a security stand point is often lacking. As a general practice, Access control policies are often added in an ad-hoc basis during the later part of the product development resulting in errors and vulnerabilities that provide a potential for exploitation. However, given the complexity and pervasiveness of today's software systems, this kind of approach might not be completely satisfactory. As such, engineering software security is essential. In order to increase the overall system security, access control requirements should be taken into account early in the software development process. An efficient Access control system design must support system evolution, such as changes in security policies, user population and their roles, and also the changes in the applications.

From the design perspective, access control policies give rise to security concerns that must be addressed in a design [1]. Prior to this view, many researchers have recommended for specifying and maintaining Access control requirements separately from the application design. This allows the Access control requirements to be analyzed clearly and the policies can be changed independently of the application. But this kind of separate policy specification has a problem associated with it-The scalability and late integration problems.

To properly address security risks and vulnerabilities without jeopardizing speed or cost, companies must bring security into the development process. This results in a tightly cohesive system with the scattering of the access control functionality across the various functional modules design [2]. This intervention of access control features with the behavior of the functional specifications makes it very difficult to understand and analyze the working of the access control policies. Moreover the access control requirements may evolve continuously. For example, Current and former employees and contractors who have or had authorized access to their organization's system and networks are familiar with internal policies, procedures, and technology and can exploit that knowledge to facilitate attacks and even collude with external attackers . Addressing this kind of pervasive access control requirements for complex information systems, using an improvised approach, would lead to a design with inconsistencies in satisfying the security constraints. This will result in security failures and violations [2].

In this paper, the access control requirements are modeled using Feature conclave Model on the basis of Feature modeling to support a systematic and planned way of integrating with the functional requirements in the design. This Model would reduce the developmental complexities and increase the effectiveness in the integration. Using the abstraction of 'Features' results in an ease of adaptation of the changes in the access control requirements and further integration with the application requirements. UML design principles are used for modeling the access control model as feature conclave model that would enhance the factor of reusability.

The paper is organized as follows. Section 2 gives an overview of current modeling techniques and its limitations. In Section 3 we present Feature conclave model and also explain the associated feature modeling method. Section 4 briefs about the formal basis of design principles for RBAC features. Section 5 demonstrates how RBAC features can be configured for an example. Section 6 concludes the paper and discusses future work.

## 2 Modeling of Access Control Requirements

The main objective of access control is to provide confidentiality of sensitive information, one of the main characteristics of security. Access control policies determine how sensitive information and computing resources are to be protected. Olden year software systems, by their very nature, have simple and generic access operations and are not concerned with the meaning of the resources they handle. On the other hand, modern Information sensitive systems and applications handle quite complex access operations due to their specific user purposes and high functionality. This complexity needs to be translated by generic models and specific access control requirements into design phase understandable processes.

Approaches for specifying and analyzing the access control requirements are based on sophisticated mathematical concepts, that formally stated, allows one to check the developed access control requirements enforce required level of protection. In practice, however, applying mathematically-based formal specification techniques can be difficult because of the high degree of mathematical skill needed. Therefore, a representation that can be analyzed without sacrificing understandability and usability is desirable. Generally the access control requirements are formally modeled to an Access control model like Mandatory Access control, Discretionary Access control or Role based Access control. Role-based access control (RBAC) models are becoming a de facto standard, greatly simplifying management and administration tasks. Role-based access control (RBAC) received considerable attention as an alternative to traditional mandatory and discretionary access control policies.

In this paper Role-Based Access control is used for modeling the access control requirements. Once requirements are presented as Role-based access control model, further modeling of RBAC is required to facilitate the smooth integration with the functional requirements. The Unified Modeling Language (UML) is considered to be the industry de-facto standard for modeling software-based systems. Use of the UML to specify RBAC policies ease the task of incorporating the policies into UML application models. The following table shows the works done in modeling the RBAC using UML and UML Profiles- the Extension mechanism of UML:

| Using UML | UML Profiles |
|---|---|
| UML class diagrams are used to illustrate the structure of RBAC in [3] and the Object Constraint Language (OCL) [4] to define RBAC constraints | A UML profile called UMLsec is proposed for modeling [8] and evaluating security aspects for distributed systems based on the multi-level security model described in [9]. |
| Object diagrams are used to visualize the RBAC constraints in [5] | SecureUML [10] is the UML profile proposed for defining security concepts based on RBAC. |
| In the work of [6] an access control model is viewed as a design pattern and uses the Gang-of-Four (GoF) pattern template [7] to describe RBAC. | Security aspects of RBAC and MAC are incorporated into UML model elements in the work proposed in [11] |

The effectiveness of the UML as a standard is predicated. Among other things, on there being a clear, precise, and pragmatic semantics for its notations. But the semi-formal semantics of UML has ambiguity and inconsistency issues. Moreover UML captures the static requirements and represents their behavior using anyone of the diagram. But the pervasive nature of the access control requirements has to be represented, so that evolving new access control requirements can be addressed. Hence there is a need for a new level of abstraction to represent the evolving access control requirements [14]. Features represent the variants- Changes.

Moreover the Access control requirements may scatter over the functional design. This will lead to many problems listed as follows: (1) Altering the access control requirement needs making changes in more number of places in the design, (2) Assessing different ways for enforcing the access control requirements as model is difficult when they crosscut a system design. These limitations can result in software that cannot be expected to protect sensitive information.Hence the impact resulting in updating an access control requirement on various functional requirements should be minimized by reducing the tightly cohesive nature of the access control requirements with the domain. This is not presented in UML kind of representations.

The above -mentioned observations have given rise to the need for developing a new modeling approach that could support the evolving access control requirements and also minimizing the impact by recognizing the integration in a step-by-step manner [13]. The new approach proposed is based on feature modeling that facilitates the reusability and comfortable composition of access control policies with the functional requirements. Reusability is advocated by creating feature models for new and evolving access control requirements by lining up the features from an available and continuously growing warehouse of features. The next section will explain the framework of Feature Conclave and feature warehouse.

## 3 Feature Conclave

Software Product Lines has emerged as a new technology to develop software product families related to a particular domain. Software product line engineering (SPL) involves developing the requirements, architecture, and component implementations for a family of systems, from which products (family members) are derived and configured. One of the multiple views of the product line is the feature modeling view. The feature model is essential for both variability management and product derivation, because it describes the product line requirements in terms of commonality and variability, as well as defining the product line dependencies. A feature is a logically cohesive piece of functionality and is present in all phases of software development. Features may occur at any level, for example high-level system requirements, architectural level, subsystem and component level. Features are basic building blocks that satisfy intuitive user formulated requirements on the software system. Features are main abstractions in design and implementation. They reflect user requirements and incrementally refine one another.

The Proposed *Feature Conclave* consists of *Feature Association rules* and also a *Feature warehouse* that houses all the features related with the domain. The *Feature Association* is based on Feature Assembly modeling [14]. This aims to bring new features as well as the previously defined features. The Feature conclave framework consists of a warehouse that houses all the features related with the domain. In the previous works of feature modeling, a Feature is viewed with its intention and also its application in developing a product. But the contribution of the feature for creating different products depends upon the requirements. Hence, it can be separated from the definition. This enables the feature definition to be independent of its applicability in various products. Therefore Features can be classified as "Concrete feature" that provides the ample definition of its capabilities and as "Optional feature", that represents the Specialization of specific features. Feature association supports reusability of features, by the usage of Feature warehouse. The Feature Association rule – indicates the level of contribution by a particular feature on a product that gives the variation. And the Feature Warehouse- 1.stores the features with annotations,2. Association relations of features and 3. Variants of features and feature dependencies. It doesn't stores the Context of Use of the particular feature. The following figure shows the Feature Conclave.

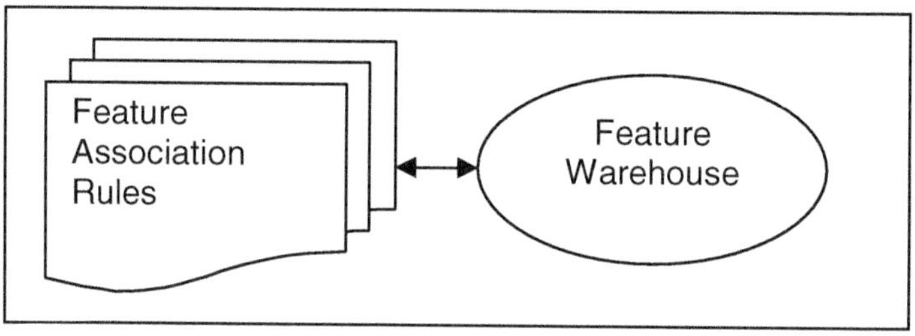


**Fig. 1.** Feature Conclave framework

### 3.1 Need for Meta-Modeling

Generally the requirements for large scale soft wares provide different scope. The requirements may scale with one another or may also produce conflicts. Hence the modeling the requirements of a particular domain, requires a multiple stand for analyzing and designing. The features of the domain may contribute differently in multiple standpoints. Hence there is a need for a Meta Model of feature that could facilitate the multiple standpoints. The Meta Model of feature will be realized according to the perspective taken by the product line at the instance. The feature warehouse is realization of this meta-modeling approach.

### 3.2 Feature Warehouse

Feature Warehouse is repository to store the feature definitions, relations between features, dependencies among features and conditions for using the features. Features in Warehouse actually represent the conceptual representations. They are stored together along with the metadata information like owners, date of creation. The relations, dependencies and conditions of the features are stored independent of the context of application. Hence the reusability of the features is possible with or without its full definition. The Concrete feature can be used with or without its descendents and options, where as the Optional feature can be used only when the options are defined. By separating the definition of the relations and its implications, higher degree of reusability is achieved. Thus the Meta-Model for the complete view of a feature association is given as follows:

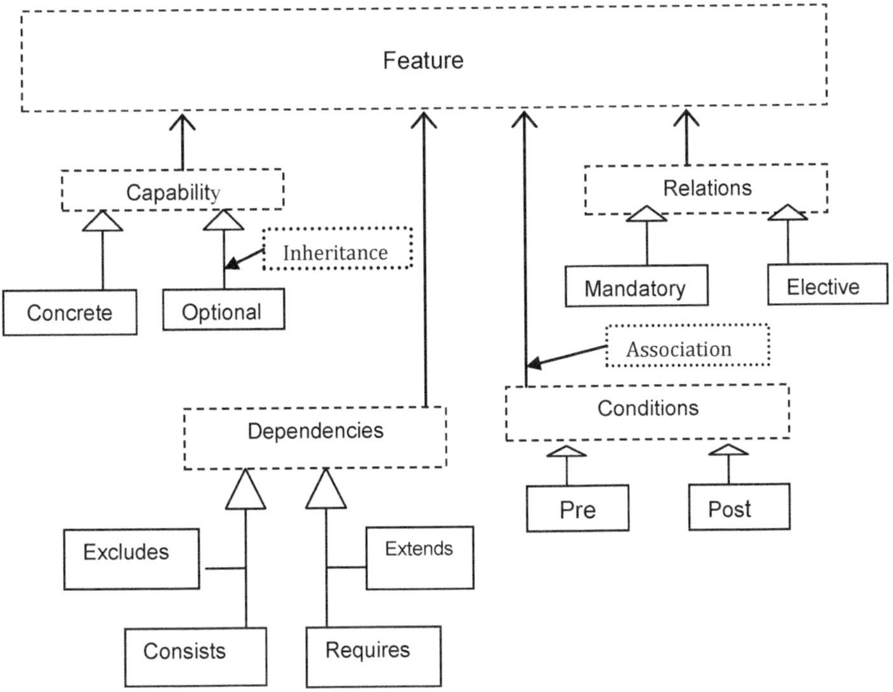


**Fig. 2.** Meta-Model of Feature Definition

### 3.2.1 Relations and Dependencies between Features

The Relations and Dependencies between features are stored as a separate property irrespective of any application in a Feature warehouse. Whenever a new product line has to be built, it's a very simple search operation through the warehouse for features that matches with the requirements. Once the selection of features are done, they need to composed together to define a product. This feature composition is supported by "Relations", either as *Mandatory* or *Optional.* A Mandatory relation depicts an "ALL" Composition that indicates the compulsory and conclusive nature. The Elective relation gives the "Choice based" composition depending upon the requirements. This Elective relation is associated with the Optional features to produce the variants. This is depicted in the following figure

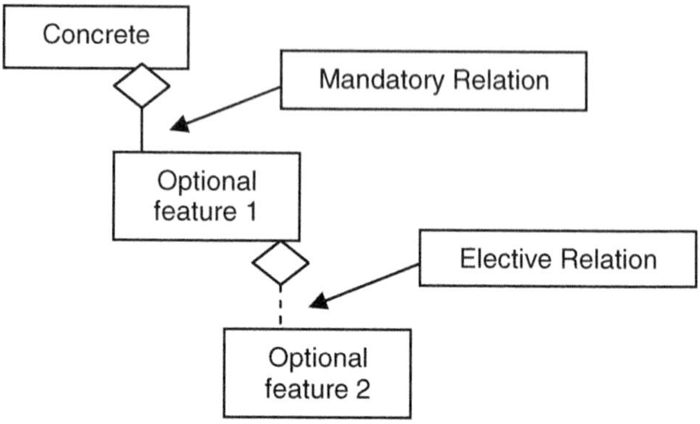


**Fig. 3.** Meta-Model of Feature Relation

In addition to the relations, the features may be defined with dependencies also. They specify how the selection of one feature has impact with other features. They also visualize the interactions between features. The different kinds of dependencies are as follows:

1. Feature A extends Feature B
2. Feature A consists Feature B
3. Feature A requires Feature B
4. Feature A excludes Feature B

#### 3.2.2 Conditions

Along with the above explained the constraints for associating a feature with another feature is also mentioned in feature definition. They are as follows:

1. Pre-Conditions for a Feature Association
2. Post-Conditions for a Feature Association.

*Pre-conditions* give the prerequisites for selecting the feature for a product design. Unless the conditions are satisfied, the corresponding feature selection will not provide its functionality. Whereas, the *Post-conditions* express the conditions needed after the feature composition. These post scenario work is very important to maintain the stability of the developing product line.

### 3.3 Feature Association Rules

The Feature Association rule indicates the level of contribution by a particular feature on a product that gives the variation. This information is a specific part of this rule, by which the specific software product line can be generated. The feature association process is referred as the feature composition in Feature modeling techniques. It is as follows:

Step -1: Analyze the requirement in the Domain → Capture the features [both mandatory and Optional] → Populate Warehouse → Create the Product Line through feature Relations and dependencies → Develop the feature Model
Step -2: New Requirement → Capture the features [both mandatory and Optional] → Check the Warehouse→If existing→ extract from warehouse → else define new features → Create feature Relations/dependencies/conditions → add the new defined features to the warehouse.

Feature Association is the process of composing the features based on feature association rules. This Association supports modeling of very large scalable software during analysis and design. The Feature association rule thus defines the process of feature association via feature definition, feature relations, dependencies and conditions.

## 4 Formulating RBAC as Feature Model

Access control requirements which tend to be pervasive can be modeled using feature modeling as the features reflect user requirements and incrementally refine one another. In this paper, RBAC is chosen, as it is an efficient and scalable access control model that governs access based on user roles and permissions. RBAC consists of group of components (Features), each providing a different access control function. From modeling perspective, The NIST RBAC standard is explored in this work.

By NIST standard [11], RBAC Core defines the minimum collection of RBAC elements, element sets, relations in order to completely achieve a role-based access control system. It includes: 1.user-role assignment 2.permission-role assignment. The Hierarchal RBAC adds relations for supporting role hierarchies. And the SOD component is used to enforce conflict of interest policies that organizations may

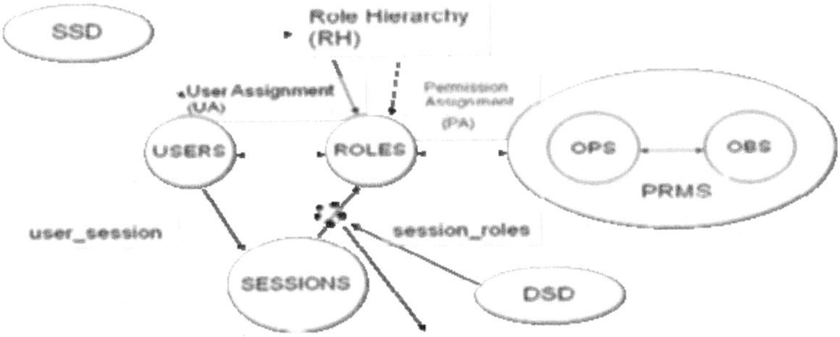


**Fig. 4.** RBAC with Core, Hierarchy and SOD Components

employ to prevent users from exceeding a reasonable level of authority for their position. The Static SOD places restrictions on sets of roles. If a user is assigned to one role, the user is prohibited from being a member of a second role. The Dynamic SOD, places constraints on the roles that can be activated within or across user's sessions. It supports each user has different levels of permission at different time. It is often referred as timely revocation of trust. The following figure shows the Core RBAC with Hierarchy and Separation of duty components.

### 4.1 RBAC – A Candidate for Feature Modelling

In this paper, the different components of RBAC are considered as a new feature to be added with the RBAC Core, which is the essential collection of RBAC elements. Hence, three features are available and they are: role hierarchy (RH), static constraints (SSD), dynamic constraints (DSD).These three features can be considered as stepwise refinement of the RBAC core feature. With this Feature oriented view, the development of RBAC models will be incremental and compositional. For example, RBAC that has Role hierarchy and dynamic constraints will be built by integrating the base model RBAC core with features RH and DSD. Therefore this approach will greatly enhance the reusability of the base model and features, as well as provide great flexibility for RBAC evolution to meet new system needs. Thus we have a nice and elegant solution to issue of addressing the evolving access control needs. RBAC feature is designed in the way that it encapsulates those and only those properties that pertain to the functions of the features [19]. . From the design guidelines, the RBAC feature has the following:

1. **Core feature – Concrete feature** consists of
   a. Users – Abstract view
      i. Roles- *Concrete*
   b. Session- *Optional*
   c. Permissions- *Optional* and Pre-conditions should be checked
      i. Operations- *Optional*
      ii. Objects- *Concrete* and Post conditions for analysis
2. Hierarchy feature – Optional Feature that requires
   a. **EITHER** General -
   b. **OR** Limited
   c. Advanced – Optional
3. SOD feature – Optional that requires
   a. Static SOD **OR**
   b. Dynamic SOD

The following figure depicts the feature association model of RBAC:

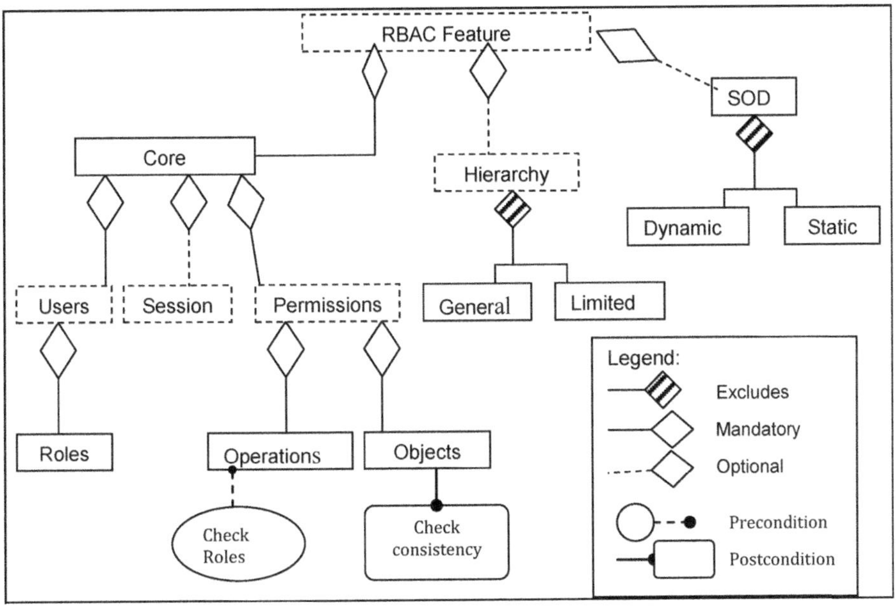


**Fig. 5.** RBAC with Core, Hierarchy and SOD Components in Feature Association model

This model can be extended with consideration of other features such as temporal access control and privacy aware policies.

# 5 Example

An example for integrating the access control requirements with the functional requirements using feature conclave model is explained in this section. Consider a Banking product line application that is supposed to have various types of services and also evolving requirements based upon the services. Let us assume the Transaction component of a Banking service. The *features identified* are:

1. Accountant, customer, Manager- Role feature
2. Pass book, Recorder, Transaction register, Cash Box- Object Feature
3. Update entries, Monitor, Cash handling – Permissions feature after checking the Roles
4. Manager is superior to Accountant – Hierarchy feature that requires General

These features are then defined in the Feature Warehouse along with the relations and dependencies. The *dependencies* are:

4. Feature Role consists of Permission features
5. Object feature requires Permission features
6. Feature Role extends Hierarchy feature

The *Pre and Post conditions* to be checked are:

7. Permission feature has to be added to the role feature before realizing with the object feature
8. Hierarchy feature has to be added before permission feature is associated with role feature.
9. Object feature should be the last feature after a role feature is associated with the permission feature and hierarchy feature.

The following figure shows the partial view of the Banking domain requirements integrated with Access control requirements.

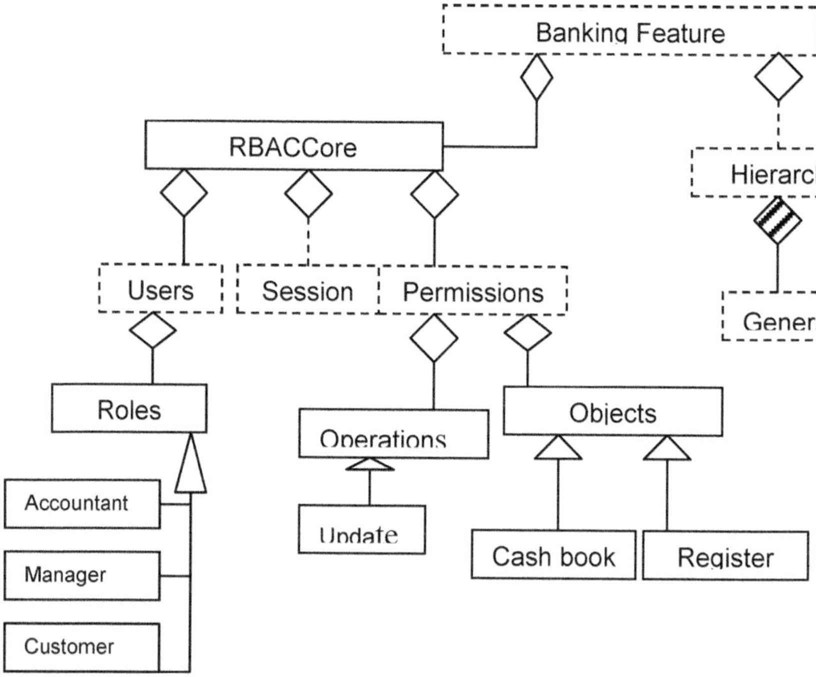


**Fig. 6.** Partial view of RBAC applied to Banking services using Feature Association model

## 5.1 Evolving Requirements

The term "Evolving requirements", can be considered as either an extension of the existing requirement or an entirely new requirement. As explained in section 3, both these situations can easily be handled by the feature conclave framework. In this example, let a new role called senior manager be introduced in the above explained banking model. This role can easily be modeled, just by extending the Manager feature and Hierarchy feature and defining a new Permission feature. Whenever this feature association has to be applied for developing a new product, for example, to support the ATM services, new features have to just define and can be combined with existing features in Warehouse in ease of manner.

## 6 Conclusion

Many related works on modeling the access control policies is done previously. This paper discusses the Access control policies for software systems that need continuous evolution of access control policies on need-basis. RBAC is chosen as the Access control model, along with static and dynamic separation of duty constraints. The Banking system is taken as case study and the composed model is produced, which is a model with access control policies integrated along with the functional requirements. This work can be further preceded by including the modeling requirements for unpredicted changes of access control policies and the verification mechanism using Propositional Logic for checking whether the composed model enforces the access control continuously.

## References

[1] Ray, I., France, R., Li, N., Georg, G.: An aspect-oriented methodology for designing secure applications. Information and Software Technology 51, 846–864 (2009), doi:10.1016/j.infsof.2008.05.004
[2] Georg, G., Ray, I., France, R.: Using Aspects to Design a Secure System. In: Proceedings of Eighth IEEE International Conference on Engineering of Complex Computer Systems 2002. IEEE Computer Society, Los Alamitos (2002)
[3] Shin, M., Ahn, G.: UML-Based Representation of Role-Based Access Control. In: Proc. of IEEE Int. Workshop on Enabling Technologies, pp. 195–200
[4] Warmer, J., Kleppe, A.: The Object Constraint Language, Getting Your Models Ready for MDA, 2nd edn. Addison Wesley, Reading (2003)
[5] Kim, D.-K., Ray, I., France, R.B., Li, N.: Modeling role-based access control using parameterized UML models. In: Wermelinger, M., Margaria-Steffen, T. (eds.) FASE 2004. LNCS, vol. 2984, pp. 180–193. Springer, Heidelberg (2004)
[6] Priebe, T., Fernandez, E., Mehlau, J., Pernul, G.: A Pattern System for Access Control. In: Proc. of Conf. on Data and Application Security, pp. 22–28
[7] Gamma, E., Helm, R., Johnson, R., Vlissides, J.: Design Patterns: Elements of Reusable Object-Oriented Software. Addison-Wesley, Reading
[8] Jürjens, J.: UMLsec: Extending UML for secure systems development. In: Jézéquel, J.-M., Hussmann, H., Cook, S. (eds.) UML 2002. LNCS, vol. 2460, pp. 412–425. Springer, Heidelberg (2002)
[9] Harrison, M., Ruzzo, W., Ullman, J.: Protection in Operating Systems. Communications of the ACM 19(8), 461–471 (1976)
[10] Lodderstedt, T., Basin, D., Doser, J.: SecureUML: A UML-based modeling language for model-driven security. In: Jézéquel, J.-M., Hussmann, H., Cook, S. (eds.) UML 2002. LNCS, vol. 2460, pp. 426–441. Springer, Heidelberg (2002)
[11] Ferraiolo, D.F., Sandhu, R., Gavrila, S., Richard Kuhn, D., Chandramouli, R.: Proposed NIST Standard for Role-Based Access Control, NIST. ACM Transactions on Information and System Security 4(3), 224–274 (2001)
[12] Fernandes, P., Werner, C., Teixeira, E.: An Approach for Feature Modeling of Context-Aware Software Product Line. Journal of Universal Computer Science 17(5), 807–829 (2011)
[13] Kim, D.-K., Lu, L., Kim, S.: A verifiable modeling approach to configurable role-based access control. In: Rosenblum, D.S., Taentzer, G. (eds.) FASE 2010. LNCS, vol. 6013, pp. 188–202. Springer, Heidelberg (2010)
[14] Zaid, L.A., Kleinermann, F., De Taylor, O.: Feature Assembly framework: Toward Scalable and reusable feature Models. In: VaMos 2011. ACM, New York (2011), doi:978-1-4503-0570-9/01/11

# Fault Diagnosis of Pneumatic Valve Using PCA and ANN Techniques

P. Subbaraj[1] and B. Kannapiran[2]

[1] Sri Nandhanam College of Engineering & Technology,
Tirupattur-635601, Vellore District, TamilNadu, India
[2] Department of Instrumentation & Control Engineering,
Arulmigu Kalasalingam college of Engineering,
Anand Nagar, Krishnankoil-626190
Srivilliputhur, Virudunagar District, Tamilnadu, India

**Abstract.** Detection and Diagnosis of faults in pneumatic valve used in cooler water spray system in cement industry is of great practical significance and paramount importance for the safe operation of the plant. In this paper the dimensionality reduction techniques such as principal component analysis (PCA) are used to reduce the input features is proposed. PCA is used to extract the primary features associated with the pneumatic valve used in cooler water spray system. The training and testing data required for the dimensionality reduction technique such as PCA model were created at normal and faulty conditions of pneumatic valve in a real time laboratory experimental setup. The performance of the developed PCA model is compared with the MLFFNN (Multilayer Feed Forward Neural Network) trained by the back propagation algorithm. From the simulation results it is observed that the performance of PCA had the best classification properties when it is compared with the performance of ANN.

**Keywords:** Fault diagnosis, Principal component analysis, Artificial Neural Network, Back propagation, Pneumatic valve.

## 1 Introduction

The cooler water spray system plays an important role in cement industry. This is an important task to reduce the temperature of the clinker at the output of the kiln. The valve is one of the most important key component used in cooler water spray system. Defects and malfunctions (simply called faults) of pneumatic valve will result in significant economic loss. There is a crucial need for checking and monitoring the condition of the pneumatic valve precisely since they are mostly subject to hazardous environments, such as severe shocks, vibration, heat, friction, dust, etc When a possible fault is detected in pneumatic valve, diagnosis is carried out to pinpoint the fault. That is to say, diagnosis is a process of locating the exact cause(s) of a fault.

Once the fault has been detected, the maintenance engineer is to identify the symptoms, analyze the symptomatic information, interpret the various error message and indications and come up with the right diagnosis of the situation in terms of

D. Nagamalai, E. Renault, M. Dhanushkodi (Eds.): CCSEIT 2011, CCIS 204, pp. 404–413, 2011.

which components may have caused the fault and the reason for the fault of the components. Since the valve has many components built inside the chamber is highly complex in nature. Fault diagnosis usually requires technical skill and experience. It also requires extensive understanding of the pneumatic valve structure and operation, and some general concepts of diagnosis. This requires an expert engineer to have a domain-specific knowledge of maintenance and know the 'inputs and-outputs' of the system [1]. In order to better equip with a non-expert to carry out the diagnosis operations, it would be wise to present an approach to define the cause–symptom relationship for quick comprehension and concise representation. Presently, many diagnosis methods have been proposed to help maintenance engineer to do diagnosis analysis. For example, expert system [2], neural network [3], soft-computing technology (fuzzy logic, rough set and genetic algorithm) [4, 5] and their integration method [6] are the popular approaches. Unfortunately, because of the bottleneck of knowledge acquisition, the application of expert system is limited. The drawback of these methods used for fault diagnosis is that the information used in the study is limited in scope and correlation of factors is neglected.

Many researchers have developed diagnostic methods by using ANNs to detect the problems of march-motors, electric motors [7], rotating machine parts [8], gears and gearboxes [9], Fisher Rosemount valves [10, 11], and compressors. But the neural network approach needs lot of data to develop the network before being put to use for real time applications. There are usually many irrelevant variables in the sampled data from real life fault diagnostic measurements. Irrelevant and redundant attributes in the input not only complicate the network structure, but also degrade the performance of the networks. By selecting only the relevant variables as input features and excluding irrelevant ones, higher performance is expected with smaller computational effort. Therefore, efficient methods for dimensionality reduction must be developed. A feasible option for solving this problem is to use feature extraction technique such as principle component analysis (PCA). PCA [12, 13] is a method based on the projection of the data into a low dimensional space that accurately characterises the state of the system. It produces a lower dimensional representation in a way that preserves the correlation structure between the process variables. This paper deals with the design and development artificial neural network based model for the fault detection in pneumatic valve used in cooler water spray system in cement industry, with PCA for reducing the input dimension.

The paper is organized as follows: in the next section, the description of the experimental system for this study is outlined. Section 3 describes the fault detection in pneumatic actuator. Sections 4 describe the development of ANN model for fault diagnosis. Sections 5 demonstrate the Principal component analysis and section 6 describe the result and discussion and finally, in Section 7, conclusions are drawn from the work.

## 2 System Description

In the cement industry the output of kiln is the clinker product which is of high temperature up to 1100 °C. It is necessary to reduce the temperature of the clinker by using cooler fans and the water spray system. The functions of the clinker cooler process are to maximize the heat recovery to the kiln process and to minimize the ultimate clinker temperature.

The measurement of clinker temperature is done by using thermocouple sensor. The direct measurement of clinker temperature is not possible on a continuous basis. Hence, the temperature of cooler vents gas temperature is measured. The sensor output is proportional to the temperature of the clinker. When the temperature of the clinker product at the outlet of kiln is up to $1100\,^{\circ}C$, then the temperature is reduced to $750\,^{\circ}C$ at the grate cooler section by using external blower fans. It is necessary to reduce the temperature of the hot gas up to $250\,^{\circ}C$ before given to the ESP (Electrostatic Precipitator) section. If the temperature of the hot gas exceeds beyond $250^{\circ}C$, then the anode and cathode plates of ESP gets damaged.

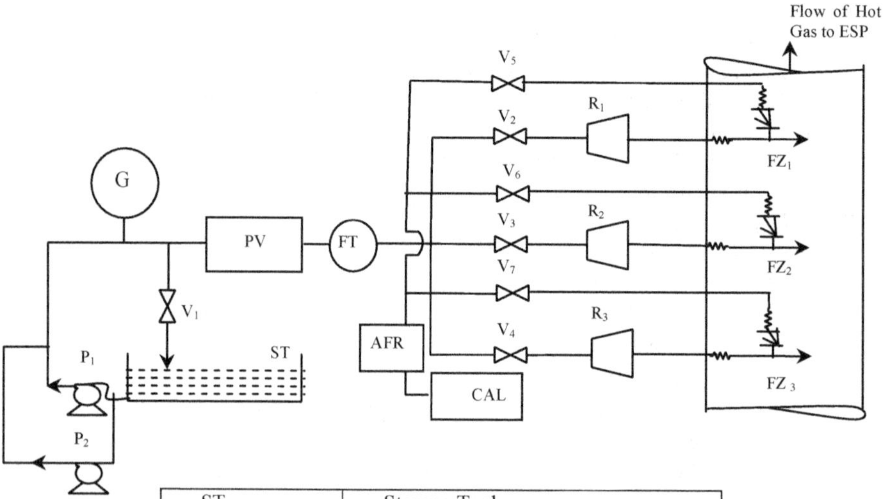


| | |
|---|---|
| ST | Storage Tank |
| $V_1$ | Return Valve |
| $V_2, V_3, V_4$ | Delivery Valve for water flow |
| $V_5, V_6, V_7$ | Delivery Valve for air flow |
| AFR | Air Filter Regulator |
| CAL | Compressed Air line from Plant |
| $P_1$-$P_2$ | Pump |
| $R_1$-$R_3$ | Flow Reducer |
| PV | Pneumatic Valve |
| FT | Flow Transmitter |
| $FZ_1$-$FZ_3$ | Flow Nozzle Banks(1-3) |
| G | Pressure gauge |
| ST | Storage Tank |

**Fig. 1.** Schematic layout of the cooler water sprays system

In order to prevent this damage to the ESP, cooler water spray system is used in cement industry. The schematic layout of the cooler water spray system setup is shown in Figure 1. The hot air recovered from this cooling process is recycled back into the kiln or preheated system to maintain the thermal energy of the hot gas. The type of valve used in the water spray system is automatic equipment designed to regulate the flow rate in a pipe system. Then the water is injected through the pneumatic valve using pump and the water reaches the flow nozzle for water spray process inside the hot gas flow duct line. The description about this system is well explained in [14]. Next section deals with fault detection in pneumatic actuator in cement industry.

## 3 Fault Detection in Pneumatic Actuator

A system that includes the capability of detecting and diagnosing faults is called the 'fault diagnosis system'. Fault detection and diagnosis are important tasks in cement industry. It deals with the timely detection, diagnosis and correction of abnormal condition of faults in the plant.

**Table 1.** List of various types of faults

| Types of fault | Name of the fault | Symbols | Name of the fault | Symbols |
|---|---|---|---|---|
| **Control valve faults** | valve clogging | F1 | external leakage ( bushing, covers, terminals) | F5 |
| | valve or valve seat erosion | F2 | internal leakage (valve tightness) | F6 |
| | valve or valve seat sedimentation | F3 | medium evaporation or critical flow | F7 |
| | increased of valve or bushing friction | F4 | - | - |
| **Pneumatic servo-motor faults** | twisted servo-motor's piston rod | F8 | servo-motor's diaphragm perforation | F10 |
| | servo-motor's housing or terminals tightness | F9 | servo-motor's spring fault | F11 |
| **Positioner faults** | electro-pneumatic transducer fault (E/P) | F12 | pressure sensor fault (PT) | F14 |
| | rod displacement sensor fault (DT) | F13 | | |
| **General faults/external faults** | Positioner supply pressure drop | F15 | fully or partly opened bypass valves | F18 |
| | increase of pressure on valve inlet or pressure drop on valve output | F16 | flow rate sensor fault (FT) | F19 |
| | pressure drop on valve at inlet or increase of pressure on valve output | F17 | - | - |

Early detection and diagnosis of plants while the plant is still operating in a controllable region can help avoid abnormal event progression and reduce productivity loss. In order to accomplish these tasks, information that reflects a change from the normal behaviour of the process has to be obtained. This is generally called symptom generation. Any method of fault diagnosis must characterize how abnormal symptoms (measurements) are related to faults (malfunctions). The major categories of faults in the pneumatic valve include Control valve faults, Pneumatic servo-motor faults, Positioner faults, General faults / External faults. The individual fault of each category along with their symbol is listed in Table 1. The sources of each fault are given in [15]. A complete description of the faults and the way they affect the valve can be found in [16].

## 4 Development of ANN Models for Fault Diagnosis

The proposed methodology for fault detection in pneumatic actuator is based on using back propagation for detecting the normal and abnormal conditions of the given parameters, which leads to various faults. The main purpose of selecting ANN as a tool is inability to form a mathematical relationship due to the nonlinearity between the inputs and the outputs, good generalization ability, fast real time operation and to perform the complicated mapping without functional relationship.

The following issues are to be addressed while developing the neural network based model for fault detection in pneumatic actuator in cooler water spray system.

### 4.1 Selection of Input and Output Variables

For the application machine learning approaches, it is important to properly select the input variables, as ANN's are supposed to learn the relationships between input and output variables on the basis of input-output pairs provided during training. In neural network based fault detection model, the input variables represent the operating state of the pneumatic actuator, and the output is the type of fault present in the valve.

### 4.2 Training Data Generation

The generation of training data is an important step in the development of ANN models. To achieve a good performance of the neural network, the training data should represent the complete range of operating conditions of the pneumatic actuator which contains all possible fault occurrences.

### 4.3 Data Normalization

If the generated data are directly fed to the network as training patterns, higher valued input variables may tend to suppress the influence of smaller ones. Data normalization compresses the range of training data between 0 and 1 or -1 to +1 depending on the type of transfer function. The input and output data are normalized using the expression,

$$X_n = (X - X_{\min})/(X_{\max} - X_{\min}) + \text{starting value} \quad (1)$$

Where $X_n$ is the normalized value of the data and $X_{min}$ and $X_{max}$ are the minimum and maximum values among all the values of the data.

## 5 Principal Component Analysis

Dimensionality reduction can be achieved either by eliminating data closely related with other data in the set, or combining data to make a smaller set of features. The feature extraction techniques used in this study are principal components analysis. Using the example of projecting data from two dimensions to one, a linear projection requires the optimum choice of projection to be a minimization of the sum-of-squares error. This is obtained first by subtracting the mean $\bar{x}$ of the data set. The covariance matrix is calculated and its eigenvectors and eigen-values are found. The eigenvectors corresponding to the M largest eigen-values are retained, and the input vectors $x^n$ are subsequently projected onto the eigenvectors to give components of the transformed vectors $z^n$ in the M-dimensional space. Retaining a subset M < d of the basis vectors $\mu_i$ so that only M coefficients $z_i$ are used allows for replacement of the remaining coefficients by constants $b_i$. This allows each x vector to be approximated by an expression of the form:

$$\bar{x} = \sum_{i-1}^{m} z_i \mu_i + \sum_{1-m+1}^{d} b_i \mu_i \tag{2}$$

Where $\mu_i$ represents a linear combination of d orthonormal vectors

## 6 Results and Discussion

This section presents the development and testing of MLFFNN and PCA models for fault detection on Pneumatic valve used in cooler water spray system. Using real time simulation on laboratory experimental setup of pneumatic valve in cooler water spray system, the required data was generated.

The parameters are collected directly from the real time system which is interfaced with personal computer using Data Acquisition (DAQ) card. The Photograph of the pneumatic actuator system interfaced with PC using DAQ is shown in Figure 2.The observations are made for normal operating condition and the procedure for observing the parameters are repeated for different faults.

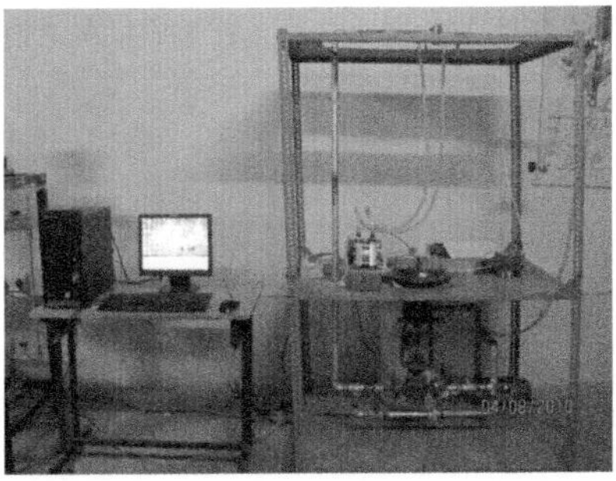

**Fig. 2.** Shows photograph of the pneumatic actuator system interfaced with PC using DAQ

## 6.1 Effect of Artificial Neural Network

The data contains 5 input features, which is given in Table 2 and one output that is labeled as either normal or as a fault, with exactly one specific fault. The total number of data generated is 1000, which contain 25% normal patterns and 75% of patterns with faults belonging to the various types of faults listed in Table 1. Among them, 750 patterns are used for training and 250 patterns are used for testing. The testing data comprises of both normal and abnormal (faulty) data, which are totally different from the training data.

**Table 2.** Name of the input features in the experimental system

| Input code | Feature Name |
|---|---|
| 1. | Differential Pressure Transmitter output(mA) |
| 2. | Temperature of water form RTD output (outlet) |
| 3. | Pressure gauge reading(Inlet Pressure) |
| 4. | Potentiometer (Stem movement) |
| 5. | Input and output power of the pump |

**Table 3.** Network performance for each fault category of control valve faults

| Type of fault | Name of the Fault | Symbol | Training pattern | Testing pattern | Efficiency |
|---|---|---|---|---|---|
| Control valve faults | Valve Clogging | F1 | 75 | 25 | 100 |
| | Valve seat erosion | F2 | 75 | 25 | 97.6 |
| | Valve seat sedimentation | F3 | 75 | 25 | 100 |
| | Bushing friction | F4 | 75 | 25 | 99.4 |
| | External leakage | F5 | 75 | 25 | 99.1 |
| | Internal leakage | F6 | 75 | 25 | 99.3 |
| | Critical flow | F7 | 75 | 25 | 99 |
| Considering all fault class in control Valve faults(total classification accuracy) | | | 75 | 25 | 98.6 |

**Table 4.** Various parameters of the ANN model

| **Description of parameters** | **values** |
|---|---|
| Number of training patterns | 75% |
| Number of testing patterns | 25% |
| Transfer function used | Logarithmic sigmoidal, Linear |
| Training Time | 12.7344 seconds |
| Mean Square Error in Training | 0.0018 |
| Mean Square Error in Testing | 0.0017 |
| Percentage of fault classification | 98.6% |

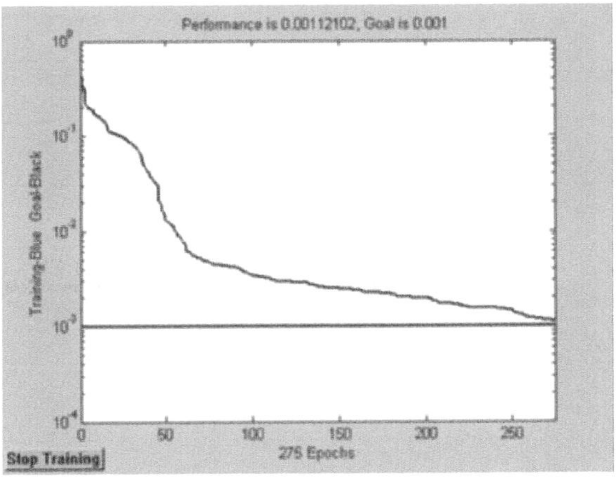


**Fig. 3.** Training Performance of the Network

After training, the generalization performance of the network is evaluated with the 250 test data that contain the combination of both normal as well as all types of fault categories. The trained neural network classified 246 data correctly, which shows an overall detection rate 98.6%. Table 3 shows the network performance for each fault category of control valve faults. The network is trained with least mean square algorithm until it reaches the mean square error of 0.001. The mean square error achieved during training is 0.0018. During testing, the mean square error achieved by the network is 0.0017. With 9×10 hidden nodes, the network took 12.7344 s to reach the error goal. This is shown in Table 4 it is found that the network has correctly classified all the data during the testing stage. This shows that the trained ANN is able to produce the correct output even for the new input.

The neural network model was trained using the back propagation algorithm with the help of MATLAB neural network toolbox. At the end of training process, the model obtained consists of the optimal weight and the bias vector. The performance of the ANN during training is shown in Figure 3.

## 6.2 Effect of Feature Extraction

The performance comparison between the various parameters of ANN and PCA model is shown in Table 5. On comparing the performance of without dimensionality reduction, it is found that the mean square error in testing and training of using PCA is incredibly low when compared with ANN and the accuracy of fault classification is high in using PCA when compared with ANN. The classification rate using the original parameters is 98.6%. The classification rate with feature extraction through PCA is 100%.

Also, the obtained results proved that the network with PCA technique performed better than ANN without PCA. The reduced performance of the classifier without feature extraction is due to the existence of irrelevant and redundant feature parameters existing in the data set. The irrelevant features burden the classifier and tend to decrease its performance. By using PCA the useful features are extracted from the original input variables.

**Table 5.** Performance comparison of ANN model with and without PCA

| Description of parameters | Without dimensionality reduction | With dimensionality reduction |
|---|---|---|
| Number of training patterns | 75% | 75% |
| Number of testing patterns | 25% | 25% |
| Functions | logarithmicsigmoidal, Linear | logarithmicsigmoidal, Linear |
| MSE in Training | 0.0018 | 9.9831 e-004 |
| MSE in Testing | 0.0017 | 8.1919 e-004 |
| Training Time | 12.7344 seconds | 11.7188 |
| Percentage of fault classified | 98.6 % | 100 |

## 7 Conclusion

This paper has presented a neural network based approach with principal component analysis based feature extraction for fault diagnosis of pneumatic actuator used in cooler water spray system in cement industry. The data required for the development of neural network model have been obtained through the laboratory grade experimental setup. Totally 19 faults from the pneumatic actuator were considered in the developed model. A key issue in neural network based approach is identifying a representative set of features from which to develop the network for a particular task. This problem has been addressed by feature extraction through principal component analysis. Simulation results show that the PCA based approach is very much effective in identifying the important key features of the network. Based on the results obtained, it is observed that the performance of the neural network model has been significantly improved by reducing the input dimension. The effectiveness of the proposed method has been demonstrated through different fault condition in pneumatic actuator system.

## References


1. Patel, S.A., Kamrani, A.K.: Intelligent decision support system for diagnosis and maintenance of automated systems. Computers Industrial Engineering 30(2), 297–319 (1996)
2. Yang, B.S., Lim, D.S., Tan, C.C.: VIBEX: an expert system for vibration fault diagnosis of rotating machinery using decision tree and decision table. Expert System with Application 28, 735–742 (2005)
3. Chen, C.Z., Mo, C.T.: A method for intelligent fault diagnosis of rotating machinery. Digital Signal Processing 14, 203–217 (2004)
4. Gayme, D., Menon, S., Ball, C., et al.: Fault detection and diagnosis in turbine engines using fuzzy logic. In: 22nd International Conference of Digital Object Identifier, vol. 24-26, pp. 341–346. North American Fuzzy Information Processing Society (2003)
5. Wang, Y.G., Liu, B., Guo, Z.B., et al.: Application of rough set neural network in fault diagnosing of test-launching control system of missiles. In: Proceedings of the Fifth World Congress on Intelligent Control and Automation, Hangzhou, PR, China, pp. 1790–1792 (2004)

6. Xiang, X.J., Shan, J.F., Fuchs, F.W.: The study of transformer failure diagnose expert system based on rough set theory. In: The Fourth International IPEMC Power Electronics and Motion Control Conference, vol. 2, pp. 533–536 (2004)
7. Bayir, R., Bay, O.F.: Serial wound starter motor faults diagnosis using artificial neural network. In: IEEE ICM 2004, Istanbul, pp. 194–199 (2004)
8. Rajakarunakaran, S., Venkumar, P., Devaraj, D., Rao, K.S.P.: Artificial neural network approach for fault detection in rotary system. Applied Soft Computing 8, 740–748 (2008)
9. Chen, D., Wang, J.W.: Classification of wavelet map patterns using multilayer neural networks for gear fault detection. Mechanical Systems and Signal Processing 16, 695–704 (2000)
10. Karpenko, M., Sepehri, N.: Neural network classifiers applied to condition monitoring of a pneumatic process valve actuator. Engineering Applications of Artificial Intelligence 15(3-4), 273–283 (2002)
11. Karpenko, M., Sepehri, N., Scuse, D.: Diagnosis of process valve actuator faults using a multilayer neural network. Control Engineering Practice 11(11), 1289–1299 (2003)
12. Chang, Y.-W., Wang, Y.-C., Liu, T., Wang, Z.-J.: Fault diagnosis of a mine hoist using PCA and SVM techniques. China Univ. Mining and Technology (18), 327–331 (2008)
13. de Moura, E.P., Souto, C.R., Silva, A.A., Irmao, M.A.S.: Evaluation of principal component analysis and neural network performance for bearing fault diagnosis from vibration signal processed by RS and DF analyses. Journal of Mechanical Systems and Signal Processing (2010)
14. Subbaraj, P., Kannapiran, B.: Artificial Neural Network Approach for Fault Detection in Pneumatic Valve in Cooler Water Spray System. Int. Journal of Comp. Applications 9(7), 0975–8887 (2010)
15. Koj, J.: The Fault Sources of Pneumatic Servo-Motor-Control Valve Assembly. In: Proceedings of the III Polish National Conference on Diagnosis of Industrial Processes, Jurata, Poland, pp. 415–419 (1998)
16. Louro, R.: Fault Diagnosis of an Industrial Actuator Valve. MSc Dissertation, Technical University of Lisbon, Portugal (2003)

# A Generalized Classification Approach for Indian Sign Language Datasets Using Instance Based Classifiers

M. Krishnaveni and V. Radha

Department of Computer Science,
Avinashilingam Deemed University for Women,
Tamilnadu, India
krishnaveni.rd@gmail.com
radharesearch@yahoo.com

**Abstract.** Sign language is the primary means of communication in the deaf community. The importance of sign language is emphasized by growing research activities and various funding agencies which deal with the technical aspects of deaf people's communication needs. Signing is primarily done with the hands. In technical terms the hands are used to describe the so-called manual parameters, which are pose, posture, position and motion. General view based sign language recognition systems extract these parameters by a single camera view because it seems to be user friendly and hardware complexity; however it needs a high accuracy classifier for classification and recognition purpose. The decision making of the system in this work employs statistical classifiers namely Navie bayes and K-NN to recognize the sign language isolated signs. Indian sign language datasets is used for the training and performance evaluation of the system.It compares the ability to deploy the K-NN and the Naïve Bayesian classifiers in solving the sign classification problem. The impact of such study may reflect the exploration for using such algorithms in other similar applications such as text classification and the development of automated systems.

**Keywords:** Sign Language, boundary based, K-NN classifier, Navie Bayes, classification accuracy.

## 1 Introduction

Sign language is important in humankind that is showing an increasing research interest in eradicating barriers faced by differently abled people in communicating and contributing to the society [1]. Automation over sign language recognition systems can greatly facilitate the vocal and the non-vocal communities which can be equivalently best and successive as speech-recognition systems [2]. Despite common misconceptions, sign languages are complete natural languages, with their own syntax and grammar. However, sign languages are not universal. As is the case in spoken language, every country has got its own sign language with high degree of grammatical variations. The sign language used in India is commonly known as Indian Sign Language (henceforth called ISL)[6].Linguistic studies on ISL were

D. Nagamalai, E. Renault, M. Dhanushkodi (Eds.): CCSEIT 2011, CCIS 204, pp. 414–422, 2011.

started around 1978 and it has been found that ISL is a complete natural language, instigated in India, having its own morphology, phonology, syntax, and grammar. ISL is not only used by the deaf people but also by the hearing parents of the deaf children, the hearing children of deaf adults and hearing deaf educators. Therefore the need to build a automation system that can associate signs to the words of spoken language, and which can further be used to learn ISL, is significant. This paper proposes a system for recognizing Indian sign language using the view based approach and the statistical classifier algorithm as a learning tool. These classifiers take the advantage of accuracy, ease of implementation and operational robustness. The paper is arranged as follows. Section 2 gives a brief overview of Indian sign language. Section 3 describes the feature extraction methodology, by using boundary detection techniques. Section 4 deals with the short description of classifiers used. Section 5 gives the performance evaluation of the classifiers using ISL datasets. Finally, section 6 summarizes the framework and future work that can be adopted.

## 2 Overview of Indian Sign Language

Indian Sign languages are rich, faceted language, and their full complexity is beyond current gesture recognition technologies[6]. The interpersonal communication problem between signer and hearing community could be resolved by building up a new communication bridge integrating components for sign[5]. Several Significant problems specific to Automatic Sign Recognition are i) Distinguishing gestures from signs ii) Context dependency (directional, verbs, inflections, etc) iii) Basic unit modeling (how do we describe them?) iii) Transitions between signs (Movements)

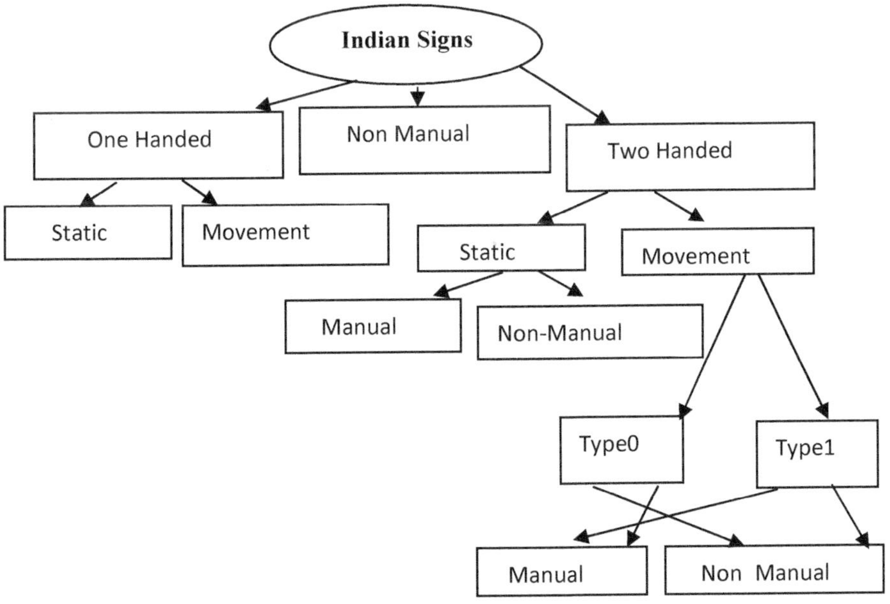


**Fig. 1.** Framework for ISL Dictionary

iv) Repetition (Cycles of movement). The ISL dictionary is been build by Sri Ramakrishna mission Vidyalaya College of Education, Coimbatore which as split the ISL into five parameters namely handshapes, locations, Orientation, Location, Movements and Facial Expression. Figure 1 depicts the framework for ISL dictionary.

## 3 Feature Extraction Methodology

The great variability in gestures and signs, both in time, size, and position, as well as interpersonal differences, makes the recognition task difficult [11]. By extracting features from image processing sequence classification can be done by a discriminative classifier. Gestures, particularly in sign language, involve significant motion of the hands[5]. Thus, in developing a sign language recognition system, it is important to model both the motion (temporal characteristics) and shape (spatial characteristics) of the hand. In this research work only the spatial characteristics of the hand are of concern. Figure 2 shows the schematic view of gesture recognition process.

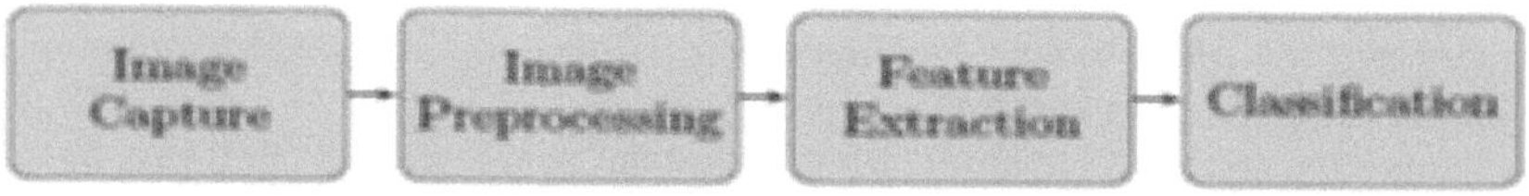


**Fig. 2.** Schematic view of gesture recognition process

Feature extraction is the process of generating a set of descriptors or characteristic attributes from a binary image. Most of the features used in existing sign language recognition systems focus only on one aspect of the signing like hand movements or facial expressions. Figure 3 shows the feature extraction concept.

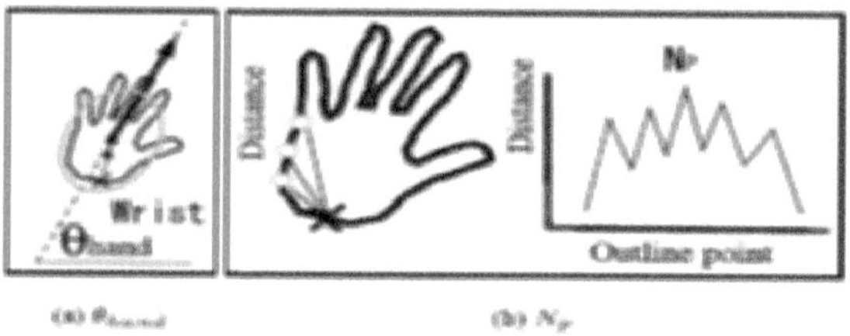


**Fig. 3.** Hand feature extraction

The strategy used for segmentation is boundary segmentation method which traces the exterior boundaries of objects, as well as boundaries of holes inside these objects, in the binary image [3]. The two important need for this method of sign extraction is First, it keeps the overall complexity of the segmentation process low. Secondly, it eliminates candidates that may produce smooth continuations, but otherwise are inconsistent with the segments in the image. This simplifies the decisions made by the segmentation process and generally leads to more accurate segmentation[7]. The feature extracted from the database used is mean intensity, area, perimeter, diameter, centroid.

**Mean Intensity:** The mean intensity $\mu$ in the selected region of interest (ROI) is given in eqn (1):

$$\mu = \frac{1}{N} \int_{x,y} ROI(x, y)\,dx\,dy \tag{1}$$

**Area:** The area of an object is a relatively noise-immune measure of object size, because every pixel in the object contributes towards the measurement. The area can be defined using eqn (2)

$$A(S) = \int\int I(x, y)\,dy\,dx \tag{2}$$

Where, I(x,y) = 1 if the pixel is within a shape,(x,y)∈ S,
0 otherwise.

**Perimeter:** The perimeter measurement gives the smooth relative boundaries and the perimeter of the region is defined as by the eqn (3)

$$P(S) = \int_t \sqrt{x^2(t) \quad + \quad y^2(t)dt} \tag{3}$$

**Diameter:** The distance around a selected region is called the circumference. The distance across a circle through the center is called the diameter. $\pi$ is the radius of the circumference of a circle.. Thus, for any circle, if you divide the circumference by the diameter, you get a value close to $\pi$. This relationship is expressed in the following eqn (4):

$$\frac{C}{d} = \pi \tag{4}$$

Where, C is circumference and d is diameter.

**Centroid:** It specifies the center of mass of the region. Centroid is the horizontal coordinate (or *x*-coordinate) of the center of mass, and the second element is the vertical coordinate (or *y*-coordinate) and it is written as given in eqn (5)

$$C = \frac{\int xg(x)dx}{\int g(x)dx} \tag{5}$$

## 4 Short Description on Classifiers

Classification (generalization) using an instance-based classifier can be a simple matter of locating the nearest neighbour in *instance space* and labelling the unknown instance with the same class label as that of the located (known) neighbour [4]. This approach is often referred to as a neighbour classifier. Classifier always tries to improve the classification rate by pushing classifiers into an optimised structure[10].

### 4.1 K-NN Classifier

The K-Nearest Neighbor classifier is a supervised learning algorithm where the result of a new instance query is classified based on majority of the K-nearest neighbor category. The purpose of this algorithm is to classify a new object based on attributes and training samples[8]. The Nearest neighbor classifier is based on learning by analogy. The training samples are described by n-dimensional numeric attributes. Each sample represents a point in an n-dimensional pattern space. More robust models can be achieved by locating *k*, where $k > 1$, neighbours and letting the majority vote decide the outcome of the class labelling. A higher value of *k* results in a smoother, less locally sensitive, function. The Nearest Neighbor classifier can be regarded as a special case of the more general K-Nearest Neighbors classifier, hereafter referred to as a K-NN classifier.

The data for KNN algorithm consists of several multivariate attributes names $X_i$ that will be used to classify the object Y. The data of KNN can have any measurement scale from ordinal, nominal, to quantitative scale, this study deals only with quantitative $X_i$ and binary (nominal) Y. All training samples are included as nearest neighbors if the distance of this training sample to the query is less than or equal to the $K^{th}$ smallest distance. In other words, the distances are sorted of all training samples to the query and determine the $K^{th}$ minimum distance. The unknown sample is assigned the most common class among its K-Nearest Neighbors. These K training samples are the closest k nearest neighbors for the unknown sample. Closeness is defined in terms of Euclidean distance, where the Euclidean between two points, $X = (x_1, x_2, \ldots x_n)$ and $Y= (y_1, y_2, \ldots y_n)$ is given in eqn (6)

$$d(X,Y) = \sqrt{\sum_{i=1}^{n}(x_i - y_i)^2} \tag{6}$$

The drawback of increasing the value of *k* is when *k* approaches *n*, where *n* is the size of the instance base, the performance of the classifier will approach that of the most straightforward statistical baseline, the assumption that all unknown instances belong to the class most frequently represented in the training data. The high degree of local sensitivity makes nearest neighbor classifiers highly susceptible to noise in the training data.

### 4.2 Naive Bayesian Classifier

A Naive Bayes classifier assigns a new observation to the most probable class, assuming the features are conditionally independent given the class value [9]. It can outperform more sophisticated classification methods by categorizing incoming objects to their appropriate class. The Naive Bayesian classifiers can handle a random number of independent variables whether continuous or categorical.

It classifies data in two steps:

1. Training step: Using the training samples, the method estimates the parameters of a probability distribution, assuming features are conditionally independent given the class.
2. Prediction step: For any unseen test sample, the method computes the posterior probability of that sample belonging to each class. The method then classifies the test sample according the largest posterior probability.

Bayes theorem used, takes the eqn as given in (7) and (8)

$$P(H|X) = P(X|H)P(H)/P(X) \tag{7}$$

It can also be expressed as

$$P(C_i|X) = \frac{P(X|C_i)P(C_i)}{P(X)} \tag{8}$$

Where P(X) is constant for all classes, only $P(X|C_i)P(C_i)$ need be maximized.

The class-conditional independence assumption greatly simplifies the training step since estimation can be done using one-dimensional class-conditional density for each feature individually. This assumption of class independence allows the Naive Bayes classifier to better estimate the parameters required for accurate classification while using less training data than many other classifiers. This makes it particularly effective for datasets containing many predictors or features. Since the build process for Naive Bayes is parallelized it can be used for both binary and multiclass classification problems.

## 5 Performance Evaluation

The proposed classifier Navie bayes is used to justify the objects using new methods to get a maximum accuracy comparing to K-NN classifier. The features are the parameters extracted from the sign images which are taken from normal camera. In each hand posture, a measure of properties like area, mean intensity, centroid, perimeter and diameter are taken; the classifier then uses these properties to determine the sign in different angles. They estimate the probability that a sign belongs to each of the target classes that is fixed. In the training phase, the training set is used to decide how the parameters must be weighted and combined in order to separate the various classes of signs. The classifier used for classification is validated using sensitivity, specificity error rate, predictive value, likelihood value, plotting the classification and misclassification error rate according to the sample datasets.

To keep track of the performance during the validation of classifiers Table 1 mentioned values are the measures taken into consideration. Conducted experiments with Indian sign language datasets shows that the Navie bayes classifier has better performance than K-NN and the current implementation for both classifiers yield good results when compared to other implementations by other researchers.

**Table 1.** Performance evaluation of NB and KNN Classifiers

| S.No | Performance Measures | NB Performance | KNN Performance |
|---|---|---|---|
| 1. | Classified rate | 1.0000 | 0.9500 |
| 2. | Sensitivity | 0.7143 | 0.5714 |
| 3. | Specificity | 0.8571 | 0.6857 |
| 4. | Error rate | 0.3429 | 0.6857 |
| 5. | Inconclusive rate | 0 | 0 |
| 6. | Positive predictive value | 0.5556 | 0.3077 |
| 7. | Negative Predictive Value | 0.9231 | 0.8636 |
| 8. | Negative likelihood | 0.3333 | 0.6316 |
| 9. | Positive Likelihood | 5.000 | 1.7778 |
| 10. | Prevalence | 0.2000 | 0.2000 |

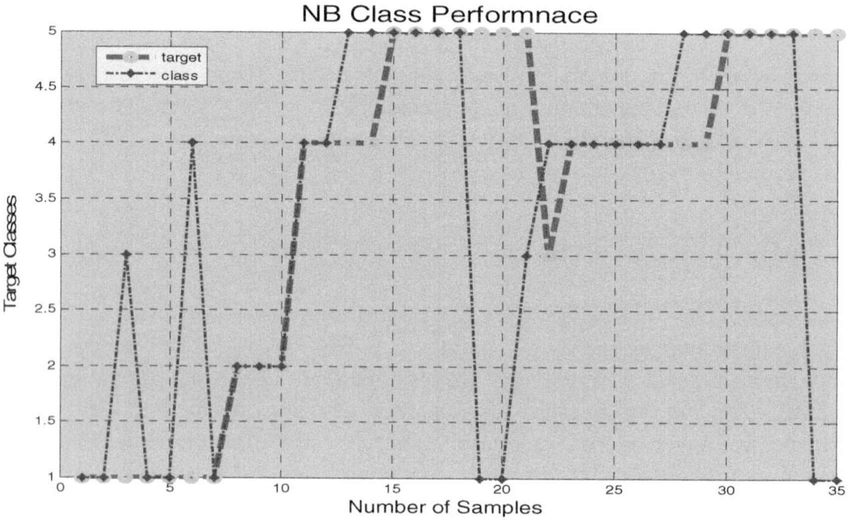


**Fig. 4.** Navie Bayes class Performance

Figure 4 and 5 depicts the classifier performance based on the datasets taken and the target class is less achieved by K-NN classifier than the proposed Navie bayes classifier.

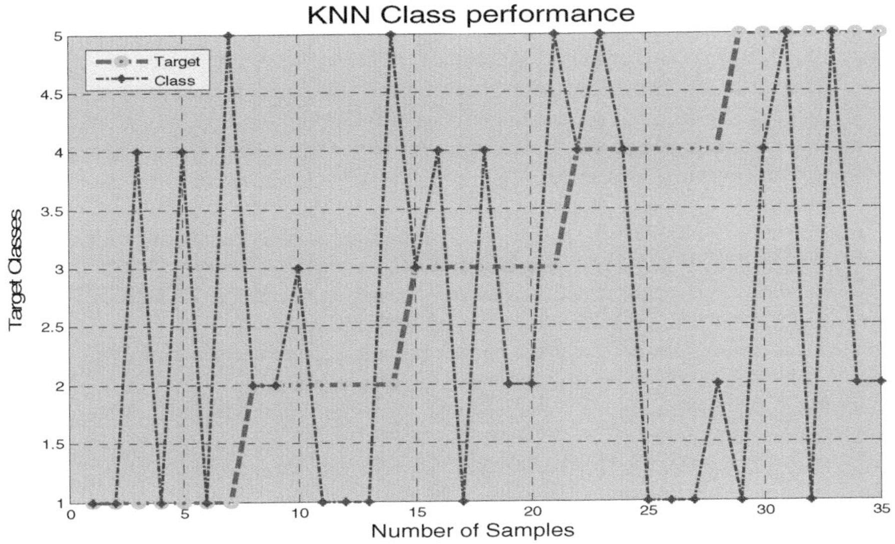


**Fig. 5.** KNN class Performance

# 6 Conclusion

The classifier methods described in this paper cover a large variety of practical applications. This help in understanding the current state of knowledge in Indian sign language research area. Additionally attempts are made to identify the main directions leading towards the standardised procedure of classifier system design. However, the results in this work are biased by the size of the database: on the one hand, the lack of training data and the large amount of singletons leads to a very difficult task. The methods used in this work are focused towards good recognition accuracy and not toward real-time performance and thus it meets the sample experimental requirements. This paper is also a contribution to the research in Indian Sign Language, which had very limited research in its automated recognition.

**Acknowledgments.** The authors thank RamaKrishna Mission Vidyalaya, International Human Resource Development Centre (IHRDC for the disabled) for providing Indian sign language dictionary.

# References

1. Phu, J.J., Tay, Y.H.: Computer Vision Based Hand Gesture Recognition Using Artificial Neural Network. Faculty of Information and Communication Technology. Universitiy Tunku Abdul Rahman (UTAR), MALAYSIA. tensaix2j@yahoo.com, tayyh@mail.utar.edu.my
2. Salleh, N.S.M., Jais, J., Mazalan, L., Ismail, R., Yussof, S., Ahmad, A., Anuar, A., Mohamad, D.: Sign Language to Voice Recognition: Hand Detection Techniques for Vision-Based Approach. Current Developments in Technology-Assisted Education (2006)

3. Kovář, J.P., Vlček, M.: Still Image Objective Segmentation Evaluation using Ground Truth. In: 5th COST 276 Workshop, pp. 9–14 (2003)
4. Elmezain, M., Al-Hamadi, A., Appenrodt, J., Michaelis, B.: A Hidden Markov Model-Based Isolated and Meaningful Hand Gesture Recognition. International Journal of Computer Systems Science and Engineering 5(2) (2009)
5. Hassanpour, R., Wong, S., Shahbahrami, A.: VisionBased Hand Gesture Recognition for Human Computer Interaction: A Review. In: IADIS International Conference Interfaces and Human Computer Interaction (2008)
6. Dasgupta, T., Shukla, S., Kumar, S.: A Multilingual Multimedia Indian Sign language Dictionary Tool. In: The 6th Workshop on Asian Language Resources (2008)
7. Haralick, R.M., Shapiro, L.G.: Survey: Image Segmentation Techniques. Computer Vision, Graphics, and Image Processing 29, 100–132 (1985)
8. Laaksonen, J., Oja, E.: Classification with learning k-nearest neighbors. In: Proceedings of ICNN 1996, USA, pp. 1480–1483 (June 1996)
9. Ruta, D., Gabrys, B.: An Overview of Classifier Fusion Methods. Computing and Information Systems, 1–10 (2000)
10. Danesh, A., Moshiri, B., Fatemi, O.: Improve text classification accuracy based on classifier fusion methods. In: Proceeding of the 10th International Conference on Information Fusion, July 9-12, pp. 1–6. IEEE Computer Society, USA (2007), doi:10.1109/ICIF.2007.4408196.
11. Xu, L., Krzyzak, A., Suen, C.Y.: Methods of Combining Multiple Classifiers and Their Applications to Handwriting Recognition. IEEE Transactions on Systems, Man, and Cybernetics 22, pt. 3, 418–435 (1992)

# Agent Based Adaptive Learning System

R. Gowri[1], S. Kanmani[2], and T.T. Sampath Kumar[3]

[1] Sri Manakula Vinayagar Engineering College,
Pondicherry India
gowrithirugnanam@gmail.com
[2] Pondicherry Engineering College Pondicherry
kanmani@pec.edu
[3] Sri Manakula Vinayagar Engineering College
samsvce@gmail.com

**Abstract.** This paper proposes the agent based intelligent, adaptive learning or tutoring system (IATS). The proposed architecture of the agent based system provides the framework for the intelligent tutoring system. We define many agents like tutoring agent, student agent, student clustering agent, resource classification agent. The system model is developed using the O-MaSE development model. The clustering is done by the combination of Evolutionary clustering methods and fuzzy c-means clustering. The classification of the resources is representation of the resources in the tree based hierarchical structure from which the adaptive resources are generated. The system provides the adaptive course materials, adaptive user interface, and adaptive examinations to the learner.

**Keywords:** Agent, Intelligent tutoring system, adaptive course content, Jade, fuzzy c-means clustering.

## 1 Introduction

Due to the increasing population of the Internet users, distance education has become a popular choice for various areas. Learning is no longer restricted to specific time slices or locations. In the e-learning environment, the design of learning materials requires more considerations for individual learners. The traditional arrangement of fixed content and learning sequence for all learners may not be suitable for all learners in the distributed environment. Intelligent tutoring systems mostly emphasize the importance of adaptive learning process. Intelligent tutoring systems are the outgrowth of the earlier computer-aided instruction models. Most of Intelligent tutoring systems are expected to be domain independent and task independent as well. Among various categories of intelligent tutoring systems, the mostly known function is to monitor students' learning process and provide suitable learning path.

Because of the adaptive characteristics of intelligent agents, researchers have tried to apply the capability of agents in the world of intelligent tutoring systems. Multi-agent systems (MAS) [1] [3] [10] bring several advantages to the implementation of

D. Nagamalai, E. Renault, M. Dhanushkodi (Eds.): CCSEIT 2011, CCIS 204, pp. 423–431, 2011.

intelligent tutoring systems since they deal well with distribution, cooperation, and integration of computer software components. The interaction protocol of the agent community would be another important issue to be considered.

Clustering is the process of grouping a data set in a way that the similarity between data within a cluster is maximized while the similarity between data of different clusters is minimized. Clustering is used for pattern recognition in image processing, and usually requires a high volume of computation. This high volume computation requires considerable amount of memory which may lead to frequent disk access, making the process inefficient. With the development of affordable high performance parallel systems, parallel algorithms may be utilized to improve performance and efficiency of such tasks. Several Clustering Algorithms are proposed for the Clustering of students [12-14] with several learner's behaviors and parameters.

## 2 Intelligent Tutoring Systems

The traditional intelligent tutoring systems model contains four components, including the domain model, the student model, the teaching model, and a learning environment or user interface. The domain model is also called expert model, which is used to determine the learning status of the student. The teaching (instructional) model is the central part of the architecture and interacts with all other models. The student model contains all learning related information used by the expert model. Learning path planning is one of the most important tasks in the intelligent tutoring systems. Y. Caseau [2] defines the routing strategies for message handling that maximize the ability of the EAI system to meet these requirements, in a self-adaptive and self-healing manner,

### 2.1 Agent-Based Solutions for ITS

In this research we propose a generic framework for the Tutoring System. We make this tutoring system as Intelligent Tutoring system by giving the adaptive property. This framework deals with the level of adaptation of the study materials to the different categories of the agents. The level of adaptation depends on the learner behavior, likeliness, the degree of learning and different learning parameters. The adaptation extends to the adaptation of the user interface for the specific learners. We define different types of agents, roles and their goals. Each agent performs a specific action based on the external parameters.

## 3 Proposed Work

In our approach the intelligent Tutoring system is modeled using the O-MaSE [11] model which specifies the agent oriented software engineering methodology. We also define agents, their roles and behaviors. These agents are independent agents; they can act independently in accordance with the environment and the system characteristics. The Similarity between the several learners are identified by the clustering of the students and analyzing their behaviors. The classification of the learning resources is done using the tree based hierarchical classification and present the resources accordingly to the student's learning capability and student's behavior. The implementation of the agent based ITS is done using the JADE [6] platform.

## 4 Agent Oriented Software Engineering

One of the major problems faced by agent-oriented software engineering is the failure to achieve a strong industry acceptance. One of the reasons hindering this acceptance is a lack of an accepted process-oriented methodology for developing agent-based systems. An interesting solution to this problem is the use of approaches that allow us to customize processes based on different types of applications and development environments. One technique that provides such a rational approach for the construction of tailored methods is Method Engineering [16].

Method Engineering is an approach by which process engineers construct processes (i.e., methodologies) from a set of method fragments instead of trying to modify a single monolithic, "one-size-fits-all" process. These fragments are generally identified by analyzing these "one-size-fits-all" processes and extracting useful tasks and techniques. The fragments are then redefined in terms of a common meta-model and are stored in a repository for later use. To create a new process, a process engineer selects appropriate method fragments from the repository and assembles them into a complete process based on project requirements[16].

### 4.1 O-MaSE Methodology

The goal of the O-MaSE [15] Process Framework is to allow process engineers to construct custom agent-oriented processes using a set of method fragments, all of which are based on a common meta-model. To achieve this, we define O-MaSE in terms of a meta-model, a set of method fragments, and a set of guidelines. The O-MaSE meta-model defines a set of analysis, design, and implementation concepts and a set of constraints between them. The method fragments define how a set of analysis and design products may be created and used within O-MaSE. Finally, guidelines define how the method fragment may be combined to create valid O-MaSE processes, which we refer to as O-MaSE compliant processes. We apply the O-MaSE methodology for developing the Intelligent Tutoring system. This has paved a way for us to create a stable agent oriented system using the O-MaSE model.

The first process we derive is appropriate for a small agent-oriented project in which reactive agents achieve goals that have been assigned at design time. Essentially, the only products required for this type of system are the system goals, agent classes, agent plans, and inter-agent protocols. This type of process leads to a rigid MAS but is very easy and fast to develop. This process may also be suitable for prototyping, where a simple and rapid process is needed.

# 5 Agents, Behaviors, Communication, Service Discovery

## 5.1 Types of Tutoring Agents

In our research, we propose several types of Agents, capabilities, roles and responsibilities. The different categories of agents are profile analysis agent, tutor agent, learner agent, classification agent, clustering agent, communication agent, service register/discover agent. The profile analysis agent and the learner agents are used to analysis the several characteristics of the learner, their likeliness, technical capabilities etc. The tutor agent performs the role of creating the course, announcement to students, analysis of student's path of learning and evaluating the learning process etc. The classification agent performs resource classification based on the hierarchical tree structure where the resources are represented in the tree like structure based on different courses. The student clustering agent helps to achieve the collaborative learning where similar learners are identified and they are allowed to communicate with each other. The communication agent performs the role of setting up the communication path between the learners and the tutors and between different types of agents.

The learner agents have three sub-types of agents. The evaluation agents perform all tests, including pre-learning tests and post-learning tests. Based on the evaluation results, the path planning agents analyze the learning profile and propose the course learning path. The problem mitigation agents have the responsibilities of learning problem analysis and assisting the learners to resolve the problems. The instructor agents can be course management agents or portfolio analysis agents. The course management agents assist the teachers to construct, arranges, and adjusts the course content. The portfolio analysis agents collect student learning portfolio and provide statistical analysis to the teacher for adjustment of course content or unit sequencing.

## 5.2 Agent Communication

Agent communication is defined by the FIPA [7] specification where the agent communicates using the Agent Communication Language (ACL). The communication architecture offers flexible and efficient messaging, where JADE creates and manages a queue of incoming ACL messages, private to each agent; agents can access their queue via a combination of several modes: blocking, polling, timeout and pattern matching based. The full FIPA communication model has been implemented and its components have been clearly distinct and fully integrated: interaction protocols, envelope, ACL, content languages, encoding schemes, ontologies and, finally, transport protocols. All agent communication is performed through message passing, where FIPA ACL is the language to represent messages.

### 5.3 Service Discovery

In our work, all the learning process is mapped into different types of services. If an agent wants to look for the available courses which is registered in the directory by the tutoring agent. All the available functionalities are mapped into services. We adopt the publisher/subscriber [4] methodology for the registering and retrieving the service details of the services which is implemented via JADE platform.

## 6 Evolving Clustering Method

Evolving clustering method [5] is a clustering algorithm which increases the number of clusters or adjusts the centers and the radius real-time dynamically accordingly to the data points. It is primarily used to solve the initial cluster center problem which is faced by FCM algorithm. In any cluster, the maximum distance between the cluster centre and data points is always less than the threshold value'd'. The threshold selection has a direct impact on the clustering numbers. Evolving clustering method is a clustering algorithm of kind of evolution, on-line and bounding by a maximum distance. It increases the number of cluster or adjusts the centers and the radius real-time dynamically as they entered sample data increasing. In any one cluster, the maximum distance between the example of cluster points and the corresponding maximum distance are less than the threshold Dthr; Dthr selection will have a direct impact on the clustering numbers. Examples of clustering process from a data stream, the whole process start clustering from an empty set. Some of the created clusters to be updated through depend on the location of the current example in the input space as well as changing the location of the cluster centers and increasing the radius of the cluster with the new examples appearance, it will no longer to be updated when its radius meet the threshold.

## 7 Fuzzy C-Means Algorithm

Fuzzy C- means [8] [9] is a method of clustering which allows one piece of data to belong to two or more clusters. It is based on the optimization of objective function. By iteratively updating the cluster centers and membership grades for each data point, Fuzzy C means moves the cluster centers to the right location within the dataset. Fuzzy C-means is a clustering algorithm that used membership degree to determine each data point belongs to a certain cluster. FCM divided the n vectors xi (i = 1, 2,..., n) into c fuzzy group, and computing the cluster center of each group, making value function of non-similarity index to achieve the minimum. FCM algorithm making each of the given data points with values between 0, 1 membership to determine its degree of belonging to various groups through fuzzy partition. And to suit the introduction of fuzzy partition, the membership matrix 'U' allowed the element value between 0, 1. The above algorithm can also initialize cluster center firstly, and run the iterative process after. The algorithm does not guarantee an optimal solution to the convergence, so its performance depends on the initial cluster centers. Therefore, Another fast algorithm can be used to determine the initial cluster centers, or start the algorithm with different initial clusters center every time, running FCM repeatedly.

## 8 Agent Based Intelligent Tutoring System Model

The system is modeled through the O-MaSE which is done using eclipse. The system model consists of the agent model diagram, domain diagram, goal diagram, and plan diagram. The different agent's services, components are explained in fig. 1.

**Fig. 1.** Agent Diagram for ITS

The plan diagram for the intelligent adaptive tutoring system is explained in the fig 2. All the functions, parameters are also explained.

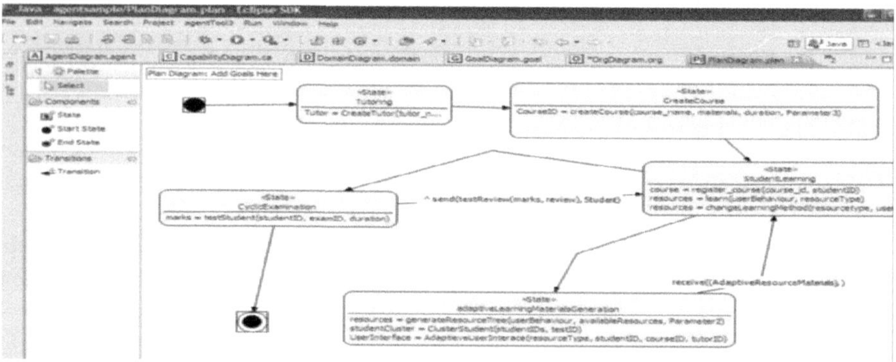

**Fig. 2.** Plan Diagram for ITS

The goal diagram describes about the various goals in the hierarchical tree representation which is explained in the figure 3. In fig. 3 the different forms of adaptation is explained. Adaptive user interface, adaptive content delivery during the course, adaptive examinations questions is divided into different sub-goals and when each sub-goal is achieved the mail goal is achieved.

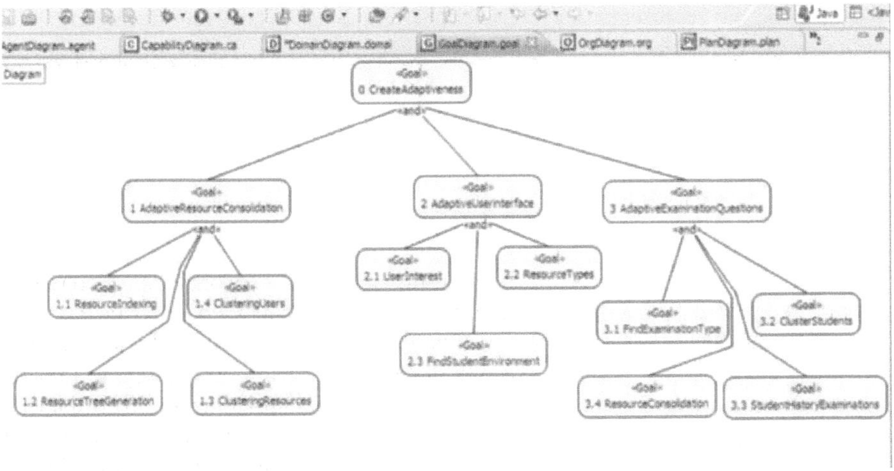


**Fig. 3.** Goal Diagram For ITS

In the fig. 4 the domain Diagram is explained in which each agent's behaviors, relationships, different objects are explained.

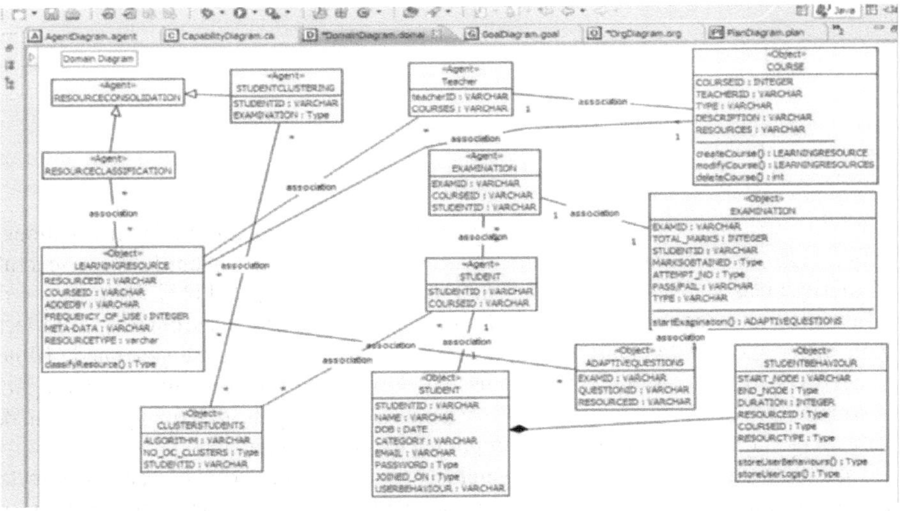


**Fig. 4.** Domain Diagram for ITS

## 9 Implementation

The system is developed with the jsp, servlets, ajax, agents developed in JADE, tomcat as web-server, mysql as database. Using the fuzzy clustering method the different types of students are identified and the different types of course contents are presented to the user according to their position in the cluster. We construct a tree like structure for the representation of the course contents. Whenever the student is in particular level of the course only the contents corresponding to the tree are only presented to the user. The resources can be classified into html, video, audio, flash or combination of those types. These contents have to be delivered adaptively to the user accordingly to the user's likeliness. The user's activity has to be monitored and in accordance to the user the adaptive user-interface is delivered by the system.

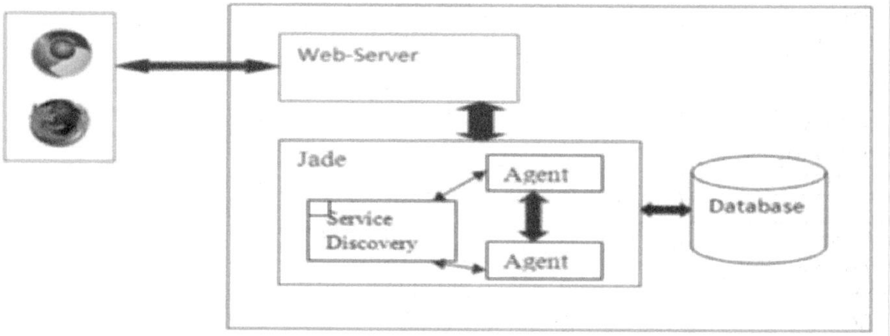


**Fig. 5.** Architecture Diagram

## 10 Conclusions and Future Work

The Intelligent Tutoring system is systematically developed applying the Agent Oriented Software Engineering approach. The modeling of the Intelligent Tutoring System is performed using the O-MaSE approach and system is analyzed using several diagrams .The adaptive behavior of the Intelligent Tutoring system is achieved by clustering of the similar students based on the likeliness, skill levels and various parameters which is performed by the ECM and FCM. The course materials are adaptively presented to the user by representing the resources in a hierarchical tree structure.

In our future work the system will be implemented in the distributed environment analyzing various parameters such as replication, fragmentation of resources consolidation of resources due to increase in the size of the course materials such as video etc and this large amount of data has to be stored and retrieved for the students in an optimum strategy.

## References


[1] Badjonski, M., Ivanovic, M., Budimac, Z.: Intelligent Tutoring System as Multiagent System. In: IEEE International Conference on Intelligent Proccssing Systems, pp. 871–875 (1997)

[2] Caseau, Y.: Self-adaptive and self-healing message passing strategies for process-oriented integration infrastructures. In: Proceedings of 11th IEEE International Conference and Workshop on the Engineering of Computer-Based Systems, pp. 506–512 (May 2004)

[3] Chang, Y.H., Lu, T.Y., Fang, R.J.: An Adaptive E-learning System Based on Intelligent agents. In: 6th Conference on WSEAS International Conference on Applied Computer Science, pp. 200–205 (2007)

[4] Eugster, P.T., Felber, P.A., Guerraoui, R., Kermarrec, A.M.: The Many Faces of Publish/Subscribe. ACM Computing Surveys (CSUR) 35(2), 114–131 (2003)

[5] Du, G.-Y., Miao, F., Tian, S.-L., Liu, Y.: A Modified Fuzzy C-means Algorithm in Remote Sensing Image Segmentation. In: International Conference on Environmental Science and Information Application Technology, ESIAT 2009, pp. 447–450 (2009)

[6] Jade-Java Agent Development Framework, `http://jade.tilab.com/`

[7] Foundation for Intelligent Physical Agents, `http://www.fipa.org/`

[8] Turksen, L.B.: Fuzzy data mining and expert system development. In: IEEE International Conference on Systems, Man, and Cybernetics (1998)

[9] Zhang, H.-Z., Chen, H., Bao, L.-X.: An improved Fuzzy C Means clustering algorithm and its application in traffic condition recognition. In: Seventh International Conference on Fuzzy Systems and Knowledge Discovery, FSKD (2010)

[10] Nargesian, F., Nazemi, E.: An Agent-Based Architecture with Centralized Management for a Distance Learning System. In: International Conference on Computational Intelligence and Multimedia Applications (2007)

[11] Jia, Y., Huang, C., Cai, H.: A Comparison Of Three Agent-Oriented Software Development Methodologies: Mase, Gaia, And Tropos. In: IEEE Youth Conference on Information, Computing and Telecommunication, YC-ICT 2009 (2009)

[12] Zhang, K., Cui, L., Wang, H., Sui, Q.: An Improvement of Matrix-based Clustering Method for Grouping Learners in E-Learning. In: 11th International Conference on Computer Supported Cooperative Work in Design (2007)

[13] Arenaz, P., Fisher, W., Kubo, C., Della-Piana, C.K., Gomez, R., Dahl, E.: CircLES: a retention program for entering students in engineering, mathematics and science. In: 29th Annual Frontiers in Education Conference (1999)

[14] Wu, F.: Apply Data Mining to Students' Choosing Teachers Under Complete Credit Hour. In: Second International Workshop on Education Technology and Computer Science, ETCS (2010)

[15] Garcia-ojeda, J.C., Deloach, S.A., Oyenan, W.H.: O-MaSE: A Customizable Approach to Developing Multiagent Development Processes. In: The 8th International Workshop on Agent Oriented Software Engineering (2007)

[16] Brinkkemper, S.: Method Engineering: Engineering of Information Systems Development Methods and Tools. Jnl. of Information and Software Technology 38(4), 275–280 (1996)

# GPGPU Implementation of Parallel Memetic Algorithm for VLSI Floorplanning Problem

Subbaraj Potti[1] and Sivakumar Pothiraj[2]

[1] Theni Kammavar Sangam College of Technology,
Theni-625 534,
Tamil Nadu, India
subbaraj_potti@yahoo.com
[2] Arulmigu Kalasalingam College of Engineering,
Krishnankoil,
Tamil Nadu-626190,
India
siva@klu.ac.in

**Abstract.** The VLSI Physical design floorplanning is the process where circuit description is converted into geometric description. This NP-Optimization problem with non-slicing blocks is very hard to solve. Memetic algorithms are used to solve NP-Optimization problems. Memetic algorithms are a family of meta-heuristic search algorithms in which a rule-based representation of Local Search (LS) is co-adapted alongside the candidate solutions within a hybrid evolutionary system. However, they may execute for a long time for difficult problems, because of several fitness evaluations. A promising approach to overcome this limitation is to parallelize these algorithms. In this paper, we propose to implement a parallel Memetic Algorithm (MA) on Graphical Processing Units. The General Purpose Graphical Processing Unit (GPGPU) is the complete parallel hardware which is best used for the Parallel Computing. The Parallel Memetic Algorithm we followed for an application perspective is a modified or hybrid genetic algorithm with extensive attributes to local search. The Parallel Memetic Algorithm operator gives a perfect exploration in the available search area and it is typical for addressing a modular assignment problem. It reduces the time efficiently for execution which is a boon for VLSI industry where we don't have to spend more weeks for Physical design to be completed. We perform experiments to compare our parallel MA with an ordinary MA and demonstrate that the former is much more effective than the latter. Both fitness evaluation and genetic operations are spread over the GPU cores, allowing for the efficient processing of very large datasets. The parallel nature of Memetic algorithm can be best harnessed with help of GPGPU and significant changes in execution time and optimization are shown for VLSI Floorplanning problem.

**Keywords:** Floorplanning, Graphical Processing Unit, Memetic algorithm, very large scale integrated circuit (VLSI).

D. Nagamalai, E. Renault, M. Dhanushkodi (Eds.): CCSEIT 2011, CCIS 204, pp. 432–441, 2011.

## 1 Introduction

The Parallel Memetic Algorithm is developed as a mathematical tool for getting solution to various problems where we need to take into account the transferring of cultural information from one generation to another. The path for finding this is realized using a sequence of steps using genetic algorithm and local search. The inspiring earlier work carried out by Maolin [1] has enabled to look for further investigation into the optimization and parallelization. The parallel Memetic Algorithm frame work will surely give its potential to a parallel computing sequence of execution and is further termed as the application boosting.

Graphics Processing Units (GPUs) are fast, highly parallel processor units. In addition to processing 3D graphics, modern GPUs can be programmed for more general purpose computation [10]. This facilitates the chance of integrating Parallel Memetic Algorithm and GPGPU to give better Platform of execution in terms of time and optimization [3]. The potential of GPGPU is carried out with help of Compute Unified Device Architecture (CUDA). The CUDA model is a collection of threads running in parallel. CUDA consists of a test of C language extensions and a runtime library that provides APIs to control the GPU[4].

Because capable hardware and software are new, there is relatively little previous work on using GPUs for evolutionary computation. For example [6] implements a evolutionary programming algorithm on a GPU, and finds that there is a 5-fold speed increase. Work by [16] expands on this, and evaluates expressions on the GPU. There, all the operations are treated as graphics operations, which make implementation difficult and limit the flexibility of the evaluations. Yu et al [17], on the other hand, implement a Genetic Algorithm on GPUs. Depending on population size, they find a speed up factor of up to 20. Here both the genetic operators and fitness evaluation are performed on the GPU [5, 7].

Memetic Algorithm covers one area of Memetic Computing, in particular dealing with areas of evolutionary algorithms that use other deterministic refinement techniques for solving optimization problems. We demonstrate a method for using the GPU as an evaluator for Memetic programming expressions and show that there is a considerable increase in speed of execution.

## 2 VLSI Floorplanning Problem

The VLSI physical design contains large number of steps which are highly complex and tentative. Floorplanning is one of them. The modular assignment problem is considered as the basic problem in Floorplanning. Consider a rectangular space where we need to arrange n modules. The search space for this is given by $n!C_n$ [1]. The rectangular space is limited with width ‘W’ and height ‘H’.

We can anticipate that the blocks must be arranged such that they should not overlap each other. Moreover even if we arrange the blocks it should give minimum wasted area. The ultimate goal is to reduce area and wire length. As a part of modular assignment with non-slicing, the efficient representation is considered as O-Tree representation. The time-complexity for the transform between an O-tree representation and its corresponding floorplan is only O(n)[14]. The representation not only covers all sub-optimal floorplan but also has a smaller search space. For a VLSI floorplanning problem of n modules, the search space of the representation is (n!)*$c_n$, where,

$$c_n = \frac{1}{2n+1}\binom{2n+1}{n} \tag{1}$$

The representation gives the geometrical relations among modules which are valuable for identifying meaningful building blocks when designing genetic operators.

A widely used global search method for VLSI floorplanning problems is genetic algorithm (GA)[2][15]. GA has been successfully applied to solve non-slicing VLSI floorplanning problems. Since the encoding scheme does not capture any topological information of the floorplans, the performance of the GA is not satisfactory. Although GA is effective in solving NP hard problems, they may execute for a long time to find better solutions, because several fitness evaluations must be performed. In this paper, we present a parallel Memetic Algorithm for a non-slicing and hard-module VLSI floorplanning problem.

## 2.1 The Problem Statement

Generally we consider a VLSI floorplanning problem as NP-hard problem. Even though floorplanning is highly tentative the assignment is carried out with some successful approaches. A module $m_i$ is a rectangular block with fixed height $h_i$ and width $w_i$, M = {$m_1$, m2, . . $m_n$} is a set of modules, and N is a net list specifying interconnections between the modules in M. A floorplan F is an assignment of M onto a plane such that no module overlaps with another. Here the cost of the floorplan is defined with an equation.

$$\text{Cost (F)} = w_1 \times \frac{Area(F)}{Area^*} + w_2 \times \frac{Wirelength(F)}{Wirelength^*} \tag{2}$$

In the above equation, Area* and Wirelength* represent the minimal area and the interconnection costs, respectively. Since we do not know their values in practice, estimated values are used[1]. The constants $w_1$ and $w_2$ are weights assigned to the area minimization objective and the interconnection minimization objective, respectively, where $0 \le w_1, w_2 \le 1$, and $w_1 + w_2 = 1$. The interconnection cost is the total wire length of all the nets, and the wire length of a net is calculated by the half perimeter of the minimal rectangle enclosing the centre of the modules that have a terminal of the net on it[13]. The minimization of the cost function is achieved by keeping the floorplan as an admissible one. A floorplan is admissible if no modules can be shifted left or bottom without moving other modules in the floorplan. In order to minimize the area, a module may be rotated 90 degrees. The area of a floorplan F, Area (F), is measured by the area of the smallest rectangle enclosing all the modules. Given M and N, the objective of the floorplanning problem is to find a floorplan F such that the cost function minimized.

# 3 Parallel Memetic Algorithm Framework

## 3.1 Representation of Genetic Population

In this section we describe the internal representation of population. If the GPU is to perform GA operations, the first thing we need to do is to represent population data in a format that is accessible by the GPU. The general idea is to store population in a set of texture maps and to exploit fragment programs to implement genetic operators. In our representation, the chromosome of each individual is sequentially divided into several segments which are distributed in a number of textures with the same position (see Fig. 1). Every segment contains four genes packed into a single pixel's RGBA channels. We call those textures population textures. Another 2D texture map, called fitness texture, is used to hold the fitness scores of each individual in the population. The mapping is done between the fitness of a particular individual and the position of the individual in the population [17]. The proposed representation enables the efficient computation of genetic operations and operators. It has several advantages: First, it naturally keeps 2D grid topology of the population. Second, for each individual, fragment programs of genetic operators only need lookup considered pixel or near neighbourhood pixels in each texture map. Thus it keeps the memory footprint of each fragment as low as possible to efficiently utilize texture cache. Third, packing of four concessive genes in one Texel makes use of the wide SIMD computations carried out by the GPU [17].

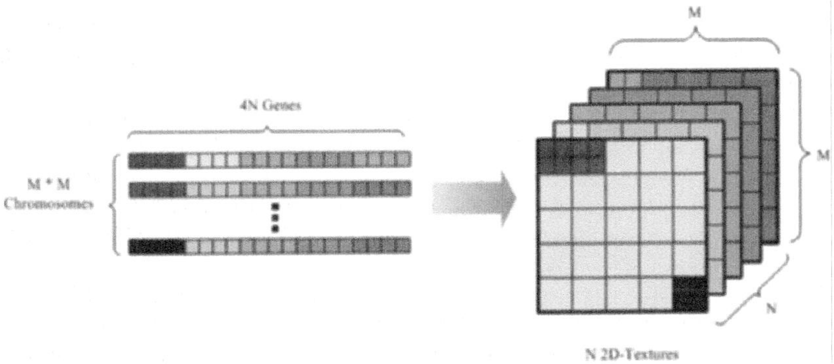


**Fig. 1 .** The representation of chromosome in a population as a set of 2D textures as shown

## 3.2 Genetic Operators

The genetic operations are defined in such a way that subtrees are transmitted and evolved through crossover operator and mutation operator. For Crossover, given two parents, both of which are admissible floorplan represented by an O-tree, the crossover operator transmits the significant structural information from two parents to child [14]. By recombining some significant structural information from two parents; it is hoped that better structural information can be created in the child.

The mutation operations randomly select a single existing individual and generate a new one by changing one or more of its genes. They are crucial to push the algorithm towards unexplored regions of the search space. The differences between the new and the existing individual change from one operation to the other. Three mutation operations have been defined: the first is the simplest one: a gene is randomly selected, and its value is changed; the second affects in a deeper way the solution: an integer k is generated, and the genes k, 2k, 3k, 4k are randomly changed; the third one radically changes the whole individual: For each gene the next implementation in the implementation list is chosen; as the list is ordered with respect to the aspect ratio of the implementations, the operator executes a rotation of all the modules.

### 3.3 Fitteness Value Evaluation

Fitness value evaluation determines the "goodness" of individuals. It is one of the core parts of GA. After each evolution, the fitness value of each individual in the current population is calculated. The result is then passed to the later stage of GA process. Each individual returns a fitness value by feeding the objective function with the genomes of the individual. This evaluation process usually consumes most of the computational time. Since no interaction between individuals is required during evaluation, the evaluation is fully parallelizable. Recall that the individuals are broken down into quadruples and stored in the tiles within the textures. The evaluation shader hence looks up the corresponding quadruple in each tile during the evaluation. The fitness values are output to an output texture of size $w \times h$, instead of $4w \times 2h$, because each individual only returns a single value[16].

### 3.4 Competition and Selection

Replacing the old generation is the last stage of each evolution. There are two major processes involved: competition and selection. GA employs a stochastic selection (soft selection) through the tournament schema. Each individual in the union set of parent and offspring population takes part in a $q$-round tournament. In each round, an opponent is randomly drawn from the union set of parents and offspring. The number of opponents defeated is recorded by the variable *win*. After the tournament, a selection process takes place and chooses the best $\mu$ individuals having the highest win values as parents for next generation.

Exploitation is the process of using information gathered to determine which searching direction is profitable. In GA, exploitation is realized by competition. Individuals in the population compete with $q$ randomly drawn opponents, where $q$ is the predefined tournament parameter. The considered individual wins if its fitness value is better than (in our case smaller than or equal to) that of the opponent. The times of winning is recorded in *win*. It tells us how good this individual is. Competition can be done either on GPU or CPU. For GPU implementation, $q$ textures of random values have to be generated by CPU and loaded to GPU memory for picking $q$ opponents within each evolution. In the shader, $q$ fitness comparisons are performed by looking up the opponent's fitness in the fitness textures obtained previously. Such look-up is achieved by regarding the random values as indices to fitness textures. The times of winning, win, is recorded and output to a texture. On the other hand, for CPU implementation, only the fitness textures of both parent and offspring population have to be transferred from GPU memory to main memory. Then the tournament is CPU performed by referring to these fitness textures. It seems that GPU implementation should be faster than the CPU one, as it parallelizes the competition. However, our experiments show that using GPU to implement the competition is slower than that of using CPU. The major bottleneck of GPU implementation is the transfer of $q$ textures towards GPU memory. It evidences the limitation of slow data transfer of current GPU. As the competition does not involve any time-consuming fitness evaluation (it only involves fitness comparison), the gain of parallelization does not compensate the loss due to data transfer [8]. Hence, we suggest to perform the tournament in CPU unless the data transfer rate is significantly improved in the future GPU.

### 3.5 Local Search Method

The local search method is based on a deterministic algorithm proposed by Guo et al. [1].Given an initial floorplan encoded in an O-tree (T, π), the local search method finds a local optimal solution through systematically examining those O-trees that can be obtained by removing a module from, and then putting it back to, the O-tree. When a module is added, its orientation may be changed if it leads to a smaller cost. The algorithm is shown below:

1) For each node m i in (T, π):
   a) Delete mi from (T, π);
   b) Insert mi in the position where we can get the smallest cost value among all possible insertion positions in (T, π) as an external node of the tree;
   c) Perform (a) and (b) on its orthogonal O-tree.
2) Output (T, π).

## 4 Experimental Results

GPUs are considered as SIMD processors, when evaluating a population of programs in parallel, the common approach so far is to implement some form of interpreter on the GPU. The interpreter is able to execute the evolved programs (typically machine code like programs represented as textures) in a pseudo-parallel manner.

The NVIDIA® Tesla™ C1060 computing processor enables the transition to energy efficient parallel computing power by bringing the performance of a small cluster to a workstation. With 240 processor cores and a standard C compiler that simplifies application development, Tesla scales to solve the world's most important computing challenges—more quickly and accurately.

The population and genetic algorithm run on the CPU, with evaluations run on the GPU. The CPU converts arrays of test cases to textures on the GPU and loads a shader program into the shader processors. The Accelerator tool kit compiles each individuals GP expression into a shader program. The program is then executed, and the resulting texture is converted back in to an array. The fitness is determined from this output array.

In this section, we experimentally evaluated the quality of the floorplan produced by our GPU based parallel Memetic algorithm on a large number of non-slicing and hard blocks that are part of the widely used MCNC circuit floorplanning benchmarks suite. All experiments were carried out on Pentium Quad core 2 processor 2.6 GHz with NVIDIA Tesla C1060 computing processor, with 8GB main memory and 512MB GPU memory.

The following parameters were used in the experiments:

- population size: $\mu$ =40, 80, 100
- tournament size: $q$= 10
- maximum number of generation: G= 100

The two objectives are mutually constrained with each other [1]. The weights for the area minimization and the interconnection minimization objectives were set to 0.8 and 0.2, respectively ($w1$ = 0.8 and $w2$ = 0.2), which means the MA focused more on minimizing the area of the floorplan. GPU based implementation was compared with software implementation running on single CPU with different benchmark circuits. Table 1 shows the best value out of 100 runs of Parallel Memetic Algorithm on CPU. Table 2 shows the best value out of 100 runs of PMA on GPGPU. Table 3 shows the statistics for the test results on the performance of the Parallel MA.

**Table 1.** Best value of 100 generation on without GPU

| MCNC Circuit | Module Count | Net | Width (mm) | Height (mm) | Floorplan Area (mm) | Total Modules Areas(mm$^2$) | Dead Area(%) | Wire (mm) | Time (s) | Cost (F) |
|---|---|---|---|---|---|---|---|---|---|---|
| ami49 | 49 | 408 | 10.6370 | 3.7150 | 39.5165 | 35.4454 | 10.3021 | 780.4 | 251 | 1097597 |
| ami33 | 33 | 123 | 1.8160 | 0.7320 | 1.3293 | 1.1564 | 13.0039 | 72.9 | 245 | 1192814 |
| Hp | 11 | 83 | 10.4830 | 1.0080 | 10.5669 | 8.8306 | 16.4314 | 175.3 | 61 | 1167594 |
| xerox | 10 | 203 | 3.2070 | 7.6440 | 24.5143 | 19.3503 | 21.0653 | 490.6 | 62 | 1248478 |
| apte | 9 | 97 | 5.3780 | 10.4530 | 56.2162 | 46.5616 | 17.1741 | 430.4 | 60 | 1170876 |

**Table 2.** Best value of 100 generation on without GPU

| MCNC Circuit | Module Count | Net | Width (mm) | Height (mm) | Floorplan Area (mm) | Total Modules Areas (mm$^2$) | Dead Area (%) | Wire (mm) | Time (s) | Cost (F) |
|---|---|---|---|---|---|---|---|---|---|---|
| ami49 | 49 | 408 | 9.9360 | 3.9850 | 39.5950 | 35.4454 | 10.4800 | 720.2 | 40 | 1040067 |
| mi33 | 33 | 123 | 1.8590 | 0.7300 | 1.3571 | 1.1564 | 14.7834 | 64.6 | 15 | 1179438 |
| Hp | 11 | 83 | 3.4400 | 2.6520 | 9.1229 | 8.8306 | 3.2040 | 163.4 | 30 | 1052091 |
| xerox | 10 | 203 | 8.8570 | 2.4380 | 21.5934 | 19.3503 | 10.3878 | 470.2 | 10 | 1137913 |
| apte | 9 | 97 | 12.2180 | 3.9850 | 48.6887 | 46.5616 | 4.3688 | 410.4 | 12 | 1051531 |

**Table 3.** Statistics for the test result on the performance of the Parallel MA

| MCNC | With GPU | | | Without GPU | | |
|---|---|---|---|---|---|---|
| Circuit | Best | Mean | Std.dev | Best | Mean | Std.dev |
| ami49 | 1040067 | 1192655 | 79400.68 | 1097597 | 1519900 | 232887.3 |
| ami33 | 1179438 | 1193769 | 22010.5 | 1192814 | 2758732 | 22649.53 |
| Hp | 1052091 | 1617788.5 | 2238.42 | 1167594 | 1873492 | 4729.34 |
| xerox | 1137913 | 2083469.4 | 10622 | 1248478 | 2513694 | 10987 |
| apte | 1051531 | 1492840 | 49203 | 1170876 | 1683491 | 87349 |

**Table 4.** Speedup for different population size for ami49

| Population Size | PMA with GPU(S) | PMA without GPU(S) | Speedup |
|---|---|---|---|
| 40 | 20 | 50 | 2.5x |
| 80 | 35 | 140 | 4x |
| 100 | 40 | 250 | 6.25x |

To study the dynamic performance of Parallel Memetic Algorithm (PMA) on GPU and PMA on CPU, we captured the dynamically changing minimal cost obtained in each run for each of the benchmarks. Fig. 3 displays the dynamic performance of the PMA on GPU and PMA on CPU for benchmark ami49. It can be seen that the dynamic performances of PMA on GPU for ami49 problem are consistent.

The execution time of PMA without GPU and PMA with GPU with population sizes of 40, 80, and 100 are shown in Fig. 2. It is interesting to notice that, the GPU based implementation is much faster than former approach. The speedup obtained is on an average from 2.5 times to 6.25 times. We see that the speedup increases as we increase the population size. When the population size is small, the speedup of GPU based implementation is marginal compared to implementation without GPU. This happens because, first time GPUs see a new piece of code, and it spends some time analyzing it and compiling various instruction sequences for faster lookup on subsequent runs. That first run will be a little slower, but for long-running computations there won't be any noticeable lag.

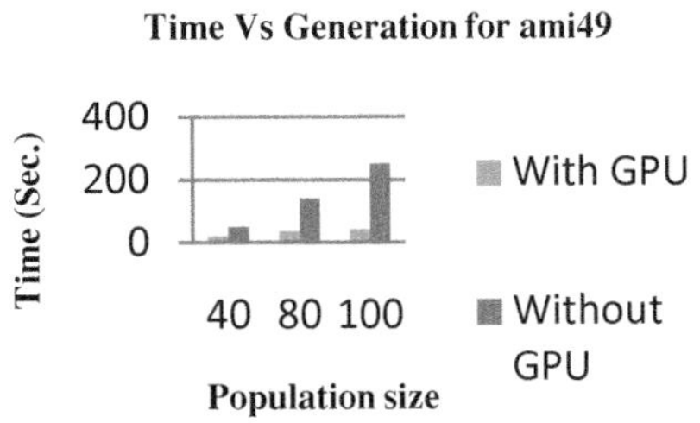


**Fig. 2.** Execution time of the PMA with GPU and without GPU

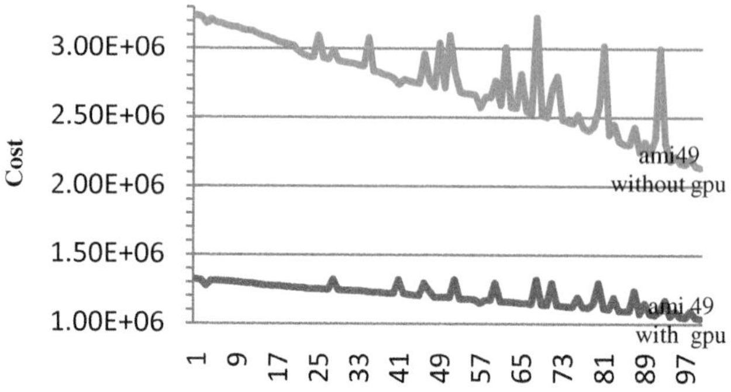


**Fig. 3.** Dynamic performance comparison of Parallel MA running on GPU and running CPU on benchmark ami49

## 5 Conclusion

In this research, an implementation of Parallel MA to solve VLSI floorplanning problem based on GPU was presented. The parallel MA on GPU is a hybrid of master-slave and fine-grained models. Competition and selection are performed by CPU while fitness evaluation, mutation, and reproduction are performed by GPU which is essentially a massively parallel machine with shared memory. GPU allows processors to communicate with any other processors directly; thus more flexible fine-grained MAs can be implemented on GPU. We have done experiments to compare our parallel MA on GPU and a sequential MA on CPU. The experimental results show that GPGPU has good accelerated performance on larger population sizes. A couple of other important factors can help to get the best performance out of our transformation to GPUs such as avoiding of excessive memory transfer, inherent parallelism and computation dependency between CPU and GPU. Note that the commercial tool can be made to run on several CPUs using a distributed option. If each of these CPUs had a GPU on board, then the GPU approach could also exploit a distributed option, and the above speedup numbers would be increased. More sophisticated cluster approaches to parallelisation will have to be examined in the future.

## References


1. Tang, M., Yao, X.: A Memetic Algorithm for VLSI Floorplanning. IEEE Trans. on Systems, Man, and Cybernetics 37, 62–68 (2007)
2. Rebaudengo, M., Reorda, M.: GALLO: A genetic algorithm for floorplan area optimization. IEEE Trans. Computer-Aided Design Integrated Circuits System 15(8), 943–951 (1996)
3. Cohoon, J., Hegde, S., Martin, W., Richards, D.: Distributed genetic algorithms for the floorplan design problem. IEEE Trans. Computer-Aided Design Integrated Circuits System 10(4), 483–492 (1991)

4. Liang, S., et al.: Design and evaluation of a parallel k-nearest neighbor algorithm on CUDA-enabled GPU. In: Proc. IEEE 2nd Symposium on Web Society (SWS), pp. 53–60. IEEE Press, Los Alamitos (2010)
5. Harding, S., Banzhaf, W.: Fast genetic programming on GPUs. In: Proceedings of the 10th European Conference on Genetic Programming, pp. 90–101 (2007)
6. Wong, M.-L., Wong, T.-T., Fok, K.-L.: Parallel evolutionary algorithms on graphics processing unit. In: IEEE Congress on Evolutionary Computation, vol. 3, pp. 2286–2293 (2005)
7. Harding, S., Banzhaf, W.: Fast Genetic Programming and Artificial Developmental Systems on GPUs,". In: Proc. 21st International Symposium High Performance Computing System and Applications (HPCS 2007), p. 2. IEEE Computer Society Press, Los Alamitos (2007)
8. Robilliard, D., Marion-Poty, V., Fonlupt, C.: Population Parallel GP on the G80 GPU. In: O'Neill, M., Vanneschi, L., Gustafson, S., Esparcia Alcázar, A.I., De Falco, I., Della Cioppa, A., Tarantino, E. (eds.) EuroGP 2008. LNCS, vol. 4971, pp. 98–109. Springer, Heidelberg (2008)
9. Jiening, W., Chufeng, Z., Jiankang, D.: Implementation of Ant Colony Algorithm based on GPU. In: Proc. of the IEEE Sixth Int. Conference on Computer Graphics, Imaging And Visualization (CGIV 2009), pp. 50–53 (2009)
10. Wu, E., Liu, Y.: Emerging technology about GPGPU. In: IEEE Asia Pacific Conference Circuits and Systems (APCCAS 2008), pp. 618–622 (2008)
11. Sutanthavibul, S., Shragowitz, E., Rosen, J.B.: An Analytical Approach to Floorplan Design and Optimization. IEEE Transactions on Computer Aided Design 10(6), 761–769 (1991)
12. Deng, Y., Mu, S.: The potential of GPUs for VLSI physical design automation. In: 9th International Conference on Solid-State and Integrated-Circuit Technology (ICSICT 2008), pp. 2272–2275 (2008)
13. Tang, X., Tian, R., Wong, M.D.F.: Minimizing wire length in floorplanning. IEEE Trans. on Computer-Aided Design of Integrated Circuits and System 25(9), 1744–1753 (2006)
14. Takahashi, Guo, T., Cheng, P.N., Yoshimura, C.K.: Floorplanning using a tree representation: a summary. IEEE Circuits and Systems Magazine 3(2), 26–29 (2003)
15. Rebaudengo, M., Reorda, M.: GALLO: A genetic algorithm for floorplan area optimization. IEEE Trans. Computer-Aided Design Integrated Circuits System 15(8), 943–951 (1996)
16. Fok, K.-L., Wong, T.-T., Wong, M.-L.: Evolutionary Computing on Consumer Graphics Hardware. IEEE Intelligent Systems 22(2), 69–78 (2007)
17. Yu, Q., Chen, C., Pan, Z.: Parallel Genetic Algorithms on Programmable Graphics Hardware. In: Wang, L., Chen, K., S. Ong, Y. (eds.) ICNC 2005. LNCS, vol. 3612, pp. 1051–1059. Springer, Heidelberg (2005)
18. Chitty, D.M.: A data parallel approach to genetic programming using programmable graphics hardware. In: Proceedings of the 9th Annual Conference on Genetic and Evolutionary Computation (GECCO 2007), vol. 2, pp. 1566–1573. ACM Press, New York (2007)
19. Tang, M., Lau, R.Y.K.: A Parallel Genetic Algorithm for Floorplan Area Optimization. In: IEEE Seventh International Conference on Intelligent Systems Design and Applications, pp. 801–806 (2007)

# Improving Cloud Security through Virtualization

B. Loganayagi[1] and S. Sujatha[2]

[1] Research Scholar (Full Time) in the Faculty of ICE
[2] Head of the Department, Department of MCA
Anna University of Technology Tiruchirappalli (Main Campus), Tiruchirappalli, India
loks20@yahoo.com, ssujatha71@yahoo.co.in

**Abstract.** Cloud Computing is accessing Services through Internet based on pay per usage model. Software as a Service (SaaS), Infrastructure as a Service (IaaS) and Platform as a Service (PaaS) are available in Cloud. Cloud based products will eliminate the need to install and manage client rich applications. Cloud Service providers are helping companies to reduce the high cost infrastructure installation and maintenance cost. The customer is charged only for the resources consumed like utility based computing. Virtualization is a technique to implement cloud computing resources such as platform, application, storage, and network. The beauty of virtualization solutions is that you can run multiple virtual machines (VM) simultaneously on a single machine. This paper provides details about cloud computing issues, how virtualization techniques can be used to overcome those issues.

**Keywords:** Cloud Computing, SaaS, IaaS, PaaS, Virtualization.

## 1 Introduction

Cloud Computing is defined by a large-scale distributed computing paradigm that is driven by economies of scale, in which a pool of abstracted, virtualized, dynamically-scalable, managed computing power, storage, platforms, and services are delivered on demand to external customers over the Internet.[1]

Cloud computing has the features of grid computing, cluster computing, service computing, and utility computing [8][9]. End user has to run one time application, like tax calculation, can use cloud application as the best choice. End users can do their work or access services with thin client, no need to invest on resources. Instead they can rent/lease the resources. More over they can choose the best available computing platform, since many companies are going to offer cloud based applications in near future, like mobile services.

Small companies, whose main business is not software related, can outsource their work to cloud providers, so that they can concentrate on such as application development only, instead of concentrating their focus on hosting environment. It is easy to scale if the application is deployed in cloud. It takes away all the risks of managing resources.

Cloud is providing preconfigured infrastructure at lower cost, which generally follows the Information Technology Infrastructure Library, can manage increased peak load capacity and moreover uses the latest technology, provide consistent

D. Nagamalai, E. Renault, M. Dhanushkodi (Eds.): CCSEIT 2011, CCIS 204, pp. 442–452, 2011.

performance that is monitored by the service provider. Dynamic allocation of the resources as and when is needed.

Furthermore cloud provides secured computing platform, and scalability won't be a problem with the use of cloud. Overall giving good performance at lower cost instead of making capital investment. The rest of the paper is organized as follows: Section 2 discusses about the need for Cloud Computing. Section 3 discusses about the Cloud providers. Section 4 discusses about the Cloud Computing and Virtualization. Section 5 discusses about the Cloud Issues and Virtualization Solutions and Section 6 discusses about the Conclusion and Future direction.

## 2 Need for Cloud Computing

Software industry is everyday changing. What you get/invest for your company/need today is going to become obsolete within a decade itself. As an end user if you need some services, instead of getting all the software, hardware straightway user can run their services in cloud, which provides necessary hardware, software support.

Cloud computing reduces capital expenditure (CAPEX) and it offers high computing at lower cost [4]. Upgrading your hardware/ software requirement also easy with the cloud, without disturbing the current work. Scalability and maintenance is easy in the case of cloud. Easily user can rent/lease the services offered by cloud computing vendors. User will be charged as pay per usage like utility based services.

## 3 Cloud Providers

Cloud computing [7] [10] [14][6]systems generally falls into three course grain categories. Infrastructure as a service (IaaS), Platform as a Service (PaaS), Software as a Service (SaaS). Many companies are offering services.

*A. Infrastructure as a Service(IaaS)*

*Infrastructure as a Service (IaaS) provisions hardware,* software, and equipments to deliver software application environments with a resource usage-based pricing model. Infrastructure can scale up and down dynamically based on application resource needs.

Typical examples are Amazon EC2 (Elastic Cloud Computing) Service and S3 (Simple Storage Service) where compute and storage infrastructures are open to public access with a utility pricing model. This basically delivers virtual machine images to the IaaS provider, instead of programs, and the Machine can contain whatever the developer want.

Example:

Full virtualization (GoGrid, Skytap), Grid computing (Sun Grid), Management (RightScale), Paravirtualization (Amazon Elastic Compute Cloud).

*B. Platform as a Service(PaaS)*

*Platform as a Service(PaaS)* offers a high-level integrated environment to build, test, and deploy custom applications. Generally, developers will need to accept some restrictions on the type of software they can write in exchange for built-in application scalability.

An example is Google's App Engine, which enables users to build Web applications on the same scalable systems that power Google applications, Web application frameworks, Python Django (Google App Engine), Ruby on Rails (Heroku), Web hosting (Mosso), Proprietary (Azure, Force.com).

*C. Software as a Service (SaaS)*

User buys a Subscription to some software product, but some or all of the data and codes resides remotely. Delivers special-purpose software that is remotely accessible by consumers through the Internet with a usage-based pricing model. In this model, applications could run entirely on the network, with the user interface living on a thin client.

Salesforce is an industry leader in providing online CRM (Customer Relationship Management) Services. Live Mesh from Microsoft allows files and folders to be shared and synchronized across multiple devices.

Identity (OAuth, OpenID), Integration (Amazon Simple Queue Service), Mapping (Google Maps, Yahoo! Maps), Payments (Amazon Flexible Payments Service, Google Checkout, PayPal), Search (Alexa, Google Custom Search, Yahoo! BOSS), Others (Amazon Mechanical Turk).

**Amazon EC2**, Developer can install anything they want in their virtual machines. EC2 gives developer lots of control, but developer still needs to install OS, Web server, Web or Application container, Database and the whole nine yards that you do today for application deployment. It provides a virtual machine, and developer is responsible for managing everything inside the VM. It offers very flexible system along with the risk of managing everything.

**Google App Engine**, here developer doesn't have any virtual machine and are not responsible for installing/managing OS, software, database. Right now developers are restricted to use Python to code the app and they need to use DataStore (persistent service) to get/put the data. Developer is responsible for developing and deploying the application. User also gets the access to Google accounts, Gmail etc. No OS, server, database headaches with the disadvantage of being restricted to the Python and DataStore. It's not for Java apps right now, but they may support more languages in the future. Google seems to be more interested in Python than Java.

**Microsoft Azure** is offering PaaS, mainly for Microsoft related technologies. Azure helps the user to develop, host and manage environment with tools for Microsoft Visual Studio to enable the creation, building, debugging, running, and packaging of scalable services.

**Force.com** is offered by Salesforce.com CRM applications and restricted to proprietary Apex language and Visual force presentation layer.Other than the listed above companies, many companies started offering cloud computing services.

## 4 Cloud Computing and Virtualization

Virtualization is a technology which exists for more than three decades. This can be used as a key technology to create cloud computing platform. Unused resources can be effectively utilized, based on the requirement easily resources can be scaled using this technology compared to physical resource scaling. Entire existing OS along with all its application also can be moved to a virtual machine or vice versa also possible.

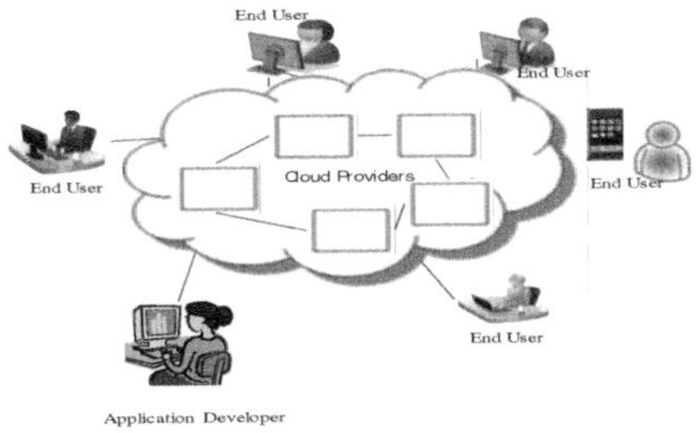


**Fig. 1.** Cloud Application deployment and Access through Cloud Providers

Cloud computing can use virtualization as a key technology, so everything can be accessed through internet. Whether resources might be platform, network, server, application or anything. Depending upon the computational need, easily it can be scaled also.

In IaaS, customers can select their infrastructure such as server, Operating System to run their application, network, and storage capacity likewise the entire necessary infrastructure to provide their service [4] by using these infrastructures as a service (Fig.2).For their application/service they are the owner at the same time maintenance of infrastructure to a third party, a trusted service provider or in other words IaaS Provider.

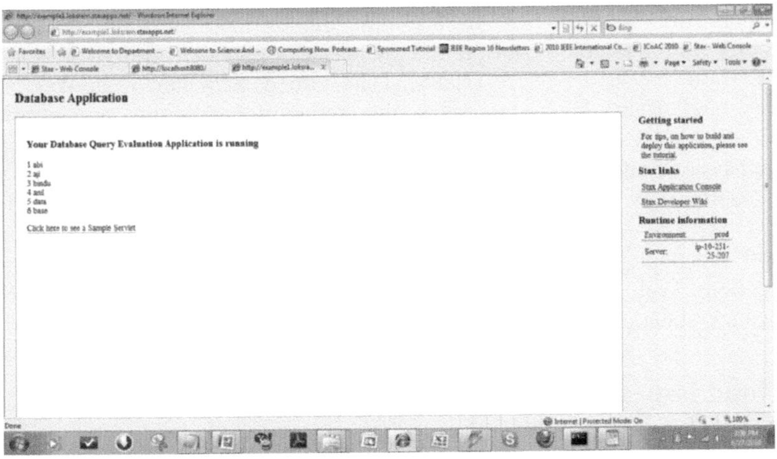


**Fig. 2.** Running Services on Cloud

Since infrastructure is accessed as a service, managing peak load also easy by scaling the resources. Customers will be charged only for their usage of resources at the same reducing capital expenditure to provide their services.

The Proposed Cloud Architecture is depicted as in fig.3. First layer consists of resources, second layer is the virtual management layer, third layer is cloud management layer, fourth layer consists of different type of web services and last one is where the end user connects with the cloud.

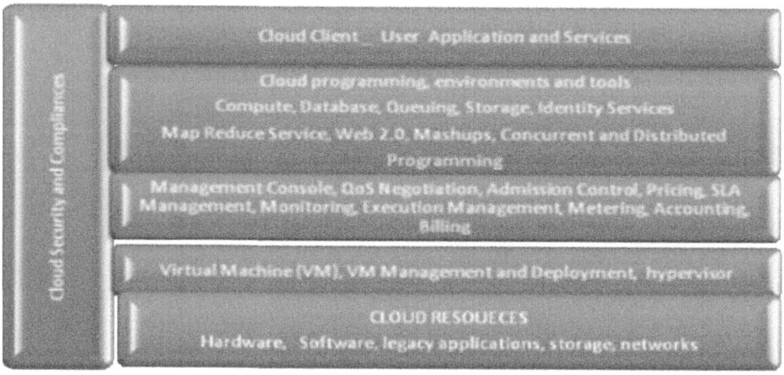


**Fig. 3.** Cloud Architecture

Cloud Resources includes hardware, software, legacy application and network resources. Virtual management layer consists of virtual machines, VM management and deployment, Xen supervisor. Cloud Management layer consists of Management Console, QoS Negotiation, Admission Control, Pricing, SLA Management, Monitoring, Execution Management, Metering, Accounting, and Billing.Fourth layer consists of Cloud Programming tools, Compute, Database, Queuing, Storage, Identity Services, Map Reduce Service, Web 2.0, Mashups, Concurrent and Distributed Programming. Topmost layer is the one, where user applications can be run. Security and Compliances applicable to all the layers.

Virtualization technologies, Middleware frameworks, Grid architecture, Load balancing, Optimal deploy configuration, Consistency models, Software as a Service (SaaS), Hardware as a Service (HaaS), Data grid & Semantic web, Web services, Security and Risk, Fault tolerance and reliability, Auditing, monitoring and scheduling, Utility computing, High-performance computing, Peer to peer computing are used as components in Cloud Computing.

To Deploy Web services, JDK™ 6 can be used. JDK™ 6 _Java EE 5 Development using GlassFish Application Server. *GlassFish,* allow application developers to develop and deploy Java EE-compliant applications such as Deploying Enterprise Applications, Web Applications, EJB module, Connector module, Life cycle Module, Application client module, Custom MBeans.

## 5 Cloud Issues and Virtualization Solution

Cloud Computing has some issues besides its nice features. The fig.4 shows the cloud challenges/issues, a survey taken by IDC in Aug 2008.

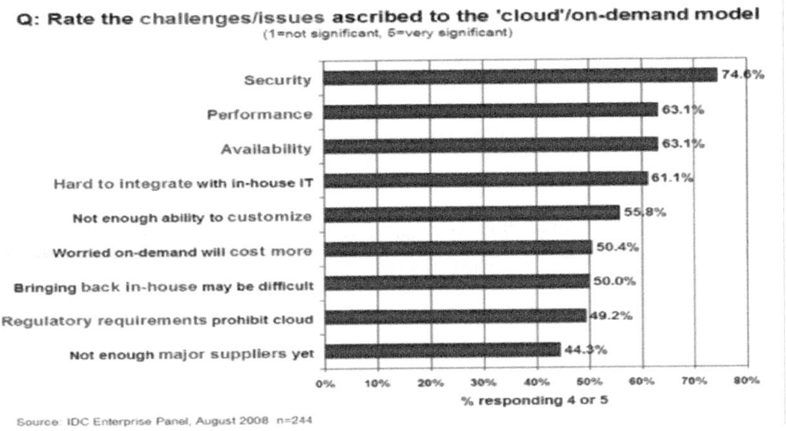


**Fig. 4.** Cloud Challenges/Issues by IDC Survey

Virtualization is not a new technique. It has emerged since from 1960s. IBM introduced this concept to increase the resource utilization. Before that Mainframe was used for one process at a time basis. After the virtualization concept, many virtual machines were introduced and each can be used to run for a separate process/application , so that effective utilization of resources were achieved.

Simply we can say virtualization is simulating hardware and/or software to run on top of another hardware and/or software. Full Virtualization deals with running one or more virtual machines directly on underlying hardware where as in Para Virtualization, hypervisor act as a intermediate layer and it will co ordinate the virtual machines and underlying host infrastructure.

Virtualization not only offers effective utilization of resources. The same feature can be used to reduce the resource cost, since in one physical machine, many virtual machines [3],[12],[13] can be run (Fig.5). Managing and monitoring these virtual machines also easier work for network administrators.

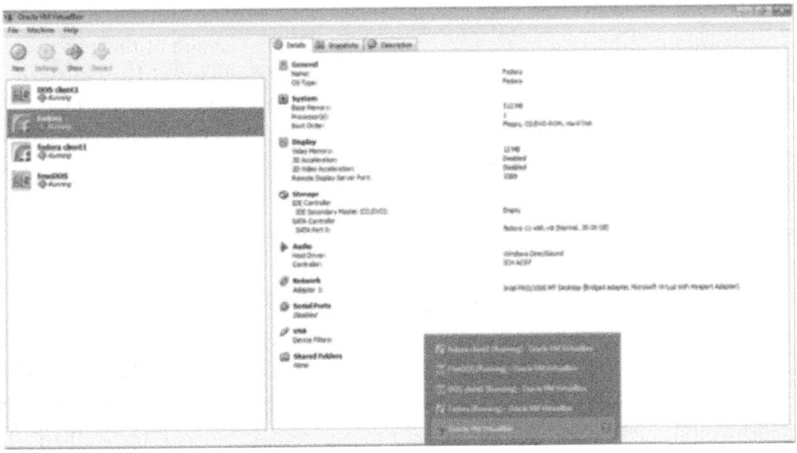

**Fig. 5.** Many VM's Running on a Single Physical Machine

Running different OS in a Single Physical Machine solves the compatibility problem (fig.5 & 6) . Entire Physical machine can be moved to VM and Vice Versa also Possible. Cloning of VM's Easy, which in turn offers the feature of reduced downtime. Fault tolerance handling of Virtual Machines are easy. For example, with the help of images taken, snapshots taken, system can be brought to its original state in case of problems [5].

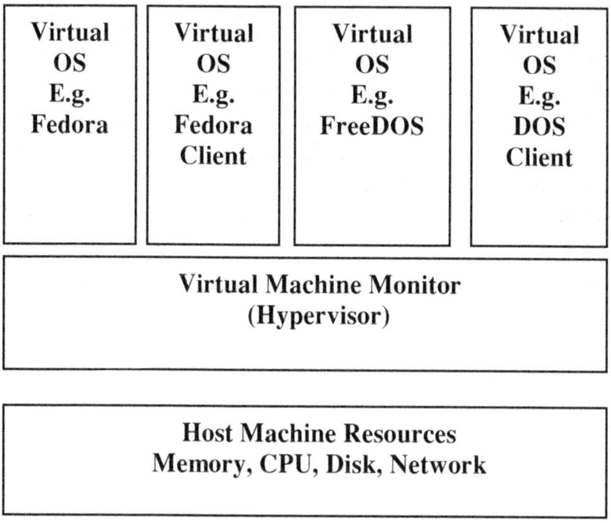


**Fig. 6.** Layered Representation of VMs running on Host machine

If suppose any VM is intruded, or any kind of error, it can be isolated separately, without affecting other VMs, since isolation is possible in virtual machines. By carefully monitoring (fig.7,8 & 9) the resource utilization like CPU, memory, network, port, Isolating the intruded Virtual Machine is possible. By taking appropriate steps to recover or it can be destroyed for security purposes.

Easily system can be rebuilt with the images taken already. Since Many VMs running on a single machine, effective utilization of resources are possible. Moving the VM from one Physical Machine to Another Physical Machine is Possible. Green Computing is achieved with the help of Virtualization technology. Some services and application may conflict with each other, that can be run on different VMs. With the help of Virtualization we can achieve Security, Reliability, Performance, Availability, Portability, Less Cost, Easy Manageability, Less downtime, etc.

Besides having nice features, single physical machine break down will bring down all the VMs running in that. By properly updating the VMs Operating system, installing and updating the firewalls in VMs like the physical machine. A single VM can be specifically used for all the incoming request whether it is genuine request or not then it can be forward.

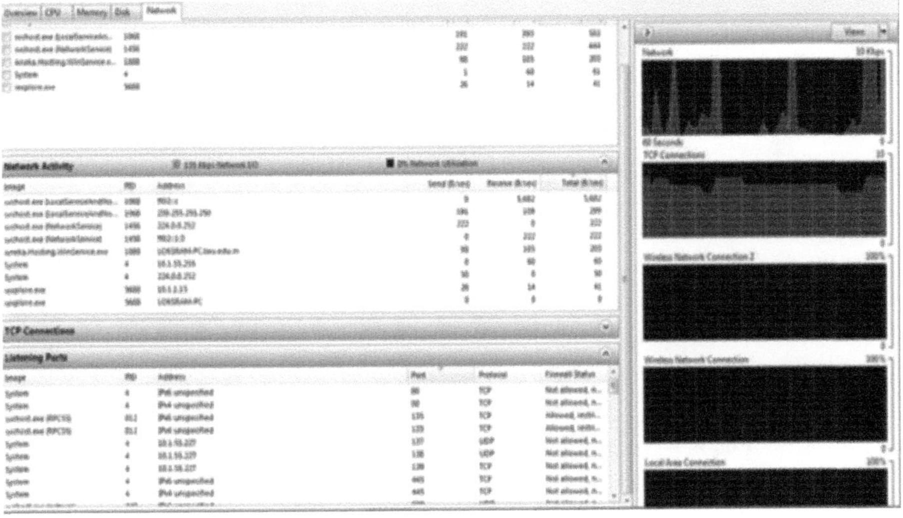

**Fig. 7.** Monitoring of Network Usage

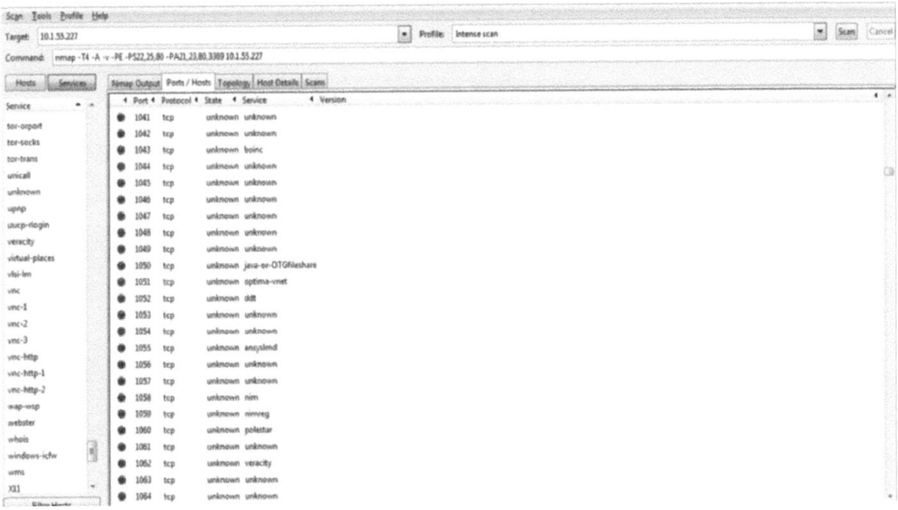


**Fig. 8.** Monitoring of Services and Ports

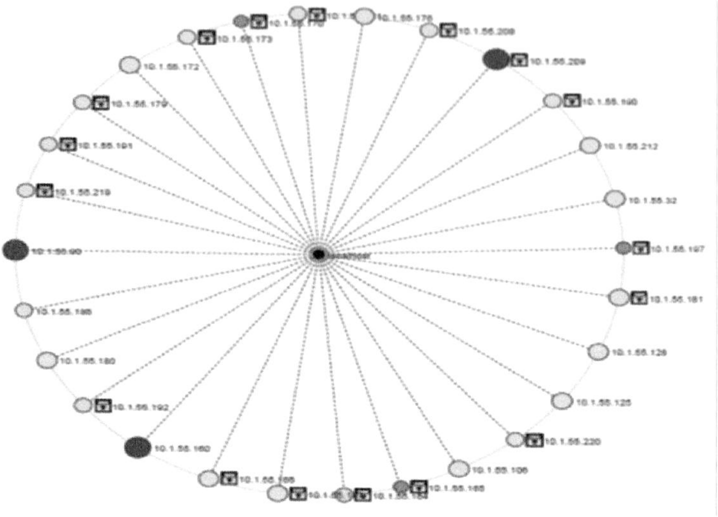

**Fig. 9.** Topology showing system running with unknown services

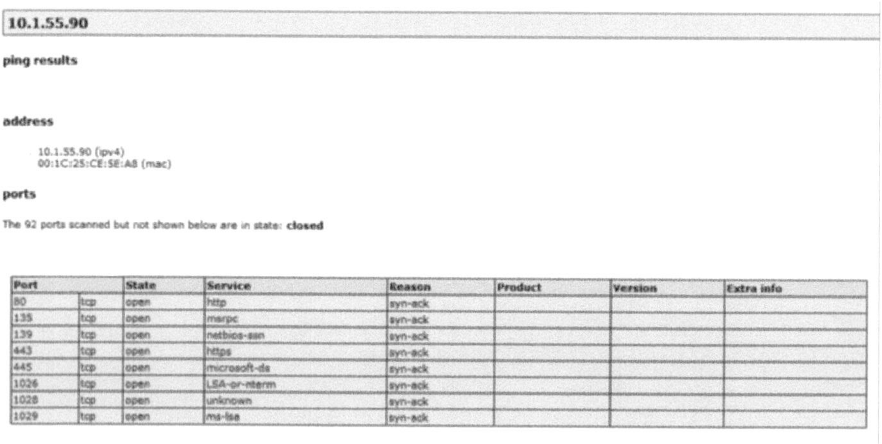

**10.1.55.90**

**ping results**

**address**

10.1.55.90 (ipv4)
00:1C:25:CE:5E:A8 (mac)

**ports**

The 92 ports scanned but not shown below are in state: **closed**

| Port | | State | Service | Reason | Product | Version | Extra info |
|---|---|---|---|---|---|---|---|
| 80 | tcp | open | http | syn-ack | | | |
| 135 | tcp | open | msrpc | syn-ack | | | |
| 139 | tcp | open | netbios-ssn | syn-ack | | | |
| 443 | tcp | open | https | syn-ack | | | |
| 445 | tcp | open | microsoft-ds | syn-ack | | | |
| 1026 | tcp | open | LSA-or-nterm | syn-ack | | | |
| 1028 | tcp | open | unknown | syn-ack | | | |
| 1029 | tcp | open | ms-lsa | syn-ack | | | |

**Fig. 10.** Report showing system running with unknown service

If there is any hidden/unknown request (Fig.9,10), along with proper resource monitoring and port scanning, intruded VM can be easily identified. Upon error, necessary steps can be taken down to overcome by isolating the intruded VM. Proper isolation is important. Since resources allocated for each machine is different, that cannot be accessed through other virtual machine. By giving restricted access/ by devising proper privilege level all the VMs to access the physical resource, security of Virtualization can be fully achieved. With Secured Virtualization technologies, cloud security can be improved.

## 6 Conclusion

Cloud computing is going to be one of the venture technology in future, even though it has some serious issues like security, privacy, social and political issues[11]. Cloud user should understand their own network, system, applications and data are moving to an unknown network which poses serious threat to security and privacy [2]. So they have to vigilantly select Cloud Service Providers by checking their security service standards and compliances.

Nowadays, Most of the vendors provide free usage to deploy their services in their cloud with certain limits like hours of usage, or hard disk space, storage or data transfer or number of end users. As a cloud user or developer, they have to choose the vendor based on their Service Level Agreements.

In future, more emphasis is going to be in the area of cloud security by defining new cryptographic methods and key management techniques, since service can be deployed anywhere in the Cloud , sometimes against the privacy policy depends upon the regional location where service is deployed that has to be taken in to consideration.

Using virtualization server, resources, network, desktop, application, operating system and storage can be virtualized. One of the major concerns in future is computing with less power. With virtualization, apart from flexibility, scalability, security, utilizing underutilized resources/idle resources, manageability, cost effective cloud computing with virtualization technology takes less power since more than one virtual machines can be run on a single physical machine. In future like IaaS, PaaS and SaaS, Anything as a Service (XaaS) is going to be possible and can be achieved through Virtualization.

## References


1. Foster, I., Zhao, Y., Raicu, I., Lu, S.: Cloud Computing and Grid Computing 360-Degree Compared. In: Grid Computing Environments Workshop, GCE 2008, November 12-16 (2008)
2. Harauz, J., Kaufman, L.M., Potter, B.: Data Security in the World of Cloud Computing. IEEE Security and Privacy 7(4), 61–64 (2009)
3. Loganayagi, B., Sujatha, S.: Creating virtual platform for cloud computing. In: 2010 IEEE International Conference on Computational Intelligence and Computing Research (ICCIC), December 28-29, pp. 1–4 (2010), doi:10.1109/ICCIC.2010.5705744
4. Loganayagi, B., Sujatha, S.: Cloud Computing in Stax Platform. In: IEEE International Conference on Computer Communication and Electrical Technology (IEEE- ICCCET 2011), March 18-19, pp. 1–5 (2011)
5. Zhao, S., Chen, K., Zheng, W.: The Application of Virtual Machines on System Security. In: 2009 Fourth ChinaGrid Annual Conference, Chinagrid, pp. 222–229 (2009)
6. Zhang, L.-J., Zhou, Q.: CCOA: Cloud Computing Open Architecture. In: IEEE International Conference on Web Services, ICWS 2009, July 6-10, pp. 607–616 (2009)
7. Viega, J.: Cloud Computing and the Common Man. Computer 42(8), 106–108 (2009)

8. Cunsolo, V.D., Distefano, S., Puliafito, A., Scarpa, M.: Volunteer Computing and Desktop Cloud: The Cloud@Home Paradigm. In: Eighth IEEE International Symposium on Network Computing and Applications, pp. 134–139 (2009)
9. Mei, L., Chan, W.K., Tse, T.H.: A Tale of Clouds: Paradigm Comparisons and Some Thoughts on Research Issues. In: IEEE Asia-Pacific Services Computing Conference, pp. 464–469 (2008)
10. Reese, G.: Cloud Application Architectures. O'Reilly, Sebastopol (2009)
11. Mather, T., Kumaraswamy, S., Latif, S.: Cloud Security and Privacy. O'Reilly, Sebastopol (2009)
12. Ritinghouse, J.W., Ransome, J.F.: Cloud Computing: Implementation Management and Security. CRC Press, Boca Raton (2010)
13. Virtualbox Information, `http://www.virtualbox.org/wiki/Documentation`
14. Cloud General Information, `http://en.wikipedia.org/wiki/Cloud_computing`

# Selection of Software Testing Technique: A Multi Criteria Decision Making Approach

Monisha Victor and Nitin Upadhyay

BITS Pilani, K.K. Birla Goa Campus,
Near NH17B Bypass road, Zuari Nagar,
Goa-403726, India
{h2009038,nitinu}@bits-goa.ac.in
http://www.bits-goa.ac.in

**Abstract.** The appropriate usage of efficient testing technique at any stage of the Software Development Life Cycle (SDLC) is still in infant stage. There are a number of testing techniques at various phases of testing. Selection of the right testing technique at any stage is one of the critical problems. The selection method should not only take the subjective knowledge but also incorporate objective knowledge for the efficient selection of testing technique as per the requirements. Selection of testing technique at every stage of SDLC depends on many factors such as resources, schedule, cost of the project, etc. Thus we formulate the selecting of testing technique as a multi criteria decision making problem and propose an efficient solution.

**Keywords:** Software Testing, SDLC, Testing Criteria, Selection of testing techniques, TOPSIS, AHP, MCDM.

## 1 Introduction

Software Testing is one of the most crucial and important phases of the Software Development Life Cycle (SDLC). The quality of the delivered product can be assured by testing the software as early in the SDLC as possible. Testing is done at every stage of the SDLC with different objective. For example, when a unit or module is implemented, the module is tested in isolation to find errors in the module [1]. Unit tested modules are integrated and the objective of integration testing is to test the integration among the unit tested modules and ensure that the newly added module does not introduce any new errors [2]. Software Testing evolved with time from testing being a part of debugging to testing requirements, design and implementation. For high confidence before deploying a system, the software has to be tested exhaustively. Since it is not feasible, the system is tested for a particular set of inputs designed based on testing techniques and checked against the expected outputs. There are numerous testing techniques of different types for testing requirements, architecture, design, code, integration of components, system [3][4]. The testing technique to be chosen at any stage depends on various factors like the software development method, test objective, cost of the project, schedule of the project,

D. Nagamalai, E. Renault, M. Dhanushkodi (Eds.): CCSEIT 2011, CCIS 204, pp. 453–462, 2011.

human resources, computational resources etc [5]. Research has been done to improve the effectiveness of testing based on reducing the number of test cases [6], experimental evaluation of testing techniques [7][8], based on data flow relationship [9], fault coverage [10], prioritizing the test cases based on certain objective function [11][12], rate at which faults are detected [13], etc. There is a need for a selection method that considers not only the subjective knowledge but also the objective knowledge and multiple criteria in selecting the best possible alternative during various stages of SDLC like unit testing, integration testing or system testing.

The approach in this paper considers the selection of the best testing technique at any phase of SDLC as a Multi Criteria Decision Making (MCDM) problem and uses Technique for Order Preference by Similarity to Ideal Solution (TOPSIS) along with Analytic Hierarchy Process (AHP) to solve it. AHP is used to determine the weights of the criteria. The weights calculated using AHP are used in TOPSIS, a technique for solving decision problems and is used to calculate the relative advantages of the alternatives. The method is based on the concept of TOPSIS that the best alternative chosen is at a shorter distance from the positive ideal solution and is the farthest from the negative ideal solution.

The methodology used to solve the problem is discussed in Section 2 and an illustrative example is given in the Section 3 of the paper.

## 2 Methodology

The selection of a testing technique can be considered as a multi criteria decision making problem. The method proposed in this paper uses TOPSIS along with AHP [14]. TOPSIS is based on the idea that the chosen alternative has the shortest distance from the positive ideal solution and is the farthest from the negative ideal solution [15]. AHP is used to calculate the weights of the criteria used for evaluation. AHP uses consistency ratio as a measure to verify the consistency of the weights obtained for the criteria. This ensures that the weights of the criteria are consistent.

Figure 1 depicts the flow chart of the methodology proposed.

The following steps need to be performed to choose the best alternative when n criteria are used to compare m alternatives.

Step 1 - Decision Matrix $[\mathbf{D}]_{m \times n}$

In this step, a decision matrix of criteria and alternatives based on the available information is developed. The value $d_{ij}$ of the matrix is the actual value of the *i*th alternative in terms of *j*th decision criteria i.e., the rating of the *i*th alternative with respect to the *j*th criteria. The order of the decision matrix is m × n.

$$D_{mxn} = \begin{pmatrix} d_{11} & d_{12} & \dots & d_{1n} \\ d_{21} & d_{22} & \dots & d_{2n} \\ \dots & \dots & \dots & \dots \\ d_{m1} & d_{m2} & \dots & d_{mn} \end{pmatrix} \tag{1}$$

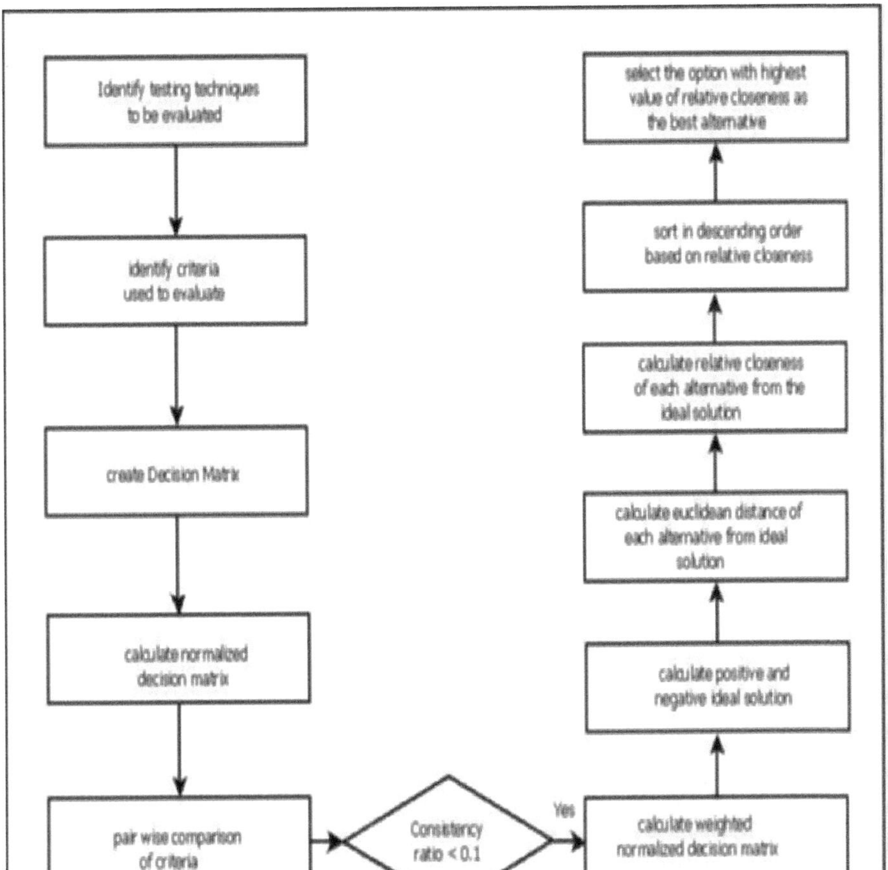


**Fig. 1.** Flow chart of the proposed methodology

Step 2 - Normalized decision Matrix **[R]** $_{m\times n}$

The decision matrix [D] is normalized so that ratings given in different scales and units to various criteria are transformed into common measurable units to make comparisons easy. The order of the normalized decision matrix is m × n. The value $r_{ij}$ of the matrix [R] is calculated using equation (2).

$$r_{ij} = d_{ij} / [\sum_{i=1}^{m}(d_{ij})^2]^{0.5} \tag{2}$$

Step 3 – Attribute Weights **[W]** $_{1\times n}$

AHP [16] is used in this step to calculate the weights of the criteria used to evaluate the alternatives. A pair wise comparison of the criteria is done and the absolute weights of the criteria are calculated using AHP. The pair wise comparison is done on a scale of 1-9 and the values along with its explanation are shown in detail in Table 1.

**Table 1.** AHP rating scale

| Value | Explanation |
|---|---|
| 1 | Equal importance |
| 3 | Moderate importance |
| 5 | Strong importance |
| 7 | Very Strong importance |
| 9 | Extreme importance |
| 2,4,6,8 | Intermediate values |

If A is the pairwise comparison matrix, the relative weights of the criteria [W] is calculated using the equation (3).

$$( A - \lambda_{max} I ) \times W = 0 \quad (3)$$

where $\lambda_{max}$ is the largest eigen value of the matrix A and I is the $n \times n$ unit matrix.

The value of consistency ratio while calculating the weights is considered as a factor which judges the consistency of the absolute weights given to the criteria. Any value less than 0.1 implies that the judgment is consistent. The consistency ratio is calculated using the equation (4).

$$CR = CI / RI \quad (4)$$

where CI is consistency index and RI is the random consistency index. The value of the consistency index is calculated using the equation (5).

$$CI = (\lambda_{max} - N) / (N - 1) \quad (5)$$

where N is the number of criteria.

The values of the random consistency index for different values of N are given in Table 2.

**Table 2.** Random consistency index

| N | 1 | 2 | 3 | 4 | 5 | 6 | 7 | 8 | 9 |
|---|---|---|---|---|---|---|---|---|---|
| RI | 0 | 0 | 0.58 | 0.9 | 1.12 | 1.24 | 1.3 | 1.41 | 1.45 |

Step 4 – Weighted Normalized Matrix $[\mathbf{V}]_{m\times n}$

The weighted normalized matrix [V] is calculated by multiplying each column of the normalized decision matrix [R] with the respective column of the [W] matrix. The value $v_{ij}$ of the matrix [V] is calculated using equation (6).

$$v_{ij} = W_j \times r_{ij} \quad (6)$$

Step 5 – Positive ideal solution $[\mathbf{I_s}^+]_{n\times 1}$ and Negative ideal solution $[\mathbf{I_s}^-]_{n\times 1}$

The positive ideal solution is calculated as the best value each criterion can attain. The negative ideal solution is calculated as the least/worst value each criterion can attain. They are produced as follows:

$$I_s^+ = \{ (\max v_{ij} / j \in \tau), (\min v_{ij} / j \in \tau') \text{ for } i=1,2,3,\ldots,m\}$$

$$= \{V_1^+, V_2^+, \ldots, V_n^+\} \quad (7)$$

$$I_s^- = \{ (\min v_{ij} / j \in \tau), (\max v_{ij} / j \in \tau') \text{ for } i=1,2,3,\ldots,m\}$$

$$= \{V_1^-, V_2^-, \ldots, V_n^-\} \quad (8)$$

Where $\tau$ is associated with the benefit or positive criteria and $\tau'$ is associated with the cost or negative criteria.

Step 6 – N dimensional Euclidean distance
The N dimensional Euclidean distance is calculated from each alternative to the positive and negative ideal solution as follows:

$$S_i^+ = \{\sum_{j=1}^{N} (V_{ij} - V_j^+) \}^{0.5}, i = 1,2,\ldots,m \quad (9)$$

and

$$S_i^- = \{\sum_{j=1}^{N} (V_{ij} - V_j^-) \}^{0.5}, i = 1,2,\ldots,m \quad (10)$$

Step 7 – Relative Closeness $[C_i^*]_{m\times 1}$
Relative closeness of each alternative with respect to the positive and negative ideal solution is calculated as follows:

$$C_i^* = S_i^- / (S_i^+ + S_i^-) \quad (11)$$

Step 8 - Ranking
The alternatives are arranged in the descending order of their relative closeness ($C_i^*$) from the ideal solutions. The alternative with the maximum value of $C_i^*$ is chosen as the best alternative.

## 3 Illustrative Example

A software to maintain the *Research Information* at a University was developed. The integration among the unit tested modules had to be tested. A list of alternatives (testing techniques) and criteria were identified and the proposed methodology was applied to find the best integration testing technique among the alternatives.

The integration testing techniques identified for testing the software are:

A = {Incremental Integration, Top-Down integration testing, Bottom-up integration testing, Sandwich, Big bang, End-to-End testing, High Frequency testing, Smoke testing}.

The set of criteria considered for evaluating and selecting the best alternative is

C = {Time when the basic functionality is exposed, Reusability of the test cases, End user view, Time of fault detection, Effort required in terms of the additional work to be done, Ease of fixing errors, Frequency of running the tests, Ease of writing the test cases, and Possibility of automating the testing}.

The following are the steps followed in finding the best alternative among the above mentioned ones. The decision matrix **[D]** which contains the rating of the *i*th alternative with respect to the *j*th criteria is shown in Table 3.

**Table 3.** Decision Matrix D

| | C1 | C2 | C3 | C4 | C5 | C6 | C7 | C8 | C9 |
|---|---|---|---|---|---|---|---|---|---|
| A1 | 3 | 4 | 1 | 3 | 4 | 5 | 3 | 4 | 3 |
| A2 | 5 | 4 | 4 | 5 | 3 | 4 | 3 | 4 | 1 |
| A3 | 1 | 1 | 1 | 2 | 3 | 4 | 3 | 5 | 1 |
| A4 | 5 | 3 | 2 | 4 | 3 | 4 | 2 | 3 | 1 |
| A5 | 5 | 2 | 4 | 4 | 3 | 1 | 1 | 3 | 1 |
| A6 | 5 | 5 | 5 | 5 | 4 | 2 | 3 | 3 | 4 |
| A7 | 3 | 3 | 1 | 5 | 4 | 5 | 5 | 3 | 5 |
| A8 | 5 | 3 | 5 | 5 | 3 | 4 | 4 | 4 | 3 |

Table 4 gives the decision matrix after normalization **[R]**.

**Table 4.** Normalized Decision Matrix **R**

| | C1 | C2 | C3 |
|---|---|---|---|
| A1 | 0.250000 | 0.423999 | 0.106000 |
| A2 | 0.416667 | 0.423999 | 0.423999 |
| A3 | 0.083333 | 0.106000 | 0.106000 |
| A4 | 0.416667 | 0.317999 | 0.212000 |
| A5 | 0.416667 | 0.212000 | 0.423999 |
| A6 | 0.416667 | 0.529999 | 0.529999 |
| A7 | 0.250000 | 0.317999 | 0.106000 |
| A8 | 0.416667 | 0.317999 | 0.529999 |

| | C4 | C5 | C6 |
|---|---|---|---|
| A1 | 0.249136 | 0.414781 | 0.458349 |
| A2 | 0.415227 | 0.311085 | 0.366679 |
| A3 | 0.166091 | 0.311085 | 0.366679 |
| A4 | 0.332182 | 0.311085 | 0.366679 |
| A5 | 0.332182 | 0.311085 | 0.091670 |
| A6 | 0.415227 | 0.414781 | 0.183340 |
| A7 | 0.415227 | 0.414781 | 0.458349 |
| A8 | 0.415227 | 0.311085 | 0.366679 |

| | C7 | C8 | C9 |
|---|---|---|---|
| A1 | 0.331295 | 0.383131 | 0.377964 |
| A2 | 0.331295 | 0.383131 | 0.125988 |
| A3 | 0.331295 | 0.478913 | 0.125988 |
| A4 | 0.220863 | 0.287348 | 0.125988 |
| A5 | 0.110432 | 0.287348 | 0.125988 |
| A6 | 0.331295 | 0.287348 | 0.503953 |
| A7 | 0.552158 | 0.287348 | 0.629941 |
| A8 | 0.441726 | 0.383131 | 0.377964 |

The weight matrix **[W]** is calculated using AHP [17] and is shown in Table 5. The consistency index calculated when finding weights of the criteria in the step 3 is 0.06 which is less than 0.1. Hence the weights calculated are consistent and can be used further.

**Table 5.** Weight Matrix **W**

| | | |
|---|---|---|
| C1 | C2 | C3 |
| 0.183400 | 0.112800 | 0.085500 |
| C4 | C5 | C6 |
| 0.056700 | 0.115000 | 0.030200 |
| C7 | C8 | C9 |
| 0.161600 | 0.200400 | 0.054300 |

Weighted Normalized Decision matrix **[V]** is calculated using the equation (6) and is shown in Table 6.

**Table 6.** Weighted Normalized Decision Matrix **V**

| | C1 | C2 | C3 |
|---|---|---|---|
| A1 | 0.045850 | 0.047827 | 0.009063 |
| A2 | 0.076417 | 0.047827 | 0.036252 |
| A3 | 0.015283 | 0.011957 | 0.009063 |
| A4 | 0.076417 | 0.035870 | 0.018126 |
| A5 | 0.076417 | 0.023914 | 0.036252 |
| A6 | 0.076417 | 0.059784 | 0.045315 |
| A7 | 0.045850 | 0.035870 | 0.009063 |
| A8 | 0.076417 | 0.035870 | 0.045315 |
| | C4 | C5 | C6 |
| A1 | 0.014126 | 0.047700 | 0.013842 |
| A2 | 0.023543 | 0.035775 | 0.011074 |
| A3 | 0.009417 | 0.035775 | 0.011074 |
| A4 | 0.018835 | 0.035775 | 0.011074 |
| A5 | 0.018835 | 0.035775 | 0.002768 |
| A6 | 0.023543 | 0.047700 | 0.005537 |
| A7 | 0.023543 | 0.047700 | 0.013842 |
| A8 | 0.023543 | 0.035775 | 0.011074 |
| | C7 | C8 | C9 |
| A1 | 0.053537 | 0.076779 | 0.020523 |
| A2 | 0.053537 | 0.076779 | 0.006841 |
| A3 | 0.053537 | 0.095974 | 0.006841 |
| A4 | 0.035691 | 0.057585 | 0.006841 |
| A5 | 0.017846 | 0.057585 | 0.006841 |
| A6 | 0.053537 | 0.057585 | 0.027365 |
| A7 | 0.089229 | 0.057585 | 0.034206 |
| A8 | 0.071383 | 0.076779 | 0.020523 |

Positive and Negative ideal solutions $[I_s^+]$ and $[I_s^-]$ are calculated from the weighted normalized decision matrix using the equation (7) and (8) and is shown in Table 7.

**Table 7.** Positive and Negative Ideal solution $I_s^+$ **and** $I_s^-$

| | C1 | C2 | C3 |
|---|---|---|---|
| $I_s^+$ | 0.076417 | 0.059784 | 0.045315 |
| $I_s^-$ | 0.015283 | 0.011957 | 0.009063 |
| | C4 | C5 | C6 |
| $I_s^+$ | 0.023543 | 0.47700 | 0.013842 |
| $I_s^-$ | 0.009417 | 0.035775 | 0.002768 |
| | C7 | C8 | C9 |
| $I_s^+$ | 0.089229 | 0.095974 | 0.034206 |
| $I_s^-$ | 0.017846 | 0.057585 | 0.006841 |

The N-dimensional Euclidean distance from each of the alternative to the Positive and Negative ideal solution is calculated using equation (9) and (10) and is shown in Table 8.

**Table 8.** Euclidean distance from Positive and Negative Ideal solution ($I_S^+$, $I_S^-$), relative closeness and ranking of the testing techniques

| | A1 | A2 | A3 | A4 |
|---|---|---|---|---|
| S+ | 0.395931 | 0.343459 | 0.486917 | 0.435656 |
| S- | 0.403377 | 0.448899 | 0.287030 | 0.360109 |
| C* | 0.504658 | 0.566536 | 0.370866 | 0.452532 |
| Rank | 5 | 3 | 8 | 6 |
| | A5 | A6 | A7 | A8 |
| S+ | 0.458015 | 0.298710 | 0.359335 | 0.298881 |
| S- | 0.331204 | 0.479840 | 0.436294 | 0.479734 |
| C* | 0.419661 | 0.616325 | 0.548363 | 0.616138 |
| Rank | 7 | 1 | 4 | 2 |

The option which has the highest value of relative closeness is *end-to-end testing technique*. The characteristics of end-to-end testing like exclusive focus on end user view, reusability of test cases [18] which are some of the criteria used to evaluate the testing techniques makes it the best choice among the available alternatives. The option with the least value of relative closeness is *Bottom-up testing technique*. Some of the characteristics of bottom-up integration testing which makes it not the best choice among the alternatives are non-reusability of the test cases, no end-user view from the beginning of the integration, effort in terms of coding drivers [2]. This shows that the methodology proposed considers all the criteria to evaluate the testing techniques and gives the appropriate result.

This method can be used in any software organization for the right selection of testing technique at any stage of the SDLC. Criteria like resources required (human or computational), previous use of a testing technique in the organization, ease of fixing the defects by the developers, training required before the use of the technique by a tester can also be used by the testing team according to the schedule, cost or resources requirement of the project in the process of decision making. Thus the method considers not only the subjective knowledge but also practical aspects of the testing techniques in choosing the best testing technique to be used.

## 4 Conclusion

Selecting the right testing technique is very critical for the success of the project. The best suitable testing technique at various phases of the software development life cycle can be chosen based on various factors. The problem of choosing the best suitable testing technique is formulated as a Multi Criteria Decision Making problem and a method using TOPSIS and AHP is proposed for the same. This method can select a testing technique from among M alternatives available based on N selection or decision criteria. An illustrative example is demonstrated to show the applicability of the proposed approach.

## References


1. Ntafos, S.C.: A comparision of some structural testing strategies. IEEE Transactions on Software Engineering 14(6), 868–874 (1988)
2. Naik, K., Tripathy, P.: Software Testing and Quality Assurance. Wiley, Chichester (2008)
3. Amar, M., Shabbir, K.: Systematic Review on Testing Aspect oriented Programs - Challenges, Techniques and Their Effectiveness. Blekinge Institute of Technology. In: MSE-2008, 17 (2008)
4. Juristo, N., Moreno, A.M., Vegas, S.: State of the Empirical Knowledge on Testing Techniques. In: Proceedings of the International Conference on Software Maintenance (1999)
5. Vegas, S.: Characterisation schema for selecting software testing techniques (2002)
6. Mahapatra, R.P., Singh, J.: Improving the Effectiveness of Software Testing through Test Case Reduction. World Academy of Science, Engineering and Technology 37, 345–350 (2008)
7. Roper, M., Miller, J., Brooks, A., Wood, M.: Towards the Experimental Evaluation of Software Testing Techniques. In: Proc. EuroSTAR 1994 (1994)
8. Frankl, P.G., Weiss, S.N.: An Experimental Comparison of the Effectiveness of Branch Testing and Data Flow Testing. IEEE Transactions on Software Engineering 19(8), 774–787 (1993)
9. Clarke, L.A., Podgurski, A., Richardson, D.J., Zeil, S.J.: A comparison of data flow path selection criteria. In: Proceedings of the Eighth International Conference on Software Engineering, pp. 244–251 (1985)
10. Basili, V.R., Selby, R.W.: Comparing the effectiveness of software testing. IEEE Transactions on Software Engineering SE-13(12), 1278–1296 (1987)
11. Rothermel, G., Untch, R.H., Chu, C., Harrold, M.J.: Test Case Prioritization: An Empirical Study. In: Proceedings of the International Conference on Software Maintenance, pp. 179–188 (1999)

12. Williams, L., Snipes, W., Meneely, A.: On Increasing System Test Effectiveness through a Test Case Prioritization Model Using Static Metrics and System Failure Data. In: Reliability Analysis of System Failure Data Workshop (2007)
13. Ma, Z., Zhao, J.: Test case prioritization based on analysis of program structure. In: Asia-Pacific Software Engineering Conference, pp. 471–478 (2008)
14. Upadhyay, N., Deshpande, B.M., Agrawal, V.P.: Integrated decision approach for COTS selection. Int. J. Decision Sciences, Risk and Management 2(3/4), 165–177 (2010)
15. Jadidi, O., Hong, T.S., Firouzi, F., Yusuff, R.M., Zulkifli, N.: TOPSIS and fuzzy multi-objective model integration for supplier selection problem. Journal of Achievements in Materials and Manufacturing Engineering 31(2), 762–769 (2008)
16. Saaty, T.L.: The Analytic Network Process. RWS Publications, Pittsburgh (1996)
17. James, T.: Whalen & Associates, LLC, `http://www.whalen.ws/`
18. Tsai, W.T., Bai, X., Paul, R., Shao, W., Agarwal, V.: End-to-end Integration Testing Design. In: Computer Software and Applications Conference, pp. 166–171 (2001)

# A Hybrid Evolutionary Algorithm for the Page Number Minimization Problem

Dharna Satsangi, Kamal Srivastava*, and Gursaran

Department of Mathematics,
Dayalbagh Educational Institute,
Agra, India
{dharna.satsangi,kamalsrivast,gursaran.db}@gmail.com

**Abstract.** In this paper we present a hybrid evolutionary algorithm (HEA) for solving the pagenumber problem. In HEA, random depth first search of the graph is used for placing the vertices on the spine and an edge embedding heuristic is designed to distribute the edges on a minimal number of pages. The results show that the algorithm achieves the optimal pagenumber for most of the standard graphs tested by us. HEA performance is also compared with the GA described by Kapoor et al. [1] on select instances of standard and random graphs. It is observed that HEA gives a better solution for most of the instances.

**Keywords:** Pagenumber problem, Hybrid Evolutionary Algorithm, Depth First Search.

## 1 Introduction

The book embedding problem consists of embedding a graph in a book with its vertices in a line along the spine of the book and its edges on the book in such a way that edges residing on the same page do not cross. The *pagenumber* of a graph $G$ is the minimum number of pages of the book into which $G$ can be embedded. This problem abstracts layout problems arising in direct interconnection networks, fault-tolerant processor arrays, fault tolerant VLSI design, sorting with parallel stacks and single row routing [1]. The problem of obtaining pagenumber is NP-complete for general graphs [1]. However, this problem has been studied for some standard graphs and their page numbers have been found. Pagenumbers of FFT (Fast Fourier Transforms), Benes, Barrel Shifter networks and t-partite graphs [2], complete graphs, grids, hypercubes, trees, X-trees, pinwheel graphs [3], planar graphs, deBruijn and for shuffle-exchange graphs ([4], [5]) are known in literature. Shahrokhi et al. [6] have given polynomial time algorithms to generate near optimal drawing of graphs on books. Berhart et al. [7], showed that the pagenumber is less than or equal to one if and only if the graph is outerplanar. Swaminathan et al. [8] have shown that bandwidth-$k$ graphs can be embedded in a book of $k$-1 pages, though it is not a tight

* Corresponding author.

D. Nagamalai, E. Renault, M. Dhanushkodi (Eds.): CCSEIT 2011, CCIS 204, pp. 463–475, 2011.

bound when considered on standard graphs with known bandwidth. Upper and lower bounds on the pagenumber of $k$-ary hypercubes are given by Bettayeb et al. [9]. In order to deal with the problem of obtaining the pagenumber for general graphs, a genetic algorithm (GA) for finding the pagenumber has been proposed by Kapoor et al. [1]. However, they have reported the results for only a few standard graphs of small sizes. In this paper a Hybrid Evolutionary Algorithm (HEA) has been designed for the pagenumber problem (PNP). In devising a solution to PNP, one has to focus on two main factors: 1) placing the vertices on the spine and 2) distributing the edges over the pages of the book. In HEA, vertex labels generated using the random depth first search (RDFS) are used for placing the vertices on the spine and a distribution heuristic motivated by the method for obtaining the pagenumber of complete graphs [3] is used for placing the edges on the pages of the book. Experimental results show that HEA outperforms genetic algorithm proposed by Kapoor et al. [1] for PNP on the graphs on which the two were compared.

In this paper, the proposed hybrid evolutionary algorithm is described in Section 2 and the implementation details are described in Section 3. In Section 4 the experiments and results obtained are described. Section 5 presents the conclusions of the work described in the paper.

## 2 Hybrid Evolutionary Algorithm for PNP

### 2.1 Notations

Let $G = (V, E)$ be an undirected graph with vertex set $V$ and edge set $E$. A labeling of the vertices of an undirected graph with $n(=|V|)$ vertices is a bijective function $\varphi: V \rightarrow [n] = \{1,2, \dots n\}$. A pair of edges $uv$ and $pq$ cross in a drawing iff $1 < \varphi(u) < \varphi(p) < \varphi(v) < \varphi(q) < n$ and both lie on the same page.

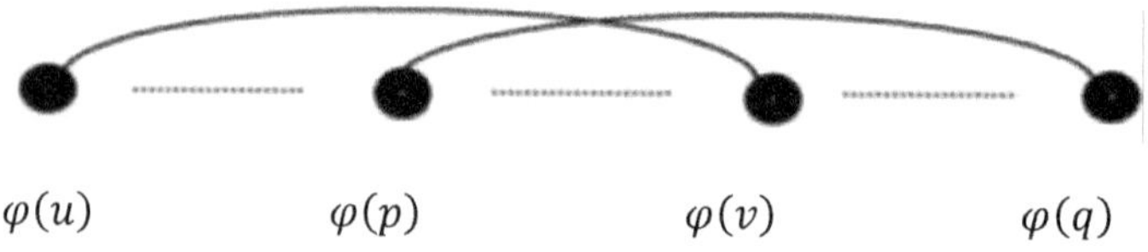


**Fig. 1.** Edge crossing condition $\varphi(u) < \varphi(p) < \varphi(v) < \varphi(q)$

### 2.2 Motivation

A depth first search (DFS) based technique for placing the vertices on the spine (a particular placement of vertices on the spine is also referred to as a *layout* of vertices in this paper) seems to be helpful in minimizing the number of crossings when edges are placed in a single page as it helps in generating layouts in which there are edges between adjacent nodes. In order to confirm this, we carried out an experiment on random connected graphs to explore the number of crossings on a single page obtained by randomly generated layouts and by DFS based layouts. For the experiments, 50 graphs of 50 vertices having an edge density 30%, where edge density of a graph with of $n$ vertices and $m$ edges is taken as $2m / n(n-1)$, were generated. Mean crossings and standard deviation for the DFS based layouts were found to be 17147.32 and 2050.042 as compared to 20943.42 and 2470.513 for the random layouts. Keeping in view the results of the experiment, we have generated an initial population of layouts for HEA using random depth first search (RDFS) where the root vertex and the neighbors of a visited vertex are chosen randomly.

PNP remains a difficult problem even when the layout is fixed [10]. Given a layout there is no guarantee that a particular embedding of edges onto the pages is minimal. Chung et al. [3] have given an edge embedding strategy that provides the optimal pagenumber for complete graphs. The edge density of complete graphs is 100% and the layout does not play a role in determining the pagenumber. Since any simple graph that is not a complete graph will have an edge density that is less than 100%, the approach proposed by Chung et al. [3] can be used to develop a strategy for distributing edges on pages. The edge embedding heuristic (EEH) has been developed by us based on the strategy proposed by Chung et al. [3] to determine a distribution of edges using minimal pages. It is described in detail in Section 3.1.

### 2.3 Solution Representation

In HEA, for a graph $G(V,E)$, $V = \{V_1, V_2, \dots, V_n\}$, each solution, $\zeta = (R, Q)$ is represented in two components:

1) An array $R$ of length $n$ that represents a labeling $\varphi$. Vertices are placed on the spine in the label sequence $1, 2, \dots, n$.
2) An $|E| \times 3$ array $Q$, where each row contains the end vertices of an edge in the first two columns and the third column stores the page index in which the edge is placed.

   The pagenumber of a solution $\zeta$ is $\max_{\forall i} Q[i][3]$ and is denoted $pg\_no(G,\zeta)$. We will use the term 'optimal pagenumber' for the optimal value of page number.

### 2.4 The Algorithm

Evolutionary algorithms (EAs) are the approximate optimization algorithms that are based on evolutionary principles. They are inspired by nature's capability to evolve living beings well adapted to their environment. At the core of EA is a population of individuals. In each iteration in the algorithm, *reproduction* operators are applied to the individuals of the current population to generate the individuals of the next generation. *Recombination* operators may also be used to recombine two or more individuals to produce new individuals. The algorithm may also use *mutation* operators that cause self-adaptation of individuals. The driving force in the EAs is the *selection* of individuals based on their fitness (which might be based on the objective function or some kind of simulation experiment). Individuals with higher fitness have a higher probability to be chosen as members of the next generation (parents) which corresponds to the principle of *survival of fittest* in natural selection. Simulated annealing (SA) proposed by Kirkpatrik et al. [11] is another powerful technique used to solve optimization problems. The algorithm is of sequential nature and converges slowly. The robustness of EAs is usually enhanced when they are hybridized with other meta heuristics such as tabu search [12], SA etc. Yip and Pao [13] proposed to incorporate the process of SA into the selection process of EA and to utilize its power to guide local search. We first give a detailed outline of the algorithm involving the hybridization of EA with SA in the context of PNP, which is followed by its pseudo code.

HEA starts with building an initial population of parents $P_0$ consisting of *pop_size* number of solutions produced by RDFS and then applying EEH to distribute the edges on the pages. The SA parameters namely, initial temperature $T_i$, final temperature $T_f$, and cooling ratio $\alpha$ are initialized. Also a non-zero positive integer $k$ is assigned to $ch\_num[i], 1 \leq i \leq pop\_size$, where $ch\_num[i]$ indicates the number of children to be produced by parent $P_t[i]$ in the $t^{th}$ generation. It means initially each parent is given an equal chance to produce children. Clearly, the total number of children to be produced by the parents $P_0$ is $k \times pop\_size$. After generating the initial population of parents $P_0$ the iterative process (steps 7 to 23) starts in which each parent $P_t[i]$ generates $ch\_num[i]$ number of children using IDFS (see section 3.3) and EEH. We refer to the population of children as $C_t$, where $C_t[i][j]$ denotes the $j^{th}$ child of parent $P_t[i]$. Now for each $i, 1 \leq i \leq pop\_size$, child with lowest pagenumber ($best\_C_t[i]$) is determined for finding the parents for the next generation. This child becomes the parent for the next generation if its pagenumber is less than that of parent; otherwise there is a probability that the child will be selected to become a parent (Steps 12 to 20). Next important step is to decide $ch\_num[i]$ for the next generation of parents $P_{t+1}$. This decision is based on the competition between parents wherein the performance of a parent is measured by comparing the pagenumber of each of its children with the $best\_pagenumber$, where $best\_pagenumber$ is the lowest pagenumber obtained so far in this process. Pseudo code for computing $ch\_num[i]$ is given in the procedure $find\_ch\_num$ at the end of this section. The iterative process is repeated until no improvement in the $best\_pagenumber$ is observed for a specified count ($cnt$) of generations or when final temperature is reached.

The pseudo code for HEA

1. Initialize $PS$, number of children to be produced by each parent of the initial generation $k$, cooling ratio $\alpha$, initial temperature $T_i$, final temperature $T_f$.
2. Set $t \leftarrow 0$; $T \leftarrow \mathrm{T_i}$;
3. **for** $i = 1$ *to* $pop\_size$
4. $\quad ch\_num[i] \leftarrow k$;
5. **endfor**
6. Generate initial population of parents $P_t[i]$ using RDFS and then apply EEH to distribute the edges on the pages
7. **Repeat**
8. $\quad$ **for** $i = 1$ *to* $pop\_size$

Apply IDFS on $P_t[i], ch\_num[i]$ times to create children $C_t[i][j], 1 \le j \le ch\_num[i]$ and then apply EEH to distribute the edges on the pages
**endfor**
Apply mutation on $k \times pop\_size \times r_m$ number of children from $C_t$.
**for** $i = 1$ *to* $pop\_size$
Find child with lowest pagenumber i.e. $best\_C_t[i]$
$\beta \leftarrow pg\_no(G, P_t[i]) - pg\_no(best\_C_t[i])$;
**if** $\beta > 0$ or $\exp(\beta/T) > \rho$ // $\rho$ is a random number lying between 0 and 1
$P_{t+1}[i] \leftarrow best\_C_t[i]$;
**else**
$P_{t+1}[i] \leftarrow P_t[i]$;
**endif**
**endfor**
Update $best\_pagenumber$
$T \leftarrow \alpha T$; $t \leftarrow t + 1$;
**Until** (stopping criteria is satisfied)

Procedure $find\_ch\_num()$

$sum \leftarrow 0$;
**for** $i = 1$ *to* $pop\_size$
$count[i] \leftarrow 0$
**for** $j = 1$ *to* $ch\_num[i]$
$\beta \leftarrow best\_pagenumber - pg\_no(G, C_t[i][j])$;
**if** $\beta > 0$ or $\exp(\beta/T) > \rho$
$count[i] \leftarrow count[i] + 1$
**endif**
**endfor**
$sum \leftarrow sum + count[i]$
**endfor**
**for** $i = 1$ *to* $pop\_size$
$ch\_num[i] \leftarrow k \times pop\_size \times count[i]) / sum)$
**endfor**

# 3 Implementation of Hybrid Evolutionary Algorithm

## 3.1 The Edge Embedding Heuristic (EEH)

Steps:

- Consider a labeling $\varphi = 1, 2, \ldots, n$ of vertices in graph $G$.
- Now for a complete graph with vertex labels $1, 2, \ldots, n$, place the edges ($x \leftrightarrow y$ denotes an edge with endpoints $x$ and $y$) of $K_n$ in a sequence $S$ according to the procedure below:

Initialize $S$ to the empty sequence;

**for** $v = 1$ **to** $\lceil n/2 \rceil$

$S += v \leftrightarrow (v+1) \leftrightarrow (v-1) \leftrightarrow (v+2) \leftrightarrow (v-2) \leftrightarrow \cdots$
$\leftrightarrow (v + \lceil n/2 \rceil - 1) \leftrightarrow (v - \lceil n/2 \rceil + 1) \leftrightarrow (v + \lceil n/2 \rceil)$

**endfor**

where += means concatenate with the sequence obtained from the previous step and addition and subtraction are modulo operators assuming labels are placed in a circle. Note that S includes all the edges in a complete graph of n vertices with some edges possibly included more than once.

- Now $E(G) \subseteq E(K_n)$. Insert edges in $E(G)$ in the same sequence as they appear in $S$ in edge matrix $Q$, without duplicating the edges.
- Place the edges in $Q$ one by one, in sequence, starting on page one using the following rule:, Assign page one to the first edge. For each $e_i$ in $Q$, $2 \leq i \leq |E|$, assign the lowest numbered page to $e_i$ if adding $e_i$ to the page does not result in a crossing. Otherwise assign a new page to $e_i$.
- Pagenumber=total number of pages assigned.

### 3.2 Initial Population

An initial population $P_0[i], i = 1 \ldots pop\_size$, consists of $pop\_size$ solutions. For each $P_0[i] = (R_i, Q_i)$, $R_i$ is generated using RDFS and EEH is used to generate $Q_i$. In this section we take an example, shown in Fig. 2a, to illustrate the components of a solution.

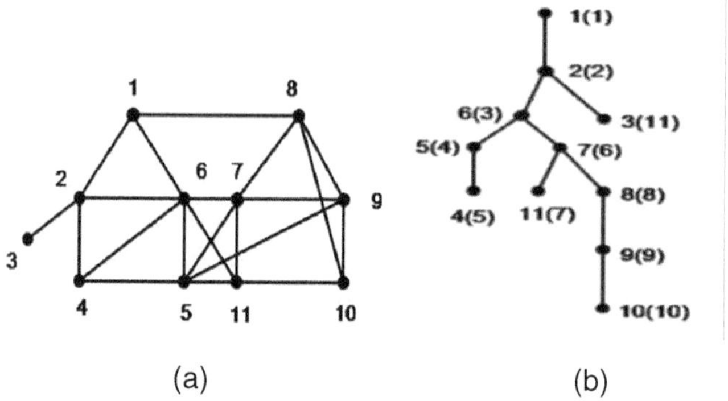


**Fig. 2.** (a) A graph and (b) its DFS tree

Starting from vertex 1, RDFS of the graph (DFS tree is shown in Fig 2(b)) provides the following labeling:

$R$=

| | | | | | | | | | | |
|---|---|---|---|---|---|---|---|---|---|---|
| 1 | 2 | 11 | 5 | 4 | 3 | 6 | 8 | 9 | 10 | 7 |

In Fig.2b, for each vertex, the number in the parenthesis corresponds to its depth first index, i.e., the label of the vertex.

EEH gives the distribution of edges in the array $Q$ as

$Q=$

| | | | | | | | | | | | | | | | | | | | | | |
|---|---|---|---|---|---|---|---|---|---|---|---|---|---|---|---|---|---|---|---|---|---|
| Edges | 1 | 2 | 5 | 7 | 7 | 2 | 1 | 7 | 5 | 2 | 10 | 8 | 8 | 4 | 4 | 6 | 1 | 9 | 5 | 5 | 6 |
| | 2 | 3 | 9 | 8 | 11 | 6 | 6 | 9 | 6 | 4 | 11 | 10 | 9 | 5 | 6 | 7 | 8 | 10 | 7 | 11 | 11 |
| Page # | 1 | 1 | 1 | 1 | 1 | 1 | 2 | 1 | 1 | 3 | 2 | 2 | 1 | 1 | 2 | 2 | 3 | 1 | 1 | 4 | 2 |

The pagenumber for this solution is 4.

### 3.3 Generating Children

We use a unary reproduction operator, intermediate depth first search (IDFS) to produce children from the parents, which is explained as follows:

Let the parent $par\_pop[i]$ and the child $child\_pop[i]$ are denoted by arrays $S\_par$ and $S\_child$ respectively. A vertex $v$ $(1 \leq v \leq n)$ is selected randomly. The label of this vertex is $S\_par[v]$. Labels of vertices with labels less than or equal to $S\_par[v]$ are copied from $S\_par$ to $S\_child$ at the respective positions. These vertices are referred to as $visited$ and remaining ones are referred to as $unvisited$. This process is shown in pseudo code below:

1. **for** $\forall u \in V$
2. **if** $S\_par[u] \leq S\_par[v]$
3. $S\_child[u] \leftarrow S\_par[u]$
4. **endif**
5. **endfor**

Now starting from the vertex $v$, DFS of the $unvisited$ vertices of the graph is carried out. The labeling of the $unvisited$ vertices starts from $S\_par[v]+1$ and the labels are stored at the respective positions of array $S\_child$. This ensures that a new obtained child solution also has a DFS based labeling. Thus IDFS helps to explore various possible DFS trees of the graph starting with the same root vertex.

Fig 3 shows a DFS tree obtained from that shown in Fig 2(b) when the RDFS is carried out after the randomly chosen vertex 6.

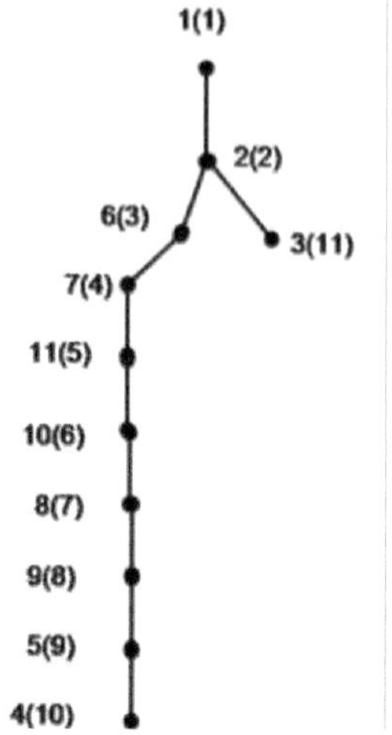


**Fig. 3.** A DFS tree obtained by applying IDFS on the tree of Fig. 2.

Application of EEH on this layout provides the following edge distribution over the pages. The pagenumber for this solution is 3.

$R=$

| | | | | | | | | | | |
|---|---|---|---|---|---|---|---|---|---|---|
| 1 | 2 | 11 | 10 | 9 | 3 | 4 | 7 | 8 | 6 | 5 |

EEH gives the distribution of edges in the array Q as

$Q=$

| | | | | | | | | | | | | | | | | | | | | | |
|---|---|---|---|---|---|---|---|---|---|---|---|---|---|---|---|---|---|---|---|---|---|
| Edges | 1 | 2 | 4 | 5 | 5 | 9 | 8 | 2 | 1 | 8 | 6 | 5 | 7 | 6 | 1 | 4 | 10 | 7 | 7 | 5 | 2 |
| | 2 | 3 | 6 | 7 | 11 | 10 | 10 | 6 | 6 | 9 | 7 | 9 | 11 | 11 | 8 | 5 | 11 | 8 | 9 | 6 | 4 |
| Page # | 1 | 1 | 1 | 1 | 1 | 1 | 1 | 1 | 2 | 1 | 1 | 1 | 1 | 2 | 2 | 1 | 1 | 3 | 3 | 1 | 1 |

### 3.4 Mutation

In HEA, *insert mutation* is used to perturb the population of children $C_r$. In this, a vertex is selected randomly and is placed at another randomly chosen position on the spine. This facilitates the entry of non-DFS solutions in the population. Mutation is applied on $k \times pop\_size \times r_m$ number of randomly selected children, where $r_m$ is mutation rate.

## 4 Experiments and Results

In this section, we first describe the classes of graphs that we use in the experiments. We then present the results of our experiments for HEA and the genetic algorithm (GA) given in Kapoor et al. [1]. The code is written in MATLAB 7.0 and the machine specifications are: Intel Core 2 Duo E8400 @ 3.00 GHz, 2.87 GB of RAM.

### 4.1 Test Sets

We use two types of graphs to test HEA: standard graphs and random connected graphs (RCGs).

**Standard graphs:** Following classes of standard graphs are used:

- *Complete graphs* ($K_n$)
- *Hypercubes* $Q_d$*:* The hypercube, of dimension $d$, is a $d$-regular graph with $2^d$ vertices and $d2^{d-1}$ edges. Each vertex is labeled by a distinct $d$-bit binary string, and two vertices are adjacent if they differ in exactly one bit.
- *Cube-connected cycles* $CC_d$*:* The Cube-connected cycle of dimension $d$ is formed from $Q_d$ by replacing each vertex $u$ with a $d$-cycle of vertices in $CC_d$ and then joining each cycle vertex of the corresponding neighbor of $u$ in $Q_d$. It has $d2^d$ vertices, $3d2^{d-1}$ edges.
- *Complete binary tree* $B(n)$: The binary tree is a tree structure in which each node has at most two child nodes. The binary tree is complete if the nodes are well balanced.
- *Pinwheel* $P(n)$: The depth-$n$ pinwheel graph, has $2n$ vertices $\{a_1, a_2, \ldots, a_n\}$ and $\{b_1, b_2, \ldots, b_n\}$ and edges connecting each pair of vertices of the form

$$a_i \leftrightarrow b_i,$$
$$a_i \leftrightarrow b_{n-i+1},$$
$$a_i \leftrightarrow a_{i+1},$$
$$b_i \leftrightarrow b_{i+1}, \quad \textit{where} \quad 1 \leq i \leq n$$

- *Star* $S_k$*:* The star graph is a complete bipartite graph $K_{1,k}$, a tree with one internal node and $k$ leaves. It has $k+1$ vertices and $k$ edges.
- *Triangulated triangle* $T_l$: The Triangulated triangle graph, is a graph whose vertices are the triples of non-negative integers summing to $l$, with an edge connecting two triples if they agree in one coordinate and differ by one in the other two coordinates.
- *Complete bipartite graphs* $K_{m,n}$: A complete bipartite graph, with $|V| = m + n$ is a simple bipartite graph such that two vertices are adjacent if and only if they are in different partite sets.
- *Shuffle Exchange* $(S(D))$: Its vertices are 0-1 vectors of length $D$. There is an edge between any vertex $(v_0, \ldots, v_{D-1})$ and vertex $(v_1, \ldots, v_{D-1}, v_0)$. Also, there is an edge between any vertex $(v_0, \ldots, v_{D-1})$ and the vertex $(v_0, \ldots, v_{D-2}, \alpha)$, $\alpha \neq v_{D-1}$.

**Random Connected Graphs (RCGs):** We generate RCGs of sizes 40 and 50, and edge densities varying from 10% to 50%. For each combination of size and edge density, 20 different graphs are randomly generated for the experiments. We use RCGs (1) to study the effect of edge density on pagenumber and (2) to compare HEA with GA.

### 4.2 GA Settings and Parameter Tuning

To compare HEA with the GA of Kapoor et al [1], we coded the GA with cyclic crossover and insert mutation operator as no specific operator is mentioned in their paper. Furthermore, as other GA parameters such as population size, mutation rate etc. are also not specified, the various parameters involved in HEA and GA are set after performing preliminary experiments on the randomly generated graphs. For GA, the size of initial population ($pop\_size$) is set as $n = |V|$ whereas for HEA, $pop\_size$=10, $k$= 10, which is the number of children generated initially, are fixed. The SA parameters, namely, initial temperature ($T_i$), final temperature ($T_f$), and cooling ratio ($\alpha$) are chosen as 1, .01 and 0.99 respectively. Mutation rate ($r_m$) of 0.2 is used in both the algorithms. In both the algorithms, the iterative process is repeated until no improvement is observed in minimum pagenumber for 50 consecutive iterations.

### 4.3 Results

The results of HEA and GA on various graphs are shown in Table 1 through Table 4. The results of standard graphs with known optimal pagenumbers are given in Table 1. HEA performs better than GA in most of the cases. The results in bold indicates the cases for which an algorithm has better performance than the other. HEA attains optimal pagenumber for most of the graphs. The pagenumber for complete graphs are all optimal as expected since EEH is based on edge distribution strategy for the optimal pagenumber of these graphs. GA does not perform well, for the hupercubes as the pagenumbers obtained by it are substantially higher than the optimal pagenumber which is $d-1$. HEA attains these values easily when tested for $d \leq 6$. Both the algorithms attain the optimal pagenumber 1 for the complete binary trees and star graphs which are obvious as these graphs are outerplanar. HEA is able to obtain optimal pagenumbers of pinwheel graphs ($P(n)$) tested for $n$ = 8, 16 and 32.

Table 2 contains the results of those classes of standard graphs for which optimal pagenumbers are not known and for which some upper bounds are available for them. The results for shuffle exchange and triangulated triangles are much less than their known upper bounds. HEA and GA both give large pagenumbers of complete bipartite graphs. The reason for this may be attributed to the fact that the *optimal* pagenumber is obtained with a non-DFS layout.

Experimental results for random graphs of sizes 40 and 50 are shown in Table 3 and Table 4 respectively. Results (Fig. 4) clearly demonstrate that HEA outperforms GA in terms of solution quality though the elapsed time of HEA is more than that of GA. The difference in the page numbers obtained by HEA and GA is more significant in the case of higher density graphs.

**Table 1.** Comparison of GA and HEA for the standard graphs with known optimal pagenumbers

| Graphs | $K_{12}$ | $K_{13}$ | $K_{14}$ | $K_{15}$ | $K_{100}$ | $K_{150}$ | $P(8)$ | $P(16)$ | $P(32)$ | $Q_3$ | $Q_4$ | $Q_5$ | $Q_6$ |
|---|---|---|---|---|---|---|---|---|---|---|---|---|---|
| GA | 7 | 7 | 8 | 9 | 86 | 135 | 2 | 3 | 6 | 2 | 3 | 7 | 13 |
| HEA | **6** | **7** | **7** | **8** | **50** | **75** | **2** | **3** | **3** | **2** | **3** | **4** | **5** |
| Optimal | 6 | 7 | 7 | 8 | 50 | 75 | 2 | 3 | 3 | 2 | 3 | 4 | 5 |

**Table 2.** Comparison of GA and HEA for the standard graphs with the upper bounds on the pagenumbers

| Graphs | $CC_3$ | $CC_4$ | $SE(4)$ | $SE(5)$ | $SE(6)$ | $SE(7)$ | $T_4$ | $T_5$ | $T_6$ | $T_7$ | $T_8$ | $K_{4,4}$ | $K_{5,5}$ | $K_{6,6}$ | $K_{7,7}$ | $K_{8,8}$ | $K_{9,9}$ | $K_{10,10}$ |
|---|---|---|---|---|---|---|---|---|---|---|---|---|---|---|---|---|---|---|
| GA | 4 | 7 | **1** | **2** | **3** | **3** | 3 | 3 | 3 | 4 | 4 | **3** | **4** | **5** | 6 | 7 | 8 | 9 |
| HEA | **3** | **5** | **1** | **2** | **3** | **3** | **3** | **3** | **2** | **3** | **4** | **4** | **5** | **6** | 7 | 8 | 9 | 10 |
| Upper Bound | 5 | 5 | 3 | 3 | 3 | 3 | 4 | 5 | 6 | 7 | 8 | 3 | 4 | 5 | 5 | 6 | 7 | 7 |

**Table 3.** Comparison of GA and HEA for the random graphs ($|V| = 40$)

| Edge density (%) | Pagenumber | | Elapsed Time (in sec.) | |
|---|---|---|---|---|
| | GA | HEA | GA | HEA |
| 10 | 6.6 | 6.6 | 147.278 | 202.65 |
| 20 | 11.2 | **10.8** | 244.8094 | 347.581 |
| 30 | 14 | **13.2** | 391.4998 | 492.0376 |
| 50 | 18.6 | **16.4** | 736.2366 | 795.784 |

**Table 4.** Comparison of GA and HEA for the random graphs ($|V| = 50$)

| Edge density (%) | Pagenumber | | Elapsed Time (in sec.) | |
|---|---|---|---|---|
| | GA | HEA | GA | HEA |
| 10 | 9.4 | **9.2** | 220.4186 | 490.2936 |
| 20 | 14.8 | **14** | 455.9404 | 845.9158 |
| 30 | 19 | **16.8** | 1.0004e+003 | 1.20776e+003 |
| 50 | 24.2 | **20.8** | 1.86306e+003 | 1.9547e+003 |

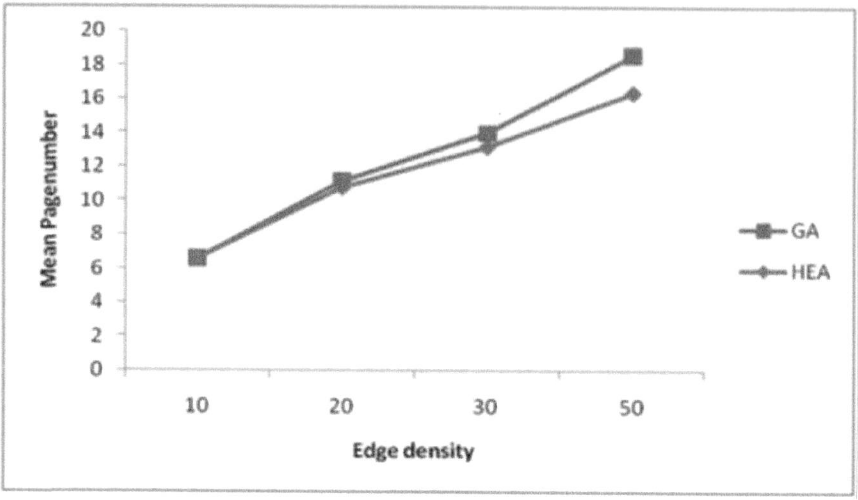


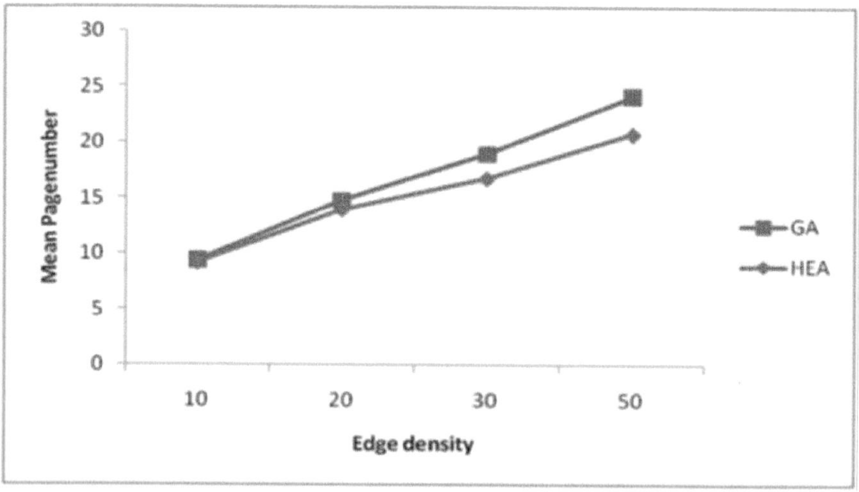


**Fig. 4.** Edge density vs Pagenumber graph for HEA and GA of sizes (a) 40 and (b) 50

## 5 Conclusion

We have presented a hybrid evolutionary algorithm for the pagenumber problem in which a random depth first search of the graph is employed for placing the vertices on the line. An edge embedding heuristic is designed to distribute the edges on a minimal number of pages. The results show that the algorithm achieves the optimal pagenumber for most of the standard graphs tested by us. HEA outperforms the GA described by Kapoor et al. [1] in terms of solution quality for most of the instances tested.

**Acknowledgement.** This work is supported by University Grants Commission (UGC), New Delhi vide letter no 36-66/2008 (SR).

## References

1. Kapoor, N., Russell, M., Stojmenovic, I.: A Genetic Algorithm for finding the Pagenumber of Interconnection Networks. J. Parallel and Distributed Computing 62, 267–283 (2002)
2. Games, R.A.: Optimal Book Embeddings of the FFT, Benes, and Barrel Shifter Networks. Algorithmica 1, 233–250 (1986)
3. Chung, F.R.K., Leighton, F.T., Rosenberg, A.L.: Embedding Graphs in Books: a Layout Problem with Applications to VLSI design. Siam J. Alg. Disc. Math. 8(1), 33–58 (1987)
4. Hasunuma, T., Shibata, Y.: Embedding deBrujin, Kautz and Shuffle-exchange Networks in Books. Discrete Applied Mathematics 78, 103–116 (1997)
5. Obrenic, B.: Embedding deBrujin and Shuffle-exchange graphs in 5 pages. In: Proc. ACM SPAA, pp. 137–146 (1991)
6. Shahrokhi, F., Shi, W.: On crossing sets, disjoint sets and Pagenumber. J. Algorithms 34(1), 40–53 (2000)
7. Bernhart, F., Kainen, P.C.: The Book Thickness of a Graph. J. Combin. Theory Ser. B 27, 320–331 (1979)
8. Swaminathan, R.P., Giriraj, D., Bhatia, D.K.: The Pagenumber of the class of Bandwidth-k Graphs is k-1. Information Processing Letters 55, 71–74 (1995)
9. Bettayeb, S., Hoelzeman, D.A.: Upper and Lower Bounds on the Pagenumber of the Book Embedding of the k-ary Hypercube. J. Digital Information Management 7(1), 31–35 (2009)
10. Garay, M.R., Johnson, D.S., Miller, G.L., Papadimitriou, C.H.: The Complexity of Coloring Circular Arcs and Chords. SIAM J. Algebraic Discrete Methods 1, 216–227 (1980)
11. Kirkpatrick, S., Gelatt, C.D., Vechhi, M.P.: Optimization by Simulated Annealing. Science 220(4598), 671–680 (1983)
12. Glover, F., Laguna, M.: Tabu Search, 2nd edn. Kluwer, Boston (1997)
13. Percy, P.C., Yip, Pao, Y.-H.: Combinatorial Optimization with use of Guided Evolutionary Simulated Annealing. IEEE Transactions on Neural Networks 6, 2 (1995)

# Trust-Enhanced Recommendation of Friends in Web Based Social Networks Using Genetic Algorithms to Learn User Preferences

Vinti Agarwal and Kamal K. Bharadwaj

School of Computer and Systems Sciences,
Jawaharlal Nehru University, New Delhi, India
{vinti.agarwal3,kbharadwaj}@gmail.co.in

**Abstract.** Web-based social networks (WBSNs) are a promising new paradigm for large scale distributed data management and collective intelligences. But the exponential growth of social networks poses a new challenge and presents opportunities for recommender systems, such as complicated nature of human to human interaction which comes into play while recommending people. Web based recommender systems (RSs) are the most notable application of the web personalization to deal with problem of information overload. In this paper, we present a Friend RS for WBSNs. Our contribution is three fold. First, we have identified appropriate attributes in a user profile and suggest suitable similarity computation formulae. Second, a real-valued Genetic algorithm is used to learn user preferences based on comparison of individual features to increase recommendation effectiveness. Finally, inorder to alleviate the sparsity problem of collaborative filtering, we have employed trust propagation techniques. Experimental results clearly demonstrate the effectiveness of our proposed schemes.

**Keywords:** Web-Based Social Networks, Friend Recommender System, Collaborative Filtering, Profile Similarity, Real-Valued Genetic Algorithm, Trust Propagation.

## 1 Introduction

Web-based Social Networks (WBSNs) are growing dramatically in size and number. The huge popularity of sites like Facebook, LinkedIn, Delicious, YouTube and others has drawn in hundreds of millions of users, and the attention of scientists and the media[1]. The public accessibility of WBSNs offers great promise for researchers interested in studying the behavior of users and how to integrate social information into applications. The Social Web therefore provides huge opportunities for recommender technology.

With the explosive growth of resources available through the Internet, information overload has become a serious concern. Web based recommender systems(RSs) are the most notable application of the web personalization to deal with problems of information and product overload[3][4]. Our Friend RS based on collaborative filtering (CF), the most widely used and commercially applied

D. Nagamalai, E. Renault, M. Dhanushkodi (Eds.): CCSEIT 2011, CCIS 204, pp. 476–485, 2011.

recommendation technique. Recommendation of friends in WBSNs poses a different set of challenges w.r.t. traditional recommending techniques such as those for movies, books and restaurants etc.

Here, we proposed a friend recommender system that suggests new friends to users in WBSNs. The main contributions of the work presented in this paper are as follows:

- Identification of appropriate attributes in a user profile to find similarity with other users in WBSNs and suggestion of suitable similarity computation schemes.
- Learning user preferences, through real-valued GA algorithm, based on individual features in an intuitive and efficient manner to increase recommendation effectiveness.
- Use of trust propagation techniques towards handling sparsity problem of collaborative filtering.

The rest of the paper is organized as follows: Section 2 summarizes related studies about social recommender systems and collaborative filtering. Section 3 introduces Friend Recommender System and set out the procedure used for profile building, similarity computation, weight learning through real-valued GA and at the end methodology used to handle sparsity problem based on trust propagation. The experimental evaluation of various scheme is presented in section 4. Finally, last section discusses the findings and presents our conclusions.

## 2 Social Recommender Systems(SRS) and Collaborative Filtering

Social Recommender Systems (SRSs) aim is to alleviate information overload over social media users by presenting the most attractive and relevant content, often using personalization techniques[4],adapted for the specific user. CF is a classical method, in we rely on recommendation about a user based on making prediction about a user preferences of a group of a user that are considered similar to this user. The most popular technique used to measure the similarity between users based on Pearson's correlations identified the list of nearest neighbors of user.

The key challenging factors that extensively affect the quality of recommendations are privacy and reciprocity [5] and attract major attention of several researchers to find better way to connect people. Other studies have incorporated social network elements into traditional CF [6],[7] but in our study we incorporate CF in WBSNs.

## 3 Proposed Friend Recommender System Framework

Recommendation systems for social networks discussed in [8] are different from other kinds of systems, since the item recommended here are rational human beings rather than goods. Thus, the 'Social' factor has to be accounted for when making a recommendation. In the following section, details of proposed friend recommender system are discussed.

### 3.1 User Profiling and Friendship Trends

In FRS a user profile is important because it records relevant information such as preferences and behavior pattern of individual users [9], [10]. Some factors are most important and strongly affect the relationship in social networks e.g. Geographic location is strongest factor affecting how relationship shapes up in social network. Presence of attributes such as hometown, city and state higher up in the testimony to that fact; relationships are more likely to develop between those belonging to same age group etc. Based on the popular WBSNs, we have identified list of 13 attributes to maintain user profile and divided into two groups i.e. Basic attributes and Behavioral attributes associated with preferences (Table 1 and Table 2).This grouping of attributes allows us to better understand the trend immerging in social networks.

**Table 1.** Basic attributes

| | |
|---|---|
| $A_1$ | Gender |
| $A_2$ | Hometown |
| $A_3$ | Language |
| $A_4$ | Relationship status |
| $A_5$ | Religion |
| $A_6$ | SES |

**Table 2.** Behavioural attributes

| | |
|---|---|
| $A_7$ | Here for |
| $A_8$ | Career interest |
| $A_9$ | Music |
| $A_{10}$ | Movies |
| $A_{11}$ | Activities/interest |
| $A_{12}$ | Books |
| $A_{13}$ | Nature |

**Similarity Measures**

One of the important steps in CF is to identify the users with similar inclinations, which leads to the formation of neighborhood set. In order to identify them several similarity measures have been proposed in the literature [11],[12],[13]. Here, we used a quantity called as weight associated with individual attributes similarity to reflect their importance while making recommendations. These weights are generated offline to every attribute for each user and determined in such a way that sum of all weights is equal to 1, i.e. $\sum w_i = 1$. Initially we considered all weights are equally distributed, indicating that user u is giving equal importance to all features.

The proposed method combines weights with other similarity of each feature to emphasize the total similarity of target user with other who shares appreciation on several features with active users. Overall similarity between target user and other user computed as in [11], aggregating all the similarities of individual attributes, which is given by

$$S(u,u_i) = w_1 * sim(A,A_{1i}) + w_2 * sim(A,A_{2i}) + \cdots .... + w_{13} * sim(A,A_{13i}) \quad (1)$$

Where

$S(.)$ – overall similarity of user u with other users $u_i$
$w_1, w_2 ........ w_{13}$ weights for each attributes
$sim(.)$ – similarity function for each attribute.

*Determination of sim() Function for.*

- Attribute $A_1$: Gender Biasedness

One plausible way of doing so is to use rating scheme assigned by user u to its friend. We keep list of persons how many user u has male and female friends and then compute male(or female) ratio by dividing total number of friends of the given user u as shown in Table 3.

The formula to compute the gender similarity for user i has in j is defined as: $sim(u_i, u_j) = 1 - \frac{|M_i - M_J| + |F_i - F_J|}{2}$ and can also be written as

$$sim(u_i, u_j) = 1 - |M_i - M_J| \text{ or } 1 - |F_i - F_J| \quad (2)$$

- Attributes $A_2$, $A_3$ ,$A_4$ and $A_5$: Geographical separation, Language , Relationship status and religion

$$sim(u_i, u_j) = \{ 1, if\ common\ else\ 0 \quad (3)$$

**Table 3.** Computation of male and female ratio

| | *Male friends* | *Female friends* | *Total friends* | *Male ratio (M)* | *Female ratio(F)* |
|---|---|---|---|---|---|
| User i | $N_{mi}$ | $N_{fi}$ | $N_i$ | $N_{mi}/ N_i = M_i$ | $N_{fi}/ N_i = F_i$ |
| User j | $N_{mj}$ | $N_{fj}$ | $N_j$ | $N_{mj}/ N_j = M_j$ | $N_{fi}/ N_j = F_j$ |

- Attribute $A_6$: Socio-Economic Status(SES)

Attribute $A_6$ could be a better dimension towards the efficient friend recommendation because for all above attributes users are provided space to identify them. Rather than omit this variable we have included it also in overall similarity computation. Here we divide the monthly income into three categories: Low, Medium and High. Then after categorization, we aggregate the difference between their class with the father's occupation to compute final similarity of SES.

- Attributes $A_7$, $A_8$, $A_9$, $A_{10}$, $A_{11}$ , $A_{12}$ and $A_{13}$: Here for, Career, Common music, Movies ,Activities, Books and Nature respectively.

While computing the similarity for these attributes we are taken into consideration the preferences order. The preferences information corresponding to each feature for a particular user is represented by a vector called user profile preference vector(UPPV).In this way UPPV of a given feature A for a user u captures the preferences given as : $UPPV_U(A)=(P_1, P_2, P_3)$.

In our experiment, we assume the maximum length(L) of UPPV vector as 3.It may vary depending on the number of preferences on common items. It could be zero if there is no common choice between two users. Here $P_i$ assumes numeric value. Consequently, we formalize the similarity of these attributes for active user a with others based on the difference of preferences. As the difference between preferences decreases similarity value gets increased.

$$sim(u_i, u_j) = \frac{1}{L} \sum_L 1 - \frac{(P_i - P_j)}{3} \quad (4)$$

The various cases generated during similarity computation based on preferences are tabulated as follows:

Case 1: if all items are common, the number of possible cases for the order of preferences between users would be C(3,3) X C(3,1) X C(3,3) X C(2,1) X C(1,1)=6.

Case 2 : Similarly ,for two common items : C(3,2) X C(3,1) X C(2,1) X C(1,1)=18.

Case 3 : And for one : C(3,1) X C(3,1)=9.

### 3.2 Learning Weight by Using Real Valued Genetic Algorithm

The above defined similarity measures treats all attributes equally, taking all $w_i$'s equal, while trying to evaluate recommendation to a user. This may not truly reflect the contribution of each attribute towards similarity. The fact that "how much" a user liked an attribute is important than he has liked it. To make it possible we proposed real valued GA method to learn these weights in an offline learning process and then used to compute the similarity value during prediction.

Traditional GA where chromosome representation technique is bit strings suffers from slow convergence due to the large chromosome structure, especially in optimization problems involving several parameters. We have used real valued GA, to learn the weights for each of the similarity measures as real numbers and therefore a floating point representation of chromosomes is most suited for our purpose. The chromosome is a set of 13 weights where the value of each weight lies between two real values which indicate maximum and minimum value as discussed in [14]. The chromosome search space for real valued GA is defined as

$$\Omega = \{ \text{w} \in \text{R}^{\text{m}} \mid w_1^{min} \leq w_1 \leq w_1^{max}, \ldots \ldots\ldots w_{13}^{min} \leq w_{13} \leq w_{13}^{max} \} \quad (5)$$

Genetic operators being used here are different from traditional operators described in[15]. GA terminates either a maximum number of generations elapsed or a desired level of fitness is reached.

#### *Fitness Function*

Fitness function validates the diversion of process towards its optimization goal by allowing best individuals to breed that lead to good recommendation [3]. Recursively the individuals that achieve a higher fitness are more likely to be selected as parents for next generation. For this, every set of weights employed in RS rerun on entire set including weights generated after crossover and mutation operators in order to compute the predicted value of rating for each user. The average difference between actual and predicted rating for all users in training set is used as fitness score for that set of weights.

$$Fitness = \frac{1}{n_R} \sum_{j=0}^{n_R} |r_j - pr_j| \quad (6)$$

### 3.3 Handling Sparsity Problem Using Trust

One of the biggest challenges of designing a collaborative filtering system is how to provide accurate recommendations with the sparse user profile. Trust has been successfully used to enhance neighborhood set in order to provide accurate recommendations [16] ,[17],[18] with sparse data. The following subsections discuss how to utilize trust value to obtain more accurate recommendation when data is highly sparse.

**Trust Computation**

Trust metric compute an estimate of what degree a user should trust on other, if are unknown to each other. It can be computed in several ways such as using similarity or rating etc. In our study we are computing trust between user a and b based on difference of rating assigned by a and b to their common friends [19]. The formula to compute trust between them is given as :

$$T(u_a, u_b) = 1 - \frac{\sum_{i=1}^{n}(r_{u_{a,i}} - r_{u_{b,i}})}{r_{\max * n}} \tag{7}$$

Here, $r_{u_a}$ is the rating assigned by a and $r_{u_b}$ assigned by b and n is the total number of common friends and $r_{max}$ is the max value of rating scale i.e. 5.

**Trust Propagation**

By formalizing trust relationship, trust between strangers can be established by analyzing path linking parties together. In this paper we use the max-min based approach described in [20] to express transitive trust relationship. Let us illustrate this with Fig.1.user a trusts b and b trusts c then it is reasonable to say that a trusts c to a certain degree but in real life trust is not always transitive [16]. However trust could not completely transfer without decay through distance. Its values reduce as path length increases. So propagation distance is a constraint in transitive relationship. As shown in Fig.1 [17] there are two paths between a and c .Hence their associated inferred trust is given by formula (14).

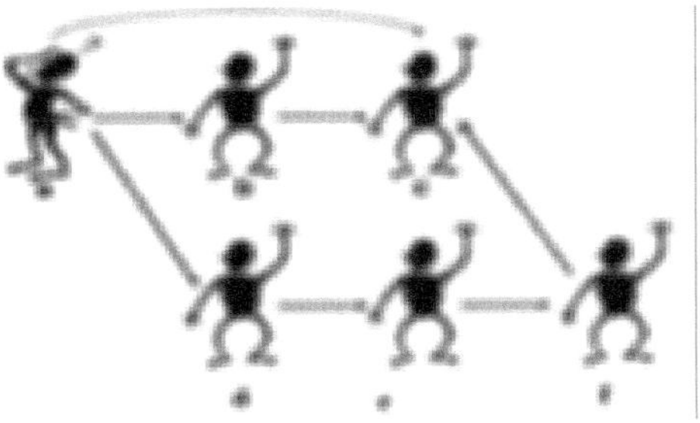

**Fig. 1.** Trust Propagation

$$T(u_a, u_c) = \max\left(\left(\min(T(u_a, u_b), T(u_b, u_c))\right), \min\left(T(u_a, u_d), T(u_d, u_e), T(u_e, u_f), T(u_f, u_c)\right)\right) \tag{8}$$

### 3.4 Main Steps of the Proposed Friends Recommender Systems

**Step 1** : Similarity computation at the attribute level.
i.e. $sim(A, A_{1i})$, $sim(A, A_{2i})$,………… $sim(A, A_{13i})$ between user u and $u_i$.
**Step 2** : Weight learning through real-valued GA.i.e. $w_1$, $w_2$…………, $w_{13}$.
**Step 3**: Generating enhanced neighborhood sets based on trust propagation.
**Step 4** : Making recommendation using CF.

## 4 Experimental Results and Discussion

### 4.1 The Dataset

To demonstrate the effectiveness of our proposed techniques experiments on simulated dataset of 20 users is conducted, which forms a small social network with each node representing people or users and ties between them denoting relationship along with a rating score from 1-5. In the first step of our experiment, we used hybrid of content and traditional CF for recommendation on both equal and learned weights. The actual ratings of an active user are randomly divided into two sets: Testing set and Training set. Training dataset used to construct neighborhood set for rating prediction. In the next experiment we demonstrate method for alleviating the sparsity problem using trust inferences and result shows that trust has potential to improve the way recommendations are made.

### 4.2 Results and Discussions

Fig. 2 illustrates the deviation of CFEW (collaborative filtering with equal weights) predicted rating score from actual rating score, compared to the deviation of CFLW (collaborative filtering with learned weights) for the same neighborhood size N=8 over 30 runs of the experiment. It is clear from the Fig. 2 that the lower value of MAE corresponds to the better performance. During the experimentation of CFLW we consider initial population size 20 where weights are real numbers and range between 0 and 30. Then 20 new individuals being produced in each generation after applying crossover and mutation operators.

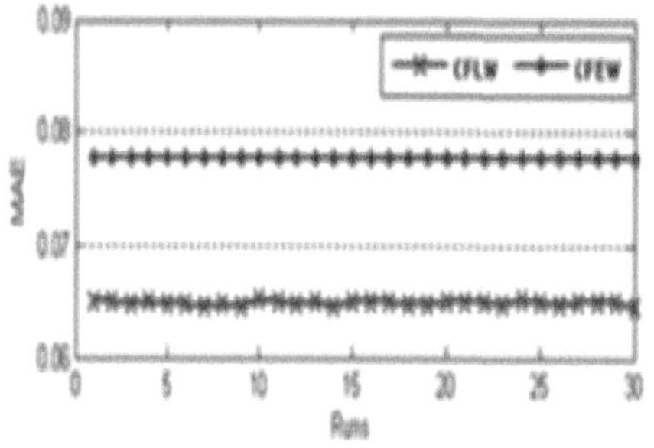


**Fig. 2.** Comparison of MAE of two neighborhood based methods: CFLW and CFEW plotted over 30 runs

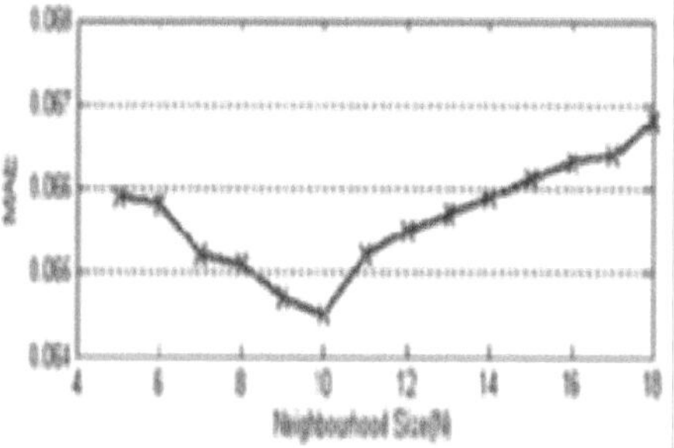


**Fig. 3.** Comparing MAE based on different neighborhood size

In our experiment GA followed by an elitist approach such that the process ends when the fitness function value for the best chromosome remains the same in 10 consecutive iterations. Fig. 3 represents the variation in MAE as neighborhood size N varies. The evaluation results show that, instead of improving the prediction results, increasing the neighborhood size might add some bad neighbors, which lead to bigger prediction error.

Table 4 presents the MAE comparison of various schemes. The results show that introduction of trust significantly improve the accuracy of recommendations. Hence, trust based rating prediction obviously reduced the deviation from actual rating score as compared to the deviation of CFEW and CFLW. MAE of trust based rating prediction is 0.0652 while that of CFEW and CFLW are 0.9417 and 0.4226 respectively.

**Table 4.** MAE for different prediction algorithms

| Prediction Algorithms | MAE(N=8) |
|---|---|
| CFEW | 0.9417 |
| CFLW | 0.4226 |
| Trust based CFEW | 0.0778 |
| Trust based CFLW | 0.0652 |

## 5 Conclusion and Future Work

We have presented a Friend Recommender System, that exploits web information to recommend relevant friends on social-networking websites. Recommendation are based on social network information using traditional CF. The method of evolving individuals weights by real-valued GA according to their preferences has significantly enhanced the performance. To deal with sparsity problem we employed a method based on trust inferences which provide additional information to CF algorithm in terms of enhanced neighborhood sets. The experimental results clearly show that the trust based CFLW outperforms rest of the schemes i.e. CFEW, CFLW and trust based CFEW on sparse datasets. The experiments in this work are performed on a simulated dataset and we plan to conduct experiments on larger dataset obtained through survey/WBSNs.

Our future work will focus on applying different techniques to compute and propagate trust [16]. It is to be seen how two level filtering based on both trust and reputation [21] and utilization of various sparsity measures [13] can be incorporated in the proposed system to further enhance its performance.

## References

1. Guy, I., Ronen, I., Wilcox, E.: Do You Know? Recommending People to Invite into Your Social Network. In: Proceedings of the 14th International Conference on Intelligent User Interfaces, IUI 2009. ACM, New York (2009)

2. Chen, J., Dugan, C., Muller, M., Guy, I.: Make New Friends, but Keep Old-Recommending People on Social Networking Sites. In: Proceedings of the 27th International Conference on Human Factors in Computing Systems, CHI 2009, pp. 201–210 (2009)
3. Al-Shamri, M.Y.H., Bharadwaj, K.K.: Fuzzy-Genetic Approach to Recommender Systems Based on a Novel Hybrid User Model. Expert Systems with Applications 35, 1386–1399 (2007)
4. Adomavicius, G., Tuzhilin, A.: Personalization Toward the next generation of recommender systems: A survey of the state-of-the-art and possible extensions. IEEE Transaction on Knowledge and Data Engineering 17(6), 734–749 (2005)
5. Garcia, R., Amatriain, X.: Weighted Content Based Methods for Recommending Connections in Online Social Networks. In: Recsys 2010, ACM, Barcelona (2010), 0-12345-67-8
6. Liu, F., Lee, H.J.: Use of Social Network Information to Enhance Collaborative Filtering Performance. Expert Systems with Applications 37(7), 4772–4778 (2009)
7. Massa, P., Avesani, P.: Trust Aware Recommender Systems. In: RecSys 2007: Proceedings of 2007 ACM Conference on Recommender Systems (2007)
8. Karkada, U.H.: Friend Recommender System for Social Networks. SI583 Term Paper, School of Information, University of Michigan (2009)
9. Hsu, W.H., King, A.L., Paradesi, M.S.R., Pydimarri, T., Weninger, T.: Collaborative and Structural Recommendation of Friends using Weblog-Based Social Network Analysis. In: AAAI Spring Symposia 2006 on Computational Approaches to Analysing Weblogs (2006)
10. Banks, L., Wu, S.F.: All Friends are NOT Created Equal: An Interaction Intensity based Approach to Privacy in Online Social Networks. In: International Conference on Computational Science and Engineering, IEEE.CSE 2009, pp. 970–974 (2009)
11. Anand, D., Bharadwaj, K.K.: Adapting User Similarity Measures for Recommender Systems:A Genetic Programming Approach. In: 2010 3rd IEEE International Conference IEEE Computer Science and Information Technology, ICCSIT (2010)
12. Anand, D., Bharadwaj, K.K.: Enhancing Acuuracy of Recommender System through Adaptive Similarity Measures Based on Hybrid Features. In: Nguyen, N.T., Le, M.T., Świątek, J. (eds.) Intelligent Information and Database Systems. LNCS, vol. 5991, pp. 1–10. Springer, Heidelberg (2010)
13. Anand, D., Bharadwaj, K.K.: Utilizing Various Sparsity Measures for Enhancing Accuracy of Collaborative Recommender Systems based on Local and Global Similarities. Expert Systems with Applications 38(5), 5101–5109 (2011)
14. Corcoran, A.L., Sen, S.: Using Real Valued Genetic Algorithm to Evolve Rule Sets for Classification. In: IEEE-CEC (1994)
15. Michalewicz, Z.: Genetic algorithms + data structures = evolution programs. AI. Springer, New York (1992)
16. Josang, A., Hayward, R., Pope, S.: Trust Network Analysis with Subjective Logic. In: Proceedings of the 29th Australasian Computer Science Conference (ACSC 2006), vol. 48 (2006)
17. Victor, P., Cock, M.D., Cornelis, C.: Recommender Systems Handbook. In: Kantor, P.B., Ricci, F., Rokach, L., Shapira, B. (eds.) Chapter in Recommender Systems Handbook, pp. 645–676. Springer, Heidelberg (2010)
18. Papagelis, M., Plexousakis, D., Kutsuras, T.: Alleviating the Sparsity Problem of Collaborative Filtering Using Trust Inferences. In: Herrmann, P., Issarny, V., Shiu, S.C.K. (eds.) iTrust 2005. LNCS, vol. 3477, pp. 224–239. Springer, Heidelberg (2005)

19. Fazeli, S., Zarghami, A., Dokoohaki, N., Matskin, M.: Mechanizing social trust-aware recommenders with T-index augmented trustworthiness. In: Katsikas, S., Lopez, J., Soriano, M. (eds.) TrustBus 2010. LNCS, vol. 6264, pp. 202–213. Springer, Heidelberg (2010)
20. Luo, H., Niu, C., Shen, R., Ullrich, C.: A Collaborative Filtering Framework Based on Both Local User Similarity and Global User Similarity. Machine Learning 72(3), 231–245 (2008)
21. Bharadwaj, K.K., Al-Shamri, M.Y.H.: Fuzzy Computational Models for Trust and Reputation Systems. Electronic Commerce Research and Applications 8(1), 37–47 (2009)

# Modeling and Verification of Server Aided Verification Protocol Using NuSMV

Vikram Saralaya, Kishore J.K., Sateesh Reddy, and Sanjay Singh

Department of Information and Communication Technology,
Manipal Institute of Technology,
Manipal University, Manipal-576104, India
sanjay.singh@manipal.edu

**Abstract.** Cryptographic algorithms are useful for signing and verifying the authenticity of sender of the message. The verifier may not have the required computational ability and relies on a powerful server to aid the verification process. The server and the illegitimate prover/signer may collaborate and try to cheat the verifier. A legitimate prover can also repudiate the message sent by himself. In this paper we model the scenario where the legitimate, cheating or repudiating prover sign's the message. The verifier then authenticates the message via an untrusted server. Specifications are written using CTL(Computational-Tree Logic). NuSMV(extension of Symbolic Model Verifier) is the tool used to verify the specifications.

## 1 Introduction

Cryptographic protocols have rarely been bug-free in its initial design. These programs are critical, for they work on the security aspect of a communication system. An organization would expect a certain degree of assurance for the designed protocol, before implementing it for their internal security. Extensive testing may expose some design flaws, but its only a best effort service. We need to formally model and verify these protocols.

Formal methods can help us detect bugs as well as improve the quality of our design. Initially a model of the protocol is developed which has all the features of its real time counterpart. The properties expected from the protocol is specified and the conformance of the model to the specification is verified. The protocol which goes through these three stages is expected to be more reliable and robust than that merely based on testing.

There has been a lot of research focused on developing better models for Server-aided verification. Girault et. al. [1] gave a clear picture about server-aided verification in theory and practice. Chow et. al. in [2] has discussed a generic pairing based server-aided verification protocol and claims it to provide better security compared with Girault et. al. in [1].

In server-aided verification there are three major entities - prover, verifier & server. The prover initially sign's a message to prove its authenticity to the verifier. Various computational intensive algorithms exist to sign and verify the

D. Nagamalai, E. Renault, M. Dhanushkodi (Eds.): CCSEIT 2011, CCIS 204, pp. 486–495, 2011.

messages. These schemes may have to be incorporated in smart cards and other low cost chips with low computational capability. Such verifiers which cannot carry out computations on its own requests some powerful but probably untrusted server to carry out major part of the computation. Wu et. al. in [3] has further worked on [1] and introduce the existential unforgeability concept. They claim their work to be more secure but computationally more intense. More work on similar lines were carried out by Wu et. al. in [4], Pfitzmann et. al. in [5] and Begiun et. al. in [6]. To the best of our knowledge, no work has been reported in literature to formally specify and verify the Server-Aided Verification Protocol. In this paper we model and verify the properties specified in [1].

The server being untrusted gives rise to two major problems, first the illegitimate prover and server can collaborate and make the verifier accept an invalid signature. The second problem is when a legitimate prover tries to repudiate the message sent by himself with the help of the server. In this paper we address these problems by formalizing the server-aided verification protocol. The properties we expect from the model, such as non-repudiation etc. are given as specifications.

NuSMV model verifier [7][8] is used to verify whether the model indeed meets its specification. The prover module is designed such that it can assume the role of a legitimate prover, cheating illegitimate prover or the one which tries to repudiate. The verifier module is modeled using two approaches slightly different from one another. The NuSMV outputs whether the prover *succeeds* or *fails* in authenticating itself to the verifier.

This paper is organized as follows. Section 2 gives an introduction to Server Aided Verification process followed in the paper. Section 3 discusses the detailed formalization and modeling of the protocol. Section 4 is about the properties any authentication protocol should satisfy and how these can be formally specified using CTL. Section 5 briefly explains about the verification results obtained. Finally section 6 concludes the paper.

## 2 Server Aided Verification Protocol Using RSA Scheme

The message signed by the prover is authenticated by the verifier with the help of the server as shown in Fig. 1. The signing can be done using any of the existing popular schemes. Let us consider the RSA(Rivest, Shamir and Adleman) scheme [9]. The functionalities of the three entities are as described below:

### 2.1 Prover/Signer

The prover can use its own private key (n,d) and sign the message. The encrypted message f(m) is used to produce the signature. A legitimate prover will produce the signature as shown below:

$$\sigma = f(m)^d \bmod n \tag{1}$$

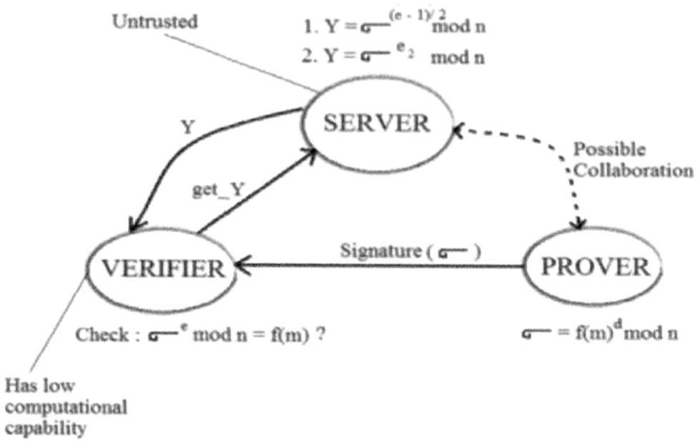


**Fig. 1.** Basic structure of Server Aided Verification protocol

If the prover tries to cheat then it will choose a value $\bar{Y}$ and generate a suitable signature using this value as shown below:

$$\sigma = f(m)^d \bar{Y}^{-2} \bmod n \tag{2}$$

If instead the prover tries to repudiate the message sent by itself, then it will generate the signature normally, but send $\bar{\sigma}$ which is co-related to $\sigma$.

## 2.2 Verifier

Verifier uses the public key (n,e) of the prover to authenticate the signature. Its job is to check whether the following equality holds:

$$\sigma^e \bmod n = f(m) \tag{3}$$

If this equality holds then the verifier can assume that the prover is indeed the one he claims to be. Due to computational constraint the verifier delegates a part of the computation to the server. The delegation can be carried out using two approaches. In the first approach the verifier asks the server to compute:

$$Y = \sigma^{(e-1)/2} \bmod n \tag{4}$$

The value of Y, so obtained from the server can be used as shown below:

$$Y^2 \sigma \bmod n = f(m) \tag{5}$$

Evaluating the above equality requires less computational resources and hence can be performed by the verifier easily. In the second approach verifier chooses two values $e_1$ and $e_2$ such that,

$$e = e_1 * e_2, e1 << e2 \tag{6}$$

Now, the server has to compute:

$$Y = \sigma^{e_2} \bmod n \tag{7}$$

The value of Y, so obtained from the server can be used as shown below:

$$Y^{e_1} \bmod n = f(m) \tag{8}$$

In section 5 we will further see the pros and cons of the two approaches with respect to the given specifications.

### 2.3 Server

A sincere server which has not collaborated with any misbehaving prover will return the value of Y as explained in the previous section. On the other hand if the server has collaborated with a cheating illegitimate prover then the precomputed value $\bar{Y}$ will be returned by the server. Finally, if the legitimate prover has collaboration with the server, then the server returns the suitable value of Y computed from the values provided by the repudiating prover : $\sigma, \bar{\sigma}$.

## 3 Modeling the Server Aided Verification Protocol

The model of the protocol involves three modules : prover module, verifier module and the server module. Each module has its own set of state variables to hold the values computed by them.

### 3.1 Prover Module

This module maintains its current state represented by the variable *prover.state*, its type of signing represented by the variable *prover.sign* and the signature of the encrypted message represented by the variable *prover.signature*. The state space of these state variables are given in Table 1.

**Table 1.** State Variables in the Prover Module

| Variable | State Space |
|---|---|
| prover.state | idle, signed, success, failure |
| prover.sign | none, legitimate, correlated, cheat |
| prover.signature | positive integer |

Initially before computing the signature it will be in the *idle* state. Once the message is signed, a transition to *signed* state takes place. The prover remains in this state until the verifier completes the authentication process. If the authentication is successful then the prover moves to *success* state else to the *failure* state.

### 3.2 Verifier Module

The current state information is held in the variable *verifier.state*. In this paper we model both the approaches discussed in section 2.2 for the verification process. The state space of the state variables are depicted in Table 2.

**Table 2.** State Variables in the Verifier Module

| Variable | State Space |
|---|---|
| verifier.state | idle, get_y1_from_server, get_y2_from_server |
| verifier.result | none, success, failure |

If the verifier decides to use the first approach then it transits to the *get_y1_from_server* state. Instead to opt for the second approach it will have to move to the *get_y2_from_server* state. Once the server returns the $Y$ value, the state of the verifier will be *verify*.

In the *verify* state the verifier has to use the value returned by the server and verify whether the prover is authentic. The verification result is stored in a separate state variable *verifier.result*.

### 3.3 Server Module

The server will remain in the *idle* state as long as the verifier does not issue a request to help it in the verification process. This is represented using the state variable *server.state*. If the verifier had chosen the first approach then the server computes the value for the variable *server.y1* else the value of *server.y2* will be set. These state variables and their state space are given in Table 3.

**Table 3.** State Variables in the Server Module

| Variable | State Space |
|---|---|
| server.state | idle, compute_y1, compute_y2, computed_y1, computed_y2 |
| server.y1 | -1 or positive integer |
| server.y2 | -1 or positive integer |

If the first approach is chosen then the value of *server.y2* will remain -1 because it will not be used by the verifier for its computation. On the other hand, if the second approach is chosen then the computed value is assigned to *server.y2* and hence *server.y1* will remain -1. In order to track the current computation taking place in server module we maintain the *compute_y1* and *compute_y2* to represent the ongoing computation. Once the computation gets over the state changes to either *computed_y1* or *computed_y2*. This change acts as an indicator for the verifier to carry out its final verification process.

### 3.4 State Machine Representation of the Protocol

The Finite State Machine(FSM) representation of protocol $\mathcal{P}$ can be written as

$$\mathcal{P} = \{V, Q, \pi, \delta, q_0\}$$

where,

V = { prover.state, prover.sign, prover.signature,
verifier.state, verifier.result,
server.state, server.y1, server.y2
}

Q = set of states in the model, with values assigned to each $v \in V$

$\pi$ : { Q $\rightarrow$ D(V) | D(V) contains range of each $v \in V$ }

$\delta$ : Transition Function

$q_0$ : Initial state $\in Q$

Initially before the prover signs the message and the verifier verifies the same, each of these has to be initialized to a stable state. From this stable state we can systematically model the signing and verification process. The initial values assigned to each of the state variables are as shown in Table 4.

**Table 4.** Initial State of the Protocol

| Variable | Value | Variable | Value | Variable | Value |
|---|---|---|---|---|---|
| prover.state | idle | verifier.state | idle | server.state | idle |
| prover.sign | none | verifier.result | none | server.y1 | -1 |
| prover.signature | -1 | | | server.y2 | -1 |

From the initial state the prover assumes the role of either a legitimate, cheating or repudiating(by generating a *correlated* signature) prover. In Fig. 2 a possible transition sequence is shown. It can be seen that once the prover signs the message and enters the *signed* state in $q_2$, the verifier has an option to choose one of the two approaches, i.e. either $q_3$ or $q_4$.

Based on the chosen approach the further computations take place and finally the verifier completes the verification and the *result* will be either *success* or *failure*. In Fig. 2, since the prover is legitimate the final state $q_9$ will have $verifier.result = success$.

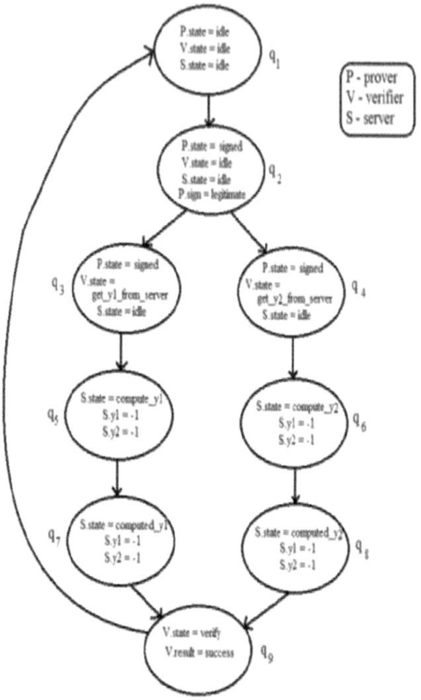


**Fig. 2.** A possible sequence of transitions from the *idle* state to final *verify* state of verifier

### 3.5 Example

In the previous sections we have defined the different state variables which are required to model the protocol. To generate and verify the signature we will have to further define some variables whose values remain constant during a complete verification cycle. Here cycle refers to the transition from *idle* state of each module to final state and back to the *idle* state again. The values we need to define are given in Table 5.

Now, let us consider $p = 2$, $q = 5$, then $n = 10$. Let public key be $e = 5$ then as per the RSA algorithm the private key is $d = 1$. Let $f(m) = 9$. The legitimate prover will generate the signature($\sigma$) as $\sigma = 9^1 \bmod 10$, i.e. $\sigma = 9$. In the rest

**Table 5.** Defined Variables in the Protocol

| Variable | Value | Variable | Value |
|---|---|---|---|
| p | prime number | f(m) | encrypted value for message |
| q | prime number | sig | $f(m)^d \bmod n$ |
| n | $p * q$ | sig_correlated | correlated value to sig |
| d | private key of prover | sig_cheat | generated using random Y value |
| e | public key of prover | | |

of the paper we represent $\sigma$ as *sig*. If the illegitimate prover wants to cheat instead, then it will randomly choose $\bar{Y}$, say $\bar{Y} = 3$ and generate the signature as sig_cheat $= 9 * 3^{-2} \bmod 10$, i.e. sig_cheat $= 1$. When the server gets request from verifier using the first approach it just sends the known value $\bar{Y} = 3$. The verifier finally computes whether $3^2 * 1 \bmod 10 = f(m)$, i.e. $9 = 9$ and hence the signature is accepted. Thus this approach fails when a illegitimate prover tries to cheat the verifier.

# 4 Specification and Model Checking of Server Aided Verification Protocol

A model is never right or wrong in itself. It either satisfies the property desired out of it or it does not. To verify this, we need to specify the properties in a way that is understandable by the system. Consider our model $\mathcal{P}$ with the states $q_i$. Now we have to check whether the model satisfies a given property $\phi$ in a particular state as shown below.

$$\mathcal{C}, s_i \models \phi$$

The property $\phi$ can be specified in a suitable logic. For our model we need to quantify over different paths rather than different states in a chosen path that change with time. To accomplish this temporal aspect we prefer the CTL. The specifications in this paper are coded using the syntax of CTL [10].

The desired properties of a good authentication protocol are :

- $\phi_1$ : Illegitimate prover should not cheat the verifier
- $\phi_2$ : Non-repudiation property has to be ensured
- $\phi_3$ : Legitimate provers signature should always succeed to prove his identity

The first property $\phi_1$ can be specified using CTL as shown below:

$$!\text{EF}(\text{prover.sign} = \text{cheat}\ \&\ \text{verifier.result} = \text{success})$$

It means if the provers intention is to cheat then the verifier must not produce the result as *success*. In order to ensure non-repudiation we need to make sure that the model will not accept a correlated signature generated by a legitimate prover. This property $\phi_2$ can be specified using CTL as shown below:

$$!\text{EF}(\text{prover.sign} = \text{correlated}\ \&\ \text{verifier.result} = \text{success})$$

The third property ensures that the model correctly authenticates a valid prover. This property $\phi_3$ can be specified using CTL as shown below:

$$\text{prover.state} = \text{legitimate} \rightarrow$$
$$\text{A}[\text{prover.state} = \text{legitimate U verifier.result} = \text{success}]$$

It states that if the prover is legitimate, then in every path(A) prover continues to remain in legitimate state until(U) verifier succeeds in its authenticating process.

## 5 Verification Results and Discussion

The properties specified in the section 4 has to be evaluated against the model developed in the section 3. To verify it we use the symbolic model verifier NuSMV. The model developed in section 3 is given as one of the inputs to the system and the properties are stated using the keyword 'SPEC'. All the three properties discussed in the section 4 has been checked in the model.

The first two properties $\phi_1$ and $\phi_2$ fails in the model and hence NuSMV produces a counterexample for the same. The property $\phi_1$ (marked as 1 inside the circle in Fig. 3) fails because in the first approach deployed by the verifier, if the server returns the precomputed value by collaborating with the prover(marked as 2 inside the circle in Fig. 3), then the verifier wrongly accepts (marked as 3 inside the circle in Fig. 3) the invalid signature. The complete state transition is shown by the screen shot in Fig. 3.

The second property $\phi_2$ fails for a similar reason. In this case though the value of $Y$ is not precomputed, the server knows $\sigma$ and the correlated value of $\sigma$. Thus when the verifier sends the request, it can easily help the prover in its attempt to repudiate the message.

The third property $\phi_3$ holds in *true* in the model. This is because when the prover is legitimate every path will lead to a state wherein verifier successfully authenticates the prover. Thus both the approaches chosen by the verifier satisfy $\phi_3$.

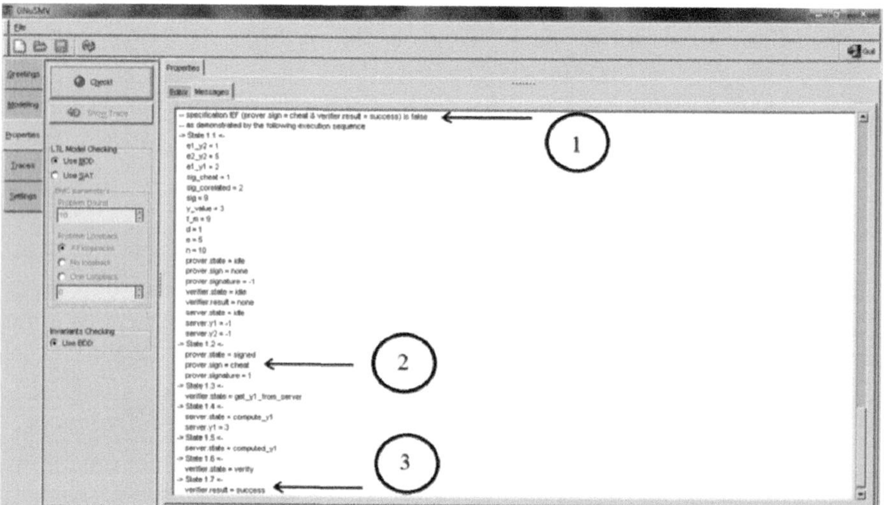


**Fig. 3.** Screenshot : Verification of Property $\phi_1$ developed in 4 which fails to hold in the model

## 6 Conclusion

Designing cryptographic algorithms and communication protocols is a complex process. Formal methods are increasingly being used in the design of these systems. It provides the stake holders a certain degree of assurance that the product is bug free. In this paper we have tried to model and verify the basic Server Aided Verification Protocol. Some of the possible loopholes in the approach were detected using NuSMV as the model verifier tool. This is just a starting work in the direction and a lot of scope for improvement exists. We considered only the prover being legitimate, cheating and repudiating. There are other attacks that the prover or the server itself can come up with. From the verifier point of view we have considered two approaches to deal with the server. Further work can be carried out to develop better approaches to delegate work to the server.

## References


1. Girault, M., Lefranc, D.: Server-aided verification: Theory and practice. In: Roy, B. (ed.) ASIACRYPT 2005. LNCS, vol. 3788, pp. 605–623. Springer, Heidelberg (2005)
2. Chow, S.S.M., Au, M.H., Susilo, W.: Server-aided signatures verification secure against collusion attack. In: Proceedings of the 6th ACM Symposium on Information, Computer and Communications Security, ASIACCS 2011, pp. 401–405. ACM, New York (2011)
3. Wu, W., Mu, Y., Susilo, W., Huang, X.: Server-aided verification signatures: Definitions and new constructions. In: Baek, J., Bao, F., Chen, K., Lai, X. (eds.) ProvSec 2008. LNCS, vol. 5324, pp. 141–155. Springer, Heidelberg (2008)
4. Wu, W., Mu, Y., Susilo, W., Huang, X.: Provably secure server-aided verification signatures. Comput. Math. Appl. 61, 1705–1723 (2011)
5. Pfitzmann, B., Waidner, M.: Attacks on protocols for server-aided rsa computation. In: Rueppel, R.A. (ed.) EUROCRYPT 1992. LNCS, vol. 658, pp. 153–162. Springer, Heidelberg (1993)
6. Begiun, P., Quisquater, J.J.: Fast server-aided rsa signatures secure against active attacks. In: Coppersmith, D. (ed.) CRYPTO 1995. LNCS, vol. 963, pp. 57–69. Springer, Heidelberg (1995)
7. Cavada, R., Cimatti, A., Jochim, C.A., Keighren, G., Olivetti, E., Pistore, M., Roveri, M., Tchaltsev, A.: Nusmv 2.4 user manual (2005)
8. Cavada, R.: Nusmv: a new symbolic model checker (2010), http://nusmv.fbk.eu/
9. Stallings, W.: Cryptography and Network Security: Principles and Practice, 5th edn. Prentice Hall, Englewood Cliffs (2010)
10. Huth, M., Ryan, M.: Logic in Computer Science, Modelling and Reasoning about Systems, 2nd edn. Cambridge University Press, Edinburgh (2005)

# Modeling and Verification of Inter Realm Authentication in Kerberos Using Symbolic Model Verifier

Punit Mundra, Madhavi Sharma, Shobhit Shukla, and Sanjay Singh

Department of Information and Communication Technology
Manipal Institute of Technology, Manipal University, Manipal-576104, India
sanjay.singh@manipal.edu

**Abstract.** In open distributed environment several users accesses the network resources on server, server will allow only authenticated users to access these resources. So it has become of prime importance for nodes communicating over a non-secure network to prove their identity to one another in a secure manner. Various authentication protocols is used for the this purpose, Kerberos protocol is one of them. Several versions of the protocol exist. The aim of this paper is modeling and verification of the Inter Realm Authentication and User to User Authentication in Kerberos protocol through NuSMV model checker. It also demonstrate that when presence of intruder in the system, and make use of service, it will generate the counter example.

## 1 Introduction

Authentication protocol are security mechanism whereby each party is assured its identity to one another. One of the authentication protocol which is commonly used is Kerberos. Kerberos was developed in the Athena Project at the Massachusetts Institute of Technology (MIT) [1]. It is a computer network authentication protocol, which allow nodes communicating over a non-secure net- work to prove their identity to one another in a secure manner. It is designed as a client/server model and it provides mutual authentication i.e. both the client and server verify each other's identity. Several versions of the protocol exist, versions 1 to 3 occurred only internally at MIT. Steve Miller and Clifford Neuman, the primary designers of Kerberos version 4, published that version in the late 1980s, although they had targeted it primarily for Project Athena.Version 5, designed by John Kohl and Clifford Neuman then appeared with the intention of overcoming the limitations and security problems of version 4.

Modeling the protocol through model checker is the best way to find the behavior of the protocol. Modeling checking is an automatic, model-based verification approach. It evaluates the all possible state and ease to find the concurrent behavior of the system. Model checking is based on temporal logic. Temporal logic expresses the dynamic aspect of the system.

For modeling the concurrent aspect Computation Tree logic (CLT) [2] is used. Computation Tree logic is basically used for Verification of hardware and communication protocol. For Modeling Kerberos, we use NuSMV model checker.

D. Nagamalai, E. Renault, M. Dhanushkodi (Eds.): CCSEIT 2011, CCIS 204, pp. 496–506, 2011.

This work has been mainly motivated by the previous work on Model Checking of Kerberos Protocol using New Symbolic Model Verifier (NuSMV) [3][4]. M. Panti et.al [3] has tried to express the basic version of Kerberos. Shreya et. al [4] also has used NuSMV for the basic version of Kerberos for authentication. They have presented the concept of freshness and which helps to find the reply attack. They have demonstrated a possible means by which the weakness of the Kerberos protocol causing the replay attack can be overcome hence making the protocol more stronger.

In this paper we have considered the kerberos protocol with two realm and try to find the reply attack. We have also demonstrated the authentication forwarding in kerberos 5. In presence of intruder NuSMV will generate the counter example. The aim of paper is to model and verify the kerberos protocol by means of NuSMV [5][6][7].

Model Checking, one of many formal verification methods [8], is an attractive and increasingly appealing alternative to simulation and testing to validate and verify systems. Given a system model and desired system properties, the Model Checker explores the full state space of the system model to check whether the given system properties are satisfied by the model. The Model Checker either verifies the given properties or generates a counter examples.

When the model checker verifies a given system property, it implies that all behaviors have been explored, and the question of adequate coverage or a missed behavior become irrelevant. There are two main advantages of using model checking compared to other formal verification methods : -

1. It is fully automatic, and
2. It provides a counter example whenever the system fails to satisfy a given property.

The remainder of the paper is organized as follows. Section 2 discusses about the theoretical background of kerberos protocol and limitation of kerberos 4. Section 3 describes how to model the Kerberos protocol in NuSMV. Section 4 shows how the authentication takes place and the results obtained in this work.

Finally conclusion has been drawn in section 5.

## 2 Theoretical Background

Kerberos provides a centralized authentication server whose function is to authenticate users to servers and servers to users. Two versions of Kerberos are in common use version 4 and version 5 [9].

### 2.1 Kerberos Version 4

This version contains mainly four entities:-

1. Client (C): - An entity which wants to make use of any service hosted on a server.
2. Server (V): - An entity which hosts different services which clients request for.

3. Authentication Server (AS): - A trusted third party which knows the passwords of all users and stores these in a centralized database. Also, the AS shares a unique secret key with each server.
4. Ticket Granting Server (TGS): - TGS, issues tickets to clients who have been authenticated to AS. Thus, the client first requests a ticket-granting ticket from the AS. The client module saves this ticket. Each time the user requires access to a new service, the client applies to the TGS, using the ticket to authenticate itself. The TGS then grants a ticket for the particular service. The client saves each service-granting ticket and uses it to authenticate its user to a server each time a particular service is requested.

The protocol also makes use of timestamps to verify that sender's clock is synchronized with that of receiver and to indicate the time at which ticket was generated, and the lifetime which informs about the lifetime of the ticket, It prevents replay after ticket has expired. The basic messages involved are shown in Fig.1.

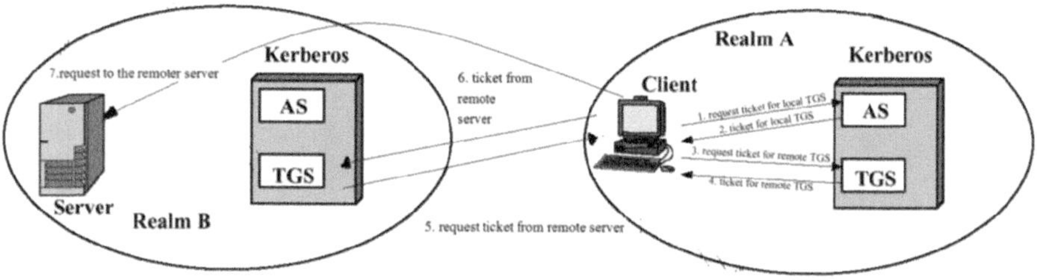


**Fig. 1.** Authentication Scenarios in Kerberos

## 2.2 Kerberos Realms

A Kerberos realm is a set of managed nodes that share the same Kerberos database. Networks of clients and servers under different administrative organizations typically constitute different realms. However, users in one realm may need access to servers in other realms, and some servers may be willing to provide service to users from other realms, provided that those users are authenticated. Kerberos provides a mechanism for supporting such inter realm authentication [9].

The Kerberos server in each inter operating realm shares a secret key with the server in the other realm. The two Kerberos servers are registered with each other. The scheme requires that the Kerberos server in one realm trust the Kerberos server in the other realm to authenticate its users. Furthermore, the participating servers in the second realm must also be willing to trust the Kerberos server in the first realm. A user wishing service on a server in another realm needs a ticket for that server.

Following are the messages exchanged in inter realm authentication:

Here AS is authentication server, CA is client in one realm A who wants to access service from server VB in second realm B TGSA is ticket granting server in realm A and TGSB is ticket granting server in realm B.

1. $- C \rightarrow AS : C \| TGS_{ID} \| TS_1$

In this message, the client requests a ticket-granting ticket by sending its identity and password to the AS, together with the identity of TGS, indicating a request to use the TGS service and a time stamp, so that the AS knows that the message is timely.

2. $- AS \rightarrow C : E(K_c, [K_{c,tgs} \| TGS_{ID} \| TS_2 \| Lifetime_2 \| Ticket_{tgs}])$, where
$- Ticket_{tgs} = E(K_{tgs}, [K_{c,tgs} \| C \| AD_c \| TGS_{ID} \| TS_2 \| Lifetime_2])$

The AS responds with a message, encrypted with a key derived from the user's password that contains the ticket requested by client. This message also contains session key $K_{c,tgs}$ to be shared between TGS and Client. The client retrieves this key.

The ticket is encrypted with TGS's key $K_{tgs}$ which is already shared with AS. The ticket contains session key $K_{c,tgs}$ which is retrieved by TGS. Like this, the session key has been securely delivered to both C and the TGS. The ticket also contains ADc which prevents the use of ticket from workstation other than one that initially requested the ticket.

3. $- C \rightarrow TGS : TGS_{rem\_ID} \| Ticket_{tgs} \| Authenticator_c$

The client sends message to TGS stating the identification of the TGS in another realm, $ticket_{tgs}$, an authenticator which includes the identity of client C, address of Client $AD_c$ and a time stamp. The authenticator is intended for use only once and has a very short lifetime. The TGS can decrypt the ticket with the key that it shares with the AS. The TGS uses the session key $K_{c,tgs}$ to decrypt the authenticator. The TGS can then check the identity and address from the authenticator with that of the ticket and with the network address of the incoming message. If all match, then the TGS is assured that the sender of the ticket is indeed the ticket's real owner.

4. $- TGS \rightarrow C : E(K_{c,tgs}[K_{c,tgs} \| TGS_{rem\_ID} \| TS_4 \| Ticket_{tgsrem}])$
$- Ticket_{tgsrem} = E(K_{tgsrem}[K_{c,tgsrem} \| C \| AD_c \| TGS_{rem\_ID} \| TS_4 \| Lifetime_4])$

The TGS then sends to C as a response to request. This message is encrypted by session key shared between TGS and client and includes a session key to be shared between Client and TGS_rem, the identification of TGS in realm B and the time stamp of the ticket.

5. $- C \rightarrow TGS_{rem} : V_{rem} \| Ticket_{tgsrem} \| Authenticator_c$

The client sends message to TGS_rem stating the identification of the requested service VB, an authenticator which includes the identity of client C, address of client $AD_c$ and a time stamp. The TGS uses the session key $K_{c,tgsrem}$ to decrypt the authenticator. The TGS can then check the identity and address from the authenticator with that of the ticket and with the network address of the incoming message. If all match, then the TGS is assured that the sender of the ticket is indeed the ticket's real owner.

6. – $TGS\_rem \rightarrow C : E(K_{c,tgsrem}[K_{c,vrem} \| V_{rem} \| TS_6 \| Ticket_v rem])$
   – $Ticket_v rem = E(K_v rem[K_{c,vrem} \| C \| AD_c \| V_{rem} \| TS_6 \| Lifetime_6])$

   The TGS_rem then sends $Ticket_v rem$ as response to C. This message is encrypted by session key shared between TGS rem and client and includes a session key to be shared between client and server V, the identification of V , and the time stamp of the ticket.

7. – $C \rightarrow V_{rem} : Ticket_v rem \| Authenticator_c$
   – $Authenticator_c = E(K_{c,vrem}[C \| AD_c \| TS_7])$

   In this message the client sends the ticket received by TGS_rem to server V. Client also sends an authenticator. The server can decrypt the ticket, recover the session key, and decrypt the authenticator.

### 2.3 Limitations of Kerberos Version 4

Kerberos 4 has limitations in two areas environmental shortcomings and technical deficiencies. The focus of this paper is on environmental shortcomings following are the environmental shortcomings of kerberos 4 [9]:-

-**Encryption system dependence:** Version 4 requires the use of DES. Ex-port restriction on DES as well as doubts about the strength of DES were thus of concern.

-**Authentication forwarding:** Version 4 does not allow credentials issued to one client to be forwarded to some other host and used by some other client. This capability would enable a client to access a server and have that server access another server on behalf of the client. For example, a client issues a request to a print server that then accesses the client's file from a file server, using the client's credentials for access. Version 5 provides this capability.

-**Inter realm authentication:** In version 4, interoperability among N realms requires on the order of N2 Kerberos-to-Kerberos relationships, as described earlier. Version 5 supports a method that requires fewer relationship.

### 2.4 Kerberos Version 5

Kerberos 5 has many improvement over kerberos version 4. Some new element are added in kerberos version 5[9] are as follows:-

-**Realm:** Indicates realm of user.
-**Options:** Used to request that certain flags be set in the returned ticket
-**Times**: Used by the client to request the following time settings in the ticket:
-**Nonce:** A random value to be repeated in message (2) to assure that the response is fresh and has not been replayed by an opponent.

The flags field included in tickets in version 5 supports expanded functionality compared to that available in version 4. There are many flags in Kerberos version 5. But for user to user authentication FORWARDABLE and FORWARDED are used.

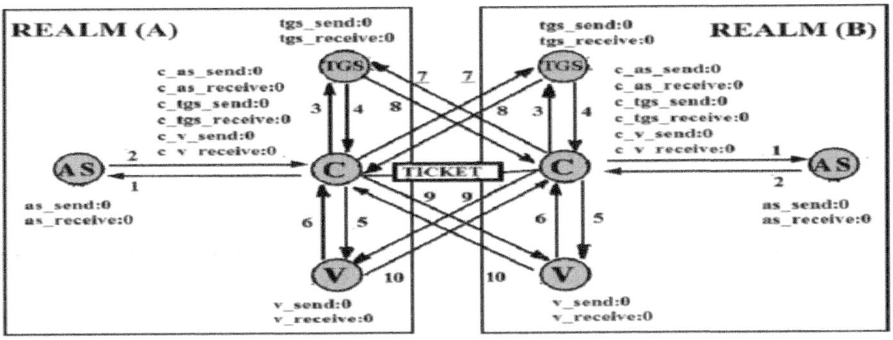


**Fig. 2.** User-to-User Authentication

**-FORWARDABLE :** Tells TGS that a new ticket-granting ticket with a different network address may be issued based on this ticket-granting ticket.

**-FORWARDED :** Indicates that this ticket has either been forwarded or was issued based on authentication involving a forwarded ticket-granting ticket.

In this paper we mainly emphasis on User to User and Inter Realm Authentication. Authentication in Kerberos happens between a client and server. The client gets a ticket for a service, and the server decrypts this ticket using its secret key. This works fine for a physically- secure server, which keeps its secret key on its local disk. But, storing the server's key on disk does not work for services that run on users' desktop machines, since no-one should keep a long-lived secret key on an insecure disk drive.

The solution to this problem is called user-to-user authentication, and it is implemented in Kerberos 5. In the user-to-user protocol, one user acts as a server, and the other user acts as a client. At the client-user's request, the server- user sends his TGT (but not his session key) to the client-user, who then gets credentials from the key distribution center(KDC), encrypted with the session keys of both TGTs. Both users can decrypt the new session-key and use it to verify each other's identity. The advantage of the User to User scheme is that the server-user exposes only his short-lived TGS session-key to theft; he keeps his long-lived secret, his password, in his biological memory. An attacker is less likely to bother to steal a short-lived server-key. However, User to User's downside is that the desktop server cannot operate autonomously; the service-operator has to refresh his TGT in order for the server to keep accepting clients' requests. Kerberos 5 has many improvement over kerberos 4, which we discuss in section II. But we focus on one of the short comings of in kerberos 4, environmental shortcomings (Inter realm Authentication). Here we also model the scenario in which one of the client can forward his authentication ticket to other client.

## 3 Modeling the Protocol

Formalizing and Modeling the inter-realm authentication, we consider the two realm and we try to _nd out the attack between the two realms. Consider the Two realm A and B.In each Realm we have considered four entity:

**Relam A**

c1 : client a1 : AS
t1 : TGS v1 : server

**Relam B**

c2 : client a2 : AS
t2 : TGS v2 : Server

These entities will communicate with each other and send message to each other as described in Section II. Each entity exchanges a set of messages with other entities in the model. Apart from entities there is one more entity Intruder (I) for the purpose of checking the protocol against any attack Each entity goes through some state during communication. Table.1 shows the following state's of entity.

**Table 1.** States Representation

| state | C | AS | TGS1 | V1 | TGS2 | V2 | state | C | AS | TGS1 | V1 | TGS2 | V2 |
|---|---|---|---|---|---|---|---|---|---|---|---|---|---|
| first state | | | | | | | next state | | | | | | |
| 1 | send1 | idle | idle | idle | idle | idle | 1 | wait | receive1 | idle | idle | idle | idle |
| 2 | wait | send1 | idle | idle | idle | idle | 2 | receive1 | wait | idle | idle | idle | idle |
| 3 | send2 | idle | idle | idle | idle | idle | 3 | wait | idle | receive1 | idle | idle | idle |
| 4 | wait | idle | send1 | idle | idle | idle | 4 | receive2 | idle | wait | idle | idle | idle |
| 5 | send3 | idle | idle | idle | idle | idle | 5 | wait | idle | idle | receive1 | idle | idle |
| 6 | wait | idle | idle | send1 | idle | idle | 6 | receive3 | idle | idle | idle | idle | idle |
| 7 | send4 | idle | idle | idle | idle | idle | 7 | wait | idle | idle | idle | receive1 | idle |
| 8 | wait | idle | idle | idle | send1 | idle | 8 | receive4 | idle | idle | idle | idle | idle |
| 9 | send4 | idle | idle | idle | idle | idle | 9 | wait | idle | idle | idle | idle | receive1 |
| 10 | wait | idle | idle | idle | idle | send1 | 10 | receive5 | idle | idle | idle | idle | idle |

- *Client(C)* : idle, send1, receive1, send2, receive2, send3, receive3
- *Authentication Server (AS)* : idle, send1, receive1
- *Ticket Granting Server (TGS)* : idle, send1, receive1
- *Server (V)* : idle, send1, receive1
- *Intruder (I)* : idle, send2, receive2, send3, receive3, eavesdrop, remove, store

Across each entity a set of counter is associated. These counter count the number of message exchanged during communication. When ever the entity send or receive the message, the corresponding counter increment. All the counter are initialize to 0, the counter will increment to its maximum number of session. counter associated with each entity are as follows.

**c_ as_ send :** counts the number of messages sent from C to AS
**c_ as_ receive :** counts the number of messages C received from AS
**c_ tgs_ send :** counts the number of messages sent from C to TGS
**c_ tgs_ receive :** counts the number of messages C received from TGS
**c_ v_ send :** counts the number of messages sent from C to V
**c_ v_ receive :** counts the number of messages C received from V
**as_send :** counts the number of messages sent from AS to C
**as_receive :** counts the number of messages received by AS from C
**tgs_send :** counts the number of messages sent from TGS to C
**tgs_receive :** counts the number of messages received by TGS from C
**v_send :** counts the number of messages sent from V to C
**v_receive :** counts the number of messages received by V from C

In the Table.2 various state of each entity is shown. Fig.2 shows how the message can transfered between the state. Number in the given figure will shows the state's of the entity which is show in Table.1.

**Table 2.** Counter

| state | c_as_s | c_as_r | c_tgs_s | c_tgs_r | c_v_s | c_v_r | as_s | as_r | tgs_s | tgs_r | v_s | v_r | tgs2_s | tgs2_r | v2_s | v2_r |
|---|---|---|---|---|---|---|---|---|---|---|---|---|---|---|---|---|
| 0 | 0 | 0 | 0 | 0 | 0 | 0 | 0 | 0 | 0 | 0 | 0 | 0 | 0 | 0 | 0 | 0 |
| 1 | 1* | 0 | 0 | 0 | 0 | 0 | 0 | 1* | 0 | 0 | 0 | 0 | 0 | 0 | 0 | 0 |
| 2 | 1 | 1* | 0 | 0 | 0 | 0 | 1* | 1 | 0 | 0 | 0 | 0 | 0 | 0 | 0 | 0 |
| 3 | 1 | 1 | 1* | 0 | 0 | 0 | 1 | 1 | 0 | 1* | 0 | 0 | 0 | 0 | 0 | 0 |
| 4 | 1 | 1 | 1 | 1* | 0 | 0 | 1 | 1 | 1* | 1 | 0 | 0 | 0 | 0 | 0 | 0 |
| 5 | 1 | 1 | 1 | 1 | 1* | 0 | 1 | 1 | 1 | 1 | 0 | 1* | 0 | 0 | 0 | 0 |
| 6 | 1 | 1 | 1 | 1 | 1 | 1* | 1 | 1 | 1 | 1 | 1* | 1 | 0 | 0 | 0 | 0 |
| 7 | 1 | 1 | 2* | 1 | 1 | 1 | 1 | 1 | 1 | 1 | 1 | 1 | 0 | 1* | 0 | 0 |
| 8 | 1 | 1 | 2 | 2* | 1 | 1 | 1 | 1 | 1 | 1 | 1 | 1 | 1* | 1 | 0 | 0 |
| 9 | 1 | 1 | 2 | 2 | 2* | 1 | 1 | 1 | 1 | 1 | 1 | 1 | 1 | 1 | 0 | 1* |
| 10 | 1 | 1 | 2 | 2 | 2 | 2* | 1 | 1 | 1 | 1 | 1 | 1 | 1 | 1 | 1* | 1 |

# 4 Authentication

## 4.1 User to User Authentication

Kerberos 5 introduced the concept of forwardable tickets. Client himself forward his ticket to other client. In kerberos 5 FORWARDABLE flag are used for user to user authentication. During the initial TGT acquisition, a client can request that the ticket be marked forwardable. Once the FORWARDABLE flag is set on a ticket, the user can use this ticket to request a new ticket. For modeling User to User authentication in NuSMV we also use the one flag with each client. If the value of flag is TRUE then user will forwarded his ticket to other when request come, other wise not. As mention in this protocol that the life time of FORWARDABLE ticket is very less. So if intruder will capture it and its life time is stale soon. For modeling this we used one counter with this flag and the value of counter is 1. For FORWARDABLE ticket server will also have one counter whose value is 1. when client1 forwarded his authenticated ticket to client2 then the value of counter is 1. When client2 forwarded ticket to server, server will check its counter value. When counter value 1 then only server will respond to that service. After granting the service server set its counter value to 0.

## 4.2 Inter Realm Authentication

For proving the authentication, corresponding property is used [3]. The corresponding property state's that :-

*If the event E occurs n times, then the event F must have occurred at least n times.*

we generalized this property by CTL specification. The correspondence property can be formalized by the means of a general schema:

$$AG(X.counter_F \geq Y.counter_E)$$

where X.counterF and Y.counterE are the counters of processes corresponding to entities X and Y. AG is the abbreviation for Always Globally CTL operator, and AGx means that the propositional formula x must be true in every possible state of the system. The authentication SPEC are as follows :-

$$\begin{aligned}
&\text{(i)} && AG(C.c_as_count_s \geq AS.as_count_r) \\
&\text{(ii)} && AG(AS.as_count_s \geq C.c_as_count_r) \\
&\text{(iii)} && AG(C.c_tgs_count_s \geq TGS.tgs_count_r) \\
&\text{(iv)} && AG(TGS.tgs_count_s \geq C.c_tgs_count_r) \\
&\text{(v)} && AG(C.c_v_count_s \geq V.v_count_r) \\
&\text{(vi)} && AG(V.v_count_s \geq C.c_v_count_r)
\end{aligned}$$

-The (i) and (ii) pair of specifications describe that if the client sends n number of protocol executions with the AS as responder, then in past the AS must have started at least n protocol executions with the client as initiator.

-The (iii) and (iv) pair of specifications describe that if the client sends n protocol executions with the TGS as responder, then in past the TGS must have started at least n protocol executions with the client as initiator.

-The (v) and (vi) pair of specifications describe that if the client sends n protocol executions with the V as responder, then in past the V must have started at least n protocol executions with the client as initiator.

**Reply Attack.** In order to detect the reply attack we make use of counter. For proving the authentication property each entity maintain the counter as in shown in Table.2. when client sends request to Authentication Server(AS),Ticket Granting Server(TGS) or Server(V) it increment its counter:

*c'as'count'send, c'tgs'count'send, c'v'count'send* respectively as shown in Table.2, and when either of these request is accepted by Authentication Server(AS),

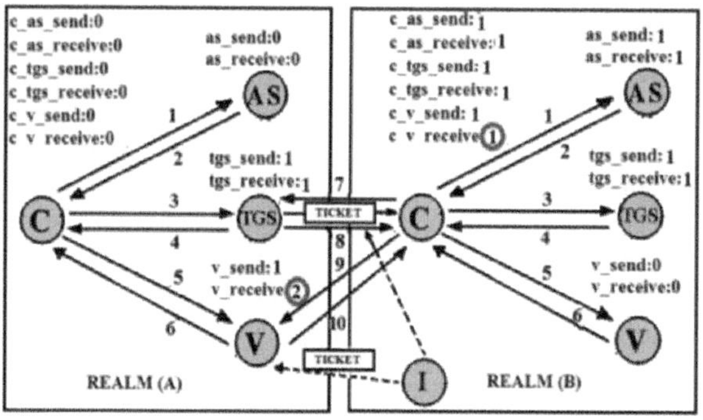


**Fig. 3.** Finding the Attack

Ticket Granting Server(TGS) or Server(V) it increment its counter:
*as_receive, tgs_ receive, v_receive*

| Loop | Step | as.network | as.state | c.network | c.state | tgs.network | tgs.state | tr2.network | tr2.state | vr2.network | vr2.state | i.state |
|---|---|---|---|---|---|---|---|---|---|---|---|---|
| | 0 | realm1 | idle | realm1 | idle | realm1 | idle | realm2 | dle | realm2 | idle | idle |
| | 1 | realm1 | idle | realm1 | send1 | realm1 | idle | realm2 | dle | realm2 | idle | idle |
| | 2 | realm1 | receive1 | realm1 | wait | realm1 | idle | realm2 | dle | realm2 | idle | idle |
| | 3 | realm1 | send1 | realm1 | wait | realm1 | idle | realm2 | dle | realm2 | idle | eavesdrop |
| | 4 | realm1 | idle | realm1 | receive1 | realm1 | idle | realm2 | dle | realm2 | idle | eavesdrop |
| | 5 | realm1 | idle | realm1 | send2 | realm1 | idle | realm2 | dle | realm2 | idle | idle |
| | 6 | realm1 | idle | realm1 | wait | realm1 | receive2 | realm2 | dle | realm2 | idle | idle |
| | 7 | realm1 | idle | realm1 | wait | realm1 | send2 | realm2 | dle | realm2 | idle | idle |
| | 8 | realm1 | idle | realm1 | receive2 | realm1 | idle | realm2 | dle | realm2 | idle | idle |
| | 9 | realm1 | idle | realm1 | send4 | realm1 | idle | realm2 | dle | realm2 | idle | eavesdrop |
| | 10 | realm1 | idle | realm1 | wait | realm1 | idle | realm2 | receive4 | realm2 | idle | eavesdrop |
| | 11 | realm1 | idle | realm1 | wait | realm1 | idle | realm2 | send4 | realm2 | idle | idle |
| | 12 | realm1 | idle | realm1 | receive4 | realm1 | idle | realm2 | dle | realm2 | idle | idle |
| | 13 | realm1 | idle | realm1 | send5 | realm1 | idle | realm2 | dle | realm2 | idle | eavesdrop |
| | 14 | realm1 | idle | realm1 | wait | realm1 | idle | realm2 | dle | realm2 | receive5 | receive5 |
| | 15 | realm1 | idle | realm1 | wait | realm1 | idle | realm2 | dle | realm2 | send5 | remove |

**Fig. 4.** Screenshot of NuSMV Result

When an Intruder (I) is present in the system, it will try to capture the ticket. When Intruder get the ticket and try to access the service on a particular server (either from Realm A or Realm B), Intruder will send the request to the server (V). In that case server will increment its receive counter. Consider the Fig.4, in which intruder will capture the packet and send it to the server, in that case counter of respective server will increment and following SPEC evaluates to FALSE.

We can also add other CTL specifications, so as to check whether Intruder (I) receives the fourth message, as given below: Consider the particular specification

$$\mathbf{SPEC\ AG\ !\ (i.state = receive5)}$$

This also returns a similar counterexample tracing the path up to the state where it becomes false and Intruder I receives the seventh message. But this specification is unrealistic since in real life scenario it is not possible to check the state of an intruder and whether it has received any messages or not.

$$\mathbf{AG}(\mathbf{C}.\boldsymbol{c\_v\_count\_send} \geq \boldsymbol{V.v\_count\_receive}).$$

Hence, the specification must hold true in all states of the system. When the present model of the protocol is run with this specification it returns false and generates a counterexample.

## 5 Conclusion

In this paper, we have tried to model and verify the Inter Realm Authentication and in Kerberos Protocol using the model checker tool NuSMV. We have used the correspondence property to obtain CTL specifications of all the security requirements of the given protocol with the help of various counters. We also demonstrate the authentication forwarding between the client. NuSMV will generate the counter example in presence of intruder. In future this method of modeling and verifying of the protocol can be extended to Pre-Authentication in Kerberos Version 5.

## References


1. Miller, S.P., Neumann, B.C., Schiller, J.I., Saltzer, J.H.: Kerberos authentication and authorization system (October 1988), `http://web.mit.edu/Saltzer/www/publications/athenaplan/e.2.1.pdf`
2. Huth, M., Ryan, M.: Logic In computer Science. Cambridge University Press, Cambridge (2004)
3. Panti, M., Spalazzi, L., Tacconi, S.: Using the nusmv model checker to verify the kerberos protocol (2007), `http://www.inform.unian.it/personale/spalazzi/repo`
4. Adyanthaya, S., Rukmangada, S., Tiwari, A., Singh, S.: Modeling freshness concept to overcome replay attack in kerberos protocol using nusmv. In: 2010 International Conference on Computer and Communication Technology (ICCCT), Allahabad, India, pp. 125–129 (September 2010)
5. Cavada, R., Cimatti, A., Jochim, C.A., Keighren, G., Olivetti, E., Pistore, M., Roveri, M., Tchaltsev, A.: Nusmv 2.4 user manual (2005), `http://nusmv.fbk.eu/NuSMV/userman/v24/nusmv.pdf`
6. Nusmv 2.4 (2011), `http://nusmv.fbk.eu/`
7. Cimatti, A., Clarke, E.M., Giunchiglia, E., Giunchiglia, F., Pistore, M., Roveri, M., Sebastiani, R., Tacchella, A.: Nusmv 2: An opensource tool for symbolic model checking. In: Brinksma, E., Larsen, K.G. (eds.) CAV 2002. LNCS, vol. 2404, pp. 268–359. Springer, Heidelberg (2002)
8. Edmund, O.G., Clarke, M., Peled, D.A.: Model Checking. MIT-PRESS, Cambridge (1999)
9. Stallings, W.: Cryptography and Network Security Principles and Practices, 5th edn. Prentice Hall, New Delhi (2010)

# A New Bluetooth Security Architecture

Mintu Philip and Asha Das

Toc H Institute of Science and Technology, Kerala, India
mintu.philip@gmail.com, asha_das_2000@rediffmail.com

**Abstract.** The paper proposes a new Bluetooth security architecture based on user authentication and chaotic image encryption. The algorithm is able to overcome man in middle attack as well as privacy for image is ensured using a fast encryption algorithm based on chaotic encryption. Chaotic encryption produces the key using a pseudorandom number generator. It is a stream cipher method.

**Keywords:** cryptography, security, chaotic, logistic map.

## 1 Introduction

The Bluetooth wireless technology provides short range wireless connectivity between common devices. Its popularity and widespread usage triggers us into giving attention to its security issues. The man-in-the-middle attack and non User Authentication are significant weaknesses in Bluetooth system specification defined at the link-level.

Chaos theory has been established since 1970s. The distinct properties of chaos, such as ergodicity, quasi-randomness, sensitivity dependence on initial conditions and system parameters, have granted chaotic dynamics as a promising alternative for the conventional cryptographic algorithms.

A new way of image encryption scheme has been proposed which utilizes a chaos-based cryptographic scheme using the logistic map. A combined image compression and encryption scheme is proposed. The model is achieved by using discrete wavelet transform and RLE for compression. Chaos based system is used for encryption. This algorithm encrypts image pixel by pixel taking consideration the values of previously encrypted pixels. This system is robust against cryptanalytic attacks. Also a simple implementation of image encryption achieves high encryption rate on general purpose computer. In mobile bandwidth constraints and power saving are needed for which the proposed algorithm is suitable.

## 2 Proposed Bluetooth Security Architecture

The proposed system consists of four modules: Bluetooth device and service discovery, Bluetooth User authentication, Image compression and chaotic image encryption. The first module discovers the Bluetooth devices within the range of the transmitting device. This is done by device discovery protocol. All the devices are

D. Nagamalai, E. Renault, M. Dhanushkodi (Eds.): CCSEIT 2011, CCIS 204, pp. 507–515, 2011.

listed and we can select the device with which communication is to be established. After selecting the device, we select the service provided by the device, which is done by the service discovery module. After communication is established, the user authentication process is started. The authentication is done using RSA and public key certificates provided by the certificate authority. If the user is authentic, the image transmission is started. The user selects the image to be transmitted and the third module will compress the image. The compression algorithm is designed based on discrete wavelet transform and run length encoding. Then the image can be encrypted using chaotic encryption technique. The cipher can then be safely sent to the channel. At the receiver side, after authentication, the cipher is decrypted using reverse chaos encryption. Then inverse DWT is applied to the bits of data to obtain the original image.

## 2.1 Bluetooth Device and Service Discovery

### 2.1.1 Device Discovery

Bluetooth splits the 2.4-GHz band into 79 channels, with the intent that all devices on a single piconet use exactly one of these channels at a time. Of these 79 channels, 32 are also used for detecting nearby devices and establishing connections. An inquiring device transmits inquiry messages on these channels, and a discoverable Bluetooth device periodically listens for and replies to the inquiry messages.

### 2.1.2 Device Discovery

The server dynamically allocates a port number. A socket is first bound and set to listening mode, before the server advertises an SDP service. The client searches for a service record and uses that information to establish a connection to the appropriate port. The concept is simple: the server advertises a service, and the client searches for and finds that service. Actual port numbers are allocated automatically.

## 2.2 User Authentication

User authentication ensures that the communicating parties are trustworthy. User is authenticated using RSA and public key certificates. The main advantage of this algorithm is that it overcomes man-in-middle attack. The steps in user authentication process are as follows:

1. Sender (A) encrypts its private key and own Bluetooth address using the certificate authority's public key.
2. A transmits the encrypted information to receiver (B).
3. B decrypts data using certificate authority's private key and receives the private key of sender.
4. B encrypts its private key and own Bluetooth address using the certificate authority's public key.
5. B transmits the encrypted information to A.
6. A decrypts data using certificate authority's private key and receives the private key of B.

7. A calculate C using equation:

$$C = KEY^{PKA} \text{ MOD } N \tag{1}$$

Where N=pq, where p and q are large prime numbers and $PK_A$ is the public key of A.

8. Send C to B.
9. B calculates the KEY from C using equation:

$$KEY = C^{KRA} \text{ MOD } N \tag{2}$$

Now, A and B communicates further using KEY.

## 2.3 Image Compression

This phase has two sub-phases: (a) discrete wavelet transforms and (b) run length encoding.

### 2.3.1 Discrete Wavelet Transform

The image is compressed using discrete wavelet transform. The advantage of DWT is that most of the coefficients become zero. DWT converts the image pixels into approximate and detail coefficients. Some of the coefficients are approximately zero which can be neglected.

### 2.3.2 Run Length Encoding

The DWT coefficients are then encoded using Run Length Encoding. RLE works by reducing the physical size of a repeating string of characters. This repeating string, called a run, is typically encoded into two bytes. The first byte represents the number of characters in the run and is called the run count. In practice, an encoded run may contain 1 to 128 or 256 characters; the run count usually contains as the number of characters minus one (a value in the range of 0 to 127 or 255). The second byte is the value of the character in the run, which is in the range of 0 to 255, and is called the run value. Thus the compressed data stream then undergoes encryption.

## 2.4 Image Encryption

The image is encrypted pixel by pixel using chaotic image encryption. The algorithm uses a coupled logistic map. It has a pseudorandom key stream generator that generates a binary stream which used in chaotic encryption. The first logistic map whose initial parameters are taken from the KEY generated during user authentication process provides the initial parameters for second logistic map.

Following are the logistic maps:

$$x_{n+1} = \mu \, x_n \, (1-x_n) \tag{3}$$

$$y_{n+1} = \mu \, y_n \, (1-y_n) \tag{4}$$

The steps involved in pseudorandom key stream generator are:

1.

$$x_{i+1} = \mu 1\ x_i\ (1-x_i) \tag{5}$$

$$y_{i+1} = \mu 1\ y_i\ (1-y_i) \tag{6}$$

2. Convert real number $x_i$ to binary equivalent$X_i$.
3. Divide $X_i$ into three parts and XOR the three parts to obtain $X_i'$.
4. Perform above steps for value of 'i' starting from 1 to n.
5. Convert $X_n$ to real value$x_n$.
6. The value $x_n$ is given as initial value μ2 to second logistic map.
7. Above steps are repeated for second logistic map.
8. The final value $y_n$ is multiplied by $10^{18}$ and is converted to binary, stored in s1.
9. Multiply the value of μ2 by $10^5$ and convert to binary and store in s2.
10. Take first 56 bits of s1 and $5^{th}$ to $15^{th}$ bits of s2 and combine it to form the key to encrypt.
11. Perform XOR operation of pixels with the key to obtain the cipher.
12. At receiver side perform XOR of cipher with the key to decrypt data.
13. Decompress the data using inverse DWT to obtain the pixels of the image.
14. Write the pixels to a new image file.

The binary sequence generated by pseudorandom key stream generator is XORed with the pixel values of the image to obtain the cipher image.

## 3 Statistical Analysis

The encrypted images should possess certain random properties in order to resist the statistical attack. Statistical analysis is done by calculating the histograms, the correlations of two adjacent pixels in the encrypted images and the correlation coefficient for several images and its corresponding encrypted images of an image database. A detail study has been undergone and the results are summarized as followings. Different images have been tested, and similar results are obtained. However, due to the page limit, only the results for the Lena.

### 3.1 Difference between the Original and the Permuted Images

*NPCR* (Number of Pixels Change Rate) is used to test the difference between the original image *P1* and the permuted one *C1*. *NPCR* stands for the number of pixel change rate. Then, if *D* is a matrix with the same size as images *P1* and *C1*, *D (i,j)*is determined as follows:

$$D(i,j) \begin{array}{l} = 0\ ,\ \text{if P1} \neq \text{c1} \\ = 1,\ \text{else} \end{array} \tag{7}$$

*NPCR* is defined by the following formula:

$$\text{NPCR} = \sum_{i=0:M-1}\sum_{j=0:N-1} D(I,j)x100/(MxN) \tag{8}$$

The NPCR of the new system is found to be 99.42.

### 3.2 Correlation Coefficients of Intra and Inter - Color -Components

To quantify the dependence between two images, Pearson's correlation coefficient is commonly used. Given by equation, this coefficient is obtained by dividing the covariance between the two images (eq. 13) by the product of their standard deviations (eq. 12 and eq. 11). E in eq.11 is the expected value operator. P1 (i, j) and C1 (i, j) are respectively the pixels gray values of the first and the second images.

$$E(x) = 1/(M \times N) \sum_{i=0:M}\sum_{j=0:N} P1(i,j) \tag{9}$$

$$D(P1) = 1/(M \times N) \sum_{i=1:M}\sum_{j=1:N} [ P1(I,j) - E(P1(I,j))]^2 \tag{10}$$

$$Cov(P1,Ci) = 1/M \times N \sum_{i=1:M}\sum_{j=1:N} [P1(i,j) - E(P1(i,j))] \times [C1(i,j) - E(C1(i,j))]$$

$$R_{p1c1} = Cov(P1,C1)/(\sqrt{D}((P1)\sqrt{D}(C1) \tag{11}$$

Following results for found for various standard images:

**Table 1.** correlation coefficient of encrypted image

| Image | Correlation coefficient |
|---|---|
| Lena | 0.00099 |
| Pepper | 0.002 |
| Cameraman | 0.0017 |

### 3.3 Distribution of Two Adjacent Pixels

Statistical analysis on large amounts of images shows that on average, 8 to 16 adjacent pixels are correlated.

In this section, some simulations are carried out to test the correlation distribution between two horizontally, vertically and diagonally adjacent pixels, in the original and permuted images. Fig. 5 shows the correlation distribution of two horizontally, vertically and diagonally adjacent pixels in the first component of the original image and the encrypted images. Then, we plot the pixel value on location (x,y+1) over the pixel value on location (x, y), location (x+1,y) over the pixel value on location (x, y) and location (x+1,y+1) over the pixel value on location (x, y).

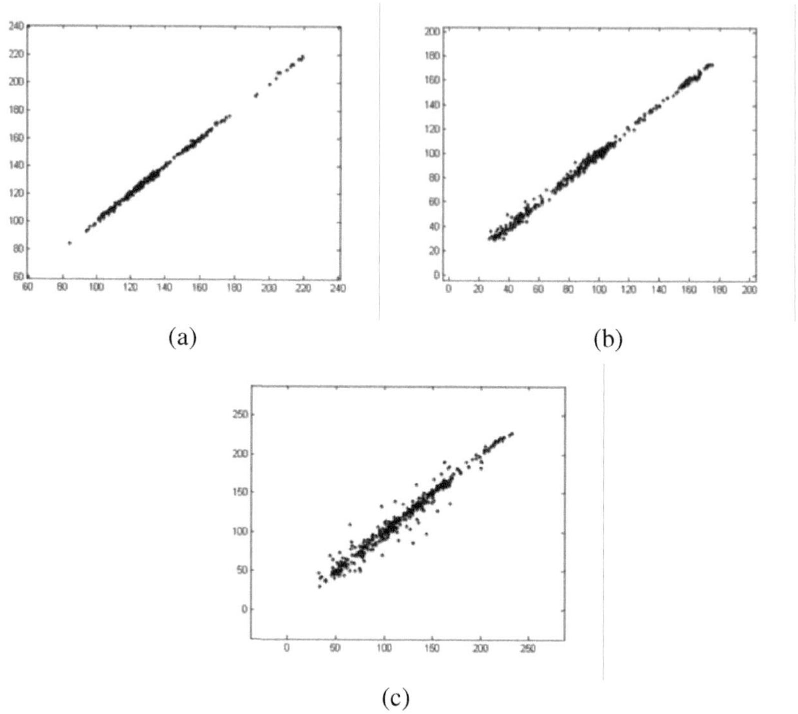


**Fig. 1.** (a) horizontal, (b) vertical and (c) diagonal correlation matrix of original image

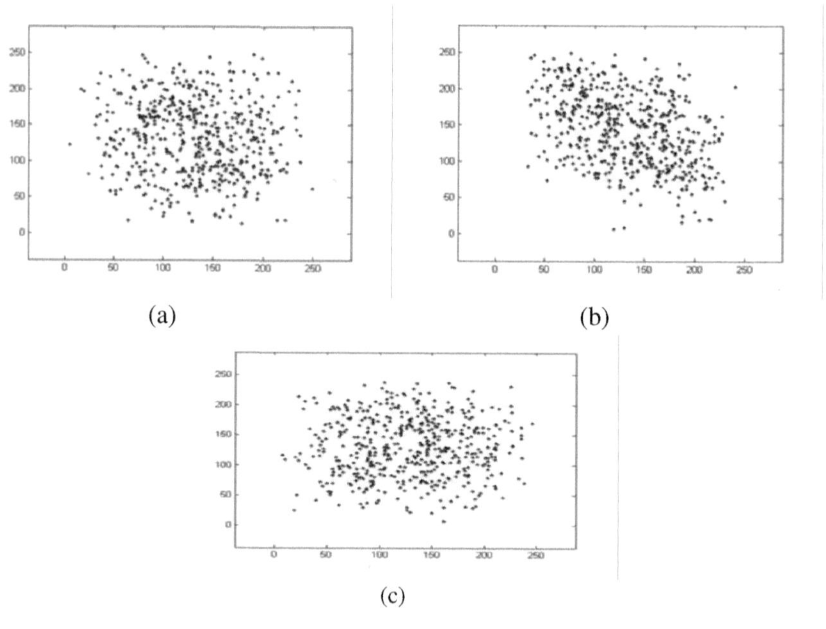


**Fig. 2.** (a) horizontal, (b) vertical and (c) diagonal correlation matrix of encrypted image

It is clear from Fig 2 that there is negligible correlation between the two adjacent pixels in the encrypted image.

### 3.4 Histogram Analysis

An image-histogram illustrates how pixels in an image are distributed by graphing the number of pixels at each color intensity level. We have calculated and analyzed the histograms of the encrypted image as well as the original colored image.

As we can see, the histogram of the encrypted image is significantly different from that of the original image.

Moreover, one can observe that the coupled map improves the uniformity of the histogram for encryption method.

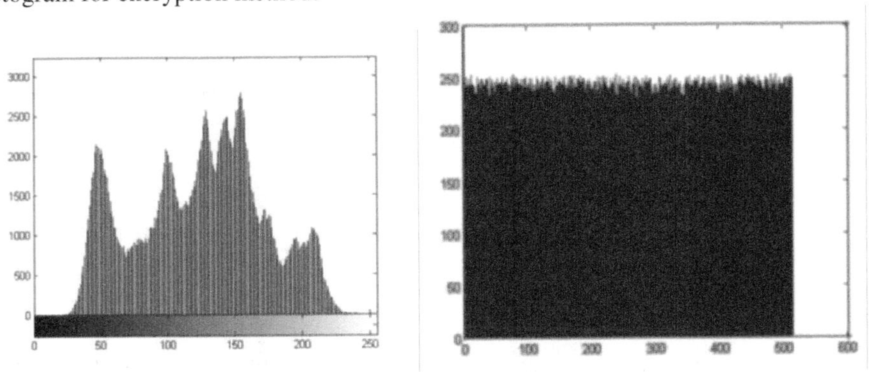


(a) (b)

**Fig. 3.** Histogram of (a) original image and (b) encrypted image

### 3.4 Information Entropy Analysis

Entropy is a statistical measure of randomness that can be used to characterize the texture of an image. It is well known that the entropy *H(m)* of a message source *m* can be calculated as :

$$H(m) = \sum_{i=0:2^N-1} p(mi) \log (1/P(mi)) \quad (12)$$

Where *p(mi)* represents the probability of message *mi*.

When an image is encrypted, its entropy should ideally be 8. If it is less than this value, there exists a certain degree of predictability which threatens its security.

**Table 2.** Information Entropy

| Image | Entropy(original) | Entropy(Encrypted) |
|---|---|---|
| Lena | 6.95 | 7.7719 |
| Pepper | 6.775 | 7.7 |
| Cameraman | 6.775 | 7.762 |

The obtained results are very close to the theoretical value. This means that information leakage in the encryption process is negligible.

## References


1. A Security Architecture Based on User Authentication of Bluetooth. International Forum on Information Technology and Applications (2009)
2. Mao, Y.B., Chen, G., Lian, S.G.: A novel fast image Encryption scheme based on the 3D chaotic baker map. Int. J. Bifurcate Chaos 14, 3613–3624 (2004)
3. Gao, H., Zhang, Y., Liang, S., Li, D.: A new chaotic algorithm for image encryption. Chaos, Solutions & Fractals 29(2), 393–399 (2006)
4. Maung, S.S., Sein, M.M.: A Fast Encryption Scheme Based on Chaotic Maps. In: GMSARN International Conference on Sustainable Development: Issues and Prospects for the GMS (2008)
5. Ahmad, M., Alam, M.S.: A New Algorithm of Encryption and Decryption of Images Using Chaotic Mapping. International Journal on Computer Science and Engineering 2(1), 46–50 (2009)
6. Wang, F., Zhang, Y., Cao, T.: Research of chaotic block cipher algorithm based on Logistic map. In: 2009 Second International Conference on Intelligent Computation Technology and Automation, pp. 678–681 (2009)
7. Yen, J.-C., Guo, J.-I.: A New Chaotic Key-Based Design for Image Encryption and Decryption. In: IEEE International Symposium on ISCAS 2000, Geneva, pp. IV-49–IV-52 (May 2000)
8. Lee, P.-H., Pei, S.-C., Chen, Y.-Y.: Generating Chaotic Stream Ciphers Using Chaotic Systems. Chinese Journal of Physics 41(6) (2003)
9. Socek, D., Shujun, L., Magliveras, S.S., Furht, B.: Short Paper: Enhanced 1-D Chaotic Key-Based Algorithm for Image Encryption. In: First International Conference on Security and Privacy for Emerging Areas in Communications Networks, pp. 406–407 (2005)
10. Rao, D., Gangadhar, K.: Modified Chaotic Key-Based Algorithm for image Encryption and Its VLSI Realization. In: International Conference on Digital Signal Processing (2007)
11. Ahmed, H.E.H., Kalash, H.M., Allah, O.S.F.: An Efficient Chaos-Based Feedback Stream Cipher (ECBFSC) for Image Encryption and Decryption. Presented at Informatical (Slovenia), pp. 121–129 (2007)
12. Liu, S., Sun, J., Xu, Z.: An Improved Image Encryption Algorithm based on Chaotic System. Journal of Computers 4(11), 1091–1100 (2009)
13. Awad, A., Saadane, A.: Efficient Chaotic Permutations for Image Encryption Algorithms. In: Proceedings of the World Congress on Engineering, vol. I (2010)
14. Zhu, A.-h., Li, L.: Improving for Chaotic Image Encryption Algorithm Based on Logistic Map. In: 2nd Conference on Environmental Science and Information Application Technology (2010)
15. Chen, G., Mao, Y., Chui, C.K.: A symmetric image encryption based on 3D chaotic maps. Chaos Solutions Fractals 21, 749–761 (2004)
16. Borujeni, S.E., Eshghi, M.: Chaotic Image Encryption Design Using Tompkins-Paige Algorithm. Hindawi Publishing Corporation, Mathematical Problems in Engineering, Article ID 762652, 22 pages (2009)
17. Salleh, M., Ibrahim, S., Isnin, I.F.: Image encryption algorithm based on chaotic mapping. Journal Teknologi, 1–12 (2003)

18. Liu, S., Sun, J., Xu, Z.: An Improved Image Encryption Algorithm based on chaotic system. Journal of Computers 4(11), 1091–1100 (2009)
19. Noura, H., El Assad, S., Vladeanu, C.: Design of a fast and robust chaos-based cryptosystem for image encryption. In: 8th International Conference on Communications (COMM), pp. 423–426 (2010)
20. Improving Pairing Mechanism in Bluetooth Security. International Journal of Recent Trends in Engineering 2(2) (November 2009); Security Overview of Bluetooth, Dave Singelée, Bart Preneel, COSIC Internal Report (June 2004)
21. Two Practical Man-In-The-Middle Attacks on Bluetooth Secure Simple Pairing and Countermeasures. IEEE Transactions on Wireless Communications 9(1) (January 2010)

# A New Trace Backing Algorithm for Maximizing Streaming Data Join

G. Hemalatha[1] and K. Thanushkodi[2]

[1] CSE Dept, Karunya University,
Coimbatore, India
Tel.: 0422-2614020
hema_latha207@yahoo.com
[2] Akshaya College of Engineering,
Coimbatore, India
thanush12@gmail.com

**Abstract.** An increasing number of database queries are executed by interactive users and applications. Since the user is waiting for the database to respond with an answer, the initial response time of producing the first results is very important. The user can process the first results while the database system efficiently completes the entire query. The state-of-art join algorithms are not ideal for this setting. Adaptive join algorithms have recently attracted a lot of attention in emerging applications where data is provided by autonomous data sources through heterogeneous network environments. The main advantage of adaptive join techniques is that they can start producing join results as soon as the first input tuples are available, thus improving pipelining by smoothing join result production and by masking source or network delays. Since the response time of the queries places a vital role in adaptive join, the join techniques like Hash Join, Sort Merge Join cannot be used because they require some prework before producing the join result. The only possible join technique that can be used in adaptive join is Nested Loop Join. In Nested Loop Join each row of the outer relation is compared with each row of the inner relation. The no. of comparisons done by the nested loop join can be reduced by using a technique called trace backing. In trace backing technique whenever a miss match occurs, the next tuple of the outer relation is compared with the mismatched inner relation tuple, instead of looping all the tuples of the inner relation.

**Keywords:** Adaptive Join, Trace Back technique, Nested Loop Join.

## 1 Introduction

Modern information processing is moving into a realm where we often need to process data that is pushed or pulled from autonomous data sources through heterogeneous networks. Adaptive query processing has emerged as an answer to the problems that arise because of the fluidity and unpredictability of data arrivals in such environments [1]. An important line of research in adaptive query processing has been towards developing join algorithms that can produce tuples “online”, from streaming,

D. Nagamalai, E. Renault, M. Dhanushkodi (Eds.): CCSEIT 2011, CCIS 204, pp. 516–522, 2011.

partially available input relations, or while waiting for one or more inputs [4], [7], [9], [12], [14], [15]. Such non-blocking join behavior can improve pipelining by smoothing or "masking" varying data arrival rates and can generate join results with high rates, thus improving performance in a variety of query processing scenarios in data integration, on-line aggregation and approximate query answering systems. Compared to traditional join algorithms ( sort-, hash or nested loops-based [13]), adaptive joins are designed to deal with some additional challenges: The input relations they use are provided by external network sources. The implication is that one has little or no control over the order or rate of arrival of tuples. Since the data source reply speed, streaming rate and streaming order, as well as network traffic and congestion levels, are unpredictable, traditional join algorithms are often unsuitable or inefficient. For example, most traditional join algorithms cannot produce results until at least one of the relations is completely available.

Waiting for one relation to arrive completely to produce results is often unacceptable. Moreover, and more importantly, in emerging data integration or online aggregation environments, a key performance metric is rapid availability of first results and a continuous rate of tuple production. Adaptive join algorithms were created in order to lift the limitations of traditional join algorithms in such environments. By being able to produce results whenever input tuples become available, adaptive join algorithms overcome situations like initial delay, slow data delivery or bursty arrival, which can affect the efficiency of the join. All existing algorithms work in three stages. During the Arriving phase, a newly arrived tuple is stored in memory and it is matched against memory-resident tuples belonging to the other relations participating in the join.

Since the allocated memory for the join operation is limited and often much smaller than the volume of the incoming data, this results in tuple migration to disk. The decision on what to flush to disk influences significantly the number of results produced during the Arriving phase. The Arriving phase is suspended when all data sources are temporarily blocked and a Reactive phase kicks in and starts joining part of the tuples that have been flushed to disk. An importance of this phase is the prompt handover to the Arriving phase as soon as any of the data sources restarts sending tuples. Each algorithm has a handover delay which depends on the minimum unit of work that needs to be completed before switching phases.

This delay has not received attention in the past, but we show that it can easily lead to input buffer overflow, lost tuples and hence incorrect results. When all sources complete the data transmission, a Cleanup phase is activated and the tuples that were not joined in the previous phases (due to flushing of tuples to disk) are brought from disk and joined. There are two new adaptive join algorithms for output rate maximization in data processing over autonomous distributed sources. The first algorithm, Double Index NEsted-loop Reactive join (DINER) is applicable for two inputs, while Multiple Index NEsted-loop Reactive join (MINER) can be used for joining an arbitrary number of input sources.

DINER follows the same overall pattern of execution discussed earlier but incorporates a series of novel techniques and ideas that make it faster, leaner (in memory use) and more adaptive than its predecessors. More specifically, the key differences between DINER and existing algorithms are (1) an intuitive flushing policy for the Arriving phase that aims at maximizing the amount of overlap of the join attribute values between memory-resident tuples of the two relations and (2) a lightweight Reactive phase that allows the algorithm to quickly move into processing tuples that were flushed to disk when both data sources block as show in Fig 1.

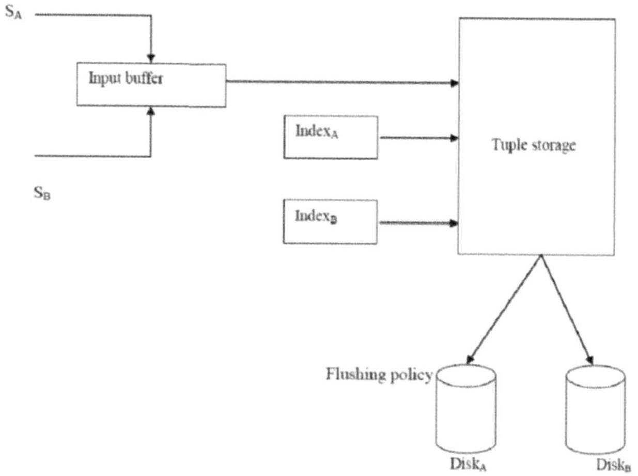


**Fig. 1.** Arriving Phase

The DINER has a higher rate of join result production and is much more adaptive to changes in the environment, including changes in the value distributions of the streamed tuples and in their arrival rates. The efficiency of the DINER algorithm can still be improved by replacing the Nested loop algorithm with a better join technique. There are many Join techniques which out performs the Nested Loop Join like Hash Join, Sort Merge Join, XJoin, Double Hash Join etc., But all these join techniques require some pre-work to produce the resulting Join tuples. As we know in adaptive environment time is a vital factor and it cannot be spend to do the pre-work. So the only possible Join technique that can be used is Nested Loop Join.

## 2 Our Contribution

In Nested loop Join each row of the outer relation is compared with each row of the inner relation which in turns results in a Cartesian product as show in Fig 2. The other joins algorithms out performs the nested loop join but still cannot be used in the steaming join. Because in all the other join algorithms some amount of preprocessing has to be done, which not possible in adaptive environment as the arrive of input tuples cannot be predicted and as soon as it is arrived it has to be join before the other set of input tuples are ready. So the only possible join technique to be used in adaptive environment is Nested Loop join.

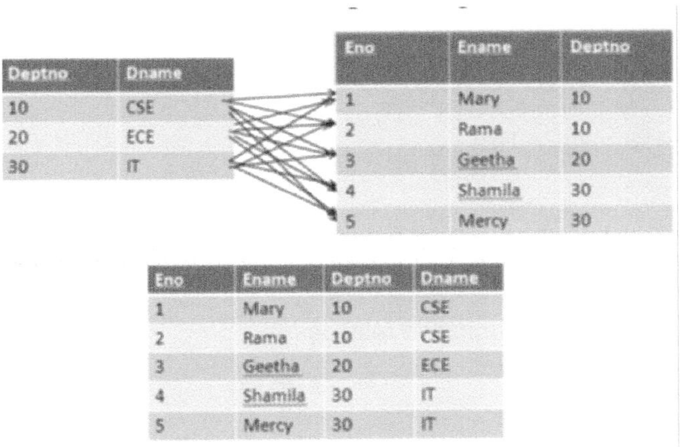


**Fig. 2.** Draw Back of Nested Loop Join

In this paper we proposed a trace backing algorithm to be used instead of nested loop join to maximize the no. of resultant tuples this in turn increased the resultant tuples produced during a time period. The proposed Trace backing algorithm [as show in the Fig 3] compares the outer row with the inner row, whenever there is a mismatch the outer row is incremented instead of inner row, the inner is incremented when there is match. This reduced the number of comparisons than it is required for Nested Loop Join.

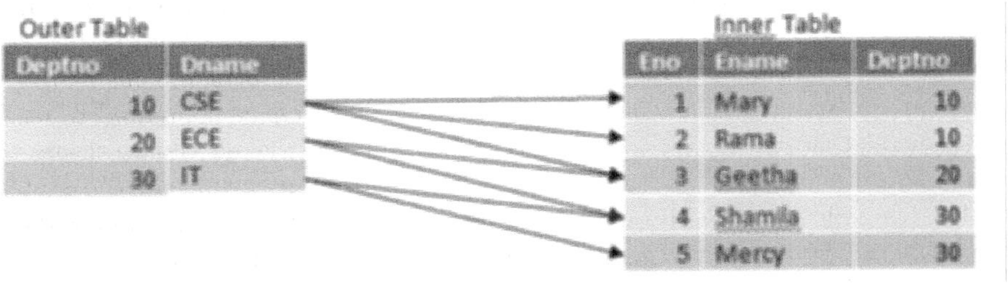


**Fig. 3.** Trace backing Technique

## 3 Algorithm for Trace Backing Technique

Let RA and RB be the two relations to be joined, the smallest relation is taken as the outer relation and the other is taken as the inner relation. The first row of the outer relation is compared with first row of the inner relation, if there is a match increment the inner pointer and compare the first row of the outer relation and the second row of the inner relation else increment the outer pointer and compare the second row of the outer relation with the first row of the inner relation. This process is repeated until all the inner relation tuples are matched.

```
Input: Tuples from relations RA and RB
Output: Joined tuples, RA >< RB

Set Ro←—smallest(RA,RB)
Set Ri←—largest(RA,RB)
PR=0;                          //     rightpointer
Pl=0;                          //     leftpointer
//n is the number of tuples in the relation
While PR <n(Ri)  do
While Pl < n(Ro) do
 // compare the tuples from two relations
 If (tRo[Pl]== tRi[PR]) then
    tRi[PR]|>< tRo[Pl]
    PR++;
 Else
    Pl++;
End if
End while
if (Pl == n(Ro)) then //check leftpointer status
leftpointer=0;
if(PR== n(Ri)) then //check rightpointer position
break;
End while
```

## 4 Performance Evaluation

The following graph shows the no. of comparisons required to perform the Nested Loop Join and Trace Back technique. The no. of comparison is greatly reduced in

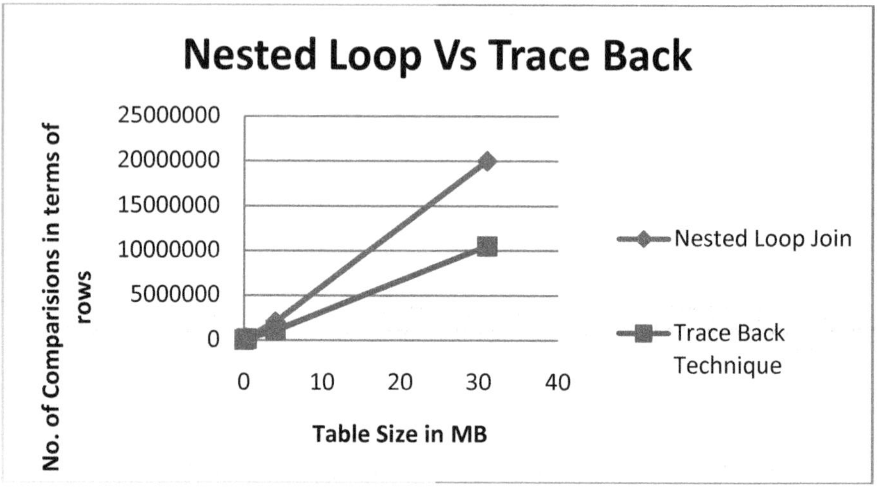


**Fig. 4.** Comparison between Nested Loop and Trace Back Join Technique

Trace Back technique. In the worst case scenario the Trace Back technique behaves similar to Nested Loop Join. In adaptive Join environment the worst case scenario occurs very rarely.

The performance of this algorithm is evaluated on the basis of throughput [the number of joined tuples per unit time] and compared with DINER. By using the trace backing algorithm the number of joined tuples is increased. For example, in 15millisecond, DINER produced 72 joined tuples, where as trace backing technique produces 108 joined tuples.

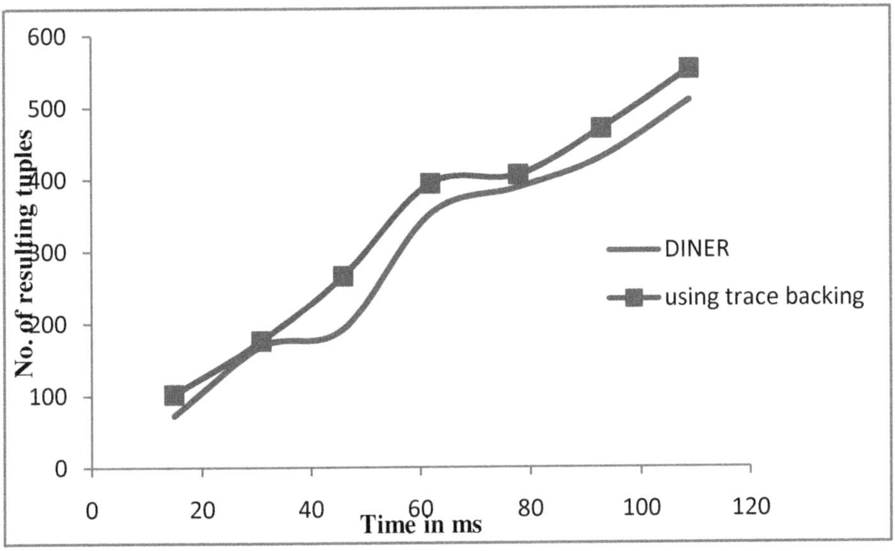


**Fig. 5.** Performance Evaluations for Double Inputs

## 5 Conclusion and Future Scope

The main advantage of adaptive join techniques is that they can start producing join results as soon as the first input tuples are available. The only Join technique that can produce the output without any preprocessing is Nested Loop Join. In Nested Loop Join each row of the outer relations are compared with the inner relation, which is same as Cartesian Join. The new technique trace backing reduced the number of comparisons when it is compared with nested loop join. By using this technique, the number of comparison are reduced which in turn increased the number of resultant tuples produced per unit of time. The trace backing technique produced 102 tuples and DINER produced 72 tuples within 15 millisecond.

In terms of future work, the flushing policy can be extended by considering more factors to specify which tuples have to be flushed. The flushing policy has an important role in improving the performance of the join operation.

## References


[1] Condie, T., Conway, N., Alvaro, P., Hellerstein, J.M., Berkeley, U.C.: Online Aggregation and Continuous Query support in MapReduce. In: Proceedings of VLDB (2004)

[2] Rahman, P., Saad, N.A., Salahi, T.: Early Hash Join: A Configurable Algorithm for the Efficient and Early Production of Join Results. In: ITSim 2008 (2008)

[3] Avnur, R., Hellerstein, J.M.: Eddies: Continuously Adaptive Query Processing. In: Proceedings of the 2000 ACM SIGMOD International Conference on Management of Data, volume 29, May 16-18, vol. 29, pp. 261–272. ACM, New York (2000)

[4] Chen, S., Gibbons, P.B., Mowry, T.C., Valentin, G.: Fractal prefeching B+trees: Optimizing both cache and disk performance. In: ACM SIGMOD International Conference on the Management of Data (June 2002)

[5] Hong, W., Stonebraker, M.: Optimization of Parallel Query Execution Plans in XPRS. Distributed and Parallel Databases 1(1), 9–32 (1993)

[6] Mihaela, B.A., Vasilis, V., Yannis, K., Antonios, D.: Adaptive Join Operators for Result Rate Optimization on Streaming Inputs. IEEE Transactions on Knowledge and Data Engineering 22(8), 1110–1125 (2010)

[7] Urhan, T., Franklin, M.J.: XJoin: A Reactively-Scheduled Pipelined Join Operator. IEEE Transactions on Knowledge and Data Engineering 23(2) (April 2000)

[8] Dittrich, J., Seeger, B., Taylor, D.: Progressive merge join: A generic and non-blocking sort-based join algorithm. In: Proceedings of VLDB (2002)

[9] Mokbel, M.F., Lu, M., Aref, W.G.: Hash-Merge Join: A Nonblocking Join Algorithm for Producing Fast and Early Join Results. In: ICDE (2004)

[10] Tao, Y., Yiu, M.L., Papadias, D., Hadjieleftheriou, M., Mamoulis, N.: RPJ: Producing Fast Join Results on Streams Through Rate-based Optimization. In: Proceedings of ACM SIGMOD Conference (2005)

[11] Hellerstein, J.M., Haas, P.J., Wang, H.J.: Online Aggregation. SIGMOD, 171–182 (1997)

[12] Haas, P.J., Hellerstein, J.M.: Symmetric Hash Joins. In: Proceedings of ACM SIGMOD (1999)

[13] Viglas, S.D., Naughton, J.F., Burger, J.: Maximizing the output rate of multi-way join queries over streaming information sources. In: VLDB 2003, pp. 285–296 (2003)

[14] Babu, S., Srivastava, U., Widom, J.: Exploiting k-Constraints to Reduce Memory Overhead in Continuous Queries Over Data Streams. In: Proceedings of the 2002 Annual ACM Symposium on Theory of Computing, pp. 370–379. ACM Press, New York (2002)

[15] Gehrke, J.: Special issue on data stream processing. IEEE Computer Society Bulletin of the Technical Communication on Data Engineering 26(1) (March 2003)

[16] Arasu, A., Babu, S., Widom, J.: The cql continuous query language: Semantic foundations and query execution. Technical report, Stanford University Database Group (October 2003)

# A Comparative Analysis of Demand Assignment Multiple Access Protocols for Wireless ATM Networks

C. Shanmuganathan and P. Raviraj

[1] Department of Computer Science &Engineering,
St.Peter's College of Engg.& Tech., Chennai, Tamilnadu, India
[2] Department of Computer Science & Engineering,
SKP Engineering College, Tiruvannamalai, Tamilnadu, India

**Abstract.** The concept of wireless ATM is now being actively considered as a potential framework for next-generation wireless communication networks capable of supporting integrated multimedia services with different QoS requirements. Several key subsystem design issues for wired ATM and wireless networks needs to be readdressed in the scope of the wireless ATM, which has the capability to extend the statistical multiplexing of wired ATM network into the wireless medium. One of the key subsystem issues is the development of appropriate medium access control (MAC) protocol. The extension of the ATM network into wireless environment faces many interesting problems. The original ATM network was designed for high speed, noiseless, and reliable channels. None of these characteristics are applicable to the wireless channel. One of the critical aspects of a wireless ATM network is the Medium or Multi Access Control (MAC) Protocol used by the Mobile station (MS) to request service from the BS, which has to consider the Quality of Service (QoS) of the specific applications. This paper analyzes recently proposed MAC protocols, particularly those of Demand Assignment Multiple Access protocols using TDMA technique with Frequency Division Duplex (FDD). It also gives performance measures of two best suited protocols for wireless network environment Distributed Queuing Request Update Multiple Access (DQRUMA) protocol and Adaptive Request Channel Multiple Access (ARCMA) protocol.

**Keywords:** MAC, DQRUMA and ARCMA.

## 1 Introduction

The concept of ATM is end-to-end communication i.e. in a Wide Area Network environment, the communication protocol will be the same i.e. ATM, and companies will no longer have to buy extra equipment (like routers or gateways) to interconnect their LANs (local area networks). Also ATM is considered to reduce the complexity of the network and improve the flexibility while providing end-to-end consideration of traffic performance [1][3].

D. Nagamalai, E. Renault, M. Dhanushkodi (Eds.): CCSEIT 2011, CCIS 204, pp. 523–533, 2011.

### 1.1 ATM Cell

ATM is a fixed size packet-based switching and multiplexing technology designed to be a connection-oriented transfer mode that supports a wide range of services. Its transmission scheme can operate at either constant or variable bit rates. The primary unit (packet) in ATM is called a cell. The ATM standard defines a fixed-size cell with a length of 53 bytes, which is composed of 5-byte header and a 48-byte payload. These fixed-sized cells reduce the complexity of ATM switches and multiplexers allowing cells to be relayed at very high speeds. The typical bit rates at the terminals are 25 Mbps, 155 Mbps, and 622 Mbps. The conversion of user data into ATM cells, and vice versa, is performed by the ATM Adaptation Layer(AAL), which resides directly above the ATM layer of the network[4][5].

## 2 Multiple Access Protocols

A Multiple access protocol is a scheme to control the access to a shared communication medium among various users. Access protocols can be grouped according to the bandwidth allocation, which can be static or dynamic, and according to the type of control mechanism implemented. Multiple access protocols can be categorized into fixed assignment, random assignment and demand assignment [2].

### 2.1 Fixed Assignment

Time-division multiple access (TDMA) and frequency-division multiple access (FDMA) are fixed assignment techniques that incorporate permanent sub channel assignment to each user.

### 2.2 Random Assignment

These techniques allocate the full channel capacity to a user for short periods, on a random basis. These packet-oriented techniques dynamically allocate the channel to a user on per-packet basis. Although there are few versions of the ALOHA protocol, in its simplest form it allows users to transmit at will [6].

## 3 Demand Assignment Multiple Access Protocols

Most DAMA protocols use time-slotted channels that are divided into frames. Depending on the transmission rate and the type of services, the channel bandwidth can be represented by a single or multiple frame(s). Each frame is divided into uplink and downlink period (channel). These periods are further divided into two sub periods or slots. They can be partitioned on a slot-by slot or period basis.

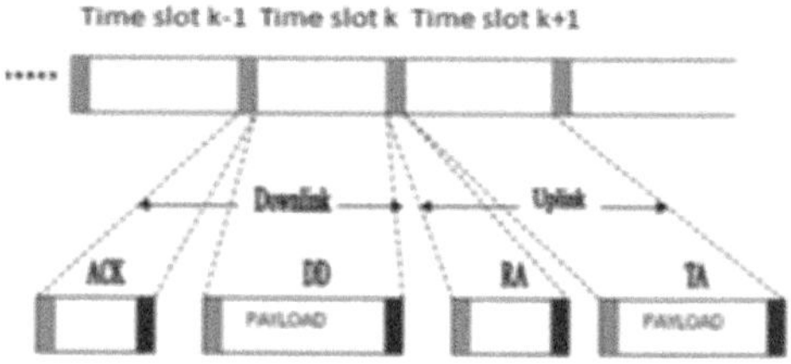


**Fig. 1.** Radio Channel Classification: (Slot-By-Slot)

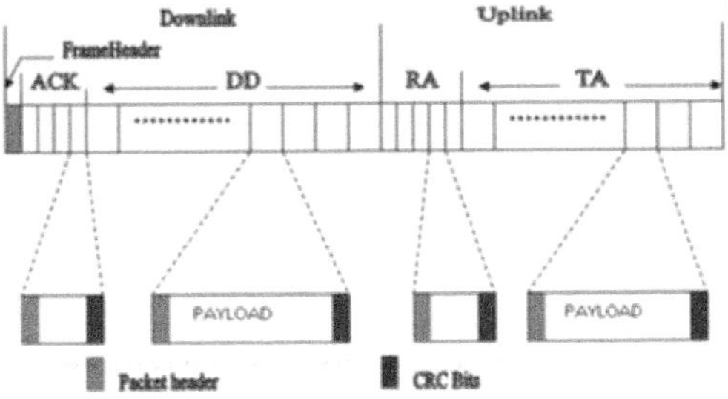


**Fig. 2.** Radio Channel Classification by Period

The uplink channel (mobile-to-base) is divided into the request access (RA) and data transmission access (TA) sub periods. On the other hand, the downlink is divided into the acknowledgement (ACK) and the data downstream (DD) sub periods.

## 4 Distributed Queuing Request Update Multiple Access Protocol

DQRUMA introduces the concept of a dynamic uplink slot where the uplink slot may be converted into a whole RA channel filled only with RA minislots. This conversion occurs when the BS senses that there is significant amount of collisions in the RA channel. It allows as many as 25 RA minislots in a single time slot where mobile can send their requests. This method drastically reduces request contention in the RA channel. Request in the RA channel use a random access protocol like slotted ALOHA. The downlink slot contains the typical DD time slot, and one or more ACK minislots, and transmission permission, Perm slot. In situations where the uplink slot is converted into a series of RA minislots, the subsequent downlink slot is converted into a series of corresponding ACK minislots.

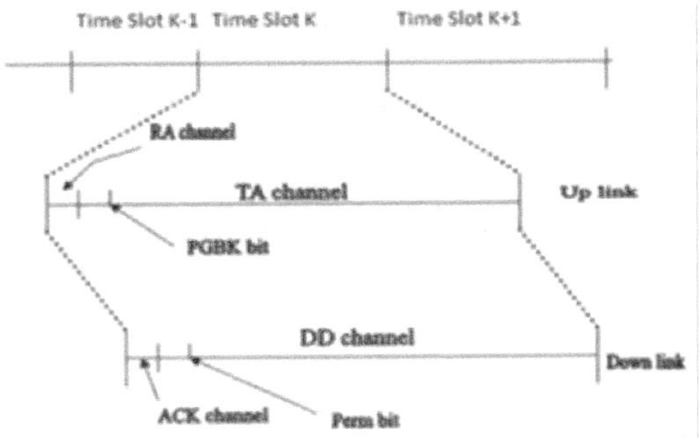


**Fig. 3.** DQRUMA Timing Diagram

DQRUMA also introduces an extra bit called piggyback (PGBK) bit in the uplink channel. Each time a mobile transmits a packet (uplink), it also includes this PGBK bit to indicate whether it has more packets in the buffer. This bit serves as a contention-free transmission request for a mobile transmitting a packet. The BS checks this bit and updates the appropriate entry in the Request Table (in the BS) accordingly. When this bit is included, a mobile does not need a request for channel access in the following time slot.

## 5 Adaptive Request Channel Multiple Access protocol

This new protocol scheme is based on the DQRUMA protocol while incorporating the periodic traffic handling of PRMA. [1].

ARCMA is essentially a demand assignment multiple access protocol with dynamic bandwidth allocation. Its basic architecture is modeled after the DQRUMA protocol. This scheme is designed to function in a cell-based wireless network with many Mobile Station (MS) communicating with the Base Station (BS) of their particular cell. Transmissions are done on a slot-by-slot basis without any frames.

### 5.1 Description of the Protocol

Since this protocol is modeled after DQRUMA, it is composed of request/acknowledgement and a permission/transmission phase. Data transmission from the BS to MS is a straight forward operation where the BS merely broadcasts the information to the entire cell. The destination MS listens to the broadcast channel and retrieves the appropriate data packets. If the transmission destination does not reside in the same cell, the BS will forward the packet to an ATM switch to be routed to the proper destination.

### 5.2 Request/Acknowledgement Phase

The request is made in the RA channel. The mobile's b-bit access ID, assigned during setup. In addition to the Access ID, the request packet also includes the type of service being requested. After every random transmission the MS needs to know if the request was successful. Since it does not detect collision by sensing the channel, the BS has to send a response to the MS indicating a successful request. When the BS receives a request from the RA channel it immediately broadcasts an acknowledgement to the MS. The BS then inserts the new request in the request table to indicate that the MS has packet(s) to transmit.

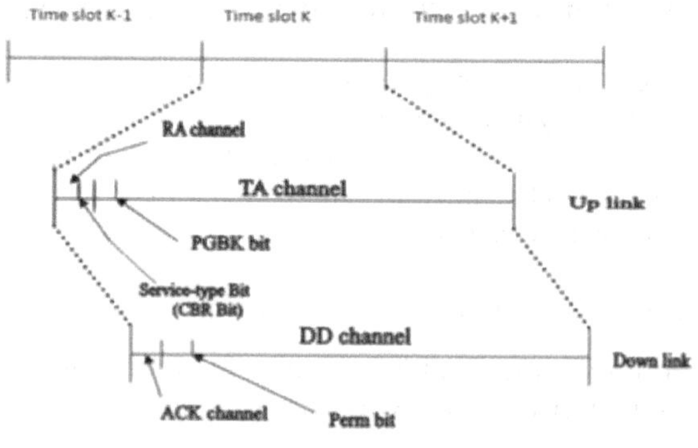


**Fig. 4.** Timing Diagram For the ARCMA Protocol

### 5.3 Permission/Transmission Phase

**General Traffic**

The BS is responsible for allocating bandwidth (time slots) to the MS by using a packet transmission policy. This protocol uses First In, First Out (FIFO) policy. The MS that makes a request first, is given permission to transmit first.

**CBR Traffic**

This connection-oriented delay-sensitive traffic is given priority in the request phase. When a CBR request arrives at the BS, it is inserted in special CBR Request Queue. The CBR Request Queue also uses FIFO policy but requests in the queue have precedence over those in the request table.

### 5.4 Adaptive Request Channel Strategy

ARCMA also implements a dynamic RA channel strategy similar to that of the DQRUMA. But ARCMA reduces the idle time slots by considering the probability of retransmission $q_r$. when the TA channel is first converted into multiple RA minislots, all new and previously unsuccessful requests are transmitted with the probability of 1. That is, all requests are sent out immediately regardless of their $q_r$. If no requests are successfully transmitted the uplink channel remains in the multiple RA mode. However, in this case, the MS's retransmit their requests according to their old $q_r$. Conversely, if successful requests were made during the multiple RA slots, the BS reverts the next time slot back to normal mode with a single RA minislot. Any remaining MS's with unsuccessful requests, retransmit their requests using their corresponding $q_r$. The following table gives an example illustrating the number of mobiles handled by each RA channel in normal and multiple RA mode.

**Table 1.** Number of Mobiles

| Mobile ID | Normal mode | | Multiple RA mode | |
|---|---|---|---|---|
| | Channel # (single channel) | # of mobiles per RA channel | Channel# (R =3) | # of mobiles per RA channel |
| 1 | 1 | 5 | 1 | 2 |
| 2 | 1 | 5 | 2 | 2 |
| 3 | 1 | 5 | 3 | 1 |
| 4 | 1 | 5 | 1 | 2 |
| 5 | 1 | 5 | 2 | 2 |

Although the all above compared protocols have their advantages and limitations when it comes to wireless data packet networks, DQRUMA and ARCMA seem to offer the most efficient scheme for wireless ATM services. But DQRUMA has one major drawback when compared to its successor, its inability to offer any distinction between different types of ATM traffic during the reservation phase.

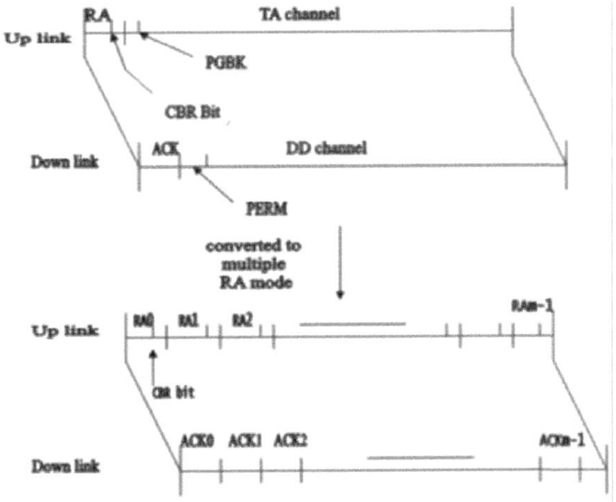


**Fig. 5.** Timeslot conversion in multiple RA mode

# 6 Performance Analysis

## 6.1 Simulated Protocols

In this simulation protocols, ARCMA and DQRUMA are compared under identical network environments and specification. DQRUMA and ARCMA have proven to be an effective multiple access protocols that can support integrated traffic in a wireless packet network. Both these protocols are implemented using the FIFO scheduling policy in the Request Table and slotted ALOHA with Binary Exponential Backoff algorithm is used as the random access scheme in the RA channel.

## 6.2 Simulation Environment

The simulation approach is modeled after the environment of a typical cellular mobile communication network. The primary characteristics of the environment are

- The network model is a typical cell in a wireless network with one BS and N active mobiles.
- MSs are buffered portable units that produce data packets according to some bursty random processes. All data packets are queued in the mobiles buffer until the BS allocates them a time slot for transmission.
- The uplink and downlink channels are simulated as separate channel. Both uplink and downlink channels are time slotted.
- The network is assumed to have fixed number of active mobiles throughout each experiment. The network traffic stays same for the duration of the experiment.
- The slotted ALOHA with BEB algorithm is used as the random access scheme in the RA channel.
- The length of the time slot is assumed the same for two protocols.
- Network delay is due to access delay and service delay. Propagation delays between mobiles and BS are assumed negligible.
- Simulations are run for 50 time slots.

## 6.3 Performance Measures

The simulation involves two different multiple access schemes performing various traffic conditions. The performance measures are:

- **Channel Throughput ($TP_c$):** $TP_c$ is defined as the ratio of the total number of transmitted packets and the total time number of time slots. That is, $TP_c = P_T/T_{TL}$, where $P_T$ is denoted as the total number of transmitted packets, and $T_{TL}$is the total number of time slots.
- **Average Transmission Delay ($D_{AVG}$):** $D_{AVG}$ is defined as the ratio of the total packet transmission delay and number of active mobiles. Hence, $D_{AVG} = D_{TL}/M$, where $D_{TL}$ is the total packet transmission delay and M is the number of active mobile.
- **Average Queue Length ($L_{AVG}$):** $L_{AVG}$ is defined as the ratio of the total number of packets in all the mobiles buffer and the number of active mobiles. Thus, $L_{AVG} = L_{TL}/M$, where $L_{TL}$ is the total number of packets in all the buffer, and M is the number of active mobiles. $L_{TL}$ is the total time taken at the end of simulation. $L_{AVG}$ is measured by number of packets [7].

## 7 Simulation Results

We present the simulation results and the performance analysis of the different multiple access strategies. We simulate protocols in the following traffic environment: with CBR traffic, non CBR traffic. As the arrival rate increases the channel throughput also increases (Fig.8) as the arrival rate increases, more packets are transmitted. Under heavy traffic condition empty request queue are caused by contention in the RA channel. Nevertheless, Tpc is very high at high arrival rates as illustrated in Fig.7.1. Figures 7.1and 7.2 show the performance in a network cell with 100 active mobiles with single packet burst and no CBR traffic.

Figure 7.3 represents the combination of the previous two figures. The exact value of the arrival rate is not a concern in our analysis. Figure 7.4 and 7.5 show the result of a simulation in an environment with no CBR traffic and single packet burst. Fig.7.4 illustrates results from a low traffic condition. We can see that as the throughput increases, so does the average delay (Davg). DQRUMA suffers from a very steep increase in delay as the throughput increases.

ARCMA follows a similar pattern, but at a much lesser rate. Fig.7.5 shows that the delay increases at a much higher rate, compared to the previous simulation, for all three scheme. From the Fig.7.6 we can see that there is an improvement in performance although there is a twice as many data packets arriving in the channel. The main reason for this improvement is the piggy-backing strategy .The second packet in every burst request accesses collision free using the PGBK bit. Figure 7.7 shows the results from a cell with mobiles with CBR traffic. The gap between their performances grows appreciably as the arrival rate increases. The increase of data arriving due to the CBR traffic causes a steep increase in the average delay for DQRUMA .The special CBR handling in ARCMA allows it take advantages of CBR traffic characteristics.

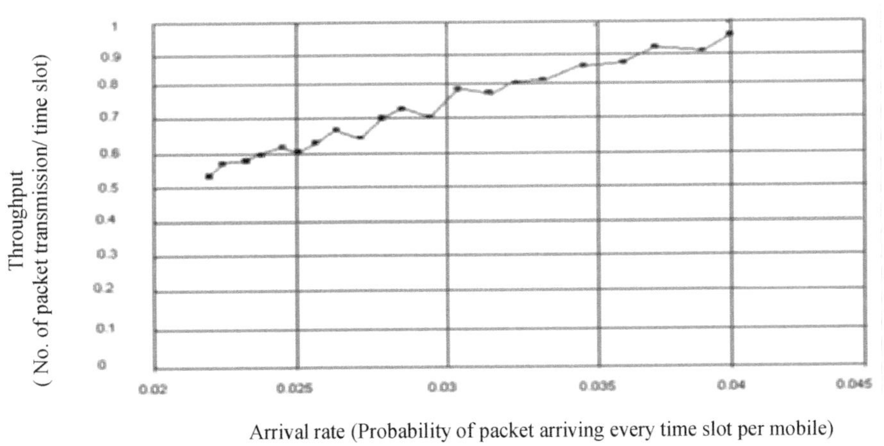


**Fig. 6.** Channel Throughput VS Arrival Rate ARCMA (M=100, CBR=0, Burst=1)

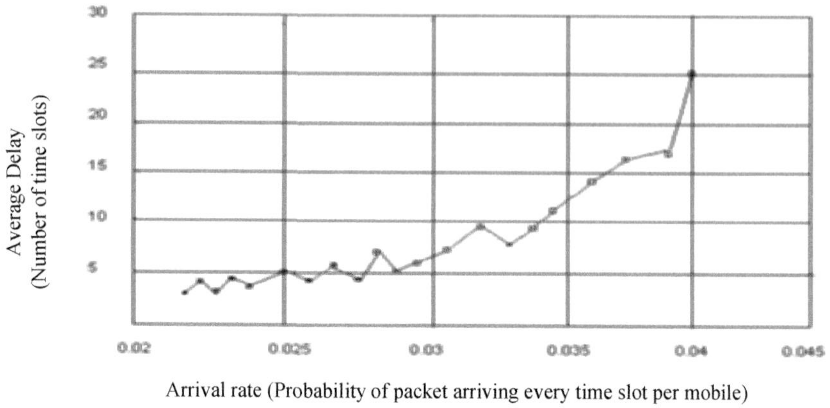


**Fig. 7.** Average delay VS Arrival rate for ARCMA (M=100, CBR=0,Burst=1)

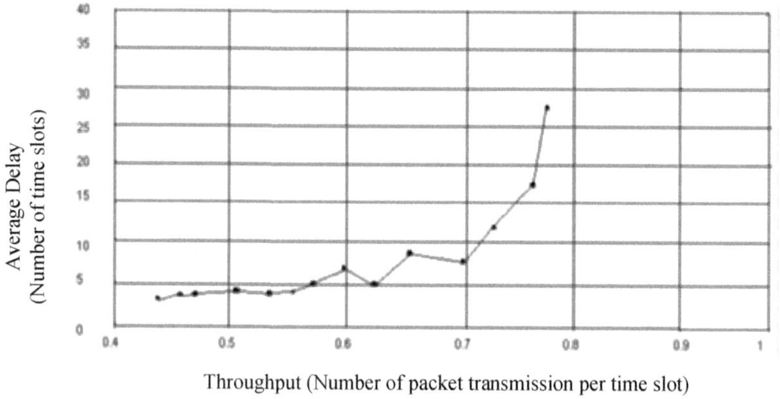


**Fig. 8.** Average Delay VS Channel Throughput for ARCMA (M=100, CBR=0, Burst=1)

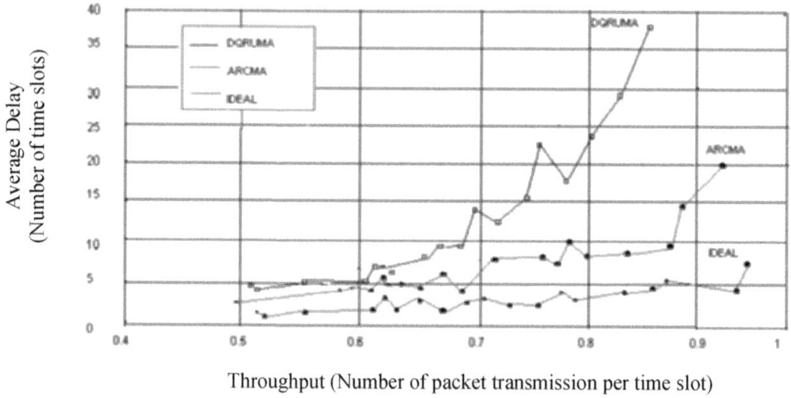


**Fig. 9.** Average Delay VS Channel Throughput for ARCMA (M=25, CBR=0, Burst=1)

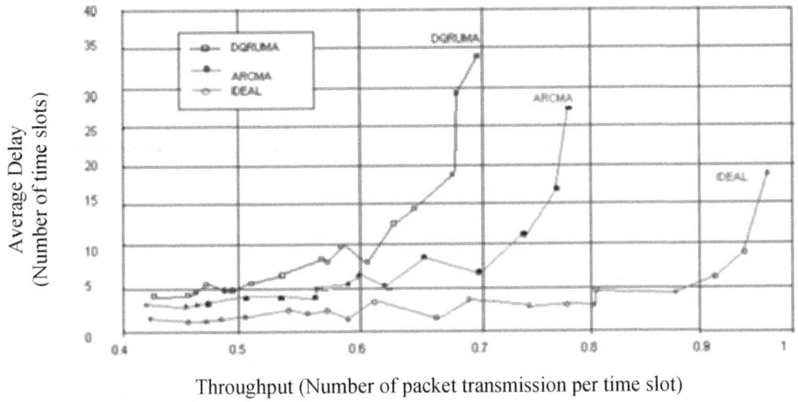


**Fig. 10.** Average Delay VS Channel Throughput for ARCMA (M=100, CBR=0, Burst=1)

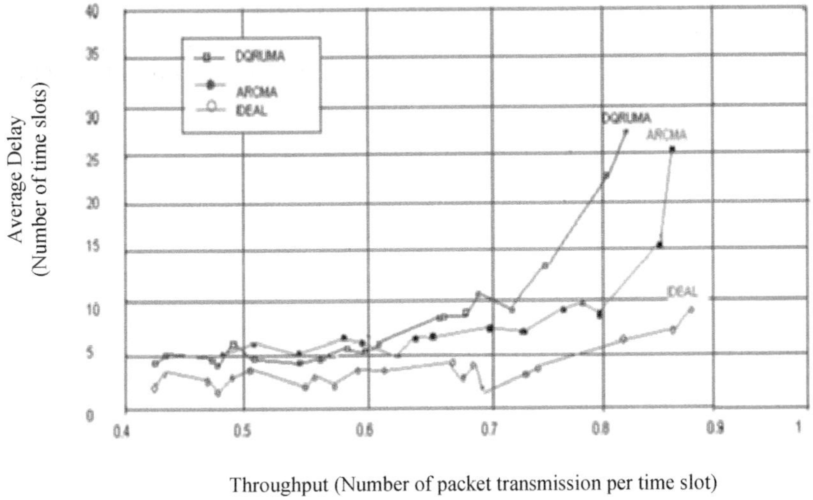


**Fig. 11.** Average Delay VS Channel Throughput for ARCMA (M=100, CBR=0, Burst=2)

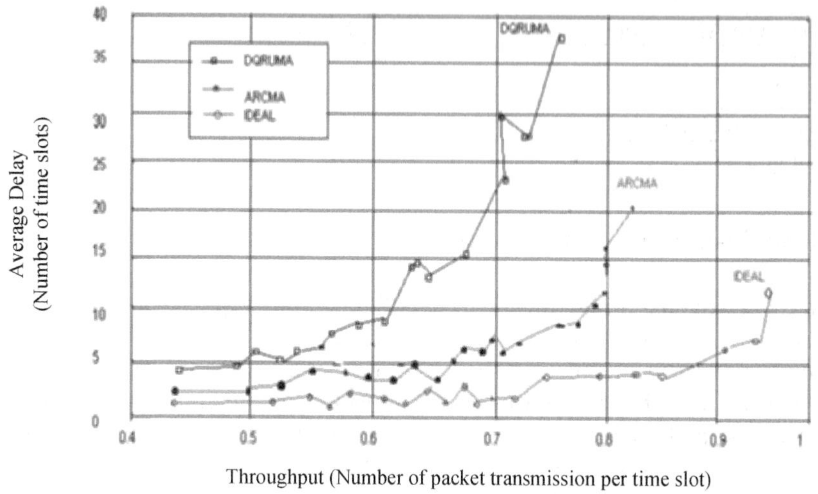


**Fig. 12.** Average Delay VS Channel Throughput for ARCMA (M=100, CBR=2%, Burst=1)

## 8 Conclusion

Wireless ATM is considered as a promising technique for future broadband wireless networks and is under research prototyping at various institutions. This paper focuses on the role of the medium access control protocol in supporting multimedia traffic and QoS. Protocols for multimedia wireless ATM are still evolving. Many protocols have been proposed, and most of them are only different in details of implementation, or variant design for specific wireless architectures. The common trend observed in these protocols is requiring the mobiles to request resources through reservations – with or without contentions. The base station can be regarded as an ATM switch that statistically multiplexes the wireless links among the mobiles. This approach will provide a way to integrate similar packet scheduling strategies on wired ATM network.

This paper presents comparison for protocols, which contain most characteristics of current available channel access protocols for wireless ATM. Although all the compared protocols have their advantages and limitations when it comes to wireless data packet networks, DQRUMA and ARCMA seem to offer the most efficient scheme for wireless ATM. Of these protocols compared only DQRUMA and ARCMA have been simulated.

The simulation results show that ARCMA performs better than DQRUMA regardless of the traffic load. ARCMA produces significantly higher channel throughput than DQRUMA because of its ability to differentiate between the services (CBR, VBR) of ATM. ARCMA brings us one step closer to designing a complete protocol suite that could be used in the wireless ATM networks.

## References


1. Hac, A., Chew, B.L.: ARCMA – adaptive request channel multiple access protocol for wireless ATM networks. International Journal of Network Management 11, 333–363 (2001), doi:10.1002/nem.411
2. Kubbar, O., Mouftah, H.T.: Multiple Access Control Protocols for Wireless ATM: Problems Definition and Design Objectives. IEEE Communications Magazine (November 1997)
3. Ayanoglu, E., Eng, K.Y., Karol, M.J.: Wireless ATM: Limits, Challenges, and Proposals. IEEE Personal Communications (August 1996)
4. Abramson, N.: Multiple Access in Wireless Digital Networks. Proceedings of the IEEE 82(9) (September 1994)
5. Handel, R., Huber, M.N., Schroder, S.: ATM networks: Concepts, Protocols, Applications
6. Tanenbaum, A.S.: Computer Networks, 3rd edn.
7. Yuan, R., Biswas, S.K., French, L.J., Li, J., Raychaudhuri, D.: Proc. "A Signaling and Control Architecture for Mobility Support in Wireless ATM Networks". In: ICC 1996, Dallas, TX, pp. 478–488 (June 1996)

# Perceptual Linear Predictive Cepstral Coefficient for Malayalam Isolated Digit Recognition

Cini Kurian and Kannan Balakrishnan

Department of Computer Applications,
Cochin University of Science & Technology,
Cochin, 682 022
cinikurian@gmail.com,
bkannan@cusat.ac.in

**Abstract.** This paper presents an optimum speaker independent isolated digit recognizer for Malayalam language. The system employs Perceptual Linear Predictive (PLP) cepstral coefficient for speech parameterization. The powerful and well accepted pattern recognition technique Hidden Markov Model is used for acoustic modeling. The training data base has the utterance of 21 speakers from the age group ranging from 20 to 40 years and the sound is recorded in the normal office environment where each speaker uttered digits zero to nine independently. The system obtained an accuracy of 99.5% with the unseen data.

**Keywords:** Speech Recognition, Malayalam, PLP.

## 1 Introduction

Humans interact with environment in several ways: sight, audio, smell and touch. Humans send out signals or information visually, auditory or through gestures [17]. Because of the increased volume data, human has to depend on machines to get the data processed. Human –computer interaction generally use keyboard and pointing devices. In fact, speech has the potential to be a better interface other than keyboard and pointing devices [12].

Keyboard a popular medium requires a certain amount of skill for effective usage. Use of mouse also requires good hand- eye coordination. Physically challenged people find it difficult to use computer. It is difficult for partially blind people to read from monitor. Moreover current computer interface assumes a certain level of literacy from the user. It expects the user to have certain level of proficiency in English apart from typing skill. Speech interface helps to resolve these issues. Speech synthesis and speech recognition together form a speech interface. Speech synthesizer converts text into speech. Speech recognizer accepts spoken words in an audio format and converts into text format [14].

Speech interface supports many valuable applications - for example, telephone directory assistance, spoken database querying for novice users, "hands busy" applications in medical line, office dictation devices, automatic voice translation into

D. Nagamalai, E. Renault, M. Dhanushkodi (Eds.): CCSEIT 2011, CCIS 204, pp. 534–541, 2011.

foreign languages etc. Speech enabled applications in public areas such as railways; airport and tourist information centers might serve customers with answers to their spoken query. Physically handicapped or elderly people might able to access services easily, since keyboard is not required. In Indian scenario, where there are about 1670 dialects of spoken form, it has greater potential. It could be a vital step in bridging the digital divide between non English speaking Indian masses and others. Since there is no standard input in Indian language, it eliminates the key board mapping of different fonts of Indian languages

ASR is a branch of Artificial Intelligence (AI) and is related with number of fields of knowledge such as acoustics, linguistics, pattern recognition etc [2]. Speech is the most complex signal to deal with since several transformations occurring at semantic, linguistic, acoustic and articulator levels. In addition to the inherent physiological complexity of the human vocal tract, physical production system also varies from one person to another [5, 6]. The utterance of a word found to be different, even when it is produced by the same speaker at different occasions. Apart from the vast inherent difference across different speakers and different dialects, the speech signal is influenced by the transducers used to capture the signal, channels used to transmit the signal and even the environment too can change the signals. The speech also changes with age, sex, and socio economic conditions, the context and the speaking rate. Hence the task of speech recognition is not easy due to many of the above constraints during recognition [13].

In most of the current speech recognition systems, the acoustic component of the recognizer is exclusively based on HMM [7,9,10]. The temporal evolution of speech is modeled by the Markov process in which each state is connected by transitions, arranged into a strict hierarchy of phones, words and sentences.

Artificial neural networks (ANN) [3,15] and support Vector machines (SVM) [2,7] are other techniques which are being applied to speech recognition problems. In ANN, temporal variation of speech can not be properly represented. SVM, being a binary static classifier, adaptation of the variability of the duration of speech utterance is very complex and confusing. SVM is a binary classifier while ASR, faces multiclass issues.

For processing speech, the signal has to be represented in some parametric form. Wide range of methods exists for parametric representation of speech signals, such as Linear Prediction coding (LPC) [10], and Mel-Frequency Cepstrum Coefficients (MFCC) [11] and Perceptual Linear Predictive (PLP) coefficient [10]. Since PLP is more adapted to human hearing, PLP parameterization technique is used in this work.

## 2 Speech Processing Using PLP

One of first decisions in any pattern recognition system is the choice of what features to use. The PLP feature converts speech signal in meaningful perceptual way through some psychoacoustic process. The various stages of this method are based on our perceptual auditory characteristics. The different stages of PLP extraction is shown in fig 1.

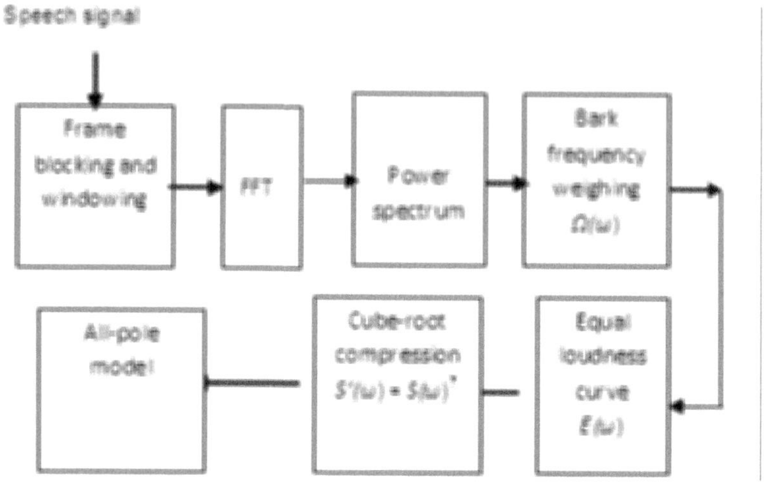


**Fig. 1.** Stages of PLP extraction

The first block in any feature extraction process is frame blocking. Here the audio signals which are basically non stationary are cut into fragments which are called frames. Then frames are passed through Hamming Window and DFT is applied over it. Then power spectrum of the speech signal is computed as follows

$$P(\omega) = Re(\ S((\omega))^2 \ + \ Im(S((\omega))^2 \qquad (1)$$

These steps are common in almost all the feature extraction process. The significant blocks of PLP analysis are as follows

### 2.1 Critical Band Integration(Bark Frequency Weighing)

The first step is a conversion from frequency to bark frequency, which is a better representation of human hearing resolution in frequency. Here power spectrum is wrapped along its frequency axis into bark frequency. In short, the speech signal is passed through some trapezoidal filters equally spaced in bark scale.

### 2.2 Equal Loudness Pre-emphasis

Different frequency components of speech spectrum are pre-emphasized by an equal - loudness curve, which is an approximation to the unequal sensitivity of human hearing at different frequencies, at about 40dB level. A filter having a specific transfer function gives the curve.

### 2.3 Intensity Loudness Conversion(Cube-Root Amplitude Compression)

Cube-root compression of the modified speech spectrum is carried out according to power law of hearing [16], which simulates the non-linear relation between the intensity of sound and its perceived loudness. Together with the psychophysical equal-loudness pre-emphasis, this operation reduces the spectral amplitude variation of the critical-band spectrum.

### 2.4 All-Pole Model

Autoregressive modeling represents the final stage of PLP analysis, and consists of approximating the spectrum by an all-pole model, by using the auto correlation method. An inverse Discrete Fourier Transformation is applied to the spectrum samples. For the M-th order all-pole model, only the first M+1 autocorrelation values are needed.

## 3 Statistical Speech Recognition and HMM

An unknown speech wave form is converted by a front-end signal processor into a sequence of acoustic vectors, $O = o1, o2, o3, \ldots$. The utterance consists of sequence of words $W = w1, w2, w3$ ----$wn$. In ASR it is required to determine the most probable word sequence, W, given the observed acoustic signal $O$. Applying Bays' rule to decompose the required probability, [11]

$$S = \arg_w \max P(W/O) = \arg_w \max(P(O/W)P(W) / P(O))$$

$$S = \arg_w \max \underbrace{P(O/W)}_{Posterior} \underbrace{P(W)}_{prior}$$

Hence a speech recognizer should have two components: $P(W)$, the prior probability, is computed by language model, while P $(O/W)$, the observation likelihood, is computed by the acoustic model. Here the acoustic modeling of the recognition unit is done using HMM.

Since HMM is a statistical model in which it is assumed to be in a Markov process with unknown parameters, the challenge is to find all the appropriate hidden parameters from the observable states. Hence it can be considered as the simplest dynamic Bayesian network [7,9]. In a regular Markov model, the state is directly visible to the observer, and therefore the state transition probabilities are the only parameters. However, in a hidden Markov model, the state is not directly visible (so-called *hidden*), but the variables influenced by the states are visible. Each transition in the state diagram of a HMM has transition probability associated with it [13, 15]. These transition probabilities are denoted by matrix A. Here A is defined as $A = a_{ij}$ where $aij = P(t_{t+1} = j \mid j = i)$, the probability of being in state j at time $t+1$, given that we were in state $i$ at time $t$. It is assumed that $a_{ij}$'s are independent of time. Each state is associated with a set of discrete symbols with an observation probability assigned to each symbol, or is associated with the set of continuous observation with a continuous observation probability density. These observation symbol probabilities are denoted by the parameter B. Here B is defined as $B = b_j(k)$, where $b_j(k) = P(v_k \text{ at } t \mid i_t = j)$, the probability of observing the symbol $v_k$, given that it is in the state $j$. The initial state probability is denoted by the matrix $\pi$, where $\pi$ is , defined as $\pi = \pi_i$ where $\pi_i = P(i_t = 1)$, the probability of being in state $t$ at $t = 1$. Using the three parameters $A$, $B$ , and $\pi$ a HMM can be compactly denoted as $\lambda = \{A, B, \pi\}$ An Example of an HMM with five states is shown in Figure 2.

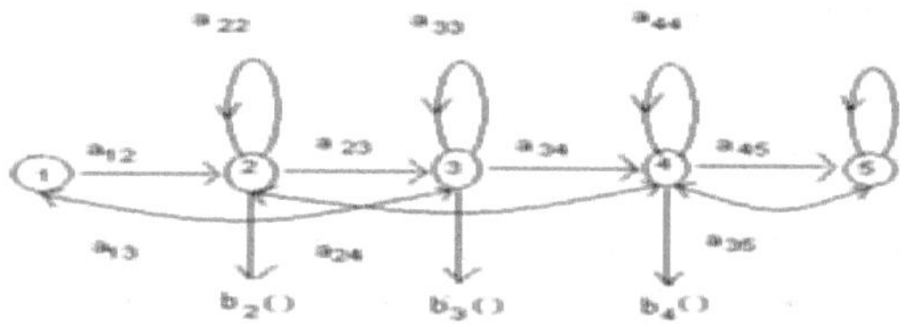

**Fig. 2.** Topology of a 5 state HMM

There are three fundamental ASR problems that can be addressed with HMM. Problem (i) is Scoring and evaluation i.e computing the likelihood of an observation sequence, given a particular HMM. This problem occurs during the recognition phase. Here for the given parameter vector sequence (observation sequence), derived from the test speech utterance, the likelihood value of each HMM is computed using forward algorithm. The symbol associated with the HMM, for which the likelihood is maximum, is identified as the recognized symbol corresponding to the input speech utterance. Problem (ii) is associated with training of the HMM for the given speech unit. Several examples of the same speech segments with different phonetic contexts are taken, and the parameters of the HMMs, $\lambda$, have been interactively refined for maximum likelihood estimation, using the Baum- Wetch algorithm [13]. Problem (iii) is associated with decoding or hidden state determination, where the best HMM state is decided.

## 4 System Development and Speech Data Base

The different steps [5] involved in the development of the proposed system is shown in figure 3.

The database contains 210 isolated spoken words recorded from 21 speakers. Each speaker uttered numbers form zero to nine separately. The recording is done in normal environment with a high quality microphone at 16kHz sampling frequency and quantized at 16 bit . The data is quantized (at 16 kHz) and digitalized (with 16 bit) and the speech data is stored in WAV format.

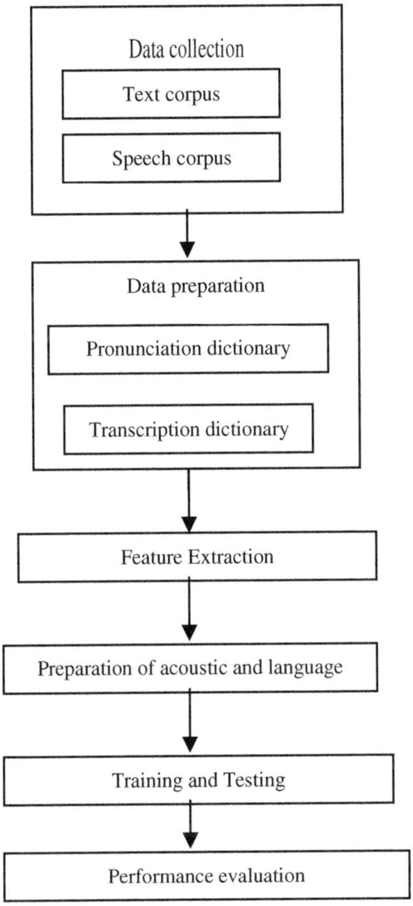


**Fig. 3.** System Development

# 5 Testing and Training

For training and testing the system, the data base is divided into three equal parts- 1, 2, 3 and the experiment is conducted in a round robin fashion. For each experiment, 2/3rd of the data is taken for training and 1/3rd of the data is used for testing. In the experiment I, part 1 and part 2 of data is given for training. Then the same trained system is taken for testing the system with part 3 of the database. In experiment II, part 1 and part 3 of the data base is taken for training and part II of the database is used for testing. In experiment III, part 2 and part 3 of the database is taken for training and tested with part 1 of the database. The result obtained from each training and testing experiment in terms of Word Accuracy, Number of words deleted, inserted, substituted are detailed in table 1.

**Table 1.** Performance Evaluation of the System with Training and Testing Data

| | TRAINING | | | | TESTING | | | |
|---|---|---|---|---|---|---|---|---|
| Experiment | Word Accuracy % | Number of Deletions | Number of Substitutions | Number of Insertions | Word Accuracy % | Number of Deletions | Number of Substitutions | Number of Insertions |
| 1 | 99.29 | 0 | 1 | 0 | 100 | 0 | 0 | 0 |
| 2 | 99.29 | 0 | 1 | 0 | 98.57 | 0 | 1 | 0 |
| 3 | 100 | 0 | 0 | 0 | 100 | 0 | 0 | 0 |

## 6 Performance Evaluation and Discussion

Word Error Rate (WER) is the standard evaluation metric for speech recognition. It is computed by SCLITE [18], a scoring and evaluating tool from National Institute of Standards and Technology (NIST). The inputs to the SCLITE are the reference text and the recognized text (output of the decoder). After each training and testing experiment, recognized text and the reference text are converted into the sclite compatible form and fed as input in sclite. Then detailed results are obtained in terms of WER, SER, and number of word deletions, insertions and substitutions. If N is the number of words in the correct transcript; S, the number of substitutions; and D, the number of Deletions, then,

$$WER = ((S + D + I)N) /100$$

Sentence Error Rate (S.E.R) = (Number of sentences with at least one word error/ total Number of sentences) * 100

The system is tested with unknown speakers. Five speakers, who are not involved in the training is selected and recognition performance of their utterance is evaluated. A word recognition accuracy of 99.5% is obtained. These results prove that the developed system is speaker independent with negligible errors.

## 7 Conclusion

The digit recognition system for Malayalam is implemented using PLP cepstral coefficients and HMM. From the results of the experiment it can be concluded that PLP and HMM are ideal candidates for spoken digit recognition. The system achieves almost cent percent (99.5) recognition accuracy.

## References


1. Ganapathiraju, A., Hamaker, J., Picone, J.: Support Vector machines for speech Recogntion. In: Proceedings of the International Conferences on Spoken Language Processing, Sdney, Australia, pp. 292–296 (November 1999)
2. Gold, B., Morgan, N.: Speech and audio signal processing. John Wiley and Sons, N.Y. (2002)
3. Behrman, Nash, L., Steck, J., Chandrashekar, V., Skinner, S.: Simulations of Quantum Neural Networks. Information Sciences 128(3-4), 257–269 (2000)
4. Gold, B., Morgan, N.: Speech and audio signal processing. John Wiley and Sons, N.Y. (2002)
5. Kurian, C., Balakrishnan, K.: Speech Recognition of Malayalam Numbers. IEEE Transaction on Nature and Biologically Inspired Computing, NaBIC 2009, 1475–1479 (2009)
6. Kurian, C., Firoz Shah, A., Balakrishnan, K.: Isolated Malayalam digit recognition using Support Vector Machines. IEEE Transaction on Communication Control and Computing Technologies, ICCCCT 2010, 692–695 (2010)
7. Burges, C.J.C.: A tutorial on support vector machines for pattern recognition. Knowledge Discovery Data Mining 2(2), 121–167 (1998)
8. Davis, S., Mermelstein, P.: Comparison of parametric representations for Monosyllabic word Recognition in continuously spoken sentences. IEEE Trans. on ASSP 28, 357–366
9. Dimov, D., Azamanov, I.: Experimental specifics of using HMM in isolated word Speech recognition. In: International Conference on Computer System and Technologies – CompSysTech 2005 (2005)
10. Felinek, F.: Statistical Methods for Speech recognition. MIT Press, Cambridge (1997)
11. Huang, X., Alex, A., Hon, H.W.: Spoken Language Processing; A Guide to Theory, Algorithm and System Development. Prentice Hall, Upper Saddle River (2001)
12. Jurasky, D., Martin, J.H.: Speech and Language Processing: An introduction to natural language Processing. In: Computational Linguistics, and Speech Recognition, 2nd edn. (2007)
13. Jyoti, Rakesh, S.: Automatic Speaker Recognition: An Approach using DWT Based Feature Extraction and Vector Quantization. IETE Technical Review 24(5), 395–402 (2007)
14. Rabiner, L., Juang, B.-H.: Fundamentals of Speech Recognition. Pearson Education, London (2008)
15. Sperduti, O., Starita, A.: Supervised Neural Networks for classification of Structures. IEEETransactions on Neural Networks 8(3), 714–735 (1997)
16. Stevens, S.S.: On the psychophysical law. Psychological Review 64(3), 153–181 (1957)
17. Grewal, S.S., Kumar, D.: Isolated word Recognition System for English language. International Journal of Information Technology and Knowledge Management 2(2), 447–450 (2010)
18. `http://www.icsi.berkeley.edu/Speech/docs/sctk-1.2/sclite.html`

# Agent Based Modeling of Individual Voting Preferences with Social Influence

Vivek Kumar Singh, Swati Basak, and Neelam Modanwal

Department of Computer Science, Banaras Hindu University,
Varanasi-221005, India
vivek@bhu.ac.in, swati.basak.bhu@gmail.com,
neelam.bhu1@gmail.com

**Abstract.** Agent Based Modeling (ABM) is usually referred to as the third way of doing modeling & simulation in contrast to the historically popular equation-based macrosimulations and microsimulations. In ABM a system is modeled as a set of autonomous agents who interact with each other and also with the environment. The agents represent various actors in the system and the environment represents agent's surroundings and the overall context. ABM has been successfully applied to model and analyze a number of complex social processes and structures, ranging from habitation patterns to spread of culture. In this paper, we present our experimental work on applying ABM to model individual voting preferences. The model explores process of formation of voting preferences, the factors governing it and its effect on the final voting pattern. We have incorporated social influence theory in our model and experimented with various settings. The paper concludes with a short discussion of the results and their relevance.

**Keywords:** Agent Based Modeling, Multi-agent Systems, Social Influence, Social Simulation, Voting Model.

## 1 Introduction

The development of social simulation over the past half century can be grouped into three broad periods: macrosimulation, microsimulation and agent-based models. Sociologists, particularly computational sociologists, first tried macrosimulations to model problems in supply-chain management, inventory control in a warehouse, urban traffic, spread of epidemics, migration and demographic patterns. Macrosimulations consist of a set of differential equations that take a holistic view of the system. However, taking the complete system as the unit of analysis had inherent limitations. Microsimulations focused on the use of individuals as the unit of analysis but continued with macro-level forecasting used in macrosimulations. Though microsimulations modeled changes to each element of the population but they did not permit individuals to directly interact or to adapt. The main focus remained forecasting macro effects of public policies that alter individual behaviours. ABM, the third methodology, takes a pure bottom-up approach and keeps the individual at the centre. It allows modeling individual actors in the population and their interactions.

D. Nagamalai, E. Renault, M. Dhanushkodi (Eds.): CCSEIT 2011, CCIS 204, pp. 542–552, 2011.

This makes it a highly suitable technique for analysis of emergent group behaviours resulting only from local interactions of individuals. Besides, there are various other advantages of this approach.

In Agent Based Modeling (ABM) a system is modeled as a set of autonomous agents, who can perceive the environment and act on the basis of some behavioural rules. The agents represent the actors in the system; environment represents the surroundings including neighbouring agents; and the behaviour rules model the interaction of the agent with other agents as well as with the environment. ABM can be used to model a number of phenomena in varied domains like Markets & Economy, Organizations, World Wide Web and Social Systems etc. Availability of fast and cheap computing power, coupled with other advances in computational sciences has paved the way for use of ABM as a preferred modeling and simulation technique. Since last few years ABM has become the frontrunner tool of the Sociologists and Psychologists who try to model social behaviours, particularly the behaviour of groups and societies. A large number of researches are now being carried out using this generative approach to model and analyze social phenomenon, such as spread of rumors, diffusion of ideas, emergence of cooperation, emergence of conventions & social norms, evolution and dynamics of spread of religions and cultures etc

This paper presents our experimental work on using ABM for understanding and analyzing voting preferences of individuals. We have modeled the process and factors shaping and affecting individual voting preferences for a particular political party. To keep the model simple and consequently analysis more accurate, we have taken individual voting preferences as '0' and '1' representing a two party political system. Every individual's final voting decision is shaped and affected by a number of factors, such as his own perception about parties, his family views & beliefs, perception of his friends, and his neighbourhood preferences. Our simulation models the role and relative weights of these factors in shaping an individual's voting preferences. We have also incorporated social influence theory in our model which states that similar persons are more likely to interact, and as a result of frequent interactions they are likely to become more similar. We experimented with multiple settings and configurations and obtained interesting results. The rest of the paper is organized as follows. Section 2 describes the ABM approach for modeling and simulation. Section 3 explains the process and factors believed to shape individual voting preferences. Section 4 presents the results of our experimental model. The paper concludes (section 5) with a short discussion of the results and their relevance.

## 2 Agent Based Modeling

Traditional modeling approaches to social systems relied on equation-based models that operate with a macro perspective. They operate on population attributes & their distributions and lack the focus on individual's role. Equation-based models, though useful for macro-scale behaviour predictions, fail to model social systems (or processes) that lack central coordination, systems that are very complex in terms of their interdependencies and systems which produce novel emergent behaviours in absence of a clear understanding of the collective phenomenon. Taking this into consideration, Axtell [1] suggested that there are three distinct uses of Agent-based computational models of social systems: (a) when numerical realizations can be proposed and solved: agent models can be used as social simulations; (b) when a mathematical model can be formulated but can only be partially solved: agent based models can be a useful tool of analysis; and (c) when mathematical models are either apparently intractable or provably insoluble: agent based modeling is perhaps the only technique available for systematic analysis.

Although technically simple, ABM is conceptually deep. Modeling a complex social process requires high conceptual clarity and analytical ability. A key issue, therefore, is to decide when to use ABM for modeling social systems. An indicative list of situations when it is better to think and model in terms of agents is: (a) when there is a natural representation of actors as agents; (b) when the interactions between the agents are complex, non-linear, discontinuous, or discrete; (c) when agents exhibit complex adaptive behaviours; (d) when the population or topology of interactions is heterogeneous; and (e) when agents have spatial component to their behaviours & interactions. ABMs in social sciences involve human agents, whose behaviours are often complex, irrational and subjective. Therefore, one needs to think carefully about the social phenomenon at hand before going for ABM. Further, the model needs to be built at the right level of description, with only the right amount of details. Unnecessarily adding complexity to a model can make the analysis difficult and the model useless.

### 2.1 Steps in ABM

Designing an Agent-based model of a social system require first identifying the purpose of the model, i.e., the potential question to be answered. The typical steps to be followed afterward can be summarized as following: (a) identifying the agents and their behaviour rules; (b) identifying agent relationships and their interaction patterns; (c) selecting an ABM platform; (d) obtaining the required relevant data; (e) validating the agent behaviour model; (f) running the model and recording the outputs; (g) analyzing the outputs with a viewpoint of linking micro-scale behaviours of the agents to macro-scale behaviour of the system; and (h) validating the model outcomes and hence the model [2].

### 2.2 ABM Computational Resources

Many rich and easy to program software platforms and toolkits for ABM are now readily available. Some of the popular modeling tools (particularly for social sciences) are: Net Logo (http://ccl.northwestern.edu/netlogo/), Repast (http://repast.sourceforge.net/), Swarm (http://www.swarm.org) and MASON (http://cs.gmu.edu/~eclab/projects/mason/). Net Logo is a multi-agent programming language and modeling environment designed in Logo programming language. It is highly suitable for modeling and exploring emergent phenomena. Repast is basically an agent-based social network modeling toolkit but has rich libraries to study dynamics of social processes as well. Swarm is a multi-agent simulation package to simulate the social and biological interactions of agents and their emergent collective behaviour. Swarm has two versions, namely Objective-C and Java versions. MASON is one of the latest Java platforms in this group. In addition to these platforms, there are many other toolkits & APIs available for modeling and visualizing social systems.

## 2.3 Benefits of ABM

Understanding social systems not only requires understanding the individuals that comprise the system but also how individuals interact with each other resulting in global behavioural patterns, which can be sufficiently novel. ABM helps researchers in investigating how large-scale macro effects arise from the micro-processes of interactions among many agents. Axelrod and Tesfatsion [3] call ABM the third way of doing science besides induction and deduction. ABM applied to social sciences take a methodical approach that could permit two important developments: (a) rigorous testing, refinement, and extension of existing theories that are difficult to evaluate using mathematical & statistical techniques; and (b) a deeper understanding of fundamental causal mechanisms in multi-agent systems. Bonabeau [4] goes a step further in proposing that ABM is a mindset and not merely a technology. He summarizes benefits of ABM over other modeling techniques in three statements: (a) ABM provides a natural description of a system; (b) ABM captures emergent phenomena; and (c) ABM is flexible.

## 2.4 Validating Agent Based Models

ABM, now an established scientific research practice, needs to incorporate a proper methodology of validation in order to verify the robustness of its findings and to truly act as a bridge between disciplines. Since simulation results of ABM are very sensitive to how agents are modeled, validation of computational model & simulation results is a critical issue in ABM [5]. Several approaches for validation of agent-based computational models have been proposed. Carley & Gasser [6] categorizes the validation approaches into three broad categories: (a) *theoretical verification*, which determines whether the model is an adequate conceptualization of the real world with the help of situation experts; (b) *external validation*, which determine whether the results from the model matches the results from the real world; and (c) *cross-model validation*, which compares the results of the model with other models. All these approaches are essentially aimed at validating the model at the macro level. Gilbert [7], on the other hand, emphasized that in order to completely validate a model it should be validated not only at macro-level but also at micro-level. Before going for a macro-level validation it is necessary to confirm that micro-level behaviours of agents are adequate representation of the actors in the system. *Econometric validation* and *companion modeling techniques* have been advocated to be suitable for empirical verification of social system models [8]. There are however still some problems in empirical validation of agent-based models [9], and validation continues to be one of the central problems of computer simulation methods including ABM [10].

### 2.5 ABM Applications in Social Science

ABM is currently being applied to model a variety of complex social phenomenon where simple local interactions generate emergent system-level behaviours. Some representative & relevant work can be found in [11], [12] & [13]. Macy & Willer [14] group most of the work on Agent-based modeling of collective behaviours in two categories: (a) *models of emergent structure* which includes works on cultural differentiation, homophilous clustering, idea diffusion, convergent behaviours and norms; and (b) *models of emergent social order* which include viability of trust, cooperation and collective action in absence of global control. Goldstone & Janssen [15] also identify three similar themes for Agent-based computational models of collective behaviour namely: (a) p*atterns and organizations* which include settlement patterns & segregation, human group behaviours and traffic patterns; (b) *social contagion* which include spread of ideas, fashions, cultures & religions; and (c) *cooperation* which include evolution of cooperation, trust & reputations and social norms & conventions.

## 3 Modeling Individual Voting Preferences

Every individual while deciding his voting preference is directly or indirectly affected by a number of factors. The final voting decision of an individual can be taken as a function of these factors. An individual makes a preference about a political party based not only on his perception about the ideology & programs advanced by it, but by several other important factors. Individual may have an inherent preference formed over a substantial period of time from his perceptions. However, family, friends, neighbourhood (representing social and religious group membership) and mass media play an important role both in shaping an individual's perception and deciding his final vote. We have modeled a population of individuals with their initial preference for a particular party, and then run the model by varying different factors to observe how (and possibly why) his initial favour changes to a particular final voting preference. Since the model deals with a two party system (used for simplicity of analysis), every individual is believed to have an initial favour for one of the two parties, represented as '0' or '1'. Individuals during course of the time interact with friends, family members and neighbours. We have also associated feature sets with every individual to represent his overall mental and behavioural makeup. The friends of an individual are then defined as those having similar feature set values as that of the individual. Family of an individual represents the agents close to him and with whom he interacts very frequently. Family members are known to have a greater influence over an individual. Neighbours are taken as relatively distant agents on the grid and are viewed to represent individuals belonging to same social and religious groups of the concerned individual. An individual is thus believed to be affected not only by his friends and family members but also by the social and religious groups he may belong. Mass media also plays role in shaping an individual's voting preference. We have devised three versions of the simulation: (i) model with social influence, (ii) a majority-based model, and (iii) model including friends, family and neighbours.

### 3.1 The Theory of Social Influence

The Social influence assumes that individuals (or agents) often imitate good behaviours through their interactions with other individuals. Its dynamics depends on familiarity of the interacting individuals, density of the neighbourhood, and popularity and spatial proximity of the other individuals. Interesting models of social influence have been proposed by Carley [16], Axelrod [17] and Axtell [18], Coleman [19], Nowak et al [20] etc. Axelrod, in his social influence model of dissemination of culture [21], focused on factors of local influence (tendency for people who interact to become more similar) and homophily (tendency to interact more frequently with similar agents). The more the agents interact, the more similar they become; and the more similar they become, the more likely they are to interact. Agents who are similar to each other interact with each other and become more similar. Axelrod's model captures culture as a group of features with multiple trait values per feature. However, the emphasis is not on the content of a specific culture but on the way in which culture is likely to emerge and spread. The simulations with varying parameters regarding grid size, number of features and number of traits per feature resulted in polarization, despite the only mechanism for change being convergence towards a neighbour. We have used the basic idea of Axelrod's culture model to represent friends of an individual. An individual's friend is believed to have behavioural feature set similar to the individual. The feature set similarity represents their like mindedness of the individuals on various issues and hence possibility of having similar voting preferences or at least being affected by each other's preferences.

### 3.2 Role of Family, Neighbourhood and Mass Media

An individual makes up his voting preference through a complex process involving non-linear feedback loop of interactions with a number of entities. Individuals interact more frequently with their family members and relatively less frequently with other persons belonging to their social and religious groups. An individual's final voting value is no doubt a function of these factors. However, what is more important to understand is that what is the extent of this effect and the by what underlying process an individual finally makes up his voting preference. We have modeled this in our second and third simulation versions. The second version is a simple model to understand the dynamics of voting preference changes due to majority effect. In this model all individuals are assigned a random initial voting preference value and then his interactions with his 8-neighbour agents is observed and analyzed. The final voting preference is a stable value that results by a process of majority sampling. Since the initial voting preference values seem to play an important role on the final outcome, we experimented with multiple initializations. This also modeled the implicit favour (or political wave) introduced by the mass media. The third version of the simulation is relatively more complex and takes into account voting preferences of individual's family and socio-cultural and religious neighbourhood. We have explored with different weights assigned to the influence of family and neighbours and observed their effect on the resultant complex set of interactions.

## 4 Experimental Setup and Results

We have designed our voting preference simulation using NetLogo [22]. The first model that we designed can be described as follows. A population of agents is created on a wrap-around agent grid of size 41 X 41. Every patch on the grid represents an individual. Every individual has certain attributes that characterize it. These attributes are represented as a feature set, which can be a 3 to 7 bit long string. Every bit in the string represents a particular trait of that attribute. The feature set values of two agents determine how similar they are. Two individuals will be similar if their feature set string contains common traits. For example, individual A (feature set: 00110) and B (feature set: 10110) are highly similar. In addition to the feature set, every agent also has a voting preference value, initialized randomly to either '0' or '1'. Agents interact with various agents in their neighbourhood. According to social influence theory, it is more likely to interact and be affected by the agents similar to it. Interactions between similar agents make them more similar, represented by a change in their voting preference so that they get similar voting preference. The model selects an agent and samples its neighbourhood. Suppose $n$ out of total $m$ agents in the neighbourhood are similar to the agent. Then the voting preference of the agent changes to the majority preference of the similar agents (n). We obtained multiple runs of the model by varying both (i) the feature size, and (ii) initial voting preference value distribution. The feature size has been varied from 3 to 7 with feature similarity threshold being varied from 2 to 3 for feature set size of 5 and 3 to 7 for feature set size of 7. The initial voting preference assignments has also been varied from 10%-90% to 50%-50% for vote values of '0' and '1'. Figures 1 and 2 show a snapshot of two runs of the model. With higher similarity threshold (in figure 1) there is less local polarization towards any vote value despite having initial favour for one vote value. Low feature set similarity however results in contiguous groups settling to same vote values.

**Fig. 1.** A snapshot of the run of feature set based simulation. The feature set size is 5 and similarity threshold is 3. In the figure on the left, the initial values of green = 757 and blue = 924 transform to final values of green = 730 and blue = 951. In the figure on the right, the initial values of green = 505 and blue = 1176 transform to final values of green = 417 and blue = 1264.

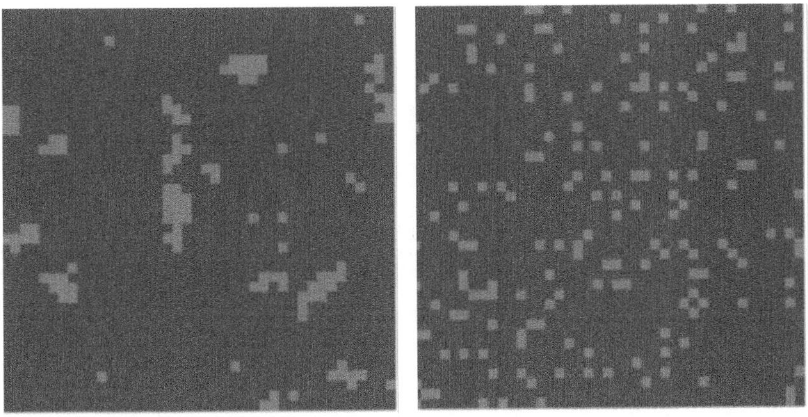

**Fig. 2.** A snapshot of the run of feature set based simulation. In the figure on the left, the feature set size is 7 and similarity threshold is 3 and the initial values of green = 505 and blue = 1176 transform to final values of green = 117 and blue = 1564. In the figure on the right, the feature set size is 7 and similarity threshold is 7 and the initial values of green = 169 and blue = 1512 transform to final values of green = 169 and blue = 1512.

In the second version of the simulation we did not use feature set concept, rather we chose to have a simple model where voting preference of an individual is simply taken as the majority vote value of its neighbours. For example, if more than 4 of the 8 neighbours of an agent prefer the vote '1', then the individual is also likely to vote '1' irrespective of his original preference. Hence the model is driven purely by the majority. At every tick, an agent sees in its neighbourhood and decides its own vote value based on the vote values of its 8-neighbours. Therefore, the majority vote value around an agent persuades him to align with it. In case of tie, that is in the 8-neighbourhood 4 agents have '1' vote value and the remaining 4 have '0' vote value, the agent's original vote value prevails. Since the final voting value of an individual depends only on the majority voting value present in the 8-neighbourhood and that the initial vote values are set randomly; we thought it would be interesting to see what happens when the initial vote values of agents have a non-uniform distribution. We run the model with multiple initial vote value distributions. The initial vote value assignments have been varied from 10% agents having '0' vote and remaining 90% having '1' vote, to many other combinations such as 20%-80%, 30%-70%, 40%-60%, 45%-55% and 50%-50% of '0' and '1' votes. A snapshot of the two interesting runs is shown in figure 3. The results demonstrate that the final voting preference of the population in the majority model depend heavily on the initial distribution of voting preferences. Whenever there is a favour (about 60% against 40%) for initial distribution of one vote value over the other (say '0' over '1'), then the final voting preference of most of the individuals becomes the favoured one ('0' vote). In case of close to equal distribution of both the vote values ('0' and '1' being less than 10% different), the final voting preference of the majority may tilt relatively favourably on either of the sides.

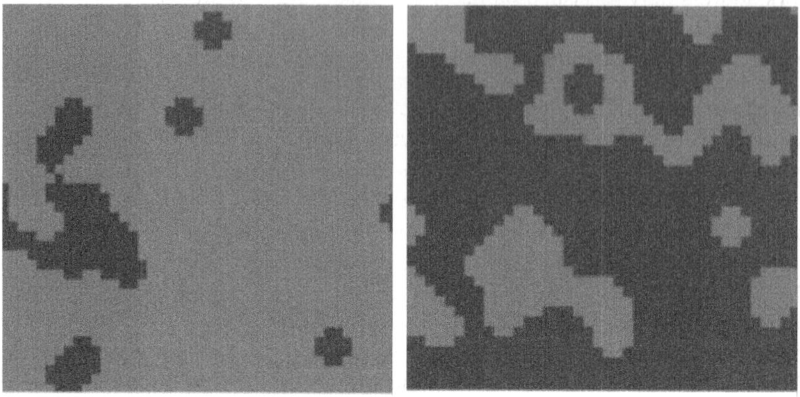

**Fig. 3.** A snapshot of the run of majority based simulation. In the figure on the left, the initial values of green = 1561 and blue = 1040 transform to final values of green = 2345 and blue = 256. In the figure on the right, the initial values of green = 1171 and blue = 1430 transform to final values of green = 659 and blue = 1942.

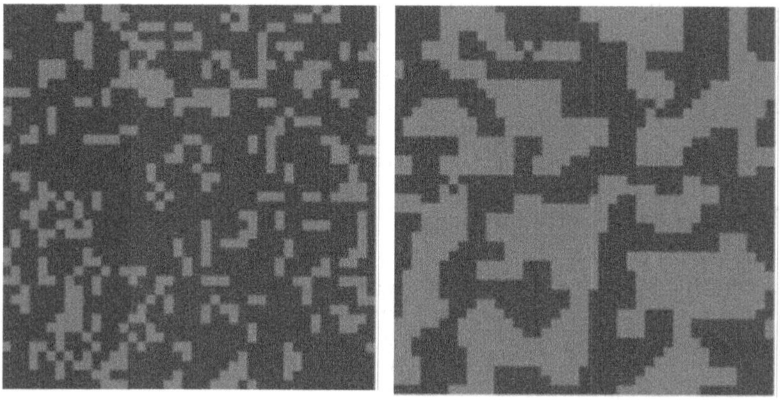

**Fig. 4.** A snapshot of the run of modified majority simulation. In the figure on the left, the initial values green = 505, blue = 1176 & weights own-vote = 80%, family-vote = 15% and neighbour-vote = 5%; results into final values green = 413, blue = 1268. In the figure on the right, the initial values green = 841, blue = 840 & weights own-vote = 40%, family-vote = 40% and neighbour-vote = 20%; results into final values green = 843, blue = 838.

The third version of the model that we explored is a modified majority model. In this version an individual's voting preference is determined not simply by the prevalent majority vote, but trough a weighted average of values of his own vote, his family members' vote and his neighbours' vote. We chose 4-neighbourhood to represent family and 8-neighbourhood to represent neighbours. Here, 8-neighbours stand to represent members of individual's socio-cultural and religious groups. Since the 8-neighbourhood also contains the 4-neighbourhood on the grid, we have chosen relatively smaller weight for neighbours. We have run the model by varying the weights in weighted average computation and also the initial vote value distribution. Figure 4 presents the results of two such runs. The results demonstrate that despite giving higher weight to the individual voting preference, the initial bias introduced in the population results to a certain degree of local polarization. Only very high value of own vote is able to sustain the other voting preference. In case of approximately equal initial voting preference also, high own-vote weight value tend to preserve most of the initial voting pattern. As the weight assignments tend to become relatively more in favour of the family and neighbourhood combine, local polarization is observed. The results in this case become somewhat similar to the majority-based version of the simulation.

## 5 Conclusion

The ABM methodology is a favorable approach to model and analyze complex social phenomena that may involve non-linear feedback loops. It has been applied successfully to model a number of social phenomena involving different social processes and organizational structures. In this paper, we have demonstrated how ABM can be applied to model and analyze the voting preference formation and resultant voting decisions of individuals in a population. The model assumes a two party system. We designed three versions of the simulation and observed the results for a large number of runs with different parameter variations. The initial voting preference distribution in the population, role of friends (likeminded individuals), affect of family members and neighbours; all produce interesting results to analyze. Social influence theory incorporated into model shows that individuals' voting decisions are affected more by those who are similar to them. However, if the similarity threshold is relatively lower, the model results in local voting preference polarizations. When the model is based only on majority, the initial distribution of voting preferences becomes the key deciding factor. These results taken together resemble to actual voting patterns observed in closely knit societies and segregations such as far off villages and tribal groups. The role of implicit favorable waves introduced by mass media can also be understood by the results of the second version of the simulation. When the weights of factors playing role in shaping an individual's voting preference are varied, the results demonstrate that high weight to individual vote is able to sustain variations in voting preferences. When the own-weight is reduced to give more favour to the family and neighbours, polarizing tendency is seen. Voting behaviour and patterns of relatively informed and well-educated persons conform to this trend. The experimental simulations attempt to model an interesting social phenomenon and present interesting results that need to be further analyzed to draw relevant inferences, analogies and explanations.

## References

1. Axtell, R.: Why agents? On the Varied Motivations for AgentCcomputing in the Social Sciences - working paper 17. Centre on Social and Economic Dynamics, Brookings Institution, Washington D.C (2000)

2. Macal, C.M., North, M.J.: Tutorial on Agent Based Modeling and Simulation. In: Kuhl, M.E., Steiger, N.M., Armstrong, F.B., Joines, J.A., Young, M. (eds.) Proceedings of the 2005 Winter Simulation Conference, The Technical Writer's Handbook, University Science, Mill Valley (1989)
3. Axelrod, R., Tesfatison, L.: A Guide for Newcomers to Agent Based Modeling in the Social Sciences. In: The Handbook of Computational Economics. Agent Based Computational Economics, vol. 2 (2005)
4. Bonabeau, E.: Agent-based Modeling: Methods and Techniques for Simulating Human Systems. Proceedings of National Academy of Science 99(3), 7280–7287 (2001)
5. Takadama, K., Kawai, T., Koyama, Y.: Micro- and Macro- level Validation in Agent-Based Simulation: Reproduction of Human-Like Behaviours and Thinking in Sequential Bargaining Game. Journal of Artificial Societies and Social Simulation 11(2), 9 (2008)
6. Carley, K.M., Gasser, L.: Computational and Organization Theory. In: Weiss, G. (ed.) Multiagent Systems - Modern Approach to Distributed Artificial Intelligence, pp. 299–330. The MIT Press, Cambridge (1999)
7. Gilbert, N.: Open Problems in Using Agent-based Models in Industrial and Labor Dynamics. In: Leombruni, R., Richiardi, M. (eds.) Industry and Labor Dynamics: the Agent Based Computational Approach, pp. 401–405. World Scientific, Singapore (2004)
8. Moss, S.: Alternative Approcahes to the Emperical Validation of Agent- Based Models. Journal of Artificial Societies and Social Simulation 11(1), 5 (2008)
9. Windrum, P., Fagiolo, G., Moneta, A.: Emperical Validation of Agent-Based Models: Alternatives and Prospects. Journal of Artificial Societies and Social Simulation 10(2), 8 (2007)
10. Kuppers, G., Lenhard, J.: Validation of Simulation: Patterns in the Social and Natural Sciences. Journal of Artificial Societies and Social Simulation 8(4) (2005)
11. Epstein, J.M.: Generative Social Sciences: Studies in Agent-based Computational Modeling. Princeton University Press, Princeton (2007)
12. Gilbert, N., Troitzsch, K.: Simulation for the Social Scientist. Open Univ. Press, Buckingham (1999)
13. Epstein, J.M., Axtell, R.: Growing Artificial Societies: Socail Science from bottom-up. MIT Press, Cambridge (1996)
14. Macy, M.W., Willer, R.: From Factors to Actors: Computational Sociology and Agent Based Modeling. Annual Review of Sociology 28, 143–166 (2002)
15. Goldstone, R.L., Janssen, M.A.: Computational Models of Collective Behaviour. Trends in Cognitive Sciences 9(9) (2005)
16. Carley, K.M.: A Theory of Group Stability. American Sociological Review 56, 331–354 (1991)
17. Axelrod, R.: The Complexity of cooperation. Princeton University Press, Princeton (1997)
18. Axtell, R., Axelrod, R., Epstein, J.M., Cohen, M.D.: Aligning Simulation Models: a case study and results. Computational and Mathematical Organization Theory 1(2), 123–141 (1996)
19. Coleman, J.S.: The Use of Electronic Computers in the Study of Social Organizations. European Journal of Sociology 6, 89–107 (1965)
20. Nowak, A., Jacek, S., Latane, B.: From Private Attitude to Public Opinion: A dynamic theory of social impact. Psychological Review 97, 362–376 (1990)
21. Axelrod, R.: The Dissemination of Culture: A model with local convergence and global polarization. Journal of Conflict Resolution 41, 203–226 (1997)
22. Wilensky, U.: NetLogo. Centre for Connected Learning and Computer-Based Modeling, Northwestern University, Evanston, IL (1999), http://ccl.northwestern.edu/netlogo

# Texture Analysis of Brain MRI and Classification with BPN for the Diagnosis of Dementia

T.R. Sivapriya[1], V. Saravanan[2], and P. Ranjit Jeba Thangaiah[3]

[1] Dept. of Computer Science, Lady Doak College, Madurai
[2] Dept. of Computer Applications, Dr.N.G.P. Institute of Technology, Coimbatore
[3] Dept. of Computer Applications, Karunya University, Coimbatore

**Abstract.** In this paper, we present an evaluation of diagnosis of dementia using texture analysis of brain MRI with wavelets and further classification by BackPropagation Network. The tests were conducted on 3D brain MRI data extracted from OASIS database. The classification is based on the following steps: First, the region of interest is extracted from the MRI images by wavelets, Gray level occurance matrix (GLCM) and Haralick features. Gabor features were characterized by the distribution of histogram of wavelet coefficients. These features were segregated into three datasets. The first dataset containing the GLCM features, the second data set has the Haralick features and the third dataset has Gabor wavelet based Haralick features. Classification was done by backpropagation network based on 3 feature vectors. From the analysis it has been found that the average efficiency of Gabor combined with Haralick features is around 97% for all types of datasets, and the average efficiency value for GLCM is 86 % and that of Haralick features was 90%. From the comparison of the average efficiency of the wavelet families, statistical features extracted from Gabor wavlets provides better efficiency than the other two methods.

**Keywords:** Texture classification, GLCM, wavelet, Gabor, Backpropagation network.

## 1 Introduction

Texture analysis is an important issue in image processing with many applications including medical imaging, remote sensing, content-based image retrieval, and document segmentation. Over the last three decades, texture analysis has been widely studied and many texture classification techniques have been proposed in the literature. Ideally texture analysis should be invariant to translation and rotation. However, most of the proposed techniques assume that texture has the same orientation, which is not always the case. In the literature, essentially two different kinds of textures like surface and visual are examined. Texture analysis, which is commonly found in the literature on image processing and computer vision [1], is expected to provide appropriate features for subsequent studies, such as content-based image retrieval, pattern recognition, and image segmentation.

Texture is an image feature that provides important characteristics for surface and object identification from image [2-4]. Texture analysis is a major component of image processing and is fundamental to many applications such as remote sensing,

D. Nagamalai, E. Renault, M. Dhanushkodi (Eds.): CCSEIT 2011, CCIS 204, pp. 553–563, 2011.

quality inspection, medical imaging, etc. It has been studied widely for over four decades. Recently, multiscale filtering methods have shown significant potential for texture description, where advantage is taken of the spatial-frequency concept to maximize the simultaneous localization of energy in both spatial and frequency domains [5]. The use of wavelet transform as a multiscale analysis for texture description was first suggested by Mallat [6]. Recent developments in the wavelet transform [7] provide good multiresolution analytical tool for texture analysis and can achieve a high accuracy rate.

Alzhemier's disease (AD) is a disease often related to ageing. The financial and social burdens of AD on the society are becoming overwhelming with the continued increase of the average life span. As such, considerable efforts have been made to develop biomarkers for early detection of AD. It is well accepted that brain atrophy is one of the most important markers of AD. Brain structures such as deep white matter, corpus callosum, hippocampus, amygdala, caudate nucleus, and thalamus are known to deteriorate in dementia subject to atrophy. Since brain ventricles are in the center of the brain, and are surrounded by the brain structures mentioned, any change in these brain structures will affect the shape of cerebral gray and white matter [8-12]. The brain volume is significantly changed in AD patients compared to healthy subjects of the same age group. Visual assessment of ventricle volume or shape change has shown to be quite reliable [13][14] in detecting AD. Zhu [15] applied Fourier analysis for image features, Ferrarini[16], Chaplot [17] applied wavelets, Selvaraj etal[18] applied Haralick components to extract features for brain MRI analysis.

The paper is organized as follows. In Section 2, we briefly review the theory of wavelets and the traditional approach to texture analysis with wavelets. The backpropagation algorithm and its architecture are reviewed in Section 3. In Section 4, the proposed methodology, feature extraction and validation are described. Experimental results and conclusion are presented in Sections 6.

## 2 Texture Analysis

### 2.1 Wavelet Transform

Given a real, mother wavelet function $\psi(t) \in L^2(R)$, which satisfies the admissibility condition,

$$C_\psi = \int_{-\infty}^{\infty} \frac{|\Psi(\omega)|^2}{|\omega|} d\omega < \infty \tag{1}$$

where $\Psi(\omega)$ is the Fourier transform of $\psi(t)$, the scaling parameter $p$ and the shifting parameter q are introduced to construct a family of basis functions or wavelets $\{\psi_{p,q}(t)\}$, $p,q \in R$, $p \neq 0$, where

$$\psi_{p,q}(t) = \frac{1}{\sqrt{|p|}}\psi(\frac{t-q}{p}). \tag{2}$$

If $\Psi(\omega)$ has sufficient decay, the admissibility condition reduces to the constraint that $\Psi(0)=0$ or

$$\int_{-\infty}^{\infty}\psi(t)dt=\Psi(0)=0 \tag{3}$$

Because the Fourier transform is zero at the origin, and the spectrum decays at high frequencies, the wavelet behaves as a bandpass filter.

The continuous wavelet transform (CWT) of a function or a signal $x(t)\in L^2(R)$ is then defined as

$$X_{CWT}(p,q)=\int_{-\infty}^{\infty}x(t)\psi_{p,q}(t)dt\,, \tag{4}$$

and the inverse wavelet transform is given by

$$x(t)=\frac{1}{C_\psi}\int_{-\infty}^{\infty}\int_{-\infty}^{\infty}X_{CWT}(p,q)\psi_{p,q}(t)\frac{dpdq}{p^2}\,. \tag{5}$$

Since $\{\psi_{p,q}(t)\}$ is a redundant basis set, it implies that, by discretizing the scaling and shifting parameters as follows:

$$p=2^k,\, q=2^k l,\ k,l\in Z\,, \tag{6}$$

an orthonormal basis $\{\psi_{k,l}(t)\}$ is obtained,

$$\psi_{k,l}(x)=2^{-k/2}\psi(2^{-k}x-l),\ k,l\in Z \tag{7}$$

which comply with the condition,

$$\int_{-\infty}^{\infty}\psi_{k,l}(t)\psi_{k',l'}(t)dt=\delta(k-k')\delta(l-l'),\forall k,k',l,l'\in Z \tag{8}$$

The wavelet transform is an appropriate tool for multiresolution analysis (MRA) [19]. In MRA, a *scaling function* $\phi(x)$ and the associated mother wavelet function $\psi(x)$, which are needed in the construction of a complete basis, must satisfy the two-scale difference equations:

$$\begin{aligned}\phi(x)&=\sqrt{2}\sum_n h(n)\phi(2x-n)\\ \psi(x)&=\sqrt{2}\sum_n g(n)\phi(2x-n)\end{aligned} \tag{9}$$

where the coefficients $h(n)$ and $g(n)$ satisfy the following:

$$g(n)=(-1)^n h(1-n) \tag{10}$$

### 2.2 Gabor Texture

Usually the Gabor features of an image M(Y) can be obtained by convolving the image with the family of Gabor filters

$$G_{u,v}(\vec{x}) = (M * \varphi_{u,v})(\vec{x}) \tag{11}$$

where $G_{u,v}(\vec{x})$ denotes resultant Gabor representation of image M at orientation u and scale v. The resultant Gabor feature set thus consists of the convolution results of an input image M(c) with all of the Gabor filters.

**Fig. 1.** Magnitude of Gabor wavelet

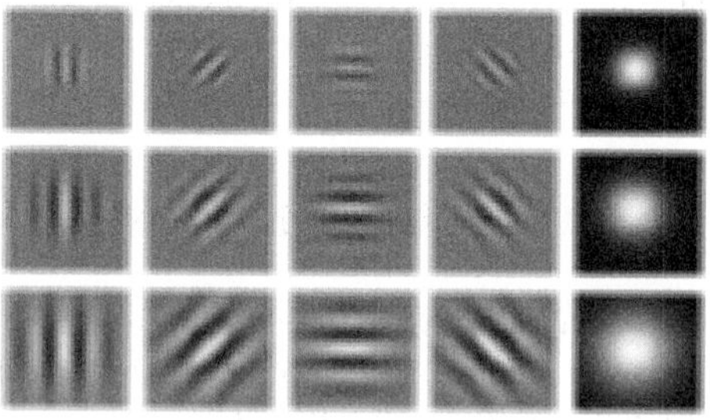

**Fig. 2.** Real part of Gabor wavelet

Figures 1 and 2 represent the magnitude and real part of Gabor wavelets.

### 2.3 GLCM Texture

One of the simplest approaches for describing texture is to use statistical moments of the intensity histogram of an image or region [20]. Using only histograms in calculation will result in measures of texture that carry only information about distribution of intensities, but not about the relative position of pixels with respect to each other in that texture. Using a statistical approach such as co-occurrence matrix will help to provide valuable information about the relative position of the neighbouring pixels in an image. Given an image I, of size N×N, the co-occurrence, matrix P can be defined as follows:

$$P(i,j)=\sum_{x=1}^{N}\sum_{y=1}^{N}\begin{cases}1, if I(x,y)=i \ and I(x+\Delta_x, y+\Delta_y)=j\\ 0, otherwise\end{cases} \tag{12}$$

Here, the offset (Δx, Δy), is specifying the distance between the pixel-of-interest and its neighbour. Note that the offset (Δx, Δy) parameterization makes the co-occurrence matrix sensitive to rotation. Choosing an offset vector, such that the rotation of the image is not equal to 180 degrees, will result in a different co-occurrence matrix for the same (rotated) image. The four matrices are used separately for classification, and then the final decision is formed by fusing the four decisions. As these matrices are symmetric, it is more convenient to use the upper or lower diagonal matrix coefficients in forming the vectors. The second-order statistical features are extracted from GLCM. In brief, a GLCM is constructed by systematically considering the relationship between pixel pairs and tabulating the frequency of various gray-level combinations within an image or within a region of interest The maximum number of gray levels considered for each image or region of interest is typically scaled down to 32 or 64 levels, rather than using the full dynamic range of, for example, 65,536 gray levels (16 bits/pixel)[21]. This quantization step is required to avoid a large number of zero-valued entries in the co-occurrence matrix.

## 2.4 Haralick Features

Haralick features are extracted from the brain MRI with the formulae given in Table 1.

**Table 1.** Haralick Features

| Property | Formula |
|---|---|
| Angular second moment | $\sum_{i=0}^{G-1}\sum_{j=0}^{G-1}[p(i,j)]^2$ |
| Correlation | $\sum_{i=0}^{G-1}\sum_{j=0}^{G-1}\frac{ijp(i,j)-\mu_x\mu_y}{\sigma_x\sigma_x}$ |
| Inertia | $\sum_{i=0}^{G-1}\sum_{j=0}^{G-1}(i-j)^2 p(i,j)$ |
| Entropy | $-\sum_{i=0}^{G-1}\sum_{j=0}^{G-1}p(i,j)\log_2[p(i,j)]$ |
| Absolute value | $\sum_{i=0}^{G-1}\sum_{j=0}^{G-1}\lvert i-j\rvert p(i,j)$ |
| Inverse difference | $\sum_{i=0}^{G-1}\sum_{j=0}^{G-1}\frac{p(i,j)}{1+(i-j)^2}$ |

# 3 Backpropagation Algorithm

The backpropagation algorithm for estimating parameters in neural networks has been the most popular in the medical literature[21-23]. One of the problems encountered by researchers utilizing the backpropagation algorithm is that low-frequency patterns (or rare patterns) may take a long training time to be recognized, because frequent patterns "dominate" the error. This problem is overcome by faster training using PSO.

A hidden layer is used for ANN's internal transformation: hidden neurons transform the "signals" from the input layer and generate signals to be sent to the next layer, so that MLP is often called the "feedforward" neural network. The input layer consists of input variables plus an input bias, which adjusts the weighted mean of the input variables. All input variables are regressed on the next layer, in the same fashion as multiple regression estimation.

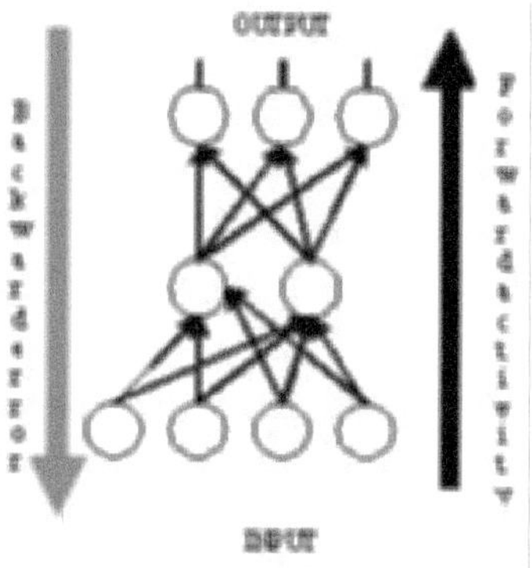

**Fig. 3.** Backpropagation Network

## 3.1 BPN Architecture Design

BPN architecture is selected based on testing and training of 7 different models. The steps are as follows:

1. Create a customized BPN model n1.
2. Train and test with new data.
3. Check for the efficiency and ability to recognize new patterns.
4. If the model n1 has comparatively higher efficiency and detects new patterns, it is selected as the ideal model for diagnosis.

# 4 Proposed Methodology

MRI images are preprocessed with Gaussian filters and normalized before feature extraction. The overall methodology is depicted in Figure 4.

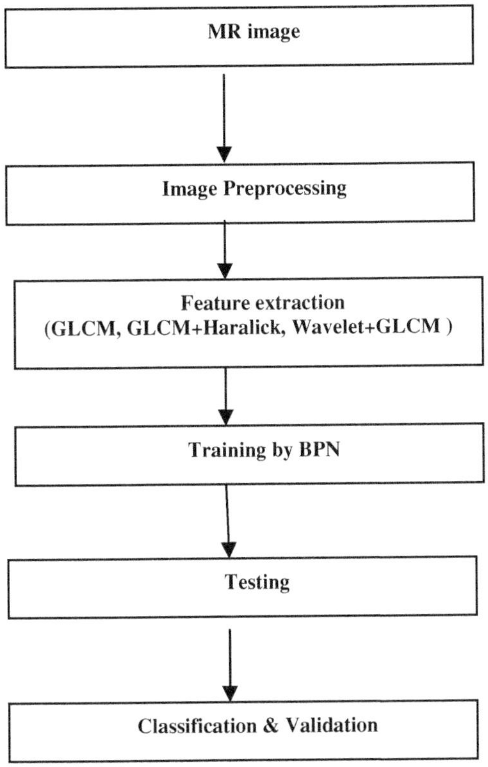


**Fig. 4.** Methodology

### 4.1 Feature Extraction Phase and Classification

For feature extraction, the following steps are done:

1. The voxel-wise texture features of image I (x, y, z) are extracted at each slice of 3D ROI by convoluting with 2D Gabor filters [24], [25]and averaging inside the ROI. The 2D Gabor filters are mathematically described at location (x, y) as $\lambda = 1/f$ is the wavelength, $\theta$ the orientation, $\gamma$ the spatial aspect ratio which determines the eccentricity of the convolution kernel.
2. A 32X32 gray level co-occurrence matrix with 32 gray are generated for the first level approximation images of each training image considered for feature extraction.
3. Haralick features listed in Table-1 is calculated for each training image.
4. For each texture feature compute the average of different angles.
5. Compute the overall mean of each feature of the entire set of images considered for feature extraction.
6. Feed features as input vectors to BPN models.
7. Test the sensitivity, specificity and accuracy of the classification based on the texture features.

### 4.2 Validation

The ability of the neural network model (a 10 X 7 X 1 structure was used) to perform the classification was examined by setting aside 20% of the patterns (or observations) as validation (or testing) data. In this cross-validation approach, the training involved repeatedly exposing the network to the remaining 80% of the patterns (training data) for several epochs, where an epoch is one complete cycle through the network for all cases. (Data were normalized before training.) A network trained in this manner is considered generalizable, in the sense that it can be used to make predictions.

## 5 Data Set

OASIS provides brain imaging data that are freely available for distribution and data analysis [26]. This data set consists of a cross-sectional collection of 416 subjects covering the adult life span aged 18 to 96 including individuals with early-stage Alzheimer's Disease (AD) . For each subject, 3 or 4 individual T1-weighted MRI scans obtained within a single imaging session are included [27]. Many studies have been done to compare manual and computerized diagnosis of dementia [28].

**Division of Data into Training and Testing**

It is easy to design an optimistically biased (low error, over-trained) algorithm. If, for example, the same data is used for model selection (i.e. optimization of C and $\gamma$ or n in our case), model training, and validation, an obvious risk is the creation of a machine which does not generalize. Such a model may be of limited effectiveness for classifying novel data. To eliminate such circular logic for this study validation is performed on independent/unseen data. The records of 50 patients with Alzhemier's disease (AD) and 50 Cognitively Normal (CN) datasets were randomly divided into two sets of 25 AD, 25 CN.

## 6 Results and Conclusion

The hybrid wavelet based BPN combined is found to effectively improve classification speed from brain MRI images. The BPN model effectively predicted the change in brain MRI in patients with Dementia as given in Figure 7. From the results of the above experiments, we can conclude that when wavelet based texture features are combined with Haralick parameters, classification is very effective and can be used for real time diagnostic decision making according to Figures 5 and 6 .

Backpropagation network classifier yields high accuracy and sensitivity which is the need of the hour in day-to-day clinical treatment. It can be observed that Gabor wavelets are an effective feature extraction method which can improve and clearly quantify the brain volume in Alzhemier's disease. Haralick features extracted in wavelet domain yields accurate results when compared with traditional GLCM feature extraction. The specificity of the BPN classifier is improved when wavelet based Haralick features are given as input to the network rather than GLCM features alone or Haralick features combined with GLCM.

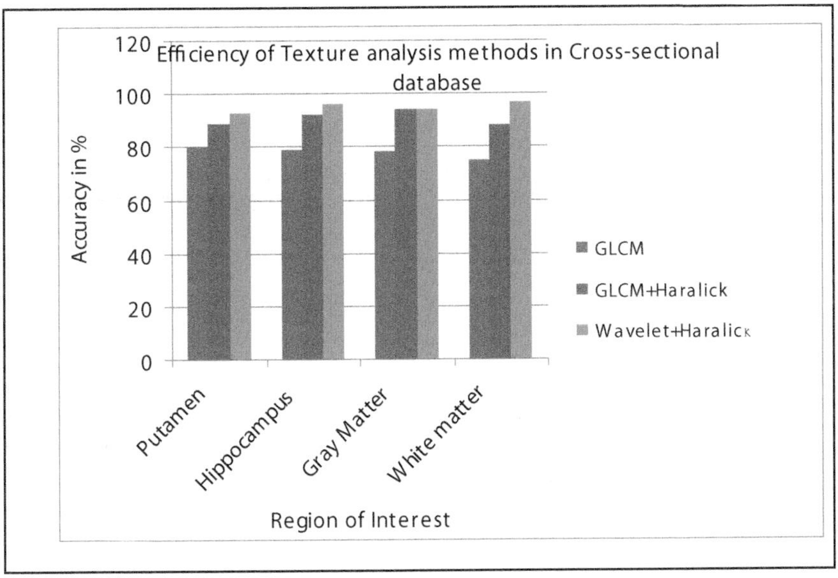


**Fig. 5.** Efficiency of Texture Analysis methods in Cross-sectional OASIS database

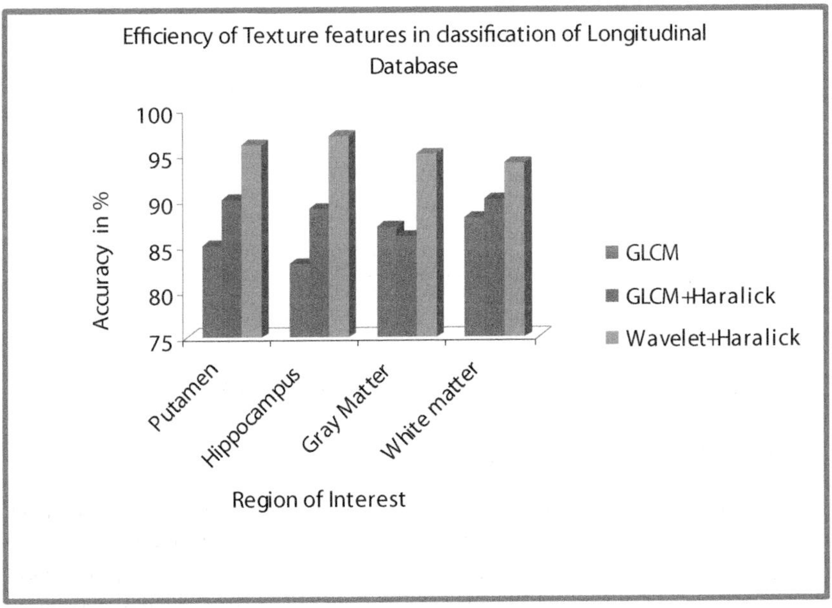


**Fig. 6.** Efficiency of Texture Analysis methods in Longitudinal OASIS database

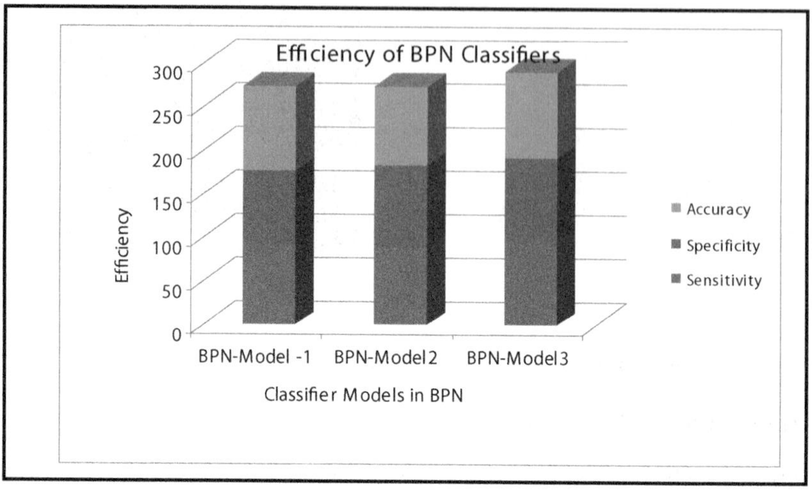


**Fig.7.** Efficiency of BPN classifiers

## References


1. Haralick, R.M., Shanmugam, K., Dinstein, I.: Textural Features for Image Classification. IEEE Transactions on Systems, Man, and Cybernetics 3(6), 610–621 (1973)
2. Smith, G., Burns, I.: Measuring texture classification algorithms. Pattern Recognition Letters 18, 1495–1501 (1997), `http://www.cssip.elec.uq.edu.au/guy/meastex/meastex.html`
3. Arivazhagan, S., Ganesan, L., Priyal, S.P.: Texture classification using Gabor wavelets based rotation invariant features. Pattern Recognition Letters 27(16), 1976–1982 (2006)
4. Vassili, Kovalev, A., Kruggel, F., Gertz, H.-J., Yves, D., von Cramon, D.: IEEE Three-Dimensional Texture Analysis of MRI Brain Datasets. Transactions on Medical Imaging 20(5) (May 2001)
5. Valkealahti, K., Oja, E.: Reduced multidimensional cooccurrence histograms in texture classification. IEEE Transactions on Pattern Analysis and Machine Intelligence 20, 90–94 (1998)
6. Mallat, S.G.: Multifrequency channel decompositions of images and wavelet models. IEEE Trans. on Acoustics Speech And Signal Processing 11(7), 674–693 (1989)
7. Gonzalez, R.C., Woods, R.E.: Digital Image Processing, 2nd edn. Prentice Hall, Englewood Cliffs (2002)
8. Kloppel, S., Stonnington, C.M., Chu, C., Draganski, B., Scahill, R.I., Rohrer, J.D., Fox, N.C., Jack Jr., C.R., Ashburner, J., Frackowiak, R.: Automatic classification of MR scans in Alzheimer's disease. Brain 131, 681–689 (2008)
9. Georgiadis, P., Cavouras, D., Kalatzis, I., et al.: Enhancing the discrimination Accuracy between metastases, gliomas and meningiomas on brain MRI by volumetric textural features and ensemble pattern recognition methods. Magn. Reson. Imaging 27, 120–130 (2009)
10. Provenzale, J.M., Mukundan, S., Barboriak, D.: Diffusion-weighted and perfusion MR imaging for brain tumor characterization and assessment of treatment response. Radiology 239, 632–649 (2006)

11. Sivapriya, T.R., Saravanan, V.: Automatic brain MRI mining using Support Vector Machine and Decision tree. CiiT International Journal of Artificial Intelligent Systems and Machine Learning 3(2), 109–116 (2011)
12. Li, G., Yang, J., Ye, C., Geng, D.: Degree prediction of malignancy in brain glioma using support vector machines. Computers in Biology and Medicine 36, 313–325 (2006) (PubMed: 16446164)
13. Mahmoud-Ghoneim, D., Toussaint, G., Constans, J., et al.: Three dimensional texture analysis in MRI: a preliminary evaluation in gliomas. Magn. Reson. Imaging 21, 983–987 (2003)
14. Snyder, A., Girton, L., Morris, J., Buckner, R.: Normative estimates of cross-sectional and longitudinal brain volume decline in aging and AD. Neurology 64, 1032–1039 (2009)
15. Zhu, H., Goodyear, B.G., Lauzon, M.L., Brown, R.A., et al.: A new local multiscale Fourier analysis for medical imaging. Med. Phys. 30, 1134–1141 (2003)
16. Ferrarini, Palm, W.M., Olofsen, H., Van Buchem, M.A., Reiber, J.H., Admriraal-Behloul, F.: Shape Differences of the brain ventricles in Alzheimer's Disease. Neuroimage (2006)
17. Chaplot, S., Patnaik, L.M., Jagannathan, N.R.: Classification of magnetic resonance brain images using wavelets as input to support vector machine and neural network. Biomedical Signal Processing and Control (2006)
18. Selvaraj, H., Thamarai Selvi, S., Selvathi, D., Gewali, L.: Brain MRI Slices Classification Using Least Squares Support Vector Machine. IC-MED 1(1), 21–33 (2007)
19. Lee, T.S.: Image Representation Using 2D Gabor Wavelets. IEEE Transaction on Pattern Analysis and Machine Intelligence 18(10), 959–971 (1996)
20. Dunn, D., Higgins, W.E.: Optimal Gabor filters for Texture Segmentation. IEEE Transaction on Image Processing 4(7), 947–964 (1995)
21. Di Luca, M., Grossi, E., Borroni, B., et al.: Artificial neural networks allow the use of simultaneous measurements of Alzheimer Disease markers for early detection of the disease. Journal of Translational Medicine 3, 30 (2005), doi:10.1186/1479-5876-3-30
22. Kyrki, V., Kamarainen, J.K., Kalviainen, H.: Simple Gabor Feature Space for Invariant Object Recognition. Pattern Recognition Letters 25(3), 311–318 (2004)
23. Laon, Z., Shen, D., Xue, Z., Karacali, B., Resnick, S.M., Davatzikos, C.: Morphological classification of brains via high-dimensional shape transformations and machine learning methods. Neuroimage 21(1), 46–57 (2004)
24. Mozer, M., Jordan, M., Petsche, T. (eds.): Advances in Neural Information Processing Systems, pp. 475–481. MIT Press, Cambridge (1997)
25. Randen, T., Husoy, J.H.: Filtering for Texture Classification: A Comparative Study. IEEE Transaction on Pattern Analysis and Machine Intelligence 21(4), 291–291 (1999)
26. Marcus, D., Wang, T., Parker, J.M., Csernansky, J., Morris, J., Buckner, R.: Open Access Series of Imaging Studies (OASIS): Cross-Sectional MRI Data in Young, Middle Aged, Nondemented and Demented Older Adults. Journal of Cognitive Neuroscience 19, 1498–1507 (2007)
27. Bagci, U., Li Bai, A.: Detecting Alzheimer Disease in Magnetic Resonance Brain Images Using Gabor Wavelets. In: IEEE 15th International Conference on Signal Processing and Communications Applications. SIU, pp. 1–4 (2007)
28. Kloppel, S., et al.: Accuracy of dementia diagnosis -a direct comparison between radiologists and a computerized method. Brain 131, 2969–2974 (2008)

# A Host Based Kernel Level Rootkit Detection Mechanism Using Clustering Technique

Jestin Joy and Anita John

Rajagiri School of Engineering & Technology, Kochi, Kerala
jestinjoy@acm.org,
anitaj@rajagiritech.ac.in

**Abstract.** Rootkits are a set of software tools used by an attacker to gain unauthorized access into a system, thereby providing him with privilege to access sensitive data, conceal its own existence and allowing him to install other malicious software. They are difficult to detect due to their elusive nature. Modern rootkit attacks mainly focus on modifying operating system kernel. Existing techniques for detection rely mainly on saving the system state before detection and comparing it with the infected state. Efficient detection is possible by properly differentiating malicious and non malicious activities taking place in a kernel. In this paper we present a novel anomaly detection method for kernel level rootkits based on k-means clustering algorithm.

## 1 Introduction

In recent years attackers employ a variety of sophisticated methods to gain access to the system. Main goal of a rootkit is to conceal the evidence of intruder activities. Most rootkits need administrative level privileges to install it in the system. Usually attacker make use of some system vulnerabilities to get administrative level privileges. An ordinary user will find it difficult to detect the presence of a rootkit, since he will not find any discrepancy in the behavior of the system, even if the system is infected by a rootkit. With the increasing use of operating system in smart phones and other embedded devices the threat posed by rootkits[4][16] are of great concern.

The first generation rootkits were easier to detect since they mainly aimed at modifying user level programs(logs, application binaries ...). Methods like checksum verification could easily detect these type of infections [11]. Rootkits first appeared in the end of 80's as method to hide log files and now they pose a serious threat to computer industry [5]. Great deal of research is going on in the area of rootkit detection. In the past, system administrators relied heavily on system utilities like ls, ps ... to find the presence of rootkits [11]. But new generation rootkits could easily hide their presence from these utilities. Kernel level rootkits modifies vital areas of an operating system like its kernel and the device drivers. Difficulty in detecting them arise out of the fact that they operate at the same security level as the operating system.

D. Nagamalai, E. Renault, M. Dhanushkodi (Eds.): CCSEIT 2011, CCIS 204, pp. 564–570, 2011.

This paper presents a technique for detecting kernel level rootkits in Linux based systems. The proposed method works based on detecting the anomaly exhibited by a malicious kernel.

## 2 Background

Generally rootkits can be classified into four: (1) Hypervisor Level rootkits [7] loads the original operating system as "guest", thereby intercepting all its activities. (2) Kernel Level rootkits modify kernel level code, so that security of the system is compromised (3) Library Level Rootkit replaces system library with malicious ones (4) Application Level rootkit [5] modifies application level binaries like ls, ps, netstat ... to give false information to the user.

### 2.1 Kernel Level Rootkit

Unlike other rootkits, kernel level rootkit modifies the kernel itself. Kernel being the lowest level of operating system makes it a good choice for the intruder to attack, since an attack on it is very difficult to detect. Also being at the kernel level provides the attacker with complete freedom to access all most all areas of an operating system.

In most cases rootkits finds its way into the kernel through Loadable Kernel Modules(LKM). LKM allows extending the functionality of the kernel, without recompiling the kernel. The code inserted using LKM provides same capability as a kernel code. Another main advantage of using LKM is that they can be added/removed on the fly. Though mainly aimed at debugging the kernel, */dev/mem* is also used by the intruders to attack the system [12]. In Linux based systems, modules can be inserted through the utilities *insmod* or *modprobe*.

Kernel level rootkits occur in different forms. They affect system call table [10], Virtual File System(VFS)(for example, by *adore−ng* rootkit), Interrupt Descriptor Table(IDT) [15] .... VFS can be thought of as a kernel subsystem which provides a unified API to user-space applications. There are certain other rootkits that employ a combination of these techniques(example *zk* rootkit [11]).

When a user level program access a system resource, it is accomplished through a system call. The user level application performs a system call, which passes the control to the kernel. The requested work is done by the kernel and result is passed back to the user level application. So system call is an important target for attackers. Kernel level rootkits attack System Call table by different mechanisms [11]. Three of them are listed below

1. System Call Table modification
   In this method, attacker replaces original system call with his own custom version. Mainly this is done by modifying the system call address by inserting malicious LKM's address. Knark [6] rootkit uses this approach.
2. System Call Target modification
   In this attack, legitimate code in the target address is modified. This type of attack does not modify system call table. It works by modifying the flow

of control in a system call. Usually a jump instruction is used to pass the control to the malicious code.

3. System Call Table Redirection
   In this method attacker replaces the call to System Call Table with his own custom version. This is usually done by overwriting the memory where System Call Table address is stored. By comparing with *System.map* file this type of attacks of could easily be detected.

*Fig.*1 show the result of a System Call redirection where, the system calls *sys_write*, *sys_open*, *sys_getdents* are replaced with modified ones; *sys_rkwrite*, *sys_rkopen*, *sys_rkgetdents*.

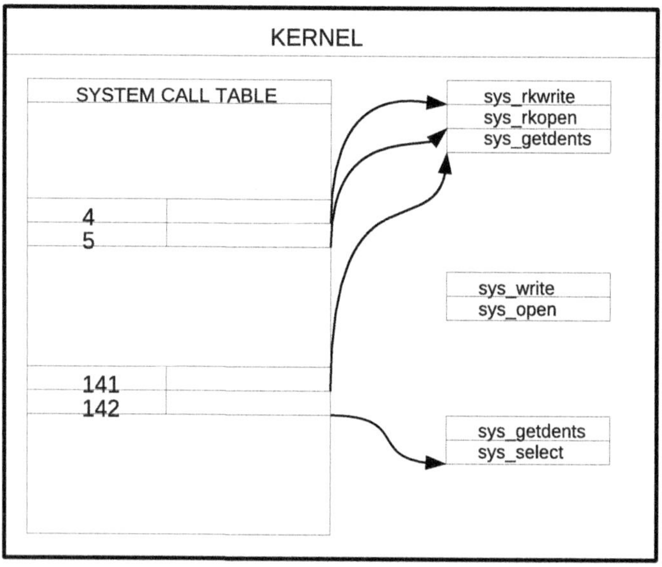


**Fig. 1.** System Call Redirection example

## 3 Detection Mechanism

Current techniques for detection mainly rely on archived copy of system files for detection [10]. But this method couldn't detect all type of attacks and also for the detection to work properly,the archived copy should be clean.

Because of its easiness and functionality attackers choose system calls as a target for modification. If kernel level rootkits are used, the method of using more than one application for getting an information doesn't work. For example we get information about the modules from both */proc/modules* and *lsmod* command. */proc* filesystem is a special filesystem that is used by the kernel to export information. Contents of this filesystem are created when they are read. Many utilities get their information form */proc/*. *lsmod* gets its information from */proc/modules*. If rootkit modifies */proc/modules* both */proc/modules* and *lsmod* command gives same information. So gathering information relying solely on user space applications doesn't help.

Cryptographic checksum based methods [11] fail when rootkits use dynamic directories to store rootkit related information. For example many rootkits rely on the */proc* directory(example *knark* rootkit) to store its contents.

Detection based on third party integrity monitors is catching up. For example Copilot[14] uses a coprocessor to check the integrity of kernel code. It uses a PCI-card to monitor the memory of the host system. Similar approach is followed by Gibraltar[3]. Like Copilot, Gibraltar executes on an observer machine, and monitors the execution of the target machine. In virtual machine[8] based rootkit detection, observer modules fetches data from the guest operating system and feds it to a rootkit detector, which can then analyze them. The need for third party mechanism makes these approaches less attractive.

Another method of importance is using cryptographically signed kernel modules. This method prevents loading unauthorized modules into the kernel space. Greg Kroah-Hartman proposed a method based on this using $RSA$ encryption to sign the modules [9]. But this method couldn't find much support from developer community because of its resistive nature.

### 3.1 Proposed Detection Mechanism

The proposed technique detects system call redirection attack which is a widely employed kernel level rootkit attack. System call redirection can be thought of as a special case of system call target modification attack. Proposed detection mechanism doesn't need prior information about the state of the system to function. The proposed system make use of statistical information from the infected system only.

An anomaly is any occurrence that is unusual, or unique. System call redirection attacks cause anomaly to the host system. System calls could be classified into either higher level system calls ( such as *sys_write*, *sys_read*) or lower level system calls ( those constituting VFS layer). Attackers modify higher layer calls since they provide easier mechanism for modifying the integrity of the system. During boot time higher level system calls are loaded first. So lower level system calls tend to have higher memory address as compared to higher layer calls.

Anomaly detection mechanism tries to find out patterns in data that do not conform to expected behavior. Creation of clusters could guide the system administrator in detecting these anomalies. Cluster based estimation can be utilized for effectively detecting the anomalies. Kernel level rootkit detection mechanism designed should need less a priori information about the host system. Cluster analysis being a unsupervised learning method acts as an efficient candidate for this purpose. Unlike other host based detection mechanisms clustering based method doesn't need prior knowledge for dividing the input data into different clusters. The proposed method uses euclidean distance as metric and variance is used as a measure of cluster scatter.

First step is to collect data, making using of the current state of the system. From this essential data are extracted. This involves memory address from system call table. Next step involves constructing clusters based on this data. Last step is analyzing the result to determine whether a rootkit attack occurred or not. The different steps involved are described in detail in the following paragraphs.

By default, Linux may access up to 4 Gigabytes of virtual memory, of which upper portion of this memory is allocated for use by the kernel (Address between $0xC0000000$ and $0xFFFFFFFF$). When a LKM based attack using new system call occurs, it gets a unusually high memory address as compared to other system calls. For the detection purpose the running kernel should be disassembled. As our aim is to detect system call related attack, we scan the system call table and the unconditional/conditional jump instructions are analyzed and their operands, which are memory addresses are collected. Along with this we collect system call addresses also. Attacker makes use of higher level system calls which are loaded earlier in memory for manipulation.

For computational purpose we convert these hexadecimal memory addresses to decimal values. We then do clustering on these decimal values and two clusters are formed with the sparse cluster representing the malicious system calls. These system calls are then checked whether they are higher level system calls. An uninfected system still shows some members in cluster two even though it is not affected by a rootkit attack. This is due to the presence of natural outliers present due to lower level system calls. Analysis of the cluster data could help to detect these false alarms.

## 4 Results and Discussion

We have evaluated the effectiveness of the proposed technique against Knark rootkit[13] and some custom implementations. Running kernel is disassembled[1] and memory addresses of system calls are retrieved for analysis. Based on this, clusters are formed to differentiate between normal and malicious system calls. Our method could detect all the system call modifications.

When tested on a clean system, some false positives are encountered. But they amount for less than 0.005% of the total analyzed instructions. When these false positives are analyzed, they turned out to be lower level system calls.

*Fig*.2 represents the distribution of system calls in a rootkit affected system. The peaks represents the modified system calls and they are detected by the clustering approach.

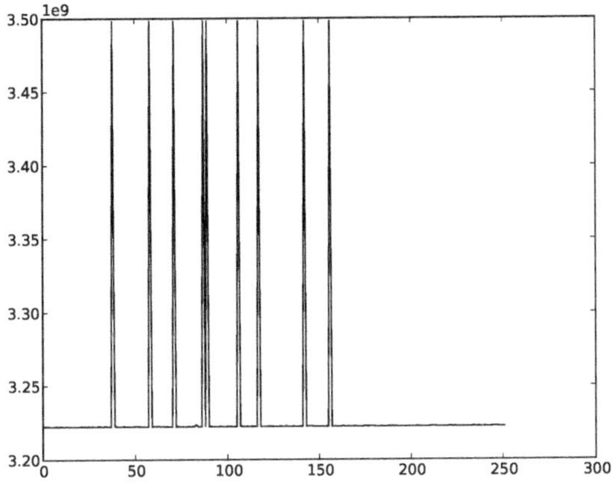


**Fig. 2.** System Call Distribution

## 5 Conclusion and Future Work

Residing in kernel makes kernel level rootkits difficult to detect. The proposed method provides an efficient host based technique for rootkit detection with reduced system overhead. Analysis could be done on a machine other than the host system. This makes the detection approach more flexible.

The presence of natural outliers is a bottleneck of this approach. Even when the system is not affected by rootkit, it will classify a very small amount of memory addresses as modified ones. This arises because of the presence of some low level system calls which are natural outliers. Right now attack based on low level system calls hasn't been found. So presence of this can easily be rejected as a false alarm.

Future work include testing the effectiveness of this method against system call table redirection attack and finding an efficient mechanism to contain the presence of false alarms in an uninfected system. As more and more news reports[2] coming out about the rootkit attack on smart phones, extending this method on smart phones could also be investigated.

## References

1. Gdb documentation (2011), `http://www.gnu.org/software/gdb/documentation/`
2. Arthur, C.: More than 50 android apps found infected with rootkit malware. Guardian Technology Blog (2011), `http://www.guardian.co.uk/technology/blog/2011/mar/02/android-market-apps-malware`
3. Baliga, A., Ganapathy, V., Iftode, L.: Detecting kernel-level rootkits using data structure invariants. IEEE Transactions on Dependable and Secure Computing 99(PrePrints) (2010)

4. Bickford, J., O'Hare, R., Baliga, A., Ganapathy, V., Iftode, L.: Rootkits on smart phones: attacks, implications and opportunities. In: Proceedings of the Eleventh Workshop on Mobile Computing Systems & Applications, pp. 49–54. ACM, New York (2010)
5. Bunten, A.: Unix and linux based rootkits techniques and countermeasures (2004)
6. Clemens, J.: Intrusion Detection FAQ: Knark: Linux Kernel Subversion (2001)
7. Desnos, A.: Detecting (and creating!) a hvm rootkit (aka bluepill-like). Journal in Computer Virology, 1–27 (2009), http://dx.doi.org/10.1007/s11416-009-0130-8
8. Garfinkel, T., Rosenblum, M.: A virtual machine introspection based architecture for intrusion detection. In: Proc. Network and Distributed Systems Security Symposium, vol. 1, pp. 253–285. Citeseer (2003)
9. Kroah-Hartman, G.: Signed kernel modules. Linux Journal (2004)
10. Levine, J., Grizzard, J., Owen, H.: A methodology to detect and characterize kernel level rootkit exploits involving redirection of the system call table. In: Proceedings Second IEEE International Information Assurance Workshop 2004, pp. 107–125. IEEE, Los Alamitos (2005)
11. Levine, J.G., Grizzard, J.B., Owen, H.L.: Detecting and categorizing kernel-level rootkits to aid future detection. IEEE Security and Privacy 4, 24 (2006), http://portal.acm.org/citation.cfm?id=1115691.1115761
12. Lineberry, A.: Malicious Code Injection via/dev/mem. Black Hat Europe (2009), http://www.blackhat.com/presentations/bh-europe-09/Lineberry/BlackHat-Europe-2009-Lineberry-code-injection-via-dev-mem.pdf
13. Miller, T.: Analysis of the KNARK Rootkit (2004)
14. Petroni Jr., N.L., Fraser, T., Molina, J., Arbaugh, W.A.: Copilot - a coprocessor-based kernel runtime integrity monitor. In: Proceedings of the 13th Conference on USENIX Security Symposium, SSYM 2004, vol. 13, p. 13. USENIX Association, Berkeley (2004), http://portal.acm.org/citation.cfm?id=1251375.1251388
15. Wang, Z., Jiang, X., Cui, W., Ning, P.: Countering kernel rootkits with lightweight hook protection. In: Proceedings of the 16th ACM Conference on Computer and Communications Security, CCS 2009, pp. 545–554. ACM, New York (2009), http://doi.acm.org/10.1145/1653662.1653728
16. Yan, Q., Li, Y., Li, T., Deng, R.: Insights into Malware Detection and Prevention on Mobile Phones. Security Technology, 242–249 (2009)

# Fuzzy Clustering for Effective Customer Relationship Management in Telecom Industry

Gayathri Asokan and S. Mohanavalli

Department of Information Technology,
Sri Sivasubramaniya Nadar College of Engineering,
Kalavakkam, Chennai, India 603 110
gayathrissnme@gmail.com, mohanas@ssn.edu.in

**Abstract.** Data mining is the process of extracting interesting patterns from data. Data mining is recently proving very effective in business decision making and is becoming a widely used strategy to improve CRM (Customer Relationship Management). CRM is the process of managing a good relationship with customer and improving the profitability of their interactions with the customer. Data mining is widely used particularly in handling large data sets as in telecom sector. Clustering is a popular mining strategy that separates those data into subsets called clusters. This research work focuses on comparing the two main approaches of clustering soft clustering and hard clustering namely the Kmeans and Fuzzy C Means (FCM) clustering algorithms on large telecom data to determine the churn ratio as a measure to enhance CRM. It is observed that FCM outperforms Kmeans in estimating churn ratio accurately and is more effective in supporting CRM.

**Keywords:** Data mining; Customer Relationship Management, Clustering, Kmeans, Fuzzy C Means.

## 1 Introduction

With the globalization of world economy and market internationalization, the market competition of the domestic Telecommunications industry has become more reasonable and increasingly fierce. The fast-growing user group, diversified operations and competition environment have put forward higher requirements for the service quality of the Telecom industry. The Telecom industry offers new products and technologies that creates a series of options and offers to customer in order to satisfy their needs and requirements. Loss of valuable customers in a telecom company is a more general problem that telecom company often faces. Retaining existing clients is the best marketing strategy to survive in industry and it is more profitable to keep and satisfy existing clients than to constantly attract new ones. Churn management, as the general concept of identifying those clients most prone to switching to another company, led to development of variety of techniques and models for churn prediction. So it is must for the telecom company to classify their customers into churn and non-churn customers [4].

D. Nagamalai, E. Renault, M. Dhanushkodi (Eds.): CCSEIT 2011, CCIS 204, pp. 571–580, 2011.

To maximize the reduction in customer loss in the face of the fierce market competition is to use the most effective and scientific method which is data mining technology, and is currently the most powerful analytical tool. Nowadays data mining is mostly used by corporate world to extract useful information from large amount of databases. Data mining technology uses the known data to find implicit business rules by establishing a mathematical model. Data mining technology based on rigorous mathematical calculations can completely change the past status that mobile enterprise can neither monitor the loss of customers nor achieve customer care after successfully acquiring customers. The application of data mining to customer relationship management can help telecom companies win new customers in the mass market, allow existing customers to create more profits, keep valuable customers, offer their customers more personalized services, etc, and the enterprise in a more favourable competitive position. Data mining technology helps telecom companies to select the right prospects of whom to focus and to identify the existing customers and the good customers who may about to leave the company [6].

CRM is one of the most important of the information technology and is an integration of business and technology process to gain the customer satisfaction and used to satisfy the need of a customer. CRM basically begins with the deep analysis of customer behavior. A telecom company that wants to create personal interaction with the customer then they must gather all the customer related data and using this data they will predict the churn customers.CRM helps telecom companies to increase the profitability of their interactions with the customer while at the same time making the interactions more friendlier through individualization. Retaining the customers is important for telecom company to identify the problem of their customers and to know about the relationship between customer and organization [1, 11].

Everyday people encounter a large amount of information. One of the vital means in dealing with these data is to classify them into set of categories or clusters. To classify the customers into churn and non-churn manually by having this large number of databases is very complex than by doing data mining algorithms technically. Classification systems are either supervised or unsupervised. We chose one of the data mining techniques that comes under unsupervised learning which groups the customer based on their similarity which is called clustering technique. Clustering is a statistical technique used to identify both minimize within group variation and maximize between group variation based on distance or dissimilarity and to find an optimal set of clusters. A large number of clustering algorithms have been developed for different type of applications. To cluster the massive amount of data in telecom industries, we proposed Kmeans clustering (hard clustering) and fuzzy clustering (soft clustering) from various type of clustering techniques. By using this both clustering techniques, we can group the telecom customers into churn and non-churn clusters, so that necessary steps will be taken before the churn customer switch to a competitor[8].

## 2 Background Work

Telecommunication companies maintain an enormous amount of information about their customers and, due to an extremely competitive environment, have great motivation for exploiting this information. For these reasons the telecommunications industry has been a leader in the use of data mining to identify customers, retain customers, and maximize the profit obtained from each customer. A key component of modeling a telecommunication customer's value is estimating how long they will remain with their current carrier. This problem is of interest in its own right since if a company can predict when a customer is likely to leave, it can take proactive steps to retain the customer. The process of a customer leaving a company is referred to as churn, and churn analysis involves building a model of customer attrition. Customer churn is a huge issue in the telecommunication industry where, until recently, telecommunication companies routinely offered large cash incentives for customers to switch carriers [13].

Data mining technique becomes gradually increasing and important tool for telecom industry. The various techniques involved in data mining is classification, estimation, clustering, association and prediction. Clustering is generally used for market segmentation and classification. Clustering separates a heterogeneous population into homogeneous subgroups or clusters so that data in each cluster share some common trait. So it is more efficient for using this clustering technique in telecom company to group their customers according to the services used by them. In paper [16], the research workers used RFM (Recency, Frequency, Money) rule to analyze the customer value. By using this RFM rule prediction of the customer characteristics can be done. Yi-Hui-Lang defined customer value is fundamental to CRM. It is a starting point of CRM to understand and measure the true value of a customer. In paper[2], the author finds the value of a customer using Kmeans clustering algorithm based on RFM analysis and assessing the value of existing customer. So by using Kmeans algorithm, author can able to classify the customers according to their similarity level[3].

In crisp clustering, each data point in the sample space assigned to only one cluster. We can able to assign each and every customer to a particular service based, in which a particular service is well used by a set of customers rather than other services. The k in the Kmeans algorithm stands for the number of cluster seeds initially provided to the algorithm. The goal of this algorithm is the assignment of each customer into one and only service. Kmeans is based on the principle which minimizes the clustering property indicator. A commonly used clustering criterion function is the minimized variance of all the same points (data or objects) to the central point in the cluster. Kmeans is a prototype-based, simple partition clustering technique which attempt to find an user specified k number of clusters. These clusters represented by their centroids, which is typically the mean of the points in the cluster [5, 9].

Hard clustering technique identifies each customer to a certain service category, with the nature of "either-or", so this type of clustering is obvious. But in the estimation of customers in telecom company is unpredictable. The usage level may vary day by day and their usage pattern is an intermediary, with the nature of "and also", so in these sample uncertain situations, soft clustering called fuzzy clustering can be used efficiently and easily. Fuzzy Clustering provides a robust and resilient method of classifying collections of large amount of (data points) customers by allowing the same customer to reside in multiple service groups (clusters) with different degrees of membership. In fuzzy clustering, a data point can have a membership in multiple clusters (each of different degree).Membership is a measure of how strongly a data point is a member of a particular cluster [14, 10].

Goal of cluster analysis is the recognition and quantification of the groupings. The quantification involves 2 processes (identifying membership of the data point and locating the center of cluster). Fuzzy clustering can able to handle extreme outliers, either very small or a higher than warranted membership degree in surrounding cluster. It provides a flexible and robust method of assigning data points into clusters. Each data point will have a degree of membership for each clusters in the range [0, 1] indicating the strength of its placement in that cluster [3, 12].

# 3 Proposed System Design

## 3.1 Kmeans Clustering

The k in the Kmeans algorithm stands for the number of cluster seeds initially provided to the algorithm. The goal of the hard Kmeans algorithm is the assignment of each data point into one and only one cluster. The cluster centers are moved according to a nearness or similarity function defining the distance from a cluster center to the data point. The crisp (or hard) Kmeans algorithm itself is fairly straightforward, involving the basic steps that are listed below [5, 7].

1 (Initialization) Select a set of k starting points. The selection may be done in a random manner or according to some heuristic.
2 (Distance Calculation) For each data point Xi,1 $\leq$ i $\leq$ n, compute its Euclidean distance to each cluster centroid Cj , 1 $\leq$ j $\leq$ k, and then find the closest cluster centroid .

$$d_{ji} = \sqrt{\Sigma \ (x_i - c_j)^2}$$

3 (Centroid Recalculation) For each 1 $\leq$ j $\leq$ k, recompute cluster centroid Cj as the average of data points assigned to it.
4 (Convergence Condition) Repeat steps 2 and 3, until convergence.

The Kmeans algorithm belongs to a family of crisp (or hard) clustering techniques that uniquely partitions the sample data points into disjoint clusters. The Euclidean distance metric measures how far away a point is from a cluster center, and thus points that share common attributes will naturally cohere into the same cluster. All is well and good if the clusters are distinctly arranged in the n-dimensional sample space. In practice (that is, in reality), however, cluster perimeters are often quite ambiguous. Data points in the sample space share attributes from one or more clusters and should thus be placed in more than one cluster (albeit with a degree of membership in one cluster that is higher or lower than the degree of membership in another).

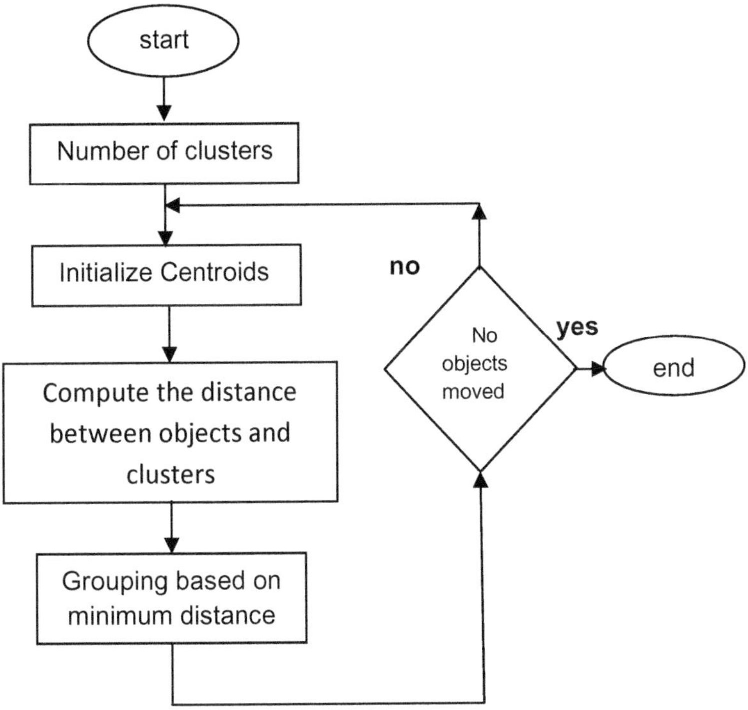


**Fig. 1.** Flow diagram of Kmeans algorithm

### 3.2 Fuzzy C Means Clustering

Based on the fuzzy set theory, firstly introduced by Zadeh in 1965. and on the concept of membership functions, the fuzzy clustering methods have been developed. In fuzzy clustering entities are allowed to belong to many clusters with different degrees of membership.The FCM Algorithm is a modified version of an original crisp cluster methodology. The idea of fuzzification parameter (m) in the range [1, n], which determines the degree of fuzziness on the cluster. When m > 1 the degree of fuzziness among points in the decision space increases, when m=1, the clouds do not overlap, but as you increase the value (m=1.25) the clouds begin to overlap and share many of the same points [5, 7].

The steps involved in FCM algorithm are as follows:

1 (Initialization) Randomly initialize the cluster center.

2 (Distance Calculation) Creating distance matrix from a point xi to each of the cluster centers to with taking the Euclidean distance between the point and the cluster center.

$$d_{ji} = \sqrt{\Sigma \ (x_i - c_j)^2}$$

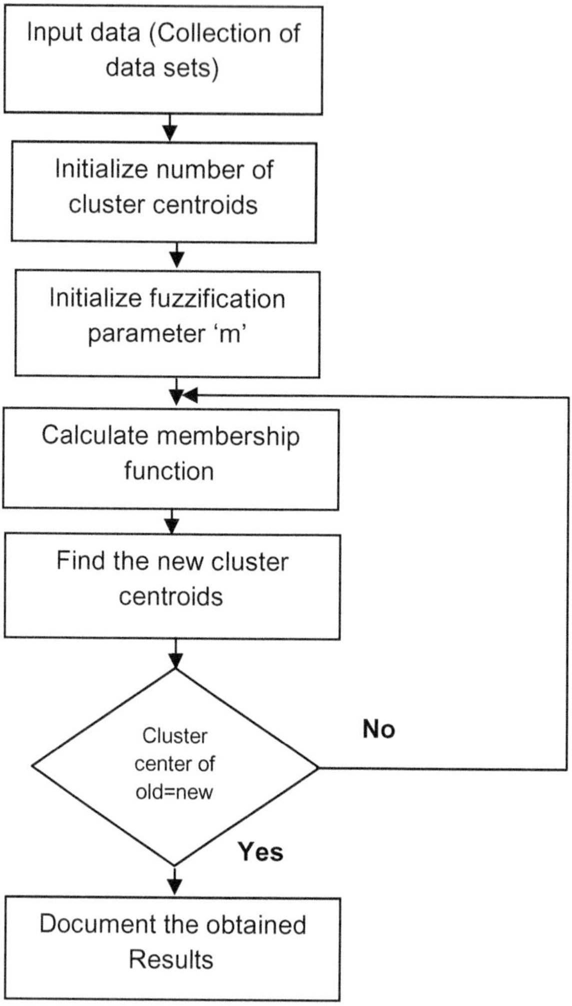


**Fig. 2.** Flow Diagram for FCM Algorithm

3 (Membership calculation) Finding Membership Calculation: takes the fractional distance from the point to the cluster center and makes this a fuzzy measurement by raising the fraction to the inverse fuzzification parameter.

This is divided by the sum of all fractional distances, thereby ensuring that the sum of all memberships is 1.

$$\mu_j(x_i) = \frac{\left(\frac{1}{d_{ji}}\right)^{\frac{1}{m-1}}}{\sum_{k=1}^{p}\left(\frac{1}{d_{ki}}\right)^{\frac{1}{m-1}}}$$

4 (Centroid Recalculation) For each $1 \leq j \leq k$, recompute cluster centroid Cj as the average of data points assigned to it.

$$C_j = \frac{\sum_i \left[\mu_j(x_i)\right]^m x_i}{\sum_i \left[\mu_j(x_i)\right]^m}$$

5 (Convergence Condition) Repeat steps 2, 3 and 4 , until convergence.

### 3.3 Advantages of Soft Clustering over Hard Clustering

When compared to Kmeans, FCM gives more accurate and efficient clustering results. Kmeans performs a crisp clustering and allows each data point to belong to only one cluster based on the Euclidean distance calculation. In FCM the fuzziness is resolved using the membership value and is more accurate than hard clustering.

## 4 Implementation and Experimental Results

The original data was retrieved from Duke University's churn management unit. The data contains 65000 records (i.e. customers) and 144 attributes. The attributes were of two types: categorical and continuous. Some of the categorical attributes such as name, age of the customer, area where the customer resides etc. seemed to be irrelevant to the problem under consideration. Hence the categorical attributes were eliminated. Some of the customer records were with lot of missing attributes and such records were ignored. Some of the continuous valued attributes like call duration, number of calls in a month, number of messages etc. which were relevant to the problem under study. Hence the customer records with the relevant attributes were taken as input data. Finally an input data set with 1171 records and the most relevant 78 attributes was obtained after applying a data filtering technique.

The preprocessed input dataset wis used in both the hard clustering (Kmeans) and soft clustering (FCM) algorithms in order to find the churn rate of customers in the telecom company accurately.The implemented results are shown as follows:

**Input given**

Number of Customers = 1171, Number of services = 78
Number of Clusters = 2,
Algorithm used = Kmeans & FCM

It was observed that Kmeans clustering with k = 2, resulted 936 customers in group A and 235 customers group B. On the other hand FCM clustering algorithm resulted in 919 customers in group A and 251 customers in group B.

The results of Kmeans algorithm was analysed and it was found that there were some misclassifications. On analyzing the results of the fuzzy c means algorithm, the misclassification was considerably reduced. The attributes which were significant in producing the maximum deviations in the churn cluster were determined.

| | A | B | C | D | E | F | G | H | I | J | K | L | M | N | O | P | Q | R | S | T | U |
|---|---|---|---|---|---|---|---|---|---|---|---|---|---|---|---|---|---|---|---|---|---|
| 1 | label | rev_Mean | mou_Mea | totmrc_M | da_Mean | rev_Rang | mou_Ran | change_m | change_re | drop_vce_ | blck_vce_ | unan_vce | plcd_vce_ | recv_vce_ | comp_vce | cc_mou_N | inonemin | mou_cvce | mou_rvce | owylis_vc | mouov |
| 2 | I1 | 23.9975 | 219.25 | 22.5 | 0.2475 | 25.99 | 244 | -157.25 | -18.9975 | 0.666667 | 0.666667 | 6.333333 | 52.33333 | 42.33333 | 45 | 0 | 18 | 90.64333 | 97.17667 | 0 | |
| 3 | I2 | 16.99 | 10.25 | 16.99 | 0 | 0 | 10 | -4.25 | 0 | 0.333333 | 0 | 2.666667 | 9 | 0.333333 | 6 | 0 | 0.333333 | 5.426667 | 0 | 0 | |
| 4 | I3 | 55.23 | 570.5 | 71.98 | 0 | 0 | 96 | 38.5 | 0 | 9.666667 | 0.666667 | 77 | 222.3333 | 94.66667 | 137 | 11.07067 | 66 | 285.2333 | 106.33 | 14.66667 | 10.81 |
| 5 | I4 | 82.275 | 1312.25 | 75 | 1.2375 | 15.42 | 362 | 156.75 | 8.145 | 52 | 7.666667 | 76 | 702 | 365 | 577.3333 | 7.753333 | 301.6667 | 369.5367 | 200.32 | 370.3333 | 243.2 |
| 6 | I5 | 17.145 | 0 | 16.99 | 0 | 0.31 | 0 | 0 | -0.155 | 0 | 0 | 0 | 0 | 0 | 0 | 0 | 0 | 0 | 0 | 0 | |
| 7 | I6 | 38.0525 | 682.5 | 52.49 | 0.2475 | 10 | 370 | 147.5 | -3.0625 | 9 | 1.666667 | 13 | 97 | 13 | 73.33333 | 5.276667 | 5 | 470.0233 | 42.19667 | 6.666667 | 6 |
| 8 | I7 | 97.3375 | 1039 | 50 | 4.95 | 46.32 | 415 | 198 | 23.2625 | 12.66667 | 3 | 176 | 533.6667 | 96 | 346.6667 | 6.25 | 45 | 460.92 | 213.8533 | 103.3333 | 54.71 |
| 9 | I8 | 31.6625 | 25.5 | 29.99 | 0.2475 | 5.7 | 83 | 59.5 | 4.0275 | 0 | 1 | 2.333333 | 6.666667 | 0 | 3.333333 | 0 | 0 | 3.8 | 0 | 0 | |
| 10 | I9 | 62.1275 | 97.5 | 65.985 | 2.475 | 30.26 | 217 | 23.5 | 6.8225 | 0 | 0.333333 | 4 | 19.33333 | 0 | 15 | 96.92 | 0 | 136.4933 | 0 | 3.666667 | 15.36 |
| 11 | I10 | 35.295 | 24.25 | 34.98 | 0 | 0.63 | 39 | 19.75 | -0.315 | 0 | 0 | 1 | 9 | 3 | 8 | 0 | 2.333333 | 5.533333 | 2.386667 | 4 | 2.926 |
| 12 | I11 | 81 | 1056.25 | 75 | 0 | 5.75 | 233 | 42.75 | 2.35 | 0 | 0 | 0 | 0 | 0 | 0 | 0 | 0 | 0 | 0 | 0 | |
| 13 | I12 | 25.225 | 2.5 | 25 | 0 | 0.45 | 8 | -2.5 | -0.225 | 0 | 0 | 0.333333 | 1.333333 | 0.333333 | 1 | 0 | 0 | 1.306667 | 1.133333 | 0.333333 | 0.023 |
| 14 | I13 | 212.515 | 1971.5 | 84.99 | 2.2275 | 283.17 | 658 | -200.5 | -106.765 | 9 | 0 | 43.66667 | 344.6667 | 271.3333 | 296 | 3.316667 | 126.3333 | 581.2967 | 718.07 | 49.33333 | 35.16 |
| 15 | I14 | 42.565 | 270.5 | 37.48 | 0.2475 | 5.46 | 369 | 201.5 | -1.785 | 3.333333 | 1.666667 | 7.666667 | 67.66667 | 25.33333 | 55.33333 | 1.026667 | 11.66667 | 117.8467 | 57.09333 | 11 | 10 |
| 16 | I15 | 63.0175 | 440.25 | 59.98 | 0 | 8.9 | 89 | -16.25 | 0.2125 | 5 | 0 | 17.33333 | 138.3333 | 53 | 118 | 8.27 | 22 | 225.0533 | 93.87667 | 0 | |
| 17 | I16 | 35.59 | 153 | 29.99 | 0 | 12.95 | 58 | 30 | 7.35 | 2 | 0.666667 | 17.33333 | 53 | 26 | 33.33333 | 0 | 14.66667 | 50.13333 | 33.10667 | 8 | 10.94 |
| 18 | I17 | 55.27 | 1212.75 | 49.99 | 0.7425 | 5.07 | 311 | 169.25 | 0.99 | 2.666667 | 6 | 43.66667 | 225 | 121.6667 | 173.6667 | 7.5 | 43 | 456.98 | 490.1267 | 50.66667 | 50.00 |
| 19 | I18 | 50.965 | 162 | 69.99 | 0 | 3.9 | 88 | -58 | -0.975 | 1.666667 | 0.333333 | 9 | 60.66667 | 14 | 50 | 0.276667 | 7 | 65.24667 | 38.37333 | 34.33333 | 17.40 |
| 20 | I19 | 172.44 | 1978 | 99.99 | 0 | 191.8 | 1797 | -1007 | -72.45 | 7.333333 | 18 | 114.3333 | 760.3333 | 157.6667 | 622 | 0.94 | 69 | 1397.513 | 515.2333 | 22.66667 | 22.08 |
| 21 | I20 | 60.185 | 1424.5 | 56.25 | 0.2475 | 74.24 | 1698 | -1134.5 | -55.185 | 23.66667 | 0 | 11.66667 | 47.33333 | 646 | 33.33333 | 0.593333 | 285.3333 | 44.04 | 939.3467 | 7.666667 | 11.76 |
| 22 | I21 | 30.205 | 360 | 29.99 | 0 | 0.6 | 240 | -105 | -0.215 | 8.666667 | 2.333333 | 19.33333 | 95.66667 | 75 | 68.33333 | 0 | 36.66667 | 103.79 | 160.8567 | 34.33333 | 33 |
| 23 | I22 | 25.4925 | 299.5 | 24.99 | 0 | 1.56 | 139 | -47.5 | -0.5025 | 2 | 0.666667 | 14.33333 | 74 | 47 | 57.66667 | 1.2 | 25.66667 | 174.25 | 77.48667 | 6.333333 | 7 |
| 24 | I23 | 98.4225 | 608 | 94.99 | 1.2375 | 33 | 768 | -563 | -12.4425 | 8.666667 | 0.333333 | 70.66667 | 273.3333 | 55 | 194.6667 | 0 | 34.33333 | 387.3467 | 57.21 | 24.66667 | 28.87 |
| 25 | I24 | 24.49 | 41.75 | 16.99 | 0 | 41.34 | 120 | 81.25 | 29.93 | 0 | 0 | 0 | 3.333333 | 0.333333 | 3.333333 | 0 | 0.333333 | 5.74 | 0 | 0 | |

fm1170r70c Sheet1 Sheet2

**Fig. 3.** Input dataset

## Misplaced Customers

44,89,184,212,273,283,320,327,354,421,429,520,736,977,1012.

| Output | | Kmeans | | | Fuzzy c means | | | | |
|---|---|---|---|---|---|---|---|---|---|
| S.No | Data point | Distance cluster 0 Group A | Distance cluster 1 Group B | Data point grouped into | Distance cluster 0 Group A | Distance cluster 1 Group B | Memb ership cluster 0 Group A | Memb ership cluster 1 Group B | Data point grouped into |
| 1 | 44 | 32407.88 | 32991.86 | Group A | 33063.22 | 30207.88 | 0.4135 | 0.5865 | Group B |
| 2 | 89 | 32249.82 | 32915.89 | Group A | 32906.74 | 30123.35 | 0.4074 | 0.5926 | Group B |
| 3 | 184 | 31340.78 | 33749.18 | Group A | 31996.78 | 30959.58 | 0.4572 | 0.5428 | Group B |
| 4 | 212 | 32001.71 | 33082.59 | Group A | 32657.69 | 30292.84 | 0.4223 | 0.5777 | Group B |
| 5 | 273 | 32461.18 | 32739.54 | Group A | 33114.35 | 29960.80 | 0.4051 | 0.5949 | Group B |
| 6 | 283 | 32051.72 | 33621.89 | Group A | 32706.56 | 30843.45 | 0.4511 | 0.5489 | Group B |
| 7 | 320 | 31226.38 | 33857.52 | Group A | 31882.68 | 31066.35 | 0.4801 | 0.5199 | Group B |
| 8 | 327 | 31442.97 | 33666.38 | Group A | 32097.67 | 30880.50 | 0.4507 | 0.5493 | Group B |
| 9 | 354 | 31507.27 | 34173.93 | Group A | 32152.51 | 31422.85 | 0.4807 | 0.5193 | Group B |
| 10 | 421 | 31226.38 | 33857.52 | Group A | 39905.88 | 23050.30 | 0.4801 | 0.5199 | Group B |
| 11 | 429 | 31442.97 | 33666.38 | Group A | 32097.67 | 30880.50 | 0.4507 | 0.5493 | Group B |
| 12 | 520 | 31463.60 | 33617.89 | Group A | 32119.92 | 30827.15 | 0.4695 | 0.5305 | Group B |
| 13 | 736 | 32280.75 | 33216.95 | Group A | 32936.43 | 30434.52 | 0.4441 | 0.5559 | Group B |
| 14 | 977 | 31166.30 | 34174.98 | Group A | 31816.38 | 31404.95 | 0.4966 | 0.5034 | Group B |
| 15 | 1012 | 32021.02 | 33168.20 | Group A | 32677.46 | 30379.07 | 0.4314 | 0.5686 | Group B |

**Comparison of both the results:**

From the comparison of clustering techniques, customer churn ratio was observed to be different as per the clustering result. FCM groups the customers based on the membership value. The membership value helps in determining the accurate churn ratio of customers moving from one group to other group. The result of soft clustering based on membership value was more accurate than Euclidean distance measure used in hard clustering because the membership value in FCM clustering resolves the fuzziness in the dataset, if existent. From the experimental results it was observed that the misclassification was reduced in FCM and it resulted in estimating the churn ratio accurately.

**Conclusion and Future Work**

It was observed that FCM gives better support to CRM rather than Kmeans in clustering telecom data. It is further proposed to improve the scalability of fuzzy algorithm to handle large real time telecom data and also to develop parallel versions of the FCM to improve the efficiency.

## References


1. Al-Mudimigh, A.S., Saleem, F., Ullah, Z., Al- Aboud, F.N.: Implementation of Data Mining Engine on CRM – Improve Customer Satisfaction. In: International Conference on Information and Communication Technologies, pp. 193–197. IEEE, Los Alamitos (2009)
2. Thomas, B., Raju, G.: A Novel Fuzzy Clustering Method for outlier Detection in Data Mining. International Journal of Recent Trends in Engineering 1(2) (May 2009)
3. Dhillon, I., Modha, D.: A Data clustering algorithm on distributed memory multiprocessors. In: Zaki, M.J., Ho, C.-T. (eds.) KDD 1999. LNCS (LNAI), vol. 1759, pp. 245–260. Springer, Heidelberg (2000)
4. Popovic, D., Basic, B.D.: Churn Prediction Model in Retail Banking Using Fuzzy C-Means Algorithm. Informatica 33 Public Sector Solutions, 243–247 (2009)
5. Cox, E.: Fuzzy modeling and genetic algorithms for Data Mining and Exploration, ch. 7, pp. 207–206
6. Wies, G.M.: Data mining in Telecommunication Industry, pp. 486–491. International Publisher of Science and Technology, IGI Global (2009)
7. Gayathri, A., Mohanavalli, S.: Enhanced Customer Relationship Management using Fuzzy Clustering. In: International Conference on Computer Science and Engineering Technology (IJCSET), vol. (7), pp. 163–167 (May 2011)
8. Gu, G., He, Y.: Research on the application of Data mining to CRM in Mobile Communication. IEEE, Los Alamitos (2010)
9. Han, J., Kamber, M.: Data Mining: Concepts and Techniques, 2nd edn. Morgan Kaufmanm, San Francisco (2006)
10. Zou, K., Wang, Z., Hu, M.: An new initialization method for fuzzy c- means algorithm. Journal of Fuzzy Optimization and Decision Making 7(4) (December 2008)
11. Levine, S.: The rise of CRM. America's Network 104(6), 34 (2000)
12. Cannon, R.L., Dave, J.V., Bezdek, J.C.: Efficient Implementation of the Fuzzy c-Means Clustering Algorithms. IEEE Transactions on Pattern Analysis and Machine Intelligence PAMI-8(2) (March 1986)

13. Chen, R.-S., Wu, R.-c., Chen, J.Y.: Data Mining Application in Customer Relationship Management of Credit Card Business. In: Proceedings of the 29th Annual International Computer Software and Applications Conference. IEEE, Los Alamitos (2005)
14. Selim, S.Z., Ismail, M.A.: Kmean type Algorithms: A Generalized convergence Theorem and characterization of Local optimality. IEEE Transactions on Pattern Analysis and Machine Intelligence 6, 81–87 (1984)
15. Rashid, T.: Classification of churn and non-churn customers for Telecom companies. International Journal of Biometrics and Bioinformatics (IJBB) 3(5)
16. Liang, Y.-H.: Combining the Kmeans and decision tree methods to promote customer value for the Automotive Maintenance Industry. In: Proceedings of IEEM. IEEE, Los Alamitos (2009)

# Dematerialized Deposits Using XSI

V. Radhaa[1,*], Surabhi Upender[1,2],
N.V.K.D. Ramesh Gorrila[1,2], and B. Vijay Kumar[1,2]

[1] Institute for Development and Research in Banking Technology,
Road #1, Castle Hills, Masab Tank, Hyderabad – 500 067 (A.P), India
Tel.: 91-040-23534981-85; Fax: 91-40-23535157
[2] Department of Computer and Information Sciences,
University of Hyderabad, Hyderabad, 500 046 (A.P), India
vradha@idrbt.ac.in, surabhi.upender@gmail.com,
gnvkdramesh@gmail.com, vijaykumar.hcu@gmail.com

**Abstract.** Traditional banking was based on paper documents like 'deposit form', 'withdrawal form' etc. The comfort level for the customer on paper based business was high as they can retain the paper and show as proof in case of disputes. The present Internet and Core banking systems which are widely in use are transaction based. One drawback of the present situation is that customer is not able to keep a copy of the transaction due to various reasons like multiple channels and lack of suitable infrastructure at client end etc. As a via media, we tried to bridge the gap and see whether we can give the comfort and trust levels of the paper document coupled with the intelligence of the technology.

In this model we propose a method to describe financial product like "Fixed Deposits" using XML technology. At present, banks give the deposits in paper form with the deposit amount, maturity date and interest rate prescribed in the document. The banks system calculates the interest and pays the customer. The paper itself can't calculate interest. It just helps the customer as a proof alone. In our model, we are trying to see whether we can embed the interest calculation algorithm into XML based deposit and give to the customer, so that customer himself can find how much he gets using tools like XSI. This will act as a good verification mechanism, instead of solely depending on banks' systems. To show case this model as a POC, we used XSD, XML and XSI (an in house tool with IDRBT).

**Keywords:** XSD, XML, Fixed Deposit, GUI , XSI.

## 1 Introduction

### Fixed Deposit

Fixed deposit is a money deposit at a banking institution that cannot be withdrawn for a certain "term" or period of time. When the term is over it can be withdrawn or it can be held for another term.

---

* Corresponding author.

D. Nagamalai, E. Renault, M. Dhanushkodi (Eds.): CCSEIT 2011, CCIS 204, pp. 581–588, 2011.

Interest rate is paid on deposit accounts by commercial banks and other depository financial institutions. Fixed interest rate is the rate which is fixed before the investment of money in the bank and cannot be changed until the maturity of the deposit is over.

**Introduction to XSI**

The XSI tool is under development in IDRBT. The XSI dynamically generates GUI for the customer from composite XML schema document. Customer can enter the data into the GUI and then XSI generates an XML document conforming to the given XSD.

**Taking input.** XSI can create new XML documents or edit existing XML documents. For new documents, XSI takes schema as input. For editing, XSI takes XML instance documents as input.

**GUI generator.** This module presents the user with GUI for the user to enter the data.

**Validator.** In this module user entered data is verified against its corresponding XSD document. Users would be alerted for appropriate corrections, in case the data is not in conformance with the XSD.

**XML document generator.** This module of XSI generates XML document by taking the XSD document as the input and user entered data.

**Save/Send the document.** The ultimate goal of XSI is to send the document to establish good communication between customer and the bank. The generated XML is either saved locally or sent through the network.

The following diagram gives the complete idea about the work flow of XSI.

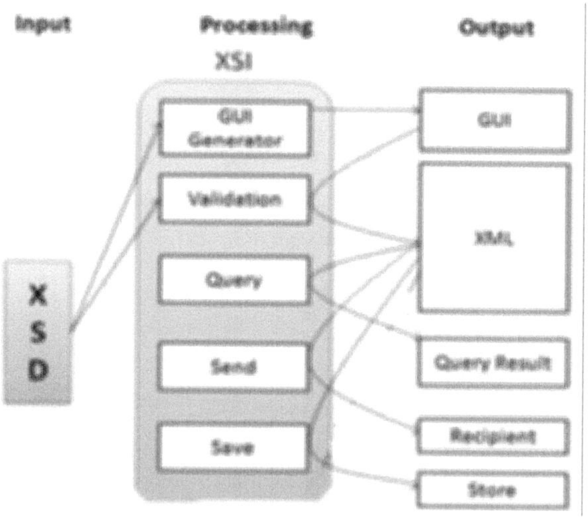


**Fig. 1.** XSI – Conceptual View

## 2 Application Overview

We developed the prototype using XML, XSD and XSI. In this section, we explain the prototype components and work flow of the application.

### 2.1 Relationship among Schema, Advertisement and Deposit Document

- There can be one fixed deposit XSD structure to define the Deposit format.
- However, the interest rates keep changing and banks advertise different rates at different points of time, based on different criteria like Principal amount, term duration etc. So, each advertisement has to be represented as an XML document.
- There can be many customers depositing money against each advertisement and so there can be many deposit instances which are again in the form of XML.

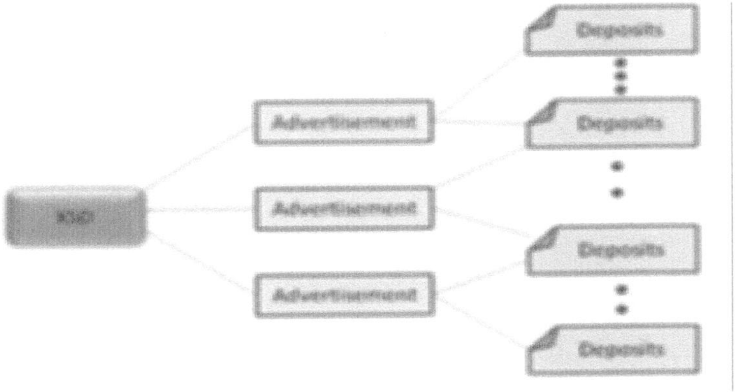

**Fig. 2.** M-M Relationship among schema, advertisement and deposit document

### 2.2 The Application Components

The application crosses three boundaries in the entire work flow. They are 1. Bank's Server, 2. Bank's Teller and, 3.Bank's Customer.

#### 2.2.1 Bank's Server

The Bank's server hosts XML and XSD files which contain all the information about bank deposit forms as well as their structure. The banks display advertisements which show FD-Fixed Deposit interests from time to time. These advertisements are coded by using XML and XSD.

#### 2.2.2 Bank's Teller

When the customer approaches the bank's teller machine, it issues the fixed deposit certificates to the customers. The bank teller contacts with the bank server to get the details in the form of XSD and XML and then fills the deposit with the customer details. Final deposit certificate is generated in the form of XML. Teller sends one copy of this XML to bank server and another to customer.

### 2.2.3 Bank's Customer

The customer receives the final deposit certificate in XML format. The customer can verify the deposit certificate at any point of time using XSI at his/her end.

## 2.3 Work Flow of the Model

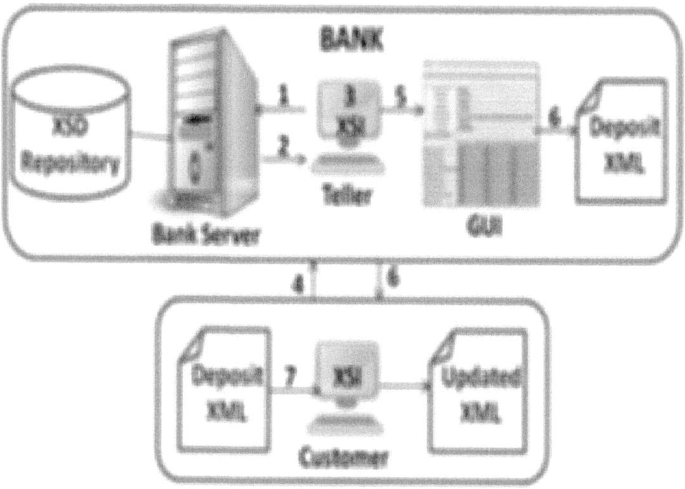


**Fig. 3.** XSI Deposit Certificate model

1. The teller sends request to bank's server to get appropriate XSD/XML advertisement for the customer requirement.

| PERIOD | INTEREST RATES ON DOMESTIC DEPOSITS (%) | | | | |
|---|---|---|---|---|---|
| DEPOSITS | Interest Rate | | | Interest Rates for Senior citizens | |
| | on Deposits Below Rs 15 lakhs | on Deposits of Rs 15 lakhs < Rs 50 Lakhs | on Deposits of Rs 50 lakhs < Rs 1 Crore | on Deposits below Rs 50 lakhs | on Deposits of Rs 50 lakhs < Rs 5 Crore |
| 9 months to less than 1 year | 6.00 | 6.00 | 6.00 | 6.75 | 6.75 |
| 1 year to less than 2 Years | 6.50 | 6.50 | 6.25 | 7.25 | 7.10 |
| 2 years to less than 3 years | 7.10 | 7.10 | 7.10 | 7.85 | 7.85 |

**Fig. 4.** Fixed deposit Advertisement

2. Bank server provides required XSD/XML advertisement to teller. These advertisements are coded by using XML and XSD.

3. The XSI generates GUI for the teller to enter customer specific details.
4. The customer approaches the bank and provides details for taking the fixed deposit at the teller. Deposit advertisement is created in GUI using deposit XSD.

```
<Deposit schemaLocation="D:\Deposit\Deposit.xsd">
    <include schemaLocation="D:\Deposit\Deposit.xsd"/>
    <Axisbank>
        <Customer_Name></Customer_Name>
        <Customer_id></Customer_id>
        <Customer_address></Customer_address>
        <Deposit_id></Deposit_id>
        <Interest_Type></Interest_Type>
        <Interest_compound></Interest_compound>
        <Principal></Principal>
          <Rate></Rate>
          <Deposit_date></Deposit_date>
        <Current_date></Current_date>
        <Maturity_date></Maturity_date>
       <Formula>( Principal * Rate * ( Maturity_date -
Deposit_date ) / 100 )</Formula>
        <Saving_InterestRate MaturityTime="days_7_to_15"
AmountRange="SI_below_15_lakhs">2.00</Saving_InterestRate>
        <Saving_InterestRate MaturityTime="days_7_to_15"
AmountRange="SI_15_to_50_lakhs">2.00</Saving_InterestRate>
                          :
                          :
       <Saving_InterestRate MaturityTime="years_2_to_3"
AmountRange="SI_15_to_50_lakhs">7.10</Saving_InterestRate>
       <Saving_InterestRate MaturityTime="years_2_to_3"
AmountRange="SI_above_50_lakhs">7.10</Saving_InterestRate>
    </Axisbank>
</Deposit>
```

**Fig. 5.** XML based advertisement conforming to deposit XSD

5. The teller enters the information in the GUI and submits.

- XSI uses the Deposit XSD to validate the information entered by the customer.
- The Deposit XSD is used to validate the information entered by the customer in the GUI. If there are any violations, customers are alerted to correct the information.

6. If the validation is successful, the teller sends the XML document to the customer and one copy is sent to the bank's server for future reference. This XML file contains all the information about the deposit money, interest rate and valid or maturity date of the deposit.
7. The customer can use XSI tool to view the XML based deposit document.

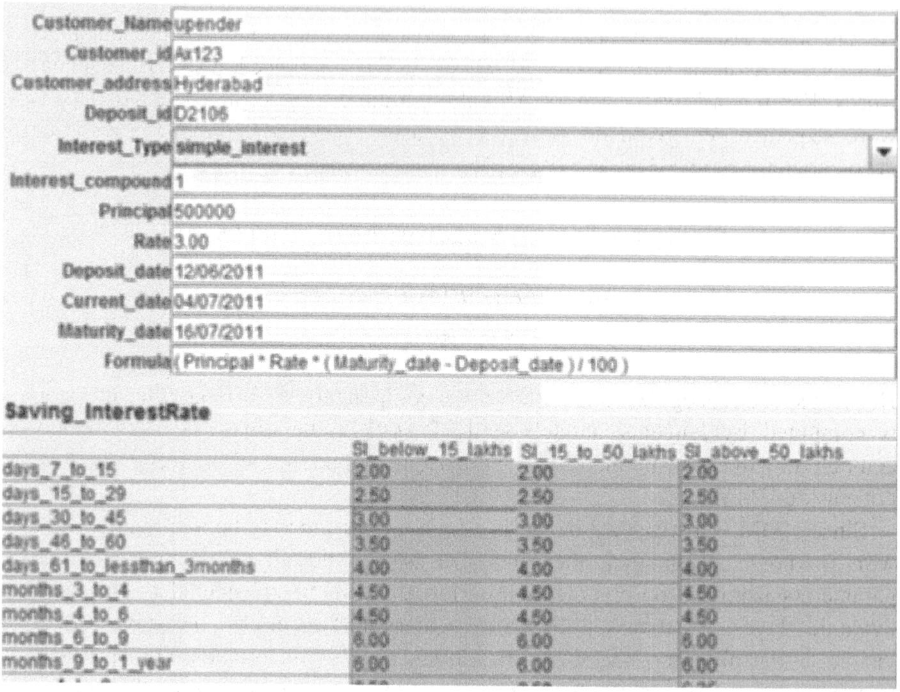


**Fig. 6.** GUI format for above XML

```
<?xml version="1.0" encoding="UTF-8"?>
<xs:schema xmlns:xs="http://www.w3.org/2001/XMLSchema" elementFormDefault="qualified">
    <xs:element name="Axisbank">
        <xs:complexType>
            <xs:sequence>
                <xs:element name="Customer_Name" type="xs:string"/>
                <xs:element name="Customer_id" type="xs:string"/>
                <xs:element name="Customer_address" type="xs:string"/>
                <xs:element name="Deposit_id" type="xs:string"/>
                <xs:element name="Interest_Type" type="xs:decimal">
                <xs:simpleType>
                        <xs:restriction base="xs:decimal">
                        <xs:enumeration value="simple_interest"/>
                        <xs:enumeration value="compound_interest"/>
                        </xs:restriction>
               </xs:simpleType>
               </xs:element>
                <xs:element name="Interest_compound" type="xs:decimal"/>
                <xs:element name="Principal" type="xs:decimal"/>
                <xs:element name="Rate" type="xs:decimal"/>
                <xs:element name="Deposit_date" type="xs:date"/>
                <xs:element name="Current_date" type="xs:date"/>
                <xs:element name="Maturity_date" type="xs:date"/>
                <xs:element name="Formula" type="xs:string"/>
               <xs:element name="Saving_InterestRate" type="InterestRate" minOccurs="30" maxOccurs="unbounded"/>
            </xs:sequence>
        </xs:complexType>
    </xs:element>

    <xs:complexType name="InterestRate">
        <xs:attribute name="MaturityTime" type="xs:string" use="required"/>
        <xs:attribute name="AmountRange" type="xs:string" use="required"/>
    </xs:complexType>
</xs:schema>
```

**Fig. 7.** XSD for Deposit form

```xml
<?xml version="1.0" encoding="UTF-8" ?>
- <Deposit schemaLocation="D:\Deposit\Deposit.xsd">
   <include schemaLocation="D:\Deposit\Deposit.xsd" />
   <advertisement schemaLocation="D:\date\Deposit.xml" />
 - <Axisbank>
     <Customer_Name>upender</Customer_Name>
     <Customer_id>Ax123</Customer_id>
     <Customer_address>Hyderabad</Customer_address>
     <Deposit_id>D2106</Deposit_id>
     <Interest_Type />
     <Interest_compound>1</Interest_compound>
     <Principal>500000</Principal>
     <Rate>3.00</Rate>
     <Deposit_date>12/06/2011</Deposit_date>
     <Current_date>04/07/2011</Current_date>
     <Maturity_date>16/07/2011</Maturity_date>
     <Formula>( Principal * Rate * ( ( Maturity_date - Deposit_date ) / 365 ) / 100 )</Formula>
     <Saving_InterestRate />
   </Axisbank>
   <Interest>1397.2602739726028</Interest>
   <Present_Value>501397.2602739726</Present_Value>
</Deposit>
```

**Fig. 8.** Final Deposit XML with customer deposit details

# 3 Advantages

## 3.1 Advantages to Bank

- Frequent changes of XML document for representing interest rates is allowed and so changes are reflected from time to time as documents just like in physical documents.
- Generates structured information contained in XML documents. The structured data is easier to handle, store, process, and analyze by application systems. Thereby, it helps organizations to make better decisions to improve efficiency, competitiveness and reduce fraud.
- Excessive processing on the server is reduced as customers need not approach the bank always. With tools like XSI, their own systems can participate in application processing.

## 3.2 Advantages to Customer

- We foresee the replacement of traditional browsers in all the financial transactions with browsers like XSI.
- The customers can calculate present deposits based on interest rates at any time.
- This tool is providing GUI, so customer can enter data in an easy way in offline mode as well, without connecting to Bank's server and he can keep all copies of his requests to bank(s) and the corresponding responses from bank(s) in a single view.
- Since the deposit is in standard XML format, it will help the customer to track his transactions in the case of banks' mergers.

– Since the Deposit Certificate is in standard XML form, new products like trading the deposits also can become a reality in future.

## 4 Conclusion

The XSI tool takes both XML documents and XSD as input and generates output in an XML format. It validates the information without human intervention. Through this tool we can provide an easy way to calculate the interest rates in the form of XML and these forms are sent to customers. So, the customer has valid proof for further verification. Bank servers will update the customer information in an easy way by changing the XML document.

## References

[1] Bray, T., Paoli, J., Sperberg-McQueen, C.M.: Extensible Markup Language, `http://www.w3.org/TR/2008/REC-xml-20081126`
[2] The World Wide Web Consortium Document Object Model (DOM), `http://www.w3.org/DOM`
[3] XSD tutorial from W3C, `http://www.w3schools.com/schema/default.asp`
[4] Radha, V., Ramakrishna, S.: XSI- A XML schema based universal customer interface. In: Enabling Technologies for Smart Appliances (ETSA), January 12-14, IEEE, Hyderabad (2005)
[5] Radha, V., Ramakrishna, S., Pradeep Kumar, N.: Generic XML Schema Definition (XSD) to GUI Translator. In: Chakraborty, G. (ed.) ICDCIT 2005. LNCS, vol. 3816, pp. 290–296. Springer, Heidelberg (2005)
[6] Radha, V., Shiva Kumar, M., Pavan Kumar, K.: SIM-Structured Information Messaging. In: CSI 2007 Annual Convention, November 28-December 1 (2007)
[7] Garvey, P.: Generating User Interfaces from composite schemas, `http://www.idealliance.org/xmlusa/03/call/xmlpapers/03-03-04.994/.03-03-04.html`
[8] Apparao, V., Byrne, S., Champion, M., Isaacs, S., Jacobs, I., Le Hors, A., Nicol, G., Robie, J., Sutor, R., Wilson, C., Wood, L.: Document Object Model (DOM) Level 1 Specification. W3C Recommendation (October 1998), `http://www.w3.org/TR/1998/REC-DOM-Level-1-19981001`
[9] Bergulund, Boag, S., Chamberlin, D., Fernandez, M.F., Kay, M., Robie, J., Simon, J.: XML Path Language (XPath) 2.0 Technical Report, W3C Working Draft (2001), `http://www.w3.org/TR/XPath20/`

# Classifying Gender from Faces Using Independent Components

Sunita Kumari and Banshidhar Majhi

National Institute of Technology, India
sunita.nitrkl@gmail.com,
bmajhi@nitrkl.ac.in

**Abstract.** Gender classification algorithms uses a number of dimensionality reduction techniques. These techniques are used to remove linear and non–linear dependency among the data. Most popular techniques are Principal Component Analysis (PCA) and Independent Component analysis (ICA). PCA makes the data uncorrelated while ICA gives independent data. In this approach, infomax ICA has been used for gender recognition and it is outperforming PCA. To perform classification using ICA features, Support Vector Machine (SVM) has been used. For larger training set SVM is performing with an accuracy of 98%. The accuracy values are obtained for change in size of testing set and the proposed system performs with an average accuracy of 96%.

**Keywords:** PCA, ICA, Gender Classification, Classifier, infomax.

## 1 Introduction

Face biometric has gained popularity because it is one of the most communicative part of social interactions. Human-Computer Interaction as well needs face recognition, gender recognition, estimation of age for passive surveillance. However, during identification of any individual the False Acceptance Rate(FAR) increases as the database size increases [5]. Further the time required increases exponentially as we need 1:N comparisons for identification. Hence, more thrust is put in any biometric system to reduce search time without compromising with accuracy. In a face recognition system, partitioning the database based on gender is an important research direction. In this paper, more emphasis is laid on classification of face database using gender. The major challenge for gender classification is two-fold, (1) to find discriminating features and (2) an optimum hyperplane for separating male and female classes. Hence most researchers thrust on these two issues while attempting gender classification using face. There has been some investigation done on classification of face database using gender.

By seeing a face image, anyone can say whether it is of a male or female but machine intelligence needs automation. This is all because the data given to machine are redundant. It needs an appropriate data/features and an optimum classifier to take such a decision. Researchers are working in both the directions equally. Geometric and appearance based features are the two ways to extract

D. Nagamalai, E. Renault, M. Dhanushkodi (Eds.): CCSEIT 2011, CCIS 204, pp. 589–598, 2011.

such keypoints from an image. The former uses an idea to modal a human face in terms of salient features like eyes, nose, mouth, etc., and the distance between the salient feature points. The latter reduces a facial image containing thousands of pixels to a handful number of pixels. The aim is to capture the distinctiveness between the two genders from a face. The geometric features are termed as *local* features and appearance features are termed as *global* features. The concept of geometric keypoints extraction has come from the work of *Leonardo Da Vinci.* He worked in the field of human anatomy and drew many pictures of human face containing different face dimensions. Appearance based keypoints are used by the researchers recently. But the work done in these fields are very contradictory. There are several methods for dimensionality reduction like Principal Component Analysis (PCA), Factor Analysis (FA), Independent Component Analysis (ICA) and many others [12]. PCA has been first applied by Kirby and Sirovich and it is shown that it is an optimal compression scheme that minimizes the mean squared error between the original and reconstructed images. Later Turk and Pentland have made it the most popular method for face recognition [13]. ICA is a generalization of PCA. This field is contradictory with respect to the performance of these algorithms. For face recognition, Bartlett et al. [2], Yuen and Lai [14], and Liu and Wechsler [8] have claim that ICA outperforms PCA while Baek et al. [1] claims that PCA outperforms ICA. Moghaddam claims that there is no significance difference between PCA and ICA [9]. It carries same effect in case of gender classification also. But all these performances varies across several factors like the problem domain, distance measure techniques and the number of basis vectors chosen. The different classifiers developed are multi–layer neural network, back–propagation neural network, decision tree, support vector machine(SVM) and many others [7] [10]. In this paper, proposition has been made with ICA as a feature extraction technique and SVM as a classifier.

The rest of the paper is organized as follows: Section 2 provides a description of the feature extraction techniques used, Section 3 discusses the classifier. Section 4 has been presented with simulation results and in last Section 5 gives conclusion.

## 2 Feature Extraction Techniques

This work has been motivated by the keen interest of researchers in face recognition using independent component analysis [2]. A common problem of biometric research is to find a suitable representation for multivariate data. Most famous transformation methods include PCA, FA, projection pursuit, etc. A new technique for linear transformation, which has been recently developed is the independent component analysis. There are two types of dependencies exist in the data (i) *linear* (ii) *non–linear*. Linear dependencies among the data can be removed by PCA and ICA removes the non–linear dependencies. PCA and ICA are unsupervised neural network based techniques which are also getting used for the purpose of dimensionality reduction. As the face image contains large redundant data inside it so there is a need to go with such a dimensionality reduction techniques. PCA decorrelates the data and ICA make the data independent. In the proposed approach, ICA has been used for the gender classification system and then it is compared with PCA.

### 2.1 ICA

ICA gives independent data and independence can be defined in terms of probability density functions (PDFs). Two variables $y1$ and $y2$ are said to be independent if joint PDF of $y1$ and $y2$ will be factorizable in terms of their marginal PDF and it can be achieved by either *minimizing mutual information* or by maximizing non-gaussianity. This paper adopts the first approach.

***Definition***: For the given set of input sample $x$, ICA finds a linear transform in such a way that

$$s = Wx$$

so that the components $s_i$ are as independent as possible. Here, $W$ is an unknown separating matrix and it needs to be determined. There are many algorithms to determine $W$ like Jade algorithm, infomax ICA, fastICA and many others [6]. Infomax ICA is used in the current approach which is proposed by Bell and Sejnosky [4]. The principle behind this is: when inputs are to be passed through a sigmoid function maximum information transfer is achieved when the sloping part of the sigmoid is optimally lined up with high density parts of the input and this can be achieved by performing stochastic gradient ascent on the entropy of the output with respect to the weight matrix $W$. It can be defined as follows:

$$\triangle W = (\mathrm{W}^T)^{-1} + \mathrm{E}(\mathrm{Y}^{'}\mathrm{X}^T)$$

Here $X$ is the input data matrix, $Y$ represents the output and can be expressed as $Y = f(S) = 1/1{+}e^{-s}$, $T$ indicates transpose and $E$ stands for expected value. The ratio between the second and first order partial derivatives of the activation function is given as: $Y_i^{'} = f_i^{''}(S_i) \,/\, f_i^{'}(S_i)$. It is shown in [3] that by maximizing the joint entropy of the output $Y$, the statistical independence between the outputs are increased or the mutual information are getting decreased.

**Preprocessing of the data.** It needs two preprocessing technique because it makes the problem of ICA estimation simpler and better conditioned.

1. *Centering*: It gives the zero–mean data like for an input vector $x$ subtract the mean $\boldsymbol{m} = Ex$ from it.
2. *Sphering/Whitening*: It is a linear transformation method which gives the whiten signal $\tilde{x}$ from an input signal x such that its components are uncorrelated and their variances equal unity. Unit variance can be achieved by making $E(\tilde{x}\tilde{x}^T) = I(identity\ matrix)$. Whitening is done by

$E\{xx^T\} = EDE^T$. Here, E is the orthogonal matrix of eigenvectors of $E\{xx^T\}$ and D is the diagonal matrix of its eigenvalues. $\tilde{x}$ is determined by

$$\tilde{x} = ED^{-1/2}E^Tx$$

**Architecture of ICA.** The goal is to find a better set of basis images so that it can easily estimate an unknown dataset. There are two ways to achieve this goal:

1. *Architecture-1/ ICA-1*: It produces a set of statistically independent basis images. The data matrix $X$ is constructed in such a way that each row represents an individual image. This architecture deals about independence of images so that images are considered as random variables and pixels as trials as shown in Fig. 1. Two images $X_i$ and $X_j$ are independent if it is not possible to predict the value of an image $X_j$ from $X_i$ by moving across pixels. A set of coefficients can be determined by the linear combination of the independent basis images from which each individual image can be easily constructed.
2. *Architecture-2/ ICA-2*: In ICA-1, the basis images were statistical independent from each other but the coefficients were not but in Architecture-2, the coefficients are statistically independent. It finds a factorial face code for the set of face images. The data matrix $X$ is constructed in such a way that each column corresponds to an individual image. This architecture deals with independence of pixels so that pixels will become random variable and images as trials. Two pixels $i$ and $j$ are said to be independent, if by moving across set of images it is not possible to predict the value of $j$ from pixel $i$ as shown in Fig. 2.

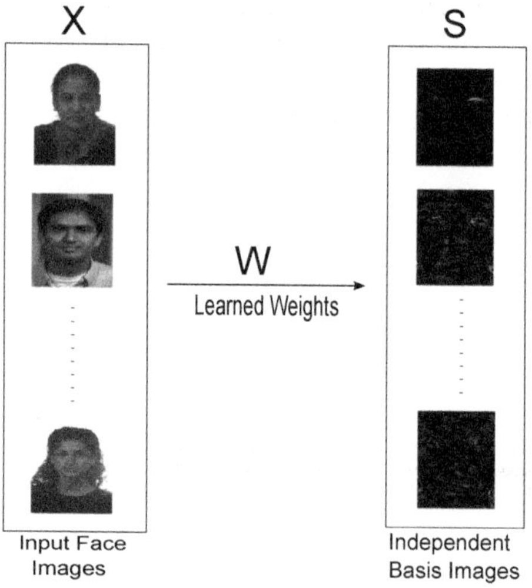


**Fig. 1.** Architecture-1

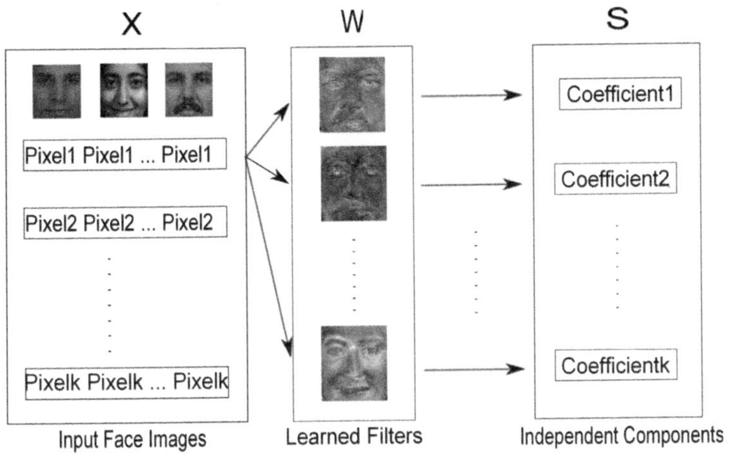


**Fig. 2.** Architecture-2

## 3 SVM

SVM is a learning algorithm for pattern recognition and it found as a better classifier than others in literature [10]. It designs a hyperplane in such a way that it will leave more space on either side so that data on both sides can move a bit more freely with less risk of causing an error [12]. A hyperplane is selected with *generalization performance* of the classifier. It is the capability of the classifier, designed using the training data set, to operate satisfactorily with an unknown dataset.

Given a training set of instance-label pairs $(X^i, y^i)$, $i=1...l$, where $X^i \in \Re^N$ and $y^n \in \{-1,1\}$.

Proposed $SVM$ is

$$f(X) = sign\langle W \cdot X\rangle + b$$

where, $W$ is the direction and $b$ is the position of the hyperplane in space. The optimal hyperplane $L$ should satisfy following criteria

$$y^i(\langle W \cdot X\rangle + b) \geq 1 \quad (1)$$

$$0 \leq y^i(\langle W \cdot X\rangle + b) < 1 \quad (2)$$

$$y^i(\langle W \cdot X\rangle + b) < 0 \quad (3)$$

Above three cases can be treated under a single type of constraint by introducing a new set of variables named as slack variable

$$y^i(\langle W \cdot X\rangle + b \geq 1 - \xi_i$$

The goal becomes

$$\text{minimize } J(w, x, \xi) = \frac{1}{2} \parallel W \parallel^2 + C\sum_{i=1}^{N} I(\xi_i)$$

$$\text{subject to } y^i(\langle W \cdot X\rangle + b) \geq 1 - \xi_i$$
$$\text{Where, } i = 1, 2 \cdots N, \xi_i > 0$$

$I(\xi_i)$ can be defined as

$$I(\xi_i) = \begin{cases} 1 & , \xi_i > 0 \\ 0 & , \xi_i = 0 \end{cases}$$

$C$ is the cost parameter and is a positive constant that controls the relative influence of above two competing terms. On simplifying above using Primal Lagrangian and dual [12], we can get the optimal hyperplane as

$$f(X) = sign(\sum_{i=1}^{l} y^i a_i K(X \cdot X^i) + b)$$

where K(· , ·) is a kernel function defined by

$$K(X^i, X^j) = exp(-\frac{\gamma||X^i - X^j||^2}{2\sigma^2}), \gamma > 0$$

Here, Radial Basis Functions ($RBF$) kernel is used because it has the property of symmetry.

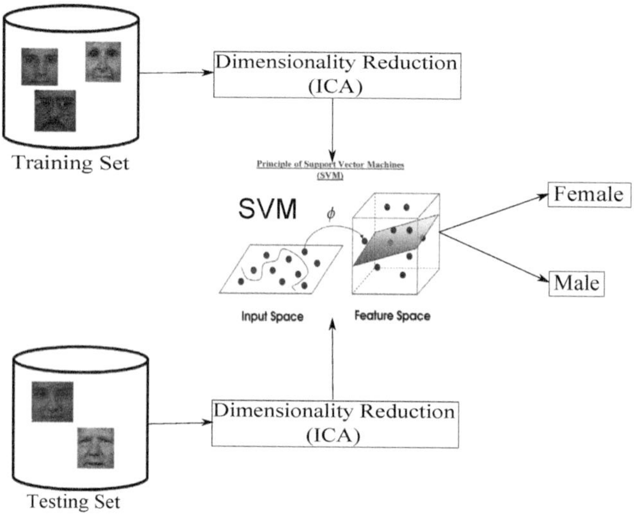


**Fig. 3.** Gender classification system

## 4 Classifier Performance

Proposed methodology for gender classification is shown in the Fig. 3. The performance of gender classification system has been evaluated for two ICA architecture using the FERET database [11](sample images from this database are shown in Fig. 5) and it is compared with existing PCA. The dataset consists of images of 500 individuals. 200 images has been randomly selected which forms the training dataset. From the remaining 300 images, four group of testing data has been made. First testset consists of 200 individuals, second has 150 individuals, third consists of 100 and fourth has 50 individual images. Performance of the system has shown on all these testset using same training set. The original size of the image was 384×256. All these images has been aligned first and then conventional triangularization approach is applied to get region of interest as shown in Fig. 6. Features has been extracted using ICA from the cropped images of size 293×241. The size of data matrix is 500×70613, where each row corresponds to one particular image. The number of independent components(ICs) got by the ICA algorithm corresponds to the dimensionality of the input. As the performance merely depends on the number of separated components so there is a need to control the number of ICs. To achieve this, ICA can be performed on linear combination of original images and this has done using PCA. The PCA baseline system uses 200 eigenvectors. Thus ICA has 200 ICs. The accuracy obtained by various combinations is shown in Fig. 4 and its associated tabular representation is given in Table: 1 .

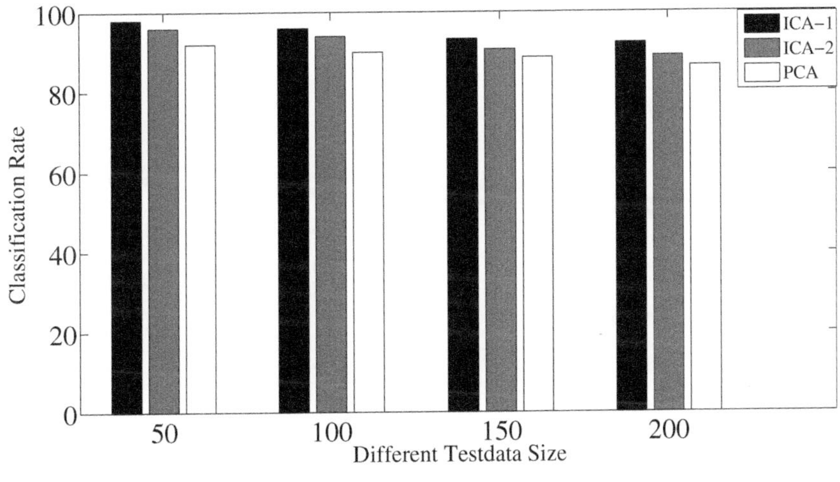


**Fig. 4.** Classification rate of genders

By comparing the difference among infomax ICA–1, ICA–2 and PCA, it is concluded that ICA–1 performs better than ICA–2 and even ICA–2 also outperforms PCA . Images obtained from PCA is shown in Fig. 7 and of ICA–1 is shown in Fig. 8 and Fig. 9 shows the faces obtained from ICA–2. Literature entails that ICA–1 and ICA–2 gives local and global features respectively. Thus, the proposed gender classifier is performing well with local features than compared to global.

**Fig. 5.** Images from FERET database

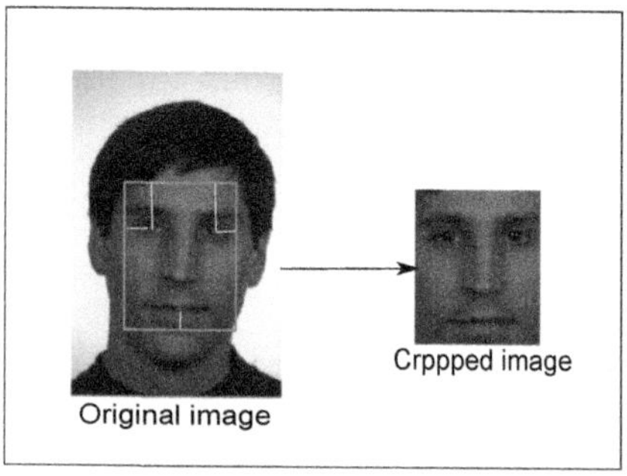


**Fig. 6.** Conventional Triangularization Approach

**Table 1.** Performance results(Accuracy)

| **Traindata size** | **Testdata size** | **ICA–1 (%)** | **ICA–2 (%)** | **PCA (%)** |
|---|---|---|---|---|
| 200 | 50 | 98 | 96 | 92 |
| 200 | 100 | 96 | 94 | 90 |
| 200 | 150 | 93.33 | 90.67 | 88.66 |
| 200 | 200 | 92.25 | 89 | 86.5. |

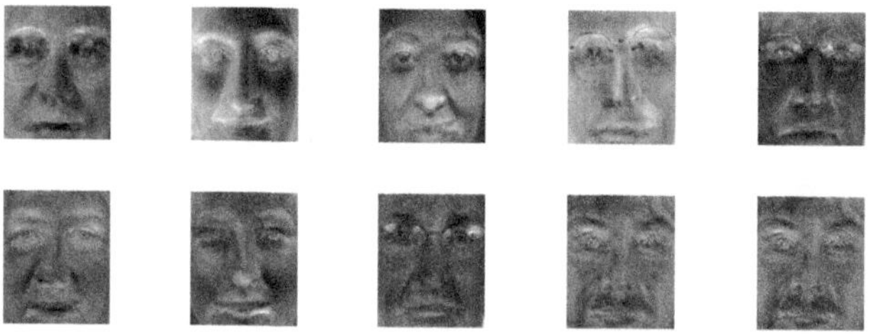

**Fig. 7.** Basis images from Principal Component Analysis

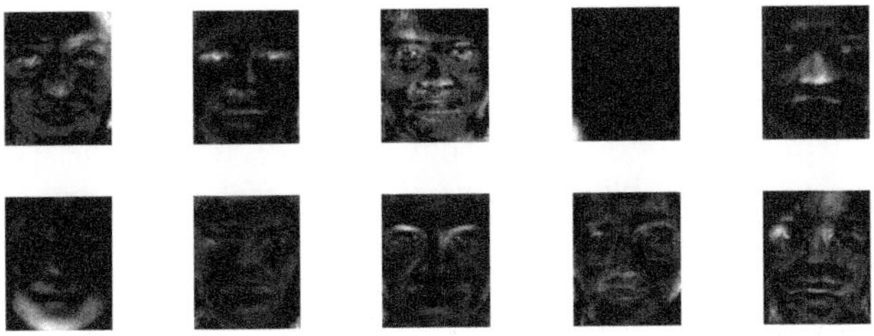

**Fig. 8.** Basis images from ICA Architecture-1: Statistical Independent Images

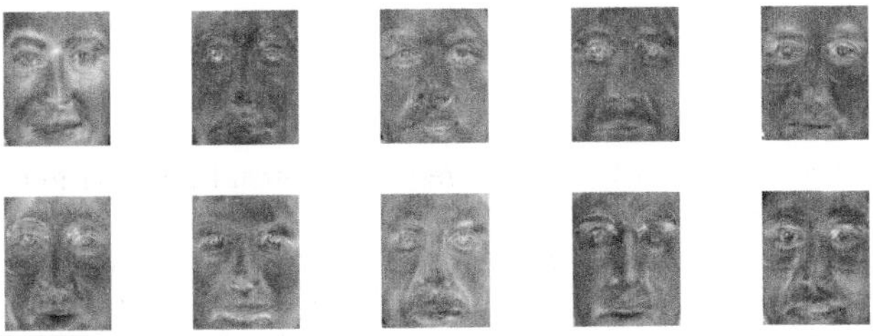

**Fig. 9.** Basis images from ICA Architecture-2: A factorial Face Code

These accuracies has been obtained using 200 basis vectors. As mentioned earlier performance also depends on basis vectors, so the variation of basis vectors on 200 training and 150 testing images is shown in Fig. 10.

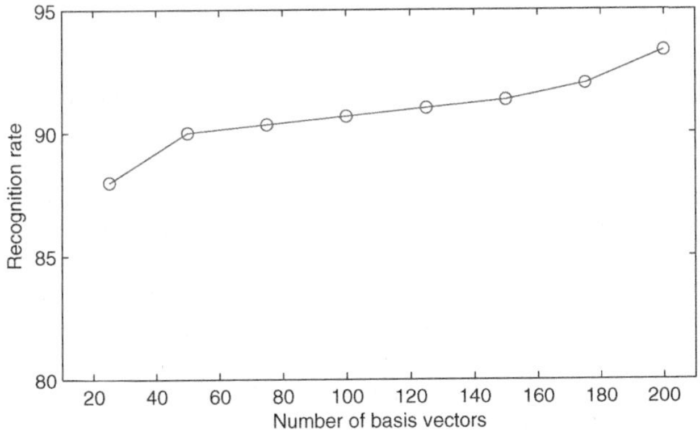


**Fig. 10.** Classification rate on varying basis vectors

## 5 Conclusions

This paper has been presented with the comparison between PCA and ICA with the problem domain, architecture, algorithm and classifier used. The performance varies with all these parameters. Among all these, The most important factor is the nature of task and then used algorithm. The dimensionality reduction techniques work better with face biometric like face identification, expression recognition, gender recognition and many others. In current application, results obtained from ICA are superior than that of the PCA. But there also, its varying with the features used, like in terms of more local and global appearance of an image. All these are dealing with the support vector machine because it has found as an optimum classifier from the literature.

## References

1. Baek, K., Draper, B.A., Beveridge, J.R., She, K.: Pca vs. ica: A comparison on the feret data set. In: Joint Conference on Information Sciences, pp. 824–827 (2002)
2. Bartlett, M., Movellan, J., Sejnowski, T.: Face recognition by independent component analysis. IEEE Transactions on Neural Networks 13(6), 1450–1464 (2002)
3. Bell, A., Sejnowski, T.J.: The 'independent components' of natural scenes are edge filters. Vision Research 37, 3327–3338 (1997)
4. Bell, A.J., Sejnowski, T.J.: An information-maximization approach to blind separation and blind deconvolution. Neural Computation 7, 1129–1159 (1995)
5. Jain, A.K., Maltoni, D., Maio, D., Prabhakar, S.: Guide to Biometrics. Springer, Heidelberg (2003)
6. Hyvrinen, A., Oja, E.: Independent component analysis: algorithms and applications. Neural Networks 13, 411–430 (2000)
7. Jain, A., Huang, J.: Integrating independent components and support vector machines for gender classification. In: Proceedings of the 17th International Conference on Pattern Recognition, ICPR 2004, vol. 3, pp. 558–561 (August 2004)
8. Liu, C., Wechsler, H.: Comparative assessment of independent component analysis (ica) for face recognition. In: International Conference on Audio and Video Based Biometric Person Authentication, pp. 22–24 (1999)
9. Moghaddam, B.: Principal manifolds and Bayesian subspaces for visual recognition. In: The Proceedings of the Seventh IEEE International Conference on Computer Vision, vol. 2, pp. 1131–1136 (1999)
10. Moghaddam, B., Yang, M.-H.: Learning gender with support faces. IEEE Transactions on Pattern Analysis and Machine Intelligence 24(5), 707–711 (2002)
11. Phillips, P.J., Moon, H., Rizvi, S.A., Rauss, P.J.: The feret evaluation methodology for face recognition algorithms. In: In (a) Example 1 (b) Example 1 (c) Right ear) (d) Mirroed Left ear
12. Sergios Theodoridis, K.K.: Pattern Recognition. Springer, Heidelberg (2006)
13. Turk, M., Pentland, A.: Eigenfaces for recognition. J. Cognitive Neuroscience 3(1), 71–86 (1991)
14. Yuen, P.C., Lai, J.H.: Independent component analysis of face images. In: IEEE Workshop on Biologically Motivated Computer Vision (2000)

# Experimental Evaluation of Fast Solution Algorithms for the Simple Motif Problem

Tarek El Falah[1,2], Thierry Lecroq[2], and Mourad Elloumi[1]

[1] Research Unit of Technologies of Information and Communication, Higher School of Sciences and Technologies of Tunis, 1008 Tunis, Tunisia
[2] University of Rouen, LITIS EA 4108, 76821 Mont-Saint-Aignan Cedex, France
Tarek.Elfalah@etu.univ-rouen.fr,
Thierry.Lecroq@univ-rouen.fr,
Mourad.Elloumi@fsegt.rnu.tn

**Abstract.** The *Simple Motif Problem* (SMP) is: given a set of strings $Y = \{y_0, y_1, \ldots, y_{n-1}\}$ built from a finite alphabet $\Sigma$, $p > 0$ an integer and $q \leq n$ a quorum, find all the simple motifs of length at most $p$ that occurs in at least $q$ strings of $Y$.

This paper presents an experimental evaluation of algorithms dealing with SMP and using a minimal forbidden pattern approach. The experiments are concerned both with running times and space consumption.

**Keywords:** algorithms, strings, motifs, SMP, complexities.

## 1 Introduction

The motif finding problem consists in finding substrings that are more or less conserved in a set of strings. This problem is a fundamental one in both Computer Science and Molecular Biology. Indeed, when the concerned strings code biological sequences, i.e., DNA, RNA and proteins, extracted motifs offer to Biologists many tracks to explore and help them to deal with many challenging problems. In the literature, several versions of the motif finding problem have been identified [4]:

- *Planted (l,d)-Motif Problem* (PMP) [2,11];
- *Extended (l,d)-Motif Problem* (ExMP) [9,14];
- *Edited Motif Problem* (EdMP) [12,13];
- *Simple Motif Problem* (SMP) [12,5].

A *simple motif* is a string built from a finite alphabet $\Sigma \cup \{?\}$ that cannot begin or end with ?, where $\Sigma$ is a set of symbols and $? \notin \Sigma$ is a wildcard symbol, it can be replaced by any symbol from $\Sigma$.

Let $Y = \{y_0, y_1, \ldots, y_{n-1}\}$ be a set of strings built from an alphabet $\Sigma$, $p > 0$ be an integer and $q \leq n$ be a quorum, the *Simple Motif Problem* (SMP) is to find all the simple motifs of length at most $p$ that occurs in at least $q$ strings of $Y$.

D. Nagamalai, E. Renault, M. Dhanushkodi (Eds.): CCSEIT 2011, CCIS 204, pp. 599–608, 2011.

Actually, concerning SMP the main existing solutions are:

- the algorithm given by Floratos and Rigoutsos [8],
- the one described by Rajasekaran *et al.* [12].
- the SMS-FORBID and SMS-H-FORBID algorithms by El Falah *et al.* [5,6].

In this paper, we focus on SMP and we propose an experimental evaluation of algorithms SMS-FORBID and SMS-H-FORBID. Both algorithms are based on a minimal forbidden pattern approach. Algorithm SMS-FORBID uses indexing structures (either suffix trees or suffix arrays) while algorithm SMS-H-FORBID uses an hash table. The experiments are concerned both with running times and space consumption. These experiments were conducted on pseudo-randomly generated data on two alphabet sizes: 4 (to simulate DNA sequences) and 20 (to simulate protein sequences).

We organize the rest of the paper as follows: In section 2, we present some useful definitions and notations. In section 3, we review algorithms SMS-FORBID and SMS-H-FORBID. Section 4 presents experimental results. In section 5, we give our conclusion and perspectives.

## 2 Preliminaries

A *simple motif* is a string built from an alphabet $\Sigma \cup \{?\}$ that cannot begin or end with ?, where $\Sigma$ is a finite set of symbols and $? \notin \Sigma$ is a wildcard symbol, it can be replaced by any symbol from $\Sigma$. Symbols of $\Sigma$ are said to be *solid* while the wildcard symbol ? is said to be *non-solid*. The length of a simple motif is the number of the symbols that constitute this motif, including the wildcard symbols.

A string of $\ell$ symbols from $\Sigma$ is called a $\ell$-mer. A string of $\ell$ symbols from $\Sigma \cup \{?\}$ is called a $\ell$-pattern. A $\ell$-pattern $z_1$ is equivalent to a $\ell$-pattern $z_2$ ($z_1 \cong z_2$), if at every position where $z_1$ and $z_2$ contains both solid symbols these symbols are equal. Formally, $z_1 \cong z_2$ if for $1 \leq i \leq \ell$: $z_1[i] = z_2[i]$ or $z_1[i] = ?$ or $z_2[i] = ?$.

A $\ell$-pattern $z_1$ is more general than a $\ell$-pattern $z_2$ if a position in $z_2$ contains the wildcard symbol implies that the same position in $z_1$ contains the wildcard symbol and if a position in $z_2$ contains a solid symbol then at the same position in $z_1$ there could be either the same symbol or a wildcard symbol. Formally $z_2[i] = ? \Rightarrow z_1[i] = ?$ and $z_2[i] = a \Rightarrow (z_1[i] = a$ or $z_1[i] = ?)$ for $1 \leq i \leq \ell$ and $a \in \Sigma$.

Let $Y = \{y_0, y_1, \ldots, y_{n-1}\}$ be a set of strings built from an alphabet $\Sigma$ and let $N = \sum_{i=0}^{n-1} |y_i|$. Let $m$ be the maximal length of the strings in $Y$. Let $\sigma$ be the size of the alphabet $\Sigma$.

## 3 Minimal Forbidden Pattern Approach

In this section, we present two algorithms SMS-FORBID and SMS-H-FORBID.

### 3.1 SMS-Forbid Algorithm

Given a set $Y$ of $n$ strings, a quorum $q \leq n$ and an integer $p$. The algorithms output the set of motifs of length at most $p$ that occurs in at least $q$ strings.

A pattern $z$ of length at most $p$ is said to be a minimal forbidden pattern if it occurs in less than $q$ strings but all its proper factors beginning and ending with a solid symbol occur in at least $q$ strings.

For each position on the input strings, we use all the windows of length $\ell$ for $3 \leq \ell \leq p$. Each window defines an $\ell$-mer. Each $\ell$-mer $x$ defines a set of $\ell$-patterns $X$. At each position of each $\ell$-pattern $z$ of $X$, the symbol of $z$ is either the symbol at the same position of $x$ or the wildcard symbol except for the first and the last symbols of $z$ that are necessarily non-wildcard symbols. Formally, $z[i] = x[i]$ or $z[i] = ?$ for $1 \leq i \leq \ell - 2$ and $z[0] = x[0]$ and $z[\ell - 1] = x[\ell - 1]$.

These $\ell$-patterns together with the generality relation form a lattice. The minimal element of the lattice is $x$ itself and the maximal element is $x[0]?^{\ell-2}x[\ell-1]$. (see example in Fig. 1: the minimal element is $x =$ CAGAT and the maximal element is C$?^{\ell-2}$T).

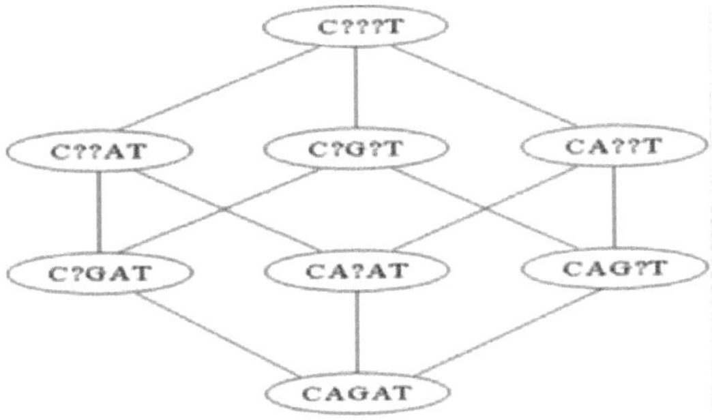


**Fig. 1.** A 5-mer $x =$ CAGAT lattice

Each node of the lattice represents an $\ell$-pattern.

The $\ell$-patterns are scanned by doing a breadth-first search of the lattice beginning from the minimal element.

When a $\ell$-pattern $z$ is considered, if it has already been output or it contains minimal forbidden patterns as factors or it is more general than an output pattern, then it is disregarded otherwise it is searched in the strings of $Y$. Then if it occurs in at least $q$ strings it is output and all its successors in the lattice are not considered since they are more general. On the contrary if it does not occur in at least $q$ strings it is added to the set of minimal forbidden patterns.

The generation of the $\ell$-patterns is performed using a breadth-first search of the lattice for the following reason. When a $\ell$-pattern is discovered all its successors in the lattice, that are more general, do not have to be considered. They are thus marked using a depth-first search of the lattice from the discovered $\ell$-pattern. During the remaining of the breadth-first search, marked $\ell$-patterns are not considered.

Contrary to the algorithm presented in [12], the new approach does not search for all the $\ell$-patterns generated from the $n$ strings of $Y$ but it begins by searching the more specific patterns i.e. the less general patterns which avoids the sorting step. Moreover it maintains a set of minimal forbidden patterns that do not occur in at least $q$ strings in order to not search for any pattern that contains a factor that has already been unsuccessfully searched. This two techniques reduce the number of patterns to be searched for. The search of one $\ell$-pattern in one string $y$ of $Y$ is done using an indexing structure of $y$ which can be done in a time proportional to $\ell$. Furthermore the new approach only outputs the more specific motifs that occur in at least $q$ strings of $Y$. This should help the biologist to identify important motifs.

**An illustrative example:** Let $Y = \{y_0 = \texttt{ACGAAGACT}, y_1 = \texttt{CGTAGCTAC}, y_2 = \texttt{CAGTACTAT}, y_3 = \texttt{ACGTACAAA}, y_4 = \texttt{CCTACTGCT}\}$ be a set of strings built from the alphabet $\Sigma = \{\texttt{A}, \texttt{C}, \texttt{G}, \texttt{T}\}$. The set of most specific simple motifs having length at most $p = 6$ and occuring in at least $q = 3$ strings of $Y$ is

$$\{\texttt{TAC}, \texttt{ACT}, \texttt{GTA}, \texttt{CTA}, \texttt{C?TA}, \texttt{CG?A}\}.$$

We will give a top-down detailed presentation of the algorithm. It builds the set *Res* of searched motifs of length at most $p$ contained in at least $q$ strings and uses a set $\mathcal{T}$ of minimal patterns that are not contained in at least $q$ strings. It scans the $n$ strings of $Y$ from $y_0$ to $y_{n-1}$. For each position of each string it considers all the windows of length $\ell$ for $3 \leq \ell \leq p$ (at the end of the strings it may be less than $p$). For each $\ell$-mer $x$ defined by each window of length $\ell$, the breadth-first search of the lattice is performed level by level.

During the breadth-first search of the lattice, when a non-marked $\ell$-pattern $x$ is considered, it is searched in the set *Res* of resulting $\ell$-motifs. For that, the set *Res* is implemented using a trie: looking if an $\ell$-pattern $x$ belongs to *Res* simply consists in spelling $x$ from the root of the trie. If $x$ belongs to *Res* then all it successors are marked using a depth-first search.

If $x$ does not belong to *Res* then the algorithm checks if it contains minimal forbidden patterns. This consists in searching for a finite set of patterns $\mathcal{T}$ with wildcard symbols in a text with wildcard symbols $x$, where a wildcard symbol in the text only matches a wildcard symbol in the set of patterns while a wildcard in the set of patterns may match any symbol in the text.

If $x$ does not contain any minimal forbidden pattern then it is searched in all the strings of $Y$ from $y_{j+1}$ to $y_{n-1}$ since it has not been considered before dealing with $y_j$ and it has at least one occurrence in $y_j$. If it occurs in at least $q$ strings, it is added to *Res* and all its successors are marked using a depth-first search. Otherwise it is added to the set $\mathcal{T}$ of minimal forbidden patterns.

The lattice is completely traversed in a breadth-first search in every cases.

Each $\ell$-pattern $x$ in the lattice is associated with an integer from 0 to $2^{\ell-2}$ whose binary representation is given by $x[1 \mathinner{.\,.} \ell-2]$ where each solid symbol is replaced by 0 and each wildcard symbol is replaced by 1. For example `ab?ba` is associated to 2 whose binary representation is 010. This enables to mark easily the nodes of the lattice.

The candidate patterns are generated by considering the strings from $y_0$ to $y_{n-1}$. When a pattern is generated from $y_j$ it occurs at least once in $y_j$ then it is searched in the following strings: $y_{j+1}, y_{j+2}, \ldots, y_{n-1}$.

This procedure considers a factorization of an $\ell$-pattern $x$ as follows:

$$x = u_0 v_0 u_1, v_1 \cdots u_{m-1} v_{m-1} u_m$$

where $u_i \in \Sigma^*$ for $0 \leq i \leq m$ and $v_j \in \{?\}^*$ for $0 \leq j \leq m-1$ (remember that an $\ell$-pattern begins and ends with a solid symbol).

Then the search of $x$ in $y$ is performed by successively searching for the $u_i$ in $y$ and merging sets of positions. Assume that $R$ contains all the positions of $u_1 \cdots u_i$ in $y$, then the procedure computes the set $T$ of all the positions of $u_{i+1}$ in $y$. The two sets are merge in order to keep only the positions of $R$ that are compatible with positions of $T$. A position $g$ of $u_i$ is compatible with a position of $T$ if there exists a position $h$ in $T$ such that $g = h + |u_1 \cdots v_i|$. The merge can be easily realized if the two sets are implemented as ordered arrays. The set of positions of $u_i$ in $y$ can be computed efficiently by using any indexing structures of $y$ (such as suffix trees or suffix arrays).

A problem remains when a more specific patterns is found after a more general pattern. Then the set of resulting patterns has to be filtered.

### 3.2 SMS-H-Forbid Algorithm

In order to easily find the candidate patterns we define a table $H$ for every couple of solid symbols and every integer $k$ from 0 to $p-3$ as follows:

$$H[a, b, k] = \{(i, j) \mid y_i[j] = a \text{ and } y_i[j + k + 2] = b\}.$$

When a candidate $\ell$-pattern is generated from position $j$ in string $y_i$, if

- it has not already been output or
- it does not contain minimal forbidden patterns as factors or
- it is not more general than an output pattern

its potential occurrences are only searched at the positions in $H[y_i[j], y_i[j+\ell-1], \ell-2]$. In practice, the elements of $H[a, b, k]$ are sorted in decreasing order of the index of the strings.

The algorithm scans the strings of $Y$ in the same order than algorithm SMS-Forbid. The breadth-first-search is performed in the same manner. The only changes appear for counting the number of strings containing an $\ell$-pattern $x$ generated from $y_j$. The occurrences of $x$ are searched using the list of pairs in $H[x[0], x[\ell-1], \ell-3]$. Furthermore those pairs $(ind, pos)$ are sorted in decreasing order thus only pairs where $ind > j$ are considered.

### 3.3 Complexities

We will now give the complexities of both algorithms SMS-FORBID and SMS-H-FORBID.

The algorithm SMS-FORBID scans all the positions of the $n$ sequences of $Y$. For each position it considers all the $\ell$-patterns defined by the corresponding $\ell$-mer for $3 \leq \ell \leq p$. The number of elements of all the corresponding lattices is bounded by $2^{p+1}$. Processing one $\ell$-pattern $x$ consists in looking if $x$ is in *Res*, checking if $x$ contains minimal forbidden patterns and searching $x$ in the $n$ sequences of $Y$. Looking if $x$ is included in *Res* can be done in $O(|x|)$ time using a trie for *Res*. Checking if $x$ contains minimal forbidden patterns consists in using an algorithm for searching a single pattern with wildcard symbols in a text with wildcard symbols for every pattern in $\mathcal{T}$. This can be done in $O(|\mathcal{T}||x|)$. The search of one $\ell$-pattern $x$ in one string $y$ of $Y$ consists in spelling all the solid factors of $x$ into the indexing structure of $y$. When a solid factor $u$ of length $|u|$ is searched in $y$, it can be done in $O(|u| \log \sigma)$ with the suffix tree of $y$ and in $O(|u| + \log |y|)$ with the suffix array and the Longest Common Prefix array of $y$ (see [3]). However the search for each solid factor can return a list of positions of size $O(|y|)$.

Thus the time complexity of this step is $O(|x| \log \sigma + p|y|)$ with suffix trees and is $O(|x| + \log |y| + p|y|)$ with suffix arrays.

The time complexity for building the indexing structures for all the $n$ sequences of $Y$ is $O(N)$. Altogether the time complexity of the algorithm SMS-FORBID is $O(N 2^p \sigma^p ((p + \log m) + pm)))$ where $m$ is the maximal length of the sequences of $Y$.

The algorithm requires to build and traverse all the lattices corresponding to $\ell$-patterns. An array of size $2^{\ell-2}$ is used to mark the nodes of each lattice. Thus the space complexity for the lattices is $O(2^p)$. The space complexity of the indexing structures for all the $n$ sequences of $Y$ is $O(N)$. In the worst case the size of *Res* and $\mathcal{T}$ is bounded by $|\Sigma|^p$. Altogether the space complexity of the algorithm SMS-FORBID is $O(N + 2^p + |\Sigma|^p)$.

A similar analysis can be done for algorithm SMS-H-FORBID. The complexities are summarized in table 1.

**Table 1.** Time and space complexities of algorithms SMS-FORBID and SMS-H-FORBID

| Algorithm | Time | Space |
|---|---|---|
| SMS-FORBID (suffix trees) | $O(N 2^p \sigma^p (p \log \sigma + pm)))$ | $O(N + 2^p + \sigma^p)$ |
| SMS-FORBID (suffix arrays) | $O(N 2^p \sigma^p ((p + \log m) + pm))$ | $O(N + 2^p + \sigma^p)$ |
| SMS-H-FORBID | $O(N 2^p \sigma^p (pm))$ | $O(\sigma^2 p + 2^p + \sigma^p)$ |

Next, we will denote by SMS-ST the SMS-FORBID algorithm using suffix trees, SMS-SA the SMS-FORBID algorithm using suffix arrays and SMS-HASH the SMS-H-FORBID algorithm.

## 4 Experimental Results

We implemented our algorithms in C. Experimental results were conducted on pseudo-randomly generated data using the KISS generator [10] on two alphabet sizes 4 (to simulate DNA sequences) and 20 (to simulate protein sequences). All the experiments were performed on a P4.2 GHz machine with 3GB memory.

We measured the computing time and the space consumption of these algorithms for different values of $m$, $n$, $p$ and $q$. All these results have been obtained through computing an average on 15 draws. Fig. 2 and 3 show the variation of computing time in function of $m$ and $n$ for $p = 7$ and $q = 10$ on DNA alphabet. The X-axis represents $m$ and the Y-axis represents the computing time $t$ in seconds. We note that we have used a nonlinear method to implement the suffix arrays for the input strings but which is fast in practice.

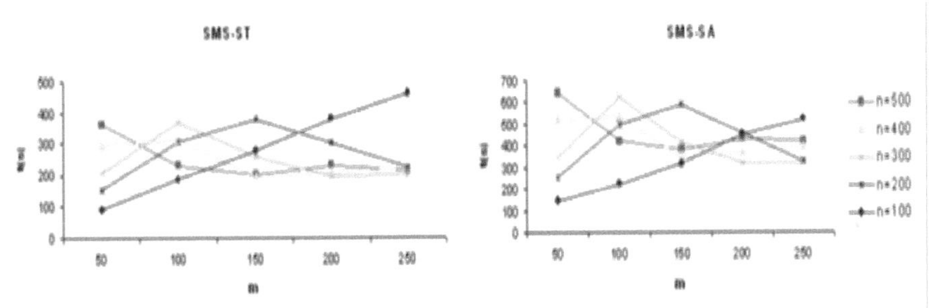


**Fig. 2.** Computing time in seconds of SMS–FORBID using Suffix Trees (left) and using Suffix Arrays (right)

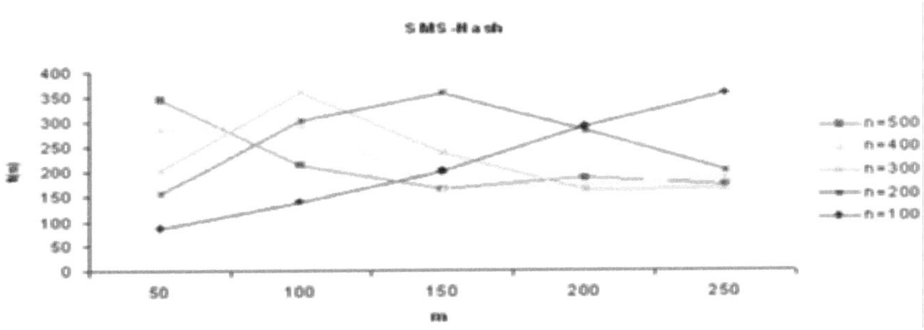


**Fig. 3.** Computing time in seconds of SMS-H-FORBID

As we can notice, SMS-H-FORBID performs better than SMS-FORBID in terms of running time since it uses a hash table instead of an indexing structure for the strings in $Y$(see Fig. 4). Indeed, the hash table used in SMS-H-FORBID makes easier the search of candidate patterns.

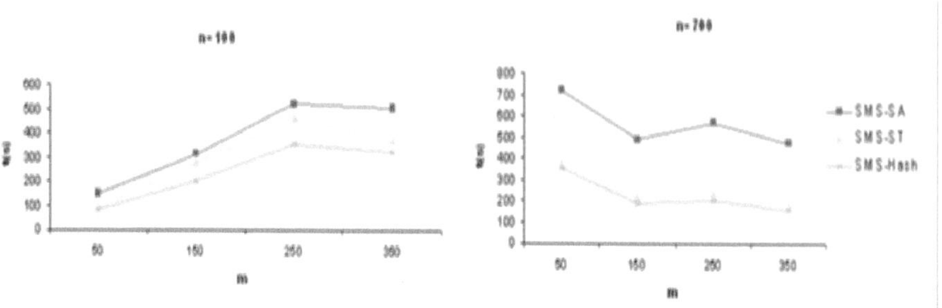


**Fig. 4.** Computing time for $n = 100$ (left) and $n = 700$ (right) with DNA alphabet

We have also computed the memory consumption in MB of the specific data structures used by each algorithm (see table 2). This table show that the SMS-FORBID algorithm using suffix arrays consumes less memory than the other algorithms.

We have employed our algorithms on various pseudo-randomly generated protein sequences (see Fig. 5). The experimental study show that SMS-H-FORBID algorithm is a faster solution to SMP when the input strings are built with protein alphabet, despite the fact that its worst case time complexity is bigger than the one of SMS-FORBID.

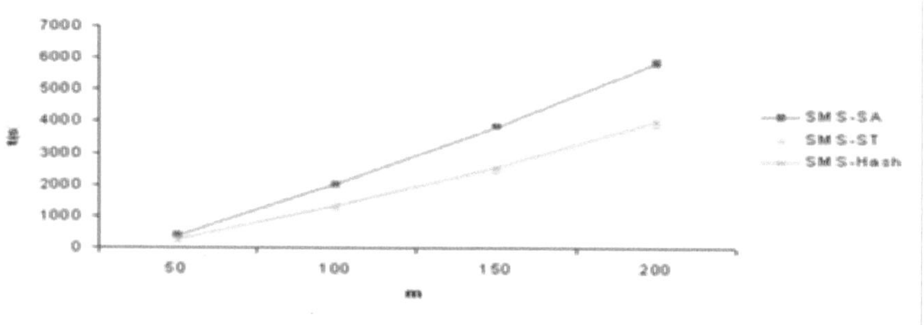


**Fig. 5.** Computing time in seconds with protein alphabet for $n = 100$

Fig. 6 shows the results obtained for SMS-H-FORBID in function of $q$ and $n$ for $m = 100$. As we can notice, the curves have a bell-like shape. Indeed, for a low value of $q$, the possibility of finding quickly the motifs is higher. When $q$ is getting closer to $n$, the number of detected minimal forbidden motifs increases. So, the possibility to reduce the computing time is also higher. It is also remarkable to note that the peak of the curves obtained with protein alphabet is for a low value of the quorum. Indeed, the size of the alphabet makes the possibility of finding simple motifs that have occurrences in at least $q$ strings decreases rapidly as $q$ increases.

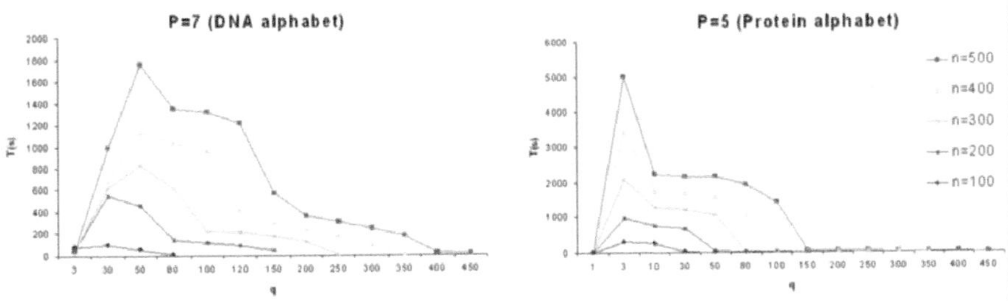


**Fig. 6.** Computing time in seconds of SMS-H-FORBID in function of $q$ and $n$ with DNA alphabet for $p = 7$ (left) and with protein alphabet for $p = 5$ (right)

**Table 2.** space consumption in MB with DNA alphabet for $n = 500$

| Algorithm \ $m$ | 50 | 150 | 250 | 350 |
|---|---|---|---|---|
| SMS-SA | 0.29 | 0.73 | 1.17 | 1.61 |
| SMS-ST | 1.6 | 4.8 | 8.02 | 11.23 |
| SMS-HASH | 0.47 | 1.5 | 2.52 | 3.55 |

## 5 Conclusion

In this paper, we have proposed an experimental evaluation of algorithms dealing with SMP. It is remarkable to note that our algorithms have the ability to identify the most specific motifs earlier in the search process, which allows less specific motifs to be pruned. By using this technique, we avoid the sorting step of SMS algorithm presented in [12] and we filter motifs respecting the generality relation between patterns. Experimental results show that our algorithms are well performing in practice. To have a more efficient solution in terms of space consumption, we are studying how to determine a new algorithm using the FM-Index [7] which is a data structure based on the Burrows-Wheeler Transform [1].

## References

1. Adjeroh, D., Bell, T., Mukherjee, A.: The Burrows-Wheeler Transform. Springer, Heidelberg (2008)
2. Chin, F.Y.L., Leung, H.C.M.: Voting algorithm for discovering long motifs. In: Proceedings of Asia-Pacific Bioinformatics Conference, pp. 261–272 (2005)
3. Crochemore, M., Hancart, C., Lecroq, T.: Algorithms on Strings. Cambridge University Press, Cambridge (2007)
4. El Falah, T., Elloumi, M., Lecroq, T.: Motif finding algorithms in biological sequences. In: Algorithms in Computational Molecular Biology: Techniques, Approaches and Applications, Wiley Book Series on Bioinformatics: Computational Techniques and Ingeneering, pp. 387–398. Wiley-Blackwell, John Wiley and Sons Ltd., New Jersey, USA (2011)

5. El Falah, T., Lecroq, T., Elloumi, M.: SMS-Forbid: an efficient algorithm for simple motif problem. In: Proceedings of the ISCA 2nd International Conference on Bioinformatics and Computational Biology, Honolulu, Hawai, pp. 121–126 (2010)
6. El Falah, T., Lecroq, T., Elloumi, M.: An efficient motif search algorithm based on a minimal forbidden patterns approach. In: Proceedings of the 5th International Conference on Practical Applications of Computational Biology and Bioinformatics. AISC, pp. 175–182. Springer, Heidelberg (2011)
7. Ferragina, P., Manzini, G.: Opportunistic data structures with applications. In: Proceedings of the 41st IEEE Symposium on Foundations of Computer Science (FOCS), Redondo Beach, CA, USA, pp. 390–398 (2000)
8. Floratos, A., Rigoutsos, I.: On the time complexity of the teiresias algorithm. Technical report, Research Report RC 21161 (94582), IBM T.J. Watson Research Center (1998)
9. Leung, H.C.M., Chin, F.Y.L.: An efficient algorithm for the extended (l,d)-motif problem, with unknown number of binding sites. In: Proceedings of the Fifth IEEE Symposium on Bioinformatics and Bioengineering (BIBE 2005), pp. 11–18 (2005)
10. Marsaglia, G., Zaman, A.: The kiss genrator. Technical report, Department of Statistics, Florida State University, Tallahassee, FL, USA (1993)
11. Rajasekaran, S., Balla, S., Huang, C.H.: Exact algorithms for planted motif challenge problems. Journal of Computational Biology 12(8), 1117–1128 (2005)
12. Rajasekaran, S., Balla, S., Huang, C.-H., Thapar, V., Gryk, M., Maciejewski, M., Schiller, M.: High-performance exact algorithms for motif search. Journal of Clinical Monitoring and Computing 19, 319–328 (2005)
13. Sagot, M.-F.: Spelling approximate repeated or common motifs using a suffix tree. In: Lucchesi, C.L., Moura, A.V. (eds.) LATIN 1998. LNCS, vol. 1380, pp. 111–127. Springer, Heidelberg (1998)
14. Styczynski, M.P., Jensen, K.L., Rigoutsos, I., Stephanopoulos, G.N.: An extension and novel solution to the (l,d)-motif challenge problem. Genome Informatics 15(2), 63–71 (2004)

# Secure Web Services Negotiation

Vakul Mohanty and Chittaranjan Hota

Birla Institute of Technology and Sciences Pilani,
Hyderabad, India
vakul.mohanty@gmail.com, hota@bits-hyderabad.ac.in

**Abstract.** The work proposed here, suggests a protocol that provides anonymity to negotiating parties and protects negotiation information without the use of encryption. The protocol is developed to ensure maximum security while causing minimum delay. Decentralization of routing and service information protects the information. The protocol makes use of K-anonymity principle to ensure security.

**Keywords:** Security, anonymity, cryptography, negotiation.

## 1 Introduction

Improvement in Internet technology and its wide reach have made Internet a valuable tool for conducting business on world wide web. Hence, more and more organizations are seeking to use web services to forward their business interests. Web services constitute of three phases: discovery, negotiation[2] and contracting. This work addresses anonymity and data security issues in the negotiation phase. In our work we device a protocol that maintains anonymity of negotiating parties and ensures that negotiation related data is kept hidden. The protocol thus not only ensures anonymity of negotiating parties but also hides the subject of negotiation, its attributes and negotiation status. A recent work [3] lists secured communication strategies based on encryption which is impractical in service sector for high latency. In this paper we take on this problem using K-anonymity [4] [5]; and the protocol is devised as a group protocol that can be well crafted into the generic architecture of webservices. In next section we propose re-engineering web service architecture to support the proposed protocol that assumes anonymity in service negotiation. The third section presents a protocol and in the next section it's performance is analysed. Then we conclude the paper with a comment.

## 2 Restructuring Web Service Architecture

A web service consists of three entities viz. service-provider, register and consumer Fig. 1 .Currently, WSDL is enabled for secure communication among services mainly by sharing keys and encrypting data. But, this could be a hindrance for mass usages for high latency. We propose to ensure security by adopting K-anonymity for communication. For that, business entities registered at UDDI are grouped into several clusters or groups. Each cluster is assigned with a group leader

D. Nagamalai, E. Renault, M. Dhanushkodi (Eds.): CCSEIT 2011, CCIS 204, pp. 609–618, 2011.

(representative) by the UDDI. The repository of conventional UDDI is augmented with additional information . The UDDI stores group information: 1. group leader's address 2. size of group 3. index of group leader.

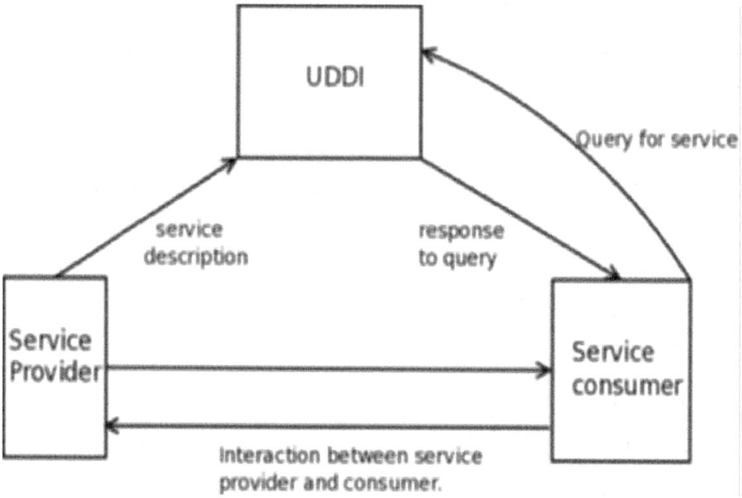


**Fig. 1.** Traditional web service architecture

### 2.1 Groups at UDDI

Initially, business entities (both service consumers and providers) register them selves at UDDI. And then, according to our scheme the entities are grouped to a predefined size say $K_{max}$. For each group a group leader is designated and the group members are uniquely indexed. The information on each group i.e its group leader and members are stored at UDDI for routing communication to and from the members via the group leader. The group leader only is aware of physical address of its members. More, the members in a group may belong to several different business domain to add diversity to group that camouflage group business domain information to external world. Next, we will discuss dynamic evolution of a group for movement i.e. joining or leaving of the group.

*Management*: The scenarios we may face after initialization of UDDI are:

1. *Entry of an entity*: As and when a new user wants to carry out business transactions it first makes a request to the UDDI. The UDDI informs one of the group leaders to include the user as its member, at the same time updates the group size. The group leader then allots the user its index hence making it a group member. Accepting an index and joining a cluster binds a user to group as a cluster member. If the new member is a service provider it updates the UDDI with all its service details. Here we would like to mention that we assume that group leaders are honest and don't participate in activities like adding users to the groups without being directed by the UDDI.

2. *New cluster formation*: In case a new group is to be created the procedure is the same as the initialization stage. The situation arises due to overgrowth of a cluster i.e when each group attains the size $K_{max.}$
3. *Exit of entity*: In case a cluster member wants to leave the cluster it informs the UDDI and its group leader. The UDDI then updates the cluster size and in case the user is a service provider its service details are also removed. The group updates itself deleting the user and its index value. By doing this the group leader removes the user from the cluster and rendering it incapable of transacting web services in habitat of the current SOA. On exit, when the population of a group is below a set minimum threshold value, the group members are directed to join different groups on dissolving the existing one.
4. *Exit of a group leader:* Suppose a group leader wants to leave a cluster, then the above procedure is carried out; in addition the UDDI choose a new group leader from among the group members and updates its data, i.e. includes the new group leader, provides a new index, updates group size. Similarly, the new group leader assumes the information, that was available at previous group leader .

The dynamic nature of the clusters effect the performance of the protocol suggested. Hence we assume that the groups engaged in negotiation do not change. We discuss the consequences of the dynamic nature of the groups later and suggest modifications that might help to overcome the problems posed.

**Table 1.** Message structures

| Message | Message structure | Path No (Fig. 2,Fig. 3) |
|---|---|---|
| ServiceRequest | < *servicename* >< *replyaddress* > < *destinationaddress* > | 1,2 |
| ServiceReply | < *Servicedetails* >< *serviceprovideraddress* > < *destinationaddress* > | 3,4 |
| Negotiation message | *{*< *attributevalues* >*}* < *negotiationstatus* > < *replyaddress* >< *destinationaddress* > | 5,6,7 8,9,10 |

## 3 Protocol

This section presents the proposed protocol designed to carry out an anonymous negotiation. The basic principle of the protocol is to generate look-real dummy information that camouflage the real so that any malicious monitor cannot differentiate the information regarding the negotiation. The identities of the parties participating in negotiation are made hidden by assigning them indexes instead of addressing their real address. Its to be noted that only UDDI knows group index and a group leader knows indexes of its members. We assume that UDDI and group leaders are trusted so identities of group members and their service informations are

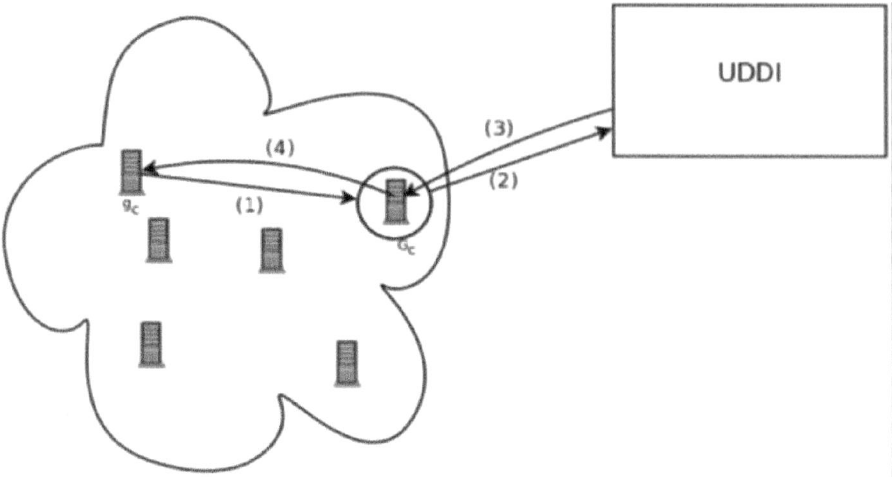


**Fig. 2.** Service Discovery phase

confidential. A negotiation message is itself buried by creating a traffic of k dummy messages from which differentiating the actual message becomes a difficult task. Further, making message structure i.e. service attributes a pre-knowledge to both the negotiating parties and avoiding usages of these attribute names at message transaction helps to hide status of negotiation to a large extent. It is to be noted that message contains only values so that a malicious monitor can not associate meaning to these values due to absence of message attribute names. This protects the subject of negotiation. Further, values in different dummy messages cast a different trends thus able to obfuscate the status of negotiation as well. Thus the proposed protocol intends to achieve 1) anonymity of negotiating parties 2) anonymity of context and 3) anonymity of negotiation status.

Assumptions made: 1. All communications with the UDDI are secure and observable. 2. all communications in the group are secure and group leaders are honest. 3. The cluster is like a closed system, they communications taking place within them are observable. The assumptions imply that negotiation between parties in the same cluster are always safe and are thus trivial. In case the negotiating parties are in different cluster we follow the proposed protocol.

Let there be a consumer $g_c$ and a provider $g_p$ belonging to their respective group leaders $G_c$ and $G_p$. A path between them is made of subpaths $g_c \rightarrow G_c$ , $G_c \rightarrow G_p$ and $G_p \rightarrow g_p$. Communications can occur on both the ways on these paths. These communications can be broadly classified as intra and inter group communications. Further, we can call messages forwarded by a group leader to a group member as a delivery message and all other communications as forwarding messages (Fig. 2, Fig. 3), from the security point of view we need to secure all inter group communications. The protocol consists of a service finding stage and then the negotiation stage. The first stage that is the service finding stage occurs once, the negotiation stage occurs several times until the negotiation process terminates. The negotiation stage [2]can be further divided into the offer making and the offer

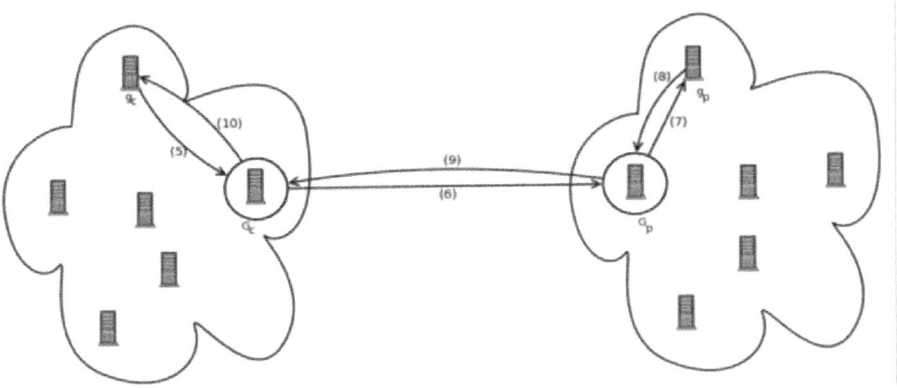


**Fig. 3.** Negotiation phase

generation stages. Offer making stage at a party involves the process by which an offer is submitted to the other party. The offer generation stage involves creating a counter offer based on a previously received offer. Now, we discuss the protocol by giving a detailed explanation for each stage. Its to be noted that here we are neither interested in offer making and counter offer generation strategies. Rather, our interest includes ensuring anonymity in negotiation.

**Service Finding Phase:** (Fig:2) Assuming a consumer $g_c$ looking for a ser-vice, it sends a ServiceRequest message to UDDI, with the service name as a message parameter. We assume that each in SOA uses a common ontology; this assumption doesn't dilute the topic in the discussion. On receiving the message, the UDDI searches its repository and replies giving the details of the best service provider through a message called ServiceReply [Table 1]. The message contains information on service provider's index i, address of the service providers group leader, service description with set of feature names and associated initial feature values < f, v >. It is to be noted that the service details are provided by the service provider when it registers at the UDDI. This process also occurs with the group leader as the intermediate node. When the service request is forwarded by the group leader it appends its index value to it, this allows the UDDI to identify that the service request is from a group member. but as communication is secure the group leader also has no idea about the contents of the messages exchanged, thus the group leader is oblivious about the nature of the service requested. Also the UDDI only entertains service requests routed through a leader. This ensures that only group members participate in the web service protocols.

**Negotiation Phase:** once serviceFinding is completed negotiation is carried by the exchange of offers and counter offers. A negotiation is shown pictorially in Fig:3.

**Offer Making:** A party that wants to make an offer sends a negotiation message M to its group leader. Along with M it also forwards a set of k dummy messages that constitute a buffer traffic to hide the negotiation message. The attributes and construction of these dummy messages is discussed in detail in a Table1. The message holds the values of the service attributes, address of the destination, reply address and negotiation status. The status is used to convey to the other party the status of the ongoing negotiation. On receiving the message the group leader accesses the destination address and forwards it to the group leader of the destination cluster.( Destination address contains address of the destination clusters group leader and the destinations index value. A reply address, of the other end of the negotiating party, also has a similar structure). When the group leader of the destination cluster receives the message it access the destination address referring to destination index; And then forwards.

**Offer generation phase:** On receiving an offer the party first checks the status field which shows whether the other party has accepted or rejected the offer, or wants to continue negotiation. If offer is accepted or rejected negotiation is terminated. If the party at other end, intends to continue negotiation then negotiation utility value is calculated. The utility value is calculated from the values it receives for various attributes in the negotiation. message. Base on utility value, it then accepts or rejects or decides to continue the negotiation [2] Once this decision is made the party makes a negotiation message by assigning appropriate values to all the fields. The message is then sent to the address that is provided in the return address field of the offer previously received. Hence the process once again enters the offer making stage. A recursion of offer making and offer generation stage compose the negotiation process. How the above protocol is able to provide data security and anonymity has been analysed in next section. An algorithm for the protocol is given at Algorithm 1 and notations used in it are described in Table 2.

**Table 2.** Notations

| | |
|---|---|
| $M_s, M_S'$ | Messages for ServiceRequest |
| $M_{SR}$ | Message for ServiceReply |
| $R_A$ | Reply address |
| $D_A$ | Destination address |
| j | Reply index |
| i | Destination index |
| $g_p$ | Service provider |
| $g_c$ | Consumer |
| C | Consumer's group leaders index |
| M | Negotiation message |
| S | Service name |
| $S_a$ | Negotiation status |
| SD | service description |
| $G_{C/P}$ | represents Consumer/Provider group leaders |

**Algorithm 1.** The routing protocol

***Service Discovery***:
At $g_c$ **do** {send($M_s$,$G_c$)}
On receive $M_s$ at($G_c$ **do** {
$M_s^{'}$= $AppendID$ ($M_s, C$) //Put consumer group leader ID
Send ($M_s^{'}, UDDI$)} // send messaeg to UDDI
On receive $M_s^{'}$ at UDDI **do** {}
**if** ValidID(C) **then**
  **do** $M_{SR}$ = SearchService($M_s^{'}$)
  send ($M_{SR}, C$) }
**end if**
On receiving ($M_{SR}$)at ($G_c$ **do** { send($M_{SR}, g_c$)}
On receiving ($M_{SR}$)at ($g_c$ **do** { SD = retrieve($M_{SR}$)}
***Negotiation phase***:
//**initialization**//
At($g_c$) **do** {
M = generateOffer(SD); send(M,$G_c$) }
//* $g_c$ *or* $g_p$ *forward negotiation message along with k dummy messages*//
//* *Initialization ends*//
On receiving (M){ At($G_c/G_P$)**do** {
**if** message is from a group member **then**
  {send(M,DA)} // to receiving group leader
**else**
  {send(M,i) } //to consumer or provider
**end if**
On receiving (M) at($g_c/g_P$)**do** {
**if** ($S_a$ = (CheckNegotiation(M)= 'rejected')) **then**
  *terminate*
  end negotiation; }
**end if**
**if** ($S_a$= (CheckNegotiation(M)= 'accepted')) **then**
  { // end negotiation and agree to provide service
  provideService($g_c$)}
**else**
  M = generateOffer(); send(M,DA) //negotiation continues
**end if**

### 3.1 Generation of Dummy Messages

When offers are being exchanged during the negotiation phase of the protocol the sender creates traffic of dummy messages to hide the negotiation message. This section deals with the generation of the messages according to the properties we define for them. When creating the dummy messages the following attributes of the dummy messages are considered:

1. The messages are all of the same length as the negotiation message and have the same number of fields. The type of the fields and their values are however different.

2. The reply address and the destination address of all the dummy messages are the same as that of the negotiation message.
3. The value in status field is immaterial and is filled with a random value.

The dummy messages are created by varying the combination of field types. If a message has F fields and we have X data types, we can form $F^X$ - 1 different dummy messages as there are X possible types for each field. A random value generator is used to generate the values required to populate the fields.

## 4 Analysis

In order to analyze the protocol we define adversary models. Using these models we predict the extent to which the protocol is able to provide security. The models are:

1. Passive monitoring: Adversary monitors and makes logs of the traffic. Traffic analysis allows the adversary to find patterns in the data exchanged and makes a guess on the data exchanged and on the communicating parties.
2. Active attacker: These adversaries are capable of modifying the messages or disrupting the proper flow of messages by adding or withholding messages.
3. Cooperative attack: Cooperative attack is an attack where two or more adversaries collect and share data among themselves to draw conclusions regarding the identity of the negotiating parties, subject of negotiation and other details.

As we are not using cryptography we hide data by decentralizing it. Service details are known only to the UDDI and the negotiating parties. The indexing structure of a group is known only to the group leader. Hence effectively no single entity has all the information regarding a negotiation process.

**Theorem 1:** The probability of identifying negotiating parties is inversely proportional to sizes of their respective groups.

The proof of the theorem follows from this discussion: Each group is indexed in range 1…$K_{max}$. Say a consumer and a provider with indices i and j respectively are to be traced from the indices of groups of provider and consumer. Assuming $P_s$ and $C_s$ are the set of members of provider and consumer groups have respectively, the probability to trace negotiating parties could be $\frac{1}{|P_s|} * \frac{1}{|C_s|}$.This can only help if an attacker can succeed compromise of the group leaders as they only have the identities of all the group members.

**Lemma 1:** The probability of tracing a negotiation message is $\frac{1}{k}$

The lemma follows directly as both the parties transact k dummy messages as discussed in previous section. And the dummy messages can be sampled from $F^X$ types, making distinguishing the real one pretty difficult.

**Theorem 2.** Proposed protocol provides both anonymity of negotiating par-ties and negotiation contents.

Theorem 1 says of possible maintenance of anonymity of negotiating parties. And Lemma 1 says the difficulty in tracing the message taking part in negotiation. Further, not providing field names of a negotiation message, for an attacker understanding of message context becomes difficult. Now we will discuss of the cases for different types of attackers. Though a passive attacker who logs transacted messages may have some time to probe but also it needs service details that are available at UDDI only. Now, such an attacker needs both group leaders as well as UDDI compromised which is highly probabilistic. Active attackers can add, subtract, substitute or modify a message in the traffic. If it either add or delete message contents, than the message format gets disturbed and this alerts the parties on possible active attack. If it changes some values in message keeping the message format intact, then negotiation process may get affected i.e either it is abandoned or false negotiation may set in. In case of the former, there is loss of secrecy except the time both have spent in negotiating. For the later case, manipulation of exact field values to force negotiation is highly unlikely as the negotiation message format is not available in the body transacted negotiation messages. Even if it happens, this can be identified in contracting stage when both the parties sign service agreement. Cooperative attackers at both group leaders can only be aware of occurrence of a negotiation but will have difficulty in identifying negotiating parties and understanding the contents of negotiation messages as service details are only known to both the negotiating parties and they are anonymous in their respective groups.

## 5 Conclusion

Our work has suggested a secure method for negotiation mostly based on anonymity. In order to gain time, we have not considered encryption. We argue that many day to day business transaction needs high response time and are also not so secret. For that kind of business purpose, the proposed protocol is workable though it needs to transact some unwanted messages. The analysis indicates that the protocol is secure to all attacks models we have discussed. However the negotiation is corrupted when an active adversary modifies the contents of the negotiating packet. Though the probability of success of the attack is low it cannot be completely avoided. Our future work is aimed at rectifying this problem. We also seek to simulate the protocol and attain experimental proof for the results we have predicted during the analysis of the protocol. We also look to make a comparative study between the protocol we have suggested and the security protocols generally used now in the business domain.

## References

1. Shannon, C.E.: Communication Theory for Secrecy Systems confidential report. A Mathematical Theory of Cryptography, September 1 (1946)
2. Mohanty, H., Kurra, R., Shymasundar, R.K.: A Framework for Web Based Negotiation. In: Janowski, T., Mohanty, H. (eds.) ICDCIT 2010. LNCS, vol. 5966, pp. 140–151. Springer, Heidelberg (2010)
3. Ren, J., Wu, J.: Survey on Anonymous Communication in Computer Networks. Computer Communication 33(4), 420–431 (2010)

4. Byun, J.-W., Kamra, A., Bertino, E., Li, N.: Efficient K-Anonymity Using Clustering Techniques. In: Kotagiri, R., Radha Krishna, P., Mohania, M., Nantajeewarawat, E. (eds.) DASFAA 2007. LNCS, vol. 4443, pp. 188–200. Springer, Heidelberg (2007)
5. Sweeney, L.: K-Anonymity: A Model For Protecting Privacy. International Journal on Uncertainty, Fuziness and Knowledge-Based Systems 10(5), 557–570 (2002)

# A Recommender System for Rural and Urban Learners

L. Arockiam[1], S. Charles[2], V. Arul Kumar[3], and P. Cijo[4]

[1] Associate Professor
[2] Assistant Professor
[3] Research Scholar
[4] Student
Department of Computer Science,
St Joseph's College, Tiruchirappalli

**Abstract.** In Educational data mining, learning process evaluation reveals the students' involvement in learning. It leads to improve the quality and success rate of the students in educational environment. Students' intellectual performance varies in learning environment and evaluated by different criteria. This study helps to find out the intelligent quotient level of the student and their performance, which are dissimilar in a classroom. Stanford Binet Intelligence Scale and Criterion Reference Model are used to evaluate the intelligence and performance of the students. Machine learning techniques are employed to find the intelligent quotient and performance of rural and urban students using students' dataset. Model based clustering technique is used to determine the similarity in the students' dataset based on the nature of the intelligent quotient and performance. Each cluster reveals the identity based on students intelligence level. The performance also categorized on descriptive modeling. Multilayer Perceptron technique classifies the intelligent level of the students and their performance. The study determines the association between intelligent level of the rural and urban students and performance through descriptive and predictive modeling. This analysis recommends the teaching community to provide right training to the rural students for their improvement of academic competence.

**Keywords:** Multilayer Perceptron, Expectation Maximization (EM) Clustering, Criterion Reference Model, Stanford Binet-Intelligence Scale, Intelligent Quotient level.

## 1 Introduction

Educational data mining is an emerging area of research in current scenario. It identifies three objectives such as educational, management and commercial. This paper deals with educational objectives, which determines the students IQ level in the college environment and their performance. Stanford-Binet Intelligence scale has been used for identifying the intelligence of the rural and urban students in post graduate course. This scale helps to measure the cognitive abilities of the same age group of rural and urban students. Criterion – Reference Model is used to categorize the performance of the students in their studies. The theory and

D. Nagamalai, E. Renault, M. Dhanushkodi (Eds.): CCSEIT 2011, CCIS 204, pp. 619–627, 2011.

programming marks can be viewed as performance. Cluster analysis is an unsupervised learning method that is used to investigate the inter-relationships among a collection of student objects, by organizing them into homogeneous clusters. EM clustering technique is used to group the student objects based on the intelligent level into classes of similar students 'objects [1]. Similarly the student performances are classified as good or fair.

The MLP is used for subcategory classification in the students' dataset. It is one of the approaches for discerning a pattern of subcategories. It estimates the useful information through mining and determines the interesting patterns of the specific category. Experimental unit variables are used for finding patterns in the students' dataset. An observational unit variable finds the subcategory of the intelligent level. Similarly it classifies the performance of students. This paper focuses on deriving association between intelligent Quotient of rural and urban students and their performance based on supervised and unsupervised learning process.

## 2 Motivation

### 2.1 Stanford Binet-Intelligence Scale

The Stanford-Binet intelligence scale is a test, which assesses intelligence and cognitive abilities of the age group between 2 and 80. It tests the intelligence Quotient level (IQL) of the student using the following criteria: general intelligence, knowledge, fluid reasoning, quantitative reasoning, visual-spatial processing, and working memory. These criteria are divided by ten subtests that include activities measuring both verbal and non-verbal intelligence. Activities include verbal absurdities, picture absurdities, verbal analogies, form patterns, procedural knowledge, sentence and word memory, position and direction, early reasoning, and quantitative reasoning. The test score is categorized as 130 and Above: Very High (VH), 120 – 129: High (H), 110 – 119: Above average (AA), 90 – 109: Average(A), 80-89: Below Average(BA), 70 – 79: Low and 60 or below: Deficient(D) [2,9]

The problem of discrimination and intelligent scale classification is mainly concerned with differentiating between g (g>=2) mutually exclusive students and with classifying intelligent level on the basis of Multivariate observations]. The mathematical problem of discrimination is to divide the observation space $R^P$ into g mutually exclusive and exhaustive regions $R_1 \ldots R_g$ so that if an observation space x falls in $R_i$ then it is classified as belonging to the population. The sub category problem is a mutually exclusive population according to their attributes and classifying an unseen instance into its sub categories using multiple features. In this study, the sub-categories are experimental unit variables and observed features are observational unit variables[3, 4].

Intelligence is characterized based on abstractness and correctness of level of intelligent the students' posses. The proposed model determines the intelligent level and is able to characterize, how the mind functions while learning.

### 2.2 Criterion-Referenced Model

A criterion-referenced model is an assessment model where the purpose is to assess the extent to which a student has achieved the goals of a course. The summative assessment is carried out aligned with specified criteria. Results are expressed in terms of how well a given student's performance matches set criteria. The standards are set before teaching takes place. The result is assigned on the basis of the standard that the student has achieved on the criteria [5].

### 2.3 Expectation Maximization (EM) Clustering

The EM algorithm is a popular iterative refinement algorithm that can be used for finding the parameter estimates. It assigns an object to the cluster with which it is most similar based on the cluster mean. Instead of assigning each object to a dedicated cluster, EM assigns each object to a cluster according to a weight representing the probability of membership. There are no strict boundaries between clusters. Therefore new means are computed based on weighted measures [6].

### 2.4 Multi-Layer Perceptron

The Multilayer Perceptron is one of the techniques for classification task. It is a network of simple neurons called perceptrons. The perceptron computes a single output from multiple real-valued inputs by forming a linear combination according to its input weights and then possibly putting the output through some nonlinear activation function [7].

## 3 DAIP Approach

In Phase I, the research questions are designed by the expert based on Stanford Binet Intelligence Scale and pretest the questionnaire. After pretesting a questionnaire, the Intelligence Scale analyze test has been conducted in Post Graduate Course. The written and practical marks are obtained from the semester exam, which are viewed as performance (good (Marks>=60) or fair (Marks<60)) [5, 8, 10] based on Criterion- Referenced Model. The students' dataset contains the test results of Stanford Binet

Intelligence scale and performance of the students in the written and practical test in programming subject.

In Phase II and III, two data mining methods are employed to find the association between the IQ level and performance. The methods are EM- clustering and Multi Layer Perceptron respectively. Each cluster reveals the identity of the students' IQ level and designates the cluster as 1) High 2) Above Average 3) Average 4) Below Average 5) Low 6) Deficient. The same method is employed for students' performance based clustering [11, 12 and 13]. The MLP technique is used to classify the Stanford Binet Intelligence level of the students. The five sigmoid nodes are used as inputs, the weights are assigned to each node, and six output layers are classified based on the category of intelligent level. According to the performance-based classification, the five sigmoid nodes are used as input and two output layers are classified based on the skills as good or fair [14]. In Phase IV, the association between intelligent level and performance of the students are analyzed using interesting measures.

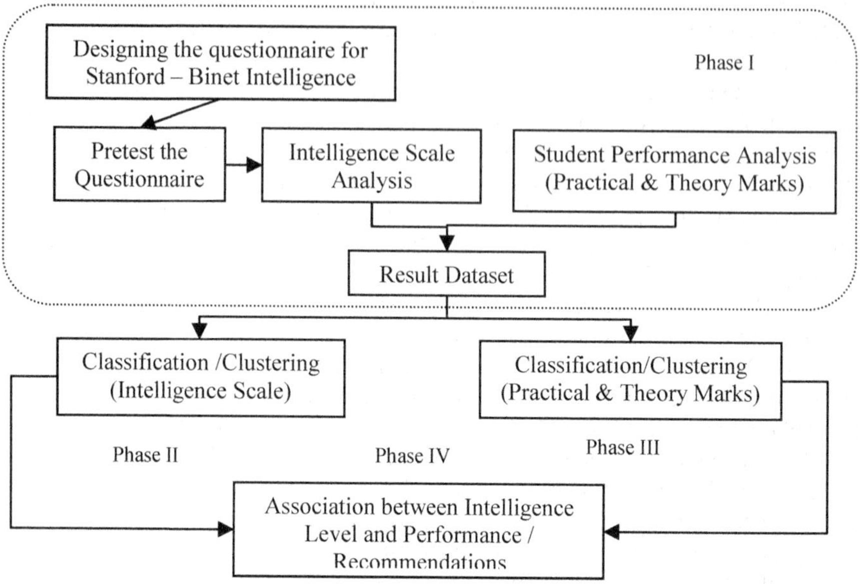


**Fig. 1.** DAIP Approach

## 4 Results and Discussions

The dataset containing 327 students' objects are used to find the association between intelligent level and their performance of the students.

**Table 1.** Cluster assignments and Intelligent Level

| C 0 | C 1 | C 2 | C 3 | C 4 | IQL |
|---|---|---|---|---|---|
| 56 | 6 | 36 | 6 | 10 | A |
| 6 | 0 | 24 | 0 | 0 | L |
| 17 | 0 | 6 | 0 | 0 | AA |
| 12 | 6 | 35 | 6 | 0 | BA |
| 36 | 24 | 12 | 29 | 0 | D |

**Table 2.** Confusion Matrix for Intelligent level

| Cross validation | | Predicted | | | | | |
|---|---|---|---|---|---|---|---|
| | | a | B | c | d | e | Intelligent Level |
| Actual | A | 114 | 0 | 0 | 0 | 0 | Average |
| | B | 0 | 30 | 0 | 0 | 0 | Low |
| | C | 0 | 0 | 23 | 0 | 0 | Above Average |
| | D | 0 | 0 | 0 | 59 | 0 | Below Average |
| | E | 0 | 0 | 0 | 0 | 101 | Deficient |

Table 1 shows that various clusters and corresponding IL are categorized. It reveals that Average IQ level of student objects are silhouetted in cluster 0, Low IQ level student objects are silhouetted in cluster 1, Above Average level student objects are silhouetted in cluster 2, Below IQ level of student objects are silhouetted in cluster 3 and Deficient IQ level of student objects are silhouetted in cluster 4. Similarly the MLP classifies the students' objects with a support of the class variable intelligent level. Table 2 reveals that the various classes of observed variables as Intelligent Level and experimental variables as demographic information of objects and theory and practical examination mark. The confusion matrix reveals that various classes using six sigmoid nodes to identify the six IQ levels.

**Table 3.** Intelligent Level Preference of Student

| **Intelligent Level** | **Response** | **Agreement %** |
|---|---|---|
| Above Average | 23 | 7 |
| Average | 114 | 35 |
| Below Average | 59 | 18 |
| Low | 30 | 9 |
| Deficient | 101 | 31 |

The intelligent level preferences and the percentage of respondents who agreed with each item are shown in Table 3. Table 3 clearly indicates that the intelligent level, which majority of Post Graduate Students posses. Approximately 7% of the students have viewed themselves as above average IQ level. 35% of the students are in average IQ level, 18% of the students are in below average IQ level, 9% of the students are in Low IQ level, 31% of the students are in Deficient, who posses with the different IQ level.

**Table 4.** Confusion Matrix for theory and practical performance

| Validation Data | | | | |
|---|---|---|---|---|
| **Actual Category** | **Predicted Category(Theory)** | | **Predicted Category(Practical)** | |
| | G | F | G | F |
| G | 154 | 0 | 303 | 0 |
| F | 0 | 173 | 4 | 20 |
| Mean absolute error | 0.0054 | | 0.151228 | |
| Root mean squared error | 0.0158 | | 0.128171 | |

Table 4 shows that the categorization of theory and practical performance of students using the experiment variables. This experiment reveals that 154 student objects are observed as good in theory subject, 173 students' objects are observed as fair in theory. Similarly the Practical marks are categorized using clustering. 303 student objects are observed as good in Practical lab, 24 students' objects are observed as fair in Practical lab.

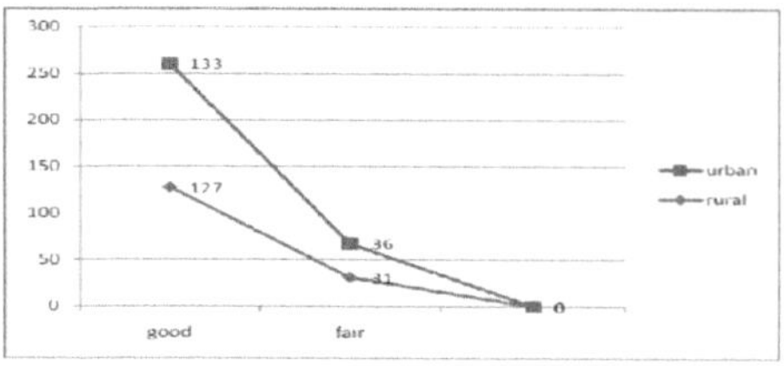


**Fig. 2.** Classification of theory performance of Rural and Urban Students

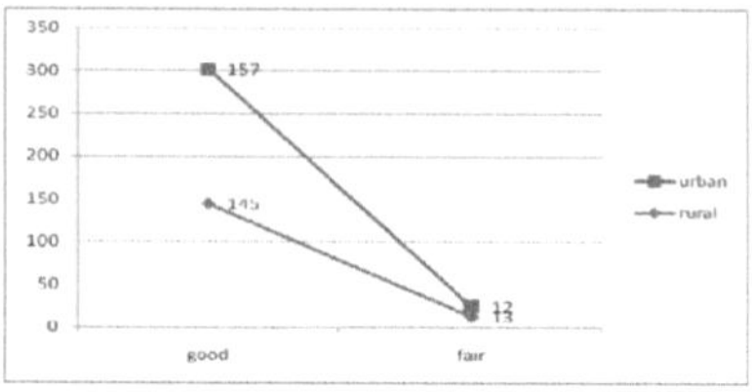


**Fig. 3.** Classification of Programming Lab Performance of Rural and Urban Students

Fig 2 reveals the classification of theory performance of rural and urban students, 127 (39%) rural students are observed as good in theory subject, 31(9%) rural students are observed as fair in theory subject, 133(41%) urban students are observed as good in theory subject, 36(11%) urban students are observed as fair in theory subject. Fig 3 reveals the classification of practical lab performance of rural and urban students, 145 (44%) rural students are observed as good in practical lab, 13 (4%) rural students are observed as fair in practical lab. 157 (48%) urban students are observed as good in practical lab, 12 (4%) urban students are observed as fair in practical lab.

**Table 5.** Intelligent Level Preference of rural and urban students

| Area/IL | AA (%) | A (%) | BA (%) | L (%) | D (%) |
|---|---|---|---|---|---|
| Urban | 5 | 22 | 6 | 2 | 18 |
| Rural | 2 | 13 | 13 | 7 | 12 |

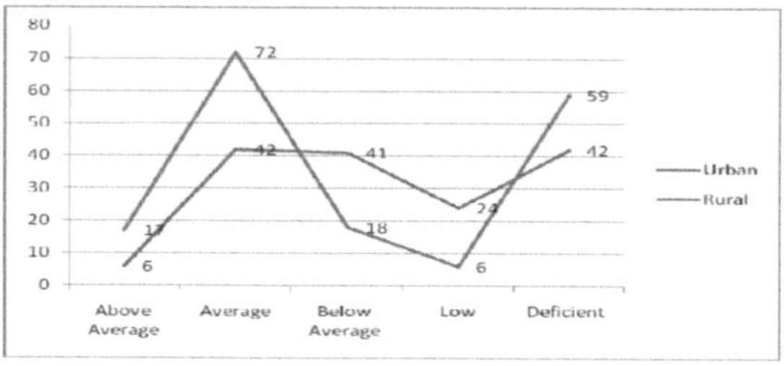


**Fig. 4.** Classification of IQ Level of the Rural and Urban Students

The urban and rural student objects are classified based on the intelligent level and Performance, 5% of urban students are in above average level, 22% of urban students are in average level, 6% of urban student are in low level, 18% of urban students are deficient level. 2% of rural student are in above average level, 13% of rural students are in average level, 13% of rural students in below average level, 7 % of rural students in Low level and 13% of rural students in Deficient level. It is evident that the urban students' intelligent levels are high compared to rural students

Fig 5 shows that the probability of students having good performance in theory and practical. The probability calibration chart is used for classification analysis to find the accuracy. The horizontal axis reveals the predicted probability of observations. The vertical axis reveals the actual probability of occurrence. The average predicted probability is calculated, which is lies between 0.62 and 0.65 and the actual probability is 0.6. The red shaded area reveals the difference between the predicted and actual probabilities. The average predicted probability error is 0.1056. The weighted probability error value is low. These error values divulge that the model is perfect.

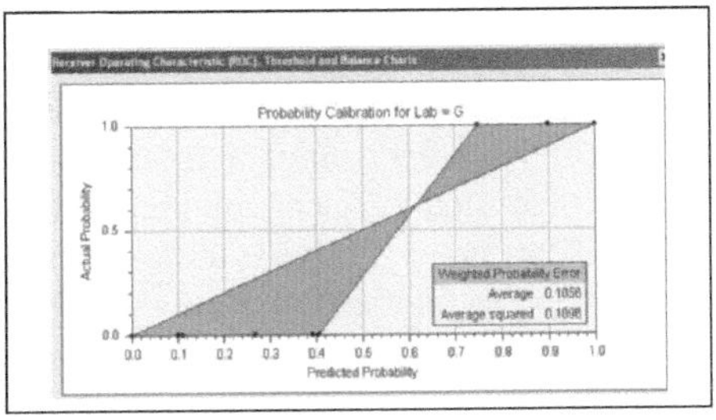


**Fig. 5.** Probability Calibration chart for Good Performance in Theory and Practical

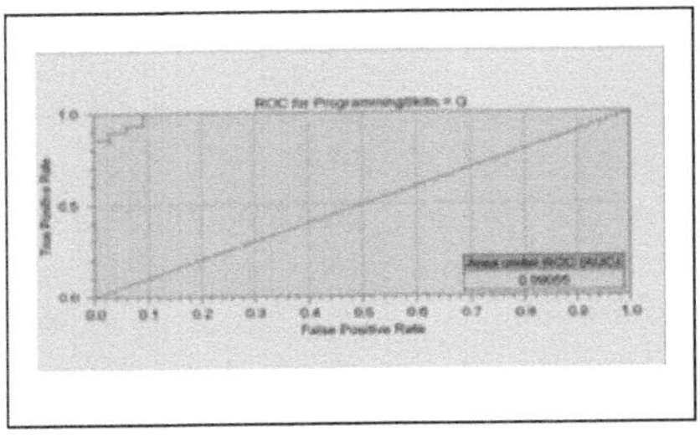

**Fig. 6.** ROC chart for Good Performance in Theory and Practical

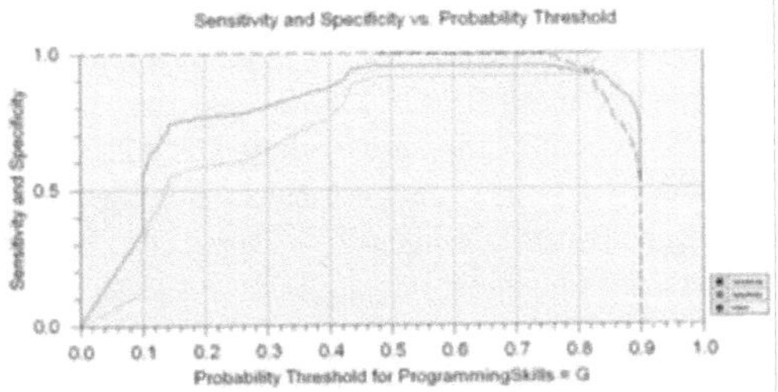


**Fig. 6.** Sensitivity and Specificity curve for Theory and Practical performance

Fig.6 shows Receiver Operating Characteristic chart and probability threshold values are elevated. It reveals that the urban students' performance is good in theory and practical. Sensitivity and Specificity curve values are equal. Fig.7 shows the sensitivity curve divulges that the urban and rural students having different IQ levels are good in Theory and Practical lab. The mapping or function can be formulated as Y = **f (X),** where X is the input and the output as Y. This function reveals the relation between X and Y. The lift measure divulges that correlated value of X and Y are 1.22, which is positively correlated. It means that the X and Y are related with each other. The cosine value of X and Y are 0.712. It reveals that, positive correlation is existed in X and Y. The observation divulges that IQ has a strong association with performance.

## 5 Conclusion

In this Paper, Model based clustering technique and Multi-Layer perceptron are used to find the association between intelligent quotient and performance. The clustering is an effective means for classifying the student objects such as Very High, High, above average, Average, Below Average, Low and Deficient and models each cluster. It reveals the identity of the IQ level of the students. The performance based clusters are observed as good and fair. The classifier reveals that, each individual student's IQ level related with the performance. More number of urban students having an average IQ level compared to rural students. Few urban students having an Above Average IQ level compared to rural students and few rural students having a below average IQ level compared to urban students. Similarly few rural students having low IQ level compared to urban students. IQ score gives us a measure of a students' academic competence such as reading, learning, problem solving, abstract reasoning and conceptual learning. This experiment reveals the academic competence of rural students and provides the recommendation to the rural learners for improving their skills. The rural learners are need to be improved their academic competence by taking enough training in various skills. The correlation analysis divulges that the Intelligent Quotient and performance have strong association.

## References

[1] Khalilian, M., Boroujeni, F.Z., Mustapha, N., Sulaiman, M.N.: K- Means Divide and Conquer Clustering. In: International Conference on Computer and Automation Engineering, pp. 306–309. IEEE Computer Society, Los Alamitos (2009)

[2] Becker, K.A.: History of the Stanford-Binet Intelligence Scales: Content and Psychometrics (Stanford-Binet Intelligence Scales, Fifth Edition Assessment Service Bulletin No. 1). Riverside Publishing, Itasca (2003)

[3] Choi, S.C., Rodin, E.Y.: Statistical Methods of Discrimination and Classification. In: Advances in Theory and Applications. Pergamon Press, Oxford (1986)

[4] Duda, R.O., Hart, P.E., Stork, D.G.: Pattern Classification, 2nd edn. John. Wiley & sons Inc., Chichester (2000)

[5] Khamis, N., Idris, S.: Issues and Solutions in Assessing Object-oriented programming Skills in the Core Education of Computer Science and Information Technology. In: 12th WSEAS International Conference on Computers, Heraklion, Greece, pp. 23–25 (2008)

[6] Wu, X., Kumar, V., Quinlan, J.R., Ghosh, J., Yang, Q., Motoda, H., McLachlan, G.J., Ng, A., Liu, B., Yu, P.S., Zhou, Z.-H., Steinbach, M., Hand, D.J., Steinberg, D.: Top 10 algorithms in data mining. Knowl. Inf. Syst. 14, 1–37 (2008)

[7] Ayers, E., Nugent, R., Dean, N.: Skill Set Profile Clustering Based on Student Capability Vectors Computed from Online Tutoring Data. In: Baker, R.S.J.D., Barnes, T., Beck, J.E. (eds.) Proceedings of 1st International Conference on Educational Data Mining, Montreal, Quebec, Canada, June 20-21, pp. 210–217 (2008)

[8] Pavlik Jr., P.I., Cen, H., Wu, L., Koedinger, K.R.: Using Item-type Performance Covariance to Improve the Skill Model of an Existing Tutor. In: Proceedings 1st International Conference on Educational Data Mining, Canada, June 20-21, pp. 77–86 (2008)
[9] Green, T.M., Jeong, D.H., Fisher, B.: Using Personality Factors to Predict Interface Learning Performance. In: 43rd Hawaii International Conference on System Sciences. IEEE Computer Society, Honolulu, HI, January 5-8, pp. 1–10. IEEE Computer Society, Los Alamitos (2010)
[10] Marshall, L., Austin, M.: The relationship between software skills and subject specific knowledge, theory and practice. Learning and Teaching Projects (March 2002)
[11] McCracken, M., Almstrum, V., Diaz, D., Guzdial, M., Hagen, D., Kollokant, Y., Lazer, C., Thomas, L.A., Utting, I., Wilusz, T.: A Luti national Study of Assessment of Programming Skills of First Year CS students. SIGC, SE Bulletin 33, 125–140 (2001)
[12] Chiu, C.: Cluster Analysis for Cognitive Diagnosis: Theory and Applications. Ph.D. Dissertation, Educational Psychology, University of Illinois at Urbana Champaign (2008)
[13] Ayers, E., Nugent, R., Dean, N.: A Comparison of Student Skill Knowledge Estimates Educational Data Mining. In: 2nd International Conference on Educational Data Mining, Cordoba, Spain, July 1-3, pp. 1–10 (2009)
[14] Nghe, N.T., Janecek, P., Haddawy, P.: A Comparative Analysis of Techniques for Predicting Academic Performance. Paper presented at 37th ASEE/IEEE Frontiers in Education Conference, Milwaukee, WI, October 10-13 (2007)
[15] Han, J., Kamber, M.: Data Mining Concepts and Techniques, 2nd edn. Morgan Kaufmann Publishers, San Francisco (2006)

# Study of Organizational Factors Affecting Usability of Online Helps with Programming Languages

Nilesh Dhannaseth[1] and Ganesh Bhutkar[2]

[1] Research Scholar,
Department of Computer Engineering,
Vishwakarma Institute of Technoogy, Pune
nileshdhannaseth@gmail.com
[2] Assistant Professor,
Department of Computer Engineering,
Vishwakarma Institute of Technoogy, Pune.
ganesh.bhutkar@vit.edu

**Abstract.** One of the most common resources for assistance to users while dealing with programming languages is online help, which is topic-oriented, procedural or reference information delivered through online software. This research paper reports observations about organizational factors with online helps for programming languages. These observations are viz. Organized Index of Content, Hyperlinks, Grouping of Items, Use of Tables / Charts / figures, Scrolling Effect, Cascading Effect, Multilingual Support and Help in Video Format. They are compiled based on authors' experience, discussions with active users and literature study about online helps. The users with programming languages can be classified as tool users, programmers, developers and architects. We find that organized index, hyperlinks, grouping of items and use of tables / charts / figures are important for all users with programming languages. These organizational factors directly affect the usability of online helps with programming languages. They should be considered by designers / developers of such online helps so that their users are encouraged in efficient usage and referencing while dealing with products / systems like programming languages.

**Keywords:** Usability, Online Helps, Organizational Factors, Programming Languages, Programmers.

## 1 Introduction

As our world is becoming more and more technology-oriented with mounting operational complexity, the volume of information generated and used is increasing tremendously. Today, it is estimated that there are approximately 150 websites for every human being on the earth [11]. Many of these websites support complex online systems, which have various advantages. The online systems are excellent for brief descriptions, quick references or even extended explanations. The online systems are quicker and also provide the ecological benefits like lesser usage and/or wastage of

D. Nagamalai, E. Renault, M. Dhanushkodi (Eds.): CCSEIT 2011, CCIS 204, pp. 628–637, 2011.

paper. The other advantages of online systems include available all time, easier transfer, access, maintenance and updation of information. While dealing with online systems, the users always require some assistance, which can be given by online consultants, e-learning systems, online forums or helps [16].

### 1.1 Online Helps of Programming Languages and Its Users

Online help is a popular system offering assistance and support to online systems. It is topic-oriented, procedural or reference information delivered through online software. It can be used to present information on a broad range of subjects. It is required with commercial, office, home, creative and/or cooperative applications, which may include military, banking, healthcare or reservation systems and products like antivirus, operating systems, web browsers or programming languages [16].

Today, hundreds of programming languages have been utilized and more languages will be developed in future for desired application development. Due to generalization as well as specialization in application development, the complexity of programming languages has increased drastically. With the continuing increase in the variety, functionality and complexity of languages with more widespread use, the usability of programming languages has become an important issue. The online helps of programming languages assist users by providing needed assistance and in acquiring necessary expertise. We have selected programming languages for study of online helps as most users learn programming languages through online helps and complain about their usability. For our study, we have selected online helps of **Visual C++ 2010, JAVA SE (Standard Edition) version 6** and **HTML version 4.01** provided by vendors - **Microsoft Corporation, Oracle-Sun** and **W3schools** respectively. Their logos are shown in figure 1.

**Fig. 1.** Selected online helps of Visual C++, HTML and JAVA

The users with programming languages differ in terms of needs, levels of expertise, usage of tools, time spent online, domain knowledge and/or work experience. These users can be classified as tool users, programmers, architects and developers as described below [6]:

**Tool users:** These users are naive or least experienced and spend more time with development tools. They have enough knowledge to look through tools like Dreamweaver, NetBeans and Ajax (Oracle.com). They also utilize GUI simply by dragging and dropping elements onto a form.

**Programmers:** They have a good understanding of coding and writing software. They rely less on tools and prefer writing programs according to the specifications determined/given. They convert software designs into a logical series of instructions.

**Architects:** Architects are well-experienced software developers. As they have domain knowledge, they can think well beyond just technical side of software development. They identify the appropriate software architecture based on the requirements and design elements contained in System Requirement Specifications (SRS). They record design in a software architecture document using use cases and the Unified Modeling Language (UML).

**Developers:** These users are expert programmers who understand that there is much more in developing software than just writing programs. Developers possess a much better perceptive of concepts and principles of software design. They think about problems in whole. They need assistance from high-end business tools like LightSwitch and Eclipse.

Online helps may be studied for various usability aspects viz. accessibility, consistency, textual elements, visual elements, readability, organization and navigation. The content of any online help is vital to a success of online system. High quality content will increase the likelihood of system utilization. But beyond providing high quality content, online helps also need to organize their content in a way that makes it accessible, readable and valuable to the users. Organizing information works well for online helps where there is variety of users. Users of online helps often experience difficulties when trying to retrieve reference topics. They have to face many problems like availability of information, textual complexity, poor layout and language support in the organization of online helps [4]. Online helps may not offer users a support in navigation through the content due to poor organization. Taking these problems into consideration, we have studied and collected several observations with organizational factors in selected online helps.

## 2 Observations about Organizational Factors in Online Helps

The observations about organizational factors in online helps are compiled based on authors' experience, discussions with active users and study of literature about online helps. Color coding is used in most of the figures in this section for differentiation among usability features. The positive features in online helps are highlighted with enclosure in rectangles using green color whereas negative features are in rectangles using red color. Also, the blue color indicates both positive and negative features.

### 2.1 Organized Index of Content

An organized index is a list of items in specifically planned order in which items have to be learnt or referred. In the case of online help, the order of topics is important for understanding of the users. A well-organized index page is sequentially ordered and is used as a navigation tool to reach the required topic of interest. Due to rapidly increasing web pages, it is necessary to include a well-organized index in online help. An organized index is located preferably in the upper-left quadrant of every web page of the online help [7]. The index usually includes the titles of the first-level header such as chapter title and often includes second-level or section titles within the chapter. The depth of details in an index depends on the availability of the structured or classified content in online help. A hierarchy containing multi-level topics is used to link entities either vertically or horizontally.

An organized index available with online help of Visual C++ is given in figure 2 and depicts a sequence of topics to be learned. Figure 2(A) shows the first-level organization of topics that represents the main content of Visual C++. Figure 2(B) shows the second-level organization of topics, which represents contents in-depth. Online help of Visual C++ uses color coding in the index structure of topics. The color used in this help is blue for unvisited links, orange for links which are in use and pink for those visited by the user. The first level topics as shown in figure 2(A) have followed 7+/-2 rule, but second level topics do not follow it [10]. The index arranged in online help of Visual C++ is in hierarchical fashion and located in upper-left quadrant of each web page.

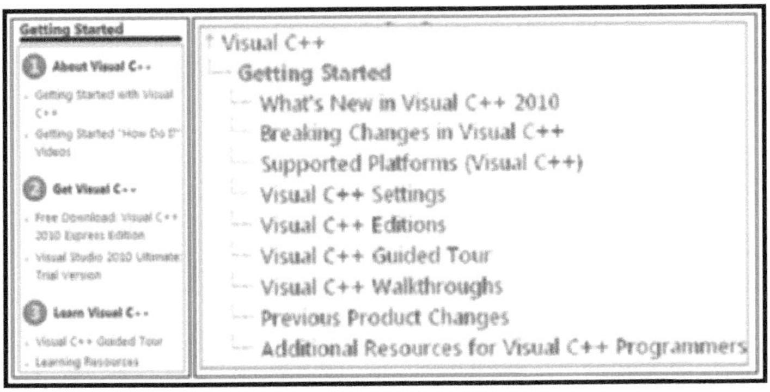


A B

**Fig. 2.** Organized index of online help for Visual C++

## 2.2 Hyperlinks

A hyperlink is an underlined word or phrase and sometimes, a graphical image or icon which has been tagged with a particular scripting language command containing the specific address of another location. By using hyperlink, the user jumps quickly from one location to another location [3]. Hyperlinks are required in online help to present a chance for user to stop reading at one place and resume somewhere else. Hyperlinks are also used to fragment or divide large information into small chunks. The concise chunks of information i.e. hyperlinks provide a limited view of large documents. Studies report that about 80% of users scan any new page and only 16% users read each word. Users spend about 12% of their time trying to locate desired information on a page [15]. So, hyperlinks are vital elements in online helps. There are four types of hyperlinks viz. text, image, bookmark and e-mail. The color used for hyperlinks are blue for unvisited links and red for visited links.

## 2.3 Grouping of Items

A grouping means all related details are placed together in order to minimize time sp-nt on searching or scanning of whole web page. With the rapidly increasing information on the web, there is a great demand for developing methods for effectively organizing the large amount of relevant details [16]. An appropriate grouping strategy can help in this regard and may lead to providing useful assistance for user in getting all information related to a topic. There are various types of grouping of items such as 'under one heading' and 'closeness by distance' [8]. Text items that share the 'same background color' typically can be seen as being related to each other. The information retrieval time is more in case of ungrouped items provided on web pages [15]. So, online helps should be designed using a suitable grouping strategy of items.

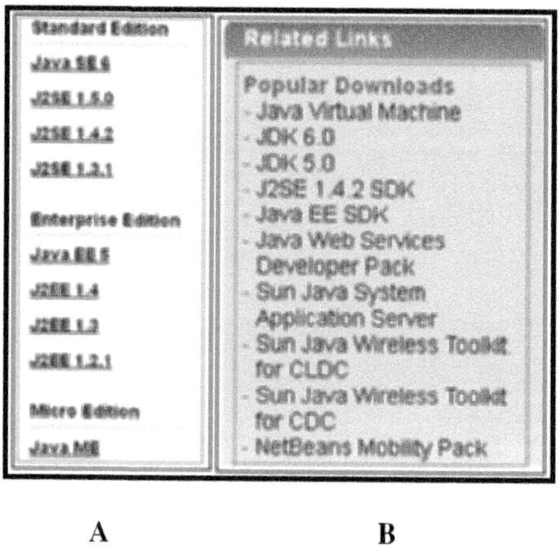


**Fig. 3.** Sample screen shots of A) Grouping of editions of JAVA help and B) Download option for various versions of Java

Figure 3 shows the groupings used in online help of JAVA. The first grouping depicted in figure 3 (A) uses 'under one heading' strategy for categorization of various editions of Java. Another strategy based on 'same background color' for grouping is shown in figure 3 (B) for popular downloads of various versions of Java.

## 2.4 Use of Tables / Charts / Figures

A table is a way of representing details like numbers, quantities or items in terms of rows and columns.

Tables are designed to structure and present details in a systematic fashion. A chart is a visual representation of numeric values, quantities or items with identified patterns. A table / chart effectively use a minimum space to communicate a large amount of details in a small space [17]. Data in tabular form takes only one third of space to represent information which is in text format [5]. According to Horton (1991), well-designed graphics used in charts help to structure data and enable the users to grasp and process information more quickly and efficiently which leads to increased information retention. There are two types of tables - unidirectional and bidirectional tables. Unidirectional table is a table in which all details displayed row wise and bidirectional table is the table whose details are depend on both rows and columns. There are chart types like bar chart, pie and line chart for representing information about online helps.

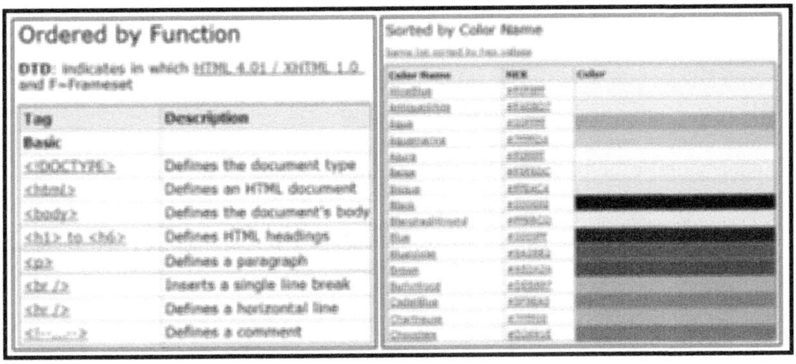


A B

**Fig. 4.** Screen shots for A) Table for functions used in HTML scripting and B) Color chart used in HTML scripting shown in online help at www.w3schools.com

Figure 4 (A) shows the unidirectional table for functions used in HTML scripting. This table is unidirectional because it has only row wise description of HTML functions. Figure 4 (B) shows the color chart that illustrates shade of colors and their indices which are used in HTML scripting.

## 2.5 Scrolling Effect

A scrolling is sliding text, image or video across a display. It does not change the layout of the text or image. But, it incrementally moves the user's view across what is apparently a larger text or image that is not wholly seen [14]. Since documents are generally too long to fit on a screen, users must frequently scroll to see other parts of the documents. There are two types of scrollings - vertical and horizontal. Sometimes, both scrollings are provided for user interaction to move viewport in two dimensions. Most of the users avoid scrolling web pages of online help and therefore, do not refer information which is present below the fold. In a study, it is found that only 10% of web users would scroll web page to see any links that are not visible in the initial display [12]. Generally, the online helps should provide very little scrolling in the web page not to miss out reference topics which are of importance.

## 2.6 Cascading Effect

A cascading effect is an effect in which users may have to click hyperlinks more times to get appropriate details. The process of cascading is really discouraging for most users [2]. In a study, it is found that when users try to find a reference topic, they are likely to quit after three clicks [13]. So, the designer of online help should design help in such a way that the users get required topic of their choice in less than three clicks. This will save valuable time of users and they don't have to jump in web pages of online help to get required details.

Figure 5 shows the graph plotted for number of clicks required to get reference topics in respective online helps for programming languages under consideration. We have taken number of clicks on Y-axis and reference topics on X-axis. The reference topics considered in graph are programming function (e.g. function to set font), bitwise operator (e.g. '&' operator) and class in language (e.g. basic I/O class). Now, we consider the example of basic input-output (I/O) class. In general, the online help of HTML requires about three clicks (average) to get basic I/O function for the users. But, in case of other helps, number of clicks required is always more than three and it may consume vital time in search/navigation of information by the users.

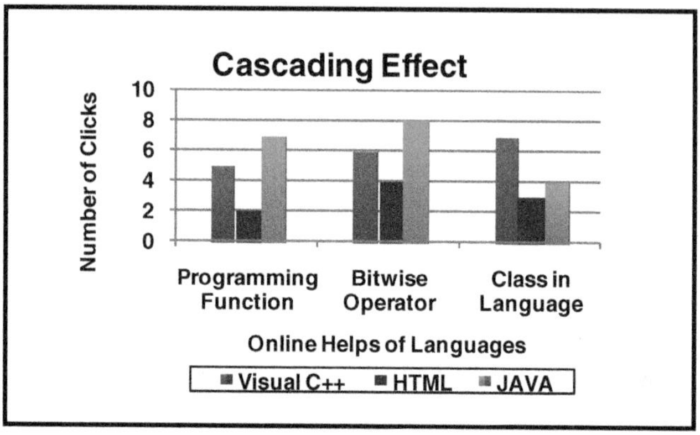


**Fig. 5.** Graph for number of clicks required to get reference topic in online helps

## 2.7 Multilingual Support

A multilingual support is a property used to mean that there exist forms that can handle several languages. An appropriate language can be selected by the user while using multilingual online help. Every online system should provide multilingual support for all widely-spoken languages like Chinese, Spanish, English, Arabic or Russian [9].

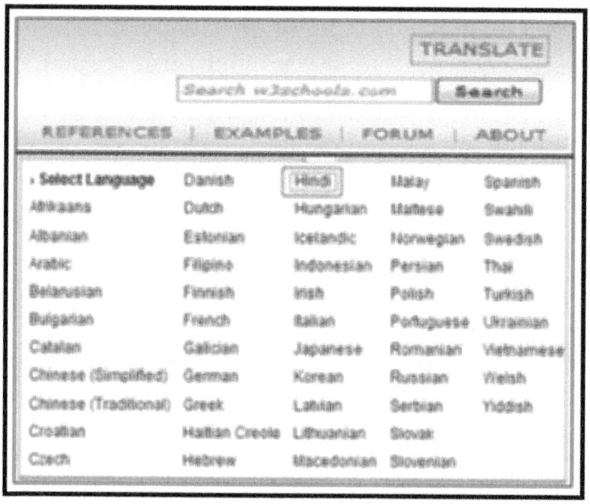


**Fig. 6.** Multilingual support in online help for HTML

It should also provide language option like Hindi, which is widely spoken in India and several countries like Mauritius, South Africa, Singapore, Nepal, and Pakistan etc. It is also desirable to provide support for major Indian languages like Bengali, Marathi, Tamil, Gujarati and Malayalam. Layout and design is an important consideration when building multilingual online helps. The problem arises due to multilingual support in the organization of web pages of online help. This include 'direction of text', 'format of things like date, time and names', 'choice of spelling convention', 'phrases and meaning which cannot be translated', 'size of text block' and 'increase in web pages of help'. The problem of multilingual support to contents affects not only users in multilingual regions like Europe or Indian subcontinent, but also all those who search information sources or databases containing material in more than one language [1]. Therefore, providing multilingual support helps making software application designs universal.

As shown in figure 6, the online help for HTML supports total 52 languages. This is an important feature. As observed, it provides multilingual support for all leading languages like Chinese, English, Arabic and Russian [9]. It also provides language option like Hindi, which is widely spoken in India. But, it fails to provide support for other major Indian languages.

## 2.8 Help in Video Format

Videos can help users to understand complex concepts and procedures that are difficult to explain with simply text and graphics. By using visual messages, users can process the information quicker which helps to encourage their learning acquisition of the reference topics. In addition, a video can help users to visualize a process or see how something works. Also, users don't have to wait for complete download of the video. With these benefits in mind, every online help should provide video support for users to convey information very easily, if needed. There are various video formats like WMV, MP4, 3GP and AVI (Fileinfo.com). Due to these formats, users have to download additional software component(s) from different websites to watch videos related to online helps. This may increase the complexity of online help for the users.

A B

**Fig. 7.** A) Video support for online help of Visual C++ and B) Additional software requires for watching videos

Figure 7 shows that the online help of Visual C++ provides help in video format to its users. The chapter 'Better Understand Using Pointers and References' is difficult to learn and complex to understand in text format. So, by using the video of same chapter, the user can grasp the topic easily. Users have to download software

'Microsoft Silverlight' to watch the videos present in this online help. But, it is not convenient thing to download additional software.

# 3 Discussion

The observations about organizational factors in online helps discussed consist of 8 organizational factors which can be classified as qualitative and quantitative. The qualitative set of factors is particularly challenging, as these factors are often described subjectively. e. g. Organized Index of Content, Scrolling Effect, Multilingual Support etc. The quantitative set of factors is classified as these factors are measurable.

A useful numerical method for calculation of *organizational index* based on organizational factors can be provided. It has two major equations and they are very easy to compute as all parameters required can be expressed in terms of natural numbers. The proposed equations/formulae are provided in equations - I & II respectively. An equation I can be used when help in video format is available; otherwise equation II can be used. The denominator part in equation II is approximate result of calculation of denominator of equation I, using studied online helps.

$$OI = \left\{ \frac{h + t + [3 * c]}{8 - \left[\frac{v}{7}\right]} \right\} \quad if\ v \neq 0$$

**Equation I.** Formula for index calculation when videos are available

$$OI = \left\{ \frac{h + t + [3 * c]}{2.75} \right\} \quad if\ v = 0$$

**Equation II.** Formula for index calculation when videos are not available

The list of parameters used in formulae are {**h** = Avg. number of hyperlinks, **t** = Number of tables / charts / figures, **c** = Avg. number of clicks, **v** = Number of videos}

The study of online helps with programming languages is carried out by considering organizational factors affecting related online helps. The equations can be used to comprehend about how well-organized the online helps are. This study may assist in highlighting organizational design deficiencies in online helps to improve usability even for tool users. Also, the proposed formulae - **OI** give a fast and useful numerical method for calculation of organizational index for online helps. These formulae also provide an opportunity for comparison among several online helps in a specific domain. These organizational factors directly affect the usability of online helps with programming languages and help web designers to develop such online helps which can encourage better interaction with their users. In future, this study can be extended to evaluate various online helps in other domains like e-commerce, e-governance or electronic gazettes.

**Acknowledgment.** We sincerely acknowledge the help and support provided by post-graduate research scholars - Mr. Tulsidas Patil and Mr. Dipak Pawar at VIT, Pune during this research work. We thank Prof. S. B. Karthick, Head, Department of Computer Engineering at VIT, Pune for his continuous encouragement to our research work.

## References


1. Biswas, S.: Multilingual Access to Information in A Networked Environment Character Encoding and Unicode Standard. In: INFLIBNET's Institutional Repository, 3rd Convention PLANNER, pp. 176–186. Assam Univ., Silchar (2005)
2. Bhutkar, G., Rajhans, N., Konkani, A., Dhore, M.: Usability Issues of User Manuals Provided with Medical Devices. The British Journal of Healthcare Computing and Information Management (2009)
3. Chaparro, B., Bernard, M., Thomasson, R.: Finding Information on the Web: Does the Amount of White Space Really Matter? In: Proceedings of the 10th Annual Usability Professionals' Association Conference, pp. 1–4 (2001)
4. Chaparro, B., Shaikh, D., Baker, J.: Reading Online Text with a Poor Layout: Is Performance Worse? Usability News, SURL, 1–5 (2005)
5. Durbin, C., Faarc, J.: Effective Use of Tables and Figures in Abstracts, Presentations, and Papers. Respiratory Care 49(10), 1233–1237 (2004)
6. Hayes, J.: Types of Programmer (2007), retrieved from `http://ezinearticles.com/?5-Types-of-Programmers—Which-Type-Are-You?&id=444143`
7. Horton, W.: Designing and Writing Online Documentation: Hypermedia for Self-Supporting Products, 2nd edn. John Wiley & Sons, Chichester (1994)
8. Katre, D., Bhutkar, G., Karmarkar, S.: Usability Heuristics and Qualitative Indicators for the Usability Evaluation of Touch Screen Ventilator Systems. In: Katre, D., Orngreen, R., Yammiyavar, P., Clemmensen, T. (eds.) HWID 2009. IFIP AICT, vol. 316, pp. 83–97. Springer, Heidelberg (2010)
9. Lewis, P.: Ethnologue: Languages of the World, 16th edn. SIL International, Dallas (2009)
10. Miller, G.: The Magic Number Seven, Plus or Minus Two: Some Limits on our Capacity for Processing Information. Psychological Review, 81–87 (1956)
11. Momento24. Number of Websites for Every Human Being (2009), retrieved from `http://momento24.com/en/2009/07/31/number-of-websites-surpasses-world-population-`
12. Nielsen, J.: Scrolling and Attention. Jakob Nielsen's Alertbox Column (2010), retrieved from `http://www.useit.com/alertbox/scrolling-attention.html`
13. Porter, J.: Testing the Three-Click Rule (2003), retrieved from `http://facstaff.elon.edu/gibson/COM322/.../Testing_the_Three_Click_Rule.pdf`
14. Sanchez, C., Wiley, J.: To Scroll or Not to Scroll: Scrolling, Working Memory Capacity, and Comprehending Complex Texts. Human Factors: The Journal of the Human Factors and Ergonomics Society 51(5), 730–738 (2009)
15. Schriver, K.: Dynamics in Document Design: Creating Text for Readers. John Wiley and Sons, Inc., Chichester (1997) ISBN: 0-471-30636-3
16. Shneiderman, B.: Designing the User Interface: Strategies for Effective Human-Computer Interaction, pp. 1–448. Addison-Wesley Longman Publishing Co., Inc., Boston (2005)
17. Stabina, R.: Quantitative Data Graphics: Best Practices of Designing Tables and Graphs for Use in Not-for-Profit Evaluation Reports. University of Oregon, Applied Information Management Program, 1–60 (2005)
18. `http://msdn.microsoft.com/en-us/visualc/default` (retrieved on December 19, 2010)
19. `http://www.oracle.com/us/technologies/java/index.html` (retrieved on January 21, 2011)
20. `http://www.w3schools.com/html/default.asp` (retrieved on January 17, 2011)

# Finding Reliable Path for Peer to Peer Application in Mobile Environment

D. Jayashree[1] and P. Yogesh[2]

[1] Research Scholar
[2] Professor
Dept. of Information Science and Technology,
Anna University, Guindy,
Chennai, India
jaishri.kannan@gmail.com,
yogesh@annauniv.edu

**Abstract.** Peer-to-Peer (P2P) as well as mobile ad hoc networks (MANETs) follows the same idea of creating a network without the help of central entities. However, P2P and MANETs operate on different network layers. A combination of both, creates new services and possibilities, but also faces several problems. As smarter mobile devices are increasingly adopted by the wider public, to access p2p applications, the question of who to trust while on the move becomes a crucial one. When operating within ad-hoc type p2p networks, a trust-based framework is the logical solution to provide ways to distinguish the most trustworthy neighbors. This promotes reliable information transfer, and encourages content and resource sharing. This paper proposes a scheme to set up such trusted paths, based upon aggregating individual peers' trust, given out by their nearest neighbor.

**Keywords:** Peer-to-peer networks, mobile network, Trust model.

## 1 Introduction

The peer to peer network differs from other distributed systems in that its member nodes undertake to provide routing support themselves. There are no dedicated router or access nodes. The implication is that trust has to be built from a neutral stand-point. One can argue that sometimes in distributed systems including ad hoc networks, concepts akin to trust may be used to decided whether to trust or not. For example, a system may define a quality of service that a peer must provide [1]. Indeed some researchers define trust as a "metaphor" that is the direct product of how interactions within a system evolve [4].

Researchers in the domain of peer to peer networking (MANET) have realised that in order to enable and sustain reliable networking, trust relationships have to be formed between interacting peers. Such relationships are acquired over time as the peer's experience of its surroundings increases.

In its simplistic form and for the purpose of this paper, trust can be seen as "an inherent property in which one autonomous component cannot completely control

D. Nagamalai, E. Renault, M. Dhanushkodi (Eds.): CCSEIT 2011, CCIS 204, pp. 638–646, 2011.

another autonomous component but which may need to rely on one another or require some cooperation from it" Recommendations generally leads to what is known as a reputation system. Such a system, operating through recommending entities, requires the presence of a centralized and trusted server to compute, store and display valid reputation information. This centralized authority is vital to such reputation systems but cannot be readily assumed to be present in the case of mobile devices. The reputation system needs to be decentralized [2, 3] such that it is accessible to entities on the move that may not have access to central repositories.

The case for peer to peer networks can be made in instances where conventional networking is not available. Conventionally, mobile devices operate by locating the nearest available trusted wireless point in order to access a known wired backbone. However, despite advances being made in lean computing and power conservation on mobile devices, these resources remain constrained and often nodes in a network find it preferable to operate on low energy in order to access local services. Furthermore, addressing unreliability through repeating retransmissions in a low energy network wastes precious bandwidth.

Superimposing trust on the p2p network therefore reinforces the transfer of information and aids in the delivery of dependable local services. One of the ways in which the success of a trust framework can be measured is via the quality of service that it can provide for members adopting it. For the purpose of this paper, the trust model presented utilizes the success rate of networking requests as one of the criteria set for achieving quality of service. Investigating quality of service itself is outside the scope of this paper and will be presented in future work. However, one of the novelties of the model is the notion that low-level network attributes and behaviour can be projected to represent higher level abstraction such as trustworthiness, given the right context. In this paper we investigate the behavior of the model with respect, to entity feedback provided for network requests, and to its ability to reflect changing circumstances, thus presenting a changing trust landscape.

In this following section 2 we have discussed the existing methods and their drawbacks. The next session explains the derivation of trust and the trust path calculation of p2p application in mobile routing. The experimental discussion have been given in the last section.

## 2 Related Work

In recent years, pee-to-peer (P2P) computing has achieved its popularity in many distributed applications, including file-sharing, digital content delivery, and P2P Grid computing. However, peer anonymity and autonomy make P2P networks quite vulnerable to attacks by selfish and malicious peers.

Damiani et al [7] proposed XRep: a reputation-based approach for evaluating the reputation of peers through distributed polling algorithm before downloading any information. The approach adopts a binary rating system and it is based on GNutella [8] query broadcasting method using TTL limit. EigenTrust [9] collects the local trust values of all peers to calculate the global trust value of a given peer. Additionally,

EigenTrust adopts a binary rating function, interpreted as positive one (representing satisfactory), or zero or negative one (representing unsatisfactory or complaint).Marti and Garcia-Molina [10] proposed a voting reputation system aiming at e-commerce applications that collects responses from other peers on a given peer. The final reputation value is calculated combining the values returned by responding peers and the requesting peer's experience with the given peer. This seems more reasonable than the model in the work by Damiani et al [7]. However, this work and the work by Kamvar et al. [9] did not explicitly distinguish transaction reputation and recommendation reputation. This may cause severe bias in reputation evaluation as a peer with good transaction reputation may have a bad recommendation reputation especially when recommending competitors.

Wang and Varadharajan [11] proposed several trust metrics for the trust evaluation in a decentralized environment (e.g., P2P network), where a trust value is a probabilistic value in the scope of [0, 1]. Prior to the interaction with an unknown Peer Px, the end-peer collects other peers' trust evaluations over Px. A method has been proposed for trust modification after a series of interactions with Px that a good value results from the accumulation of constant good behavior leading to a series of constant good trust values. In another work by Wang and Varadharajan [4], the temporal dimension is taken into account in the trust evaluation wherein fresh interactions are weighted more than old ones.

Yu et al. [12] proposed the method of exponential averaging taking into account a series of interactions of the requesting peer itself. It is similar to the work by Wang and Varadharajan. In addition, Wang and Varadharajan [13] proposed Trust, a model for trust evaluation considering recommendation trust. Trust also includes a method to measure the recommendation trust, based on the requesting peer's interaction experience with the target peer and the recommendations of responding peers in multiple rounds. Lin et al. [14] proposed a method of reputation-based trust evaluation in service-oriented environments based on a novel architecture of distributed trust management brokers.

## 3 Trust Based Mobile Peer-to-Peer Protocol

In this paper we proposed a trust mechanism in MPP (Mobile Peer-to-Peer Protocol) [5] that include both trust calculation and finding shortest trust path. MPP allows a realistic operation of Peer-to-Peer services over ad hoc networks. It spans from the network layer and tries to reuse existing protocols as far as possible. Additionally the network layer needs knowledge about the trustworthiness of these peers. Providing a trusted path for resource sharing can help the peers to obtain trustworthy acquaintance of neighbours without the fear of losing the packets to malicious peers.

To meet the requirements of finding trust path within the mobile network, trust is extended on EDSR. The general structure of EDSR has been referred in EDSR protocol [5]. EDSR provides the basis for the realization of trusted P2P applications in MANETs. The following section explain how trust path is formed and the calculation of trust on each peer.

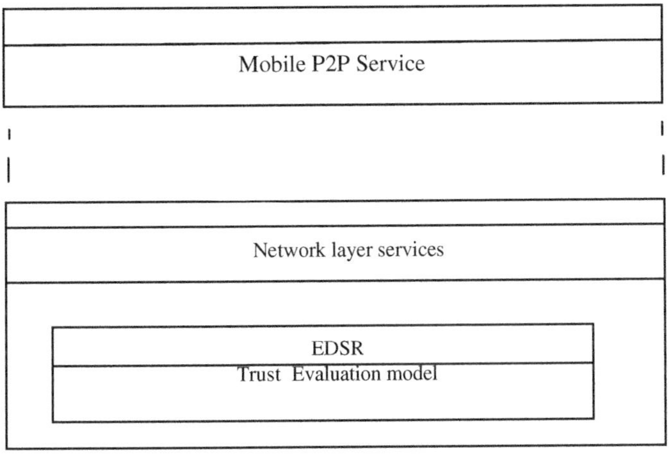


**Fig. 1.** Trust Architecture of Mobile peer

This figure shows where the trust evaluation model of trust has been placed in MPP architecture.

### 3.1 Trust Evaluation

In a Mobile P2P network, some peers join or leave the network frequently, which leads to the dynamic topology changes. As a result, a trust management system needs to frequently update trust ratings, which in turn can increase the communication overhead. Here we have present an approach, which can easily adopted by any routing algorithm without much overhead to find a trust path.

Trust relationships are asymmetric. This means that if a node A's trust in a node B in performing an action is 0.5 (where 0.5 is the trust metric), it does not automatically imply that node B has the same amount of trust about node A. Each node holds an individual trust relationship, defined by a trust metric, with its immediate neighbours (on a one-hop basis only). All nodes are trustors, with their respective neighbours being their trustees. The trust metric evolves over time due to two factors: firstly it may rise or fall depending on new data received by the trustor and secondly, it may undergo a period of natural "decay" over time if data regarding a certain entity is not received over time. The "decay" promotes cooperation within the network, as nodes are eager to gain a good reputation. All nodes are observed by their peers in relation to their success or failure in carrying out this specific action. The trust metric is therefore a measure of the reputation of a given node.

This section describes the procedure of the proposed trust ratings aggregation scheme, This scheme takes into account the trust ratings based on direct and indirect interactions. In a scenario with two possible paths, this proposed algorithm considers the best path based on confidence, $\Psi$. The value of $\Psi \in [0,1]$ is calculated by individual peers based on the number of positive and negative interactions with a peer. We use $\Psi$ for trust ratings aggregation from witness interactions only. Since it is possible that a peer can act maliciously and provide false trust ratings, the value of $\Psi$ can determine the behavior of peer in recent interactions. A newly joined peer can send a trust request to the network. Peers that receive the trust request choose to send back their trust lists. When the new peer receives this reply, its initial local trust list is established for its own neighbors. It is possible that a trustworthy peer gets disconnected due to power shortage or physical location change in the MANET environment, for this reason, each entry in the t_list has a ttl variable indicating time to live. If ttl exceeds a threshold and no connection can be established for a node, its trust ratings are removed from the t_list. If z is a direct peer and rating of x in the local t_list is higher than threshold then z can be trusted.

If "z" a neighbor peer has poor trust ratings, then a second opinion about recommendations is sought from other direct neighbors. Any direct neighbor that can give recommendations about y, while having a higher trust rating compared to z is considered and the local t_list is updated accordingly. The trustworthiness of recommendations from a distant node (witness) will be considered if and only if the value of confidence $\Psi$ is greater than a predefined constant. This is because multiple ratings are received over multiple hops that can be modified by malicious peers in a given period.

The weighted average of trustable witness recommendations ($\Psi$ > constant) provides balanced trust ratings. In witness interactions, the value of $\Psi$ is multiplied to the calculated trust in order to obtain the witness trust interaction value. If the ttl for a trust rating stored in the local t_list expires, an update request is made to node z to provide the latest t_list. If no reply is received, the node is assumed to have been disconnected, any subsequent trust ratings are therefore removed from the local t_list. The value of confidence $\Psi$ is determined by the behavior of a peer in terms of interactions. If the number of completed interactions $\alpha$ is larger than incomplete interactions $\beta$, the value of $\Psi$ is increased. Similarly, as a consequence to a large number of incomplete transactions this value is decreased. The value of $\Psi$ is given by $(\alpha - \beta) / \alpha$. Frequent disconnections due to mobility in mobile networks cause disruptions in communication and therefore large number of incomplete transactions.

Although the effect of frequent disconnections on the confidence value due to mobility is essential to the quality of trust ratings, it is beyond the scope of this paper.

### 3.2 Path Trust Calculation

Path trust is the trust value associated with the path. This value is defined as the weighted average of the trust values of the nodes in the path. As trust is considered to be asymmetric mutual trust between the nodes is are used. Hop count also plays a prominent role in the selection of the path since the larger the number of nodes, more is the delay in the network and chances of information modification also increases.

To search for trusted data provider EDSR offers a similar message like the Gnutella query-message. At the end of each message, the current node stores its own IP address and the trusted value of the next hop peer. Thus, the route of the reply message is predetermined.

Request packet: Request packet header is modified by adding a F-trust value. Hence, that it now contains the following fields: source IP address, destination IP address, a sequence number and F-trust. Each broadcast packet is modified to include the trust value of the node from which packet is received. So when Source A broadcasts a request packet and node B receives it, it updates the F-trust (forward trust) value such as

$$\text{F-trust} = \text{F-trust} + {}_{A}T^{B} \text{ ( trust value of A for B)} \quad (1)$$

At the destination, F-trust contains information about the trust of all nodes involved in the path.

Reply Packet: Reply packet is modified to accommodate the F-trust and B-trust values. Each node updates it by including the trust value of the node from which it received the packet. So when node X receives reply from node Y, it updates B-trust as:

$$\text{B-trust} = \text{B-trust} + {}_{X}T^{Y} \text{ ( trust value of X forY )} \quad (2)$$

When the Reply packet reaches the source node, the most secure path is selected by it. It calculates the path trust based on the trust values F-trust and B-trust received in the Reply packet and the number of nodes in the path. The path selected is the one which has the maximum path trust.
Trust value of $i^{th}$ path is:

$$\text{Trust Path value} = [(\text{F-trust} + \text{B-trust}) / 2\ ] * Wt_i \quad (3)$$

Where Wt of $i^{th}$ path is $Wt_i$ =

$$[1/\text{no. of nodes in ith path}] \Big/ \sum_{i=1}^{n} \frac{1}{\text{no of nodes in path } i} \quad (4)$$

n is the total number of path discovered from s to d. $Wt_i$ is the weight assigned to the $i^{th}$ path. : trust path value is the trust value of the $i^{th}$ path. : trust path value is the trust value of the path selected as the most trust-worthy path.

As in the case of search requests, the structure of a reply is based on the structure of a ROUTE REPLY (RREP) of EDSR. Additionally it holds the the transmitting peer id, destination peer id and the trust value for the next hop peer. Otherwise all the other packet details have been maintained along with their trust value.

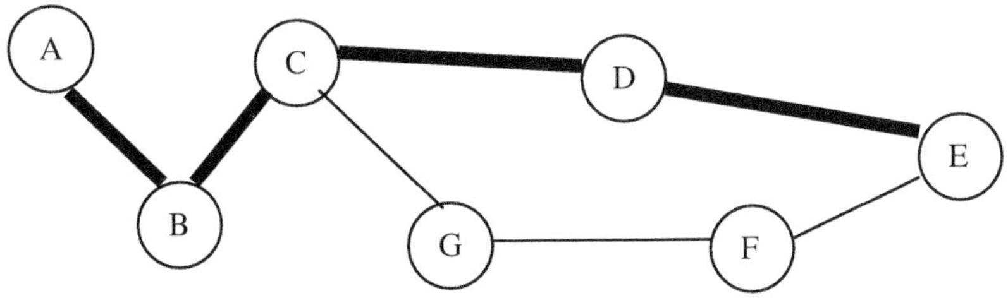


**Fig. 2.** Path Trust Calculation from peer A to E

The above network model explains the Path Trust value calculation from node A to E through nodes A, B, C, D, and E.

Trust Path Value $_{A\text{-}E}$

$$= [\{ {}_{A}T^{B} + {}_{B}T^{C} + {}_{C}T^{D} + {}_{D}T^{E} + {}_{E}T^{D} + {}_{D}T^{C} + {}_{C}T^{B} + {}_{B}T^{A} \} / 2 ] * Wt_i \quad (5)$$

Trust Path from A to E contains mutual trust information of all the nodes involved in the path from A to E.

## 4 Performance Evaluation

In this section, we discussed the performance on simulation using the following parameters defined here.

To simulate the trust path calculation in a mobile ad-hoc network environment, NS2 v2.28, a network simulation tool, is used. EDSR protocol is adopted as a routing protocol, with 1500m1500m in the size of network area. The simulation is carried out with a change in the number of peers ranging from 20 up to 100. A mobile scenario for mobile peers where pause time on average is 2 seconds with average velocity and maximum velocity varying 3 to 20 m/sec and 5 to 30 m/sec, respectively, is designed for evaluating their performance. Table 1 shows the values of parameters used in a simulation.

On top of that, when having found out applicable objects, the number of physical hops for paths to the objects is reduced, which means that, when an object requester searches applicable objects, it can quickly find them without going through various peers and hops count of peers are relatively less from source peer to destination peer.

**Table 1.** Simulation Parameters

| Parameter | Value |
|---|---|
| Number of Peers | 20 to 100 |
| Network size | 1500m x 1500m |
| Mean query generate time(secs) | 5 |
| Routing protocol | DSDV, AODV |
| Movement average speed (m/s) | 3 to 20 |
| Movement maximum speed (m/s) | 5 to 30 |
| Channel | Channel / Wireless Channel |
| Propagation | Propagation / TwoRayGround |
| Network interface | Phy / WirelessPhy |
| Mac | Mac / 802_11 |

The simulation was carried out as required with varying loads of traffic through various cycles. Figure 3 pertains to the traffic through one node \only within a network. Data was analyzed at the end of each cycle.

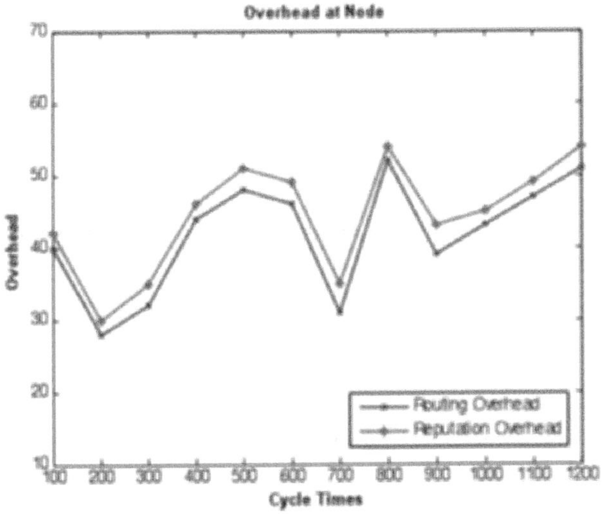


**Fig. 3.** The overheads generated at single peer

As far as the node is concerned, it can be seen in Figure 3 that the overhead added by the trust mode by virtue of its reputation mechanism is not appreciably higher than the outing overhead.

## 5 Conclusion

This paper presents a new trust path-finding scheme for mobile P2P networks. This relies on direct trust ratings and witness recommendations from reliable peers to determine trust ratings for a node using confidence value. Simulation results demonstrate that the overall performance of this scheme is reliable and have marginal increment in their overheads for detecting trusted peers in P2P mobile networks.

In the future, we should do more work to research the discrimination of the malicious actions and deploy the framework in a real system.

## References


1. Poslad, S.: Ubiquitous Computing: Smart, Devices, Environments and Interactions, pp. 309–311. Wiley, Chichester (2009) ISBN: 978-0-470-03560-3
2. Chakravorty, R., Agarwal, S., Banerjee, S., Pratt, I.: MoB: A mobile bazaar for wide-area wireless services. In: Proceedings of MobiCom 2005, pp. 228–242 (2005); Quercia, D., Hailes, S., Capra, L.: MobiRate: Making Mobile Raters Stick to their Word. In: Proceedings of UbiComp 2008, Seoul, Korea (2008)
3. Quercia, D., Hailes, S., Capra, L.: B-trust: Bayesian Trust Framework for Pervasive Computing. In: Stølen, K., Winsborough, W.H., Martinelli, F., Massacci, F. (eds.) iTrust 2006. LNCS, vol. 3986, pp. 298–312. Springer, Heidelberg (2006)

4. Wen, D., Huaimin, W., Yan, J., et al.: A recommendation-based peer-to-peer trust model. Journal of Software 15(4), 571–583 (2004)
5. Schollmeier, R., Gruber, I., Niethammer, F.: Protocol for Peer-to-Peer Networking inMobile Environments. In: Proc. 12th International Conference on Computer Communications and Networks (ICCCN 2003), October 20-22, pp. 121–127. IEEE, Los Alamitos (2003)
6. Damiani, E., di Vimercati, S.D.C., Paraboschi, S., Samarati, P., Violante, F.: A reputation based approach for choosing reliable resources in peer to peer networks. In: Proceedings of ACM CCS 2002, p. 207 (November 2002)
7. GNutella, `http://www.GNutella.com/`
8. Kamvar, S.D., Schlosser, M.T., Garcia-Molina, H.: The eigentrust algorithm for reputation management in p2p networks. In: Proceedings of the 12th International WWW Conference, Budapest, Hungary (May 2003)
9. Marti, S., Garcia-Molina, H.: Limited reputation sharing in p2p systems. In: Proceedings of ACM EC 2004, New York, USA, pp. 91–101 (May 2004)
10. Wang, Y., Varadharajan, V.: Interaction trust evaluation in decentralized environments. In: Bauknecht, K., Bichler, M., Pröll, B. (eds.) EC-Web 2004. LNCS, vol. 3182, pp. 144–153. Springer, Heidelberg (2004)
11. Yu, B., Singh, M.P., Sycara, K.: Developing trust in largescale peer-to-peer systems. In: Proceedings of 2004 IEEE First Symposium on Multi-Agent Security and Survivability, pp. 1–10 (August 2004)
12. Wang, Y., Varadharajan, V.: Trust2: Developing trust in peer-to-peer environments. In: Proceedings of 2005 IEEE International Conference on Services Computing (SCC 2005), Orlando, Florida, USA, pp. 24–31 (2005)
13. Lin, K.J., Lu, H., Yu, T., Tai, C.E.: A reputation and trust management broker framework for web applications. In: Proceedings of The 2005 IEEE International Conference on e-Technology, e-Commerce and e-Service (EEE 2005), pp. 262–269 (2005)

# Text Mining for Interpreting Gene

K. Prabavathy[1] and P. Sumathi[2]

[1] Department of Computer Science,
PSG College of Arts and Science, Coimbatore-14
Tamil Nadu, India
Tel.: 9942262728
praba_bud@yahoo.co.in
[2] Department of Computer Science,
Chikkanna Govt. Arts College , Tirupur-02
Tamil Nadu, India
Tel.: 9150588820
sumathirajes@hotmail.com

**Abstract.** Text mining provides methods to retrieve and extract information contained in free-text automatically. The advances in molecular biology techniques have given rise to high throughput biology that produces tremendous data and related publications in the past decades. Here the efficient constituents of Co-reference resolution incorporating Natural Language Processing (NLP) to interpret Gene expression are focused. It may overcome the challenges and limitations of text mining in biological data for resolving unsolved problems and this paper describes a new phase of text mining process to uncover interesting term correlations, genomic term identification in curation process , identification of biological relations and help the biologists in their analysis of complex problems. The new phase of text mining process is organized in four tasks of namely *use NLP, find correlated terms, Co-reference resolution and built a structured database.*

**Keywords:** Bio-NLP, Correlated terms, Co-reference resolution, Gene, Structured database.

## 1 Introduction

Data mining refers to automated search for meaningful patterns of data stored in large databases distributed over the web. Text mining is a young interdisciplinary area that draws on information retrieval, data mining, machine learning, statistics and computational linguistics. Text mining generally refers to the process of extracting the interesting and non-trivial information and knowledge from unstructured text.

The Data mining tools require input data to be fairly homogenous like those found in conventional database (numbers, characters, etc.). Text mining unlike Data mining works on unformatted data and semi-structured data. A real text mining discovers new pieces of knowledge, from approaches that find overall trends in textual data.

D. Nagamalai, E. Renault, M. Dhanushkodi (Eds.): CCSEIT 2011, CCIS 204, pp. 647–653, 2011.

MEDLINE searches crossed 82.07 million being a challenge with approximately 10% growth every year. This in turn strongly depends on efficiency of database interoperability and data mining integration. The huge amount of data of this post genomic era includes genomic, proteomic and literature data. Bioinformatics has emerged as one and only one solution to handling such complex data.

## 2 Applications of Text Mining

The text mining is considered as an art than to a science, but contributes very much to the Concept driven science or conceptual biology. Text mining is one of the tools now used in fields like Astrophysics, Sequence analysis, Drug Discovery, Gene identification and disease relation, disease identification, protein interaction studies, to study protein - protein interactions, protein – gene interactions, metabolomics and even in the nascent field of Systems biology.

Figure 1 shows the Applications of BioNLP and Text mining .The major topics are represented by the inner circle of seven approaches, and the corresponding applications are given in the outer layers of boxes.

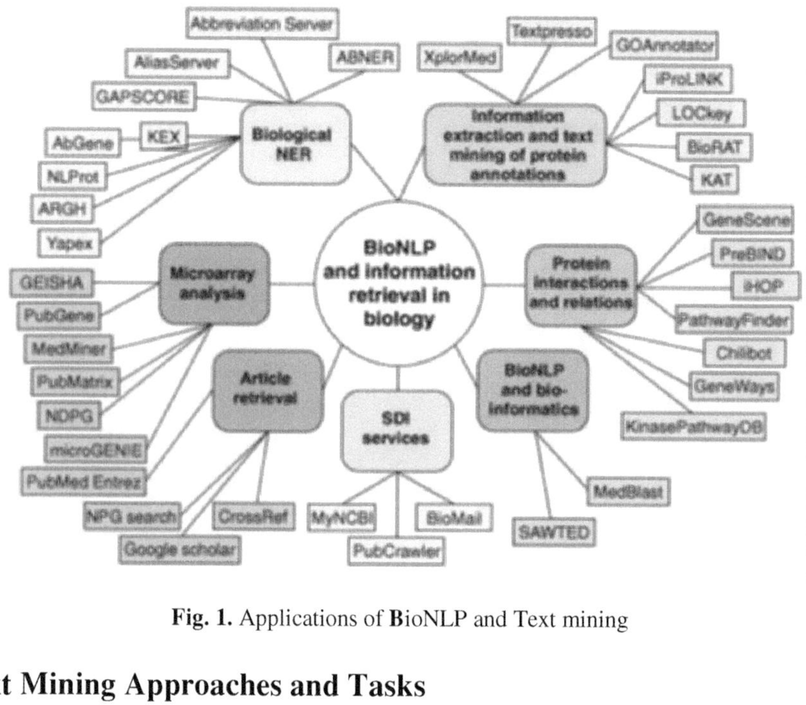


**Fig. 1.** Applications of BioNLP and Text mining

## Text Mining Approaches and Tasks

The text mining is a process of finding, organizing and analyzing information. It creates the new information that is not obvious in a collection of documents. New information is defined as a pattern, trend, or relationship that can't be easily gleaned by reading individual documents. The two approaches of text mining are Bag of Words that looks for features.

Natural Language Process (NLP) looks for Text analysis by using Syntax and semantics

## 3 Natural Language Processing(NLP)

A variety of different systems is available to do natural language processing, but no system can cope with the complexity of natural language in a spoken language. NLP is based on the use of computers to process language, and it includes techniques developed to provide the basic methodology required for automatically extracting relevant functional information from unstructured data, such as scientific publications. Information retrieval and NLP systems are soon likely to become important not only for extracting information but also for assisting in various aspects of research such as the discovery of new facts, the interpretation of findings, and the design of experiments. NLP based systems provide solutions on chunks of sentences. The main advantage of NLP based approaches can be found in the fact that sentence analysis can be shaped following a hand-crafted linguistically model

Figure 2 represents the NLP and its Steps. The three main tasks of Text mining are Information Retrieval, Semantics and Information Extraction.

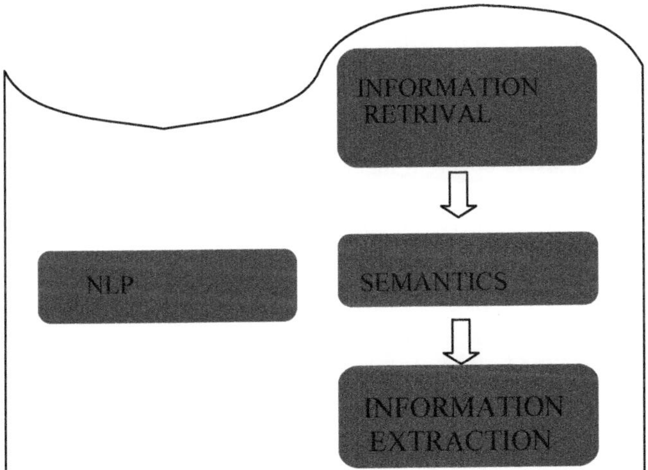


**Fig. 2.** NLP and its Steps

**Information Retrieval -** It is an automatic process of extracting files based on Query based approach or Document based approach. Document based approach retrieves documents of similar kinds, whereas, query based approach retrieves documents that contain certain combination of user defined words or key words.

**Semantics -** The biomedical texts are often ambiguous inclusive of words or terms that may have number of synonyms, like, NF-kβ or NF- KB, gene AGP1 or YCC5, etc. Vector models and other methods are blind enough and very much automatic to sense such similarity and differences. Semantics identifies the definition to such terms from the text.

**Information Extraction -** This automated methods look for facts in the text that may represent combination of terms or patterns. Selection also considers sequence of the words and takes into account their morphological changes through the text. Information extraction is the process of using this information to extract meaning from a text and regular expressions to produce a structured, machine readable file that contains essential information.

### 3.1 Related Work

Text mining is a system for phrase extraction and proximity analysis. Text analysis as a major new frontier in bioinformatics, whose smashing success in the area of gene sequence analysis is based, after all, on nothing more than algorithms for finding and comparing patterns in the four-letter language of DNA. Most of the efforts have been focused on a few specific problems although that there are many other interesting areas of potential applications.

Bob Futrelle has organized a large "bio-NLP" information network and enunciated a radical vision by using coherent theory based on the biological nature of language. Swanson's, Kostoff's, Tanabe's, and Chowdhury's reliance on human expertise and manual filtering interfaces between biomedical text mining and the cognitive tradition in IR. K. Bretonnel Cohen , Lawrence Hunter articulated an in-depth treatment of statistical approaches to biomedical text mining.

## Proposed Phase - Gene Interpretation by Text Mining

Text is "unstructured, amorphous, and difficult to deal with" but also "the most common vehicle for formal exchange of information." The way to represent the huge amounts of unstructured text in a compact form is difficult and can do by designed phase of text mining process. Figure 3 describes the tasks of designed phase that provides the basis for the biologists in their analysis of complex problems. The tasks that are undertaken in this paper are

- Use NLP
- Find out the correlated terms
- Co-reference Resolution
- A Structured database

### 3.2 Use NLP

The first step of NLP approach retrieves the Genes from open access database; the second step apply semantics for identifying the physical gene structure and by third step finalizing the characterization of Gene are done. By this the text is indexed into terms or words that looks for particular keyword or pattern as specified. It helps researchers to identify the natural behavioral of gene and to find Gene- Disease relationships.

### 3.3 Finding Out the Correlated Terms

Define the concepts by the sets of correlated terms rather than by raw term which can quantify the relations between documents or portions of text. Correlated terms usually having a more specific meaning are taken and predict the subsequences of diseases. The parameters were set to discover a thousand combinations of correlated terms. These sets of terms were further added to the basic index and applied to the information retrieval problem. The extraction of annotations of genes and proteins automatically and for detecting protein-protein interactions and regulatory pathways can be done with correlated terms. Also the extraction of associations between proteins and their functional features that pose an interesting biological problem are found out from the co-related terms.

**Co-reference Resolution**

Two noun phrases referring to the same entity are said to co-reference. Co- reference resolution is important for many text mining tasks:

- Information extraction
- Summarization
- First story detection

Summarization of Functional genomics can be done by Co-reference resolution of Gene. A number of text-mining methods have been developed for summarizing groups of genes.

**Structured database**

A Biological database can be built from the extracted information. So with this, Gene expression data (co-regulated genes) along with co-occurrence information can be predicted which a major resource to study various Gene -Diseases Relationship. Proteins and genes are characterized within biological databases through unique identifiers; each identifier is associated with its corresponding protein or nucleotide sequence and functional descriptions.

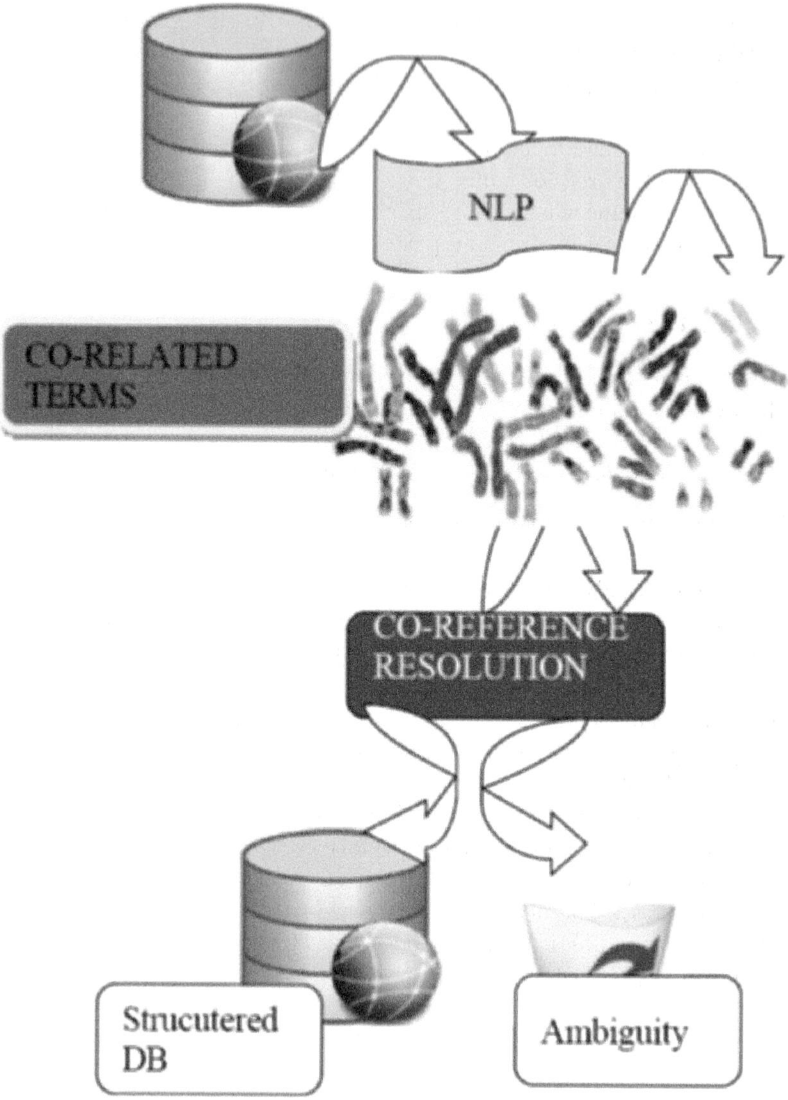


**Fig. 3.** Tasks of Designed phase

## 4 Conclusion

This paper describes a new text mining process along with NLP to uncover interesting term correlations and Gene- Disease relationship. This algorithm comprising these above said steps identifies combinations of terms that optimize an objective function, which is the cornerstone of the process. A scientist now may use this information from the database to know why the genes were co-regulated, what disease may be associated? , what symptom, or the identification of a gene in conjunction with a mutation. etc., these can also be linked with other studies like NMR data of metabolites (metabolomic data) to define causes of disease. Finding meaningful concepts within a huge collection of corpuses in a reasonable timeframe is a difficult task to accomplish and that can be accomplished in future.

## References

1. Cohen, K.B., Hunter, L.: Getting started in text mining. PLoS Comput. Biol. 4, e20 (2008)
2. Rzhetsky, A., Seringhaus, M., Gerstein, M.B.: Getting started in text mining: part two. PLoS Comput. Biol. 5, e1000411 (2009)
3. Friedman, C., Kra, P., Yu, H., Krauthammer, M., Rzhetsky, A.: GENIES: a natural-language processing system for the extraction of molecular pathways from journal articles. Bioinformatics 17(suppl.1), S74–S82 (2001)
4. Perez-Iratxeta, C., Wjst, M., Bork, P., Andrade, M.: G2D: a tool for mining genes associated with disease. BMC Genet. 6, 45 (2008)
5. Koussounadis, A., Redfern, O.C., Jones, D.: Improving classification in protein structure databases using text mining. BMC Bioinformatics 10, 129 (2009)
6. Krallinger, M., Valencia, A.: Text-mining and information-retrieval services for molecular biology. Genome Biol. 6, 224 (2005)
7. Shatkay, H., Höglund, A., Brady, S., Blum, T., Dönnes, P., et al.: SherLoc: high-accuracy prediction of protein sub cellular localization by integrating text and protein sequence data. Bioinformatics 23, 1410–1417 (2007)
8. Natarajan, J., Berrar, D., Dubitzky, W., Hack, C., Zhang, Y., et al.: Text mining of full-text journal articles combined with gene expression analysis reveals a relationship between sphingosine-1-phosphate and invasiveness of a glioblastoma cell line. BMC Bioinformatics 7, 373 (2006)
9. Galvez, C.: Knowledge management for biomedical literature: the function of text-mining technologies in life-science Research
10. Editorial, K.J.: Briefings in Bioinformatics. (3), 220–221 (2005)

# Improved Input-Delayed Dynamic Neural Network for GPS/INS Integration

M. Malleswaran[1], A. Saravanaselvan[2], and V. Vaidehi[3]

[1] Dept of Electronics and Communication Engineering
Anna University of Technology Tirunelveli, Tirunelveli, India
[2] Dept of Electronics and Communication Engineering
National Engineering College, Kovilpatti
[3] Dept of Computer Technology
MIT Campus, Anna University, Chennai, India
malleshl971@yahoo.com,
selvansaravanan@yahoomail.com,
vaidehivijay@gmail.com

**Abstract.** At present, real time navigation systems rely on kalman filtering for integrating global positioning system (GPS) and inertial navigation system (INS) in order to provide feasible solution for reliable navigation. But there exist some inadequacies related to the stochastic error models of inertial sensors, requirement of prior knowledge in fusing data and long design time. Moreover, recently artificial intelligence (AI) techniques based on static neural networks were suggested in place of kalman filter. But there also exist drawbacks like inadequate training during GPS absence, less reliability and more training time. To solve this problem, a new dynamic architecture based on Input-delayed model has been developed. This paper, therefore proposes a scheme based on Improved Input-delayed dynamic neural network (IIDDNN) to model the samples of INS values with respect to GPS values as target vectors (Normal mode). This model is capable of providing a reliable and accurate solution especially for GPS absence (prediction mode). The prediction mode output obtained using IIDDNN is compared with the performance of normal mode of IIDDNN. The proposed IIDDNN model is evaluated using test data of fixed trajectory path for an aircraft vehicle.

**Keywords:** Global Positioning System, Inertial Navigation System, Artificial Intelligence, Improved Input-Delayed Dynamic Neural Network, Static Neural Networks.

## I. Introduction

### *A. Motivation*

Nowadays land vehicles are equipped with Global Positioning System (GPS) to provide accurate position and velocity information [14-17]. However, there are several situations where GPS experience satellite signal blockage or deterioration of accuracy (due to multipath effects and clock bias error). Therefore, GPS is usually

D. Nagamalai, E. Renault, M. Dhanushkodi (Eds.): CCSEIT 2011, CCIS 204, pp. 654–672, 2011.

combined with Inertial Navigation System (INS)[18,19]. The integration of GPS and INS, therefore, provides a navigation system that has superior performance in comparison with either a GPS or an INS stand-alone system.

***B. Navigation using GPS***

The last two decades have shown an increasing trend in the use of positioning and navigation technologies in land vehicle applications. Most of the current vehicular navigation systems rely on the global positioning system (GPS) that is capable of providing accurate measurements the GPS needs at least four satellites with good geometry. In addition, there must be a direct line of sight between the GPS antenna and those satellites. Unfortunately, this is not always possible since a GPS signal may be lost when driving around obstacles (through urban areas, overpasses or tunnels on highways, tree lined streets etc) or when operating in poor weather conditions. Moreover, the GPS signal is purposefully jammed while an aircraft vehicle on fly. The satellite signal blockage results in deterioration of the overall position accuracy [1, 4].

***C. Navigation using INS***

Inertial Navigation System (INS) is a self-contained system. It incorporates three orthogonal accelerometers and three orthogonal gyroscopes, which measure three linear accelerations and three angular rates respectively. A set of mathematical transformations and integrations with respect to time, known as the mechanization equations are applied to the raw measurements from the INS sensors to determine position, velocity and altitude information. Unfortunately, the INS cannot replace the GPS or operate on a stand-alone basis. During the mechanization procedure, the accuracy of INS position components deteriorates with time due to the integral of the sensor errors that exhibit considerable long-term growth. These errors include white noise, correlated random noise, bias instability and angle random walk. The errors are stochastic and wide sense stationary process in nature and can cause a significant degradation in the INS performance over a long period of operation [10, 11, and 12]

***D. Existing INS/GPS data fusion techniques***

In order to overcome the problems associated with the operation of GPS and INS on their own, the two systems are often paired together so that the drawbacks associated with each system are eliminated. The GPS/INS data fusion is commonly performed real time using a kalman filter (KF) [1,4]. This method requires a dynamic model of both INS and GPS errors, a stochastic model of the inertial sensor errors and a priori information about the covariance of the data provided by both systems. Data fusion employing a KF has been widely used and is considered the benchmark for INS/GPS integration.

There are however, several considerable drawbacks to its use [2, 5, and 11]. These include the following: (1) the necessity of accurate stochastic modeling, which may not be possible in the case of low cost and tactical grade sensors; (2) the requirement for a priori information of the system and measurement covariance matrices for each new sensor, which could be challenging to accurately determine; (3) relatively poor accuracy during long GPS absence; (4) the weak observability of some of the error states that may lead to unstable estimates of other error states; and (5) the necessity to tune the parameters of the stochastic model and a priori information for each new sensor system. The above drawbacks can be overcome by using intelligent networks or Artificial Neural Networks (ANN) [2, 4]. The IIDDNN is found to have optimized architecture when compared to static neural networks in the prediction mode. [3, 11, 12]

## 2 Literature Survey

More recently, several techniques based on Artificial Intelligence (AI) have been proposed to replace KF in order to eliminate some of its inadequacies [1, 2, 13]. The major inadequacy related to the utilization of KF for GPS/INS integration is the necessity to have a predefined accurate stochastic model for each of the sensor errors [3, 12]. Furthermore, prior information about the covariance values of both INS and GPS data as well as the statistical properties (i.e. the variance and the correlation time) of each sensor system has to be known accurately [3]. Several AI – based GPS/INS architectures using Multi-Layer Perceptron Neural Networks (MLPNN) [2], Radial Basis Function Neural Networks (RBFNN) [5] and Adaptive Neuron-Fuzzy Inference Systems (ANFIS) were reported for GPS/INS integration [6–7]. The main idea behind all of these methods is to mimic the latest vehicle dynamics by training the AI module during the availability of the GPS signals. An empirical model that processes the INS output (position or velocity) and provides the corresponding INS position or velocity error can be established during the normal (training) procedure. In addition, some of the recently published results showed that KF might outperform some of the AI–based techniques during the first few seconds of the GPS absence [1,7]. This is due to the inaccurate linearized dynamic model of INS errors utilized by KF. Such linearized model operates adequately only in the short-term of GPS absence. In case of long GPS absence, the ignored non-linear and nonstationary parts of the errors grow to large values and deteriorate the overall positioning accuracy. The negative impact of the inappropriate stochastic models of inertial sensors also appears in cases of long GPS absence [4–6]. In this research, we aim at developing an AI based GPS/INS integration based on IIDDNN model and thus providing a better positioning accuracy during the absence of GPS signal. Such technique combines the advantages of IIDDNN model in representing the sequential process in the time varying input pattern (INS position).

## 3 Improved Input Delayed Dynamic Neural Network

The INS dynamic error model used by KF is a linearized one. In addition the stochastic error model for the sensor errors is also linearized in the form of 1st order difference equations. Such state space model is required for the operation of KF. The non-linear and the non stationary parts of the INS errors can be modeled using extended kalman filter, particle filter and unscented kalman filter but they all required complex modelling process. Moreover, INS position error will grow to large value using kalman filter thus deteriorating the positioning accuracy, especially in the long term GPS absence. This leads to large position errors in case of relatively long GPS outages. That is the reason for opting Artificial Neural Network (ANN) to find the solution for navigation problem in an aircraft vehicle. The non-linear complex AI – based modeling capabilities is therefore suggested in this study.

Dynamic networks are generally more powerful than static networks (although they may be somewhat more difficult to train) [7,20]. Because dynamic networks have memory, they can be trained to learn sequential or time-varying patterns. In fact, in order to predict temporal patterns, an ANN requires two distinct components: a memory and an associator. The memory holds the relevant past information, and the associator uses the memory to predict future events. The associator is simply a static multi layer perceptron neural network (MLPNN) and the memory is generated by a time delay unit (or shift register) that constitutes the tapped delay line. Traditional MLPNN is a static and memory less network that is effective for complex non linear static mapping. A common method is to consider a sequence of time varying input patterns. This approach is suitable to find accurate position of an aircraft vehicle at the time of GPS absence. Here, the network is provided with a static memory that is specifically expert knowledge-dependent. A major limitation of the IIDDNN model is the increase of computational complexity since the input layer has to have a number of neurons equal to the number of past samples. As a result, both the complexity and training time will dramatically increase.

### *A. Proposed Architecture description*

In this case, the static neural network is transformed into a dynamic neural network by incorporating a memory and associator units at the input layer of input delay neural network. This input delayed model is trained to learn sequential or time varying patterns of the INS samples. The memory holds past samples of INS values and the associator uses the memory to predict future events. The model configuration of Improved Input-delayed neural network is shown in figure1. The case shown in figure1 considers a tapped delay line that involves the 'd' most recent inputs. For a case of 'd' delay elements, the network processes time varying pattern of the INS samples applied as inputs. This dynamic model neural network considers a cluster inputs for the training process whereas the static neural networks produces the future outputs based on current samples position alone. In other words, we can say that the static neural networks output does not considers the past events. Because static neural networks are memory less topology that is effective for complex non linear static mapping. In fact, INS position error prediction is a procedure where previous errors have to be seriously considered. A fixed number of past events are selected and introduced to the input layer of the network. Therefore, the network is involved with a static memory that is specifically expert knowledge-dependent. This has a beneficial effect on the prediction accuracy especially in the case of GPS absence (prediction mode). Furthermore, the IDNN is trained with the New dynamic back propagation learning algorithm [11, 12].

The training criteria of the network involves New dynamic back propagation learning algorithm based on input delayed model of IIDDNN. During the time of GPS absence, the path selector unit is switch over into prediction mode from the normal mode for training INS values. In this case, IIDDNN model gives the INS position error value and this value is added with the current samples of INS position so as to arrive the new target values. The path selector unit allows GPS signal as target values to the IIDDNN model in normal mode whereas in the prediction mode

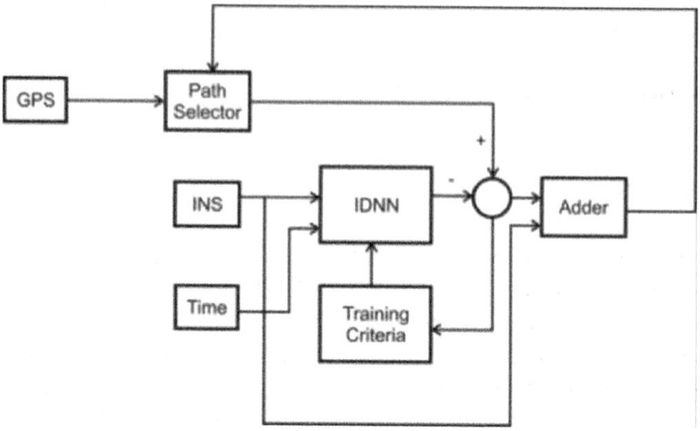


**Fig. 1.** System configuration of proposed scheme

the GPS signal does not exist. In this case, the adder unit which adds the INS position error value to the current samples of INS position to determine the new target value for training the IIDDNN model. At that time, the path selector unit allows the adder output as target values to the IIDDNN model to carry out the prediction mode training process.

The IIDDNN system model processes the input delay elements as x(t), x(t -1), x(t - 2), ... and x(t - d), where d is known as the tapped delay line memory length.

The input signal $X_i$ (t) to the input layer of IIDDNN model shown in figure 1 is mathematically represented as

$$X_i (t) = \sum_{L=0}^{d} W_i(k)\, x(t - L) + \alpha_i$$

Wi(k) – Synaptic weight for neuron i at input layer of IIDDNN
$\alpha_i$ – bias value between input node to hidden node
d – Tapped delay line memory length

The non linear activation function used in the hidden layer of IIDDNN is given as

$$H_i = S\left( \sum_{L=0}^{d} W_i(k)\, x(t - L) + \alpha_i\right)$$

$S_i$ – Hidden layer output
H(.) – Sigmoidal activation function
$S(X_i (t)) = (1 - e^{-Xi (t)}) / (1+e^{-Xi (t)})$

The linear activation function used in the output layer of IIDDNN is represented as

$$Y_j(t) = F\left( \sum_{i=1}^{N} W_{ij} H_i + \beta_j\right)$$

Yj(t) – neuron output j
$\beta_j$ – bias value between hidden node to output node
$W_{ij}$ – Weight factor between hidden node to output node
L(.) – Transfer function of the output neuron j

$$C_j(t) = C\left( F\left( \sum_{i=1}^{N} W_{ij} H_i + \beta_j \right) \right)$$

C(.) – Comparator function of the trained output neuron j and input neuron i.

***B. Training mode***

Along each of the East, North and vertical directions, the INS position ($P_{INS}$) and the time(t) are the inputs to one of the IIDDNN modules while the error in the corresponding INS position ($dp_{INS}$) is the module output. During the availability of the GPS signal, the IIDDNN module operates in the normal mode. In order to train the network, the INS position error provided at the output of this IIDDNN module should be compared to a certain target or a desired response. In this case, the target (or true) INS position error ($dp_{INS/GPS}$) is the difference between the INS original position and the corresponding GPS position. The difference between IIDDNN module output ($dp_{INS}$) and the true position error $dp_{INS/GPS}$ is the estimation error (D (dp)) of the IIDDNN module. In order to minimize this error, the IIDDNN module is trained to adjust the IIDDNN parameters that are continuously updated according to the least square criterion until reaching certain the minimal Mean Square Error (MSE). The training procedure continues and is repeated for all GPS/INS data windows until a GPS absence is detected.

***C. Prediction mode***

When the satellite signal was blocked (during GPS absence), the system was switched into the prediction mode where the IIDDNN module is used to process the INS position $P_{INS}$ at the input and predict the corresponding position error ($dp_{INS}$) using the latest IIDDNN parameters obtained before losing the satellite signals. The estimated error for past INS samples along with current INS values is then applied into a comparator network in order to predict the new GPS position ($P_{GPSC}$). In order to utilize the IIDDNN based module in real-time GPS/INS integration, a time varying pattern with memory size (d) is selected. A number of samples equal to memory size of INS position component and the corresponding GPS position are collected. The prediction procedure of the IIDDNN parameters starts after collecting the dth sample of both INS and GPS position components. The patterns of INS and GPS position components are used to train the IIDDNN module to reflect the latest vehicle dynamics and the INS error trend. The IIDDNN module is trained until a certain number of MSE is reached. This procedure is repeated after collecting completely new patterns of INS and GPS of a size equal to the memory size. To provide a complete navigation solution for a aircraft vehicle, the longitude, latitude and altitude components are considered and applied to the IIDDNN module to estimate the MSE.

***D. New dynamic back-propagation learning algorithm***

- The weight is initialized as small random values
- The Input signal $x_i$ is given to each input unit and the input unit transmits the signal to all hidden layers.
- Each hidden unit sums its weighted input signals as

$$z_{-inj} = \sum_{L=0}^{d} W_i(k)\, x(t - L) + \alpha_i$$

and the activation function is applied for training that input signal.

$$zj = f(z_{-inj})$$

- The hidden units output is given as input to output unit. Each output unit sums it weighted input signals as

$$y_{-ink} = F\left( \sum_{i=1}^{N} W_{jk}\, H_i + \beta_j \right)$$

and the activation function to calculate the output signals

$$y_k = f(y_{-ink})$$

- Comparator function of the trained output neuron j and input neuron i is given as

$$C_j\,(t) = C\left( F\left( \sum_{i=1}^{N} W_{jk}\, H_i + \beta_j \right) \right)$$

- Each output unit receives a GPS target pattern. According to the received GPS signal corresponding error information is calculated.

$$\delta_k = (t_k - y_k) * f(y_{ink})$$

- Each hidden unit receives the feedback from output unit and sums its delta inputs from the units in the layer above

$$\delta_{-inj} = \sum_{k=1}^{m} \delta_j * w_{jk}$$

The error information is calculated as

$$\delta_j = \delta_{-inj} * f(z_{-inj})$$

- Each output unit updates its bias and weights
- The weight correction term is given by

$$\Delta w_{jk} = \alpha\, \delta_k\, z_j$$

- And the bias correction term is given by

$$\Delta\beta_j = \alpha\,\delta_k$$

Therefore

$$W_{jk(new)} = w_{jk(old)} + \Delta w_{jk}$$

$$\beta_{j(new)} = \beta_{j(old)} + \Delta\beta_j$$

- Each hidden unit updates its bias and weights
  The weight correction term

$$\Delta\, W_i(k) = \alpha\, \delta_j\, x_i$$

The bias correction term

$$\Delta\, \alpha_i = \alpha\, \delta_j$$

Therefore

$$W_i(k)_{(new)} = W_i(k)_{(old)} + \Delta\, W_i(k)$$

$$\alpha_{i\,(new)} = \alpha_{i\,(old)} + \Delta\, \alpha_i$$

- After the errors are minimized, the testing conditions are stopped.

***E. Network over-fitting***

Network over-fitting is a machine-learning problem that has been investigated by many researchers. Network over-fitting usually occurs when the network captures the internal local patterns of the training data set rather than recognizing the global patterns of the data sets. The knowledge base that is extracted from the training data set is therefore not general. As a consequence, it is important to realize that the specification of the training samples is a critical factor in producing a neural network which is capable of making the correct responses. Two procedures have been evaluated to overcome the problem of over-fitting namely, early stopping and regularization.

***F. Early stopping***

In general, the aim of early stopping is to mimic the prediction of future individuals form the whole. This will be achieved in case that the training data fully represents the sample space and each left out individual can lie anywhere in this space. Large samples and small dimensionality generally satisfy these requirements. During the IIDDNN training stage, the module performs the function of understanding the input and output mapping. To ensure the generalization of the IIDDNN model a hold out kind of early stopping criterion was chosen due to its suitability for real time implementation. This approach allows IIDDNN to read just its parameters to detect the dynamics of the data in a transition interval before moving to the prediction stage. Furthermore, it avoids the possibility of model over-fitting for the training data.

***G. Regularization***

The second method utilized to avoid the over-fitting problem and to optimize the IIDDNN model is the regularization technique. This is known to be a very desirable procedure when the scaled conjugate gradient descent method is adopted for training. The regularization technique involves modifying the performance function of mean square error (MSE) which is normally chosen to be the sum of squares of the error between the network output and the desired response on the training data set. [9]

## 4 Experimental Setup

In this experiment, the training mode and the prediction mode of IIDDNN was performed. The normal mode and prediction mode training were performed using bipolar sigmoidal activation function (non linear activation function). During the presence of GPS signal, the proposed system relies on GPS position information to train the IIDDNN model. During the normal mode, the IIDDNN model is trained to mimic the latest vehicle dynamic, determine the INS position error, and correct the corresponding INS position component. The data is processed as follows: first, the INS and GPS signals are taken as input vector and target vector respectively. The input vector is send as one of the input signal to input node during the training mode. The value of the weight is initially considered to be a small value and bias is chosen close to the first input value. The learning rate of this experiment is also chosen accordingly. The INS position is then used as the input to IIDDNN, and is trained until it nearly equals the target GPS position. During the absence of GPS signal, the proposed system operates in the prediction mode where the IIDDNN model predicts the corresponding INS position error based on the knowledge stored during the normal mode procedures. This predicted INS position error is then removed from the corresponding INS position component to obtain the corrected INS position. [8, 21]

## 5 Simulation Results and Tables

The trajectory path shown in figure 2 is taken into consideration and the New Dynamic Back Propagation Learning Algorithm (NDBPLA) for IIDDNN is implemented using matlab. The position and time value obtained after INS mechanization are taken as input vectors. The position values obtained from GPS using GPS receiver is taken as the target vector. Training was performed for latitude, longitude and altitude components. The actual output i.e., corrected INS value for all the three components were found using NDBPLA. From the output obtained during the final epoch of IIDDNN model is shown in Fig. 7,8,9,10,11 and 12 for normal mode and predicted mode. The mean square error (RMSE) and performance index (PI) values were calculated for each and every epoch. The formula for the RMSE and PI are given as

$$\mathrm{RMSE} = \sqrt{\frac{\sum_{i=1}^{m}(y_i - ay_i)^2}{m}}$$

$$\mathbf{PI} = \frac{\sqrt{\sum_{i=1}^{m}(y_i - ay_i)^2}}{\sum_{i=1}^{m}|y_i|}$$

Where $y_i$ is the target output, $ay_i$ is the computed output, m is the number of data patterns and $|\mathbf{y}_i|$ is modulus of target output. The performance of the IIDDNN model is given in table 7. The less RMSE and PI values of prediction mode IIDDNN gives superior performance than normal mode IIDDNN model.

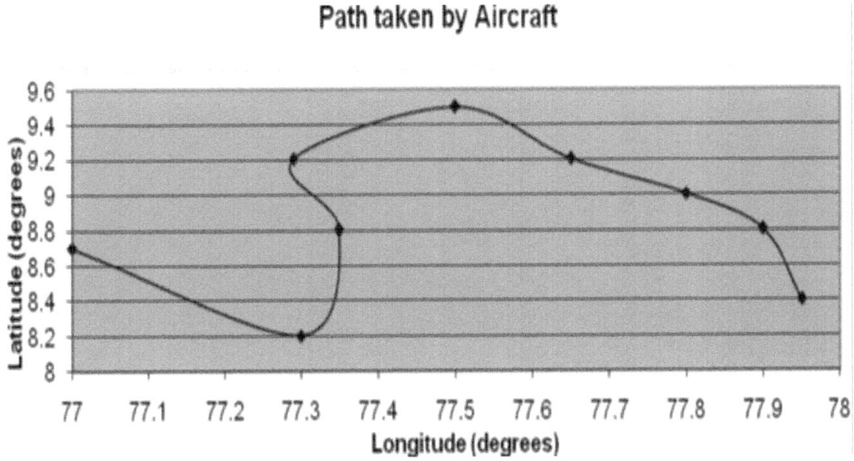


**Fig. 2.** Trajectory path

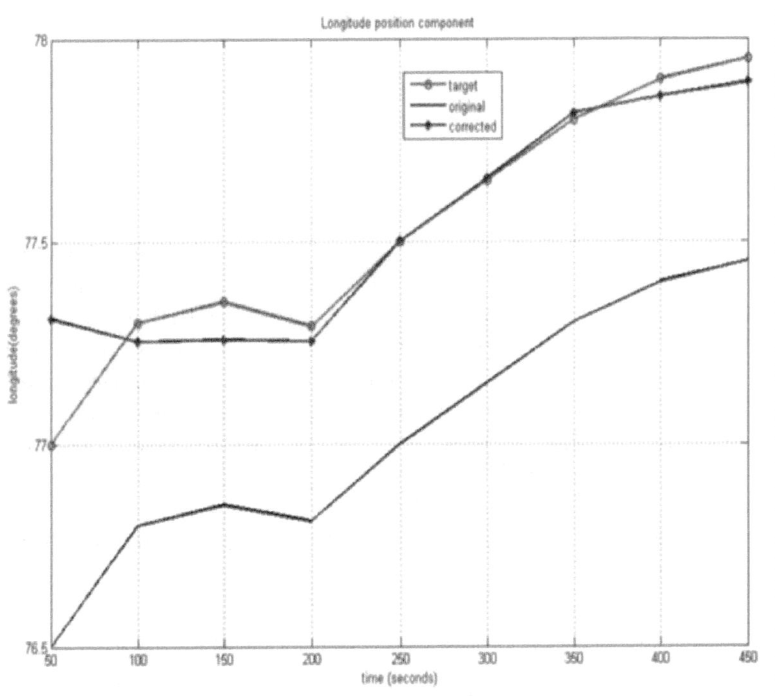


**Fig. 3.** Output of longitude component in Normal mode when EPOCH=4

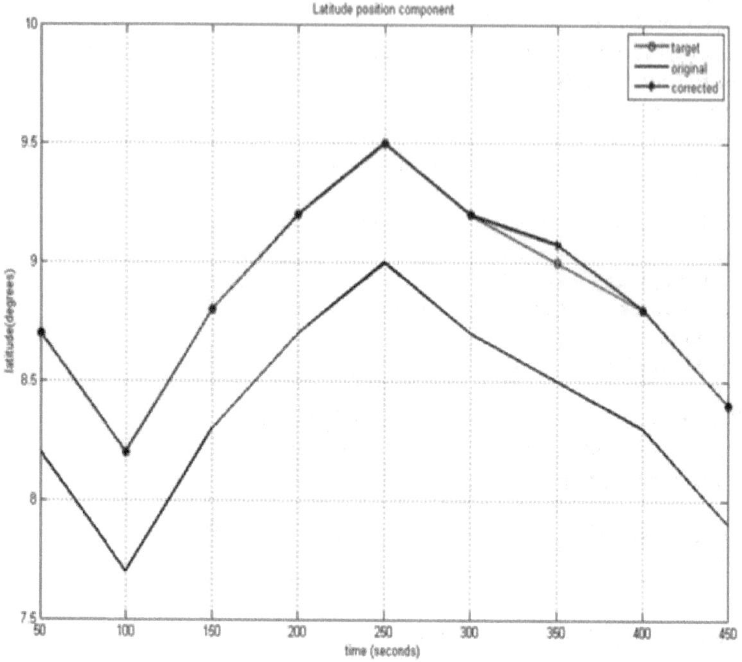


**Fig. 4.** Output of latitude component in Normal mode when EPOCH=4

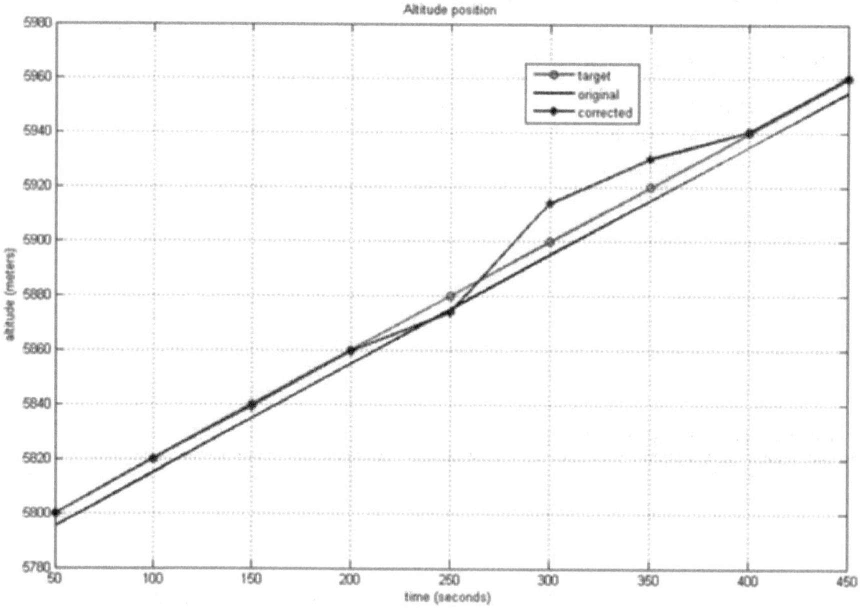


**Fig. 5.** Output of altitude component in Normal mode when EPOCH=4

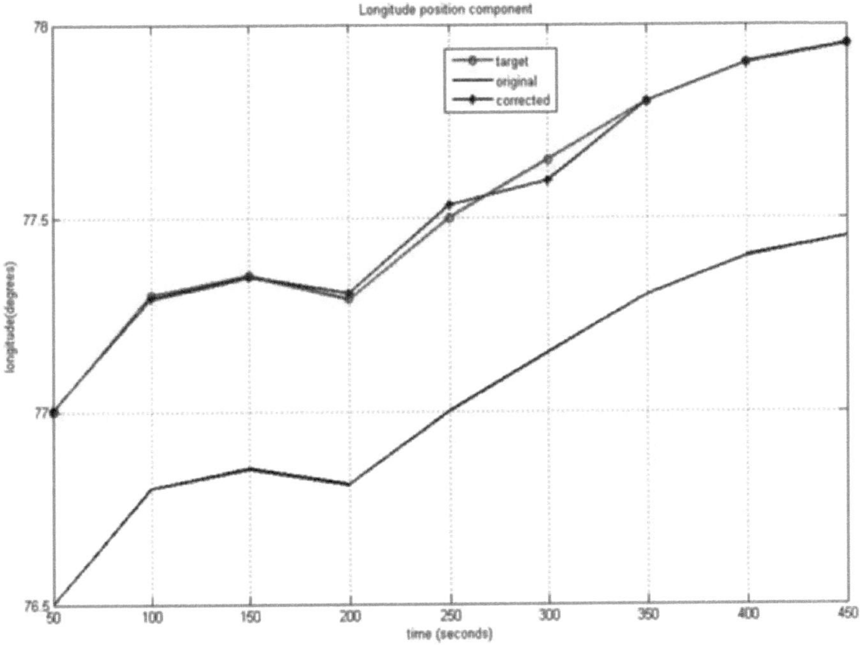


**Fig. 6.** Output of longitude component in Normal mode when EPOCH=7

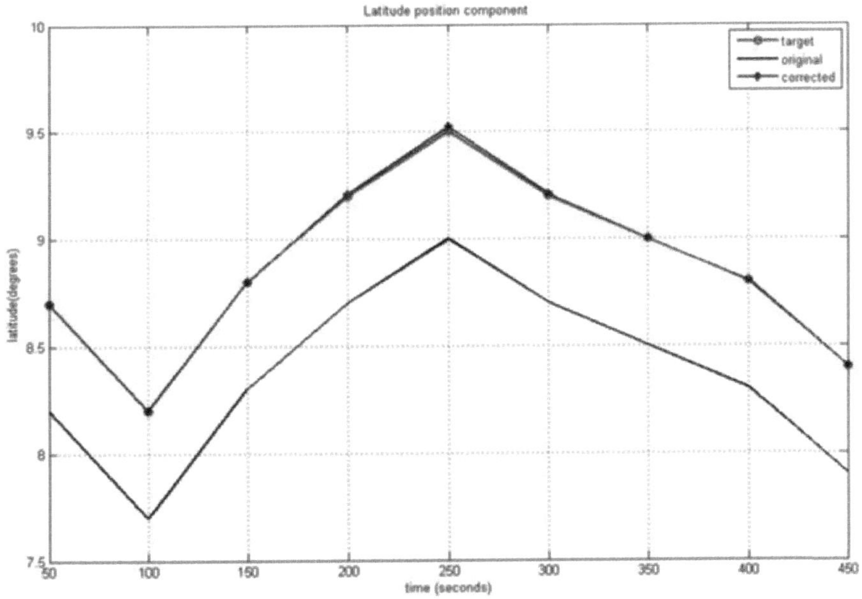


**Fig. 7.** Output of latitude component in Normal mode when EPOCH=7

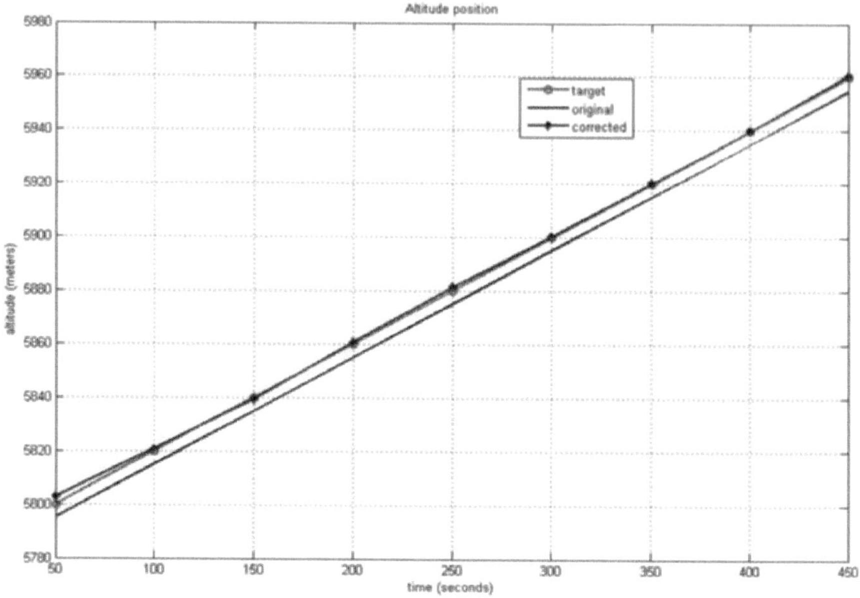


**Fig. 8.** Output of altitude component in Normal mode when EPOCH=7

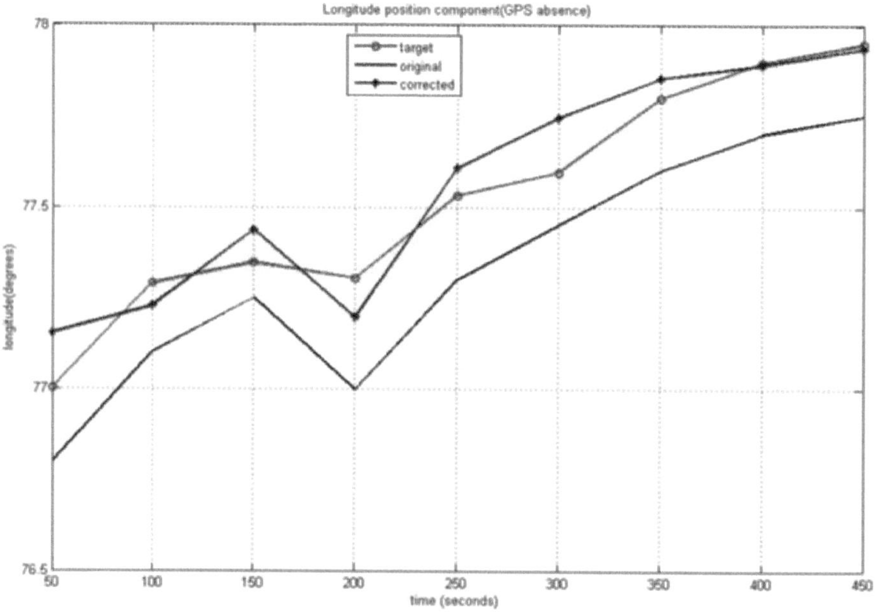


**Fig. 9.** Output of longitude component in predicted mode when EPOCH=3

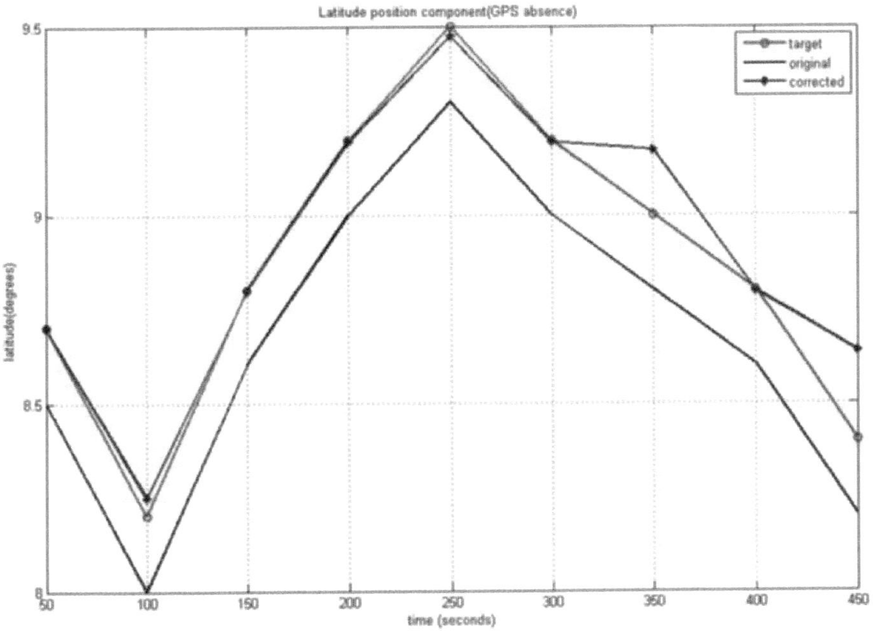


**Fig. 10.** Output of latitude component in predicted mode when EPOCH=3

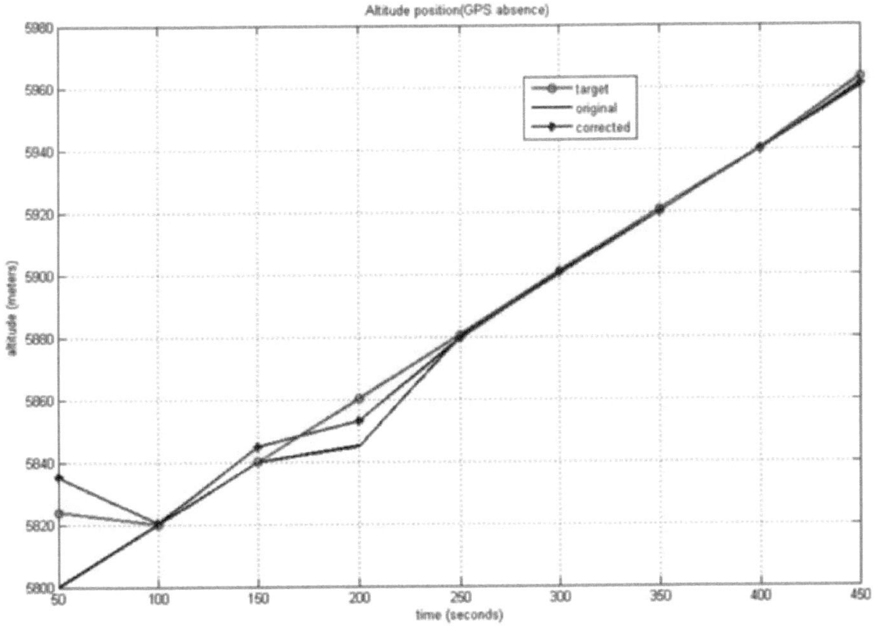


**Fig. 11.** Output of altitude component in predicted mode when EPOCH=3

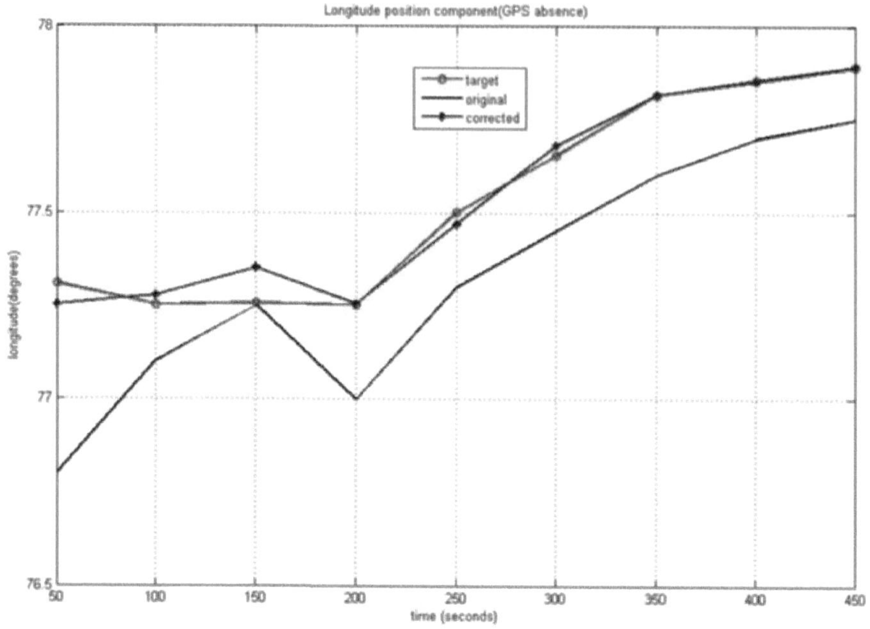


**Fig. 12.** Output of longitude component in predicted mode when EPOCH=5

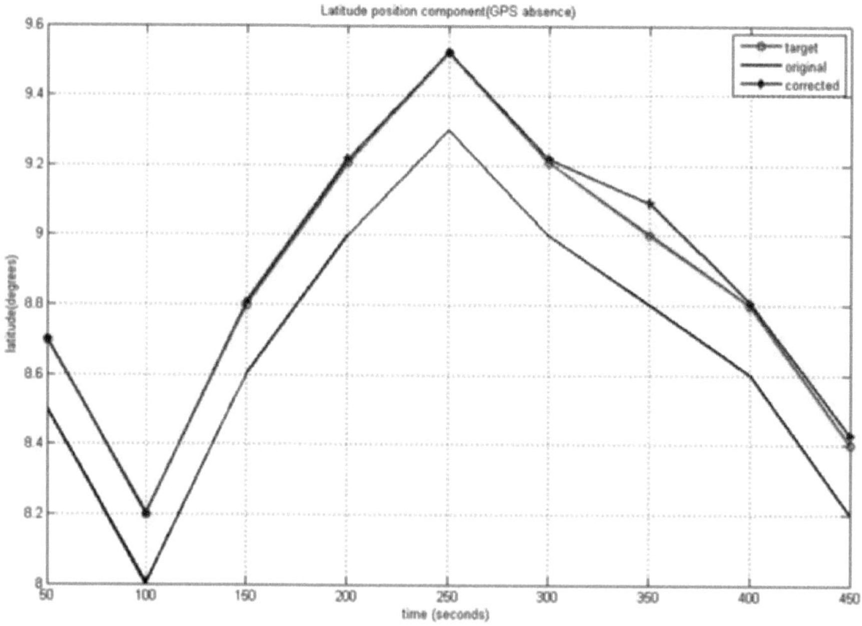


**Fig. 13.** Output of latitude component in predicted mode when EPOCH=5

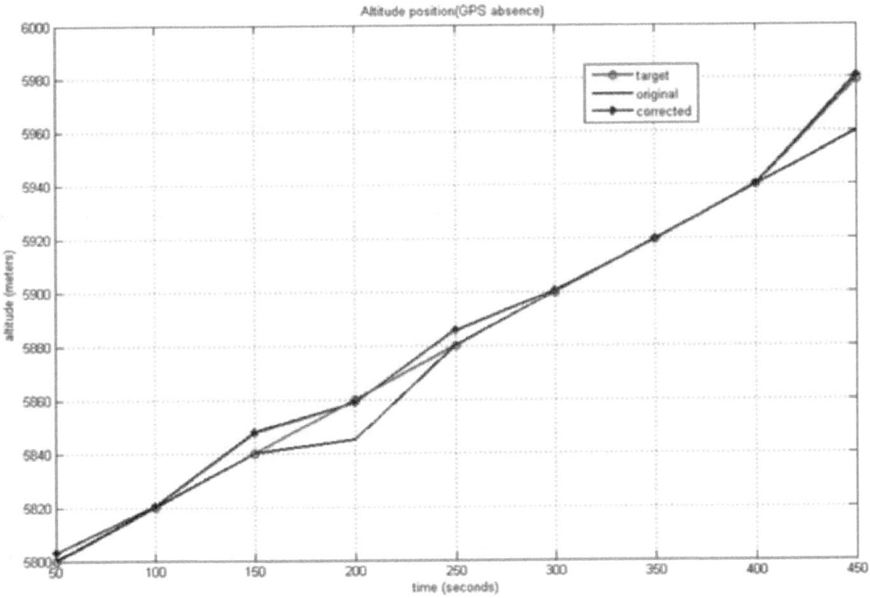


**Fig. 14.** Output of altitude component in predicted mode when EPOCH=5

**Table 1.** Mean square error of IIDDNN model in Normal mode

| Epoch no | Longitude (MSE) | Latitude (MSE) | Altitude (MSE) |
|---|---|---|---|
| 1 | 120.50 | 118.46 | 150.34 |
| 2 | 504.54 | 367.45 | 290.89 |
| 3 | 258.99 | 233.89 | 180.46 |
| 4 | 69.80 | 100.48 | 97.90 |
| 5 | 15.52 | 40.45 | 50.67 |
| 6 | 7.65 | 10.35 | 15.78 |
| 7 | 1.89 | 1.76 | 1.78 |

**Table 2.** Weight updates in output layer during each epoch of IIDDNN training in Normal mode

| Epoch no | Weight updates in output layer | | | | |
|---|---|---|---|---|---|
| | W1 | W2 | W3 | W4 | W5 |
| 1 | 0.5210 | 0.9053 | 1.5226 | 2.4269 | 3.6583 |
| 2 | 0.4652 | 0.4676 | 0.3761 | 0.4449 | 1.5453 |
| 3 | 0.4124 | 0.0731 | -0.6298 | -1.4743 | -1.5273 |
| 4 | 0.3623 | -0.2782 | -1.4401 | -2.7918 | -3.8410 |
| 5 | 0.3148 | -0.5878 | -2.0438 | -3.3882 | -3.8983 |
| 6 | 0.2700 | -0.8581 | -2.4651 | -3.5756 | -3.8813 |
| 7 | 0.2275 | -1.0922 | -2.7448 | -3.6239 | -3.8865 |

**Table 3.** Weight updates in hidden layer during each epoch of IIDDNN training in Normal mode

| Epoch no | Weight updates in hidden layer | | | | |
|---|---|---|---|---|---|
| | W1 | W2 | W3 | W4 | W5 |
| 1 | -1.5500 | -1.5500 | -1.5500 | -1.5500 | -1.5500 |
| 2 | -1.5500 | -1.5500 | -1.5500 | -1.5499 | -1.5496 |
| 3 | -1.5500 | -1.5500 | -1.5500 | -1.5502 | -1.5530 |
| 4 | -1.5500 | -1.5500 | -1.5500 | -1.5510 | -1.5729 |
| 5 | -1.5500 | -1.5500 | -1.5502 | -1.5616 | -1.9903 |
| 6 | -1.5500 | -1.5500 | -1.5513 | -1.7068 | -10.13 |
| 7 | -1.5500 | -1.5500 | -1.5593 | -3.7926 | -168.11 |

**Table 4.** Mean square error of IIDDNN model in Predicted Mode

| Epoch no | Longitude (MSE) | Latitude (MSE) | Altitude (MSE) |
|---|---|---|---|
| 1 | 160.40 | 304.55 | 267.81 |
| 2 | 128.44 | 220.78 | 139.30 |
| 3 | 44.82 | 110.55 | 49.87 |
| 4 | 22.75 | 30.67 | 28.62 |
| 5 | 1.12 | 0.98 | 0.87 |

**Table 5.** Weight updates in output layer during each epoch of IIDDNN training in Predicted mode

| Epoch no | Weight updates in output layer | | | | |
|---|---|---|---|---|---|
| | W1 | W2 | W3 | W4 | W5 |
| 1 | 1.1648 | 0.2738 | -1.0394 | -2.0166 | -13.1677 |
| 2 | 0.5898 | 0.3632 | 0.6924 | 1.7445 | 27.6166 |
| 3 | 0.0245 | 0.4394 | 1.6642 | 2.5573 | 101.7480 |
| 4 | -0.1676 | 0.4738 | 2.0029 | 2.9893 | 156.6126 |
| 5 | -0.3428 | 0.5328 | 2.8862 | 3.4526 | 182.4536 |

**Table 6.** Weight updates in hidden layer during each epoch of IIDDNN training in Predicted mode

| Epoch no | Weight updates in hidden layer | | | | |
|---|---|---|---|---|---|
| | W1 | W2 | W3 | W4 | W5 |
| 1 | -1.5500 | -1.5500 | -1.5500 | -1.5500 | -1.5502 |
| 2 | -1.5500 | -1.5500 | -1.5500 | -1.5501 | -1.5615 |
| 3 | -1.5500 | -1.5500 | -1.5500 | -1.5495 | 0.0412 |
| 4 | -1.5500 | -1.5500 | -1.5500 | -1.5517 | -285.39 |
| 5 | -1.5500 | -1.5500 | -1.5500 | -1.6243 | -328.24 |

**Table 7.** Comparison between Normal mode and Prediction mode of IIDDNN model

| Position components | Normal mode training of IIDDNN model | | | | Prediction mode training of IIDDNN model | | | |
|---|---|---|---|---|---|---|---|---|
| | Number of epochs | RMSE | Performance index | Execution Time(sec) | Number of epochs | RMSE | Performance index | Execution Time(sec) |
| Longitude | 7 | 0.5428 | 0.0289 | 7.3 | 5 | 0.4567 | 0.03456 | 7.1 |
| Latitude | 7 | 0.5267 | 0.0278 | 8.6 | 5 | 0.4879 | 0.00321 | 8.3 |
| Altitude | 7 | 0.5098 | 0.0293 | 8.8 | 5 | 0.4673 | 0.00036 | 8.4 |

## 5 Conclusion

From the results and statistical analysis, the prediction mode training of IIDDNN model shows superior performance in Root Mean Square Error, Performance index and Execution time with less number of epochs when compared to the normal mode training of IIDDNN model. So, IIDDNN is optimized neural network for GPS/INS Integration in terms of error accuracy and training time even during the signal blockages in GPS. This model relies on input delay elements at the input layer so that the output INS position error is modeled based on the present and past samples of the corresponding INS position. The impact of different delay elements present at the input layer improves the overall positioning accuracy. So, this novel approach is proposed to predict the accurate position of the moving vehicle particularly in GPS absence.

## References


1. Noureldin, A., Karamat, T., Eberts, M., El-Shafie, A.: Performance enhancement of MEMS based INS/GPS integration for low cost navigation applications. IEEE Trans. Veh. Technol. 58(3), 1077–1096 (2009)
2. El-Sheimy, N., Chiang, K.-W., Noureldin, A.: Developing a low cost MEMS IMU/ GPS integration scheme using constructive neural networks. IEEE Trans. Aerosp. Electron. Syst. 44(2), 582–594 (2008)
3. Malleswaran, M., Vaidehi, V., Angel Deborah, S., Manjula, S.: Integration of INS and GPS using Radial basis function Neural Networks for Vehicular Navigation. IEEE Xplore (2010), 978-1-4244-7815-6
4. Noureldin, A., El-Shafie, A., Taha, M.R.: Optimizing neuro-fuzzy modules for data fusion of vehicular navigation systems using temporal cross-validation. Eng. Appl. Artif. Intell. 20(1), 49–61 (2007)
5. Sharaf, R., Noureldin, A., Osman, A., El-Sheimy, N.: Online INS/GPS integration with a radial basis function neural network. IEEE Aerosp. Electron. Syst. 20(3), 8–14 (2005)
6. Nassar, S., Noureldin, A., El-Sheimy, N.: Improving positioning accuracy during kinematic DGPS outage periods using SINS/DGPS integration and sins data de-noising. Surv. Rev. 37(292), 426–438 (2004)
7. Hiliuta, A., Landry, R., Gagnon, F.: Fuzzy corrections in a GPS/INS hybrid navigation system. IEEE Trans. Aerosp. Electron. Syst. 2(2), 591–600 (2004)

8. Noureldin, A., Osman, A., El-Sheimy, N.: A neuro-wavelet method for multi-sensor system integration for vehicular navigation. J. Meas. Sci. Technol., Lond., UK 15(2), 404–412 (2004)
9. Noureldin, A., Irvine-Halliday, D., Mintchev, M.P.: Accuracy limitations of FOG-based continuous measurement-while-drilling surveying instruments for horizontal wells. IEEE Trans. Instrum. Meas. 51(6), 1177–1191 (2002)
10. Hirokawa, R., Nakakuki, K., Sato, K., Ishihara, R.: Threading the maze GPS/INS landmark sensing and obstacle avoidance. GPS World, 20–26 (2004)
11. Malleswaran, M., Vaidehi, V., Angel Deborah, S., Manjula, S.: Comparison of RBF and BPN Neural Networks applied in INS and GPS Integration for Vehicular Navigation. International Journal of Electronics Engineering research 2(5), 753–763 (2010)
12. Malleswaran, M., Vaidehi, V., Angel Deborah, S.: Data fusion using MLFFNN for land vehicle application. International Journal of Engineering Science and Technology 2(12), 7676–7690 (2010)
13. Dissanayake, G., Sukkarieh, S.: The aiding of a low-cost strapdown inertial measurement unit using vehicle model constraints for land vehicle applications. IEEE Trans. Robot. Automat. 17(5), 731–747 (2001)
14. Sharaf, R., Tarbouchi, M., El-Shafie, A., Noureldin, A.: Real-time implementation of INS/GPS data fusion utilizing adaptive neuro-fuzzy inference system. In: Proceedings of the Institute of Navigation National Technical Meeting – ION NTM 2005, San Diego, CA, January 24-26 (2005)
15. Reda Taha, M.M., Noureldin, A., El-Sheimy, N.: Improving INS/GPS positioning accuracy during GPS outages using fuzzy Logic. In: Proceedings of the 16th International Technical Meeting of the Satellite Division of the Institute of Navigation ION GPS/GNSS 2003, September 9-12 Oregon Convention Center, Portland (2003)
16. Lobo, J., Lucas, P., Dias, J., Traca de Almeida, A.: Inertial navigation system for mobile land vehicles. In: Proceedings of the IEEE International Symposium on Industrial Electronics, ISIE 1995, vol. 2, pp. 843–848 (July 1995)
17. Mynbaev, D.K.: Errors of an inertial navigation unit caused by ring laser gyro errors. In: Proceedings of the IEEE Position Location and Navigation Symposium, pp. 833–838 (1994)
18. El-Sheimy, N., Schwarz, K.P.: Integrating differential GPS receivers with an inertial navigation system (INS) and CCD cameras for a mobile GIS data collection system. In: Proceedings of the International Society for Photogrammetry and Remote Sensing Conference, Ottawa, Canada, pp. 241–248 (1994)
19. Wong, R.V.C., Schwarz, K.P., Cannon, M.E.: High-accuracy kinematic positioning by GPS-INS. Navig.: J. Inst. Navig. 35, 275–287 (Summer 1988)
20. Sivanandam, S.N., Sumathi, S., Deepa, S.N.: Introduction to Neural Networks using Matlab 6.0. Tata McGraw-Hill Publication, New York (2002)
21. Bishop, C.M.: Neural Networks for Pattern Recognition. Oxford University Press, New York (1995)

# Application of Artificial Neural Network for the Prediction of Groundwater Level in Hard Rock Region

M. Kavitha Mayilvaganan[1] and K.B. Naidu[2]

[1] Research Scholar
[2] Professor
Department of Mathematics,
Sathyabama University, Chennai
m.kavithaa@gmail.com

**Abstract.** Artificial neural network has been shown to be an efficient tool for non-parametric modeling of data in a variety of different contexts where the output is the nonlinear function of inputs. Neural networks are the preferred tool for many predictive data mining applications because of their power, flexibility and ease of use. In the present study Feed-Forward Network based Artificial Neural Network (ANN) model is used as a method to predict the groundwater levels. The ANN model was trained using back propagated algorithm with two hidden layer, and with logsig activation function. The models are evaluated using three statistical performance criteria namely Mean Average Error (MAE), Root Mean Squared Error (RMSE) and Regression coefficient ($R^2$). The results of the study clearly showed that ANN can be used to predict groundwater level in a hard rock region with reasonably good accuracy even in case of limited data situation.

**Keywords:** Artificial Neural Networks Model, Back Propagation Algorithm, MATLAB, Observation Well, Ground Water level.

## 1 Introduction

In this paper, groundwater level has been estimated for the hard rock aquifer system of a small region of Tamilnadu named Thurinjapuram watershed using the method of ANN. It is an information processing paradigm that is inspired by the way biological nervous systems, such as the brain processes information. It determines the relationship between inputs and outputs of physical systems by a network of interconnecting nodes adjusted by connecting weights based on the training samples, and extracts patterns and detects trends that are too complex to be noticed by either humans or other computational techniques. Neural network takes a different approach to problem solving than that of conventional computers. It has remarkable ability to derive meanings from complicated and imprecise data. It has an ability to learn and apply the knowledge based on the data given for training or initial experience [1].Moreover, it validates the learning model with a chosen validation set from the input data to achieve a desired level of accuracy. The accuracy could be maximized by repeating the training process with different topological model [2]. Recently, ANN

D. Nagamalai, E. Renault, and M. Dhanushkodi (Eds.): CCSEIT 2011, CCIS 204, pp. 673–682, 2011.

has been used for forecasting groundwater level successfully [3-5]. Thus proper design and implementation of ANN models could provide an efficient tool in water resource management and forecasting. Moreover, ANN could be employed to predict various aspects in hydrology such as rainfall-runoff model, precipitation forecasting and water quality model [6].

## 2 Artificial Neural Network (ANN)

ANN based methods are data analysis methods and algorithms loosely based on nervous systems of humans and animals. In this study, feed forward network based neural network is used for the input data set to make the future predictions. In this network, the information moves in only forward direction from the input nodes to the output nodes, through the hidden nodes. There are no cycles or loops in the network. Input nodes are the resultant where identified weight functions (depending on the input data set) are applied to the inputs. Based on the topology it traverses through multiple intermediate weight functions before concluding to a resultant output (represented as output node). All these intermediate weighted resultants are known as hidden nodes and the cumulative application of weight function decides the number of hidden layers (depending on the topology)

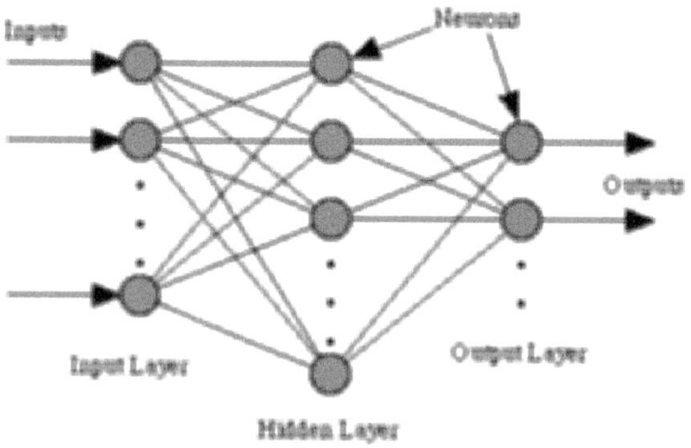


**Fig. 1.** Structure of feed-forward ANN

In this paper, Levenberg–Marquardt algorithm (LMA) has been used to train the ANN and it has surfaced as the best model resulting high level of accuracy and efficiency with the given input data. [7]. Next is Training Parameters Selection. In this phase, training algorithm's parameters like quick propagation coefficient, learning rate, momentum and a condition to stop the network training process, mean square error and minimum improvement has been decided. Error is the kind of conditions based on which training process are said to be reached at the permissible level for forecasting. In this phase the ANN validates its' authenticity forecasting from the training subset and comparing it with the validation subset. Once it achieves this, it certifies itself for forecasting.

## 3 Study Area and Data

The Thurinjapuram watershed covers geographical area of 151.38 sq. km and is located in between 12°12'58" and 12° 21'11" North latitudes and 78°59'45" and 79°9'28" East longitudes (Fig. 2.) It is mainly situated in Thiruvannamalai district of Tamilnadu, India. It is mainly located in Thurinjapuram block and partially falls into two other blocks (Chengam and Thiruvannamalai). Thurinjalar is one of the major tributaries of Ponnaiyar Major River originating from Kavuttimalai reserve forest in Chengam Taluk of Tiruvannamalai district. Thurinjalar River, which is the major stream draining the area, exhibits only sporadic flow during the rainy season. The drainage characteristics are very good. Bedrock is peninsular gneiss of Archean age. The Thurinjapuram area can be classified as "hard rock terrain". The predominant soil types in this river basin are Entiso, Inceptisols, Vertisol and Alfisols. The soil in this minor basin is observed to have good infiltration characteristics. Hence groundwater recharge is possible in this area. The economy of the Thurinjapuram sub watershed depends mainly on agriculture.

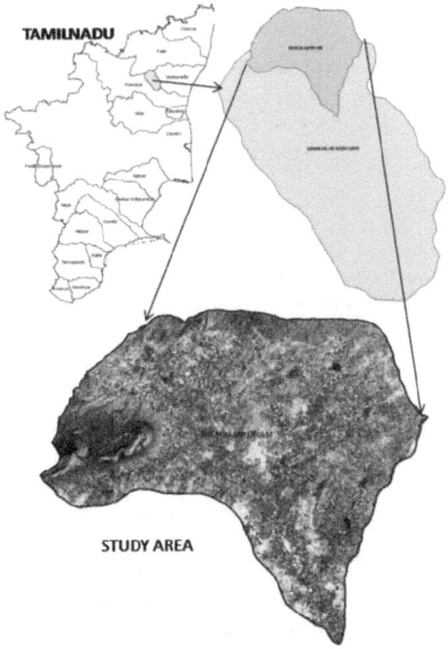


**Fig. 2.** Study area

The input data used for water level prediction are monthly Rainfall and Ground water (level in the observation well) data of Thurinjapuram watershed in Tamilnadu, India, and one month ahead groundwater level as output. For the present study monthly water level data for three observation wells (23112, 23142, and 23143) during 1985 to 2008 has been collected from Thiruvannamalai Groundwater subdivision. In the same period monthly Rainfall data were collected from Kilnatchipattu Raingauge station.

# 4 Methodology

There is no fixed theory for the selection of a particular network type or various internal parameters within it. The various parameters, which affect the performance of network, are number of hidden nodes, number of hidden layers, learning rate and momentum factor. The ANN architecture consist of an input layer with two nodes, a hidden layer with varying nodes and one output layer with single node, thus leading to a multi input single output network. The input data are the rainfall the ground water levels during time 't', 't-1' and 't-2' etc and output data is ground water level during time 't+1'. The data length of 276 data sets has been divided into 70% for training and 30% for testing. Thus the ANN model was trained using 192 randomly selected data sets while the testing set belonged to remaining data sets. The closeness of predictions with the actual observations was judged quantitatively by drawing a scatter plot and a time series comparison plot and quantitatively by evaluating three error measures, namely, Regression coefficient, mean average error and root mean square error.

# 5 ANN Model

Neural network with single hidden layer was trained with different number of neurons. Its regression percentage and performance values were given in Table1.

## 5.1 The Effect of Different Number of Neurons Per Layer

At 32 neurons per layer the data had a good enough fit, with an increase of neurons above this level having no real effect on the prediction anymore, see Table 1. In order to get the best fit and resolution a range of 30 to a 35 were found to be optimal number of neurons in two layer neural networks.

**Table 1.** Training results for two layer neural networks with different number of neurons.

| Number of neurons | Linear regression percentage | Performance value |
|---|---|---|
| 10 | 48 | 0.01796 |
| 15 | 63 | 0.01094 |
| 20 | 77 | 0.00701 |
| 25 | 82 | 0.00660 |
| 30 | 86 | 0.00581 |
| **32** | **88** | **0.00508** |
| 35 | 85 | 0.00683 |
| 40 | 79 | 0.00695 |
| 50 | 73 | 0.00723 |
| 60 | 68 | 0.00755 |

## 5.2 Number of Layers in the Neural Network

The influence of the number of layers was also investigated. These results indicate that the 3 layer network was better than the 2 layer network, while 4 and more layers gave no results. This allows for neural networks to be constructed for 3 layer neural networks, since 2 layer networks have less resolution than 3 layer neural networks. Table 2.presents the results of the number of layers.

**Table 2.** Result of the ANN for different number of layers

| Number of Layers | Linear regression percentage (Reg %) | Performance value |
|---|---|---|
| 2 | 88 | 0.00508 |
| **3** | **94** | **0.00581** |
| 4 | No Fit | No Fit |

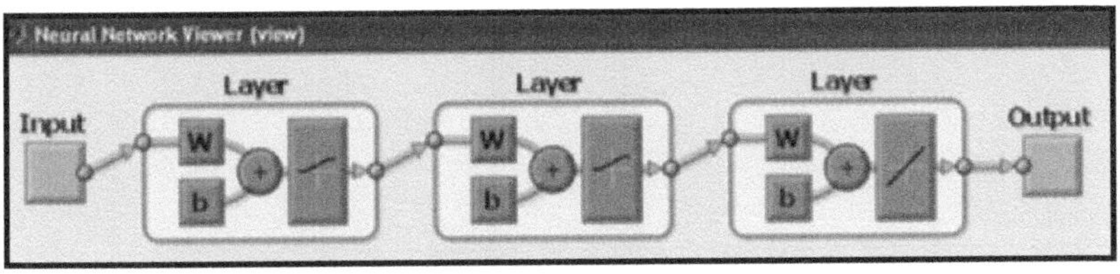


**Fig. 3.** Feed forward neural network with two hidden layers

### 5.3 The Effect of Transfer Function on Data Estimation

Three layer networks was constructed with different transfer functions contained within it to determine the stability of the neural network. The results are given below in Table 3.

**Table 3.** Result of ANN for different activation function

| No of neurons | Activation fun | RMSE | Reg % | performance |
|---|---|---|---|---|
| 12-20-1 | Tansig | 0.5832 | 94 | 0.00581 |
| **12-20-1** | **Logsig** | **0.4665** | **96** | **0.00463** |

The difference can be observed from the average fit value in Table 3, in which the three layer log sigmoid neural network performed the best (Fig3).

Different neural network architectures were developed in order to establish a relationship between the input and output. All the networks were of the feed forward type. The network architectures were trained by varying the number of hidden layer and then by varying number of neurons in each hidden layer.

The momentum factor and learning rate were also fixed by trial and error process. Momentum factor of 0.9 with learning rate of 0.6 were found better for all networks. Also for the given data log sigmoid as an activation function performed better than the other activation functions. Number of training epochs given were 2000 to 3000 early stopping methods was applied, and error goal was set to be 0.001( Fig 4(a)(b)).

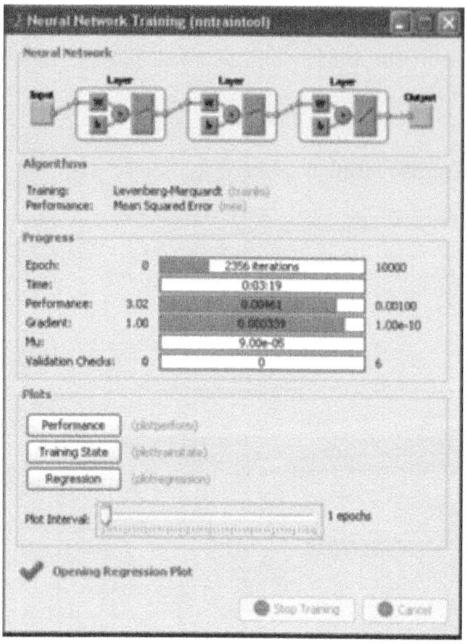


Fig. 4(a). Neural network training tool

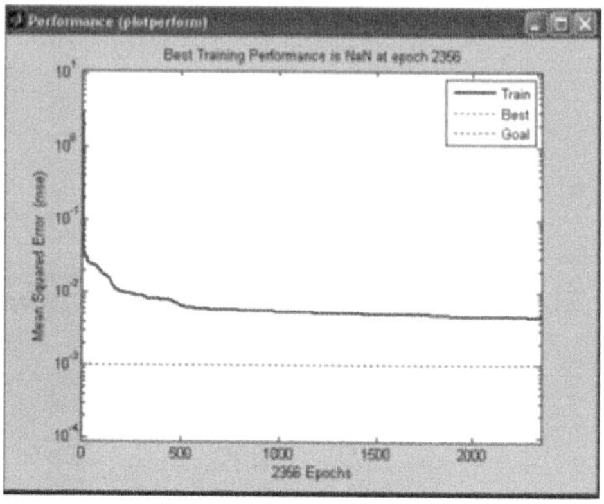


Fig. 4(b). Performance plot

# 6 ANN vs. Linear Regression (LR)

As a benchmark for ANN performance, LR was also conducted on the same data sets with the identical input and output variables. Pearson's correlation coefficient test (2-tailed) was used to test correlations between rainfall and groundwater level . Water level was estimated by a linear curve estimation regression model of SPSS 13.0. All the statistical analyses were conducted with SPSS 13.0 for Windows. A comparison in predictive accuracy between the two methods is shown in Fig 5.

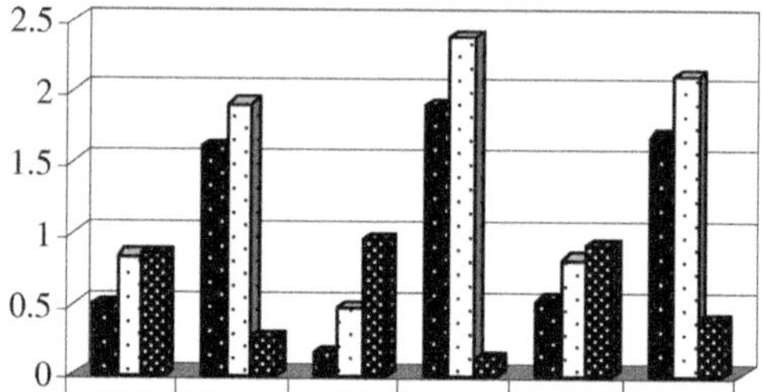


| | ANN | LR | ANN | LR | ANN | LR |
|---|---|---|---|---|---|---|
| | 23112 | | 23141 | | 23143 | |
| MAE | 0.5 | 1.6 | 0.17 | 1.9 | 0.53 | 1.68 |
| RMSE | 0.84 | 1.91 | 0.47 | 2.38 | 0.82 | 2.11 |
| $R^2$ | 0.86 | 0.27 | 0.96 | 0.11 | 0.91 | 0.39 |

Fig. 5. Comparison of ANN and LR error for all the three wells

## 7 Results and Discussion

A quantitative evaluation of the results was made by deriving the correlation coefficient for various network architectures. Large numbers of ANN models has been developed based upon type of network, network architecture, training algorithm, transfer function, momentum rate, training and testing data set etc. among these models a model of two hidden layer with 12 and 20 nodes performed better .

High $R^2$ value, small MAE and RMSE values of ANN shows that the model fits the data well. Analysis of data in randomized sets ( Fig 5) clearly showed that ANN model is best fit for predicting the groundwater level than the LR model. Since statistical regression is problematic if there is vagueness in the relationship between the independent (input) and the dependent variables (output).

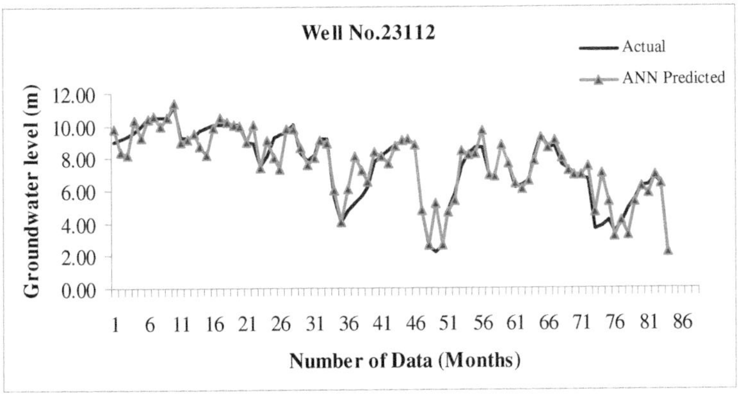


(a)

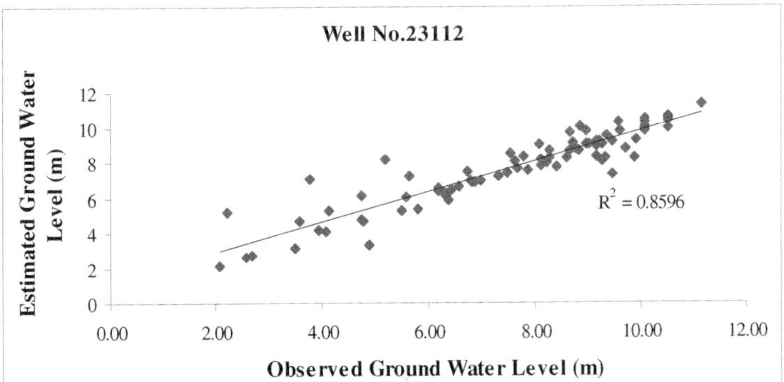


(b)

**Fig. 6**(a), (c), (e) Comparison of predicted and actual values of groundwater level.6 (b), (d), (f) Scatter Plot between Observed and Predicted groundwater level of the wells

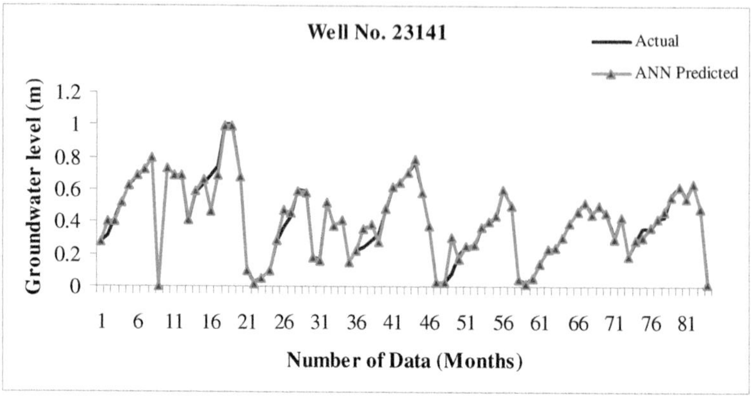


(c)

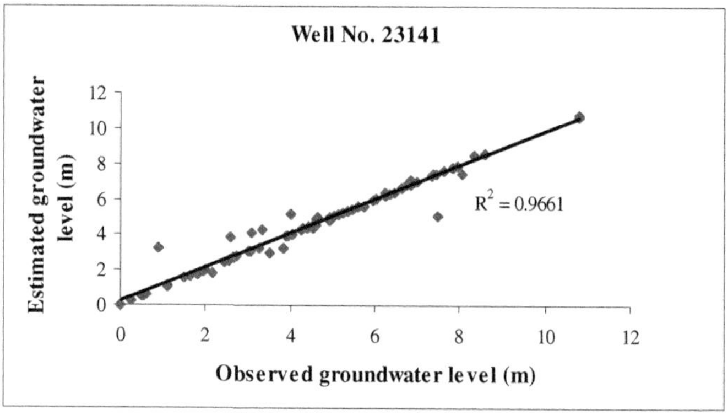


(d)

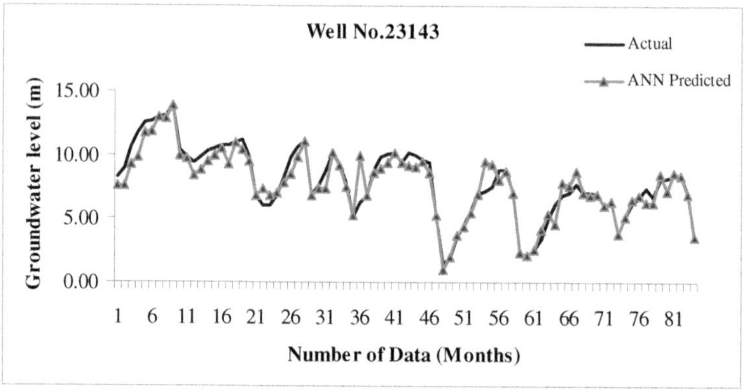


(e)

**Fig. 6.** (*Continued*)

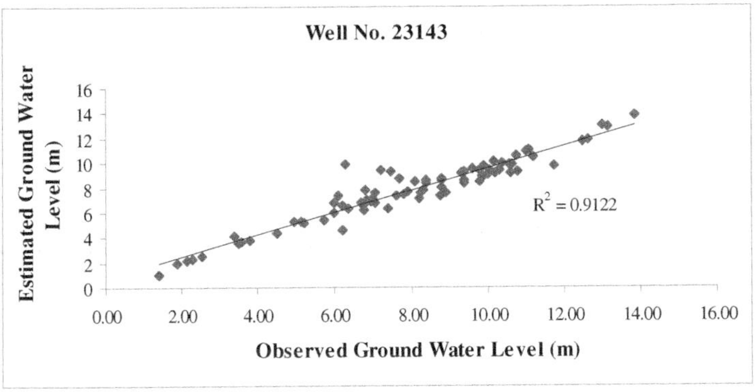


(f)

**Fig. 6.** (*Continued*)

Further, the data were analyzed separately for each independent well point to have a clear comparison of the mean observed and estimated water levels. Here also, the results exhibited similar trend as in the case of set-wise data analysis, i.e. ANN model was found to be the best fit for predicting groundwater levels at the Thurinjapuram watershed. The performance trend of the model, individually for each of the three wells along with observed and simulated values indicated in Fig 6 showing that the difference between observed and simulated values of groundwater levels with more accuracy and stability over conventional methods. Scatter plot between observed and predicted groundwater level is also given for the goodness of fit.

## 8 Conclusion

In this study, a better forecasting model using ANN has been developed for predicting monthly groundwater level fluctuations in the Thurinjapuram watershed. As evident from Fig 6 there have been few deviations of the predicted groundwater level value from the actual. In conventional regression analysis, deviations between observed and estimated values are due to indefiniteness of structure of a system or imprecise observations. As rainfall and groundwater level predication involves tremendous amount of imprecision and uncertainty, soft computing technique might warrant the ideal prediction model. On the other hand, the proposed prediction models based on soft computing are easy to implement and produces desirable mapping function by training on the given dataset. The model requires information only on the input variables for generating forecasts.

ANN configurations were tested in terms of optimum results for a prediction horizon of 84 months. The most suitable configuration for this task proved to be a 12-20-1 feed forward network trained with the Levenberg–Marquardt method as it showed the most accurate predictions of the decreasing groundwater levels. From the results of the study it can also be inferred that the Levenberg–Marquardt algorithm is more appropriate for this problem. In general, the results of the case study are satisfactory and demonstrate that neural networks can be a useful prediction tool in the area of groundwater hydrology.

**Acknowledgement.** The First author gratefully acknowledges the support given by DST Govt. of India, under women scientist scheme.

## References

1. Caulibaly, P., Anctil, F., Bobee, B.: Multivariate reservoir inflow forecasting using temporal neural networks. J. Hydrol. Eng. (9-10), 367–376 (2001)
2. Prochazka, A.: Seasonal time neural networks and –series prediction by artificial neural networks. In: Fifth International Conference, Digital Object Identifier, vol. 440(7-9), pp. 36–41 (1997)
3. Coppola, E.J., Anthony, J.R., Poulton, M., Szidarovszky, V.W.: A neural network model for predicting aquifer level elevations. Groundwater 43(2), 231–241 (2005)
4. Nayak, P.C., Satyaji, U.R., Sudheer, K.P.: Groundwater level forecasting in shallow aquifer using artificial neural network approach. Water Resour. Manage. 20, 77–90 (2006)
5. Daliakopoulos, I.N., Coulibaly, P., Tsanis, K.I.: Groundwater level forecasting using artificial neural network. J. Hydrol. 229, 229–240 (2005)
6. Govindaraju, R.S., Ramachandra, R.A.: Artificial neural networks in hydrology. Kluwer, Netherlands (2000)
7. Banerjee, P., Prasad, R.K., Singh, V.S.: Forecasting of groundwater level in hard rock region using artificial neural network. Environ. Geol. 58, 1239–1246 (2009)

# Energy Efficient and Fault Tolerant GPSR in Ad Hoc Wireless Network

Jyotsana Jaiswal and Pabitra Mohan Khilar

Department of Computer Science and Engineering,
National Institute of Technology Rourkela, India
jyotsanaa.jaiswal@gmail.com, pmkhilar@nitrkl.ac.in

**Abstract.** Routing in wireless network is a key research area which establishes path between source and destination node pairs. In this paper, we have designed and evaluated an energy-efficient and fault tolerant Greedy perimeter Stateless routing (EFGPSR) protocol for wireless Ad hoc network. The proposed protocol is divided into four phases: Fault testing phase, Planarization phase, Energy efficient greedy forwarding phase and Energy efficient perimeter forwarding phase. In fault testing phase, all nodes come to know about their fault free neighbours. Next is planarization phase which is subset of perimeter forwarding phase and can be done reactively or proactively. Next are energy efficient greedy forwarding and perimeter forwarding phases. Both these phases try to maintain balance between the metrics to choose the next hop (i.e. distance from destination in greedy forwarding phase and minimum angle w.r.t the line connecting the forwarding node and destination in perimeter forwarding phase) and selection of node having highest energy among the neighbouring node to extend network lifetime. Evaluation and comparison of GPSR and EFGPSR is done through NS-2 simulator. Simulation shows that EFGPSR performs better in terms of increasing the network lifetime, successful packet delivery ratio with insignificant increase in number of hop count.

**Keywords:** Wireless network, geographic routing, fault tolerance, GPSR, planarization.

## 1 Introduction

In recent years, wireless and mobile communication technologies have grown rapidly. As wireless network evolves there is a trend towards decentralized, deployable and self organizing networks.

Routing plays a vital role in establishing a communication links between various nodes. Since last few years, geographic routing (GR) protocols, also known as georouting or position based routing for wireless networks has gained a significant attention. The idea behind GR protocol is that the source node sends a packet to destination node using the geographic location of destination instead of using network address. The essential requirements of GR are a) each node should be capable of determining its own location (geographic coordinates) and

D. Nagamalai, E. Renault, M. Dhanushkodi (Eds.): CCSEIT 2011, CCIS 204, pp. 683–692, 2011.

b) source should be aware of the location of destination. The main advantage of GR over traditional routing protocol is that each node require to maintain only location of itself and its neighbours for its functioning. However, in traditional routing protocols for wireless networks (e.g. AODV [1], DSDV [2]), nodes usually have to keep significant amount of routing information. Greedy Perimeter Stateless Routing (GPSR) [3] is a well known and most commonly used position-based routing protocol for wireless networks. In GPSR, source includes the location information of destination in the header of every packet. If the destination is not directly reachable, the source starts with greedy forwarding, i.e., the source node forwards the data packet to the neighbour that is closest to the destination in the coordinate space. Such greedy forwarding is repeated at the intermediate node, until the destination is reached. However, GPSR itself suffers from few drawbacks. Firstly, greedy forwarding over geographic coordinates may not be optimal due to unawareness of connectivity information of the network. Secondly, packet may get stuck in local minimum condition (a forwarding node could not find a neighbour that lies closer to the destination than itself) for sparse networks. To deal with above stated local minimum problem, the nodes switch to perimeter forwarding. In perimeter forwarding, the packet is forwarded to that neighbouring node which comes first in a planar sub graph of the network , when the line connecting the forwarding node and destination is rotated in the counter clockwise direction. The location of forwarding node where perimeter forwarding starts is recorded in the header of the data packet. Greedy forwarding is resumed when the data packet reaches a forwarding node which can find a neighbour node whose distance is smaller than the distance between the destination node and the node at which perimeter forwarding begun. The probability of finding a route between source destination node pairs is very high.

Most existing designs of wireless network routing protocols are based on the assumption that every node in the network is fault free. However, such assumption usually does not hold in realistic environments. Various kinds of faults such as crash fault, transient fault, Byzantine faults can occur in a node. The occurrence of faults affects the routing process therefore fault tolerance is of increasing importance in applications where it is essential to maintain efficient routing. In routing, fault tolerance helps in controlling the overhead which is there due to the faulty node and thus helps in considering the reliable routes. Many fault tolerance routing protocols have been proposed in wireless network. In Wireless sensor network(WSN), fault tolerance mechanism in routing protocol has been classified into two schemes, retransmission and replication. In MANET, works that focus on fault tolerant routing problem are very less. Wireless network comprises of large number of energy constraint node, some node may run out of energy and die and shorten the lifetime of network. Therefore, many routing protocols in MANET and WSN have been proposed with the aim to reduce energy consumption. GPER protocol was proposed to provide power efficient geographic routing in WNS. Traditionally, greedy forwarding and perimeter forwarding consider only single metrics i.e. minimum distance from destination and minimum angle in counter clockwise direction with respect to the line connecting the forwarding node and destination. These parameters does not take into account the energy conservation for evaluating the routing performance. In this paper, an energy efficient and fault tolerant routing algorithm has been proposed. The rest of the paper is organized as follows: The system models are described in Section 2. The energy-efficient and fault tolerant GPSR routing protocol is presented in Section 3. Section 4 presents the simulation result and comparison with other existing protocol. Section 5 conclude the paper and lists the future works.

## 2 Preliminaries

### 2.1 System, Fault and Energy Model

**System Model.** The system is composed of homogeneous $n$ nodes, each having a unique id. All nodes have similar computing, storage resources and identical communication range. A set of nodes with circular radio range $r$, can be seen as a graph: each node is a vertex, and edge $(n, m)$ exists between nodes $n$ and $m$ if the distance between $n$ and $m$, $d(n, m) \leq r$. Source node knows the location of final destination.

**Fault Model.** Wireless network consists of many nodes which may be either fault free or faulty. The following assumptions have been made about faulty nodes:

1. No new faults occur during the execution of the routing protocol i.e. faults are permanent (a faulty node remains faulty until it is repaired and/or replaced).
2. Faulty nodes can be either hard faulty or soft faulty.

Nodes become hard faulty due to two main reasons. First, a node may be damaged during deployment or immediately after that. Second, depletion of battery power. When a node becomes hard faulty it does not participate in any further communication. Soft faulty nodes are more subtle than hard faulty nodes, since a soft-faulted node continues to communicate with the other node in the system although with altered specifications i.e., the faulty nodes may produce some random results instead of expected results. In this work, both hard-faulted and soft-faulted nodes have been considered.

The proposed routing protocol apply the following testing model given in Table 1 to check whether the node is faulty or fault free. The testing model uses the test task to check the validity of the node. The test task is to find out two's complement of five bit number. Tester node knows the result of the test task. Tester node broadcast the test task along with its id and location information. On receiving the test task, all the neighbouring node unicast the result of the test task along with their id and location information. As shown in the table of test model, five cases are observed. First case, the status of the tester node and tested node are both fault free then test result is zero. This means that there is match between the expected result of the tester node and the actual result returned by the tested node. Second case, the tester node is fault free and the tested node is soft faulty then test result is 1. This means there is a mismatch between the expected result and actual result. Third case, the tester node is fault free or soft faulty and the tested node is hard faulty then test result is NULL because a hard faulty node can receive a beacon test message but cannot send reply to it. Fourth case, the tester node is soft faulty and tested node is soft faulty or fault free then test result is 1. Here also there is mismatch between the expected result and actual result. Fifth case, the tester node is hard faulty and tested node is also faulty (soft or hard) or fault free then test result is NULL because a hard faulty node cannot send a beacon test message.

When the test result is 0, the node is considered as fault free, otherwise faulty. The test model given in Table 1 is used to select fault free nodes while establishing the paths between the nodes pair.

**Table 1.** Test model

| Status of Tester node | Status of Tested node | Test Result |
|---|---|---|
| Fault free | Fault free | 0 |
| Fault free | Soft faulty | 1 |
| Fault free | Hard faulty | NULL |
| Soft faulty | Hard faulty | NULL |
| Soft faulty | Soft faulty | 1 |
| Soft faulty | Fault free | 1 |
| Hard faulty | Soft faulty | NULL |
| Hard faulty | Hard faulty | NULL |
| Hard faulty | Fault free | NULL |

## 2.2 Energy Model

In wireless network energy consumption at each node can be due to transmission and reception of message. The energy model proposed by Heinzelman et al [6] to transmit an $n$-bit message over a distance $d$ or receive an $n$-bit message are as follows:

$$E_{TX}(m,d) = E_{TX-elec}(m) + E_{TX-amp}(m,d) \tag{1}$$

$$= m * E_elec + m * \varepsilon_{amp} * d^2 \tag{2}$$

$$E_{RX}(m) = E_{RX-elec}(m) = m * E_{elec} \tag{3}$$

Table 2 gives related parameters and their definations.

**Table 2.** Radio parameters

| Parameter | Definition | Unit |
|---|---|---|
| $E_{elec}$ | Energy dissipation rate to run the radio | $50nJ/bit$ |
| $\varepsilon_{amp}$ | Energy dissipation rate to run transmit amplifier | $100pJ/bit/m^2$ |

## 3 The Proposed Algorithm

The proposed algorithm consists of four different but dependent phases which are as given below:

### 3.1 Fault Testing Phase

In this phase, all nodes come to know about the location information and residual energy information of their fault free neighbours at regular interval. Each tester node broadcast a beacon-test message containing its own id (e.g. IP address), location information and test task. Information regarding the test task in given in 2.1. Each tester node knows the result of the test task. On receiving the beacon-test message, all the neighbouring node execute the test task and unicast the reply message. The reply message contains the result of the test task, their unique Id, location information and residual energy information of the node. In this phase, tester node checks whether the neighbouring nodes present are fault free or faulty (i.e., soft or hard) by comparing the result of test task using the test model given in Table 1. Each node maintains the location information and residual energy information of all fault free neighbouring nodes in their range.

### 3.2 Planarization Phase

After fault testing phase, planarization phase starts. This phase can be done reactively or proactively. In proactive version, each node begins the planarization phase immediately after the fault testing phase. In reactive version, each node begins the planarization phase when local minimum condition occur. The main reason for using the planarization phase is that the right-hand rule does not work properly on full connected graphs with crossing edge. The right hand rule is used by perimeter forwarding phase. Each node should run the planarization algorithm in a distributed fashion. The Relative Neighbourhood Graph (RNG) and Gabriel Graph (GG) are two well known planar graph [7]. RNG planarization has been considered in this paper. During this phase one important property to be taken care is that, while removing edges from the graph to reduce it to RNG must not disconnect the graph.

### 3.3 Energy Efficient Greedy Forwarding Phase (EEGF)

In this phase, a forwarding node $F$ uses full graph in choosing next hop. The optimal choice of next hop considers two metrics first minimum distance from destination($D$) and second the energy. Algorithm 1 explains the pseudo code for the proposed energy efficient greedy forwarding used at a forwarding node. In traditional GPSR, the distance between forwarding node as well as its neighbour with respect to destination is calculated. The forwarding node selects that neighbour as the next hop that lies closest to the destination. If the forwarding node $F$ could not find a neighbour node that lies closer to the destination than itself, then node switches to perimeter forwarding. However in EEGF, forwarding node first form a set of Selected-Neighbour($F$) which is a subset of the Neighbour-list($F$). Each node $I \varepsilon$ Selected-Neighbour($F$), if the distance between the neighbour node $I$ and the destination node $D$ is less than the distance between the forwarding node $F$ and $D$. After this, $Cost(I)$ of every node $I \varepsilon$ Selected-Neighbour(F) is calculated. $Cost(I)$ is defined as

$$Cost(I) = \frac{ResidualEnergy(I)}{Distance(F, I)^2} \tag{4}$$

where, $ResidualEnergy(I)$ is the $AvailableEnergy(I)$ divided by $InitialEnergy$ $(I)$ and $Distance(F, I)$ is the difference in the distance between $F$ And $D$ and the distance between $I$ and $D$ divided by the distance between $F$ and $D$. The reason for calculating this cost is that relation between energy and distance are inversely proportional to each other because from energy balance point of view, the node with more residual energy should be selected as the next hop and from distance point of view, the node which is closer to the destination should be selected as the next hop. Thus, $Cost(I)$ balances the two metrics energy and distance and helps in finding the optimal next hop. The forwarding node selects that neighbour $I \varepsilon$ Selected-Neighbour($F$) as next hop, whose $Cost(I)$ is maximum. If the forwarding node $F$ could not find a neighbour node that lies closer to the destination than itself, that is Selected-Neighbour($F$) is empty then node switches to perimeter forwarding.

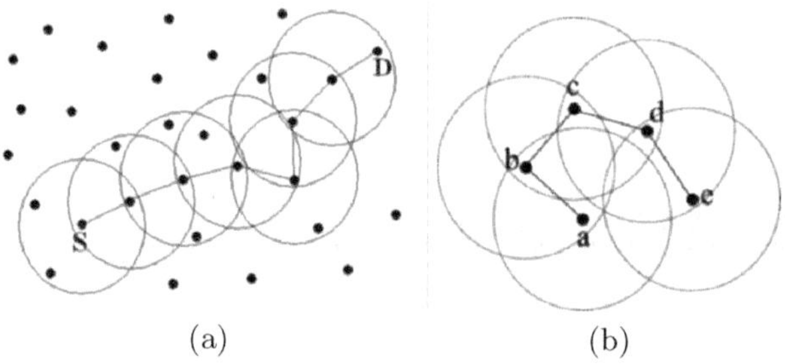


**Fig. 1.** (a) Greedy forwarding and (b) Local minima condition

**Require:** Destination $D$, Forwarding node $F$, Neighbour-list($F$)
**Ensure:** Next-Node // if Energy efficient greedy forwarding is successful
Switch to perimeter forwarding // if Energy efficient greedy forwarding is not successful
**Intialization:** Next-Node=NULL; Maximum-Cost=0.0;
Selected-Neighbour(F)=0
1. *Dist(F-D)* $= \sqrt{(x_F - x_D)^2 + (y_F - y_D)^2}$
2. **while** neighbour-node $I \varepsilon$ Neighbour-list($N$) **do**
3. *Dist(I-D)* $= \sqrt{(x_I - x_D)^2 + (y_I - y_D)^2}$
4. **if** Distance(I-D) < Dist(F-D) **then**
5. Selected-Neighbour($F$) $\rightarrow$ Selected-Neighbour($F$)$\cup$($I$)
6. **end if**
7. **end while**
8. **while** neighbour-node $I \varepsilon$ Selected-Neighbour-list($N$) **do**
9. $ResidualEnergy(I) = \frac{AvailableEnergy(I)}{InitialEnergy(I)}$
10. $Distance(F, I) = \frac{Dist(F-D) - Dist(I-D)}{Dist(F-D)}$
$Cost(I) = \frac{ResidualEnergy}{Distance(F,I)^2}$
11. **if** Maximum-Cost < Cost($I$) **then**
12. $Maximum - Cost = Cost(I)$
13. $Next - Node \leftarrow I$
14. **end if**
15. **end while**

**Algorithm 1.** Energy-Efficient Greedy Forwarding Algorithm

### 3.4 Energy Efficient Perimeter Forwarding

Energy efficient perimeter forwarding uses right hand rule/left hand rule as given below:

**Right Hand Rule.** The right-hand rule to traverse a graph is shown in Figure2 (a). This rule states that when arriving at node $b$ from node $a$, the next edge traversed is the next one sequentially counterclockwise about $b$ from edge $(a, b)$. On graphs with edges that cross, the right-hand rule may take a degenerate tour of edges that does not trace the correct path as shown in figure 2 (b). Now on removing the crossing edges right hand rule traces the correct path (Figure 2 (c)). In this context planarization comes into picture. The left hand rule also works in a similar fashion the only difference is that the next edge traversed is in clockwise direction

**Energy Efficient Perimeter Forwarding.** In this phase, complete EFGPSR is described which combines fault testing phase, energy efficient greedy forwarding phase and energy efficient perimeter forwarding phase. Greedy forwarding phase is executed on the full network graph whereas energy efficient perimeter forwarding phase on the planarized network graph where greedy forwarding

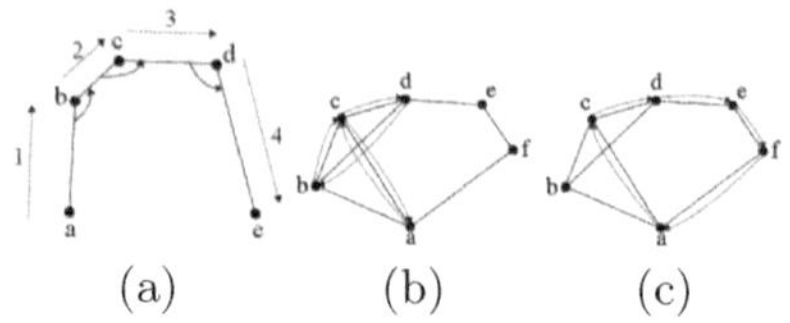


**Fig. 2.** (a) Right Hand Rule (RHR), (b) RHR with crossing edge, (c) RHR without crossing edge

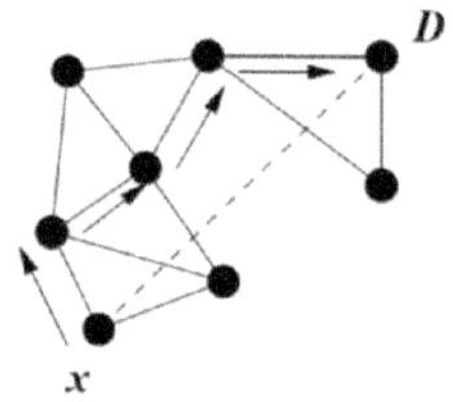


**Fig. 3.** Perimeter forwarding example

fails. A flag is used in the packet header of EFGPSR which indicate whether the packet is in greedy mode or perimeter mode. Initially, source node marks all data packets as greedy-mode. On receiving greedy-mode packet, a forwarding node uses the energy efficient greedy forwarding algorithm to find the optimal next hop. If no optimal node is found i.e. greedy forwarding fails, the node marks the packet into perimeter mode. In traditional perimeter forwarding as shown in figure 3 the packet is forwarded to that neighbour node that comes first in a planar sub graph of the network , when the line connecting the forwarding node and destination(D) is rotated in the counter clockwise direction. The location of forwarding node where perimeter forwarding starts is recorded in the header of the data packet. Greedy forwarding is resumed when the data packet reaches a forwarding node or a node which can find a neighbour node whose distance is smaller than the distance between the destination node and the node at which perimeter forwarding begun. However, in energy efficient perimeter forwarding, the forwarding node select the next hop using the following algorithm:

## 4 Simulations and Results

In this paper, the performance of GPSR and EFGPSR is evaluated based on NS-2 network simulator. The network dimension used for simulation is 1000×1000meter square. The transmission range of each node is assumed to be 250m. For traffic source 20 traffic flow originated by 20 sending nodes has been considered. The MAC layer protocol used is IEEE 802.11. Simulation results are shown in figure 4 and figure 5.

**Require:** Forwarding node $F$, Neighbouring-node-list containing *index*, *angle* and *energy* of each node
**Ensure:** Next-Node
**Intialization:** Next-Node=NULL; Temp=0; Max=Max-energy; Min=Min-energy; Mid=Mid-energy
Sort the Neighbouring-node-list in increasing order of angle
for all Neighbouring-node
Temp=First nodes after sorting
**if** Temp[energy] < Mid and Temp[energy] >= Max **then**
    Select node index
    break
**end if**
End for

**Algorithm 2.** Energy-Efficient Perimeter Forwarding Algorithm

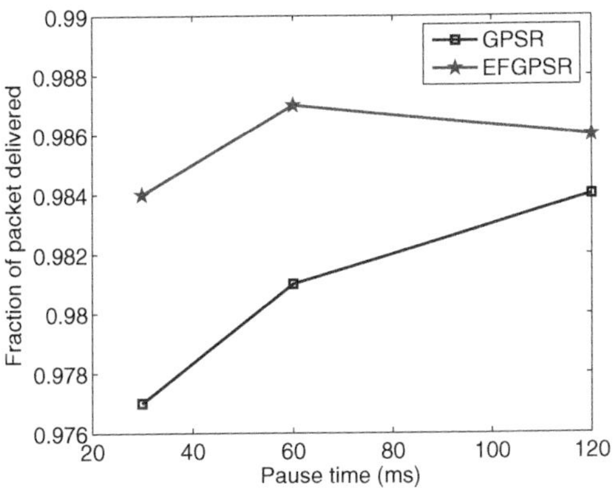


**Fig. 4.** Packet Delivery Success Rate. GPSR and EFGPSR with becon interval 1. 50 nodes

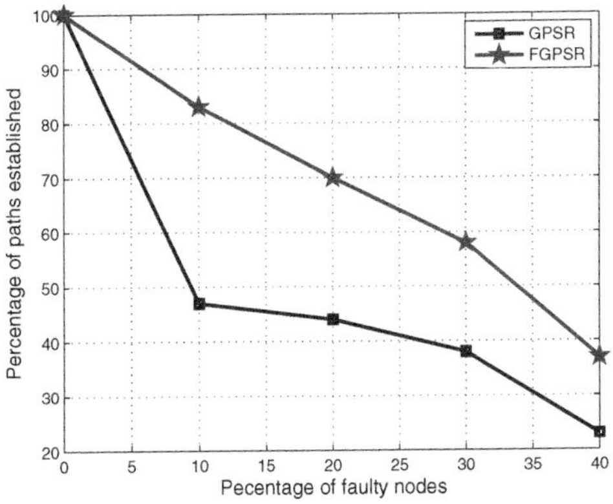


**Fig. 5.** GPSR vs EFGPSR

## 5 Conclusions

In this paper, fault tolerant GPSR routing algorithm has been proposed. EFGPSR and GPSR have been simulated for the 1000×1000-sized static networks of 50 node. 20 traffic flow originated by 20 sending nodes has been considered. The result carried out shows that EFGPSR provides higher no. of path between source and destination out of the total no. of paths (i.e., 20) than GPSR due to the fact that GPSR does not check the faulty node which causes long detouring paths and may not establish path to the destination shown in figure 5. And also EFGPSR try to maintain balance between the metrics to choose the next hop (i.e. distance from destination in greedy forwarding phase and minimum angle with respect to the line connecting the forwarding node and destination in perimeter forwarding phase) and selection of node having highest energy among the neighbouring node to extend network lifetime. Evaluation and comparison of GPSR and EFGPSR shows that EFGPSR performs better in terms of increasing the network lifetime, successful packet delivery ratio with insignificant increase in number of hop count.

## References

1. Perkins, C., Belding-Royer, E., Das, S.: Ad hoc On-Demand Distance Vector (AODV) Routing, RFC Editor (2003)
2. Charles, E., Perkins, B.P.: Highly dynamic Destination-Sequenced Distance-Vector routing (DSDV) for mobile computers. SIGCOMM Comput. Commun. Rev. 24, 234–244 (1994)
3. Karp, B., Kung, H.: GPSR: greedy perimeter stateless routing for wireless networks. In: Proceedings of the 6th Annual International Conference on Mobile Computing and Networking, Boston, MA, USA, pp. 243–254 (August 2000)
4. Wu, S., Candan, K.: GPER: Geographic Power Efficient Routing in Sensor Network. In: Proceeding of Int'l Conference Network Protocols (ICNP), pp. 161–172. IEEE, Los Alamitos (2004)
5. Melodia, T., Pompili, D., Akyildiz, I.: Optimal Local Topology Knowledge for Energy Efficient Geographical Routing in Sensor Network. In: Proceeding of INFOCOM. IEEE, Los Alamitos (2004)
6. Heinzelman, W., Chandrakasan, A., Balakrishnan, H.: Energy-Efficient Communication Protocol for Wireless Microsensor Networks. In: Proceedings of the 33rd Hawaii International Conference on System Sciences, vol. 8. IEEE Computer Society, Hawai (2000)
7. Agarwal, P., Matausek, J.: Relative neighborhood graphs in three dimensions. In: Proceedings of the Third Annual ACM-SIAM Symposium on Discrete Algorithms, pp. 58–65. Society for Industrial and Applied Mathematics, Philadelphia (1992)

# An IDE for Developing Mobile Applications for Smart Phones

Sakila Banu Ali and Nisar Hundewale

College of Computers and Information Technology,
Taif University,
Saudi Arabia
shakeelamjb@gmail.com,
nisar@computer.org
http://cit.tu.edu.sa

**Abstract.** Mobile Devices have been gaining increasing acceptance for developing Multimedia-rich applications. Everyday new Technologies in Mobile Devices have been developed by developers and are competing with the Previous in Market. Nowadays developing applications for Smart phones is common today and nearly thousands of new applications are coming to the Market every day. Software firms have also developed many programming platforms and tools for the developers to write programs for Mobile Devices. But so far there is no Specific IDE developed to create mobile application easily by just Drag and Drop method to make even the non-programmers to develop application for the smart phones.

This paper presents an IDE for developing Mobile Application without Programming for Smart phones. The IDE allows non-programmer to develop a mobile application for smart phones efficiently. The IDE has navigation control tools in iconic format that can be used to incorporate within the application by drag and drop method and directly installed to the smart phones as applications. With the help of the GUI Builders and ADT plug-in the IDE is created and tested against the Android Mobile Phones.

**Keywords:** IDE-Integrated Development Environment, GUI-Graphical User Interface, ADT-Android Development Tool.

## 1 Introduction

The Mobile Applications are classified into the following types. They are Communications, Games, Multimedia and productivity. Almost all developed applications come into this one category. Mobile devices are gaining acceptance for multimedia applications more as they are attracting the mobile users mostly. Based on the users Area of Interests Developers are providing applications for smart phones.

D. Nagamalai, E. Renault, M. Dhanushkodi (Eds.): CCSEIT 2011, CCIS 204, pp. 693–703, 2011.

For example, if the user is more interested in learning Islamic topics, then the developers has to collect the useful pages from various Islamic sources based on the users interests and the pages have to be wrapped in one unit to present it as a Mobile Application for the smart phones.

This paper provides an IDE that allows non-programmers to develop such applications without the knowledge of Programming language. The IDE has tools in Iconic format needed to wrap the pages. The UI presented in IDE are designed for Drag and Drop method. Thus an application can be developed within minutes.

This IDE is mainly designed for the non-programmer who will develop mobile application without the knowledge of API and programming language. This IDE is first aimed to develop applications for android mobile phone developers as android is open source and hence applications created can be easily installed on android mobile devices as well as the developers has complete access to use android tools. In addition, android phones are competing today with apples iPhones in market, day by day android mobile users are increasing.

The paper is structured as follows. The proposed IDE is developed using Eclipse with Android plug-in. Eclipse IDE is used here because android can be used as a plug-in to the eclipse IDE hence the created applications using the proposed IDE can be tested right there on android emulator in eclipse. This paper first presents the design of the proposed IDE. The IDE is designed using Jigloo GUI designer as Jigloo can also be used as a plug-in to the eclipse. Using Jigloo GUI designer the IDE is designed first with all necessary Icons with drag and drop functions. Programs are written in java for each icon in the IDE to perform the actions. The created IDE is tested with various applications and the results are shown below. The application thus created with this IDE is tested again with android emulator in eclipse. A table is provided at last to show the time consuming for different applications using this IDE.

## 1.1 PROPOSED IDE

The IDE is divided into the following. They are Widgets, Main View and Slider View. The Main View is the Heart of this IDE where the Dragged Icons and Slides are placed to this Main View. The slider view is shown in the Right Side of the IDE. The slides will be distributed in the Slider View once the file has been chosen from the File option. The Slides are dragged to the Main View by selecting the Slides from the Slider View and dragging it to the Main View.

The widget is shown in the left side of the IDE. The widget has the icons arranged vertically. The icons are dragged to the main view and are placed in the selected slides.

Once the icons are dropped into the slide, the input box is shown to enter the slide number based on the icon selected. Finally, the slides are consolidated and converted to .apk files to install into the android mobile phones. The IDE is divided into three views.

I. Icon View
II. Main View
III.Slider View

**I.Icon View:** The Icons like Previous, Next, Home and Scroll Bar are placed in the Icon View. These icons are programmed with Drag and Drop functionalities. The Icon is dragged to the main view to place on the selected slide. Once the icon is dropped on the slide in the Main View, input box will be shown to enter the Next or Previous Slide Number based on the Icon selected.

**II.Main View:** The Main View is the heart of the IDE where the slides and Icons are Dragged and Dropped to this Main View. The slides are dragged to the Main View to place the icon on the Slide. The slides with the icons are then consolidated by the Slide Number which the user enters while placing the icons on the slide. The consolidated slides are then saved with the .apk file type and are send to the Android Emulator to test the developed application and installed to the Android Device.

**III.Slider View:** The Slides are placed vertically one by one in the Slider View. Once the file is selected from the File Menu, the entire slides from the selected file are distributed to the Slider View. The slides are arranged vertically with the slide number.

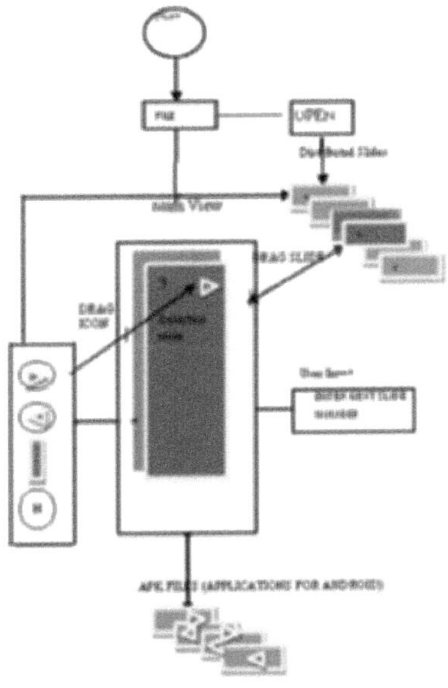

**Fig. 1.** Architecture of the Proposed IDE:

## 2 Tools Used to Develop IDE

The tools used for this IDE are the following;

I. Eclipse
II. Jigloo GUI Builder
III. Android Development Tools
IV. Android Emulator to test the developed Application

### 2.1 Eclipse

Nowadays Eclipse IDE is used by all software developers to develop applications. Eclipse is easy to use and any number of plug-ins can be added to the Eclipse IDE. Moreover, eclipse is an open source IDE and includes a number of unique features such as code refactoring, automatic code updates. In its entirety, the Eclipse Platform contains the functionality required to build an IDE. However, the Eclipse Platform is itself a composition of components; by using a subset of these components, it is possible to build arbitrary applications. The Eclipse Rich Client Platform (RCP) is one such subset of components. Figure 2 shows a representation of some of the components in the Eclipse Platform and highlights the subset that makes up the RCP (in reality there are a great many more components).

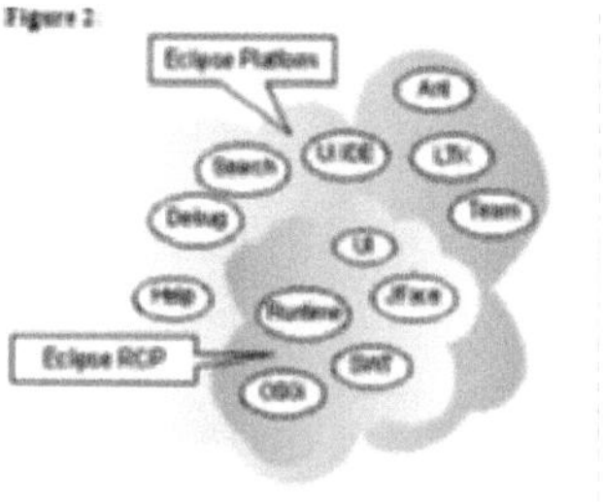


**Fig. 2.** Eclipse

### 2.2 Jigloo GUI Builder

Jigloo is a visual Java GUI builder. Jigloo is an Eclipse plug-in that enables to build sophisticated graphical user interfaces (GUIs) rapidly that run on the Java platform. It can be used to build Swing-based applications and Standard Widget Toolkit (SWT)-based applications[11]. Jigloo is an easy-to-use visual editor, and is easy to create the UI for desktop application quickly.

Desktop Applications can be easily developed by Jigloo GUI Builder. The main advantage of Jigloo is that it can be used as a Plug in for Eclipse to design the UI for the Application. With Jigloo, it is easy to build UIs targeting Swing or SWT.

**Things Possible with Jigloo:** Jigloo is a classic WYSIWYG editor for building Swing and SWT GUIs. To develop a graphical, desktop application in Java, Jigloo is a great option. Jigloo supports round-tripping. That means not only to make visual changes that cause the underlying code to change but also directly edit the code and see the changes that causes in the GUI. This makes Jigloo a great choice. Jigloo was built with Swing/SWT compatibility in mind. It supports many layout options supported by either technology. Not only to use either technology for building a new GUI but it can even convert between Swing and SWT using Jigloo.

## 2.3 Android Development Tools

Android Development Tools (ADT) is a plug-in for the Eclipse IDE that is designed to give you a powerful, integrated environment to build Android applications. ADT extends the capabilities of Eclipse to let quickly set up new Android projects, create an application UI, add components based on the Android Framework API, and debugs the applications using the Android SDK tools, and even export signed (or unsigned) .apk files in order to distribute the application to the smart phones.

Developing in Eclipse with ADT is highly recommended and is the fastest way to get started. With the guided project setup, as well as tools integration, custom XML editors, and debug output pane, ADT gives an incredible boost in developing Android applications.

## 2.4 Android Emulator

Android is a software stack for mobile devices that includes an operating system, middleware and key Applications. The Android SDK provides the tools and APIs necessary to begin developing applications on the Android platform using the Java programming language. Android emulator is a device that appears like Android phones, where the application developed can be tested right there without the need of external android device.

**Android Application Framework:** By providing an open development platform, Android offers developers the ability to build extremely rich and innovative applications. Developers are free to take advantage of the device hardware, access location information, run background services, set alarms, add notifications to the status bar, and much, much more. Developers have full access to the same framework APIs used by the core applications. The application architecture is designed to simplify the reuse of components; any application can publish its capabilities and any other application may then make use of those capabilities. This same mechanism allows components to be replaced by the user.

**Android Runtime:** Android includes a set of core libraries that provides most of the functionality available in the core libraries of the Java programming language. Every Android application runs in its own process, with its own instance of the Dalvik virtual machine. Dalvik has been written so that a device can run multiple VMs efficiently. The Dalvik VM executes files in the Dalvik Executable (.dex) format which is optimized for minimal memory footprint. The VM is register-based, and runs classes compiled by a Java language compiler that have been transformed into the .dex format by the included "dx" tool. The Dalvik VM relies on the Linux kernel for underlying functionality such as threading and low-level.

## 3 Image of IDE

The left side panel shows the Icon View, the Middle panel shows the Main View and the Right Panel shows the Slider View. The Previous, Next, Scroll view and the Home Icons are arranged vertically in the Icon View. The Main View is the place where the selected slide and the Icons reside. The Slider View has the Slides that are arranged one by one vertically.

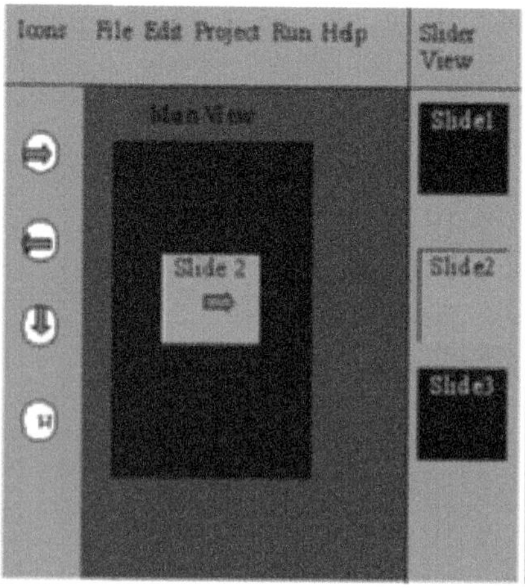


**Fig. 3.** Image Of IDE

### 3.1 Program Structure

The IDE is developed using Jigloo GUI Builder. Jigloo is an Eclipse Plugin. Jigloo can be installed using the eclipse powerful automatic updates.

The Structure of the Program is shown from the following figure

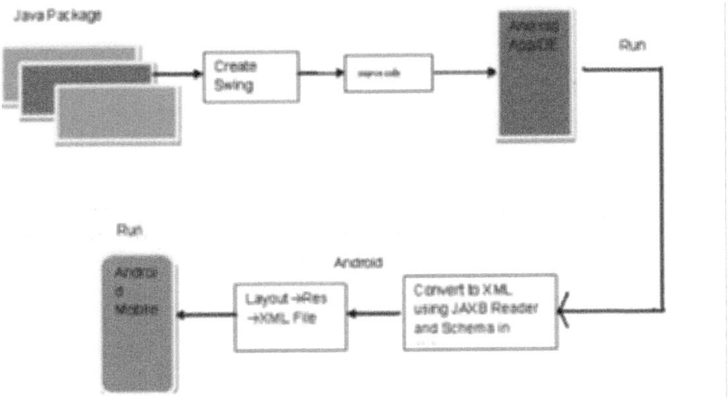


**Fig. 4.** Program Structure

**Eclipse with Jigloo Plugin to Create the Application IDE.** Steps Followed in Creating the IDE:

1. The IDE is developed as a package using Jigloo as a plugin in eclipse. Package Name: org.developerworks.jiglooIDE
2. A Swing Application is created in the Package Swing Application Name: NewIDE.java
3. The NewIDE.java contains the main source code required to build the IDE *Sample Source Code for NewIDE.java:*

```
NewIDE.java\\
{
jButton1 = new JButton();
Sliderpanel.add (jButton1);
jButton1.setText ("slide1");
jButton1.setBounds(31, 26, 73, 92);
---------
---------
}
```

4. The Xml Code of the IDE is generated using Jigloo Code Parser. JAXB is used to read and write the XML data for the created IDE[12]. *Example*

   *NewIDE Schema*

```
<dw:workflow xmlns:dw="org:developerworks:newIde"
xmlns:xsi="http://www.w3.org/2001/XMLSchema-instance"
xsi:schemaLocation="org:developerworks:newIde
workflow.xsd ">
<complexType name="AppIDE">
<sequence>
<element name="slide1" type="dw:priority"/>
```

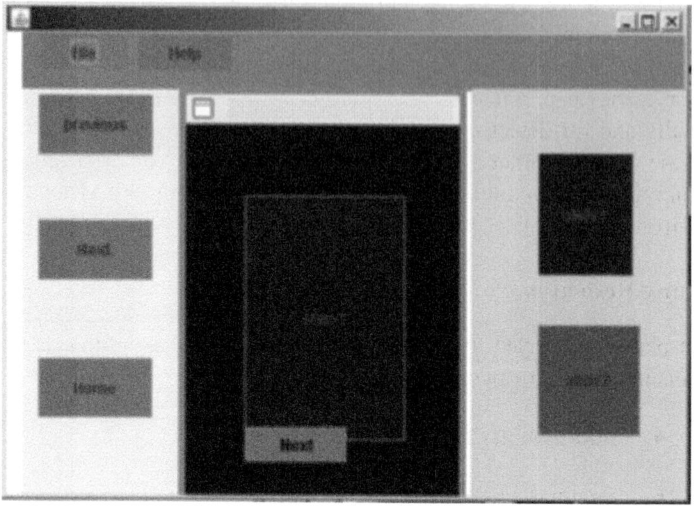

**Fig. 5.** Result:Design of NewIde.java

```
<element name="slide2" type=" dw:priority "/>
<element name="previous" type=" dw:priority " minOccurs="0"/>
<element name="next" type="string"/>
</sequence>
</complexType>
------
------
------
>
```

5. The XML code for the design of the IDE is copied and saved in a new file called AppIDE.xml

### 3.2 Eclipse with ADT Plugin(Android) to Test the Application

Steps Followed to Run the Developed AppIDE in Android Project.

1. ADT Plug in is used with Eclipse to test the Application.
2. A New Android Test Project is Created.
3. The created XML file AppIDE.xml is imported to the Android -¿res-¿Layout Folder
4. Applications are created using this ADT plugin in Eclipse using the new Developed IDE AppIDE.xml file.
5. Once the Application are created with ADT plugin it will create the .APK files automatically.
6. The Application .APK files are installed to the Android Mobile Phones.

Thus Website Application for Android Mobile Devices is developed in few minutes and the Experimental Results are shown below.

### 3.3 Image of the Created IDE in Jigloo

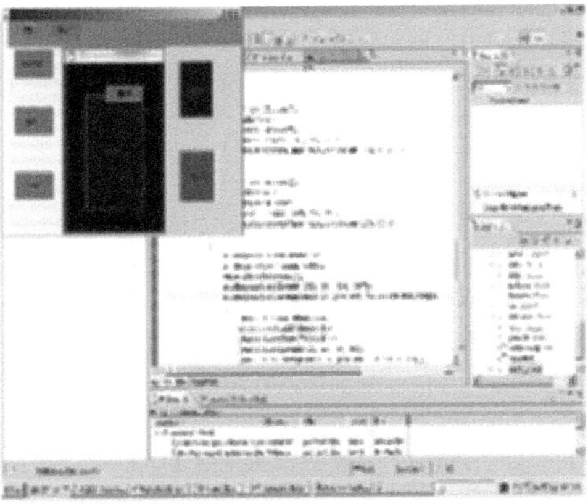

**Fig. 6.** Image of Created IDE in Jigloo

## 4 Snapshot of IDE Using Java

### 4.1 Experimental Setup and Results

The IDE is first experimented for Islamic Sources. Various Islamic Sites are collected for Dhua, Historic Places and Prophets Sayings etc. Selected Pages from

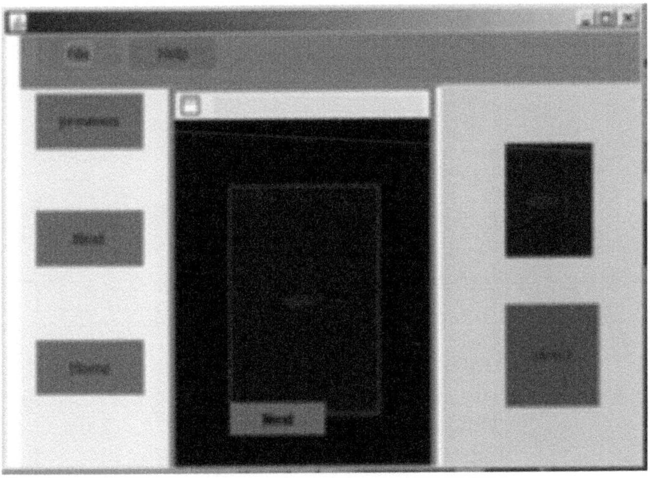

**Fig. 7.** Snapshot of IDE

various sites are collected for different types of application like audio files, PowerPoint and PDF. Audio files are tested by collecting Dhuas from various sources and developed into mobile application as .apk files using the IDE .Applications to install into the Android Phones are experimented with the Created IDE and the Results are Shown Below.

### 4.2 Use Case 1: Development Using PowerPoint

The below figure shows three different kinds of applications that fall into the same category. For Example, the PDF, the web pages and the word documents are treated here as a slides. These applications will disperse the contents as the slides and the slides are consolidated using this IDE. The slides are then converted to .APK files to install into the smart phones. The Android OS will accept only .APK files to install into the smart phones. For this the file type is converted to .APK extensions. This is done by the Android plug-in that is plugged into the eclipse IDE. The xml source code is then placed into the Android Plug-in into the Eclipse IDE to obtain the Target file Extension .APK. The applications that use similar file types are shown below. Some Applications are tested with the developed IDE and the time required to create the application using this IDE is described below in the Table.

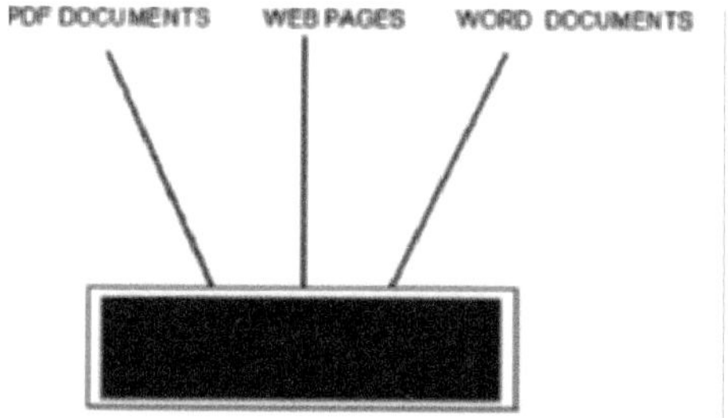


**Fig. 8.** Common Files

**Table 1.** TEST CASES

| Application Tested | | | | |
|---|---|---|---|---|
| SerialNo | TypeOfApplication | TypeOfFiles | NoOfPages | TimeTaken |
| 1 | Dhua | AudioFiles | 10 | 5 Minutes |
| 2 | Historic Places | PowerPoint | 10 | 4 Minutes |
| 3 | Quran | PDF | 30 | 10 Minutes |
| 4 | Prophet sayings | AudioFiles | 15 | 6 Minutes |
| 5 | Prophets Life | PowerPoint | 20 | 6 Minutes |

### 4.3 Conclusion and Future Work

This IDE is a useful tool for Developing Mobile applications easily and the main advantage of this IDE is even a non-programmer who is programming for first time can use this IDE to develop applications for mobile devices. Currently the IDE has limited controls and in future more controls will be added to this IDE to make more efficient for mobile application developers. Using this IDE Mobile Applications can be created more easily with Drag and Drop functionalities.

**Future Work:** The IDE currently has limited icons that are necessary to develop application, and in future more icons will be added to develop advance applications for android smart phones. More Functionalities will be added in future in order to save the created application in different files. Currently the IDE is designed for Android Smart Phones and in future it will be extended to develop applications for all types of smart phones.

## References


1. Bloom, S., Book, M., Gruhn, V., Hrushchak, R., Kohler, A.: Write Once Run Anywhere. A survey of Mobile Runtime Environments. In: Wu, S., Yang, L.T., Xu, T.L. (eds.) GPC 2008. LNCS, vol. 5036, pp. 132–137. Springer, Heidelberg (2008)
2. Kenteris, M., Gavalas, D., Economou, D.: An Innovative Mobile EI (2009)
3. Kelly, M.D.: Getting started with Jigloo, a GUI builder for Eclipse
4. Holzer, A., Ondrus, J.: Trends in Mobile Application Development. In: Proceedings of the 2nd International Conference Mobile Wireless Middleware, Operating Systems and Applications (Mobileware 2009), pp. 55–64 (2009)
5. Android-An Open Handset Alliance Project, `http://code.google.com/android`
6. J2ME Polish, `http://www.j2mepolish.org` (retrieved December 13, 2009)
7. JavaFX Mobile, `http://javafx.com/` (retrieved December 13, 2009)
8. Java ME, `http://java.sun.com/javame` (retrieved December 13, 2009)
9. JForm Designer, `http://www.formdev.com/jformdesigner`
10. Android- An Open Handset Alliance Project, `http://code.google.com/android`
11. Jigloo GUI Builder, `http://www.ibm.com/developerworks/opensource/tutorials`
12. Jigloo SWT/Swing GUI Builder For Eclipse and Webshphere, `http://www.cloudgarden.com/jigloo/`
13. A Guide to Install APK apps On Google Android phones, `http://www.talkandroid.com/guides/install-apk-files-on-android/`

# Function Approximation Using SVM with FCM and Slope Based Partition

Amera Almas and Nisar Hundewale

College of Computers and Information Technology,
Taif University, Saudi Arabia
almasamera14@gmail.com, nisar@computer.org
http://cit.tu.edu.sa

**Abstract.** For a large training data set, the memory space required to store the kernel matrix is a difficult task for the solution of QP problems for SVR. Support Vector Regression (SVR) is used to approximate the function. So we are proposing the slope based partition algorithm, which automatically evolves the partitions based on change in slope. In this paper we are improving the performance of function approximation with SVM by preprocessing the dataset with Fuzzy c-Mean clustering (FCM) and slope based partition methods. Using FCM and slope, we are portioning the data, and for every partitioned dataset we are finding the function approximation, and aggregating the result. The results indicate that Root Mean Square Error (RMSE) has been reduced with the partitioned data, compare to the processing entire dataset, with the increase in performance approximately 40%

**Keywords:** Support Vector Regression (SVR), Function approximation, FCM, slope based partition, partition of data sets.

## 1 Introduction

During the last decade a most of the researchers are using Support Vector Machines (SVM) to solve pattern classification and function approximation problems. Function approximation or regression problems, unlike classification problems, have continuous output variable. SVM is a kernel based tool, can be applied to both classification and Regression problems. In classification problems ( [Burges, 1998], [Osuna et. al, 1997], [Pontil and Verri, 1997]) SVM finds a maximal margin hyper plane that separates two classes and increases the performance in accuracy. SVR aims to find a linear regressor through a given set of data points, where the points may be in the linear space or in a higher dimensional feature space. So the main objective is to find the hyper plane with small norm which minimizes the distances from data points to the hyperplane. The SVR method was proposed for estimating function approximation for the regression data sets, constructing multidimensional splines, and solving linear operator equations [Vapnik,1995]. The need for an approximating function often arises in scientific and economic problems. One possibility is that one has a finite set of data points and wants to determine the underlying functional form.

D. Nagamalai, E. Renault, M. Dhanushkodi (Eds.): CCSEIT 2011, CCIS 204, pp. 704–714, 2011.

Recently Fuzzy Functions and SVMs have been proposed to solve the regression estimation problems Fuzzy logic system and SVM is used to be powerful tool for modeling nonlinear functions. Fuzzy logic system design is mainly neural fuzzy networks. One major disadvantage of using neural fuzzy networks is that network learning is based empirical risk minimization which does not account for small test error. Whereas SVR learning formulation is based on the principle of structural risk minimization and attempts to minimize a bound on the generalization error. The SVR learning approach also exhibits desirable properties such as generalization and over-fitting prevention capabilities. Many studies have used SVRs to deal with regression estimation problem due to their high generalization ability. SVRis a novel method for tackling the problems of function approximation and regression estimation based on the statistical learning theory. [Osuna et. al, 1997] suggested optimizing only a fixed-size subset (working set) of the training data each time, while the variables corresponding to the other patterns are frozen. Going to the extreme, the sequential minimal optimization (SMO) algorithm [Platt, 1999] breaks the original QP into a series of smallest possible QPs, each involving only two variables.

As proved by [Chia et. al, 2010] Fuzzy function along with SVR gives better performance . In this paper we are to deal mainly with dataset which is partitioned to according to the logic used in slope based and FCM. First the given system domain is fuzzy partitioned into c clusters using fuzzy c-means clustering (FCM) algorithm. Then, one regression function is calculated to model the behavior of each partition. In the second approach, the dataset is partitioned based on maximum change of slope and for each partition, regression function is calculated.

## 2 Literature Survey

The T-FR methods represent the fuzzy regression approaches proposed by Tanaka et al. [Tanaka et. al, 1982], where the regression coefficients of input variables are fuzzy numbers. Fuzzy Functions and SVMs have been proposed to solve the regression estimation problems [Asli et. al, 2007]. Fuzzy clustering based Neural Fuzzy Network with SVR has been proposed to differentiate the performance in noisy and clean training data [Chia et. al, 2010].

The organization of this paper is as follows. Section 3 introduces Support Vector Regression, Section 4 introduces partition methods, FCM, and slope based partition, Section 5 Experiments Section and Section 6 Results and Discussion. In the Results section, we display results obtained from the application of the proposed slope based partition and FCM on operating system dataset.

## 3 Support Vector Regression

The Support Vector method can also be applied to the case of regression, maintaining all the main features that characterise the maximal margin algorithm: a non-linear function is learned by a linear learning machine in a kernel-induced feature space while the capacity of the system is controlled by a parameter that does not depend on the dimensionality of the space. As in the classification case the learning algorithm minimises a convex functional and its solution is sparse. We are given a training set $\{(x_i, y_i)\}_{i=1}^{n} \subset R^d \times R$, where $x_i$ and $y_i$ are the input and output, respectively, of the $i^{th}$ pair. Support vector regression ( [Smola and Scholkopf,2004] is a kernel method that performs nonlinear regression based on the kernel trick. Essentially, each input $x_i \in R^d$ is mapped implicitly via a nonlinear feature map $\varphi(x)$ to some kernel-induced feature space $F$ where linear regression is performed. In the case of linear functions $f$ taking the form

$$f(x) = \sum_{i=1}^{n} w_i x_i + b \quad with \quad w \in R, b \in R. \tag{1}$$

One way to ensure minimum $\varepsilon$ deviation is to minimize the norm, i,e $||w||^2 = \sum_{i=1}^{n} W_i^T W_i$ The problem can be written as a convex optimization problem: minimize $\frac{1}{2}||w||^2$ subject to

$$subject \quad to \quad y_i - \sum_{i=1}^{n} W_i x_i - b \leq \varepsilon, \sum_{i=1}^{n} W_i x_i + b - y_i \leq \varepsilon \tag{2}$$

The assumption in (2) is that such a function $f$ actually exists that approximates all pairs $(x_i, y_i)$ with $\varepsilon$ precision. In the case where constraints are infeasible, slack variables , $\xi_i, \xi_i^*$ were introduced. This case is called soft margin formulation [Bennet and Mangasarian, 1992] and is described by the following problem;

$$Minimize \qquad \frac{1}{2}||w||^2 + C\sum_{i=1}^{l}(\xi_i + \xi_i^*) \tag{3}$$

subject to

$$y_i - \sum_{i=1}^{n} W_i x_i - b \leq \varepsilon + \xi_i \tag{4}$$

$$\sum_{i=1}^{n} W_i x_i + b - y_i \leq \varepsilon + \xi_i^* \tag{5}$$

$$\xi_i, \xi_i^* \geq 0 \tag{6}$$

The constant C ¿ 0 determines the amount up to which deviations larger than $\varepsilon$ are tolerated. This is called -insensitive loss function $|\xi|_\varepsilon$ and is described by

$$|\xi|_\varepsilon = 0 \qquad if \quad |\xi| \leq \varepsilon, \tag{7}$$

$$|\xi|_\varepsilon = |\xi| - \varepsilon \qquad otherwise, \tag{8}$$

Figure 1 depicts this $\varepsilon$ -insensitive loss function graphically. Only the points inside the shaded region contribute to the prediction accuracy, whereas the points outside the shaded region contribute to the error, as the deviations are penalized

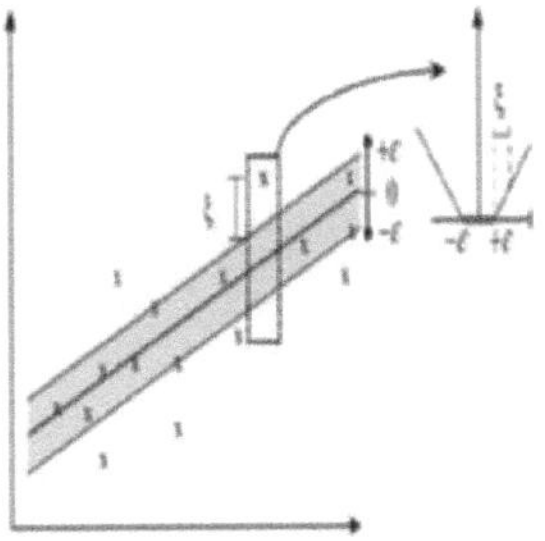

**Fig. 1.** The soft margin loss setting for a linear SVR

in a linear fashion. It is observed that in most cases the optimization problem can be solved more easily in its dual formulation. Hence, Lagrange multipliers are used to get the dual formulation as described in [Fletcher, 1989], which is as follows;

$$L = \frac{1}{2}||w||^2 + C\sum_{i=1}^{l}(\xi_i + \xi_i^*) - \sum_{i=1}^{l}(\eta_i\xi_i + \eta_i^*\xi_i^*)$$
$$-\sum_{i=1}^{l}\alpha_i(\varepsilon + \xi_i - y_i + \sum_{i=1}^{n} w_i x_i + b)$$
$$-\sum_{i=1}^{l}\alpha_i^*(\varepsilon + \xi_i - y_i + \sum_{i=1}^{n} w_i x_i + b) - \sum_{i=1}^{l}(\eta_i\xi_i + \eta_i^*\xi_i^*) \tag{9}$$

Here L is the Lagrangian and $\eta_i, \eta_i^*, \alpha_i, \alpha_i^*$ are Lagrange multipliers. Hence the dual variables in (4) have to satisfy positivity constraints i.e.

$$\eta_i, \eta_i^*, \alpha_i, \alpha_i^* \geq 0 \tag{10}$$

According to the saddle point condition the partial derivatives of L with respect to the primal variables $(w, b, \xi_i, \xi_i^*)$ have to vanish for optimality. Therefore, we get

$$\frac{\partial L}{\partial b} = \sum_{i=0}^{l}(\alpha_i^* - \alpha_i) = 0, \tag{11}$$

$$\frac{\partial L}{\partial w} = w - \sum_{i=1}^{l}(\alpha_i - \alpha_i^*)x_i = 0, \tag{12}$$

$$\frac{\partial L}{\partial \xi_i} = C - \alpha_i - \eta_i = 0, \tag{13}$$

$$\frac{\partial L}{\partial \xi_i^*} = C - \alpha_i^* - \eta_i^* = 0, \tag{14}$$

Substituting (11), (12), (13) and (14) into (9) yields the dual optimization problem;

$$maximize \; -\frac{1}{2}\sum_{i,j=1}^{l}(\alpha_i-\alpha_i^*)(\alpha_j-\alpha_j^*)\sum_{i,j=1}^{n}x_ix_j-\varepsilon\sum_{i=1}^{l}(\alpha_i-\alpha_i^*)$$
$$+\sum_{i=1}^{l}y_i(\alpha_i-\alpha_i^*) \quad (15)$$

subject to

$$\sum_{i=1}^{l}(\alpha_i-\alpha_i^*)=0 \quad and \quad \alpha_i,\alpha_i^*\in[0,C]. \quad (16)$$

To make the SV algorithm nonlinear all training patterns $x_i$ are mapped $\phi : x \to f$ into some feature space $f$ as described in ( [Aizerman et al, 1964], [Nilsson, 1965]) and then using the standard SV regression a linear hyperplane is constructed the feature space. Using the trick of kernel functions [Cortes and Vapnik, 1995] following QP problem is formulated.

$$Maximize \quad -\frac{1}{2}\sum_{i,j=1}^{l}(\alpha_i-\alpha_i^*)(\alpha_j-\alpha_j^*)K(x_i,x_j), \quad (17)$$

$$-\varepsilon\sum_{i=1}^{l}(\alpha_i-\alpha_i^*)+\sum_{i=1}^{l}y_i(\alpha_i-\alpha_i^*) \quad (18)$$

subject to

$$\sum_{i=1}^{l}(\alpha_i-\alpha_i^*)=0 \quad and \quad \alpha_i,\alpha_i^*\in[0,C]. \quad (19)$$

The optimal solution obtained

$$w=\sum_{i=1}^{l}(\alpha_i-\alpha_i^*)\phi(x_i) \quad and \quad f(x)=\sum_{i=1}^{l}(\alpha_i-\alpha_i^*)K(x_i,x)+b, \quad (20)$$

where K(.,.) is a kernel function. Usually more than one kernel used in the literature to map the input space into feature space [Christianini and Shawe, 2000].

## 4 Data Partition

### 4.1 Based on Fuzzy Clustering Means(FCM)

In Fuzzy clustering, each point belongs to one or more cluster, whereas in fuzzy logic, the point belongs to only one cluster. Thus, points which are outlier of cluster, has less membership value than the points which lies near the center of cluster. For each point $x$ we have a coefficient giving the membership value of the $k$th cluster is $\mu_k(x)$. Usually, the sum of those coefficients for any given $x$ is defined to be 1:

$$\forall x \left[ \sum_{k=1}^{num.clusters} \mu_k(x) = 1 \right] \tag{21}$$

The centroid of a cluster is the mean of all points, weighted by their membership value of belonging to the cluster:

$$center_k = \frac{\sum_x \mu_k(x)^m x}{\sum_x \mu_k(x)^m} \tag{22}$$

The membership value is related to the inverse of the distance to the cluster center:

$$\mu_k(x) = \frac{1}{d(center_k, x)}, \tag{23}$$

then the coefficients are normalized and fuzzyfied with a real parameter $m > 1$ so that their sum is 1. So

For m equal to 2, this is equivalent to normalizing the coefficient linearly to make their sum 1. When m is close to 1, then cluster center closest to the point is given much more weight than the others, and the algorithm is similar to k-means. [WIKI]. Clustering is then obtained by considering the points which has more membership value than the other clusters. The mean shift algorithm does not have prior knowledge about the shape of the clusters, and the number of classes can be determined based on the number of clusters.

### 4.2 Data Partition Based on Slope

For a large training data set, the memory space required to store the kernel matrix becomes very challenging for the solution of QP problems for SVR. Thus, the traditional optimization algorithms, such as Newton, cannot be directly used. In this section, we propose an unsupervised partition algorithm, through which fixed no. of partitions are automatically generated based on the slope of data set, based upon the random change of slope. The resultant partitions of the algorithm are then used for SVR training. Preprocessing of data includes taking

---

preprocess the data p=$(x_i, y_i)_{i=1}^{N}$
Find $Z = \Delta(y)$
sort $z$ in ascending order
select top $m$ values
select Minimum length of the partition
For $j = 1tom$
For$k = z[1]toz[m]$
$p_j = (x_i, y_i)_{i=k-1}^{k}$
For each partition $p_i$ apply SVR and find RMSE
Calculate the aggregate RMSE

---

care of missing values, noise, etc. Also, fetching the set of target values from the data into the program is also a part of that pre-processing. The main idea is to partition the data into multiple small sections. The section limits are defined by any great change in the data. The secondary condition is to make partitions of greater than a or equal to a minimum length in order not to end up in very badly sized subsets.

To find a greatest amount of change in the data, first the difference of the data is found. This difference data is also called as slope (of data). Now, the maximum value in this slope is actually the greatest change in the original data. There might be multiple points where the data has same highest amount of change. The program is set to find all those positions in the data where the slope is the highest. From these found positions, the program checks the points that are so apart from each other (and boundaries) that they can make a partition of the minimum required size or more. The positions that cannot make partitions large enough are ignored. If the required number of partitions is not found, this process has to be redone. In order to make next iteration possible, the highest value is made to a lowest or Not-a-Number value in the same iteration. That makes the next highest value as the highest value. Then, this process is repeated until either the data is divided into a set number of partitions (or more) or to a point where there are no more divisions possible.

A dataset may contain a slope which has more common highest values than the number of required partitions. Also, in a certain iteration the number of partitions made might be much more than the remaining number of partitions required. Hence, it is certainly possible that the above process ends up giving more number of partitions than required. But it should also be noted that iteration will not take place unless there is a need of partition. So, any extra partition generated has to be from the last iteration. Now, all these new partitions has same peak value. At this juncture, the program is set to skip the partitions from the last iterations which had minimum size. Or in other words, the partitions which give larger size of partitions, if removed, are skipped. This way, the data is divided into the exact number of partitions. . In FCM, every point belongs to all the clusters, but we have considered only those points in a cluster which has the maximum value of membership compare to other clusters. After finding the partitions, each partition is used to train SVR separately and the performance is evaluated in RMSE. And all the partitions RMSE values are aggregated.

## 5 Experiment Section

For the experiments we have considered data set from the website [DACS] reliability datasets, SS1A, SS1B, SS1C, SS2, SS3,SS4. The Software Reliability Dataset was compiled by John Musa of Bell Telephone Laboratories. These datasets contains t failure interval data to assist software managers in monitoring test status and predicting schedules and to assist software researchers in validating software reliability models. These models are applied in the discipline of Software Reliability Engineering. The dataset consists of software failure data for operating system. This data set consists of 3 attributes, called Failure Number, Failure Interval, Day of Failure. For these datasets we have applied the FCM and slope based algorithms and portioned the dataset in to five datasets. Later each subset of dataset is processed through LibSVM to get the performance in RMSE. While training the SVR we have chosen the RBF kernel as it gives the better performance than the other kernels. In this we are doing experiments with the values Cost (c) value is 6.4 and epsilon (p) is 0.026, as it is proved by Shawon et. al., that the performance of trined SVM model is better than the SVM trained with other combination of values [Shaowu et.al, 2007].

The figure 2 shows the results of this process in a graphical form. This graph has multiple scale for the original data and its slope are completely different in magnitude. The left vertical axis has the the scale for original data magnitude. The right one is for the difference or slope of the data. And partitions are shown with vertical bars across the data. Also shown are the skipped partitions, if extra partitions were found. It also shows the graph representing the max peaks of SS1A data set, in which the lines representing with pink color has been selected for the partition and the blue color represents the skipped partition, as in this the partition length color represents the skipped partition, as in this the partition length is less than 4.

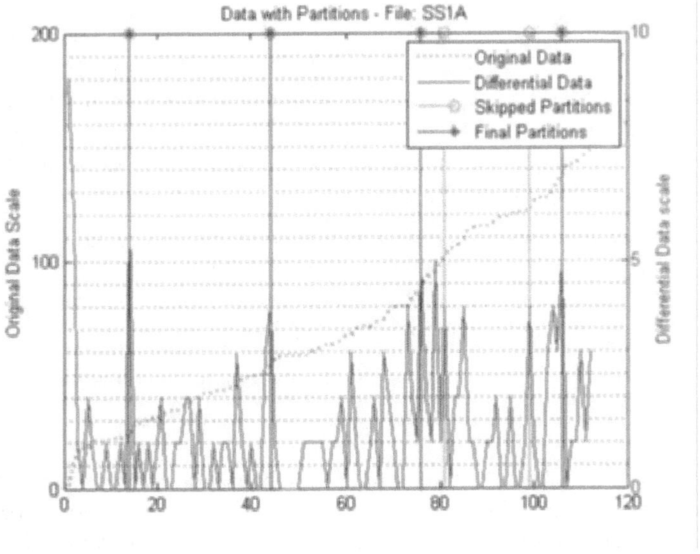


**Fig. 2.** Graph Based on Slope Partition

Fig 3 explains the proposed approach. First the dataset is processed through the FCM and slope based partition then each sub dataset is passed to train SVR through LibSVM measuring the performance in RMSE.

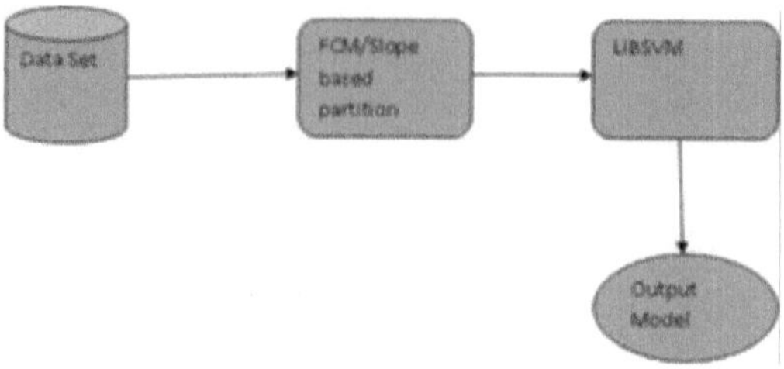


**Fig. 3.** Data Flow Diagram

## 6 Results and Discussion

The results are shown in the Tables 1 and Table2. In Table 1, it is representing the results of the datasets before and after applying slope based partition. From the table it is observed that SVR error rate is reduced when it is trained with partitioned data. Considering the SS1A dataset we made 5 partitions, based up on the max peaks, which is the change in slope. All the 5 partitions are used to train data individually, passed through LibSVM using RBF kernel, by setting the parameter. And the SVR model is tested by other partitions. The performance is measured in RMSE, and tabulated the RMSE rate on the original data and the aggregated RMSE of partitioned data. This table shows that the results are better than the original data. Similarly, Table 2 represents the results of the datasets for training SVR with the original data and datasets after partitioning. From the table we can conclude that in all datasets we are getting the better performance, after partitioning the data and aggregating the RMSE, than considering complete data.

**Table 1.** Shows the results of slope based partition

| Dataset | RMSE(original Data) | RMSE(after partitioning) | % of performance improved |
|---|---|---|---|
| SS1A | 17.85 | 7.3766 | 58.67 |
| SS1B | 90.1762 | 51.8399 | 42.51 |
| SS1C | 60.9547 | 29.7321 | 51.22 |
| SS2 | 44.8908 | 24.4178 | 45.61 |
| SS3 | 63.2017 | 48.4788 | 23.3 |
| SS4 | 38.1437 | 30.3019 | 20.56 |
| Over all average performance= | 40.31 | | |

**Table 2.** Shows the results based on FCM

| Dataset | RMSE(original Data) | RMSE(after partitioning) | % of performance improved |
|---|---|---|---|
| SS1A | 17.85 | 10.2024 | 42.84 |
| SS1B | 90.1762 | 52.7324 | 41.52 |
| SS1C | 60.9547 | 26.5313 | 56.47 |
| SS2 | 44.8908 | 19.6897 | 56.14 |
| SS3 | 63.2017 | 30.2127 | 52.20 |
| SS4 | 38.1437 | 23.0987 | 39.44 |
| Over all average performance= | 48.1017 | | |

## 7 Conclusions and Future Work

In this paper we have proposed new approach called Slope based partition, and compared the results with the FCM partitions, for training SVR. The proposed method reduces model complexity by first grouping similar components with same slope into a partition and then training SVR and getting local function approximation. Which is the simple slope logic based partition, which reduces the complexity, and also SVR model simplification. We have considered the local maxima, and restricted the no. of partitions for all data sets. This is the limitation of our algorithm. We could not handle the stopping criteria. In FCM algorithm we need to mention the no. of clusters has to be initialized. While training data set should be known in prior, which is used to train SVR in both the algorithms, which is the limitation of both the algorithms. Considering original data training SVR we get good functional approximation. But preprocessing the data with FCM and slope based partition and training SVR, overall finding the aggregating the RMSE gives reduced error rate. So from our algorithmic approach, we can conclude that both FCM and slope based partitions are comparable. The study about joining the results of the subsets is crucial. The effect of partitioning on a data which has periodical slope should also be studied. Also, the effectiveness of this technique on a data with non-increasing values should also be critically studied. This research also calls for the study of simpler selection of peaks based on thresholds rather than analysis.

## References

[Aizerman et al, 1964] Aizerman, M.A., Braverman, E.M., Rozonoer, L.I.: Theoretical foundations of the potential function method in pattern recognition learning. Automation and Remote Control 25, 821–837 (1964)

[Asli et. al, 2007] Celikyilmaz, A., Turksen, B.I.: Fuzzy functions with support vector machines. Science Direct, Information Sciences 177, 5163–5177 (2007)

[Bennet and Mangasarian, 1992] Bennet, K.P., Mangasarian, O.L.: Robust linear programming discrimination of two linearly inseparable sets. Optimization Methods and Software 1, 23–34 (1992)

[Burges, 1998] Burges, C.J.C.: A tutorial on support vector machines for pattern recognition. Data Mining and Knowledge Discovery 2(2), 121–167 (1998)

[Chia et. al, 2010] Juang, C.-F., Hsieh, C.-D., Hong, J.-L.: Fuzzy Clustering-Based Neural Fuzzy Network, with Support Vector Regression. IEEE, Los Alamitos (2010)

[Cortes and Vapnik, 1995] Cortes, C., Vapnik, V.: Support vector networks. Machine Learning 20, 273–297 (1995)

[Christianini and Shawe, 2000] Cristianini, N., Shawe-Taylor, J.: An introduction to support vector machines and other kernel-based learning methods. Cambridge University Press, New York (2000)

[Fletcher, 1989] Fletcher, R.: Practical Methods of Optimization. John Wiley and Sons, New York (1989)

[Nilsson, 1965] Nilsson, N.J.: Learning machines: Foundations of trainable pattern classifying systems. McGraw-Hill, New York (1965)

[Osuna et. al, 1997] Osuna, E.E., Freund, R., Girosi, F.: Support vector machines: Training and applications. Technical report, Massachusetts Institute of Technology, Artificial Intelligence Laboratory, AI Memo No. 1602 (1997)
[Pontil and Verri, 1997] Pontil, M., Verri, A.: Properties of support vector machines. Technical Report, Massachusetts Institute of Technology, Boston MA, USA (1997)
[Platt, 1999] Platt, J.C.: Fast training of support vector machines using sequential minimal optimization. In: Schölkopf, B., Burges, C., Smola, A. (eds.) Advances in Kernel Methods - Support Vector Learning, pp. 185–208. MIT Press, Cambridge (1999)
[Shaowu et.al, 2007] Zhou, S., Wu, L., Yuan, X., Tan, W.: Parameter Selection of SVM for Function Approximation Based on Differential Evolution (2007)
[Smola and Scholkopf,2004] Smola, A.J., Schölkopf, B.: A tutorial on support vector regression. Statistics and Computing 14(3), 199–222 (2004)
[Tanaka et. al, 1982] Tanaka, H., Uejima, S., Asai, K.: Linear regression analysis with fuzzy model. IEEE Transactions on Systems, Man, Cybernetics SMC-2, 903–907 (1982)
[Vapnik,1995] Vapnik, V.N.: The nature of statistical learning theory. Springer-Verlag, New York Inc., New York (1995)
[DACS] www.thedacs.com
[WIKI] http://en.wikipedia.org/wiki/Clusteranalysis

# A PIC-Controller Based Doubly Fed Induction Generator (DFIG) for Wind Energy Conversion Systems

K. Sureshkumar[1], P. Vijaya kumar[2], R. Dhandayuthabani[3], and A. Sakthivel[4]

[1] Department of EEE, Velammal Engineering College,
Chennai-600066
sureshbarath@gmail.com
[2] Department of EEE, Karpagam College of Engineering,
Coimbatore-641 032
vijay_p6@hotmail.com
[3] Product Development Engineer, Automotive Infotronics, Chennai
dhandayuthabanir@yahoo.co.in
[4] Department of EEE, Dr. Mahalingam College of Engg & Tech,
Pollachi-641003
ersakthi@yahoo.com

**Abstract.** The design details of a PIC-controller based Doubly Fed Induction Generator (DFIG) for wind power generation is described. The proposed PIC-DFIG not only provides a low cost Solution to today's electric power needs, but is also aimed at delivering better power quality, making it ideally suited for wind energy conversion systems operating at variable wind speeds and connected to variable and often, dynamic utility loads. Power quality improvement is achieved by cancellation of the largest and the most annoying harmonics in the supply grid. Additionally, power factor correction and adaptive response to variable reactive loading are also addressed by this design. The proposed design involves two current-controlled power converters connected through a slip ring assembly. The rotor control is achieved using a commercially available PIC controller (model 16F877A) employing a position sensorless, field oriented control algorithm. Some preliminary simulation efforts and modeling scenarios are also presented.

**Keywords:** PIC-controller, doubly-fed induction generator, wind energy, harmonic compensation, power factor improvement.

## 1 Introduction

In view of the rapid depletion of fossil fuels and the increasing global concern for clean environment, considerable efforts are being spent on developing green, renewable energy sources. Of such resources, wind energy offers a sustainable, low-cost and environmentally friendly solution to energy needs. However, the intermittent and variable nature of wind currents offers many challenges to tapping the wind power effectively. Sometimes a Variable Speed Power Generation (VSPG) system is used to harness wind energy, where a generator parameter, such as the wind turbine speed, is dynamically adjusted to optimize the output power as the ambient wind

D. Nagamalai, E. Renault, M. Dhanushkodi (Eds.): CCSEIT 2011, CCIS 204, pp. 715–723, 2011.

speed changes. Fixed speed generating systems are also used widely in wind power generation and owe their popularity to a few advantages such as low cost, simplicity of design, the ability to control both active and reactive currents independently and so on. The system we propose, the Doubly Fed Induction Generator (DFIG) is one such fixed speed system. The optimization of output power in this system is achieved through regulating the power flow in the two generator windings, stator and rotor. Since the rotor side winding handles only the slip power, which is easier to control, often it is enough to control only the rotor side flow in order to regulate the overall flow and obtain an improved power rating. We propose the use of PIC controllers to regulate such rotor side converter flows because they offer the following advantages among others: (1) Low cost (2) Off the shelf units easily available (3) The Assembly Language coding results in fast response and therefore more dynamic control (4) multi-functionality and flexibility, leading to the development of customizable optimization algorithms. Once the rotor parameters' ranges are determined and control algorithm is fine-tuned, we may be able to implement the design using commercially available PLC units. in place of PIC controllers for easier end-user programming, if needed. We will discuss some concepts on which we base such control algorithms and propose how best to generate and optimize wind power.

First, in addition to wind power conversion, it is also important to improve the power quality of the generated power, especially given the wide variety of electrical appliances with stringent supply quality requirements that are being used by today's utility consumers. Power quality, in terms of harmonic current distortion in the distributed power and power factor, may also be subject to government regulations and compliance can be quite challenging. The DFIG design described in this paper addresses these concerns [1-6] as well and is capable of compensating for the most prevalent harmonic current components in the grid, in addition to its ability to generate effective wind power. The proposed design also corrects power factor and offers reactive power control. The current sensing in the rotor is done using a position sensorless, field oriented control [10] method and is used in regulating the rotor circuit dynamically via PIC controllers.

As a prelude to actual implementation of the design, we offer a simulation of the design and some scenarios using MATLAB. In particular, we provide some sampling of PIC programming to give a flavor of the controller programming.

## 2 Architecture of DFIG Power Grid System

The proposed DFIG design and a section of the generator-grid interface are shown schematically in Fig. 1. The exact circuitry of the controllers, the power converters and associated electronics, details of the load etc. are not shown for brevity.

It must be noted that this design affords independent control of active (wind) power generation, reactive (load) power compensation and grid harmonics distortion cancellation. The two power converters used, one for the stator and one for the rotor, are the current-controlled voltage source type [10]. The stator flux field orientation is sensed in a position-less manner and is used to regulate flow in rotor side windings. This way, both the fundamental as well as the harmonics are sensed. The exact level of harmonic currents to be used in command for the rotor side is obtained either by estimating the harmonics current theoretically based on the externally connected load or by actually sensing the current in the load. In general there are two major power

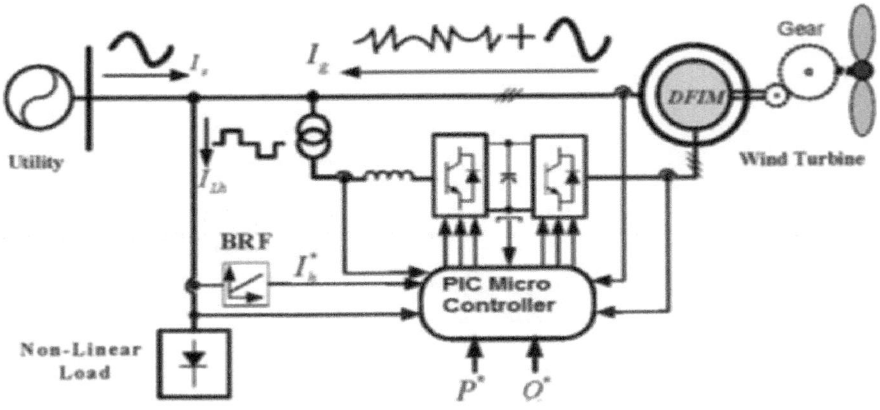


**Fig. 1.** Block diagram of DFIG-POWER grid system

components in this rotor side regulation algorithm, one due to the active process of wind energy generation and associated variability due to wind speed and other characteristics of wind turbine and another harmonics component due to the utility load and grid harmonics. By suitable controller commands, these components are dynamically translated into the appropriate rotor side excitation. This in turn results in high quality power output with little harmonic distortion and adjusted for reactive load. i.e., better power factor.

## 3 Design Details

### 3.1 Rotor Side Design

In our design, the non linear load currents and phase currents of the line are sampled and converted into d-q co-ordinate values by using Parks transformation. Based on the quadrature and direct rotor current, the gate signal for the back to back ac/dc/ac converter is given.

In order to have a common frame of reference for both the stator and the rotor equations and also to reduce the complexity of the voltage equations, one can use a rotating set of d-q axes attached to the stator flux[9] which rotates at an angular frequency of θ. The flux linkages of stator and rotor in the Excitation reference frame will be:

$$\lambda^e_{qs} = L_s\, i^e_{qs} + L_m\, i^e_{qr}, \tag{1}$$

$$\lambda^e_{ds} = L_s\, i^e_{ds} + L_m\, i^e_{dr} \tag{2}$$

$$\lambda^e_{qr} = L_r\, i^e_{qr} + L_m\, i^e_{qs}, \tag{3}$$

$$\lambda^e_{dr} = L_r\, i^e_{dr} + L_m\, i^e_{ds} \tag{4}$$

Rotor voltages in the *d–q* excitation reference frame can be shown as

$$v^e_{qr} = r_r\, i^e_{qr} + (\omega e - \omega r)\, \lambda^e_{dr} + P\, \lambda^e_{qr} \quad (5)$$

$$v^e_{dr} = r_r\, i^e_{dr} - (\omega e - \omega r)\, \lambda^e_{qr} + P\, \lambda^e_{dr} \quad (6)$$

If the stator flux vector is aligned with the *d*-axis in the synchronously rotating reference frame, then $\lambda^e qs = 0$, and the rotor currents components can be shown as

$$i^e_{qr} = -(L_s/L_m)\, i^e_{qs} \quad (7)$$

$$i^e_{dr} = -(L_s/L_m)\, i^e_{ds} + (1/L_m)\, \lambda^e_{ds} \quad (8)$$

The rotor voltage equations in the excitation reference frame

$$v^e_{qr} = r_r i^e_{qr} + \omega_{slip}\sigma L_r i^e_{dr} + \omega_{slip}(L_m/L_s)\lambda^e_{ds} + \sigma L_r (d/dt) i^e_{qr} \quad (9)$$

$$v^e_{dr} = r_r i^e_{dr} - \omega_{slip}\sigma L_r i^e_{qr} + \sigma L_r (d/dt) i^e_{dr} + (L_m/L_s)\,(d/dt)\lambda^e_{ds} \quad (10)$$

where,

$$\omega_{slip} = \omega e - \omega r \quad (11)$$

The quadrature and direct rotor current demand can be derived from the active and reactive power references *P**, *Q**, and harmonic currents needed for compensation of nonlinear load current. The variables $i^{e*}_{qr}$ and $i^{e*}_{dr}$ can be derived as follows:

$$i^{e*}_{qr} = i^{e*}_{qr\text{–active}} + i^{e*}_{qrh} \quad (12)$$

$$i^{e*}_{dr} = i^{e*}_{dr\text{–reactive}} + i^{e*}_{drh}. \quad (13)$$

where,

$$i^{e*}_{qrh} = -\,(L_s/L_m).\ i^e_{qLh} \quad (14)$$

$$i^{e*}_{drh} = -\,(L_s/L_m) i^e_{dLh} + (1/L_m)\lambda^e_{ds} \quad (15)$$

### 3.2 Direct and Quadrature Rotor Current Command Generation

The block diagram of quadrature and direct rotor currents commands generation is shown in Fig. 3. In this method, wind velocity is measured, and if it is higher than the cut-in value, the system goes to current mode control. Active and reactive power references are being generated in respect to the wind speed and turbine characteristics. This will generate the active power command I**qr*-active. If the wind turbine power falls below Pmin, the speed control mode is activated, and using speed regulator, the active power command is obtained.

In this topology, while direct axis is used to control the reactive power and the quadrature axis current component is used to regulate the active power, ac signal portion of the quadrature and the direct rotor current components are set to control harmonics compensation of the grid. In the proposed algorithm, the decoupled control of active power, reactive power, and harmonic compensation are achieved.

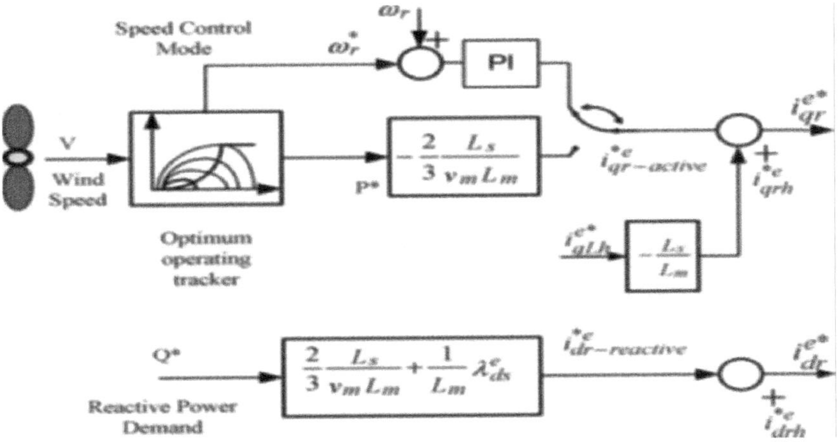


**Fig. 2.** Direct and quadrature rotor current command generation

The active and reactive powers of the stator are:

$$P{*}s=-3/2(v_m(L_m/L_s)i{*}^{e}_{qr\text{-}active}) \tag{16}$$

$$Q{*}s=-3/2v_m(\lambda^{e}_{ds}-L_m i{*}^{e}_{dr\text{-}reactive}/L_s) \tag{17}$$

### 3.3 Position – Sensorless Rotor Position Detection

Proper rotor current orientation [7] with respect to the stator flux requires transformations between the stator and the rotor coordinates. It means that information of the instantaneous rotor position is necessary to implement correct field orientation. The following method is based on simple trigonometric computations3 Therefore the estimation is possible instantaneously, and the control on the fly is enforceable without any appreciable transient. The stator flux depends on the stator voltage and frequency of excitation.

The *d–q* components of the rotor current in the stationary reference frame can be obtained from

$$i^{s}_{dr} = (1/L_m)(\int(v^{s}_{ds}- r_s\, i^{s}_{ds})d-(L_m + L_{ls})i^{s}_{ds}) \tag{18}$$

$$i^{s}_{qr} = (1/L_m)(\int(v^{s}_{qs} - r_s\, i^{s}_{qs})dt - (L_m+L_{ls})i^{s}_{qs}) \tag{19}$$

If we project these currents into two-phase orthogonal reference frame fixed on the rotor we can write,

$$i_r = \sqrt{(i^{s\ 2}_{dr} + i^{s\ 2}_{qr})} \tag{20}$$

$$i_r = \sqrt{(i^{r\ 2}_{dr} + i^{r\ 2}_{qr})} \tag{21}$$

$$\theta 1 = \tan^{-1}(i^{s}_{qr}/i^{s}_{dr}) \tag{23}$$

$$\theta 2 = \tan^{-1}(i^{r}_{qr}/i^{r}_{dr}) \tag{24}$$

The rotor current vectors in the orthogonal *d–q* stationary and rotary reference frames are illustrated in Fig. 4. It can be inferred that the rotor current vector in the stationary reference frame rotates with synchronous speed.

However, in the rotor reference frame, this vector rotates with slip frequency. It is obvious that the rotor position angle can be calculated from

$$\theta_r = \theta_1 - \theta_2 \tag{25}$$

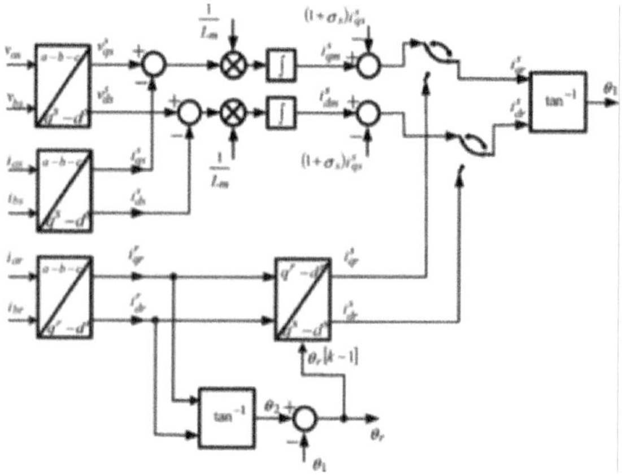

**Fig. 3.** Sensorless rotor position calculator

## 4 Simulation Results

We have simulated some scenarios using our model.

**Case Study 1. Grid status without non linear load**

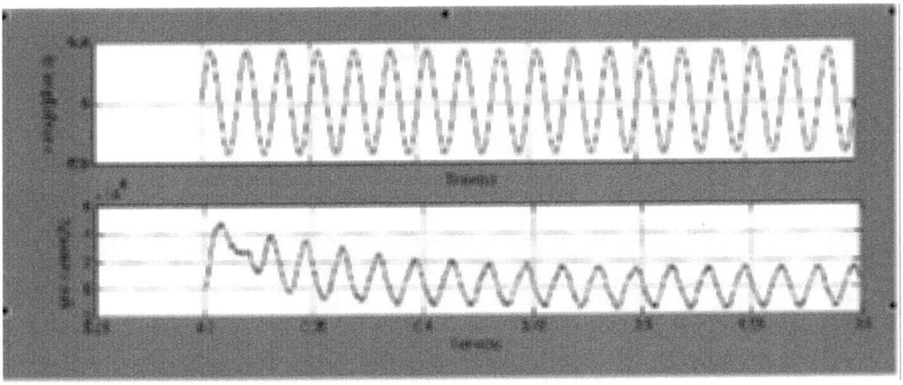

**Fig. 4.** Case study 1. Grid voltage and current.

**Case Study 2. Grid status with non linear load**

Nonlinear load current is generated by a diode rectifier feeding a variable resistor. The magnitude and phase of the non-linear load current are varied to evaluate the performance of the proposed Method. After connecting the non linear load across the grid, the Total harmonic distortion of current is 29.73%.

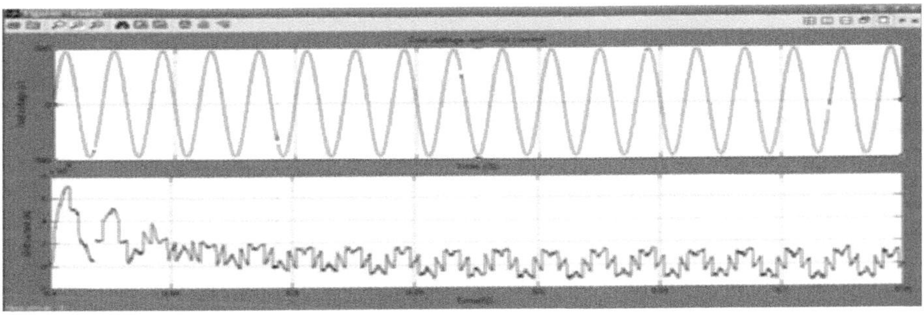

**Fig. 5.** Case study 2. Grid voltage and Grid current with connecting non linear load

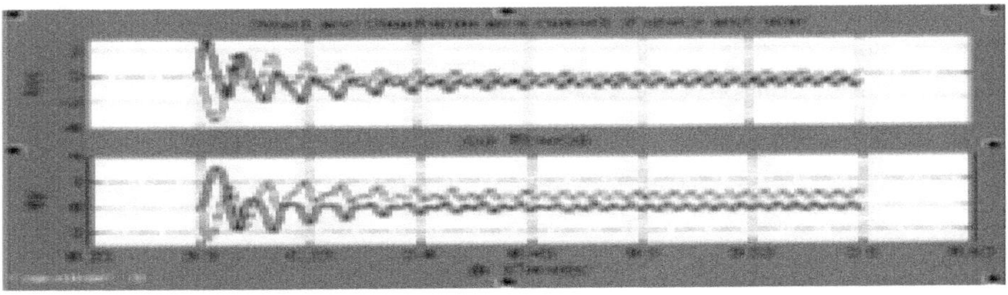

**Fig. 6.** Case study 2. (a)Direct and Quadrature axis current of stator. (b)Direct and Quadrature axis current of Rotor without controller.

**Case Study 3. Grid status with controller**

The controller has been developed to cancel harmonics needed for non linear load by varying the stator flux based on the command gate signal. After connecting the controller, the total harmonic distortion is reduced (THD = 3.07%).

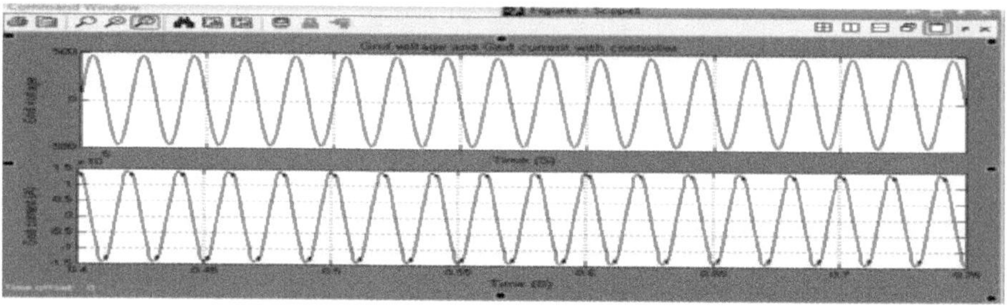

**Fig. 7.** Case study 3 .Grid voltage and current with connecting non linear load with controller.

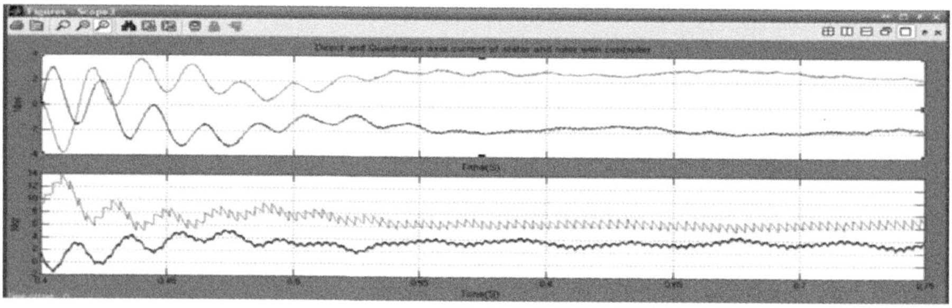

**Fig. 8.** Case study 3. (a)Direct and Quadrature axis current of stator. (b)Direct and Quadrature axis current of Rotor with controller.

## 5 Conclusion

The design and modeling of a PIC controller-based DFIG-Wind power system is described. The power generation is not only optimized for maximum conversion of wind energy, but is also capable of producing high quality power with dynamic cancellation of grid harmonics. The design algorithm is based upon controlling rotor current flows and in the control of reactive current components due to external utility load. The use of PIC controller in the rotor side control affords a simple, low cost way of implementing the proposed algorithm and consequent generation of high quality wind power. Such a system is well suited for low to medium scale wind power generation units.

## References

1. Abolhassani, T., Niazi, P., Toliyat, H.A., Enjeti, P.: Integrated Doubly Fed Electric Alternator/ActiveFilter (IDEA), a Viable Power Quality Solution, for Wind Energy Conversion System. Proc. IEEE Trans. Energy Conversion 23(2), 642–650 (2008)

2. Abolhassani, M.T., Niazi, P., Toliyat, H.A., Enjeti, P.: A sensoreless Integrated doubly-fed electric alternator/active filter (IDEA) for variable speed wind energy system. In: Proc. IEEE Ind. Appl. Soc., vol. 1, pp. 507–514 (2003)
3. Datta, R., Ranganathan, V.T.: A Simple Position-Sensorless Algorithm for Rotor-Side Field-Oriented Control of Wound-Rotor Induction Machine. IEEE Transaction on Industrial Electronics 48(4) (August 2001)
4. Xu, L., Cheng, W.: Torque and reactive power control of a doubly fed induction machine by position sensorless scheme. IEEE Trans. Ind. Applicat. 31, 636–642 (1995)
5. Akagi, H.: New Trends in Active Filters for Power Conditioning. IEEE Transactions on Industry Applications 32(6), 1312–1322 (1996)
6. Pena, R., Clare, J.C., Asher, G.M.: Doubly fed induction generator using back-to-back PWM converters and its application to vari-able-speed wind- energy generation. Proc. Inst. Elect. Eng., pt. B 143(3), 231–241 (1996)
7. Datta, R.: Rotor side control of grid-connected wound rotor Induction machine and its application to wind power generation. Ph.D. dissertation, Dept. Elect. Eng., Indian Inst. Science, Bangalore, India (February 2000)
8. Akagi, H., Sato, H.: Control and Performance of a Doubly-Fed Induction Machine Intended For a flywheel Energy Storage System. IEE Transaction on Power Electronics 17(1) (January 2002)
9. Hopfensperger, B., et al.: Stator-flux-oriented control of a doubly-fed induction machine with and without position encoder. IEE Proc. Electr., Power Appl. 147(4) (July 2000)
10. Morel, L., Godfroid, H., Mirzaian, A., Kauffmann, J.M.: Double-Fed induction machine: converter optimization and field oriented control without position sensor. Proc. Inst. Elect. Eng., pt. B 145(4), 360–368 (1998)
11. Akagi, H., Sato, H.: Control and performance of a doubly-fed Induction Machine intended for a flywheel energy storage system. IEE Trans. Power Electron. 17, 109–116 (2002)

# Packet Scheduling Scheme with Enhanced Quality of Service for Mobile WiMAX

V. Sampath Kumar, C. Kalyana Chakravarthy, and V.R.L. Priyanka

M.V.G.R.College of Engineering, Vizianagaram
vatti.sampath@gmail.com, kalyan@mvgrce.edu.in, priya_mvgr@yahoo.com

**Abstract.** A new packet scheduling scheme for Mobile WiMAX network is proposed. The network architecture includes one BS and several MSs which are in different locations. The uplink traffic from Mobile Stations (MS's) that belong to various classes to the Base Station (BS) is scheduled.It is assumed that the MS's request the BS for bandwidth by a Bandwidth request and the BS station grants the bandwidth. The available slots for scheduling of uplink traffic is done based on the predefined priorities of the traffic classes. The BS then allocates the bandwidth to each MS based on the class it belongs and distance from the BS. All the mobile WiMAX parameters like transmit power, receive gain, transmit gain, mobility have been considered for simulation. Results indicate a clear reduction in delay and increase in throughput for different classes of flows compared to the existing Quality of Service based 2-Tier Ad-Hoc Scheduling Scheme (2T-AHSS).

## 1 Introduction

The WiMAX family of standards (802.16) concentrates on two types of usage models a fixed WiMAX usage model and a mobile WiMAX usage model. The basic element that differentiates these systems is the ground speed at which the systems are designed to manage. Based on mobility, wireless access systems are designed to operate on the move without any disruption of service; wireless access can be divided into three classes; stationary, pedestrian and vehicular.

A mobile WiMAX network access system is one that can address the vehicular class, whereas the fixed WiMAX serves the stationary and pedestrian classes. This raises a question about the nomadic wireless access system, which is referred to as a system that works as a fixed WiMAX network access system but can change its location.

Mobile WiMAX systems offer scalability in both radio access technology and network architecture, thus providing a great deal of flexibility in network deployment options and service offerings. Some of the salient features supported by Mobile WiMAX are high data rates, Quality of Service, Scalability, Security and Mobility.

Scheduling is the main component of the MAC layer that helps assure QoS to various service classes. Scheduling can be done for uplink and downlink.

D. Nagamalai, E. Renault, M. Dhanushkodi (Eds.): CCSEIT 2011, CCIS 204, pp. 724–734, 2011.

The scheduling algorithms developed in WiMAX Networks are vendor dependent. Every algorithm has its own advantages and disadvantages depending on what scenario it is being used. Hence selecting the appropriate algorithm for scheduling the packets for the downlink channel plays a key role [1]. Example is the Round Robin Scheduler [2] that fairly assigns allocation one by one to all connections. With packet-based allocation, stations with larger packets have an unfair advantage. Moreover, RR may be non-work conserving in the sense that the allocation is still made for connections that may have nothing to transmit; since RR cannot assure QoS for different service classes. The Weighted Round Robin algorithm has been applied for WiMAX scheduling [3]. A WRR algorithm assigns weight to each SS and the bandwidth is then allocated according to the weights. The weights can be used to adjust for the throughput and delay requirements. The WRR is an extension of the Round Robin algorithm. It is a work-conserving algorithm in that it will continue allocating bandwidth to the SSs as long as they have backlogged packets. Though the above algorithms are effective for fixed WiMAX where the data rates at the subscriber stations change less frequently, In Mobile WiMAX the rates at which the mobile stations transmit the data are changed at every instant. We describe a new efficient scheme for mobile station downlink scheduling that meets the challenges faced by a Mobile WiMAX network. A study of an optimized scheduling scheme in WiMAX networks has been done which includes allocation of resources to the mobile stations and scheduling the traffic flows according to their QoS parameters[4], channel aware scheduling algorithms. The goal of the work is to achieve the key performance objectives such as complete QoS requirements, fair to all services and using resources efficiently.

In this study, we propose a new packet scheduling scheme for Mobile WiMAX network. The network architecture includes one BS and several MSs which are in different locations. We schedule the uplink traffic from Mobile Stations (MS's) that belong to various classes to the Base Station(BS). Here we assume that the MS's request the BS for bandwidth by a Bandwidth request and the BS station grants the bandwidth. The available slots for scheduling of uplink traffic is done based on the predefined priorities of the traffic classes. The BS then allocates the bandwidth to each MS based on the class it belongs and distance from the BS.

In our work, we have considered all the mobile wimax parameters like transmit power, receive gain, transmit gain, mobility, restricting the no of users to a limit of 20 mobile station each requesting single type of service class at a time.

## 2 Previous Work

Our algorithm is an extension to the 2-Tier Ad-Hoc Scheduling Scheme (2T-AHSS)[5] that makes the bandwidth allocation for the TDD mode of operation of the IEEE 802.16 standard, and supports all types of service flows as indicated by the standard. 2T-AHSS is fair, efficient, and granular; and it follows the IEEE 802.16 variable transmission rate and signaling mechanism rules and regulations through dynamic adjustments of the bandwidth allocation. 2T-AHSS operates in two stages - Ad-Hoc STAGE and Dynamic Bandwidth Allocating Priority Queue (DBPAQ) for scheduling different classes of Traffic.

The first stage involves the process of identifying the traffic class of the incoming requests by the MSs and applying corresponding packet scheduling algorithms for each class. Table below gives the different traffic classes and their corresponding scheduling algorithms.

**Table 1.** Traffic classes and their corresponding algorithms

| Service Type | Scheduling Algorithm Used |
|---|---|
| UGS traffic | EDF Scheduling |
| ertPS traffic | Weighted Fair Queuing |
| rtPS traffic | Weighted Fair Queuing |
| nrtPS traffic | RoundRobin Scheduling |
| BE traffic | FIFO Scheduling |

In stage 2, As all the classes are given a predefined share of total bandwidth , previously a strict priority algorithm is used for allocating the bandwidth for the packets in AQL (Active Queue List). As for the entire scheduling process, we first perform the bandwidth allocation for various classes of traffic flows in a priority manner, and then perform the admission control for each type of traffic flows, respectively. The relevant traffic scheduling algorithm is described in the following algorithm.

1. Traffic Scheduling at each MSi.At the beginning of each frame ft Perform the bandwidth allocation in order for UGS(ertPS), rtPS, nrtPS and BE, respectively
2. Get a transmission rate Ri(t)for each MSi;
3. Perform Admission Control for each type as follows

**a** For a UGS or ertPS session ii with request of rii
   If (rii ¿ Ri(t ))
   Accept flow ii; Ri (t ) ? Ri (t ) ?rii ;
   Else Reject flow ii

**b** For an rtPS flow ij with request of rij
   If (rij ¿ Ri(t ))
   Accept flow ij; Ri (t ) ? Ri (t ) ?rij ;
   Else If (flow ij is within its delay deadline d max(ij))
   Put flow ij into the rtPS waiting queue;

**c** For a nrtPS flow ik with a minimal request of min (ik)
   If ( min (ik) ¿ Ri (t ))
   Accept flow ij; Ri (t ) ? Ri (t ) ? min (ik);
   Else Put ik into the nrtPS waiting queue;

**d** For a BE flow il
If ( Ri(t) ¿ 0)
Allocate a fair share of Ri (t ) to il;
Else Put il into the BE waiting queue;

4. End Process

## 3 Proposed Work - Enhanced DBAPQ for Improved QoS

In the above algorithm we deal with all the traffic flows in a strict priority manner according to their service urgency status, and then schedule them according to the available resource (bandwidth) Ri (t ).

However, it has a disadvantage of starvation as least priority classes get least chance for bandwidth allocation.

In the Enhanced DBAPQ scheme, the bandwidth to each service class is allocated in a dynamic way instead of allocating in static percentages [6]. Here we introduce a new value Bandwidth Remaining (BWR). At start of scheduling, this value is same as bandwidth of base station ie 100.

As the mobility in the mobile WiMAX is most vital factor, we need to allocate the bandwidths dynamically based on the distance of the MS node. If the MS node is far away from BS, then its actual data rates are lesser than sending data rates since as distance increases, the path losses of the signal will also increase during the transmission. So while considering mobility, we also consider the data rates of the mobile station at different distances. But because of this , the MS nodes far away suffer with low bandwidths and nearer MSs will have higher bandwidths which causes the bandwidth allocation useless and may also exceeds the actual request than actually it needed as per the requesting service class. So we calculate the distance of the MS node and find out the bandwidth it can actually receive at that distance. If that bandwidth is more than what it actually needed, then remaining bandwidth is re-allocated to the MS nodes which are suffering from lower bandwidths.

The bandwidth based on the distance from the base station that is calculated from the formula[7]:

$$SNR = \left(\frac{P * G_t * G_r}{KTB_\eta}\right)\left(\frac{\lambda}{4\prod}\right)^2\left(\frac{1}{d}\right)^n \quad (1)$$

Where Pt=transmit power
Gt=antenna transmit
Gr=antenna gain
K= Boltzmann constant
T=temperature
B= Bandwidth
d = distance of MS
SNR=Signal Noise Ratio

Thus our proposed algorithm helps in allocating the bandwidth dynamically not only based on class but also the distance of the MS. As the nodes far from the BS suffer from improper service they are allocated an equal share of bandwidth same as those of nearer MSs. Thus it results in higher throughput of MSs which are far from the base station same as those nearer to BS. The proposed algorithm also makes use of unutilized bandwidth of a particular class and distributes it to the remaining classes which are requesting for bandwidth in that particular frame.

For the purpose of Dynamic Bandwidth allocation, we introduce a new value Bandwidth Remaining (BWR). At start of scheduling, this value is same as bandwidth of base station ie 100% of BS bandwidth. Based on the type of service class, we allocate the dynamic % of BWR to each service class.

The proposed scheduling scheme is able to send the data packets to the base station at higher data rates with lesser delay compared to the existing scheme. By using the bandwidth re-allocation technique it is possible to grant bandwidth to all users preventing improper service and starvation.

Table below indicates the allocation of bandwidths considered for different service classes.

**Table 2.** Allocated bandwidth on the frame, per service class

| Service Type | Default Bandwidth(%) |
|---|---|
| UGS traffic | 45% of BWR |
| ertPS traffic | 36.5% of BWR |
| rtPS traffic | 57.2% of BWR |
| nrtPS traffic | 75% of BWR |
| BE traffic | 5% |

## 4 Simulation Results

To simulate our proposed algorithm, we have developed a simulator. The simulator provides options to the end users to create as many Mobile stations and set their properties and global properties accordingly.

As the distance varies the data rates to that mobile station varies. So as the station moves far from the base station, the data rates decreases. Since the resources in wireless networks are limited, another important consideration is how to maximize the total system throughput. One of the best ways to increase throughput is to reduce delay between the success transmissions, which is the actual transmitted data not including the overhead and lost packets. The bandwidth request is indicated in number of bytes. If we allocate the bandwidths dynamically as per the request of the service classes, appropriate bandwidths are allocated and the services flows are scheduled in lesser time reducing the delays between the rounds. Since the time taken is reduced, the throughput is increased. Our scheduling scheme using Dynamic Bandwidth allocation clearly shows reduced delay compared to the existing scheme within a given Frame time.

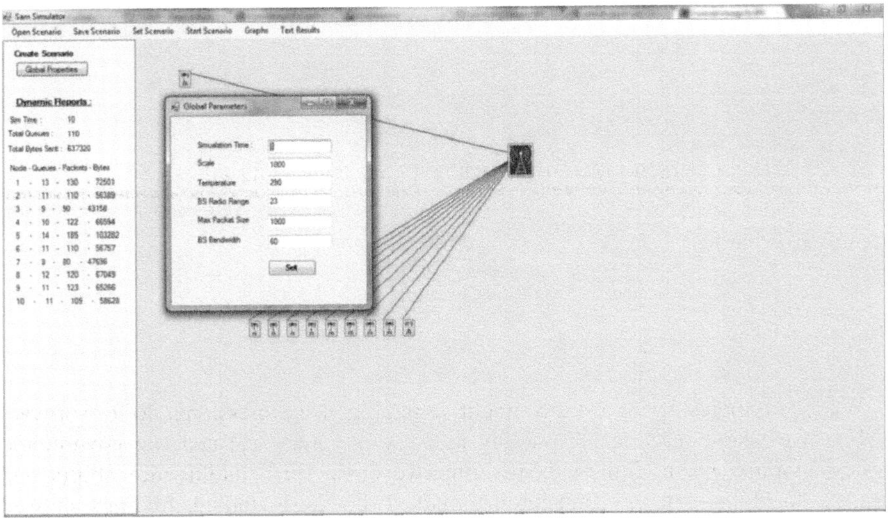


**Fig. 1.** Setting Global Parameters

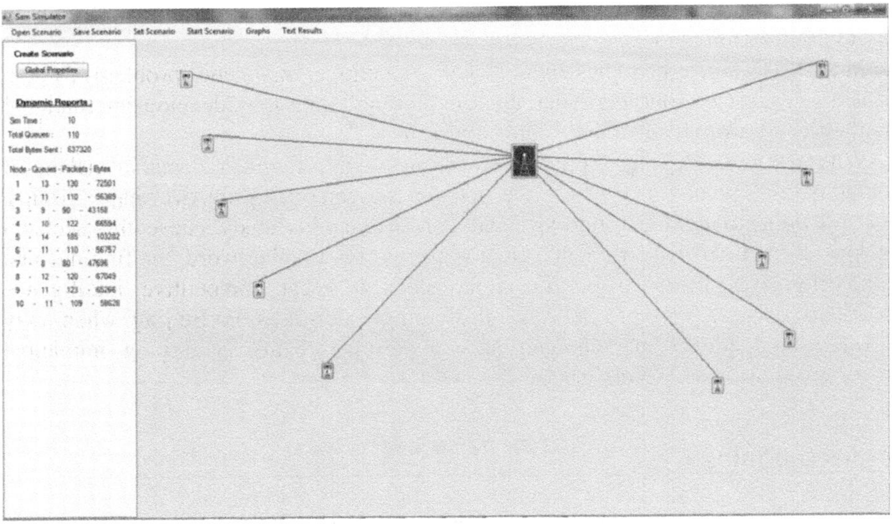


**Fig. 2.** Setting Properties for Mobile Stations (MS)

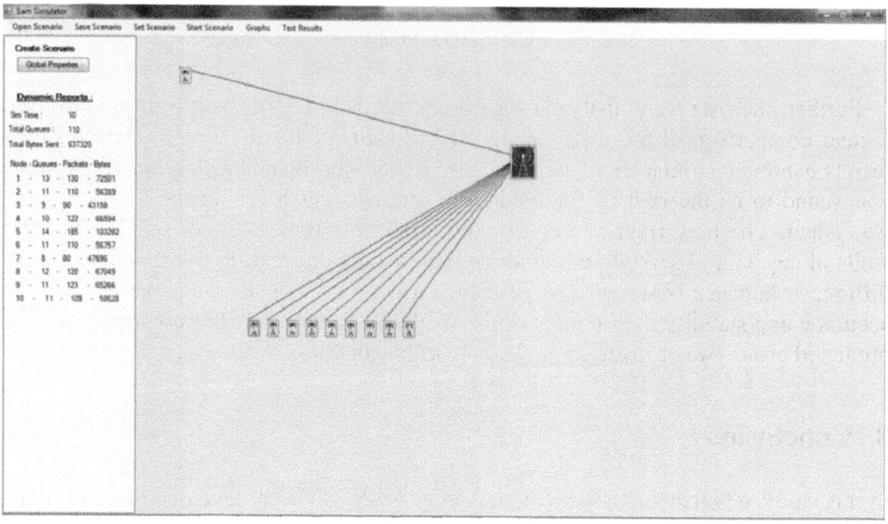

**Fig. 3.** Mobile Station statistics of the packet generation

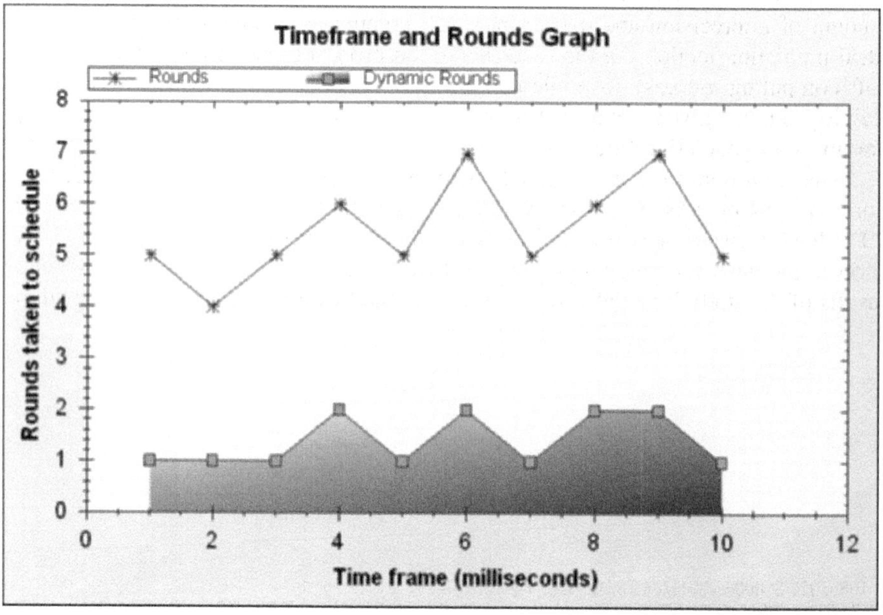


**Fig. 4.** Time Frame Vs Rounds taken to schedule

Also the comparison of Data Rates for various classes of traffic with increase in distance has been measured.

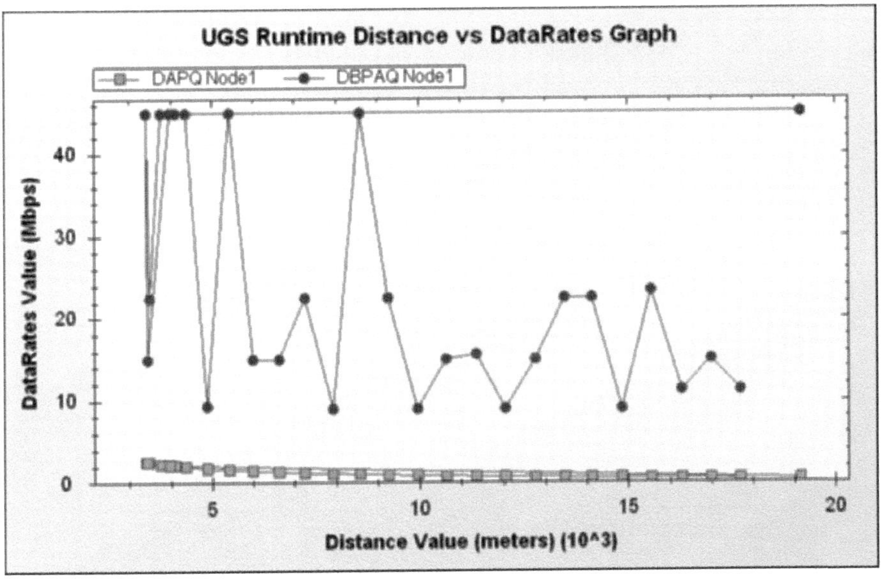


**Fig. 5.** Distance Vs Data rates for UGS Packets

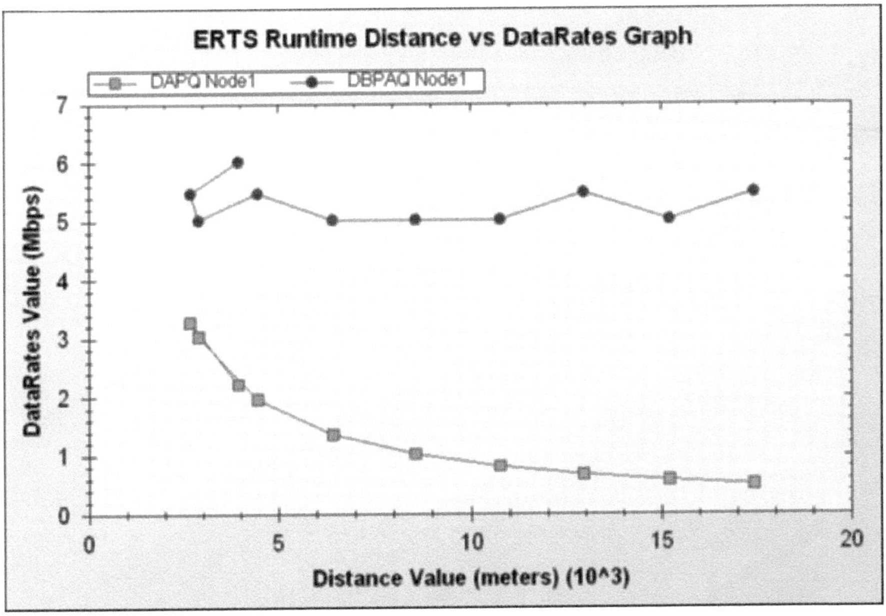


**Fig. 6.** Distance Vs Data rates for ERTPS Packets

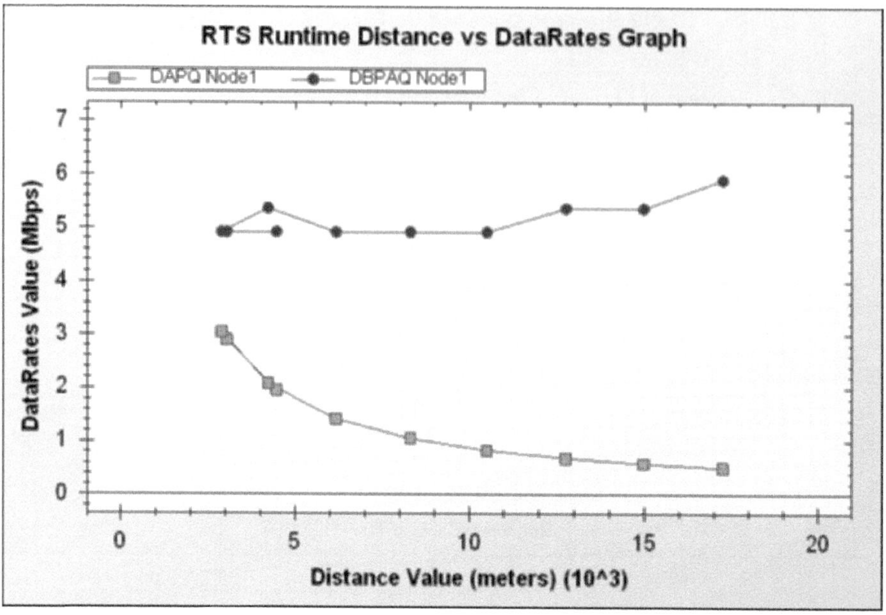


**Fig. 7.** Distance Vs Data rates for RTPS Packets

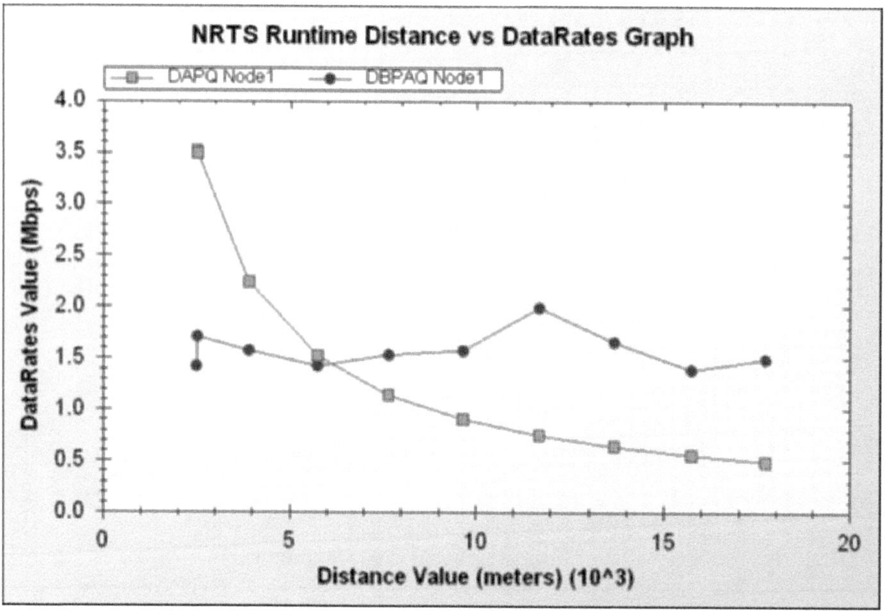


**Fig. 8.** Distance Vs Data rates for NRTPS Packets

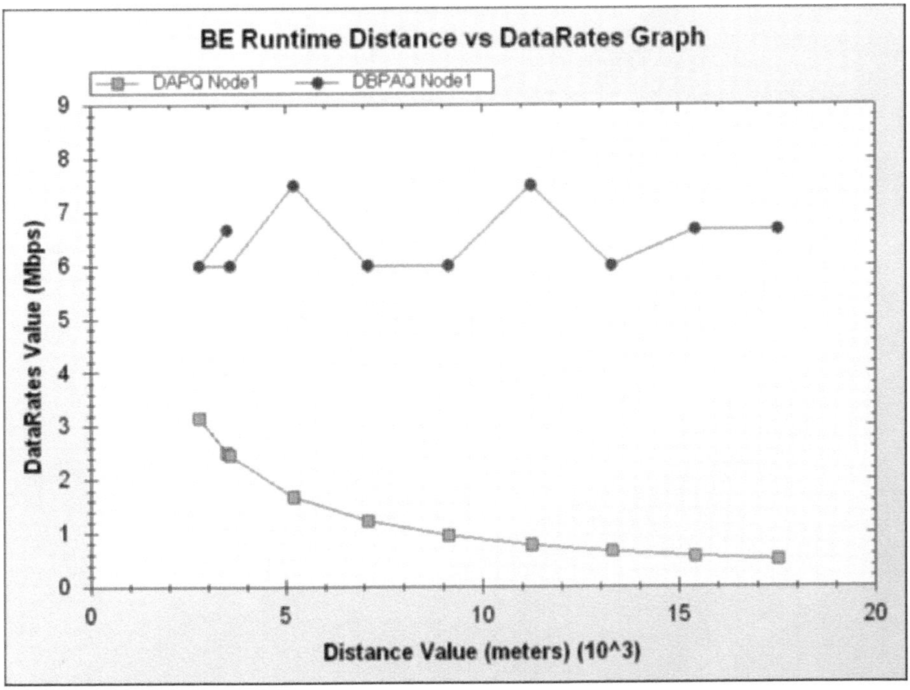


**Fig. 9.** Distance Vs Data rates for BE Packets

## 5 Conclusion

In a mobile WiMAX scenario, The same mobile subscriber station may have different service requirements at different times in the WiMAX network.Scheduling strategies that can improve network throughput, maintain relative fairness, and lower delay over the existing scheduling schemes such as the round robin and weighted round robin algorithms are desirable.In this project, a packet scheduling scheme considering the QOS in mobile WiMAX network is proposed. In Mobile WiMAX the rates at which the mobile stations transmit the data are changed at every instant. Even in this condition the proposed scheduling scheme is able to send the data packets to the base station with better bandwidth utilization and lesser delay compared to the existing scheme.

## References

1. Lu, S., Bharghavan, V.: Fair Scheduling in Wireless Packet Networks. IEEE/ACM Trans. Networking 7 (1999)
2. So-In, C., Jain, R., Al-Tamim, I.A.: Generalized Weighted Fairness and its support in Deficit Round Robin with Fragmentation in IEEE 802.16 WiMAX. In: The 2nd International Conference on Computer and Automation Engineering (ICCAE), Singapore, vol. 1, pp. 784–788 (2010)

3. Katevenis, M., Sidiropoulos, S., Courcoubetis, C.: Weighted Round-Robin Cell Multiplexing in a General-Purpose ATM Switch Chip. IEEE J. Selected Areas in Comm. 9(8), 1265–1279 (1991)
4. Mobile WiMAX QoS Classes, http://www.wimaxforum.org/technology/downloads/Mobile_WiMAX_Part1_Overview_and_Performance.pdf
5. Esmailpour, A., Nasser, N.: Packet Scheduling Scheme with Quality of Service Support for Mobile WiMAX Networks. In: The 9th IEEE International Workshop on Wireless Local Networks (WLN 2009), Zürich, Switzerland, October 20-23 (2009)
6. Vaidya, N.H., Bahl, P., Gupta, S.: Distributed fair scheduling in a Wireless LAN. IEEE Trans. Mobile Comput. 4, 616–629 (2005)
7. Tran, M., Zaggoulos, G., Nix, A., Doufexi, A.: Mobile wimax: Performance Analysis and Comparision with Experimental Results. IEEE, Los Alamitos (2008)

# Author Index

GPSR Compliance

*The European Union's (EU) General Product Safety Regulation (GPSR) is a set of rules that requires consumer products to be safe and our obligations to ensure this.*

*If you have any concerns about our products, you can contact us on ProductSafety@springernature.com*

In case Publisher is established outside the EU, the EU authorized representative is:

Springer Nature Customer Service Center GmbH
Europaplatz 3
69115 Heidelberg, Germany

**Batch number: 10838023**

Printed by Printforce, the Netherlands